Springer Water

The book series Springer Water comprises a broad portfolio of multi- and interdisciplinary scientific books, aiming at researchers, students, and everyone interested in water-related science. The series includes peer-reviewed monographs, edited volumes, textbooks, and conference proceedings. Its volumes combine all kinds of water-related research areas, such as: the movement, distribution and quality of freshwater; water resources; the quality and pollution of water and its influence on health; the water industry including drinking water, wastewater, and desalination services and technologies; water history; as well as water management and the governmental, political, developmental, and ethical aspects of water.

Rahul Kumar · Renu Yadav · R. K. Gupta
Editors

Sustainability in Aquaculture and Fisheries

Springer

Editors
Rahul Kumar
Department of Zoology
Chaudhary Charan Singh Haryana
Agricultural University
College of Agriculture, Bawal
Rewari, Haryana, India

R. K. Gupta
Department of Zoology
College of Basic Sciences and Humanities
Chaudhary Charan Singh Haryana
Agricultural University
Hisar, Haryana, India

Renu Yadav
Department of Zoology, College of Basic
Sciences and Humanities
Chaudhary Charan Singh Haryana
Agricultural University
Hisar, Haryana, India

Department of Zoology
Desh Bandhu Gupta Government College
Panipat, Haryana, India

ISSN 2364-6934 ISSN 2364-8198 (electronic)
Springer Water
ISBN 978-3-032-16604-3 ISBN 978-3-032-16605-0 (eBook)
https://doi.org/10.1007/978-3-032-16605-0

This Springer imprint is published by the registered company Springer Nature Switzerland AG
The registered company address is: Gewerbestrasse 11, 6330 Cham, Switzerland

Contents

Chapter 1
Capitalizing Beneficial Bacteria for Sustainable Aquaculture

Parul Puri

Abstract Aquatic systems are growingly acknowledged for their multi-pronged contributions in improving food security, nutrition, mitigation of poverty, generation of employment with revenue and resource creation at a low environmental imprint. Aquaculture systems are strongly under influence of astute balance between helpful and antagonistic microbial strains. Moreover, inherent bacterial community present in culture environment in association to aquaculture species and processes is dynamically influenced by farming controls and variables. Beneficial potential of bacterial microbials can be utilized to enhance growth, survival, immunoprophylactic and therapeutic effects, remitting disease incidences in culture environment. Harnessing capabilities of beneficial strains and their products can hence enhance multitude of aquaculture practices, including waste water treatment, reutilisation; bio-mitigation; feed quality; fish welfare and consequent improvement in overall productivity. Additionally, maintenance, propagation of remedial heterotrophic; phototrophic bacterial strains such as autotrophic nitrifiers in innovative aquaculture approaches, biofloc; integrated multitrophic and recirculating systems, is essential for sustainable expansion of aquaculture.

Keywords Aquaculture · Sustainability · Probiotics · Effective microorganisms · Synbiotic · Biofloc · IMTA · RAS

1.1 Introduction

Fish and fish products are source of essential minerals, fatty acids, proteins catering to global nutrition demands and food security (Puri et al. 2022). Aquaculture the organized farming (= production) of finfish, shellfish and aquatic plants in controlled environment has witnessed an overall high in the year 2022, with worldwide yield of aquatic products, aquatic animals, algae up to 223 million tonnes (MT), refer

P. Puri (✉)
Assistant Professor, Department of Zoology, Maitreyi College, University of Delhi, Bapu Dham Complex, Chanakyapuri, Delhi-110021, India
e-mail: parulpuri@maitreyi.du.ac.in

R. Kumar et al. (eds.), *Sustainability in Aquaculture and Fisheries*,
Springer Water, https://doi.org/10.1007/978-3-032-16605-0_1

Table 1.1 Aquaculture contribution to the global production of top five aquatic animal species and total production in 2022 (*Source* FAO 2024)

Top five aquatic animal species (2022)	Aquaculture production (MT)
Whiteleg shrimp (*Penaeus vannamei*)	**6.8**
Cupped oysters nei (*Crassostrea* spp.)	**6.2**
Grass carp (*Ctenopharyngodon idellus*)	**6.2**
Nile tilapia (*Oreochromis niloticus*)	**5.3**
Silver carp (*Hypophthalmichthys molitrix*)	**5.1**
Total aquatic animal production[a]	**94.41**
Finfish	61.56
Crustaceans	12.75
Molluscs	18.91
Other aquatic animals	1.18
Algae (microalgae, seaweeds)[a]	36.50
Total aquaculture—all species[a]	**130.94**

[a] Inland, marine, coastal aquaculture

Table 1.1. Table 1.1 describes worldwide aquaculture and aquatic animal production including finfish, molluscans, crustaceans, algae and other aquatic animals from marine, coastal and inland sources estimating top five culture species yields in 2022. Aquaculture and fisheries have contributed to 15% of animal protein needs of global population, higher up to 50% specifically in developing nations of Asia and Africa (FAO 2024). Aquaculture alone has contributed to over 57% of aquatic animal products for direct human consumption projecting a high nearing 62% by 2030 (World Bank 2013; FAO 2024). Besides steady uprise, intensive aquaculture development at higher stocking densities has raised array of environmental concerns including overuse of resources, dependence on commercial feed, waste generation, rapid spread of virulence and disease outbreaks (Tovar et al. 2000; Kumar et al. 2022a, b; Thompson et al. 2024). Aquaculture growth and fisheries development deem sustainable efforts in wake of adding more and nutritious food to human platter. In this ambit, the implementation of sustainable aquaculture practices will provide foothold for long standing aquaculture expansion, and to achieve social economic and environmental sustainability.

1.1.1 Predisposition of Aquaculture Environment: Threats to Bacterial Beneficence

Prevailing physical, chemical, biological, technical stressors inherent to culture practices (refer Table 1.2) largely impact animal welfare, health and survival (Folke and Kautsky 1992). These alterations primarily effect gut microbial ecology, impairing healthy microbial balance causing dysbiosis. Gastrointestinal tract (GIT) of organisms and culture ecosystem per se can be inhabited by beneficial and opportunistic bacterial strains (De Schryver and Vadstein 2014). Aquaculture health and well-being is a complex interplay of inherent (= autochthonous) healthy gut microbiota, and extrinsic (= allochthonous) water, food acquired microbial balance (Tarnecki et al. 2017; Ringø and Birkbeck 1999). Microbes present in culture system can be of types beneficial, neutral or harmful (pathogenic) types (Zhou et al. 2009). Beneficial strains can cater change of neutral strains to beneficial ones, together these strains have capacity to revive ecosystems by reducing pollution and support growth needs of culture species. Stable, slow growing (K-strategist) microbiota are characteristic of beneficial, healthy microbiome while r-strategic, rapid growth is found in opportunistic strains prone to pathogenic incidences and transfers (Puri et al. 2022; Bakke et al. 2017; Urmila et al. 2025; Attramadal et al. 2014). Various works have explored effects of stressors on shift in microbial dynamics and their impact on farming environment some as listed below.

1.1.1.1 Abiotic Stressors

Existent water quality parameters such as temperature, pH, dissolved oxygen (DO), ammonia, nitrite, nitrate, at a range lower than optimal can be a major abiotic stressor in aquafarming environment.

Physical Stress

Temperature and Salinity Stress

Farming temperature affects growth and body weight by altering the metabolic rate of fish depending on type of species investigated. Temperature effects have been studied extensively in Nile tilapia (*Oreochromis niloticus*) (Abdel-Tawwab and Wafeek 2017), channel catfish (*Ictalurus punctatus*) (Buentello et al. 2000), sea bass (*Dicentrarchus labrax*) (Besson et al. 2016). Fish growth and survival are important criteria in aquaculture productivity and output. In general, increase in water temperature correlates to increment in fish growth although this increase affects survival differently. Increase in temperature above 16 °C decreased survival of freshwater crayfish (*Paranephrops zealandicus*) although growth rates increased with water temperature at maximum specific rates reaching 0.57 (Hammond et al. 2006). White shrimp (*Penaeus vannamei*) juveniles show ideal growth at temperature range 25–30 °C with less effect of salinity at 20–50%, survival is best at 20–30 °C at water salinity requirement above 20%. Together growth and survival optima for *P. vannamei* correspond at 28–30 °C and salinities within 33–40% akin their prevalent natural existence (Ponce-Palafox et al. 1997; Sharma et al. 2024). Yuan et al. (2017) confirmed similar temperature range 28–30 °C for optimum growth and molting of mitten crab (*Eriocheir sinensis*) juvenile. Mean low temperature at 18 °C had negative effect on

Table 1.2 Response of aquaculture species to aquafarming stressors

Classification	Type of Stressors	Aquaculture species	Stress response	References
1. Abiotic stressors A. Physical stress	Temperature	Channel catfish (*Ictalurus punctatus*)	FE↑, FI↓, WG↓ (at low temperatures 15.7 °C)	Buentello et al. (2000)
		Chinese mitten crab (*Eriocheir sinensis*)	Survival↓ (at high temperature 22°, 28°, 30 °C), GR↓ (at low temperature 18°, 22 °C)	Yuan et al. (2017)
	Salinity	Black sea bass (*Centropristis striata*) juvenile	Growth↓, mortality↑, FCR↑ at 10% salinity	Cotton et al. (2003)
		Pacific white shrimp (*Litopenaeus vannamei*)	*V. parahaemolyticus, owensii campbellii incidence* at low 15% salinity	Bauer et al. (2021)
B. Chemical stress	Low dissolved oxygen (= hypoxia)	Nile tilapia (*Oreochromis niloticus*)	Liver and muscle glycogen ↓, β-oxidation↑, LDH↑	Li et al. (2018)
		Large yellow croaker (*Larimichthys crocea*)	Oxidative stress index (ROS↑, SOD↑, CAT↑, MDA↑, GSH↑); Apoptotic factor (bax↑, caspase-3↑, caspase-9 mRNA↑)	Liu et al. (2024a, b)
		Chinese mitten crabs (*E. sinensis*)	LDH↑, SOD↑, CAT↑, MDA↑, upregulation of immune-genes (MyD88, ALF1, Relish, Crustin)	Jiang et al. (2023)
		Pacific white shrimp (*L. vannamei*)	*V. parahaemolyticus incidence*↑	He et al. (2025)
		Darkbarbel catfish (*Pelteobagrus vachelli*)	Pathogenic bacteria *Edwardsiella*↑, beneficial bacteria *Clostridium*↓, tissue necrosis↑ (upregulated *p53, Caspase-3, 9, Bcl-2, Bax)*	Zheng et al. (2021)
	pH	Mummichog (*Fundulus heteroclitus)*	*V. parahaemolyticus incidence*↑	Boleza et al. (2001)

(continued)

Table 1.2 (continued)

Classification	Type of Stressors	Aquaculture species	Stress response	References
		Giant freshwater prawn (*Macrobrachium rosenbergii*)	*Feeding*↓, *growth*↓, *survival*↓	Chen and Chen (2003)
		Yellowtail kingfish (*Seriola lalandi*)	*Survival*↓, *growth*↓, FCR	Abbink et al. (2012)
	Chemical treatments-antibiotics	Zebrafish	*A. hydrophila* infection↑, innate immunity↓, gut microbiota dysbiosis	He et al. (2017)
		Channel catfish (*Ictalurus punctatus*)	Microbiota richness and diversity↓ (beneficial Firmicutes↓, Bacteroidetes↓; pathogenic Proteobacteria↑)	Abdelhamed et al. (2019)
		Channel catfish (*I. punctatus*)	Pathogenic bacteria↑	Wang et al. (2019)
	Feed composition, quality, availability	Nile tilapia (*O. niloticus*)	Growth↓, GLU↑, AST↑, ALT↑, creatinine↑, NH_3 excretion↑, COR↑	Abdel-Tawwab (2016)
		Dolly varden char (*Salvelinus malma*)	Growth↓, mortality under hypoxia stress↑, HSI↓, serum LYZ↓, WG↓, serum AST↑, ALT↑, MDA↑, TAC↑	Guo et al. (2018)
	Nitrogenous, other metabolic wastes	Juvenile turbot (*Scophthalmus maximus*)	Growth↓, FI↓, mortality↑	Person-Le Ruyet et al. (1997)
		Juvenile white shrimp (*Litopenaeus vannamei*)	Mortality↑	Li et al. (2007)
	Heavy metals	Rainbow trout (*O. mykiss*)	Mortality↑	Davidson et al. (2009)
		European seabass (*Dicentrarchus labrax*)	Growth↓, Cr, Mn, Co, Ni, Cu, As, thallium (Tl), Cd accumulation in liver and muscle	Deviller et al. (2005)

(continued)

Table 1.2 (continued)

Classification	Type of Stressors	Aquaculture species	Stress response	References
C. Technical stress	Handling	Eurasian perch (*Perca fluviatilis*)	Plasma COR↑, GLU↑, SSI↓, spleen LYZ activity↓, spleen total Ig's↑	Milla et al. (2010)
	Transportation	Pearl gentian grouper (*Epinephelus fuscoguttatus* female x *Epinephelus lanceolatus* male)	TAN↑, TFAA↑, SOD↑, CAT↑, GSH-PX↑, MDA↑, LDH↑, AST↑, ALT↑, plasma COR↑, GLU↑, pH↓, DO↓, survival↓, at 15 °C/120 rpm; 25 °C/70, 120 rpm; mucus secretion, gill lamellar curvature, swollen epithelial cells	Fang et al. (2023)
		Rainbow trout (*O. mykiss*)	Plasma COR↑, GLU↑, tight junction gene (claudin 7, 8d, 12)↑, mucin gene↑, antimicrobial peptides↓ (DB1, DB2, cathelicidin 1, 2, hepcidin), cytokines (TGF-β1a, β1b)↑	Tacchi et al. (2015)
		Rainbow trout (*O. mykiss*)	MDA↑, SOD↑, CAT↑, TNF-α↑, IL-1 β↑, IL-8↑, Hsp70↑, NOS↑, *ACP*↑, *AKP*↑	Ren et al. (2022)
2. Biotic Stressors	Stocking density	Juvenile turbot (*Scophthalmus maximus*)	BWI↓, SGR↓, FCR↓, plasma COR↑	Liu et al. (2017)
		Chinese sturgeon (*Acipenser sinensis*)	Serum COR↑, GLU↑, lactate↑, ACTH↑, GST↑, LYZ↓, IgM↓, AKP↓, ACP↓, HAMP1↓, TNF-α↓, IL 1β↓, IL 8↓, GH↓, IGF-I↓	Long et al. (2019)
	Pathogens	Hybrid red tilapia (*Oreochromis* sp.)	Disease incidence↑, mortality↑	Dong et al. (2019)
		Yellow catfish (*Tachysurus fulvidraco*)	Necrosis↑, inflammation↑, immune response↑, disease incidence↑, mortality↑	Hongsen et al. (2024)

Abbreviation—ACP: acid phosphatase; ACTH: adrenocorticotropic hormone; AKP: alkaline phosphatase; AST: aspartate aminotransferase; ALT: alanine aminotransferase; BWI: body weight increase; CAT: catalase; COR: cortisol; FCR: feed conversion ratio; FE: feed efficiency; FI: feed intake; GLU: glucose; GR: growth rate; GSH: glutathione; GSH-PX: glutathione peroxidase; GST: glutathione-s-transferase; HAMP 1: hepcidin antimicrobial peptide1; HIS: hepatosomatic index; IL: interleukin; LDH: lactate dehydrogenase; LYZ: Lysozyme; MDA: malondialdehyde; NOS: nitric oxide synthase; ROS: reactive oxygen species; SOD: superoxide dismutase; SSI: spleno-somatic index; TAC: total antioxidant capacity; TAN: total ammonia nitrogen; TFAA: total free amino acid; TNF-α: tumor necrosis factor-alpha; WG: weight gain economic value of growth rate of sea bass (*Dicentrarchus labrax*), reduced stocking density per batch decreasing overall output (Besson et al. 2016). Temperature fluctuations can also transgress microbial dynamics favouring pathogenic outbreaks and disease incidences by inducing changes in non-pathogenic forms to acquire pathogenic traits involving formation of flagellar processes, biofilm development, adhesion capability, under direct or indirect temperature controls (Cascarano et al. 2021). Hence temperature variations can affect host physiology, susceptibility and virulence shifts in farming scenario.

Effects of salinity are variable for different life history stages of fish. Increasing salinity stress above 2.5 parts per trillion and time of exposure to high salinities decreased embryo hatching, induced larval abnormalities and augmented fry mortality in freshwater pabda (*Ompok pabda*) (Alam et al. 2020). Cotton et al. (2003) studied 12 week effects of various salinities (10, 20, 30%) and temperature (15, 20, 25, 30 °C) on juvenile sea bass (Table 1.2). Salt concentration at 20, 30% with farming temperatures at 25 °C gave best growth parameters of weight gain (WG), relative growth rate (RGR), specific growth rate (SGR), daily weight gain (DWG) and feed conversion ratio (FCR). Significant negative effects on these parameters were found at 10%, 15 °C of salinity and temperature respectively. Renaud and Parry (1994) investigated salinity effects on biochemical constitution fatty acid profile of marine microalgae, *Isochrysis* sp., *Nannochloropsis oculata, Nitzschia frustulum.* Salinity change demonstrated significant variability in lipid profile, total fatty acid content of microalgae with lesser effect on total protein, chlorophyll a and carbohydrate values. Salinity can alter bacterial microflora composition in recirculating aquaculture systems (RAS) affecting denitrification potential (von Ahnen et al. 2019; Bakke et al. 2017). High salinities can induce microbial abundance of autotrophic denitrifiers but decreasing gross denitrification process in RAS systems. Lowering salinity can cause pathogenic switching among *Vibrio* species, increasing virulence augmenting genes occurrence in shrimp aquaculture (Bauer et al. 2021).

Chemical Stress

Hypoxia (= Low Dissolved Oxygen)

Dissolved oxygen (DO), the readily available form of oxygen to organisms, is a crucial factor limiting fish growth in aquaculture environment (Wang et al. 2023b; Claireaux and Chabot 2016; Shefali et al. 2021). Large variation in DO content across farming areas can affect growth, reproduction and survival of aquaculture species. Solubility of oxygen in water is 33 time lesser compared to that in air, with diffusivity even lower (3 lakh times low), making small fluctuations in oxygen values render perilous effects on aquatic life (Verberk et al. 2011). Hypoxia is a subthreshold saturation of oxygen, a state at which DO values are below 2.8 mg/L (Wu 2002). In aquaculture, low DO values of the range between 0.28 to 4 mg/L results in hypoxic condition (Zhan et al. 2023; Saha et al. 2022). Hypoxic stress (DO value under 4 mg/L) is related to growth retardation, histopathological developments, oxidative stress, endocrine disruption, immuno-suppression and disease outbreaks, leading to fish mortality (Wu 2002; Ruan et al. 2020). Hypoxia can cause alterations in sex hormones, vitellogenin production; affect gonadal development; spawning, hatching rate and larval survival (Saha et al. 2022). Hypoxia loss are irreversible to fish and shellfish. Sudden, short-term hypoxia up to hours lead to acute effects, whereas sustained long-term drop in minimal oxygen required can have chronic effects. Among crustaceans shrimps are highly susceptible to hypoxia stress (He et al. 2025). Liu et al. (2024b) studied 96 h acute hypoxia stress DO 2 ± 0.1 mg/L on oxidative state, apoptotic rate and intestinal microflora of large yellow croaker (*Larimichthys crocea*). Increased prevalence of opportunistic, pathogenic bacteria belonging to genera *Mycoplasma, Staphylococcus* and *Pseudomonas* indicated gut microbial dysbiosis. Hypoxia induced changes in gut microflora (Firmicutes and Bacteroidetes for energy utilisation) and intestinal histology (decrease in goblet cell number, height and width of intestinal villi) correlated to increased disease risk. Similar duration (96 h) hypoxic stress at DO 3.0 ± 0.1 mg/L was evaluated for changes in gut flora, immune responsiveness, oxidative state and apoptotic stress in 180 mitten crabs (*E. sinensis*) by Jiang et al. (2023). Pro-apoptotic gene expression of Caspase 3, 8 and BAX was upregulated with increased apoptotic rate in direct effect to increased oxidative stress. Overall negative impact of hypoxia was on compromised immune response and altered gut flora affecting hepatopancreas-intestinal health. Variable response to hypoxic conditions is found among different fish species. Wu et al. (2016) demonstrated wide tolerance range of juvenile turbots (*Scophthalmus maximus*) to varying DO concentrations. Respiratory rates, blood pH, haemoglobin and haematocrit values increased under hypoxia (1.2, 3.5 mg/L DO), with reverse trend of decrease in respiration rate under mild hyperoxia (11.14 mg/L DO) condition. Gill physiology was damaged after prolonged hyperoxia incidence, with no apparent hypoxic gill damage to fish.

pH

Difference in pH can have effects on growth and survival of aquaculture species. pH of seawater ranges between 7.85 to 8.5 (Abbink et al. 2012). Aquafarming systems rely on pH optima between 6.5 and 8.5 (Timmons et al. 2002). Reported optimal range of temperature and pH for the growth of giant freshwater prawn (*Macrobrachium rosenbergii*) are 29° to 31 °C and 7 to 8.5, respectively (Chen and Chen 2003). At low pH (= 6.8) exposure for 56 days juvenile *M. rosenbergii* exhibited significant decrease in moulting and body weight. Additionally, rate of feeding reduced at pH of 6.8 and below. Variation to pH tolerance has been found among different life history stages with increased tolerance to low pH in order adult > juvenile > fingerlings being reported for Nile Tilapia (*O. niloticus*) (Mustapha and Atolagbe 2018). Visible structural and behaviour changes included erratic swim, skin erosion, fin bleed, impaired feeding. Acid stress had direct effect on survival with mortality resulting from blood acidosis, epidermal erosion and inhibition of ion regulation. According to Mota et al. (2018) low aquatic pH can enhance fish growth preventing accumulation of toxic ammonia release by fishes across its conversion to less toxic ammonium ions as well decrease parasitic incidences and disease load. In extensive aquaculture high alkalinity can improve carbon availability for photosynthesis, augmenting buffering action in semi-intensive and intensive systems (Somridhivej and Boyd 2017). Concurrent lowering of pH at high carbon dioxide content (= hypercapnia) may potentiate negative effects of low DO values (Boleza et al. 2001). Hypoxia alone has hazardous effects on immune mechanism of mummichog (*Fundulus heteroclitus*), while hypercapnic hypoxia (pO_2 15 torr, pCO_2 8 torr, pH 7) favours establishment of opportunistic pathogens by compromising bactericidal action of phagocytic anterior head kidney cells.

Chemical Treatments—Antibiotic Antimicrobials

Antimicrobials are veterinary drugs generally used to inhibit the growth of pathogenic microorganisms. Their use, overuse, persistence and slow degradation can cause hazardous build up imposing risk to biota and environment. Use of antibiotics in aquaculture world-over find measures in disease prophylaxis (control prevention), therapy (treatment), metaphylaxis (mass treatment against outbreaks) and enhancing growth (Quesada et al. 2013). Injudicious antibiotic use and chemical therapeutic treatments translate inharmonious side effects on fish physiology beside their long trailing impacts on the environment (Rasul and Majumdar 2017; Chen et al. 2022). β-lactams, quinolones, sulfonamides, tetracyclines, macrolides are prominent antibiotics applied for disease control, prevention and treatment in susceptible aquafarming conditions. Quinolone and sulfonamide are few of the most prevalently persistent aquaculture use antibiotics (Johansson et al. 2014; Song et al. 2016). Limbu et al. (2018) investigated 12 week chronic exposure effects of bacteriostatic antibiotics oxytetracycline (oxy) and sulfamethoxazole (sulf) in Nile tilapia (*O. niloticus*) and their incidental human effects on consumption. *O. niloticus* were treated with low concentration antibiotic supplemented to water (oxy 420 ng per litre and sulf 260 ng per litre) and allowed aquaculture doses of antibiotic supplementation to diet (oxy 80 mg kg^{-1} and sulf 100 mg kg^{-1}, daily). Both water and dietary antibiotic treatment impaired fish growth, nutrient digestibility, leading to dysregulated intestinal microflora, liver damage, compromising immune health. Allowed aquaculture concentrations can impose human health risk, bringing forth need to assess antibiotic use in aquaculture management. Feed is an important contributor of antibiotic to freshwater aquaculture imposing contamination risk to algae, as sedimentary residuals and potent human hazard (Hu et al. 2024). Indiscreet antibiotic use, selection pressures in intensive aquafarming environments, has led to interference competition among bacteria with development of increased virulence and antibiotic resistance mechanisms (ARMs) among pathogens, as well as arise of newer pathogens some including *Plesiomonas shigelloides, Stenotrophomonas maltophilia, Streptococcus iniae* (Sundberg et al. 2016; Puri et al. 2022, 2023; Llewellyn et al. 2014). ARMs associated with virulence prevalence genes of *Aeromonas hydrophila* has led to large scale disease outbreak of tilapia, channel catfish, common and grass carp in freshwater aquaculture of Northern Vietnam (Nhinh et al. 2021). 74.7% of bacterial isolates had multidrug resistance values above 0.2, relating it to antibiotic effects. Recent outbreak of pathogenic *S. iniae* strain 2022SI08 has led to mass mortality of yellow catfish (*Tachysurus fulvidraco*) in aquaculture farm of China during year 2022 (Hongsen et al. 2024).

Feed—Composition, Quality, Availability

Feed is the largest contributor to intensive aquaculture development and growth. Feed composition and ingredient quality determine the acceptability, palatability, digestibility, utilisation and interference characteristics of feed to fish physiology (Glencross et al. 2007; Glencross 2009). Fishmeal based diets have long-standing contribution to aquaculture nutrition yet, its scarcity, expense and environmental impacts have shifted focus to alternative plant-based feeds. Although many plant-based ingredients (such as legume seeds, rapeseeds, oilseed cake) have limited utilisation in fish feed due to presence of antinutritional substances known to hamper host physiology. Saponins, phytates, tannins, complex carbohydrates, gossypol and phorbol esters are key antinutritional substances (ANS) (Francis et al. 2001). Presence of ANS in aquafeeds negatively influence the protein digestibility and organoleptic properties, as well as the bioavailability of other nutrients. ANS such as phytate in plant-based diets hinder calcium availabilities to fish (Puri 2025). Feed inclusion with rapeseed protein source can impair protein digestibility, lowering availability of phosphorus and other bivalent, trivalent minerals (Kaiser et al. 2021b). Dietary phytate and rapeseed protein at high inclusion rates lowered feed intake and growth performance of rainbow trout (*O. mykiss*) (Kaiser et al. 2021a, 2021b). Additionally ANS such as protease inhibitors, phytates, contribute to soluble, and insoluble waste accumulation (through saponins, polyphenols); concurrently increasing aquaculture nitrogen (N) and phosphorous (P) loads (Kokou and Fountoulaki 2018).

Nitrogenous and Other Metabolic Wastes Accumulation

Waste products generated from culture practices pose risk to environment and thwarts aquaculture sustainability. Aquaculture intensification requires increased feed input to raise production causing relatable feed waste. Unutilized feed and fish metabolites are source of N and P based waste accumulation (Dauda et al. 2019). Dietary input of 75% N and 69–86% dietary P is released in water (Lazzari and Baldisserotto 2008; Luo et al. 2020). Nitrogenous wastes in aquaculture systems predominantly comprise ammonia, nitrites and nitrates. Ammonia is primary nitrogen waste and it can be decomposed to nitrite-N and nitrate-N by bacterial action (Luo et al. 2020). Together sum of unionised (NH_3-N) and ionised ammonia-N (NH_4^+-N) comprise total ammonia nitrogen (TAN). Both water pH and temperature determine ammonia toxicity by forming NH_3 from NH_4^+ form of ammonia toxic to fish (Tomasso 1994). Ammonia and nitrite toxicity is known to impair growth, immune function, disease resistance and fish survival. Recommended ammonia below 1 mg/L and nitrite levels less than 0.5 mg/L are considered safe in culture systems (Ajani et al. 2011) though NH_3 at concentrations 0.025 mg N/L is highly toxic to most aquatic animals (Luo et al. 2020). Person-Le Ruyet et al. (1997) reported chronic ammonia toxicity effects on growth and survival of turbot. Li et al. (2007) studied effects of salinity on ammonia toxicity to white shrimp (*L. vannamei*). At low salinity (3%) resultant increment in ammonia toxicity caused energy expense of juvenile shrimp to maintain osmoregulation in susceptible culture conditions. Maintaining low possible level of ammonia is important requirement for aquaculture. Moreover, nitrate-N and phosphorous build up is related to eutrophication in culture systems and by runoff to neighbouring natural aquatic habitats (Jahan et al. 2003; Chen et al. 2024; Sharma et al. 2025). Nutrient loading of ammonium and phosphate (as inorganic phosphorous) can largely shift microbial community dynamics favouring proliferation of opportunistic strains (Olsen et al. 2017). Hornick and Buschmann (2018) observed changes in bacterial community diversity, structure and function in salmon aquaculture. Gammaproteobacteria and various taxa (such as *Shewanella, Pseudoalteromonas, Psychrobacter*) carrying antimicrobial resistance were significantly augmented at aquaculture sites.

Heavy Metal

Metals with relatively higher density and high toxicity at low levels are termed as heavy metals (Huseen and Mohammed 2019; Yadav et al. 2023). Heavy metal concentration (HMC) in aquaculture sediments are known to effect food and environment safety with aquafeed protein as a larger contributor of HMC in culture sediments. Major contribution of heavy metal deposition in culture environment (44.3%) came from agricultural source including aquaculture feed compared to those from natural source (19.05%) and industrial runoff (36.66%) (Yadav et al. 2022a, b; Liu et al. 2024a). Unused feed in aquaculture wastewater contains starch, crude fibre and feed proteins with high chelating potential to heavy metals (Hu et al. 2010). Copper and zinc are appended as preservatives, additives in fish feed or as anti-fouling agents (Farmaki et al. 2014). Heavy metal accumulation increase with decrease in water replacement rates in RAS, with demonstration of increments in iron (Fe), manganese (Mn), As, nickel (Ni) and zinc (Zn) concentrations in low water exchange recirculating systems (Martins et al. 2011). HMC was found to be higher in liver in comparison to in muscle deposition of Nile tilapia (*O. niloticus*) except for As. According to Martins et al. (2011), increments in heavy metals in RAS water do not correlate to similar deposition in fish tissues. However, van Bussel et al. (2014) in their work demonstrated direct association between aquatic HMC deposition to whole body metal accumulation in juvenile turbot (*Psetta maxima*). Exposure to dissolved metals displayed decreased dry matter content of whole body for juvenile turbot, with non-significant effect on either growth, feed conversion or performance. In similar work studying potential heavy metal effect on rainbow trout high mortality related to chronic copper toxicity was established (Davidson et al. 2009). Martins et al. (2009) related accumulation of Arsenic (As) and copper (Cu) in RAS to impair embryonic and larval development of common carp (*Cyprinus carpio*) resulting in large scale mortality.

Technical Stress

Handling and Transportation

Handling and transportation are essential activities in aquaculture. Fang et al. (2023) studied during and post-transportation effects of temperature and vibration frequency for juvenile pearl gentian grouper (*Epinephelus fuscoguttatus* female x *Epinephelus lanceolatus* male). Deterioration in meat quality with protein depletion was found at high temperature and high frequency transport. Additionally during and post-transportation effects were noticed in abnormal gill histological changes; increased oxidative, hematological stress response and decline in water quality. Optimal conditions for transport of pearl gentian grouper were determined to be at water temperature 15 °C and vibration frequency of 70 rpm. Primary response to transportation and handling stress are related to increase in plasma cortisol and blood glucose levels with increment in energy metabolism (Davis 2006; Acerete et al. 2004). The propensity of these physiological responses is related to intensity and duration of stressor. Transportation caused three times cortisol rise at a value 146 ng/ml and 1.5 times increase in glucose levels. While severe handling stress released cortisol at values 121 ng/ml correlated to stressor, within initial 2 h (Acerete et al. 2004). Transportation can cause oxidative stress response in rainbow trout (*O. mykiss*) with increased production of stress markers such as malondialdehyde (MDA), catalase (CAT), superoxide dismutase (SOD), Hsp 70 and immune upregulation of IL-1 β, IL-8, TNF-α in gill; IL-1 β, IL-8 in muscle. These changes can set in within short span of transportation of about 2 h (Ren et al. 2022; Fang et al. 2023). Transportation can increase susceptibility of fish to pathogenic bacteria by dysregulating skin homeostasis (Tacchi et al. 2015). According to Tacchi et al. (2015) transportation stress can upregulate anti-inflammatory cytokines, down-regulate skin antimicrobial peptides, reshaping bacterial community dynamics of skin mucosa in *O. mykiss* increasing susceptibility to pathogenic *Vibrio anguillarum.* Milla et al. (2010) studied acute handling stress of 1 min dewatering on Eurasian perch (*Perca fluviatilis*). Their work demonstrated downregulation of splenic antimicrobial defences such as lowered spleen lysozyme and hepcidin gene activity in subduing innate pathogenic safeguards bringing forth hazardous effect of short term handling stress on fish physiology.

1.1.2 Biotic Stressors

Stocking Density

Stocking density is the rearing density determined as organisms managed per unit farming area or volume. In aquaculture too high or too low stocking densities are known to negatively impact growth, physiology, welfare/productivity (Saraiva et al. 2022) Overcrowding due to higher stocking density in RAS negatively affected growth of turbot (*Scophthalmus maximus*) with lower final weight, body weight,

SGR, FCR, coefficient of variation for weight, concentration of growth hormone (GH) post 120 day stocking at densities 14.27 kg m^{-2} (Liu et al. 2017). Although effect of stocking density on plasma GLU, Cl^-, protein level and fish survival were non-significant. In intensive RAS system high stocking density at 12.68 kg m^{-2} similarly downregulated growth of Chinese sturgeon (*Acipenser sinensis*), decreasing immune responsiveness, serum GH, insulin-like growth factor-I (IGF-I), SOD, GSH-PX with increase in stress responses of heat shock protein (Hsp) 70, MDA, serum lactate, glucose and cortisol levels (Long et al. 2019). The work substantiates oxidative damage, stress effects due to overly rearing density, with optimal suggested RAS farming densities for *A. sinensis* between 4.8 and 8.99 kg m^{-2}. Roy et al. (2021) demonstrated that long term low density stocking (3.76 kg m^{-3}, at a rate 9 fish per tank) can decrease survival in juvenile rainbow trout (*O. mykiss*) with mortality reaching up to 50%, upsurge in plasma cortisol owing to chronic stress, with non-significant acute stress responses for lactate and lysozyme. Alongside physiological stress, low dopamine and serotonin levels confirmed psychological stress owing to low density stocking effects on depressed social behaviour of tilapia. Low rearing density thus alike overcrowding has overall suppressive effects on immune response, physiology in intensifying disease susceptibility among culture organisms.

Pathogens

Aquaculture dynamics in itself promote evolution of highly virulent pathogens (Kennedy et al. 2016; Pulkkinen et al. 2010). Evolution of pathogens and development of virulence are closely associated to emergence of disease in aquaculture practices. Bacteria, viruses, fungi and parasites are primary cause of infections in fish and shellfish in aquaculture environment. Pathogenic bacteria are the main cause for illness and disease imposed mortality in aquaculture and capture fisheries (Senthamarai et al. 2023; Austin and Austin 2007). Overcrowding (high density stocking), handling and transport can induce stress situations leading to incidences of bacterial infections, one of the commonest cause of disease outbreaks in aquaculture (Rupp et al. 2019). Stress influenced disease incidences are demonstrated in their work on RAS culture of European perch (*Perca fluviatilis*). *Aeromonas* sp., *A. hydrophila, salmonicida* and *caviae* were most recurrent bacterial isolates from diseased perch causing deep skin ulcerations, muscle fiber necrosis and systemic infection. Gill epithelial hyperplasia and necrosis was caused by *Flavobacterium* sp. with secondary infection from *Saprolegnia* sp. High density stocking, improper feed ratio and inadequate exchange of water can make intensive culture systems such as RAS prone to attacks from opportunistic bacteria (Masser et al. 1999; Martins et al. 2009). Bacterial pathogens are prominent cause of vibriosis (*V. anguillarum, V. alginolyticus*) in many marine fishes; botulism (*Clostridium botulinum*) and furunculosis (*A. salmonicida*) in salmonids; gill rot, edwardsiellosis (*Edwardsiella ictaluri*) in tilapia; gill diseases (*Flavobacterium* sp.) in turbot; mycobacteriosis (*Mycobacterium* sp.); hemorrhagic septicaemia (*Aeromonas hydrophila*) in fresh water fishes; septicaemia (*Vagocococcus salmoninarum*) in Atlantic salmon, meningoencephalitis (*Streptococcus difficilis*) in carp, trout; streptococcosis (*Streptococcus iniae*) in rainbow trout (Irshath et al. 2023; Puri et al. 2022; Dong et al.

2019). *Vibrio* spp. *parahaemolyticus, fluvialis, alginolyticus* are common cause of stress related infections in commercial crabs (Hariharan and Amadi 2016). The proficiency of opportunistic virulent strains to attach and colonise aquaculture substrates explains their capacity to from biofilms (Yu et al. 2021). Cai and Arias (2017) demonstrated that common bacterial pathogens *A. hydrophila, Flavobacterium columnare, E. ictaluri* had the potential of colony formation on surfaces and farming facilities making biofilm as pathogenic reservoirs, in fish farms. Biofilm environment supports long term bacterial survival preventing predation, desiccation, helping surpass antibiotic treatments. Additionally, many of these fish bacterial pathogens are known to be passed through fish consumption and handlings to humans, likely relating to health hazards (Ghittino et al. 2003; Hernroth and Baden 2018).

1.2 Beneficial Bacteria in Aquaculture Development

Maintenance of healthy bacterial community and their ecological function is vital to confer sustainability to aquaculture, as is evident from the prelisted investigations examining effect of stressors on microbiota and concomitant aquaculture welfare. Prominence of bacterial microflora in aquatic environment, their interaction to host physiology, alongside vagaries of culture environment has led to evolution of opportunistic bacterial strains, pathogenic occurrences and disease outbreaks limiting aquaculture productivity (Olafsen 2001). Beneficial bacteria consist friendly microflora with potential to contribute health and welfare to systems they reside into (Zhou et al. 2009). These bacteria are known for their probiotic ('pro' meaning *'in favour of*; 'biosis' for '*vitality*') effects hence are also designated as 'probiotics'. Standard FAO/WHO (2002) definition on 'probiotics' states them as live microbials whose adequate ministration bestows health benefits to the host harbouring them. Considering aquaculture environment, Merrifield et al. (2010) proposed term probiotic as, viable, non-viable, or even part of the bacteria that when added to the water or feed can render benefits to the host or to its environment. Probiotics in aquaculture comprise Gram-positive, Gram-negative bacteria (including genus *Bacillus*, lactic acid bacteria LABs of genera *Lactobacillus* and *Bifidobacterium*); yeasts (*Rhodotorula, Saccharomyces, Candida, Pichia*); unicellular algae (*Phaeodactylum*) (Ghosh et al. 2023; Carballo et al. 2019; Ringø et al. 2018). Additionally recent applications focus on probiotic use of fungus (*Aspergillus, Geotrichium* spp.) in improving growth and immunological indices of rohu (*Labeo rohita*) and tilapia (*O. niloticus*) (Amir et al. 2019; Dawood et al. 2020)

Use of beneficial microbials 'probiotics' in aquafarming management can curtail increasing emergence of antibiotic-resistance and pathogenic bacteria in aquaculture. Probiotic bacteria utilise various mechanisms against virulent opportunistic strains based on direct interference and indirect controls (de Bruijn et al. 2017). Direct effects involve probiont tactics such as competitive/niche exclusion preventing pathogenic establishment; colonisation impedance; and antipathy using antimicrobial peptides, quorum quenching (exhausting pathogenic quorum sensing), antipathetic compounds bacteriocins, siderophores (iron chelators), lysozymes, proteolytic enzymes, hydrogen peroxides. Indirect effects include processes to strengthen host physiology such as production of short-chain fatty acids (SCFAs) such as acetate, butyrate, propionate with immunoprophylactic and health promoting effects. Figure 1.1 details probiotic routes, pathogenic combats and strategies for accomplishing fish welfare and aquaculture health. Besides, their use in biofloc technology (BFT); integrated multi-trophic aquaculture (IMTA); recirculating aquaculture systems (RAS); beneficial bacteria contribute towards enhancing aquaculture sustainability; alleviating need for antibiotics or chemical treatment; waste management and remediation. Nearly 90% of bacterial microbiota of marine and freshwater fish comprise the phyla Proteobacteria, Fusobacteria, and Firmicutes, and to a smaller extent Bacteroidetes, Actinobacteria, and Verrucomicrobia (Hennersdorf et al. 2016; Estruch et al. 2015; Tarnecki et al. 2017). Bacterial genera *Bacillus, Lactobacillus,* and *Bifidobacteria* are the most recognized beneficial bacteria in fed aquaculture management.

Application of beneficial bacteria towards sustainable aquaculture development encompasses their use in augmenting feed characteristics, quality; host health, aquaculture productivity; water quality parameters; mitigating effects of abiotic, biotic stressors; support sustainable technologies such as BFT, IMTA-RAS systems as well as help waste water (WW) treatment to support circular bioeconomy (Fig. 1.2). Recent applications utilising potential of beneficial bacteria in aquaculture are further elaborated in Table 1.3. Nutrition and feeding are key themes in sustainable aquaculture discourse, besides, recycling and reutilisation of waste by-products are pillars for circular bioeconomy (Puri 2024, 2025). Additionally, sustainable aquafarming endeavours facilitate animal welfare, health and wellbeing of natural and culture environment.

1.2.1 Beneficial Bacteria in Feed Supplementation

Intensive aquaculture systems largely depend on quality feed for fish welfare and sustainable produce. Feed is the primary contributor to achieve economic, social and environmental viability of farming procedures owing to its cost and quality impact. Probiotic LAB, *Lactobacillus brevis* has been investigated for nutrition value improvement of Atlantic salmon (*Salmo salar*) diet (Refstie et al. 2005), refer Table 1.3. *L. brevis* fermented soybean meal eliminated ANFs and dietary allergens improving digestibility and fish growth. *Pediococcus acidilactici* as commercial probiotic Bactocell in Atlantic salmon (*Salmo salar*) diet could diminish inflammatory effects of ANF in plant-based soybean diet (Nordvi et al. 2023). Dietary supplementation of Bactocell at 0.03% mediated immune gene suppression and

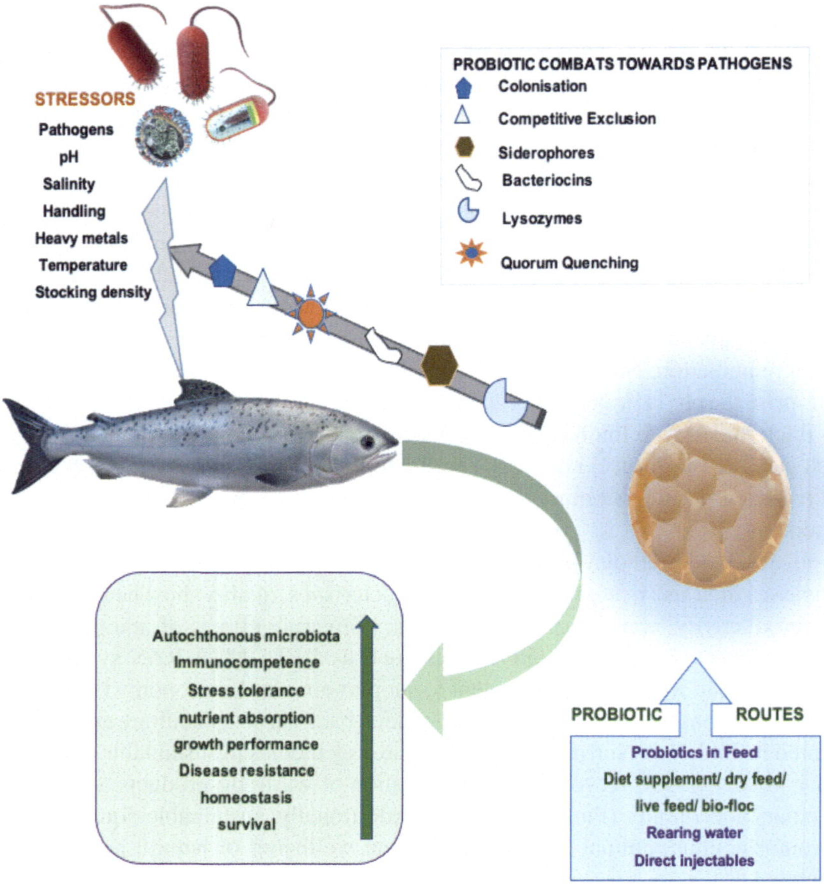

Fig. 1.1 Probiotic combats, routes and strategies for host welfare and health in aquaculture (*Source* Author)

transmembrane proteins enhancement of type IV collagen alleviating host gut homeostasis. Utiswannakul et al. (2011) demonstrated probiotic potential of host associated probiotic (HAP) *Bacillus subtilis* BP11 in tiger shrimp *Penaeus monodon.* BP11 supplemented diet improved growth, survival and immunoprophylaxis against *Vibrio* incidences. Additionally viability of BP11 during ambient and cold temperature storage improved shelf-life of stored feed rations. Mixed bacterial strains in pacu (*Piaractus mesopotamicus*) diet mitigated effects of aflatoxin (AFB1) exerting positive effects on liver health (Godoy et al. 2024). *Lactiplantibacillus pentosus* restored intestinal integrity by improving intestinal microbiota, of AFB_1 inflicted *P. vannamei,* enhancing immune-responsiveness, survival and growth (Khalafalla et al. 2022).

Fig. 1.2 Use of beneficial bacteria for sustainable aquaculture development (*Source* Author)

1.2.2 Beneficial Bacteria in Heavy Metal Decontamination

Microbial use is a choicest method to detoxify and alleviate heavy metal (HM) accumulation in cultured fish species owing to their safety, bio-efficacy, ease of use, without any residual effects (Giri et al. 2024; Arun et al. 2021). Microbe-based HM removal is dependent on metabolic and non-metabolic mechanisms (Chandrashekharaiah et al. 2022). Enzymatic reduction through probionts oxidise HMs into less toxic forms, additionally bacterial strains have capacity to accumulate HM on their cell wall, macromolecules by biosorption, as well form metal complexes, causing clearance from surrounding environments. Chandrashekharaiah et al. (2022) investigated bacteria-algal synergism in bioremediation of lead (Pb^{2+}), cadmium (Cd^{2+}) intoxicated WW. Consortium consisting bacterial probiont *Bacillus* spp., *Micrococcus* sp. and algal counterpart *Scenedesmus acutus*, *Chlorella pyrenoidosa*, were eco-friendly, sustainable and green alternatives for alleviation and reutilisation of WW. Probionts were effective against high respective concentrations (200

Table 1.3 Recent applications utilising potential of beneficial bacteria in aquaculture

Beneficial bacteria (concentration)	Host species, aquaculture application	Beneficial effects	Probiotic approach/ Probiotic route	References
Pediococcus acidilactici (5.23×10^5 CFU g^{-1})	Atlantic salmon (*Salmo salar*)	Gut health↑, immunoprophylaxis↑	Probiotic (commercial-Bactocell)/ diet addition	Nordvi et al. (2023)
Bacillus subtilis BP11 (10^9 CFU L^{-1})	Black tiger shrimp larvae (*Penaeus monodon*)	FCR↑, growth↑, survival↑, immunity to *V. harveyi* 639	Probiotic/culture water addition	Utiswannakul et al. (2011)
B. subtilis E20 (10^9 CFU L^{-1})	White shrimp larvae, (*Litopenaeus vannamei*)	Survival↑, immunity↑, growth↑	Probiotic/culture water	Liu et al. (2010)
Limosilactobacillus fermentum GMNL-93, *Lactiplantibacillus plantarum* GMNL-141, *Lacticaseibacillus casei* GMNL-277 *Bifidobacterium lactis* GMNL-550 (each 5×10^6 CFU g^{-1})	Juvenile Asian seabass (*Lates calcarifer*)	Growth↑, intestinal health↑, muscle amino acid composition↑, disease resistance↑	Probiotic mixture (commercial—GM-Lac)/diet addition	Wang et al. (2023a)
B. subtilis, Bifidobacterium bifidum, Enterococcus faecium, Lactobacillus acidophilus, L. casei, L. lactis, L Plantarum, P. acidilactici (each at 5.0×10^{12} CFU kg^{-1})	Pacu (*Piaractus mesopotamicus*)	AFB1 mitigation	Probiotic mixture (commercial-ANTIMICOX MA)/ diet addition	Godoy et al. (2024)
B. cereus, B. amyloliquefaciens, Pseudomonas stutzeri	Tilapia (*O. mossambicus*), waste-water treatment	Survival↑, toxic NH_3↓, nitrite↓, nitrate↓	Probiotic mixture/culture water addition	John et al. (2020)
Sphingomonas sp. LPN080	White shrimp (*L. vannamei*), waste-water treatment	Toxic NH_3↓, nitrite↓, *Vibrio* sp.↓	Probiotic/culture water addition	Yun et al. (2019)

(continued)

Table 1.3 (continued)

Beneficial bacteria (concentration)	Host species, aquaculture application	Beneficial effects	Probiotic approach/ Probiotic route	References
L. acidophilus ATCC 4356 (10^6 CFU mL^{-1} kg^{-1} diet)	Thinlip grey mullet (*Liza ramada*)	FBW↑, WG↑, SGR↑	Probiotic/diet addition	Khalafalla et al. (2022)
Bacillus sp., *Micrococcus* sp., *Scenedesmus acutus, Chlorella pyrenoidosa*	HM decontamination	Detoxification of Pb^{+2}, Cd^{+2}, remediation of waste-water and reutilisation	Probiotic/water addition	Chandrashekharaiah et al. (2022)
L. fermentum 1744 (ATCC 14,931) (10^7 CFU g^{-1}) + lactulose (10 g Kg^{-1} of diet)	Rainbow trout (*O. mykiss*), HM decontamination	FCR↑, HM removal↑	Synbiotic (encapsulated)/diet addition	Madreseh et al. (2019)
L. plantarum CCFM8661 (10^8 CFU g^{-1})	Nile tilapia (*O. niloticus*), HM decontamination	Survival↑, innate immunity↑, Pb toxicity and tissue accumulation↓, digestive enzyme↑, δ-aminolevulinic acid dehydratase activity↑	Probiotic/diet addition	Zhai et al. (2017)
Nitrosomonas (1.62 kg ha^{-1}) + *Nitrobacter* (0.82 kg ha^{-1})	*Pangasius sutchi; Catla catla; L. rohita,* water quality treatment	*Pseudomonas*↓, *Nitrosomonas*↑, *Nitrobacter*↑, THB↑, NH_3↓, phosphate↓, nitrite↓	Probiotic (commercial)/pond water	Padmavathi et al. (2012)
Rhodovulum sulfidophilum (10^5 CFU mL^{-1})	Milkfish (*Chanos chanos*); Asian tiger shrimp (*P. monodon*); Asian hard clam (*Meretrix lusoria*); seaweed (*Gracilaria* sp.), IMTA-RAS	Pathogenic *Vibrio*↓, *Escherichia*↓, cyanobacteria↓, algae↓, NH_3↓, P stable, beneficial nitrifiers↑, acidobacteria↑	Probiotic/pond water	Chang et al. (2019)

(continued)

Table 1.3 (continued)

Beneficial bacteria (concentration)	Host species, aquaculture application	Beneficial effects	Probiotic approach/ Probiotic route	References
Phaeobacter sp. 4UAC3 (10^7 CFU mL^{-1}) enriched *Ulva ohnoi*	Turbot (*Scophthalmus maximus*) larvae, IMTA-RAS	Mortality↓	Probiotic/culture water addition	Pintado et al. (2023)
B. subtilis, L. plantarum, L. rhamnosus, L. acidophilus, L. delbrueckii	Nile tilapia (*O. niloticus*), Biofloc	Immune parameter (SOD↑, CAT↑, AKP↑, MYP↑, RB↑), mortality↓, water quality↑	Probiotic (single and multi-species)/culture water addition	Mohammadi et al. (2021)
B. velezensis strain AP193 (1×10^7 CFU g^{-1} of diet), *B. subtilis* (3.6×10^4 CFU g^{-1}, 7.2×10^4 CFU g^{-1})	Nile tilapia (*O. niloticus*), Biofloc	FCR↑	Probiotic/diet addition	Oliveira et al. (2025)

Abbreviation—AFB: aflatoxin B; MYP: Myeloperoxidase; RB: Respiratory burst; THB: total heterotrophic bacteria

+ 1.5 ppm) of Pb + Cd. Adjunctly, treated water reported better growth, biomass, chlorophyll content of remediated plants. Paulina et al. (2014) in their study recognised 11.6% copper removal potential of Gram negative bacterium *Pantoea agglomerans* from contaminated marine sediments utilising atomic absorption spectrometric analysis. Functional feed supplementation of *L. plantarum* improved growth, survival of Pb exposed tilapia. Moreover, probiotic *L. plantarum* could efficiently subside Pb build-up in fish tissues (including gill, muscle, brain, kidney, liver and gonads), lowered HM-stimulated oxidative distress, empowering host innate immune mechanisms (Zhai et al. 2017). Madreseh et al. (2019) demonstrated synergising effect of probiotic *L. fermentum* with prebiotic (non-digestible probiotic substrate) lactulose fed three times daily for 56 days to trout. Encapsulation of probiotic in synbiotic combination (= 'probiotic' + 'prebiotic as probiotic substrate') precluded toxic impacts of heavy and trace metals (Pb, Cd, Zn, Ni) in trout tissue, augmenting overall growth. Figure 1.3 discusses common synbiotic approach in aquaculture (for detail refer Puri et al. 2022).

1.2.3 Beneficial Bacteria in Waste-Water Treatment and Reutilisation

Bacterial consortium can be utilised for bioremediation of culture waste-water (WW) reducing harmful ammonia, nitrite concentrations (John et al. 2020; Nisha et al. 2025) helping foster WW—reutilisation and circular bioeconomy aiming to reduce; reuse; restore, using biological systems (Das et al. 2023). Bioaugmentation potential of bacterial cumulations can ablate pollutants and harmful toxins restoring sustainability of culture ecosystem and increase fish survival. Recently, anaerobic ammonium oxidising 'anammox' bacterial isolates *Candidatus* (Brocadia, Kuenenia, Anammoxoglobus) capable of on-site remediation of polluted culture sediments have been recognized from aquaculture sites (Shen et al. 2016). 'Anammox' can concomitantly remove both ammonia and nitrite, lowering sludge formation. 'Anammox', as compared to denitrifying bacteria, are chemolithoautotrophs with slow growth potential. They are important source of nitrite-N and ammoniacal-N removal from freshwater aquaculture. According to Yun et al. (2019) 'anammox' such as *Sphingomonas* sp. endow safe, efficient, alternative to support environmentally sustainable aquafarming models.

1.2.4 Beneficial Bacteria in Water Quality Improvement: IMTA-RAS, BFT

Prowess of beneficial bacteria has been extended for quality improvement of culture water. Padmavathi et al. (2012) studied probiotic abilities of autotrophic nitrifying

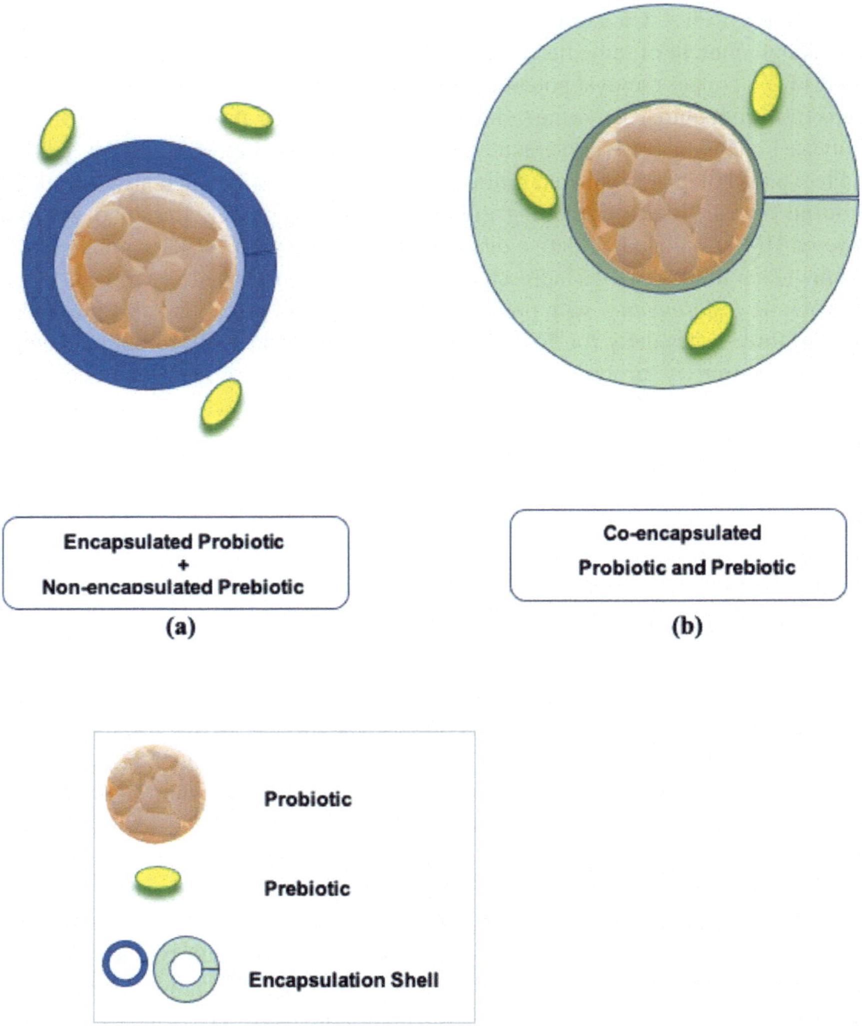

Fig. 1.3 Encapsulation strategy of probiotics in synbiotic combination for aquaculture (*Source* Author)

bacteria *Nitrosomonas* and *Nitrobacter* sp. Their addition to culture ponds improved probiotic nitrifying and heterotrophic bacteria counts, decreasing pathogenic *Pseudomonas*, with low ammonia, nitrite, phosphorus loads, improving water quality. Innovative culture systems IMTA, RAS, BFT ameliorate water quality for sustainable aquaculture (Thompson et al. 2024; Pintado et al. 2023). BFT due to stable community structure prevents pathogenic growth of bacteria in culture set-up. These techniques have become benchmark of resilient, profitable, environmentally sustainable alternatives for long-standing aquaculture expanse (Granada et al. 2015). RAS can optimise water quality parameters (for pH, salinity and temperature) as per needs of culture species. Efficient RAS design can utilize lesser water (90–99% low) compared to other aquaculture approach, benefitting farming sustainability (Martins et al. 2010). Utilising photosynthetic *Rhodovulum sulfidophilum* in innovative multitrophic RAS (MRAS) for milkfish augmented beneficial bacterial species of ammonium-oxidizers, nitrite-oxidizers, denitrifiers and Acidobacteria in water as well as sediments. IMTA and BFT offer possibility of feed, faecal waste reutilisation from one species to serve fertilization and nutrition of other cultured organisms (Oliveira et al. 2025). Mohammadi et al. (2021) studied role of single species (SS) and multi-species (MLS) probiotics in BFT. SS, *L. plantarum* treated tilapia yielded better growth, weight gain, digestive enzyme activity, while MLS (*B. subtilis*; *L. plantarum, rhamnosus, acidophilus, delbrueckii*) showed better impact on immune parameters and antioxidant system. It was concluded that probiotic strain in mixture had superior effects for tilapia farming in a BF system. Oliveira et al. (2025) used probiotics coated diet for Nile tilapia, unlike Mohammadi et al. (2021), their work found no effect of feed coated forms on fish growth although *B. subtilis* at high and low concentration enhanced survival, water quality parameters, bacterial community of water and fish excreta.

1.2.5 Beneficial Bacteria Against Pathogens and Antibiotic Treatment

In a 90 day feed trial El-Rhman et al. (2009), investigated in vitro and in vivo activity of host-derived probiotic supplemented feed for Nile tilapia. Both *Micrococcus luteus* and *Pseudomonas* sp. produced in vitro beneficial effect against *A. hydrophila* challenge. Only *M. luteus* produced in vivo effect in improving feed utilisation, fish growth, and survival. Putative probiotics are tested for their in vitro antimicrobial potential by production of inhibition zones against pathogenic strains. Midhun et al. (2018) investigated antipathogenic potential of cellular preparations of *B. coagulans* MTCC 9872. Bacterial cell-free extract filtrate showed high probiotic impact compared to separations from whole cell (WC) or heat killed WC in production of inhibition zones. According to Antony et al. (2011) probiotic activity is based on production of antimicrobial peptides (AMPs). In their study on white spot virus inflicted tiger shrimp, *Penaeus monodon*, it was found that use of probiotic *Bacillus* MCCB101, *Micrococcus* MCCB104 singly and in combination upregulated the AMP gene expression. While best response was obtained with *Bacillus* strain towards AMP expression, increasing the fish survival. Above works explicate that bacterial probionts have immunoregulatory effects on culture organisms besides enhancing growth, survival and health. Probiotics have been also been investigated for their beneficial effects against pathogenic incidence of *Aeromonas* spp. (Fečkaninová et al. 2017) in salmonids; *A. hydrophila, Pseudomonas fluorescens, S. iniae* in*O. niloticus* (Aly et al. 2008) administering both prophylactic and biotherapeutic effects.

1.2.6 *Effective Microorganisms in Aquaculture*

Peculiar group of naturally existing beneficial microbials utilised as co-culture of probiotic organisms is called 'Effective Microorganisms' (EM); a concept proposed by Higa and Parr (1994) and Balogun et al. (2016), with remarkable ability to remediate, restore, and conserve culture systems. EM consist microbial composites of LABs, photosynthetic bacteria, yeast and actinomyces. Though the initial concept of EMs was proposed to develop sustainable soil and plant agriculture systems; applications of EM use in aquaculture are growing. EM in aquaculture can bioremediate WW by HM removal and preventing eutrophication (Stephane et al. 2016; Kumar and Singh 2022).

EM helps ameliorate growth, reproductive rate, fish and water quality, controlling disease outbreaks in aquaculture hence are suitable probiotic alternative to chemical antibiotic treatment (Abdel-Aziz et al. 2020, 2021; Gao et al. 2020). EM in aquaculture improvement has been investigated fervently in recent applications on tilapia. Their inclusion in GIFT tilapia diet (Xu et al. 2021) promoted growth, immunity, and food intake capacity improving aquaculture productivity.

1.3 Conclusion

Sustainable aquaculture growth and welfare rely on judicious development of its economic, social, and environmental paradigms. Biosecurity; disease mitigation, control; aquaculture health management, world-over, can help attain long withstanding sustainable aquaculture goals. In this view utilizing potential of helpful bacterial microflora ('autochthonous bacteria') inherent to aquatic species and culture systems ('allochthonous') can outnumber virulence episodes and disease incidence inflicted externalities on culture organisms. Besides alleviating pathogenic episodes, probiotic bacteria can ablate toxic effects of heavy metals, chemical treatments, waste pollutants (such as ammonia, nitrite, N, P) improving water quality. Beneficial microbiota can augment host health by improving gut homeostasis, histomorphology, zootechnical parameters contributing to growth, immunity, survival and productivity. By enhancing feed parameters such as organoleptic properties, feed acceptance, removal of ANF, shelf-life of storage feed, probiotics can improve palatability and utilisation of diet by host. At a significant level probiotics can mitigate effects of stressors (abiotic and biotic) inherent to culture environment. Harnessing potential of bacterial beneficence can assure long term food security, alleviating poverty, with related socio-economic benefits for population world-over, thereby escalating sustainable aquaculture development and welfare.

1.4 Future Prospects

Bacterial beneficence in aquaculture extends besides use as single or multi-strain probiotics. Synbiotic combination and encapsulated forms of probiotics are known to prolong bacterial sojourn in host thus enhancing beneficial effects on growth, immune-responsiveness, overall health and survival of culture species. Future developments focussing on identification of putative probiotic strains, specifically host associated probiotics (HAPs), probiotic products, and metabolites can mitigate disease outbreaks expanding bacterial benefits to culture environment. Applications based on utilisation of EM's, probiotics are growing, dealing with addressal of issues of waste water treatment, reutilisation; bio-mitigation; feed quality; fish welfare and consequent improvement in overall productivity.

References

Abbink W, Garcia AB, Roques JAC, Partridge GJ, Kloet K, Schneider O (2012) The effect of temperature and pH on the growth and physiological response of juvenile yellowtail kingfish Seriola lalandi in recirculating aquaculture systems. Aquaculture 330–333:130–135. https://doi.org/10.1016/j.aquaculture.2011.11.043

Abdel-Aziz M, Bessat M, Fadel A, Elblehi S (2020) Responses of dietary supplementation of probiotic effective microorganisms (EMs) in *Oreochromis niloticus* on growth, hematological, intestinal histopathological, and antiparasitic activities. Aquacult Int 28:947–963. https://doi.org/10.1007/s10499-019-00505-z

Abdel-Aziz MF, Abdel-Tawwab YA, Sadek MF, Yones AMM (2021) Evaluation of use effective microorganisms (EM) with different feeding strategies on growth performance, body chemical composition and economic efficiency of monosex Nile tilapia *Oreochromis niloticus* juveniles. Aquat Living Resour 34:21. https://doi.org/10.1051/alr/2021021

Abdelhamed H, Ozdemir O, Waldbieser G, Perkins A, Lawrence M, Karsi A (2019) Effects of florfenicol feeding on diversity and composition of the intestinal microbiota of channel catfish (*Ictalurus punctatus*). Aquacult Res 50:3663–3672. https://doi.org/10.1111/are.14325

Abdel-Tawwab M (2016) Effect of feed availability on susceptibility of Nile tilapia, *Oreochromis niloticus* (L.) to environmental zinc toxicity: growth performance, biochemical response, and zinc bioaccumulation. Aquaculture 464:309–315

Abdel-Tawwab M, Wafeek M (2017) Fluctuations in water temperature affected waterborne cadmium toxicity: hematology, anaerobic glucose pathway, and oxidative stress status of Nile tilapia, *Oreochromis niloticus* (L.). Aquaculture 477:106–111. https://doi.org/10.1016/j.aquaculture.2017.05.007

Acerete L, Balasch JC, Espinosa E, Josa A, Tort L (2004) Physiological responses in Eurasian perch (*Perca fluviatilis*, L.) subjected to stress by transport and handling. Aquaculture 237:167–178

Ajani EK, Akinwole AO, Ayodele IA (2011) Fundamentals of fish farming in Nigeria. Walecrown Ventures, Ibadan, Nigeria, pp 89–105

Alam MR, Sharmin S, Islam SMM, Alam MA, Ehiguese FO, Pattadar SN, Shahjahan M (2020) Salinity intrusion affects early development of freshwater aquaculture species pabda, *Ompok pabda*. Aquacult Rep 18:100476. https://doi.org/10.1016/j.aqrep.2020.100476

Aly SM, Ahmed YAG, Ghareeb AAA, Mohamed MF (2008) Studies on *Bacillus subtilis* and *Lactobacillus acidophilus*, as potential probiotics, on the immune response and resistance of Tilapia nilotica (*Oreochromis niloticus*) to challenge infections. Fish Shellfish Immunol 25(1–2):128–136. https://doi.org/10.1016/j.fsi.2008.03.013

Amir I, Zuberi A, Kamran M, Imran M, Murtaza MH (2019) Evaluation of commercial application of dietary encapsulated probiotic (Geotrichum candidum QAUGC01): effect on growth and immunological indices of rohu (Labeo rohita, Hamilton 1822) in semi-intensive culture system. Fish Shellfish Immunol 95:464–472

Antony SP, Singh ISB, Jose RM, Kumar PRA, Philip R (2011) Antimicrobial peptide gene expression in tiger shrimp, *Penaeus monodon* in response to gram-positive bacterial probionts and white spot virus challenge. Aquaculture 316(1–4):6–12. https://doi.org/10.1016/j.aquaculture.2011.03.025

Arun KB, Madhavan A, Sindhu R, Emmanual S, Binod P, Pugazhendhi A, Sirohi R, Reshmy R, Awasthi MK, Gnansounou E, Pandey A (2021) Probiotics and gut microbiome—prospects and challenges in remediating heavy metal toxicity. J Hazard Mater 420:126676. https://doi.org/10.1016/j.jhazmat.2021.126676

Attramadal KJK, Truong TMH, Bakke I, Skjermo J, Olsen J, Vadstein O (2014) RAS and microbial maturation as tools for K-selection of microbial communities improve survival in cod larvae. Aquaculture 432:483–490. https://doi.org/10.1016/j.aquaculture.2014.05.052

Austin B, Austin DA (2007) Bacterial fish pathogens: disease in farmed and wild fish, 4th edn. Springer, Praxis Publishing, Chichester, UK

Bakke I, Åm AL, Kolarevic J, Ytrestøyl T, Vadstein O, Attramadal KJK, Terjesen BF (2017) Microbial community dynamics in semi-commercial RAS for production of Atlantic salmon post-smolts at different salinities. Aquacult Eng 78(Part A):42–49. https://doi.org/10.1016/j.aquaeng.2016.10.002

Balogun RB, Ogbu JU, Umeokechukwu EC, Kalejaiye-Matti RBC (2016) Effective microorganisms (EM) as sustainable components in organic farming: principles, applications and validity. In: Nandwani D (ed) Organic farming for sustainable agriculture. Sustainable development and biodiversity, vol 9. Springer, Cham. https://doi.org/10.1007/978-3-319-26803-3_12

Bauer J, Teitge F, Neffe L, Adamek M, Jung A, Peppler C, Steinhagen D, Jung-Schroers V (2021) Impact of a reduced water salinity on the composition of *Vibrio* spp. in recirculating aquaculture systems for Pacific white shrimp (*Litopenaeus vannamei*) and its possible risks for shrimp health and food safety. J Fish Dis 44(1):89–105. https://doi.org/10.1111/jfd.13270

Besson M, Vandeputte M, van Arendonk JAM, Aubin J, de Boer IJM, Quillet E, Komen H (2016) Influence of water temperature on the economic value of growth rate in fish farming: the case of sea bass (*Dicentrarchus labrax*) cage farming in the Mediterranean. Aquaculture 462:47–55. https://doi.org/10.1016/j.aquaculture.2016.04.030

Boleza KA, Burnett LE, Burnett KG (2001) Hypercapnic hypoxia compromises bactericidal activity of fish anterior kidney cells against opportunistic environmental pathogens. Fish Shellfish Immunol 11(7):593–610. https://doi.org/10.1006/fsim.2001.0339

Buentello JA, Gatlin DM, Neill WH (2000) Effects of water temperature and dissolved oxygen on daily feed consumption, feed utilization and growth of channel catfish (*Ictalurus punctatus*). Aquaculture 182(3–4):339–352. https://doi.org/10.1016/S0044-8486(99)00274-4

Cai W, Arias CR (2017) Biofilm formation on aquaculture substrates by selected bacterial fish pathogens. J Aquat Anim Health 29:95–104. https://doi.org/10.1080/08997659.2017.1290711

Carballo C, Pinto PIS, Mateus AP, Berbel C, Guerreiro CC, Martinez-Blanch JF, Codoñer FM, Mantecon L, Power DM, Manchado M (2019) Yeast β-glucans and microalgal extracts modulate the immune response and gut microbiome in Senegalese sole (*Solea senegalensis*). Fish Shellfish Immunol 92:31–39

Cascarano MC, Stavrakidis-Zachou O, Mladineo I, Thompson KD, Papandroulakis N, Katharios P (2021) Mediterranean aquaculture in a changing climate: temperature effects on pathogens and diseases of three farmed fish species. Pathogens 10:1205. https://doi.org/10.3390/pathogens10091205

Chandrashekharaiah PS, Gupte Y, Sarkar P, Prasad S, Sanyal D, Dasgupta S, Banik A (2022) Algae-bacterial aquaculture can enhance heavy metals (Pb^{2+} and Cd^{2+}) remediation and water re-use efficiency of synthetic streams. Resour Conserv Recycl 180:106211. https://doi.org/10.1016/j.resconrec.2022.106211

Chang B-V, Liao C-S, Chang Y-T, Chao W-L, Yeh S-L, Kuo D-L, Yang C-W (2019) Investigation of a farm-scale multitrophic recirculating aquaculture system with the addition of *Rhodovulum sulfidophilum* for milkfish (*Chanos chanos*) coastal aquaculture. Sustainability 11(7):1880. https://doi.org/10.3390/su11071880

Chen SM, Chen JC (2003) Effects of pH on survival, growth, molting and feeding of giant freshwater prawn *Macrobrachium rosenbergii*. Aquaculture 218(1–4):613–623. https://doi.org/10.1016/S0044-8486(02)00265-X

Chen J, Huang L, Wang Q, Zeng H, Xu J, Chen Z (2022) Antibiotics in aquaculture ponds from Guilin, South of China: occurrence, distribution, and health risk assessment. Environ Res 204(Part B):112084. https://doi.org/10.1016/j.envres.2021.112084

Chen J, Liu X, Chen J, Jin H, Wang T, Zhu W, Li L (2024) Underestimated nutrient from aquaculture ponds to Lake Eutrophication: a case study on Taihu Lake Basin. J Hydrol 630:130749. https://doi.org/10.1016/j.jhydrol.2024.130749

Claireaux G, Chabot D (2016) Responses by fishes to environmental hypoxia: integration through Fry's concept of aerobic metabolic scope. J Fish Biol 88(1):232–251

Cotton CF, Walker RL, Recicar TC (2003) Effects of temperature and salinity on growth of juvenile black sea bass, with implications for aquaculture. N Am J Aquacult 65:330–338

Das SK, Mondal B, Sarkar UK, Das BK, Borah S (2023) Understanding and approaches towards circular bio-economy of wastewater reuse in fisheries and aquaculture in India: an overview. Rev Aquacult 15(3):1100–1114. https://doi.org/10.1111/raq.12758

Dauda AK, Ajadi A, Tola-Fabunmi AS, Akinwole AO (2019) Waste production in aquaculture: sources, components and managements in different culture systems. Aquacult Fisher 4(3):81–88. https://doi.org/10.1016/j.aaf.2018.10.002

Davidson J, Good C, Welsh C, Brazil B, Summerfelt S (2009) Heavy metal and waste metabolite accumulation and their potential effect on rainbow trout performance in a replicated water reuse system operated at low or high system flushing rates. Aquacult Eng 41(2):136–145. https://doi.org/10.1016/j.aquaeng.2009.04.001

Davis KB (2006) Management of physiological stress in finfish aquaculture. N Am J Aquacult 68(2):116–121. https://doi.org/10.1577/A05-007.1

Dawood MA, Eweedah NM, Moustafa EM, Farahat EM (2020) Probiotic effects of *Aspergillus oryzae* on the oxidative status, heat shock protein, and immune related gene expression of Nile tilapia (Oreochromis niloticus) under hypoxia challenge. Aquaculture 520:734669

de Bruijn I, Li Y, Wiegertjes GF and Raaijmakers JM (2017) Exploring fish microbial communities to mitigate emerging diseases in aquaculture. FEMS Microbiol Ecol 94:fix161

De Schryver P, Vadstein O (2014) Ecological theory as a foundation to control pathogenic invasion in aquaculture. ISME J 8(12):2360–2368. https://doi.org/10.1038/ismej.2014.84

Deviller G, Palluel O, Aliaume C, Asanthi H, Sanchez W, Franco Nava MA, Blancheton J-P, Casellas C (2005) Impact assessment of various rearing systems on fish health using multibiomarker response and metal accumulation. Ecotoxicol Environ Saf 61(1):89–97. https://doi.org/10.1016/j.ecoenv.2004.07.011

Dong HT, Senapin S, Jeamkunakorn C, Nguyen VV, Nguyen NT, Rodkhum C, Khunrae P, Rattanarojpong T (2019) Natural occurrence of edwardsiellosis caused by edwardsiella ictaluri in farmed hybrid red tilapia (Oreochromis Sp.) in Southeast Asia. Aquaculture 499:17–23. https://doi.org/10.1016/j.aquaculture.2018.09.007

El-Rhman AMA, Khattab YAE, Shalaby AME (2009) *Micrococcus luteus* and *Pseudomonas* species as probiotics for promoting the growth performance and health of Nile tilapia, Oreochromis Niloticus. Fish Shellfish Immunol 27(2):175–180. https://doi.org/10.1016/j.fsi.2009.03.020

Estruch G, Collado M, Peñaranda D, Vidal AT, Cerdá MJ, Martínez GP, Martinez-Llorens S (2015) Impact of fishmeal replacement in diets for gilthead sea bream (Sparus aurata) on the gastrointestinal microbiota determined by pyrosequencing the 16S rRNA gene. PLoS One 10(8):e0136389. https://doi.org/10.1371/journal.pone.0136389

Fang D, Mei J, Xie J, Qiu W (2023) The effects of transport stress (temperature and vibration) on blood biochemical parameters, oxidative stress, and gill histomorphology of pearl gentian groupers. Fishes 8:218. https://doi.org/10.3390/fishes8040218

FAO (2024) The state of world fisheries and aquaculture 2024. Blue Transformation in Action, Rome. https://doi.org/10.4060/cd0683en

FAO/WHO (2002) Report of a joint FAO/WHO expert consultation on guidelines for the evaluation of probiotics in food. World Health Organization and Food and Agriculture Organization of the United Nations, London, Ontario, Canada

Farmaki EG, Thomaidis NS, Pasias IN, Baulard C, Papaharisis L, Efstathiou CE (2014) Environmental impact of intensive aquaculture: investigation on the accumulation of metals and nutrients in marine sediments of Greece. Sci Total Environ 485–486:554–562. https://doi.org/10.1016/j.scitotenv.2014.03.125

Fečkaninová A, Koščová J, Mudroňová D, Popelka P, Toropilová J (2017) The use of probiotic bacteria against Aeromonas infections in salmonid aquaculture. Aquaculture 469:1–8. https://doi.org/10.1016/j.aquaculture.2016.11.042

Folke C, Kautsky N (1992) Aquaculture with its environment: prospects for sustainability. Ocean Coast Manag 17(1):5–24. https://doi.org/10.1016/0964-5691(92)90059-T

Francis G, Makkar HPS, Becker K (2001) Antinutritional factors present in plant-derived alternate fish feed ingredients and their effects in fish. Aquaculture 199(3–4):197–227. https://doi.org/10.1016/S0044-8486(01)00526-9

Gao J, Tai X, Shao N, Sun Y, Nie Z, Wang Y, Li Q, Xu P, Xu G (2020) Effects of effective microorganisms on the growth performance, nutritional composition and flavour quality of the pond-cultured *Eriocheir sinensis*. Aquacult Res 52(3). https://doi.org/10.1111/are.14941

Ghittino C, Latini M, Agnetti F, Panzieri C, Lauro L, Ciappelloni R, Petracca G (2003) Emerging pathologies in aquaculture: effects on production and food safety. Vet Res Commun 27(Suppl. 1):471–479

Ghosh K, Harikrishnan R, Mukhopadhyay A, Ringø E (2023) Fungi and actinobacteria: alternative probiotics for sustainable aquaculture. Fishes 8(12):575. https://doi.org/10.3390/fishes8120575

Giri SS, Kim HJ, Jung WJ, Lee SB, Joo SJ, Gupta SK, Park SC (2024) Probiotics in addressing heavy metal toxicities in fish farming: current progress and perspective. Ecotoxicol Environ Saf 282:116755. https://doi.org/10.1016/j.ecoenv.2024.116755

Glencross B (2009) Ingredient evaluation in aquaculture: digestibility, utilisation and other key nutritional parameters. In: Burnell G, Allan G (eds) Woodhead Publishing series in food science, technology and nutrition, new technologies in aquaculture. Woodhead Publishing, pp 387–416. https://doi.org/10.1533/9781845696474.3.387

Glencross BD, Booth M, Allan GL (2007) A feed is only as good as its ingredients—a review of ingredient evaluation strategies for aquaculture feeds. Aquacult Nutr 13(1):1353–5773. https://doi.org/10.1111/j.1365-2095.2007.00450.x

Godoy AC, Ziemniczak HM, Fantini-Hoag L, da Silva WV, Ferreira ACV, Saturnino KC, Neu DH, Gandra JR, Pereira UP, Honorato CA (2024) The effects of probiotic-based additives on aflatoxin intoxication in *Piaractus mesopotamicus*: a study of liver histology and metabolic performance. Vet Res Commun 48:2281–2294. https://doi.org/10.1007/s11259-024-10409-w

Granada L, Sousa N, Lopes S, Lemos MLF (2015) Is integrated multitrophic aquaculture the solution to the sectors' major challenges?—A review. Rev Aquacult 8(3):283–300

Guo Z, Cui J, Li M, Liu H, Zhang M, Meng F, Shi G, Wang R, He X, Zhao Y (2018) Effect of feeding frequency on growth performance, antioxidant status, immune response and resistance to hypoxia stress challenge on juvenile dolly varden char *Salvelinus malma*. Aquaculture 486:197–201

Hammond KS, Hollows JW, Townsend CR, Lokman PM (2006) Effects of temperature and water calcium concentration on growth, survival and moulting of freshwater crayfish, *Paranephrops zealandicus*. Aquaculture 251(2–4):271–279. https://doi.org/10.1016/j.aquaculture.2005.05.032

Hariharan H, Amadi V (2016) Shellfish as reservoirs of bacterial pathogens. J Coastal Life Med 4(4):253–258. https://doi.org/10.12980/jclm.4.2016J6-13

He H, Huang S, Weng S, He J, Li C (2025) Acute hypoxia stress decreases the resistance of *Litopenaeus vannamei* to *Vibrio parahaemolyticus* infection. Aquaculture 599:742131. https://doi.org/10.1016/j.aquaculture.2025.742131

He S, Wang Q, Li S, Ran C, Guo X, Zhang Z, Zhou Z (2017) Antibiotic growth promoter olaquindox increases pathogen susceptibility in fish by inducing gut microbiota dysbiosis. Sci China Life Sci 60(11):1260–1270. https://doi.org/10.1007/s11427-016-9072-6

Hennersdorf P, Kleinertz S, Theisen S, Abdul-Aziz MA, Mrotzek G, Palm HW, Saluz HP (2016) Microbial diversity and parasitic load in tropical fish of different environmental conditions. PLoS One 11(3): e0151594. https://doi.org/10.1371/journal.pone.0151594

Hernroth BE, Baden SP (2018) Alteration of host-pathogen interactions in the wake of climate change—increasing risk for shellfish associated infections? Environ Res 161:425–438

Higa T, Parr JF (1994) Beneficial and effective microorganisms for a sustainable agriculture and environment. https://www.em-pars.com/wp-content/uploads/2018/05/7-beneficial-and effective-microorganisms-for-a-sustainab-.pdf

Hongsen X, Nengbin Z, Yiling C, Huamei Y, Meiqin Z, Eakapol W, Qianrong L, Rui W (2024) Pathogenicity of *Streptococcus iniae* causing mass mortalities of yellow catfish (*Tachysurus fulvidraco*) and its induced host immune response. Front Microbiol 15:1374688. https://doi.org/10.3389/fmicb.2024.1374688

Hornick KM, Buschmann AH (2018) Insights into the diversity and metabolic function of bacterial communities in sediments from Chilean salmon aquaculture sites. Ann Microbiol 68:63–77. https://doi.org/10.1007/s13213-017-1317-8

Hu G, Huang S, Chen H, Wang F (2010) Binding of four heavy metals to hemicelluloses from rice bran. Food Res Int 43(1):203–206. https://doi.org/10.1016/j.foodres.2009.09.029

Hu H, Qi M, He P, Chen X, Li Z, Cheng H (2024) Occurrence and risk assessment of quinolones and sulfonamides in freshwater aquaculture ponds in Northeast Zhejiang, China. Sci Total Environ 953:176066. https://doi.org/10.1016/j.scitotenv.2024.176066

Huseen HM, Mohammed AJ (2019) Heavy metals causing toxicity in fishes. J Phys Conf Ser 1294:062028. https://doi.org/10.1088/1742-6596/1294/6/062028

Irshath AA, Rajan AP, Vimal S, Prabhakaran VS, Ganesan R (2023) Bacterial pathogenesis in various fish diseases: recent advances and specific challenges in vaccine development. Vaccines 11(2):470. https://doi.org/10.3390/vaccines11020470

Jahan P, Watanabe T, Kiron I, Satoh SH (2003) Improved carp diets based on plant protein sources reduce environmental phosphorus loading. Fish Sci 69:219–225

Jiang S, Zhang W, Qian X, Ji J, Ning X, Zhu F, Yin S, Zhang K (2023) Effects of hypoxia and reoxygenation on apoptosis, oxidative stress, immune response and gut microbiota of Chinese mitten crab, *Eriocheir sinensis*. Aquat Toxicol 260:106556. https://doi.org/10.1016/j.aquatox.2023.106556

Johansson CH, Janmar L, Backhaus T (2014) Toxicity of ciprofloxacin and sulfamethoxazole to marine periphytic algae and bacteria. Aquat Toxicol 156:248–258. https://doi.org/10.1016/j.aquatox.2014.08.015

John EM, Krishnapriya K, Sankar TV (2020) Treatment of ammonia and nitrite in aquaculture wastewater by an assembled bacterial consortium. Aquaculture 526:735390. https://doi.org/10.1016/j.aquaculture.2020.735390

Kaiser F, Harloff HJ, Tressel R-P, Kock T, Schulz C (2021a) Effects of highly purified rapeseed protein isolate as fishmeal alternative on nutrient digestibility and growth performance in diets fed to rainbow trout (*Oncorhynchus mykiss*). Aquacult Nutr. https://doi.org/10.1111/anu.13273

Kaiser F, Harloff HJ, Tressel R-P, Steffens D, Seibel H, Jung C, Schulz C (2021b) Effects of supplemented anti-nutritive substances from rapeseed on growth and health performance of rainbow trout (*Oncorhynchus mykiss*). Aquaculture 536:736422. https://doi.org/10.1016/j.aquaculture.2021.736422

Kennedy DA, Kurath G, Brito IL, Purcell MK, Read AF, Winton JR, Wargo AR (2016) Potential drivers of virulence evolution in aquaculture. Evol Appl 344–354

Khalafalla MM, Zayed NFA, Amer AA, Soliman AA, Zaineldin AI, Gewaily MS, Hassan AM, Doan HV, Tapingkae W, Dawood MAO (2022) Dietary *Lactobacillus acidophilus* ATCC 4356 relieves the impacts of aflatoxin B_1 toxicity on the growth performance, hepatorenal functions, and antioxidative capacity of thinlip grey mullet (*Liza ramada*) (Risso 1826). Probiot Antimicrob Proteins 14:189–203. https://doi.org/10.1007/s12602-021-09888-z

Kokou F, Fountoulaki E (2018) Aquaculture waste production associated with antinutrient presence in common fish feed plant ingredients. Aquaculture 495:295–310. https://doi.org/10.1016/j.aquaculture.2018.06.003

Kumar R, Singh D (2022) Insecticide-tolerant bacterial population in Eisenia fetida's gut and vermicast exposed to chlorantraniliprole and fipronil. Appl Biol Res 24(3):273–279. https://doi.org/10.5958/0974-4517.2022.00033.7

Kumar R, Gupta R, Dahiya T, Yadav R, Saifi R, Yodha K, Kumari B, Singh P, Shefali (2022a) Surveillance of *Heteropneustes fossilis* (Stinging Catfish) diseases and health management practices in Haryana. Appl Ecol Environ Sci 10(4):261–267. https://doi.org/10.12691/aees-10-4-12

Kumar R, Gupta RK, Yadav R, Saifi R, Yodha K, Pooja (2022b) *Eisenia fetida* as protein source for growth enhancement of *Heteropneustes fossilis*. Egypt J Aquat Biol Fisheries 26(2):577–588. https://doi.org/10.21608/ejabf.2022.234553

Lazzari R, Baldisserotto B (2008) Nitrogen and phosphorus waste in fish farming. Bol Inst Pesca 34(4):591–600

Li E, Chen L, Zeng C, Chen X, Yu N, Lai Q, Qin JG (2007) Growth, body composition, respiration and ambient ammonia nitrogen tolerance of the juvenile white shrimp, *Litopenaeus vannamei*, at different salinities. Aquaculture 265(1–4):385–390. https://doi.org/10.1016/j.aquaculture.2007.02.018

Li M, Wang X, Qi C, Li E, Du Z, Qin JG, Chen L (2018) Metabolic response of Nile tilapia (*Oreochromis niloticus*) to acute and chronic hypoxia stress. Aquaculture 495:187–195. https://doi.org/10.1016/j.aquaculture.2018.05.031

Limbu SM, Zhou L, Sun SX, Zhang ML, Du ZY (2018) Chronic exposure to low environmental concentrations and legal aquaculture doses of antibiotics cause systemic adverse effects in Nile tilapia and provoke differential human health risk. Environ Int 115:205–219. https://doi.org/10.1016/j.envint.2018.03.034

Liu B, Jia R, Zhao K, Wang G, Lei J, Huang B (2017) Stocking density effects on growth and stress response of juvenile turbot (*Scophthalmus maximus*) reared in land-based recirculating aquaculture system. Acta Oceanol Sin 36:31–38. https://doi.org/10.1007/s13131-017-0976-4

Liu K-F, Chiu C-H, Shiu Y-L, Cheng W, Liu C-H (2010) Effects of the probiotic, Bacillus subtilis E20, on the survival, development, stress tolerance, and immune status of white shrimp, Litopenaeus Vannamei larvae. Fish Shellfish Immunol 28(5–6):837–844. https://doi.org/10.1016/j.fsi.2010.01.012

Liu S, Wu K, Yao L, Li Y, Chen R, Zhang L, Wu Z, Zhou Q (2024a) Characteristics and correlation analysis of heavy metal distribution in China's freshwater aquaculture pond sediments. Sci Total Environ 931:172909. https://doi.org/10.1016/j.scitotenv.2024.172909

Liu Y, Wang J, Ding J, Zhang Y, Hou C, Shen W, Wu X, Zhu J (2024b) Effects of hypoxia stress on oxidative stress, apoptosis and microorganisms in the intestine of large yellow croaker (*Larimichthys crocea*). Aquaculture 581:740444. https://doi.org/10.1016/j.aquaculture.2023.740444

Llewellyn MS, Boutin S, Hoseinifar SH, Derome N (2014) Teleost microbiomes: the state of the art in their characterisation, manipulation and importance in aquaculture and fisheries. Front Microbiol 5:207

Long L, Zhang H, Ni Q, Liu H, Wu F, Wang X (2019) Effects of stocking density on growth, stress, and immune responses of juvenile Chinese sturgeon (*Acipenser sinensis*) in a recirculating aquaculture system. Comp Biochem Physiol C Toxicol Pharmacol 219:25–34. https://doi.org/10.1016/j.cbpc.2019.02.002

Luo G, Xu J, Meng H (2020) Nitrate accumulation in biofloc aquaculture systems. Aquaculture 520:734675. https://doi.org/10.1016/j.aquaculture.2019.734675

Madreseh S, Ghaisari HR, Hosseinzadeh S (2019) Effect of lyophilized, encapsulated *Lactobacillus fermentum* and lactulose feeding on growth performance, heavy metals, and trace element residues in rainbow trout (*Oncorhynchus mykiss*) tissues. Probiot Antimicrob Proteins 11:1257–1263. https://doi.org/10.1007/s12602-018-9487-7

Martins CI, Pistrin MG, Ende SS et al (2009) The accumulation of substances in recirculating aquaculture systems (RAS) affects embryonic and larval development in common carp *Cyprinus carpio*. Aquaculture 291:65–73. https://doi.org/10.1016/j.aquaculture.2009.03.001

Martins CIM, Eding EH, Verdegem MCJ, Heinsbroek LTN, Schneider O, Blancheton JP, d'Orbcastel ER, Verreth JAJ (2010) New developments in recirculating aquaculture systems in Europe: a perspective on environmental sustainability. Aquacult Eng 43(3):83–93

Martins CIM, Eding EH, Verreth JAJ (2011) The effect of recirculating aquaculture systems on the concentrations of heavy metals in culture water and tissues of Nile tilapia *Oreochromis niloticus*. Food Chem 126(3):1001–1005. https://doi.org/10.1016/j.foodchem.2010.11.108

Masser MP, Rakocy J, Losordo T (1999) Recirculating aquaculture tank production systems: management of recirculating system, Agriculture Do (ed). SRAC Publication

Merrifield DL, Dimitroglou A, Foey A, Davies SJ, Baker RTM, Bøgwald J, Castex M, Ringø E (2010) The current status and future focus of probiotic and prebiotic applications for salmonids. Aquaculture 302:1–18

Midhun SJ, Neethu S, Vysakh A, Radhakrishnan EK, Jyothis M (2018) Antagonism against fish pathogens by cellular components/preparations of *Bacillus coagulans* (MTCC-9872) and it's in vitro probiotic characterisation. Curr Microbiol 75:1174–1181

Milla S, Mathieu C, Wang N, Lambert S, Nadzialek S, Massart S, Henrotte E, Douxfils J, Mélard C, Mandiki SN, Kestemont P (2010) Spleen immune status is affected after acute handling stress but not regulated by cortisol in Eurasian perch, *Perca fluviatilis*. Fish Shellfish Immunol 28:931–941

Mohammadi G, Rafiee G, Tavabe KR, Abdel-Latif HMR, Dawood MAO (2021) The enrichment of diet with beneficial bacteria (single- or multi- strain) in biofloc system enhanced the water quality, growth performance, immune responses, and disease resistance of Nile tilapia (*Oreochromis niloticus*). Aquaculture 539:736640. https://doi.org/10.1016/j.aquaculture.2021.736640

Mota VC, Hop J, Sampaio LA, Heinsbroek LTN, Verdegem MCJ, Eding EH, Verreth JAJ (2018) The effect of low pH on physiology, stress status and growth performance of turbot (*Psetta maxima* L.) cultured in recirculating aquaculture systems. Aquacult Res 49:3456–3467. https://doi.org/10.1111/are.13812

Mustapha MK, Atolagbe SD (2018) Tolerance level of different life stages of Nile tilapia *Oreochromis niloticus* (Linnaeus, 1758) to low pH and acidified waters. J Basic Appl Zool 79:46. https://doi.org/10.1186/s41936-018-0061-3

Nhinh DT, Le DV, Van KV, Huong Giang NT, Dang LT, Hoai TD (2021) Prevalence, virulence gene distribution and alarming the multidrug resistance of *Aeromonas hydrophila* associated with disease outbreaks in freshwater aquaculture. Antibiotics 10(5):532. https://doi.org/10.3390/antibiotics10050532

Nisha A, Rani S, Kumar S, Kumar R, Kumar A (2025) Recent technologies and strategies for bioremediation of toxic and hazardous waste. In: Dubey KK, Shah MP, Yadav A (eds) Challenges and sustainable solutions in bioremediation. CRC Press, pp 1–16. https://doi.org/10.1201/9781003407317-1

Nordvi MF, Løvmo SD, Whatmore P, Sundh H, Sigholt T, Olsen RE (2023) Low intestinal inflammation model (HP48) in Atlantic salmon (*Salmo salar*) and inflammatory mitigation by Bactocell. Aquaculture 563:738920

Olafsen JA (2001) Interactions between fish larvae and bacteria in marine aquaculture. Aquaculture 200:223–247

Oliveira BPN, Padeniya U, Bledsoe JW, Davis DA, Liles MR, Hussain AS, Wells DE, Bruce TJ (2025) Evaluation of probiotic effects on the growth performance and microbiome of Nile tilapia

(*Oreochromis niloticus*) in a high-density biofloc system. Aquacult Nutr 868806. https://doi.org/10.1155/anu/5868806

Olsen LM, Hernández KL, Ardelan MV, Iriarte JL, Bizsel KC, Olsen Y (2017) Responses in bacterial community structure to waste nutrients from aquaculture: an in situ microcosm experiment in a Chilean fjord. Aquacult Environ Interact 9:21–32. https://doi.org/10.3354/aei00212

Padmavathi P, Sunitha K, Veeraiah K (2012) Efficacy of probiotics in improving water quality and bacterial flora in fish ponds. Afr J Microbiol Res 6(49):7471–7478. https://doi.org/10.5897/AJMR12.496

Paulina P, Corsini G, Mario T, González A (2014). *Pantoea agglomerans* an agent to remove residual copper from aquaculture activity. Adv Mater Res 945–949:3479–3482. https://www.scientific.net/AMR.945-949.3479

Person-Le Ruyet J, Galland R, Le Roux A, Chartois H (1997) Chronic ammonia toxicity in juvenile turbot (*Scophthalmus maximus*). Aquaculture 154(2):155–171. https://doi.org/10.1016/S0044-8486(97)00052-5

Pintado J, Olmo GD, Guinebert T, Ruiz P, Nappi J, Thomas T, Egan S, Masaló I, Cremades J (2023) Manipulating the *Ulva* holobiont: co-culturing *Ulva ohnoi* with *Phaeobacter* bacteria as a strategy for disease control in fish-macroalgae IMTA-RAS aquaculture. J Appl Phycol 35:2017–2029. https://doi.org/10.1007/s10811-023-02986-1

Ponce-Palafox J, Martinez-Palacios CA, Ross LG (1997). The effects of salinity and temperature on the growth and survival rates of juvenile white shrimp, *Penaeus vannamei*, Boone, 1931. Aquaculture 157(1–2):107–115. https://doi.org/10.1016/S0044-8486(97)00148-8

Pulkkinen K, Suomalainen L-R, Read AF, Ebert D, Rintamäki P, Valtonen ET (2010) Intensive fish farming and the evolution of pathogen virulence: the case of columnaris disease in Finland. Proc R Soc B Biol Sci 277:593–600

Puri P (2024) Formulating sustainable feeds: understanding nutrient and energy requirements in finfish aquaculture. Academicia Int Multidiscip Res J 14(12):71–80. https://doi.org/10.5958/2249-7137.2024.00026.3

Puri P (2025) Biochemical and microbiological assessments for nutritional profile of aquafeeds under storage conditions: natural antioxidants and biotherapeutic augmentation strategy. Indian J Nat Sci 15(88):90036–90047

Puri P, Puri P (2024) Bioresource potential of duckweed in aquaculture practices: an updated bibliometric analysis through Scopus database. Academicia Int Multidiscip Res J 14(11):15–28. https://doi.org/10.5958/2249-7137.2024.00018.8

Puri P, Sharma JG, Singh R (2022) Biotherapeutic microbial supplementation for ameliorating fish health: developing trends in probiotics, prebiotics, and synbiotics use in finfish aquaculture. Anim Health Res Rev 23(2):113–135. https://doi.org/10.1017/S1466252321000165

Puri P, Singh R, Sharma J (2023) Micro-/bio-/nano-/syn-encapsulations and co-treatments of bioactive microbial feed supplementation in augmenting finfish health and aquaculture nutrition: a review. Benefic Microbes 14(3):281–302. https://doi.org/10.3920/BM2022.0087

Quesada SP, Paschoal JAR, Reyes FGR (2013) Considerations on the aquaculture development and on the use of veterinary drugs: special issue for fluoroquinolones—a review. J Food Sci 78(9):R1321-1333

Rasul MG, Majumdar BC (2017) Abuse of antibiotics in aquaculture and it's effects on human, aquatic animal and environment. Saudi J Life Sci 2(3):81–88. https://doi.org/10.21276/haya

Refstie S, Sahlström S, Bråthen E, Baeverfjord G, Krogedal P (2005) Lactic acid fermentation eliminates indigestible carbohydrates and antinutritional factors in soybean meal for Atlantic salmon (*Salmo salar*). Aquaculture 246(1–4):331–345. https://doi.org/10.1016/j.aquaculture.2005.01.001

Ren Y, Men X, Yu Y, Li B, Zhou Y, Zhao C (2022) Effects of transportation stress on antioxidation, immunity capacity and hypoxia tolerance of rainbow trout (*Oncorhynchus mykiss*). Aquacult Rep 100940

Renaud SM, Parry DL (1994) Microalgae for use in tropical aquaculture II: effect of salinity on growth, gross chemical composition and fatty acid composition of three species of marine microalgae. J Appl Phycol 6:347–356. https://doi.org/10.1007/BF02181949

Ringø E, Birkbeck TH (1999) Intestinal microflora of fish larvae and fry. Aquacult Res 30:73–93

Ringø E, Hoseinifar SH, Ghosh K, Doan HV, Beck BR, Song SK (2018) Lactic acid bacteria in finfish—an update. Front Microbiol 9:1818. https://doi.org/10.3389/fmicb.2018.01818

Roy J, Terrier F, Marchand M, Herman A, Heraud C, Surget A, Lanuque A, Sandres F, Marandel L (2021) Effects of low stocking densities on zootechnical parameters and physiological responses of rainbow trout (*Oncorhynchus mykiss*) Juveniles. Biology 10(10):1040. https://doi.org/10.3390/biology10101040

Ruan W, Ji WW, Zheng L, Yu D, Fang H (2020) On hypoxia stress in fish and its nutritional regulation and response. Mar Fisher 42(6):751–761

Rupp M, Knüsel R, Sindilariu PD, Schmidt-Posthaus H (2019) Identification of important pathogens in European perch (Perca fluviatilis) culture in recirculating aquaculture systems. Aquacult Int 27:1045–1053. https://doi.org/10.1007/s10499-019-00382-6

Saha N, Koner D, Sharma R (2022) Environmental hypoxia: a threat to the gonadal development and reproduction in bony fishes. Aquacult Fisher 7(5):572–582. https://doi.org/10.1016/j.aaf.2022.02.002

Saraiva JL, Rachinas-Lopes P, Arechavala-Lopez P (2022) Finding the "golden stocking density": a balance between fish welfare and farmers' perspectives. Front Vet Sci 9:930221. https://doi.org/10.3389/fvets.2022.930221

Senthamarai MD, Rajan MR, Bharathi PV (2023) Current risks of microbial infections in fish and their prevention methods: a review. Microb Pathog 185:106400. https://doi.org/10.1016/j.micpath.2023.106400

Sharma MMM, Kapoor D, Rohilla R, Loyal A, Sharma P, Husen A (2024) Role of essential minerals and rhizosphere in plant under changing environmental conditions. In: Husen A (ed) Essential minerals in plant-soil systems. Elsevier, pp 235–252. (ISBN: 978-0-443-16082-0). https://doi.org/10.1016/B978-0-443-16082-0.00002-3

Sharma P, Kapoor D, Kumar R, Husen A (2025) Rhizosphere and osmolytes coordination in plant systems under adverse situations. In: Husen A (ed) Roles of osmolytes in changing environment. Academic Press, pp 101–117

Shefali, Kumar, R., Sankhla, M.S., Kumar, R., and Swaroop S. Sonone (2021) Impact of pesticide toxicity in aquatic environment. Biointerface Res Appl Chem 11(3):10131–10140. https://doi.org/10.33263/BRIAC113.1013110140

Shen L, Hs W, Zq G, Ruan Yj XuXh, Li J, Sj M, Zheng P (2016) Evidence for anaerobic ammonium oxidation process in freshwater sediments of aquaculture ponds. Environ Sci Pollut Res 23:1344–1352. https://doi.org/10.1007/s11356-015-5356-z

Somridhivej B, Boyd CE (2017) Likely effects of the increasing alkalinity of inland waters on aquaculture. J World Aquacult Soc 48(3):496–502. https://doi.org/10.1111/jwas.12405

Song C, Zhang C, Fan L, Qiu L, Wu W, Meng S, Hu G, Kamira B, Chen J (2016) Occurrence of antibiotics and their impacts to primary productivity in fishponds around Tai Lake, China. Chemosphere 161:127–135. https://doi.org/10.1016/j.chemosphere.2016.07.009

Stephane C, Marion T, Wayman S, Steinberg C, Guillemin JP (2016) Bioherbicides: dead in the water? A review of the existing products for integrated weed management. Crop Prot 87:44–49. https://doi.org/10.1016/j.cropro.2016.04.016

Sundberg L-R, Ketola T, Laanto E, Kinnula H, Bamford JKH, Penttinen R, Mappes J (2016) Intensive aquaculture selects for increased virulence and interference competition in bacteria. Proc R Soc B Biol Sci 283:20153069. https://doi.org/10.1098/rspb.2015.3069

Tacchi L, Lowrey L, Musharrafieh R, Crossey K, Larragoite ET, Salinas I (2015) Effects of transportation stress and addition of salt to transport water on the skin mucosal homeostasis of rainbow trout (*Oncorhynchus mykiss*). Aquaculture 435:120–127. https://doi.org/10.1016/j.aquaculture.2014.09.027

Tarnecki AM, Burgos FA, Ray CL, Arias CR (2017) Fish intestinal microbiome: diversity and symbiosis unravelled by metagenomics. J Appl Microbiol 123:2–17

Thompson CC, Wasielesky W, Landuci F, Lima MS, Bacha L, Perazzolo LM, Lourenço-Marques C, Soares F, Pousão-Ferreira P, Hanson L, Gomez-Gil B, Thompson M, Varasteh T, Silva TA, Swings J, Zhang XH, de Souza W, Thompson FL (2024) Understanding the role of microbes in health and disease of farmed aquatic organisms. Mar Life Sci Technol 6:579–609. https://doi.org/10.1007/s42995-024-00248-8

Timmons MB, Ebeling JM, Wheaton FW, Summerfelt ST, Vinci BJ (2002) Recirculating aquaculture systems, 2nd edn. Northeast Reg. Aquaculture Center Publ. No. 01-002

Tomasso JR (1994) Toxicity of nitrogenous wastes to aquaculture animals. Rev Fish Sci 2(4):291–314. https://doi.org/10.1080/10641269409388560

Tovar A, Moreno C, Mánuel-Vez MP, García-Vargas M (2000) Environmental impacts of intensive aquaculture in marine waters. Water Res 34(1):334–342. https://doi.org/10.1016/S0043-1354(99)00102-5

Urmila, Gupta RK, Sharma S, Kumar R (2025) Earthworms: the environmentally preferred bioengineers for managing heavy metals. In: Dey S, Bhattacharya S (eds) Biotechnological interventions in the removal of emerging pollutants. Springer Nature Singapore, Singapore, pp 365–384

Utiswannakul P, Sangchai S, Rengpipat S (2011) Enhanced growth of black tiger shrimp *Penaeus monodon* by dietary supplementation with *Bacillus* (BP11) as a probiotic. J Aquat Res Dev 3(4):2155–9546

van Bussel CGJ, Schroeder JP, Mahlmann L, Schulz C (2014) Aquatic accumulation of dietary metals (Fe, Zn, Cu Co, Mn) in recirculating aquaculture systems (RAS) changes body composition but not performance and health of juvenile turbot (*Psetta maxima*). Aquacult Eng 61:35–42. https://doi.org/10.1016/j.aquaeng.2014.05.003

Verberk WCEP, Bilton DT, Calosi P, Spicer JI (2011) Oxygen supply in aquatic ectotherms: partial pressure and solubility together explain biodiversity and size patterns. Ecology 92(8):1565–1572. https://doi.org/10.1890/10-2369.1

von Ahnen M, Aalto SL, Suurnäkki S, Tiirola M, Pedersen PB (2019) Salinity affects nitrate removal and microbial composition of denitrifying woodchip bioreactors treating recirculating aquaculture system effluents. Aquaculture 504:182–189. https://doi.org/10.1016/j.aquaculture.2019.01.068

Wang E, Yuan Z, Wang K, Gao D, Zhanjiang L, Lilesa M (2019) Consumption of florfenicol-medicated feed alters the composition of the channel catfish intestinal microbiota including enriching the relative abundance of opportunistic pathogens. Aquaculture 501:111–118. https://doi.org/10.1016/j.aquaculture.2018.11.019

Wang Y-C, Lin H-Y, Chang P-S (2023a) Evaluation of probiotic potentiality of GM-Lac (Lactobacillus and Bifidobacterium) in juvenile Asian seabass Lates calcarifer. Aquacult Rep 30:101615. https://doi.org/10.1016/j.aqrep.2023.101615

Wang Z, Pu D, Zheng J, Li P, Lü H, Wei X, Li M, Li D, Gao L (2023b) Hypoxia-induced physiological responses in fish: from organism to tissue to molecular levels. Ecotoxicol Environ Saf 267:115609. https://doi.org/10.1016/j.ecoenv.2023.115609

World Bank (2013) Fish to 2030—prospects for fisheries and aquaculture. Report number 83177-GLB. https://www.fao.org/4/i3640e/i3640e.pdf

Wu RSS (2002) Hypoxia: From molecular responses to ecosystem responses. Mar Pollut Bull 45(1–12):35–45. https://doi.org/10.1016/S0025-326X(02)00061-9

Wu Z, You F, Wen A, Ma D, Zhang P (2016) Physiological and morphological effects of severe hypoxia, hypoxia and hyperoxia in juvenile turbot (*Scophthalmus maximus* L.). Aquacult Res 47(1):219–227. https://doi.org/10.1111/are.12483

Xu W, Mawolo PW, Gao J, Chu L, Wang Y, Nie Z, Song L, Shao N, Gao J, Xu P, Xu G (2021) Effects of supplemental effective microorganisms in feed on the growth, immunity, and appetite regulation in juvenile GIFT tilapia. Aquacult Rep 19:100577

Yadav R, Gupta RK, Kumar R, Kaur T, Moun R (2022a) Biochemical response of earthworm, *Eisenia fetida* to heavy metals toxicity. J Appl Nat Sci 14(3):990–998

Yadav R, Gupta RK, Kumar R, Kaur T (2022b) Assessment of toxicity of lead and nickel on the biochemical and immunological parameters of earthworm, *Eudrilus Eugeniae*. Agric Sci Digest 44(5):922–929

Yadav R, Kumar R, Gupta RK, Kaur T, Yodha K, Kour A, Kaur S, Rajput A (2023) Heavy metal toxicity in earthworms and its environmental implications: a Review. Environ Adv 100374

Yu E-M, Li Z-C, Li Z-F, Wang G-J, Xia Y, Zhang K, Tian J-J, Gong W-B, Xie J (2021) New insights into the relationships between bacterial dynamics and water quality of aquaculture systems supplemented with carbon source and biofilm substratum. Microorganisms 9(10):2168. https://doi.org/10.3390/microorganisms9102168

Yuan Q, Wang Q, Zhang T, Li Z, Liu J (2017) Effects of water temperature on growth, feeding and molting of juvenile Chinese mitten crab *Eriocheir sinensis*. Aquaculture 468(1):169–174. https://doi.org/10.1016/j.aquaculture.2016.10.007

Yun L, Yu Z, Li Y, Luo P, Jiang X, Tian, Ding X (2019) Ammonia nitrogen and nitrite removal by a heterotrophic Sphingomonas sp. strain LPN080 and its potential application in aquaculture. Aquaculture 500:477–484. https://doi.org/10.1016/j.aquaculture.2018.10.054

Zhai Q, Wang H, Tian F, Zhao J, Zhang H, Chen W (2017) Dietary *Lactobacillus plantarum* supplementation decreases tissue lead accumulation and alleviates lead toxicity in Nile tilapia (*Oreochromis niloticus*). Aquacult Res 48:5094–5103. https://doi.org/10.1111/are.13326

Zhan Y, Ning B, Sun J, Chang Y (2023) Living in a hypoxic world: a review of the impacts of hypoxia on aquaculture. Mar Pollut Bull 194(A):115207. https://doi.org/10.1016/j.marpolbul.2023.115207

Zheng X, Fu DY, Cheng JH, Tang RY, Chu MX, Chu P, Wang T, Yin SW (2021) Effects of hypoxic stress and recovery on oxidative stress, apoptosis, and intestinal microorganisms in *Pelteobagrus vachelli*. Aquaculture 543. https://doi.org/10.1016/j.aquaculture.2021.736945

Zhou Q, Li K, Jun X, Bo L (2009) Role and functions of beneficial microorganisms in sustainable aquaculture. Biores Technol 100(16):3780–3786. https://doi.org/10.1016/j.biortech.2008.12.037

Chapter 2
Biofloc Farming: Transforming Aquaculture with Sustainable Practices

Parul Puri

Abstract Developing aquaculture systems keeping in pace with demand of food sector is a propitious task. Besides resilience and biosecurity are important issues in course of sustainable aquaculture expansion. Biofloc farming is eco-friendly, state-of-art culture environment with nearly self-sustained potential of reutilisation and remediation of generated waste. Moreover biofloc technology has become benchmark sustainable technique for intensive farming to improve productivity with fewer parameters to optimize at minimal water exchange requirement. Biofloc comprises beneficial heterotrophic-autotrophic microbial consortium capable of ameliorating health of farmed species and aquaculture welfare. Improved growth, feed conversion rates, immunity, survival, biosecurity, disease resistance, stress mitigation, and pliability towards integration to sustainable techniques are some merits fundamental to biofloc farming approach. Present chapter describes role of biofloc farming practices in attaining sustainability goals considering environmental, social, economic benefits for aquaculture welfare and growth.

Keywords Heterotrophic bacteria · Biofloc · In situ · Emergy · Circular bioeconomy · Biosecurity

2.1 Introduction

Intensification of aquaculture require incremental feed provisions to improve productivity, this causes residual accumulation of unutilised feed, waste metabolites, organic matter, resulting in depletion of water quality, farming yields, and welfare (Dauda et al. 2019). Biofloc systems are sustainable, eco-friendly, state-of-art culture environment initially developed to solve water quality issues and are applied extensively to aquaculture settings (Avnimelech 2007; De Schryver et al. 2008; Bossier and

P. Puri (✉)
Assistant Professor, Department of Zoology, Maitreyi College, University of Delhi, Bapu Dham Complex, Chanakyapuri, Delhi 110021, India
e-mail: parulpuri@maitreyi.du.ac.in

R. Kumar et al. (eds.), *Sustainability in Aquaculture and Fisheries*,
Springer Water, https://doi.org/10.1007/978-3-032-16605-0_2

Ekasari 2017). Biofloc is a co-cultured microbial consortium consisting of aggregations of bacteria (heterotrophic, autotrophic, cyanobacteria), microalgae, yeasts, protozoans and fungi. These biological flocs can be resided by rotifers and zooplanktons like copepods, cladocerans, ostracods, nematodes (Raza et al. 2024). Biofloc technology (BFT) is alternatively termed as zero-exchange autotrophic heterotrophic system, microbial-floc system, active suspension, active sludge, heterotrophic pond, suspended bacteria based system, green soup, suspended growth system (Avnimelech 2007; Dauda et al. 2019). BFT allows growth of microbial biomass based on utilisation of excrements and unused feed thus driving concurrent removal of waste from culture settings (De Schryver et al. 2008). Identification and regulation of the microbial community structure as well as composition is a significant measure to improve and optimize BFT (Wei et al. 2020). BFT system dominated by photoautotrophic bacteria and algae including cyanobacteria, chlorophytes typically display green hue; while systems of heterotrophic bacterial dominance appear brown largely due to presence of carotenoid pigments (Pimentel et al. 2023). Microbial community colour index (MCCI) provides an effective measure to depict transition of sunlight exposed green-algal-biofloc to a brown heterotrophic bacterial system on the basis of increase in feed load in culture environment (Kirk 2010; Hargreaves 2013). Changes in colour of biological flocculation from green to brown can help predict type of microbial dominance in farming ecosystem. Bioflocs can be of type in-situ and bioflocs (Menaga et al. 2019; Shefali et al. 2021). In-situ flocs are co-cultured along with cultured species in farming environment. Ex-situ flocs are grown in controlled environment using bioreactors. Adjustable microbial community structure and controllable nutritional profiles are some advantages of ex-situ biofloc systems (Kuhn et al. 2012; Kumar et al. 2022b). Deng et al. (2022) compared impact of in-situ biofloc and ex-situ viable, non-viable flocs on microbial community and growth of Nile tilapia. Growth increment of tilapia was found in in-situ biofloc systems, with relative abundance of beneficial gut microflora. Lack of growth promoting effects of ex-situ flocs was related to loss of active biological component owing to processing effects. Due to its minimal water exchange requirement, waste reuse potential, and as a potential harbour to beneficial microbial community capable of sustaining improved fish growth, survival, feed utilisation, BFT is fast becoming a benchmark sustainable alternative among intensive farming approaches (McCusker et al. 2023; Kumar et al. 2022c; Castilla-Gavilan et al. 2025; Emerenciano et al. 2025). Upcoming sections elaborates on fundamentals of BFT and utilisation of its potential towards sustainable aquaculture transformation.

2.1.1 Essential Requisites for BFT Optimisation

2.1.1.1 C:N Ratio, Carbon Source, Aeration

Carbon to nitrogen ratio (C:N), aeration, dissolved oxygen content are important variables for biofloc dynamics hence their appropriate optimisation is an essential input for managing effective BFT conditions (Hargreaves 2013). Proficient mixing and aeration help maintain active biofloc in suspended water column, controlling water quality. Aeration improves oxygen concentration of water for bacterial, fish proliferation and growth (Ogello et al. 2021). Turbulent mixing prevents anoxic dead zones that may be lethal for the cultured organisms due to extensive ammonia accumulation. Ammonia concentration in biofloc is dependent on the C:N ratio of feed and inputs (Deswati et al., 2023; Luo et al., 2020). Most commercial feeds with a protein concentration ranging between 30–35%, provide low C:N ratio up to 10:1. Improving the C:N inputs to 15:1 favours the heterotrophic growth and ammonia control (Avnimelech 1999; Li et al. 2018; Hargreaves 2013). Higher C:N ratio can be achieved by including carbon source and by reducing protein content of feed. Choice of the carbon sources is hence significant to BFT. Both nature, type of carbon source influence constitution and structure of biofloc microbial community thereby governing control over pathogenic proliferation and water quality (Li et al. 2018; Wei et al. 2020). Different works utilising effective C:N ratio and carbon source for beneficial BFT aquaculture turnouts are elaborated in Table 2.1. While numerous works describe glucose and molasses as carbon source for biofloc fertilization (Monroy-Dosta et al. 2025; El-Naga et al. 2024; Li et al. 2022; Braga et al. 2023; Gaona et al. 2016; Krummenauer et al. 2014); use of soy pellet, wheat flour, ragi, sucrose and corn starch have also been explored (Castilla-Gavilan et al. 2025; Panigrahi et al. 2022; Mahalakshmi et al. 2024; Miao et al. 2020). Recently biodegradable polymers (BPs) including poly-hydroxybutyrate-hydroxyvalerate (PHB-HV) and poly-butylene-succinate (PBS) have been incorporated with longan powder blends in biroflocs (Li et al. 2018). BPs have advantage of high nitrate- and nitrite-nitrogen ablation efficiency and steady carbon supply with less management needs. Organic carbon source (OCS) apart from affecting microbial dynamics of bioflocs, was found to influence configuration of fish gut microflora. OCS help stimulate microbial metabolism investing conversion of nutrient source, organic waste, nitrogen into assimilated microbial protein. Microbial community differences based on variations in carbon source utilised in bioflocs were described in their work by Wei et al. (2020). It was found that both density and diversity of microbes are dependent on type of carbon sources provided for development of microbial flocs. Among three carbon source studied glycerol flocs had highest Shannon diversity index of 5.83 compared to that of glucose (5.64) and starch (4.66). *Proteobacteria* and *Bacteroidetes* comprised major phyla (more than 70%) for all developed biological flocs. *Bacteroidetes* represented class Alpha, Beta, Delta, and Gamma-proteobacteria comprising important nitrogen cycle participating and denitrifying bacterial strains. While *Cyanobacteria* was most abundant in starch based flocs, absence of pathogenic *Vibrio* in these flocs confirmed role of carbon source on absence of pathogenic incidence in BFT. Largely, management of microbial ecology, appropriate C:N ratios and sludge control are significant measures to avoid pathogenic episode in BFT (Khanjani et al. 2024).

Table 2.1 Advantages of biofloc technology to aquaculture performance

Farmed species	C:N ratio	Carbon source	Positive effects	References
Nile tilapia (*Oreochromis niloticus*)	20:1	Molasses	↑Nocturnal digestive enzyme activity	Monroy-Dosta et al. (2025)
Grass carp (*Ctenopharyngodon Idella*)	15:1	Glucose monohydrate	↑Relative abundance of beneficial bacteria, ↑weight gain, ↑length	Liu et al. (2024)
Whiteleg shrimp (*Litopenaeus vannamei*)	10:1	Molasses	↑SR, ↑feed intake, ↑growth	El-Naga et al. (2024)
Red tilapia (*Oreochromis* spp.)	15:1	Molasses	↑Growth, ↑immune response, ↑antioxidant activity, ↑digestive enzymes, ↑reproductive performance	Sallam et al. (2024)
Pacu (*Piaractus mesopotamicus*)	20:1	Corn:wheat bran:molasses (1:1:2)	↑Growth performance	Sgnaulin et al. (2021)
Nile tilapia (*O. niloticus*) juveniles	15:1	Sugar cane molasses powder	↑Growth performance, ↑health	Durigon et al. (2020)
Whiteleg shrimp (*L. vannamei*)	6:1	Molasses	↑Survival, ↑productivity, ↓FCR	Gaona et al. (2016)
Tench (*Tinca tinca*), Mullet (*Mugil cephalus*)	20:1	Glucose (anhydrous)	↑CF	Vinatea et al. (2018)
Minnow (*Rhynchocypris lagowskii*)	20:1	Glucose	↑Growth performance, alleviate effects of soybean meal diet	Li et al. (2022)
Giant freshwater prawn (*Macrobrachium rosenbergii*)	1:10	Glucose, sucrose, molasses	↑Antioxidant activity,↑antibacterial response, ↑survival from pathogen challenge	Miao et al. (2020)
Giant freshwater prawn(*M. rosenbergii*)	6:1	Molasses	↑Growth performance, ↑NH_3 tolerance, ↓water use	Braga et al. (2023)
Lagoon sand shrimp (*Gammarus insensibilis*)	16:1	Corn flour	↑Water quality, ↑survival, ↑nutritional content	Castilla-Gavilan et al. (2025)
Pacific white shrimp (*Penaeus vannamei*)	15:1	Molasses: rice bran-wheat flour- ragi (3:1)	↑Yields, ↑immune response, ↑proteomic changes, ↑survival, ↑growth, ↑water quality	Panigrahi et al. (2022)

(continued)

Table 2.1 (continued)

Farmed species	C:N ratio	Carbon source	Positive effects	References
Peacock cichlid (*Aulonocara* sp.)	20:1	Soy pellets	↑Stocking density, ↑hematological parameters (Hb, RBCs, WBCs, Ht), ↑immune parameters (RBA, MYP, SP), ↑RPS to *Aeromonas veronii* challenge	Mahalakshmi et al. (2024)
Crucian carp (*Carassius carassius*)	15:1	Wheat flour	↑Gonadal performance, ↑carotenoid content, ↑survival	Nazir et al. (2025)
Speckled shrimp (*Metapenaeus monoceros*)	15:1	Corn starch, glycerine	↑Growth performance, ↓FCR, ↑SGR, ↑TBC	Kaya et al. (2019)

Abbreviations—CF: Condition factor; FCR: feed conversion ratio; Ht: hematocrit; Hb: hemoglobin; MYP: myeloperoxidase; RBA: respiratory burst activity; RBCs: red blood cells; RPS: relative percentage survival; SGR: specific growth rate; SP: Serum protein; SR: Survival rate; TBC: Total bacterial count; WBCs: white blood cells

2.1.1.2 Biofloc Supplementation: Probiotics, Prebiotics, Synbiotic Feed Additives

BFT supplements fish with macro, micronutrients; health stimulating bioactive compounds including vitamin C, carotenoids, poly-β-hydroxybutyrate (PBHB) and related probiotic benefits (García-Montero et al. 2021). Intake of microbial biofloc offers non-specific immune augmentation through release of microbe associated molecular patterns (MAMPs) including peptidoglycans, ß-glucans, lipopolysaccharides (LPS) conferring survival value to host (Ekasari et al. 2019). According to Li et al. (2024), in view of constancy of yield traditional bioflocs propagating from foregoing flocs are less efficient when compared to bioflocs frequently provisioned with probiotic supplementation. Hence, functional supplementation of probiotics help improve efficiency of BFT systems. Probiotic supplementation to biofloc microbials in BFT can potentially synergise beneficial effects on lowering pathogenic incidences, suppressing propagation of harmful bacteria in culture environment, as well by improving host physiology (Bakar et al. 2021; Ramasubburayan et al. 2025). Prebiotics are non-digestible probiotic substrates capable of increasing abundance of probionts in host gut (Gibson et al. 2010; Bindels et al. 2015). Probiotic and prebiotic mixture have synergistic and novel effects in combination as 'synbiotics' (Puri et al. 2022). Romano (2021) has related OCS addition to bacterial bioflocs akin supplementation of prebiotic (OCS) to beneficial heterotrophic bacteria (probionts) in flocs. Relative merits of supplementation of biofloc with probiotics, prebiotic additives, synbiotics, to aquaculture. Supplementation with functional additives (probiotics, prebiotics, synbiotics), artificial feed, fertilization of biofloc is known to augment growth, survival, immune parameters and water quality as shown in Table 2.2 (Hwihy et al. 2025; Panigrahi et al. 2022; refer Table 2.2).

2.2 BFT for Aquafarming Sustainability

BFT in aquafarming environment has several advantages such as low investment cost, or need of any reactor, minimal exchange requirement of water in comparison to use of sustainable techniques such as recirculating aquaculture systems (RAS).

Biofloc systems can enhance fish production ranging between 9 and 27% compared to conventional flow-through systems or RAS (Luo et al. 2014; Hisano et al. 2019). Additionally, BFT can inherently improve production by retaining high stocking densities of fish and resource utilisation efficiency supporting circular economy goals (Emerenciano et al. 2025; Vaishali et al. 2024). Improved growth, feed conversion rates (FCR), feed utilisation, disease resistance, stress mitigation, survival and integration to sustainable techniques are some merits integral to BFT approach (Padeniya et al. 2022; Kaya et al. 2019). Figure 2.1 elaborates on essential inputs and role of BFT in fulfilling economic, social and environmental objectives towards aquaculture sustainability.

2.2.1 BFT in Improving Aquaculture Welfare and Productivity

BFT is an efficient intensive culture system supporting high stocking densities for improved productivities. Mahalakshmi et al. (2024) applied BFT to improve yields of ornamental fish peacock cichlid (*Aulonocara* sp.). Higher stocking density were supported at 285.50 fish per cubic metre. Pathogenic *Aeromonas veronii* challenge could be overcome with lower infectivity in fish reared in biofloc systems with resulting survivals up to $93.33 \pm 3.33\%$. According to Nazir et al. (2025) it is essential to select appropriate stocking density for each fish species at each life stage in a given BFT environment. Crucian carp (*Carassius carassius*) cultured in BFT at high density stocking demonstrated augmentation in absolute, relative fecundity values, with advanced gonadal maturation and performance. Additionally, increment in biofloc stocking density at 2 kg m^{-3} improved skin carotenoid content and overall fish survival (Nazir et al. 2025). Functional addition of probiotic *Pseudoalteromonas piscicida* to biofloc system conferred resistance to Pacific shrimp (*P. vannamei)* against severe *Vibrio parahaemolyticus* infection at a test density of 10^7 CFU mL^{-1} (Gustilatov et al. 2024). Improved hepatopancreatic protection, immune response and survival were benefits of biofloc probiotic augmentation. BFT can support high productivity nearly at negligible water renewal (Gaona et al. 2016) by adjusting hydric flow rates for suspended solid removal. Sludge clearance at flow rate 1750 L

Table 2.2 Use of probiotics, prebiotics and synbiotics in biofloc systems and their relative merits to aquaculture

Farmed species	Probiotics, prebiotic (effective conc.)	Probiotic modalities (combination, route, feeding rate)	Merits	References
Giant freshwater prawn (*Macrobrachium rosenbergii*)	*Bacillus subtilis* (10^9 CFU ml^{-1})	Water addition, 2% BW/ twice daily	↑Water quality, ↑intestinal beneficial bacteria, ↑immunity, ↑digestion	Qiu et al. (2023)
Whiteleg shrimp (*L. vannamei*)	Pond toss®—*B. subtilis, licheniformis amyloliquefaciens, pumilus* (10^9 CFU g^{-1}); LactéolFort®—*Lactobacillus delbruekii, L. fermentum* (each 10^{13} CFU g^{-1})	Lyophilized commercial probiotic powder with maize oil sprayed in diet, 2% BW/twice daily	↑SGR, ↑SR (87.35–90.21%), ↑WG	El-Naga et al. (2024)
Brown shrimp (*Penaeus aztecus*)	**Synbiotic** = [**probiotic (Olvit)**—*L. plantarum, L. acidophilus, L. salivarius, B. subtilis* (10^9 CFU 100 g^{-1} feed) + **prebiotic**-MOS (0.2 g 100 g^{-1} feed)]	Synbiotic = [Commercial probiotic (Olvit) + prebiotic-MOS]; spirulina (*Arthrospira platensis*) at 0.2 g 100 g^{-1} feed, 6% BW/twice daily	↑THC, ↑B cells, ↑WG, ↑FBW, ↑PER, ↓FCR	Kaya (2025)
Pacific White Shrimp (*Penaeus vannamei*)	*Bacillus tequilensis* AP BFT3 (10^9 CFU ml^{-1})	Water addition, 5% BW/ daily	↑BW, ↑immune response, ↑survival, ↑growth, ↑water quality	Panigrahi et al. (2022)
Red tilapia (*Oreochromis* spp.)	*Lactobacillus plantarum* (10^7 cell ml^{-1})	Water addition, satiation/ thrice daily	↑FBW, ↑WG, ↑SGR, ↑FI, ↑hematology, ↑biochemical parameters, ↑reproductive performance, ↑survival	Sallam et al. (2024)

(continued)

Table 2.2 (continued)

Farmed species	Probiotics, prebiotic (effective conc.)	Probiotic modalities (combination, route, feeding rate)	Merits	References
Nile tilapia (*O. niloticus*)	*B. velezensis* AP193 (1×10^7 CFU g^{-1}); *B. subtilis* (BiOWiSH Feedbuilder Syn3 - 3.6×10^4 CFU g^{-1}, 7.2×10^4 CFU g^{-1})	Top-coated to feed, 4% BW/4 weeks	↓FCR, ↑fecal bacteria composition	Oliveira et al. (2025)
Whiteleg shrimp (*L. vannamei*)	*Bacillus* spp., *Enterococcus* spp., *Thiobacillus* spp., *Paracoccus* spp. (AquaStar® Pond @ 1×10^6 CFU L^{-1}); *Bacillus* spp., *Enterococcus* spp., *Lactobacillus* spp. (AquaStar®Growout—3×10^6 CFU g^{-1} feed)	Multi-strain probiotic; feed, water addition; 10% BW/thrice daily	↓FCR, ↑growth, ↑survival	Krummenauer et al. (2014)
Red tilapia (*Oreochromis* sp.)	Aquablend™, Aquaprotec™, Nutrimax™	Commercial probiotic, feed addition, 8% daily	↑Feed efficiency, ↑immune response, ↑antioxidant response	Bañuelos-Vargas et al. (2021)
Chinese shrimp, (*Fennerpenaeus chinensis*)	*Bacillus* sp, *Lactobacillus* sp, *Rhodobactor* sp. (1×10^7 CFU ml^{-1})	Biofloc addition	↑Growth, ↑immune response, ↓antioxidant stress	Kim et al. (2015)
Triangle sail mussels (*Hyriopsis cumingii*)	*B. subtilis* GFP strain (106 CFU L^{-1})	Biofloc addition	↑Digestion, ↑metabolism, ↑immune response, ↓pathogenic bacteria *Aeromonas, Pseudomonas*, ↑organ physiology, ↑water quality	Tang et al. (2024)
Pacific white shrimp (*L. vannamei*)	*Rhodospirillum rubrum, B. subtilis, Providencia rettgeri* (10^{11} CFU L^{-1})	Water addition	↑Water quality, ↑growth, ↑immune response (TLR, PrPO, hsp70), ↑stress resistance, ↓*Vibrio parahaemolyticus,*	Li et al. (2024)

(continued)

Table 2.2 (continued)

Farmed species	Probiotics, prebiotic (effective conc.)	Probiotic modalities (combination, route, feeding rate)	Merits	References
Asian stinging catfish (*H. fossilis*)	*Bacillus* spp., *Lactobacillus* spp., *Saccharomyces* spp. (24×10^9 CFU g^{-1} @25 g L^{-1})	Commercial probiotic (Virvac Probiotics V5), water addition	↑Growth, ↑survival, ↑biomass, ↑beneficial bacteria (Firmicutes and Actinobacteria), ↑stress response, ↑hematology, ↑immunity, ↑antioxidant response	Mali et al. (2024)
Butter catfish (*Ompok bimaculatus*	Cibafloc™	Commercial probiotic, water addition	↑Water quality, ↑growth, ↑digestive enzymes (AMY, PRO, LIP), ↑feed utilization, ↑survival	Behera et al. (2025)
Whiteleg shrimp (*L. vannamei*)	*Lactobacillus pentosus* BD6 (10^{10} cfu kg^{-1})	Feed addition, 3% BW/ twice daily	↑Antibacterial activity, ↑growth performance, ↑immunity, ↓mortality, ↑disease resistance	Chiu et al. (2021)
Hybrid grouper (*Epinephelus fuscoguttatus* ♀ × *Epinephelus polyphekadion*♂) juveniles	*B. licheniformis* (0.375, 0.75, 1.50 g m^{-3})	Commercial probiotic (VLPPro® SB538), water addition	↑Water quality, ↑growth performance (↑WG, ↑SGR, ↓FCR), ↑intestine health	Huang et al. (2023)

(continued)

Table 2.2 (continued)

Farmed species	Probiotics, prebiotic (effective conc.)	Probiotic modalities (combination, route, feeding rate)	Merits	References
Pacific white shrimp (*Penaeus vannamei*)	*Pseudoalteromonas piscicida* 1Ub (10^6 CFU mL^{-1})	Biofloc addition	↑Survival, ↑immune response (THC, PA, RBA, PO, SOD)	Gustilatov et al. (2024)
Nile tilapia (*O. niloticus*)	**Synbiotic + biofloc meal = Probiotic kg^{-1}feed** [*L. Plantarum* (10 ml, 1.5×10^9 cob g^{-1}, *Saccharomyces boulardii* (250 mg, 5×10^9 cob g^{-1}), *L. acidophilus* (25 mg, 5×10^9 cob g^{-1})], **prebiotic** MOS (2 g kg^{-1} feed)] + **biofloc meal** (3 g kg^{-1} feed)	Feed addition, satiation/ twice daily	↑Gut bacterial count, ↑FBW, ↑WG, ↓FCR, ↑body composition, ↑tissue histomorphology	Hersi et al. (2023)

Abbreviations—AMY: amylase; BW: body weight; CFU: Colony forming units; cob: CFU per unit biofloc; FBW: Final body weight; FCR: feed conversion ratio; FI: feed intake; GFP: green fluorescent protein; HSP: heat shock protein; LIP: lipase; MOS: mannan oligosaccharide; PA: phagocytic activity; PER: Protein efficiency ratio; PO: phenoloxidase; PrPO: prophenoloxidase; PRO: protease; RBA: respiratory burst activity; SGR: specific growth rate; SOD: superoxide dismutase; SR: Survival rate; THC: Total hemocyte count; TLR: toll-like receptor; WG: weight gain

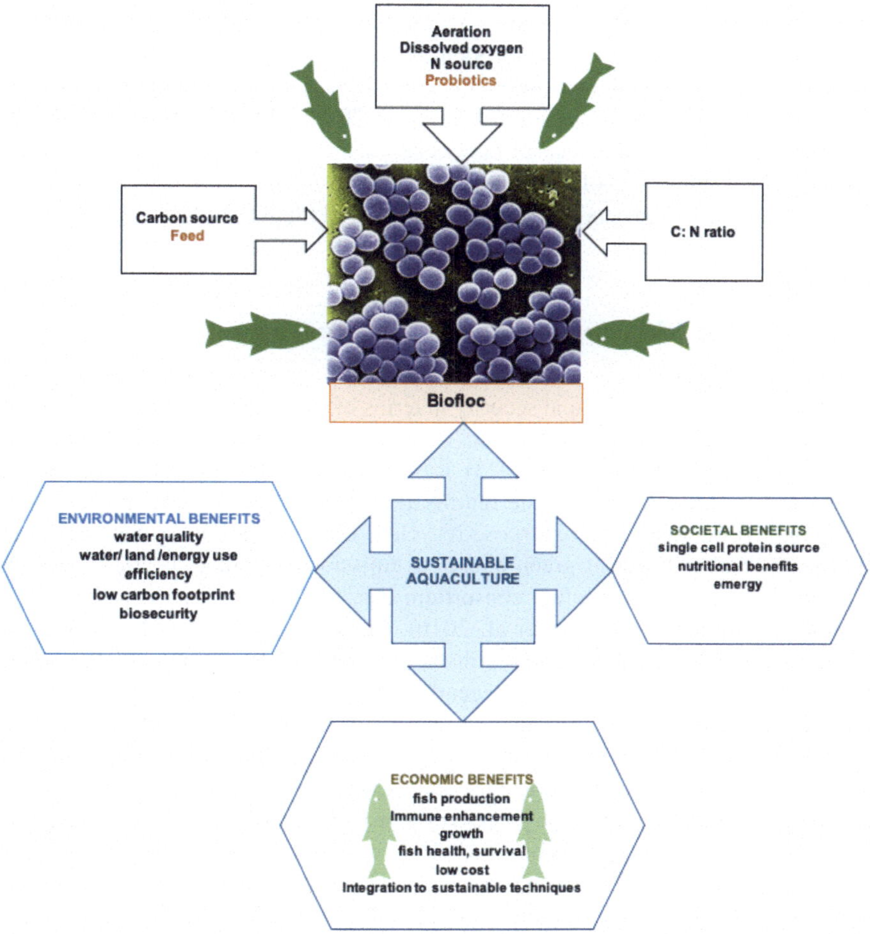

Fig. 2.1 Role of biofloc technology in sustainable aquaculture paradigm

per hour augmented survival, FCR and shrimp yields. Monroy-Dosta et al. (2025) described rhythmic increment in nocturnal periodicity of digestive enzyme (trypsin, lipase and leucine aminopeptidase) activity of tilapia grown in BFT. It is suggested to synchronise artificial feeding regimes of Nile tilapia to adjust maximum photoperiod response during dark phase despite continuous availability of biofloc feed biomass. BFT system was developed by Kaya et al. (2019) for study of individual effect of OCS corn starch, glycerine, and in combination with prebiotic mannan oligosaccharides (MOS) on speckled shrimp (*Metapenaeus monoceros*). Best shrimp growth in terms of weight gain, FCR, was obtained with corn starch MOS combination in biofloc. Effective concentration of MOS were found at 3 g per kg shrimp diet. Multistrain probiotic has significant control over *V. parahaemolyticus* infliction to *L. vannamei* improving shrimp growth rate and survival in BFT farming environment

(Krummenauer et al. 2014). Addition of *Bacillus subtilis* and effective microbials comprising yeast, actinomyces, autotrophic and lactic acid bacteria in the BFT system has ameliorating effect on the water body, digestive system and immunoresponsiveness of *Macrobrachium rosenbergii* (Qiu et al. 2023). *Bacillus* sp. in the biofloc can improve relative abundance of beneficial bacteria in the intestinal tract of *M. rosenbergii.* Incorporation of BFT under high salinity stress conditions can optimize growth, health, and reproductive performance for red tilapia (Sallam et al. 2024).

2.2.2 BFT in Biosecurity and Circular Economy

BFT can contribute to biological security in terms of sustainable production, disease mitigation, environment resilience and water quality improvement (Behera et al. 2025; Li et al. 2024; Huang et al. 2023). Resource use potential of BFT (feed, land, water, energy efficiency and waste reutilisation) aligns with circular bioeconomy exemplars of "reduce-reuse-repurpose-recycle". Biofloc microbials can sequester atmospheric carbon lowering green-house gas emissions (Ogello et al. 2021). Certain phytoplankton species in biofloc consortium can act as carbon sinks to be stored in biomass production (Manan et al. 2016). Carbon, nitrogen (feed and floc) in BFT system can be circulated and reutilised to provide primary production as aquaculture yield (Zhang et al. 2024). Recently Liu et al. (2025) reported biosecurity potential and nutrient removal capacity of novel ultrafiltration linked marine BFT systems Vieira et al. (2021) analysed emergy values and microbial composition to estimate sustainability, food safety of farming systems utilising probiotic and BFT for shrimp production. Emergy is the process of transformation of energy into product or service outcome. Compared to traditional flow-through technologies super-intensive BFT system produced 400 shrimps per square metre area with best renewability, environment load ratio and emergy sustainability values (at 14.79%, 5.76, 0.20 respectively).

2.2.3 Biofloc Meal: Nutritional Source of Bioprotein and for Fishmeal Substitution

Dried floc material after harvesting is a valuable source of single cell protein (SCP) (Ogello et al. 2021) and can be utilised as 'biofloc meal' (BM) for partial replacement of costly source of protein such as fishmeal (FM) and soybean meal (Dantas et al. 2016). Along with natural wet floc biomass, dried BM is a promising bioprotein containing high content of essential amino acids, minerals, vitamins and carbohydrates, alternative feed ingredient to FM and feed supplement in aquafeeds (Khanjani et al. 2023). Protein content of BM ranges between 12–50%, with high lipid and carbohydrate values reaching up to 13% and 36.4% respectively (Emerenciano et al.

2013). BM collected from catfish culture was utilised as feed constituent for Pacific shrimp (*L. vannamei*) diet (Ekasari et al. 2019). Daily feeding rate was kept at 5% shrimp biomass at feeding frequency four times daily. 30% FM replaced BM diet lowered FCR, improving digestibility of protein, fat, phosphorous; specific growth rate and shrimp survival. Better phosphorous (P) digestibility of BM-based diet was due to microbial phytase dependent conversion of non-digestible P in phytic acid from plant-based feed ingredients, to digestible P for the cultured organism. According to Wei et al. (2024) 40% substitution of FM with BM in common carp (*Cyprinus carpio*) diet enhanced growth, immunity and metabolism of dietary protein. Although BM substitution higher at 80% had adverse effect on growth and hepatopancreatic health. Works of Wei et al. (2024) and Ekasari et al. (2019) confirm that variations in the digestibility and nutritional quality of BM are dependent on their source of bioflocs. Hersi et al. (2023) studied merits of BM together with synbiotic incorporation of probiotic *Lactobacillus plantarum, L. acidophilus, Saccharomyces boulardii*, and prebiotic mannan oligosaccharides (MOS) to tilapia diet. BM and synbiotic substantiated as effective strategy to mitigate adverse effects of salinity on tilapia growth, improving gut microbiota, tissue histomorphology and whole body composition. Lunda et al. (2020) investigated suitability of BM as novel protein source for red swamp crayfish (*Procambarus clarkii*) diet. Feed inclusion levels ranging from 33–66% improved crayfish growth. Since higher inclusion levels of BM above 66% compromised fish growth, confirming similar observations in *C. carpio*, thus indicating limited substitution of BM in aquafeeds for best possible effects on physiological parameters of farmed species.

2.2.4 BFT Integration to Other Sustainable Aquaculture Techniques

BFT is amenable to integration with other sustainable approaches to augment sustained aquaculture growth and development. Legarda et al. (2019) in their work investigated integrated RAS of mullet and shrimp using BFT. Integrated shrimp mullet biofloc increased productivity by 11.9% and P retention by 16.8% compared to shrimp monoculture. The haematological and immune parameters in BFT was comparable for fish species maintaining fish health and shrimp growth. At commercial level BFT proved as cost effective approach. Zimmermann et al. (2023) explains integration of biofloc to RAS, with recirculation in multi-compartment biofloc as bio-RAS system. Comparing to traditional single biofloc BFT, bio-RAS reduces waste load of one compartment by sludge reuse in other compartments maximising process, productivity, health and resource use efficiency. Taufik et al. (2024) amalgamated clear water RAS to BFT with probiotic intervention. This system benefits post-larval shrimp survival, is amenable to biosecurity, improving shrimp production at minimal space requirement. Newer techniques such as Flocponics integrate BFT in aquaculture to conventional aquaponics for soil less production of plants (Pinho et al. 2022).

2.3 BFT Limitations

Bioflocs lack essential vitamins, minerals, amino acids, omega n3 fatty acids such as eicosapentaenoic and docosahexaenoic acid (EPA and DHA) vital for growth and survival of fish and marine shrimp (Crab 2010; Amaya et al. 2007). Nutrient compensation in biofloc hence require addition of artificial diets, thus making total replacement of additional feeds less possible. Requirement of proper aeration can increase cost to benefit ratio of BFT making it less supportable in terms of energy expense (Minaz et al. 2024). Besides the inherent amenability of BFT to feed reutilisation require proper management in terms of quality of supplemental feed. Channa et al. (2024) in their work reported problem of feed derived severe heavy metal (HM) accumulation in biofloc cultured Nile tilapia. Commercial feed rendered HM concentrations were significantly high in both culture water and fish tissues posing potential risk to environment, fish, and consumer health. Among fish tissue, liver had highest tendency of copper (Cu) and lead (Pb) bioaccumulation, followed by gills. While cadmium (Cd) and chromium (Cr) values were higher in tilapia gills. Another limitation to BFT is the problem of controlling composition of microbial community dynamics to optimize water quality and aquaculture wellness. This require apposite identification of microbials and their community role in biofloc (Wei et al. 2020).

2.4 Conclusions

Intensification of aquaculture has increased burden of waste management, increasing threat of pathogenic outbreak compromising aquaculture and fish health. BFT is an environment friendly, sustainability driven, self-nurturing approach with capacity to reuse, recycle, remediate generated culture waste (feed and metabolic products) inherent to farming system. Bioflocs can be source of nutrition and quality protein, bioactives, vitamins, carotenoids to culture species. The advantage of BFT lies in its inherent potential to renew, revitalise, repurpose products into services. Harnessing beneficial interactions of microbials for biomass production in culture ecosystem is one of the biggest strength of BFT. As a sustainable aquaculture initiative use of fertilization, probionts, prebiotics and synbiotics can augment BFT beneficence. Careful management and optimisation of heterotrophic community dynamics; technological variables (aeration, flow rate); apposite C:N ratio (with carbon source addition) can offer improved yields per unit arable area. Bio flocculants can feat carbon sequestration for biomass production, improving carbon footprint. Nutrient reclamation, with efficient land, water, energy use help propagate circular food production, supporting sustainable development goals (SDGs) for aquaculture. Conjunction to other sustainable systems (including RAS, aquaponics, hydroponics) increases the variety of culture services adding food security for consumers.

2.5 Future Prospects

While BFT has direct potential of harnessing sustainable aquaculture development, optimisation and meticulous monitoring of BFT parameters for dissolved oxygen, water quality, C:N ratio, microbial community growth can be supported by conjugation to automated techniques. This approach can be fostered using Artificial Intelligence (AI) driven systems supporting real time monitoring and intermediations as well as intelligent systems supported through Internet of Things (IoT). Use of such modern data driven methodologies and approaches can help in easy predictions for necessary, timely intervention, augmenting biofloc system performance and outputs. Additionally, future prospects of BFT expansion can focus on cultivation of efficient carbon sources for fulfilling effective fertilisation needs.

References

Amaya EA, Davis DA, Rouse DB (2007) Replacement of fish meal in practical diets for the Pacific white shrimp (*Litopenaeus vannamei*) reared under pond conditions. Aquaculture 262:393–401

Avnimelech Y (1999) Carbon/nitrogen ratio as a control element in aquaculture systems. Aquaculture 176:227–235

Avnimelech Y (2007) Feeding with microbial flocs by tilapia in minimal discharge bio-flocs technology ponds. Aquaculture 264(1–4):140–147. https://doi.org/10.1016/j.aquaculture.2006.11.025

Bakar SNHA, Hasan HA, Abdullah SRS, Kasan NA, Muhamad MH, Kurniawan SB (2021) A review of the production process of bacteria-based polymeric flocculants. J Water Process Eng 40:101915. https://doi.org/10.1016/j.jwpe.2021.101915

Bañuelos-Vargas I, Rodríguez-Montes GAO, Martínez-Montaño E, Pérez-Jiménez A, Mendoza-Gamboa OA, Estrada-Godínez JA, Hernández C (2021) Antioxidant and immune response of juvenile red tilapia (*Oreochromis* sp) cultured at different densities in sea water with biofloc plus probiotics. Aquaculture 544:737112. https://doi.org/10.1016/j.aquaculture.2021.737112

Behera S, Das PC, Felix N, Ferosekhan S, Swain HS, Kumari R, Athithan S, Padmavathy P (2025) Effect of different carbon supplements on growth performance and digestive enzyme activities of butter catfish (*Ompok bimaculatus* Bloch, 1794) in biofloc system. Aquaculture 603:742384. https://doi.org/10.1016/j.aquaculture.2025.742384

Bindels LB, Delzenne NM, Cani PD, Walter J (2015) Towards a more comprehensive concept for prebiotics. Nat Rev Gastroenterol Hepatol 12:303–310

Bossier P, Ekasari J (2017) Biofloc technology application in aquaculture to support sustainable development goals. Microb Biotechnol 10(5):1012–1016. https://doi.org/10.1111/1751-7915.12836

Braga IFM, Araújo MT, Brito LO, Correia ES (2023) Influence of BFT and water exchange systems on growth, ammonia tolerance, and water footprint in Macrobrachium rosenbergii nursery in intensive systems. Aquacult Int 31:1775–1788. https://doi.org/10.1007/s10499-023-01046-2

Castilla-Gavilan M, Guerra-García JM, Hachero-Cruzado I (2025) Integrating Gammarus insensibilis in biofloc systems: a sustainable approach to nutrient enrichment and waste valorisation in aquaculture. Aquaculture 597:741922. https://doi.org/10.1016/j.aquaculture.2024.741922

Channa MJ, Ayub H, Ujan JA, Habib SS, Ullah M, Attaullah S, Khayyam K, Khan K (2024) Human health risk assessment due to the incidence of heavy metals in different commercial feeds used for the culturing of biofloc fish (Nile tilapia: *Oreochromis niloticus*). Biol Trace Elem Res 202(4):1741–1751. https://doi.org/10.1007/s12011-023-03767-0

Chiu ST, Chu TW, Simangunsong T, Ballantyne R, Chiu CS, Liu CH (2021) Probiotic, Lactobacillus pentosus BD6 boost the growth and health status of white shrimp, Litopenaeus vannamei via oral administration. Fish Shellfish Immunol 117:124–135. https://doi.org/10.1016/j.fsi.2021.07.024

Crab R (2010) Bioflocs technology: an integrated system for the removal of nutrients and simultaneous production of feed in aquaculture. PhD thesis, Ghent University, pp 178

Dantas Jr EM, Valle BCS, Brito CMS, Calazans NKF, Peixoto SRM, Soares RB (2016) Partial replacement of fishmeal with biofloc meal in the diet of postlarvae of the Pacific white shrimp *Litopenaeus vannamei*. Aquacult Nutr 22:335–342

Dauda AB, Ajadi A, Tola-Fabunmi AS, Akinwole AO (2019) Waste production in aquaculture: sources, components and managements in different culture systems. Aquacult Fisher 4(3):81–88. https://doi.org/10.1016/j.aaf.2018.10.002

Deng Y, Borewicz K, van Loo J, Olabarrieta MZ, Kokou F, Sipkema D, Verdegem MCJ (2022) In-situ biofloc affects the core prokaryotes community composition in gut and enhances growth of Nile Tilapia (*Oreochromis niloticus*). Microb Ecol 84:879–892. https://doi.org/10.1007/s00248-021-01880-y

De Schryver P, Crab R, Defoirdt T, Boon N, Verstraete W (2008) The basics of bio-flocs technology: the added value for aquaculture. Aquaculture 277(3–4):125–137. https://doi.org/10.1016/j.aquaculture.2008.02.019

Deswati, Tetra ON, Yusuf SYY, Suparno, Pardi H (2023) Dynamics and fluctuations of ammonia, nitrite and nitrate in the utilization of tilapia cultivation waste in Aquaponics-NFT (nutrient film technique) based on biofloc. AACL Bioflux 16(3):1254–1265. http://www.bioflux.com.ro/aacl

Durigon EG, Lazzari R, Uczay J, Lopes DLA, Jerônimo GT, Sgnaulin T, Emerenciano MGC (2020) Biofloc technology (BFT): adjusting the levels of digestible protein and digestible energy in diets of Nile tilapia juveniles raised in brackish water. Aquacult Fisher 5(1):42–51. https://doi.org/10.1016/j.aaf.2019.07.001

Ekasari J, Suprayudi MA, Elas P, Senja RK (2019) The digestibility of biofloc meal from African catfish culture medium as a feed raw material for Pacific white shrimp. Jurnal Akuakultur Indonesia 18(1):1–8. https://doi.org/10.19027/jai.18.1.1-8

El-Naga WA, Sharawy ZZ, Tahoun AA, El-Ashram A, Suloma A (2024) Probiotics enhance the zootechnical, biochemical, and water quality aspects of culturing the whiteleg shrimp, *Litopenaeus vannamei* in the biofloc system. Egypt J Aquat Biol Fisher 28(4):1303–1314

Emerenciano M, Gaxiola G, Cuzon G (2013) Biofloc technology (BFT): a review for aquaculture application and animal food industry. In: Matovic MD (ed) Biomass now—cultivation and utilization. IntechOpen, pp 301–328. https://doi.org/10.5772/53902

Emerenciano MGC, Khanjani MH, Sharifinia M, Miranda-Baeza A (2025) Could biofloc technology (bft) pave the way toward a more sustainable aquaculture in line with the circular economy? Aquacult Res 2025:1020045. https://doi.org/10.1155/are/1020045

Gaona CAP, Serra FP, Furtado PS, Poersch LH, Wasielesky W Jr (2016) Biofloc management with different flow rates for solids removal in the *Litopenaeus vannamei* BFT culture system. Aquacult Int 24:1263–1275. https://doi.org/10.1007/s10499-016-9983-2

García-Montero C, Fraile-Martínez O, Gomez-Lahoz AM, Pekarek L, Castellanos AJ, Noguerales-Fraguas F, Ortega MA (2021) Nutritional components in Western diet versus Mediterranean diet at the gut microbiota–immune system interplay. Implications for health and disease. Nutrients 13(2):699

Gibson GR, Scott KP, Rastall RA, Tuohy KM, Hotchkiss A, Dubert-Ferrandon A, Gareau M, Murphy EF, Saulnier D, Loh G, Macfarlane S, Delzenne N, Ringel Y, Kozianowski G, Dickmann

R, Lenoir-Wijnkoop I, Walker C, Buddington R (2010) Dietary prebiotics: current status and new definition. Food Sci Technol Bull Funct Foods 7:1–19

Gustilatov M, Widanarni W, Ekasari J, Julyantoro PGS, Waturangi DE (2024) Biofloc system supplemented by *Pseudoalteromonas piscicida* 1Ub protects the Pacific white shrimp *Penaeus vannamei* from *Vibrio parahaemolyticus* infection. Aquacult Fisher 9:967–974. https://doi.org/10.1016/j.aaf.2023.05.003

Hargreaves JA (2013) Biofloc production systems for aquaculture. Southern regional aquaculture center, SRAC Publication No. 4503

Hersi MA, Genc E, Pipilos A, Keskin E (2023) Effects of dietary synbiotics and biofloc meal on the growth, tissue histomorphology, whole-body composition and intestinal microbiota profile of Nile tilapia (Oreochromis niloticus) cultured at different salinities. Aquaculture 570:739391. https://doi.org/10.1016/j.aquaculture.2023.739391

Hisano H, Barbosa PTL, Hayd LA, Mattioli CC (2019) Evaluation of Nile tilapia in monoculture and polyculture with giant freshwater prawn in biofloc technology system and in recirculation aquaculture system. Int Aquati Res 11:335–346. https://doi.org/10.1007/s40071-019-00242-2

Huang J, Amenyogbe E, Wen Z, Ou G, Li Y, Jiang X, Xie R, Chen G (2023) Effect of Bacillus licheniformis probiotic on the culture of hybrid grouper (♀ Epinephelus fuscoguttatus × ♂ Epinephelus polyphekadion). Aquacult Rep 33:101798. https://doi.org/10.1016/j.aqrep.2023.101798

Hwihy HM, Zeina AF, Husein MSA, El-Tabakh MAM (2025) Biofloc biosecurity: a revolutionary paradigm in augmenting aquaculture health and resilience against *Aeromonas hydrophila*. Aquacult Fisher 10(4):608–616. https://doi.org/10.1016/j.aaf.2024.03.001

Kaya D (2025) Improvement of brown shrimp (*Penaeus aztecus*) culture parameters through dietary enriched synbiotic in a biofloc system. Aquacult Int 33:232. https://doi.org/10.1007/s10499-025-01909-w

Kaya D, Genc MA, Aktas M, Eroldogan OT, Aydin FG, Genc E (2019). Effects of biofloc technology (BFT) on growth of speckled shrimp (*Metapenaeus monoceros*). Tarım Bilimleri Dergisi J Agric Sci 25:491–497. https://doi.org/10.15832/ankutbd.441745

Khanjani MM, Mozanzadeh MT, Sharifinia M, Emerenciano MGC (2023) Biofloc: a sustainable dietary supplement, nutritional value and functional properties. Aquaculture 562:738757. https://doi.org/10.1016/j.aquaculture.2022.738757

Khanjani MM, Sharifinia M, Emerenciano MGC (2024) Biofloc technology (BFT) in aquaculture: what goes right, what goes wrong? A scientific-based snapshot. Aquacult Nutr 2024:7496572. https://doi.org/10.1155/2024/7496572

Kim M-S, Min EY, Kim J-H, Koo J-K, Kang J-C (2015) Growth performance and immunological and antioxidant status of Chinese shrimp, *Fennerpenaeus chinensis* reared in bio-floc culture system using probiotics. Fish Shellfish Immunol 47(1):141–146. https://doi.org/10.1016/j.fsi.2015.08.027

Kirk KR (2010) Modeling microbial and nutrient dynamics in zero-discharge aquaculture systems. https://tigerprints.clemson.edu/all_dissertations/648

Krummenauer D, Poersch L, Romano LA, Lara GR, Encarnação P, Wasielesky W (2014) The effect of probiotics in a *Litopenaeus vannamei* biofloc culture system infected with *Vibrio parahaemolyticus*. J Appl Aquacult 26:370–379. https://doi.org/10.1080/10454438.2014.965575

Kuhn DD, Lawrence AL, Patnaik S, Boardman GD (2012) In-situ and ex-situ biofloc technology for Shrimp Culture. In: Proceedings of the 9th international conference on recirculating aquaculture. Roanoke, Virginia

Kumar R, Gupta RK, Yadav R, Saifi R, Yodha K, Pooja (2022b) *Eisenia fetida* as protein source for growth enhancement of *Heteropneustes fossilis*. Egypt J Aquat Biol Fisher 26(2):577–588. https://doi.org/10.21608/ejabf.2022.234553

Kumar R, Yadav R, Kumar, Gupta R, Pooja (2022c) A sustainable way for fish health management by replacement of chemical and drugs by earthworm. In: Environmental degradation in Asia:

land degradation, environmental contamination, and human activities. Springer International Publishing, Cham, pp. 329–352. https://doi.org/10.1007/978-3-031-12112-8_16
Legarda EC, Poli MA, Martins MA, Pereira SA, Martins ML, Machado C, de Lorenzo MA, Vieira FN (2019) Integrated recirculating aquaculture system for mullet and shrimp using biofloc technology. Aquaculture 512:734308. https://doi.org/10.1016/j.aquaculture.2019.734308
Li J, Liu G, Li C, Deng Y, Tadda MA, Lan L, Zhu S, Liu D (2018) Effects of different solid carbon sources on water quality, biofloc quality and gut microbiota of Nile tilapia (*Oreochromis niloticus*) larvae. Aquaculture 495:919–931. https://doi.org/10.1016/j.aquaculture.2018.06.078
Li L, Yang ZY, Qu ZH, Zhu R, Li DL, Wang HT, Wei XF, Shang GJ, Wu LF (2022) Alleviative effect of biofloc technology (BFT) on extruded soybean meal (ESBM)-induced growth inhibition and intestinal barrier dysfunction in *Rhynchocypris lagowskii*. Aquaculture 561:738677. https://doi.org/10.1016/j.aquaculture.2022.738677
Li Z, Du Q, Jiao T, Zhu Z, Wan X, Ju C, Liu H, Li Q (2024) Effects of probiotic supplementary bioflocs (*Rhodospirillum rubrum, Bacillus subtilis, Providencia rettgeri*) on growth, immunity, and water quality in cultures of Pacific white shrimp (Litopenaeus vannamei). Aquaculture 591:741141. https://doi.org/10.1016/j.aquaculture.2024.741141
Liu Y, He Y, Jia X, Fan Y, Yuan L, Shen Y, Xu X, Li J (2024) Effects of biological flocculation technology (BFT) on water quality dynamics and immune response of grass carp (*Ctenopharyngodon idella*). Aquacult Fisher 9:921–928
Liu W, Yang Y, Lu Z, Li X, Ning M, He H (2025) Innovative approaches for enhanced nutrient (N and P) removal and biosecurity in marine recirculating aquaculture systems using biofloc and ultrafiltration combined systems. Sep Purif Technol 356(Part A):129766. https://doi.org/10.1016/j.seppur.2024.129766
Lunda R, Roy K, Dvorak P, Kouba A, Mraz J (2020) Recycling biofloc waste as novel protein source for crayfish with special reference to crayfish nutritional standards and growth trajectory. Sci Rep 10:19607. https://doi.org/10.1038/s41598-020-76692-0
Luo G, Gao Q, Wang C, Liu W, Sun D, Li L, Tan H (2014) Growth, digestive activity, welfare, and partial cost-efectiveness of genetically improved farmed tilapia (Oreochromis niloticus) cultured in a recirculating aquaculture system and an indoor biofLoc system. Aquaculture 422–423:1–7. https://doi.org/10.1016/j.aquaculture.2013.11.02333
Luo G, Xu J, Meng H (2020) Nitrate accumulation in biofloc aquaculture systems. Aquaculture 520:734675. https://doi.org/10.1016/j.aquaculture.2019.734675
Mahalakshmi S, Antony C, Ravaneswaran K, Uma A, Lingam RSS (2024) Influence of biofloc and stocking density on physiological responses and disease resistance in peacock cichlid (*Aulonocara* sp). Aquaculture 592:741168. https://doi.org/10.1016/j.aquaculture.2024.741168
Mali P, Ghosh S, Dash G, Chowdhury S (2024) Influence of various carbon and nitrogen ratios in a jaggery-based biofloc rearing system on the Asian stinging catfish *Heteropneustes fossilis*. Aquacult Eng 107:102469. https://doi.org/10.1016/j.aquaeng.2024.102469
Manan H, Julia H, Moh Z, Kasan NA, Suratman S, Ikhwanuddin M (2016) Study on carbon sinks by classified biofloc phytoplankton from marine shrimp pond water. AACL Bioflux 4
McCusker S, Warberg MB, Davies SJ, Valente CS, Johnson MP, Cooney R, Wan AHL (2023) Biofloc technology as part of a sustainable aquaculture system: a review on the status and innovations for its expansion. Aquacult Fish Fisher 3:331–352. https://doi.org/10.1002/aff2.108
Menaga M, Felix S, Charulatha M, Gopalakannan A, Panigrahi A (2019) Effect of in-situ and ex-situ biofloc on immune response of genetically improved farmed Tilapia. Fish Shellfish Immunol 92:698–705. https://doi.org/10.1016/j.fsi.2019.06.031
Miao S, Hu J, Wan W, Han B, Zhou Y, Xin Z, Sun L (2020) Biofloc technology with addition of different carbon sources altered the antibacterial and antioxidant response in *Macrobrachium rosenbergii* to acute stress. Aquaculture 525:735280. https://doi.org/10.1016/j.aquaculture.2020.735280
Minaz M, Yazici IS, Sevgili H, Aydin I (2024) Biofloc technology in aquaculture: advantages and disadvantages from social and applicability perspectives—a review. Ann Anim Sci 24(2):307–319. https://doi.org/10.2478/aoas-2023-0043

Monroy-Dosta MDC, Becerril-Cortés D, Lazo JP, Mena-López A, Negrete-Redondo P, Nogueda-Torres E, Navarro-Guillén C, Mata-Sotres JA (2025) Effect of biofloc culture on the daily rhythmicity of the activity and expression of digestive enzymes in tilapia, *Oreochromis niloticus*. Aquacult Nutr 617425. https://doi.org/10.1155/anu/6617425

Nazir U, Rather MA, Chesti A, Wani SM, Zamir F, Bhat BA, Qadiri SSN, Farooq S, Aijaz A (2025) Effect of stocking density on gonadal development, pigmentation and survival of *Carassius carassius* (Crucian carp) in a biofloc system. Aquaculture 595:741477. https://doi.org/10.1016/j.aquaculture.2024.741477

Ogello EO, Outa NO, Obiero KO, Kyule DN, Munguti JM (2021) The prospects of biofloc technology (BFT) for sustainable aquaculture development. Sci Afr 14:e01053. https://doi.org/10.1016/j.sciaf.2021.e01053

Oliveira BPN, Padeniya U, Bledsoe JW, Davis DA, Liles MR, Hussain AS, Wells DE, Bruce TJ (2025) Evaluation of probiotic effects on the growth performance and microbiome of Nile Tilapia (*Oreochromis niloticus*) in a high-density Biofloc system aquaculture nutrition. 2025:5868806. https://doi.org/10.1155/anu/5868806

Padeniya U, Davis DA, Wells DE, Bruce TJ (2022) Microbial interactions, growth, and health of aquatic species in biofloc systems. Water 14:4019. https://doi.org/10.3390/w14244019

Panigrahi A, Esakkiraj P, Saranya C, Das RR, Sundaram M, Sudheer NS, Biju IF, Jayanthi M (2022) A Biofloc-based aquaculture system bio-augmented with probiotic bacteria *Bacillus tequilensis* AP BFT3 improves culture environment production performances, and proteomic changes in *Penaeus vannamei*. Probiot Antimicrob Proteins 14:277–287. https://doi.org/10.1007/s12602-022-09926-4

Pimentel OGLF, Amado AM, They NH (2023) Biofloc colors as an assessment tool for water quality in shrimp farming with BFT systems. Aquacult Eng 101:102321. https://doi.org/10.1016/j.aquaeng.2023.102321

Pinho SM, de Lima JP, David LH, Emerenciano MGC, Goddek S, Verdegem MCJ, Keesman KJ, Portella MC (2022) FLOCponics: the integration of biofloc technology with plant production. Rev Aquacult 14:647–675. https://doi.org/10.1111/raq.12617

Puri P, Sharma JG, Singh R (2022) Biotherapeutic microbial supplementation for ameliorating fish health: developing trends in probiotics, prebiotics, and synbiotics use in finfish aquaculture. Anim Health Res Rev 23(2):113–135. https://doi.org/10.1017/S1466252321000165

Qiu Z, Xu Q, Li S, Zheng D, Zhang R, Zhao J, Wang T (2023) Effects of probiotics on the water quality, growth performance, immunity, digestion, and intestinal flora of giant freshwater prawn (*Macrobrachium rosenbergii*) in the Biofloc culture system. Water 15(6):1211. https://doi.org/10.3390/w15061211

Ramasubburayan R, Prakash S, Immanuel G, Mubarakali D, Rajakumar G, Thirumurugan D, Palavesam A (2025) The transformative role of prebiotics, probiotics, and microbiomes in biofloc systems for sustainable aquaculture: a comprehensive review. Rev Aquacult 17(1):e13000. https://doi.org/10.1111/raq.13000

Raza B, Zheng Z, Yang W (2024) A review on biofloc system technology, history, types, and future economical perceptions in aquaculture. Animals 14(10):1489. https://doi.org/10.3390/ani14101489

Romano N (2021) Probiotics, prebiotics, biofloc systems, and other biocontrol regimens in fish and shellfish aquaculture. In: Kibenge FSB, Baldisserotto B, Chong RSM (eds) Aquaculture pharmacology. Academic Press, pp 219–242. https://doi.org/10.1016/B978-0-12-821339-1.00003-9

Sallam GR, Shehata AI, El Basuini MF, Habib YJ, Henish S, Rahman ANA, Hassan YM, Fayed WM, El-Sayed AFM, Aly HA (2024) Integrated biofloc technology in red tilapia aquaculture: salinity-dependent effects on water quality, parental stock physiology, reproduction, and immune responses. Aquacult Int 32:8731–8761. https://doi.org/10.1007/s10499-024-01588-z

Sgnaulin T, Pinho SM, Durigon EG, Thomas MC, de Mello GL, Emerenciano MGC (2021) Culture of pacu *Piaractus mesopotamicus* in biofloc technology (BFT): insights on dietary protein sparing and stomach content. Aquacult Int 29:2319–2335. https://doi.org/10.1007/s10499-021-00748-9

Shefali, Kumar R, Sankhla MS, Kumar R, Sonone SS (2021) Impact of pesticide toxicity in aquatic environment. Biointerface Res Appl Chem 11(3):10131–10140

Tang X, Wang T, Yang Q, Zheng S, Ma S, Yao W, Li Y, Wu Z (2024) Potential benefits of *Bacillus subtilis*-mediated bioflocs in supplementary feeding on triangle sail mussels *Hyriopsis cumingii*: a pilot study of fluorescence-labeled floc-forming bacteria. Aquacult Rep 34:101904. https://doi.org/10.1016/j.aqrep.2023.101904

Taufik M, Ismail TIT, Manan H, Ikhwanuddin M, Salam AIA, Rahim AIA, Ishak AN, Kamaruzzan AS, Draman AS, Kasan NA (2024) Synergistic effects of Recirculating Aquaculture System (RAS) with combination of clear water, probiotic and biofloc technology: a review. Aquacult Fisher 9(6):883–892. https://doi.org/10.1016/j.aaf.2023.07.006

Vaishali MA, Holeyappa SA, Khairnar SO, Barik S, Tyagi A, Surasani VKR (2024) Growth performance, health status and flesh quality of striped catfish (*Pangasianodon hypophthalmus*) reared in variable stocking densities in biofloc system. Aquaculture 590:741047. https://doi.org/10.1016/j.aquaculture.2024.741047

Vieira JL, Nunes LS, de Menezes FGR, de Mendonça KV, de Sousa OV (2021) An integrated approach to analyzing the effect of biofloc and probiotic technologies on sustainability and food safety in shrimp farming systems. J Clean Prod 318:128618. https://doi.org/10.1016/j.jclepro.2021.128618

Vinatea L, Malpartida J, Carbó R, Andree KB, Gisbert E, Estévez A (2018) A comparison of recirculation aquaculture systems versus biofloc technology culture system for on-growing of fry of Tinca tinca (Cyprinidae) and fry of grey *Mugil cephalus* (Mugilidae). Aquaculture 482:155–161. https://doi.org/10.1016/j.aquaculture.2017.09.041

Wei X-F, Meng S-T, Wang Y-T, Li L, Zhu R, Li D-L, Liu S-Y, Wu L-F (2024) Effects of replacing fish meal with biofloc meal on growth performance, nutrients metabolism, immune response and intestinal microbiota of common carp (*Cyprinus carpio*). Aquaculture 591:741124

Wei Y-F, Wang A-L, Liao S-A (2020) Effect of different carbon sources on microbial community structure and composition of ex-situ biofloc formation. Aquaculture 515:734492. https://doi.org/10.1016/j.aquaculture.2019.734492

Zhang Z, Liu H, Jin J, Zhu X, Han D, Xie S (2024) Towards a low-carbon footprint: current status and prospects for aquaculture. Water Biol Secur 3(4):100290. https://doi.org/10.1016/j.watbs.2024.100290

Zimmermann S, Kiessling A, Zhang J (2023) The future of intensive tilapia production and the circular bioeconomy without effluents: Biofloc technology, recirculation aquaculture systems, bio-RAS, partitioned aquaculture systems and integrated multitrophic aquaculture. Rev Aquacult 15(S1):22–31

Chapter 3
Concepts and Practical Applications of Probiotics, Prebiotics and Synbiotics in Aquaculture

Jiun-Yan Loh, Chou-Min Chong, Chen-Fei Low, and Yan Chen

Abstract Aquaculture represents a rapidly growing segment of global animal production. However, aquaculture diseases and outbreaks pose significant challenges for this sector. The rampant use of pharmaceuticals within fish and shellfish hatcheries and farms has given rise to antimicrobial-resistant (AMR) pathogens, which potentially threaten biosecurity in the aquatic environment. This chapter covers sustainable proactive measures, particularly the application of biological products such as pro-, pre-, and synbiotics. Probiotics refer to live microbes that colonize or proliferate within their host, subsequently producing natural antimicrobial biocompounds that could stimulate vital growth and immune parameters of their host. These beneficial impacts of probiotics are achieved through multiple mechanisms of action, for example, by competitive exclusion of the host's intestinal mucosa tissue, producing compounds that inhibit microbes, promoting nutrient absorption, exhibiting antiviral activity, and modulating the immune system. As for prebiotics, they serve as energy sources for probiotics residing within the gastrointestinal tract of the host. Prebiotics refer to non-digestible fibers, for example, fructooligosaccharides (FOS), mannanoligosaccharides (MOS), and inulin that help in the development of beneficial gut microorganisms, which in turn improve growth, survival, and immune function. Prebiotics help in the modulation of gut microbiota and its morphology, although some prebiotics may inadvertently stimulate pathogenic bacteria. The addition of

J.-Y. Loh (✉)
Tropical Futures Institute, James Cook University, Sims Drive, Singapore
e-mail: james.loh@jcu.edu.au

C.-M. Chong
International Institute of Aquaculture and Aquatic Sciences (I-AQUAS), Universiti Putra Malaysia, Negeri Sembilan, Port Dickson, Malaysia
e-mail: choumin@upm.edu.my

C.-F. Low
Institute of Systems Biology, Universiti Kebangsaan Malaysia, Bangi, Selangor, Malaysia
e-mail: low@ukm.edu.my

Y. Chen
Yazhou Bay Innovation Institute, Hainan Tropical Ocean University, Sanya, PR China
e-mail: cy969398hn@126.com

R. Kumar et al. (eds.), *Sustainability in Aquaculture and Fisheries*,
Springer Water, https://doi.org/10.1007/978-3-032-16605-0_3

probiotics to prebiotics further enhances these benefits, and this combinatory treatment is termed synbiotics. The chapter refers to their use in finfish and shellfish and emphasizes changes in feed efficiency, disease susceptibility, and gut health. Different methods of administration are discussed, such as bioencapsulation in live feed, incorporation into feed pellets, and direct supplementation into culture water. Enhanced molecular methodologies like next-generation sequencing allow for the study of probiotic colonization, providing perspectives into the selection of probiotics. More research is essential to optimize these sustainable approaches for better aquatic animal welfare and production efficiency.

Keywords Antimicrobial · Immunity · Prebiotics · Probiotics · Synbiotics

3.1 Introduction

Aquaculture is rapidly expanding within the animal production industry, which frequently faces challenges associated with disease outbreaks. Such occurrences often cause significant economic losses for aquaculture operators. Fish diseases are attributable to various pathogens, including bacterial, fungal, parasitic, and viral infections, with pathogenic bacteria being the most common culprits. In freshwater aquaculture, notable bacterial pathogens include *Aeromonas* sp., *Edwardsiella* sp., *Streptococcus* sp., *Pseudomonas* sp., and *Vibrio* sp., while marine fish diseases are primarily associated with *Vibrio* species. These bacterial infections typically manifest with similar early-stage symptoms, including fin rot, exophthalmia, and skin ulcers, leading to increased mortality rates and stunted growth in severe cases.

To address these challenges, chemotherapeutic agents such as antibiotics are often used to manage bacterial infections. While they can effectively control the spread of bacterial diseases and eliminate pathogenic bacteria, their widespread use presents significant biosecurity risks to the environment (Wibawati et al. 2024). The use antibiotics has been constrained due to the development of resistant bacterial strains in cultivated organisms, even though these medications are valuable for managing diseases, particularly during critical infection periods (Lazado et al. 2015; Manjeet et al. 2022; Ulkhaq et al. 2024).

Antibiotics have proven effective in controlling fish diseases. The common antibiotics used in aquaculture include aminoglycosides, aztreonam, cephalosporins, ciprofloxacin, imipenem, nitrofurantoin, quinolones, penicillin, sulfamethoxazole, and trimethoprim, chloramphenicol, gentamicin, norfloxacin, oxytetracycline, cefazolin, and aztreonam (Clark et al. 1991; Sahoo and Mukherjee, 2002; Zhang et al. 2005; Zhu et al. 2006; Loh and Ting 2017; Ulkhaq et al. 2024). However, prolonged exposure of aquatic organisms to antibiotics can lead to the development of genes conferring antibiotic resistance within fish. These antibiotic resistance genes can be passed across either through direct inheritance (vertically) or lateral transfer (horizontally), throughout the broader microbial populations in the environment, which may include pathogens that affect human health (Rico et al. 2017; Kumar et al. 2022a).

Environmental microbes can acquire resistance through inheritance (intrinsic resistance) or lateral gene transfer. Intrinsic resistance is an ancient survival mechanism in bacteria, with many species naturally evolving or inheriting traits that help them withstand antimicrobial compounds (Munita and Arias 2016; MacGowan and Macnaughton, 2017). Acquired resistance, on the other hand, results from horizontal gene transfer among bacteria (Romero et al. 2012). Research conducted by Jin Jun and colleagues in 2004 highlighted four distinct genes responsible for tetracycline resistance, notably tet-A, tet-D, tet-B, and tet-G within the bacterium *E. tarda*, these genes are present in mobile plasmids. This discovery underscores the possibility that the widespread application of antibiotics in disease managing and preventing could cause bacteria resistant to the drugs (Jin Jun et al. 2004).

Administration of antibiotics in aquaculture differs markedly from practices used in terrestrial animal farming. In land-based operations, antibiotics are typically delivered orally to animals such as poultry and cattle. Conversely, in aquaculture, antibiotics may be incorporated into fish feed or directly added to culture water to prevent or manage bacterial infections. Such practices can exert considerable pressure on aquatic environments and cultured species. Moreover, the application of diluted antibiotics could accelerate the widespread emergence of antibiotic resistance genes among aquatic microbes. Therefore, alternative preventive and control strategies, such as probiotics, prebiotics, immunostimulants, phytotherapeutics, or vaccines, should be adopted to reduce antibiotic usage in fish farming operations (Loh and Ting 2017).

3.2 Concept of Probiotics

In the nineteenth century, microbiologist Louis Pasteur discovered that lactic acid bacteria (LAB) fermentation could prevent food spoilage (Berche 2012). In 1899, pediatrician Henri Tessier reported that *Bifidobacterium bifidum*, isolated from the fecal samples of breastfed infants, could reduce the risk of diarrhea in infants. Probiotics were first described by Russian biologist Élie Metchnikoff in 1908 (Lauzon et al. 2014). In 1974, Richard B. Parker introduced the term ‘probiotics,’ combining the Latin ‘pro’ (for) and Greek ‘bios’ (life) to signify beneficial microorganisms. Initially, probiotics were defined as live microorganisms in the digestive system that support general well-being of the host (Parker, 1974). In 2001, this concept was refined by the FAO/WHO as “live microbes that, when administered in sufficient quantities, may confer various health benefits to their host.“ For over a century, probiotics have been used as functional products containing beneficial microorganisms aimed at enhancing human health.

In the modern era, the definition of probiotics has evolved to encompass various therapeutic applications, including bioaugmentation, bioremediation, and biocontrol (Lauzon et al. 2014). Bioaugmentation or bioremediation refers to beneficial microorganisms could improve water quality by degrading organic wastes and pollutants (Moriarty 1997; 1998; Gatesoupe 1999). In contrast, when beneficial bacteria are used to inhibit pathogens and maintain gastrointestinal health, this application also termed as biocontrol (Maeda et al. 1997; Moriarty 1998).

In natural ecosystems, the mucosal surfaces of most fish act as a primary defense against pathogenic invasion and provide a habitat for diverse microbial communities (Hansen and Olafsen 1999). These microbes colonize and thrive on mucosal organs such as gills, skin, and the alimentary canal, creating unique microbial profiles for every individual (Kiron 2015). These profiles are significantly influenced by various biotic and abiotic factors, including age, nutrient availability, feeding habits, seasonal food sources, interactions with other organisms, and external environmental conditions such as soil, water, and biochemical availability (Bevins and Salzman 2011). Resident microbes can be either beneficial or pathogenic; some facultatively pathogenic microorganisms coexist with the indigenous bacterial community in fish (Roberts 2012; Sudheesh et al. 2012). Potentially pathogenic bacteria may harm the host when their populations exceed normal levels or when immunity of the hosts is compromised due to stress (Denev et al. 2009; Ringø et al. 2010).

Intestinal microflora represents a highly complex microbial ecosystem, with the gastrointestinal (GI) tract acting as the greatest example of an ecological niche for diverse microorganisms. The composition and abundance of these microbes are significantly influenced by the external environments to which fish are exposed (Loh and Ting 2017). The GI tract of fish can host up to 10^8 heterotrophic organisms and approximately 10^5 anaerobes per gram of tissue. Notably, the microbial density within the GI tract often exceeds that in the aquatic environment (Austin 2006). Additionally, the species and diversity of gut microflora in aquatic organisms differ markedly from those in terrestrial animals, primarily belonging to bacterial phyla such as Actinobacteria, Proteobacteria and Firmicutes (Sanchez et al. 2012; Loh and Ting 2017; Kumar et al. 2025). Common bacterial genera identified in marine and freshwater organisms include *Aeromonas*, *Bacteroides*, *Edwardsiella*, *Flavobacterium*, as well as species from *Debaryomyces* and *Saccharomyces* (Cahill 1990; Gatesoupe 1999; Nayak 2010; Lauzon et al. 2014; Loh and Ting 2017; Urmila et al. 2025). Some of these gut microbiotas may be harnessed in developing probiotics for aquaculture. Interest in probiotic development dates to the early 1960s when dietary probiotic supplementation became available for farm animals to enhance immunity (Tannock 2003). However, the practical application of probiotics has only started two decades ago in the aquaculture industry (Gatesoupe 1999; Kiron 2015). Rigorous screening of beneficial bacteria is essential for developing aquaculture probiotics to ensure quality and confirm the reproducibility of their observed beneficial effects (Loh and Ting 2017). The pathway for probiotic development in aquaculture is shown Fig. 3.1.

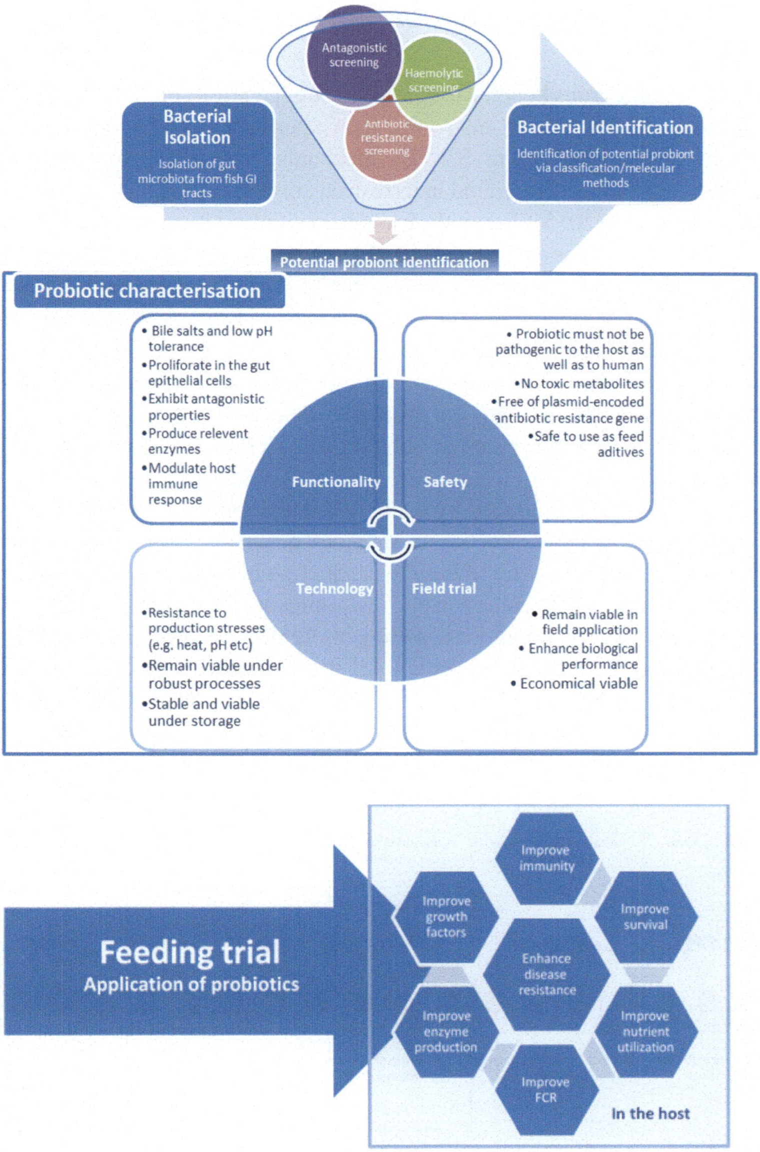

Fig. 3.1 The pathway for probiotic development in aquaculture started with in vitro selection of beneficial bacteria through screening process, includes haemolytic and antibiotic-resistance analyses. A candidate probiont must then be thoroughly evaluated for its functionality, safety, and technological aspects, followed by field testing. Critical assessments during in vivo trials include the candidate's ability to colonize gut epithelium, enhancements of digestibility, nutrient utilization efficacy, and immunomodulatory or immunostimulatory properties. Commercialization of a probiont candidate should only proceed after multiple batches of the bacteria have been tested and confirmed

3.2.1 Probiotics Applications in Fish and Shellfish

Probiotics offer numerous benefits to the host through various biological approaches. Research indicates that these microorganisms perform multiple anti pathogenic mechanisms as illustrated in Fig. 3.2. The mechanisms by which probiotics operate in aquaculture resemble those found in terrestrial animals, a perspective supported by studies such as Kiron (2015). Typically, probiotics employ several strategies to benefit the host (Fig. 3.2). These strategies include competing for adhesion sites, thereby preventing pathogenic organisms from attaching to the host's tissues and producing inhibitory compounds that inhibit harmful microbes. Probiotics also compete for essential nutrients, depriving pathogens of the resources they need to thrive, while enhancing the host's immune response through immunomodulation. Additionally, some probiotic bacteria have antiviral properties that provide protection against viral infections. Collectively, these actions highlight the crucial role probiotics play in promoting host health.

Chemical substances produced by bacteria that exhibit the capability to suppress the growth of pathogens are known as antagonistic or inhibitory compounds. The administration of probiotics stimulates the production of these biocompounds, which help prevent pathogen proliferation in the host's intestine. For example, LAB produce inhibitory compounds that suppress pathogens (either or both Gram-positive and Gram-negative bacteria) in fish. Specific LAB strains, particularly *Lactococcus lactis* subsp. *lactis* CF4MRS (Loh et al. 2014, 2017, 2018; Sharma et al. 2024), *Pediococcus acidilactici*, and *Lactobacillus fermentum* (Chong et al. 2019), have been shown to inhibit fish diseases caused by *E. tarda*, *Escherichia coli*, *Pseudomonas fluorescens*, *P.*

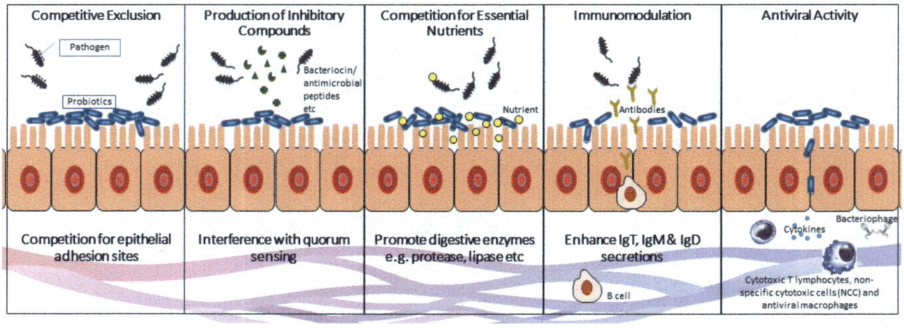

Fig. 3.2 Competitive exclusion occurs in the gut environment as probiotics proliferate and colonize the gut epithelium, thereby preventing pathogen adhesion. The production of inhibitory substances involves probiotics generating chemical compounds that disrupt bacterial quorum sensing and inhibit pathogen growth. Probiotics also compete for essential nutrients, rendering these nutrients unavailable to pathogens, while some probiotics produce enzymes and growth factors that enhance feed digestibility, leading to improved growth performance. Immunomodulation is facilitated by the presence of probiotics, which can enhance macrophage activity and antibody levels against pathogenic invasions. Additionally, probiotic strains exhibit antiviral activity, bolstering immunity and resistance against viral diseases

aeruginosa, *Klebsiella pneumoniae*, *Serratia marcescens*, and *A. hydrophila*. *L. sakei* BK19, also shown promising result in reducing streptococcosis through enhancing macrophage phagocytic as well as peroxidase and lysozyme activities in the kidney of grouper fish (Harikrishnan et al. 2010). Fečkaninová et al. (2017) found that LAB could also mitigate *Aeromonas* infections in salmonids. Numerous studies have focused on LAB and the chemical inhibitory substances they produce. The primary inhibitory substances from LAB are derived from organic acids, such as lactic acid, which can eliminate unwanted bacteria by reducing pH levels in the host's GI tract (Huang et al. 2014; Loh et al. 2017). For instance, lactic acid by *P. pentosaceus* LAB4012 inhibited the growth of *V. anguillarum*. Hydrogen ions from lactic acid was claimed crucial to this inhibitory effect (Huang et al. 2014). Other chemical inhibitory substances produced by LAB include nisin, D-isomers of amino acids, hydrogen peroxide (H_2O_2), diacetyl (2,3-butanedione), acetaldehyde, reuterin, and some bacteriocins (Montville et al. 1995; Cintas et al. 2001; Loh et al. 2017; Yadav et al. 2022, 2023). Bacteriocins are referred as small antimicrobial peptides (AMPs) naturally synthesized by bacteria, including LAB (Drider et al. 2006), which exhibit resistance to its own produced bacterial toxins (Montville et al. 1995). Some bacteriocins may include autolysins, enzymes that induce bacterial autolysis through damaging and destructing of pathogen's cell membranes (Bruno and Montville 1993; Montville et al. 1995; Abee 1995; Moll et al. 1998). Studies have shown that bacteriocins produced by LAB, such as *L. acidophilus*, demonstrate significant antimicrobial activity towards important fish diseases such as *A. hydrophila* and *Streptococcus agalactiae* (Villamil et al. 2014). Nisin, a natural biocompound derived from *L. lactis* subsp. *lactis* exhibited strong antimicrobial activity against *Vibrio* infections in European seabass, *Dicentrarchus labrax* (Touraki et al. 2013). Certain chemical compounds produced by specific LAB strains (primarily from *Bifidobacterium* sp., *Lactobacillus* sp., and *Bacillus cereus*) can disrupt bacterial quorum sensing, interfering with the sensory molecules used in pathogen communication through enzymatic reactions or the production of autoinducer antagonists, also known as quorum quenching molecules (Brown 2011).

The symbiotic relationship between probiotics and their host's GI tract significantly contributes to gut microbial diversity. These beneficial bacteria enhance fish digestion, growth, and stress tolerance, and are vital in regulating the host's nutritional factors (Wang et al. 2008; Loh et al. 2014; Lazado et al. 2015). These gut microbiotas require essential nutrients, e.g., carbon, oxygen, and minerals, for growth and metabolism. Specific probiotic strains, including *Agrobacterium* sp., *Clostridium* sp., and *Staphylococcus* sp., are recognized for their roles in regulating nutritional factors by providing essential nutrients to the host (Sakata 1990; Ringø et al. 1995; Kumar et al. 2022b, 2022c). These enzyme-producing bacteria, for instance, *Bacillus*, could significantly improve digestive processes in aquatic animals (Ray et al. 2012). These bacteria facilitate food digestion by producing exogenous enzymes, which could further stimulate endogenous enzyme production (enzymes secreted by the host) to enhance growth performance in fish and shrimp (Zhou et al. 2009a; 2009b; Shefali et al. 2021; Ziaei-Nejad et al. 2006; Loh and Ting 2016). Fish and shrimp, being monogastric organisms, have a limited capacity to digest complex or fibrous feedstuffs found in artificial diets (Rønnestad et al. 2003). The exogenous enzymes secreted by probiotics can facilitate the digestion of artificial diets, particularly during the weaning period when animals transition from larval to adult stages. As shown in many studies, probiotics could enhance growth and digestive activities in various aquatic animals. For example, *B. coagulans* has shown promising improvements in digestive enzyme activity in Pacific whiteleg shrimp, *Penaeus vannamei*, larvae (Zhou et al. 2009a). Additionally, probiotics, for instances, *Lactobacillus plantarum*, *L. sakei*, *L. lactis*, *E. faecium*, and *B. subtilis* have evidence to improve feed utilization efficiency, physiological performance, and immunity in grouper fish (Sun et al. 2010; Reyes-Becerril et al. 2014).

Probiotics employ competitive strategies on adhesion sites or competitive exclusion as a critical strategy to inhibit or displace pathogens (Ouwehand et al. 2000; Juntunen et al. 2001). Figure 3.2 depicts the mechanism involves competition for essential nutrients and the adhesion space on intestinal mucosal surfaces in the GI tract (Bandyopadhyay and Mohapatra 2009). Probiotics often gain a competitive advantage by producing antimicrobial substances, which could alter the gut environment to make it less favourable to competitors (Rettger 1905; Bandyopadhyay and Mohapatra 2009; Kumar et al. 2024; Loh and Ting 2016). Bacterial competition is a natural ecological phenomenon. The pre-emptive colonization of beneficial bacteria through probiotics, especially during early larviculture stages or before disease onset, may prevent opportunistic pathogens from colonizing the host GI tract and reduce the risk of fish diseases later (Tinh et al. 2008). Literatures have well-documented the importance of probiotics on cultured species. Take a few examples, Iwashita et al. (2015) demonstrated that probiotic mixture of *B. subtilis*, a common brewer's yeast—*Saccharomyces cerevisiae*, and *Aspergillus oryzae* enhanced the survival rates of tilapia when challenged with the pathogens, i.e., *A. hydrophila* and *S. iniae*. The application of *B. subtilis* BT23 resulted in enhanced survival rates of tiger shrimp against vibriosis in a study conducted by Vaseeharan and Ramasamy (2003). In a snook culture, *Bacillus* sp. showed promising results in larval survival while reducing the population of suspected pathogenic bacteria within the GI tracts of *Centropomus undecimalis* (Kennedy et al. 1998).

Like other higher vertebrates, including mammals, fish immunity is modulated by innate (nonspecific), and acquired (specific) immunity. Innate immune system is activated by humoral, and complement factors, or by the cellular responses and antimicrobial peptides, while the acquired immune system requires antibody activation and the establishment of immunological memory to respond effectively to bacterial or antigenic challenges. Numerous studies have indicated that probiotics can regulate bacterial and viral infections. For instance, metabolites of *Pediococcus pentosaceus* LAB 4012 exhibited significant anti-*V. anguillarum* activity by elevating erythrocyte levels and promoting leukocyte proliferation, leading to enhanced phagocytic activity in grouper (Huang et al. 2014). When *E. coioides* larvae were administered *B. clausii* DE5 and *B. pumilus* SE5 bioencapsulated livefeed copepods, it also demonstrated promising results in suppressing *Vibrio* sp. and improving growth performance of fish (Sun et al. 2013). Other probiotics, e.g., *Moraxella* sp., reported to mitigate poliovirus infestations in marine environments (Girones et al. 1989). Certain *Vibrio* strains, such as NICA 1030 and NICA 1031, exhibited strong antiviral activities in infectious hematopoietic necrosis virus (IHNV) (Direkbusarakom et al. 1998). Balcázar et al. (2007) reported several controversial probiotic strains, including *Aeromonas* sp. and *Vibrio* sp., could suppress IHNV in laboratory experiments. Although probiotics show potential as prophylactic agents for managing viral diseases in aquaculture, the immunomodulatory mechanisms underlying their antiviral effects remain inadequately understood. More studies are needed to investigate the complex physiological modulations induced by probiotics at the cellular level (Table 3.1).

3.3 Concepts of Prebiotics

Prebiotics are non-digestible carbon sources such as dietary fibers that improve immune functions and growth factors through promoting beneficial microflora in the gut (Ganguly et al. 2010, 2013). It is essential to differentiate prebiotics from immunostimulants, which are often classified as drugs or nutrients that activate immune components and enhance the immune system. In contrast, prebiotics are food sources metabolized by specific microorganisms, resulting in alterations in the intestinal microbial community that favour beneficial bacteria. These non-digestible fibers fall under the carbohydrate category based on molecular weight or degree of polymerization.

Prebiotics can be classified for human and animal consumption into monosaccharides, polysaccharides, and oligosaccharides (Loh et al. 2020). In aquaculture, oligosaccharides are the commonly used prebiotics for both finfish and shellfish. Specific oligosaccharides utilized in finfish include trans-galactooligosaccharides (TOS), lactulose, and inulin (Gibson et al. 2004; Roberfroid 2007; Ringø et al. 2014; Sharma et al. 2025). In contrast to finfish, prebiotic applications in shellfish farming are significantly less documented. Only a few types are commonly utilized, including FOS, MOS, XOS, IMO, oligofructose, scFOS, and polysaccharide inulin (Daniel and Hoseinifar 2014).

These fibers serve as carbon and energy sources, supporting the establishment of bacteria and probiotics in animal's GI tract environment (Delgado et al. 2011). Although many research interests are currently focused on developing prebiotics for aquaculture, there are still significant research gaps concerning their effects on fish and shellfish metabolism and immunomodulation. Gibson et al. (2004) outline several key criteria for an effective prebiotic. A good prebiotic should resist gastric acidity, avoid hydrolysis by digestive enzymes, and prevent gastrointestinal absorption. Meanwhile, the prebiotics should be able to ferment and convert to an energy source by the intestinal microbiota. An effective prebiotic also should be able to selectively promote the growth or biological activity of beneficial intestinal bacteria, subsequently enhancing health and well-being of its host.

Table 3.1 Common probiotics and their functions in finfish and shellfish, specifically in Asian aquaculture

Aquatic species	Probiotics	Strain [†]	Function(s)	References
Tilapia *Oreochromis niloticus* (Nile tilapia)	*B. pumilus*	Organic Green™	Enhance immunity and survival rate. Protect host from *A. hydrophila* infection	Aly et al. (2008a)
	L. acidophilus	*n/s*	Enhance immunity and protect against *S. iniae* and *P. fluorescens*	Aly et al. (2008b)
	L. rhamnosus	*n/s*	Control against *edwardsiellosis*	Pirarat et al. (2006)
	S. faecium	*n/s*	As a potential growth promoter	Lara-Flores et al. (2003)
	L. lactis	RQ516	Increase in final weight, daily weight gains, serum protein content, and globulin; enhance immune responses	Zhou et al. (2010b)
	P. acidilactici, P. pentosaceus, L. plantarum	*n/s*	High antagonistic effects on *P. mirabilis, K. pneumoniae*, and *S. aureus*; produces lipase, β-galactosidase	Coulibaly et al. (2023)
Catfish *Ictalurus punctatus* (Channel catfish)	*B. subtilis*	AP79	Decrease mortality rate due to *E. ictaluri*	Ran et al. (2012)
Clarias gariepinus (African catfish)	*L. acidophilus*	*n/s*	Enhance fish health, survival, better feed efficiency, and blood parameters	Al-Dohail et al. (2009)
Snakehead *Channa striata* (Striped snakehead)	*L. acidophilus*	*n/s*	Improve growth, immunity; protect host from *A. hydrophila* infection	Talpur et al. (2014)
Carp *Cyprinus carpio* (Common carp)	*S. faecium*	M74	Enhance overall weight gain, feed efficiency, intestinal microflora, and reduce gut microbiome from Enterobacteriaceae group of pathogens	Bogut et al. (1998)

(continued)

Table 3.1 (continued)

Aquatic species	Probiotics	Strain †	Function(s)	References
Catla catla (South Asian carp)	*B. circulans*	PB7	Enhance survival rate and protect host from *A. hydrophila* infection	Bandyopadhyay and Mohapatra (2009)
Labeo rohita (Indian major carp/Rohu)	*B. subtilis*	*n/s*	Improve overall survival rate, weight gain, blood biochemical constituents; protect host from *A. hydrophila* infection	Kumar et al. (2006)
Freshwater Prawn *Macrobrachium rosenbergii* (Freshwater giant prawn)	*L. sporogenes*	*n/s*	Improve survival, growth, and blood biochemical constituents	Seenivasan et al. (2012)
	Micrococcus	*MCCB 104*	Inhibit *V. alginolyticus*	Jayaprakash et al. (2005)
Sea bass *Dicentrarchus labrax* (European sea bass)	*Brevibacillus brevis*	*n/s*	Control against vibriosis. Improve growth performances	Mahdhi et al. (2012)
	Vagococcus fluvialis	*n/s*	Control against *V. anguillarum* infection; colonize GI tracts effectively	Sorroza et al. (2012)
	B. subtilis *L. plantarum*	*n/s*	Both probiotics protect *Artemia* nauplii against *V. anguillarum*	Touraki et al. (2012)
Grouper *Epinephelus coioide* (Orange-spotted grouper)	*B. pumilus*	SE5	Suppress pathogen *S. saprophyticus* and enhance immunity	Yang et al. (2014); Sun et al. (2010, (2011a)
	Psychrobacter sp. *L. lactis* *Enterococcus faecium*	*n/s* MM1 MM4	Improve feed utilization efficiency, digestive enzyme production, and general immunity	Sun et al. (2011b, 2012)
	Pediococcus pentosaceus	LAB4012	Inhibit *V. anguillarum;* enhance growth performance, erythrocyte counts; elevate leukocytes and phagocytic activities	Huang et al. (2014)

(continued)

Table 3.1 (continued)

Aquatic species	Probiotics	Strain †	Function(s)	References
	B. subtilis *Saccharomyces cerevisiae* *Lactobacillus plantarum*	E20 P13 *n/s*	Improve growth performance, and immune resistance through upregulation of innate and humoral immune responses. Control against infections caused by *Streptococcus* sp. and iridovirus	Chiu et al. (2008); Son et al. (2009); Liu et al. (2012)
	B. clausii	DE5	Improve overall feed utilization efficiency, immunity, and overall growth	Sun et al. (2010)
E. bruneus (Kelp grouper)	*L. sakei*	BK19	Protect host against streptococcosis e.g. *S. iniae* and *S. parauberis*; enhance innate immune responses	Harikrishnan et al. (2010)
	B. clausii *B. pumilus*	DE5 SE5	Increase body length and weight Suppress *Aliivibrio fischeri*, and vibriosis	Sun et al. (2013)
Mycteroperca rosacea (Leopard grouper)	*L. sakei*	5–4	Enhance immune responses. The combinatory treatment with inulin could enhance growth and immune responses e.g. IgM expression	Reyes-Becerril et al. (2014)
	Debaryomyces hansenii	CBS8339	Enhance immunity, growth performance, and antioxidant activity; control against *A. hydrophila*	Reyes-Becerril et al. (2011)
Hybrid (*E. lanceolatus* ♂ × *E. fuscoguttatus* ♀)	*B. cereus* *L. acidophilus* *C. butyricum*	BC-01 LAG01 CBG01	Improve growth, host's digestive enzyme activities, and immunity responses	He et al. (2017)
	Bacillus sp.	*B. velezensis* PGSAK01 *B. stercoris* PGSAK05 *B. velezensis* PGSAK17 *B. subtilis* PGSAK19	Improve immunity, overall growth, blood biochemical constituents, gut microbial composition; upregulate immune gene expressions	Amoah et al. (2023); Amoah et al. (2024)
Shrimp/Prawn *Litopenaeus vannamei* (Pacific white shrimp)	*Arthrobacter* sp.	XE-7	Enhance several immune responses, e.g. serum phenoloxidase (PO) activity, hemocyte counts, respiratory burst activity, and phagocytosis	Li et al. (2008)

(continued)

Table 3.1 (continued)

Aquatic species	Probiotics	Strain [†]	Function(s)	References
	B. subtilis	L10/G1	Improve growth performances and control against *V. harveyi* through enhanced immune responses	Zokaeifar et al. (2012)
	B. subtilis	E20	Enhance stress tolerance levels and immunity of postlarvae	Liu et al. (2010)
	Bacillus OJ	PB	Improve immunity, host's intestinal microflora, and control against white spot syndrome virus (WSSV)	Li et al. (2009)
Penaeus monodon (Tiger prawn)	*Bacillus* sp.	P64	Enhance the immunity indices; protect host from *Vibrio* sp. infection	Gullian et al. (2004)
	Bacillus sp.	PV11/PV12	Improve survival rate and control against *Vibrio* sp.	Vijaya Baskar and Kannan (2009)
	Streptomyces sp.	*n/s*	Enhance growth parameters	Das et al. (2006)
	E. faecium *S. phocae*	MC13 PI80	Protect the post larvae against *V. parahaemolyticus*	Swain et al. (2009)
	L. acidophilus	04	Improve survival rate of juvenile P. monodon, control against *V. alginolyticus*	Sivakumar et al. (2012)

[†] *n/s:* not specified

3.3.1 Prebiotics Applications in Fish and Shellfish

As discussed in probiotic research, prebiotics in fish and shellfish primarily focus on improving growth, survival, and disease resistance (Table 3.2). Many studies explore multiple areas, such as enhancing immune responses (Bakke-McKellep et al. 2007; Sang et al. 2009; Zhang et al. 2011; Hoseinifar et al. 2011a; b), altering the gastrointestinal microbiome (Burr et al. 2005; Sang and Fotedar 2010; Zhang et al. 2011; Ortiz et al. 2012), and boosting digestive enzyme production (Wang et al. 2010; Loh and Ting 2016). Most prebiotic developments are evaluated based on several criteria, including their influences on the host's growth, survival performance, feed utilization efficiency, gut microflora composition and morphology, cellular damage within the gut, resistance to invasion of pathogenic bacteria, and the host's innate immune responses.

Figure 3.3 illustrates the overall benefits of prebiotics for finfish and shellfish. Unlike probiotics, the application of prebiotics is simpler in aquaculture, as they are non-living substances. Figure 3.4 shows the various methods for administering pre- and probiotics commonly used in aquaculture, including both direct and indirect supplementation. When incorporating probiotics into diets or supplementing them directly in culture water, it is crucial to consider bacterial viability, particularly regarding their storage and processing to maintain their beneficial effects on the host. In contrast, prebiotic application does not involve biotic factors. Prebiotic supplementation can be achieved through livefeed bioencapsulation. As noted by Daniels et al. (2010), bioencapsulating *Artemia* with the prebiotic mannan-oligosaccharides (MOS) at a concentration of 12 mg L^{-1}, either as a single treatment or alongside the commercial probiotic suspension *Bacillus* sp. at 100 mg L^{-1}, could promote European lobster (*Homarus gammarus*) growth performance and survival rates. Furthermore, administering MOS alone significantly altered the intestinal microbiota of the lobster.

Over the past few decades, studies have documented the effects of prebiotics on the physiological improvements of fish and crustaceans (Burr et al. 2005; Sang et al. 2009; Zhang et al. 2011; Hoseinifar et al. 2011a, b). The inclusion of FOS in dietary formulations shown promising results particularly on the growth of turbot larvae (Mahious et al. 2006), red drum (Zhou et al. 2010a), and Pacific whiteleg shrimp (Zhou et al. 2007). Additionally, the application of MOS, either alone or in combination with probiotics such as *Bacillus* sp., has improved the quality of European lobster larvae (Daniels et al. 2010). MOS supplementation has also demonstrated positive effects on the growth of other crustaceans such as western king prawn, tiger prawn and freshwater crayfish (Hai and Fotedar, 2009; Genc et al. 2007a; Mazlum et al. 2011). Enhancement in growth performances were similarly observed in European sea bass (Torrecillas et al. 2007) and rainbow trout (Rodrigues-Estrada et al. 2008).

Table 3.2 List of prebiotics and their functions in aquaculture

Aquatic species	Prebiotics	Function(s)	References
Carp *Cyprinus carpio* (Common/Jian carp)	MOS	Improve non-specific immune system	Zhou and Li (2004); Staykov et al. (2005)
	Immunogen®	Enhance carcass protein contents, hematological, blood serum, control against *A. hydrophila*	Ebrahimi et al. (2011)
Labeo rohita (Rohu)	MOS	Improve survival after challenging with *A. hydrophila*, promote growth and survival in fingerlings with 1% MOS in diet	Andrew et al. (2009)
Carassius auratus gibelio (Crucian carp)	XOS	Improve growth parameters and enhance enzymatic activity	Xu et al. (2009)
Ctenopharyngodon idellus (Grass carp)	Inulin	Reduce disease susceptibility and protect host from *A. hydrophila* and *E. tarda* infections	Wang and Wang (1997)
Catfish *Ictalurus punctatus* (Channel catfish)	Bio-MOS™	Improve white blood cell (WBC) counts with 2 g kg^{-1} Bio-MOS™	Welker et al. (2007)
	MOS	Improve survival and SGR with 2 g kg^{-1} MOS	Hernández et al. (2012)
Clarias gariepinus (African catfish)	AXOS	Improve gut health through modulation of hindgut microbial population	Rurangwa et al. (2008)
Tilapia *Oreochromis niloticus* (Nile tilapia)	FOS	Improve survival, growth, hematology, non-specific immunity, and innate immune responses	He et al. (2003); Panase et al. (2023)
	MOS	Improve overall immunity, reduce daily feed consumption, improve weight and length gain, and protect against *Streptococcus* sp.	Shelby et al. (2009); Sado et al. (2008)
	Inulin	Improve SGR, survival and protect host from *A. hydrophila* infection through enhancement of lysozyme activity	Ibrahem et al. (2010)
	Grobiotic®-A	Improve weight gain, SGR, survival, and resistance against *A. hydrophila*, enhance neutrophil oxidative production, lower feed conversion ratio (FCR)	Zheng et al. (2011)

(continued)

Table 3.2 (continued)

Aquatic species	Prebiotics	Function(s)	References
O. niloticus × *O. aureus* (Hybrid tilapia)	MOS	Increase height of intestinal villi and thickness of mucosal layer with 0.4% MOS; however, MOS does not improve fish growth and immune system	Yuji-Sado et al. (2015)
		Increase body protein contents; improve survival rate, and non-specific immune responses	Genc et al. (2007b); He et al. (2003)
	scFOS	Improve overall growth rate, feed consumption (lower FCR), survival, and modulate gut microbiome	Zhou et al. (2009b)
O. aureus (Blue tilapia)	Inulin	Protect host from *A. hydrophila* and *E. tarda* infections	Wang and Wang (1997)
Shrimp/Prawn *L. vannamei* (Pacific whiteleg shrimp)	Inulin	Increase the phenoloxidase activity	Luna-González et al. (2012)
	scFOS	Enhance hemocyte respiratory burst with supplementation of 0.025 to 0.8% scFOS	Li et al. (2007)
	Grobiotic®-A	Improve overall survival rate of shrimps in low-salinity (2 ppt) water	Gatlin et al. (2006)
	Previda®	Improve microbial community in shrimp's GIT; enhance immune responses, and blood biochemical constituents, e.g. hemolymph protein	Anuta et al. (2014); Anuta et al. (2021)
Penaeus monodon (Tiger prawn)	MOS	Improve growth, feed utilization efficiency, and gut microbiotas; enhance disease resistance against WSSV	Andrino et al. (2014)
Sea bass *Morone chrysops* x *M. saxatilis* (Hybrid striped bass)	Grobiotic®-AE	Enhance feed utilization efficiency, however not on the growth performance; Improve survival rate when challenged with *S. marinum* and *M. marinum*	Li and Gatlin (2004); Li and Galin (2005)
Dicentrarchus labrax (European sea bass)	MOS	Enhance immune system, and increase disease resistance with supplementation of 0.4% MOS	Torrecillas et al. (2007)
Grouper *Epinephelus fuscoguttatus* x *E. lanceolatus* (Hybrid)	MOS	Improve growth, blood biochemical constituents, liver immunity, and intestinal health; control against *V. harveyi*	Amoah et al. (2023)

(continued)

Table 3.2 (continued)

Aquatic species	Prebiotics	Function(s)	References
Atlantic salmon *Salmo salar*	Inulin	Increase mass of gastrointestinal tract; however, it does not affect absorptive capacity	Refstie et al. (2006)
	MOS	Improve feed utilization efficiency, and survival	Grisdale-Helland et al. (2008)
	GOS	Promote nitrogen and reduce energy losses through non-fecal nitrogenous excretion	Grisdale-Helland et al. (2008)
Turbot *Psetta maxima*	FOS	Increase growth performances of turbot larvae with < 20 g kg^{-1} FOS	Mahious et al. (2006)
	Inulin	Alter gastrointestinal microflora, and improve *Bacillus* species by 14% while reducing *Vibrio* spp.	Mahious et al. (2006)
Trout *Oncorhynchus mykiss* (Rainbow trout)	MOS	Improve feed efficiency, body protein concentration, and survival with 2 g kg^{-1} MOS	Staykov et al. (2007); Yilmaz et al (2007); Grisdale-Helland et al. (2008)
Sea bream *Pagrus major* (Red sea bream)	Lactosucrose	Enhance the integrity of the intestinal epithelial cells in the host	Kihara et al. (1995)
Arctic charr (*Salvelinus alpinus*)	Inulin	Enhance bacterial colonization in the hindgut, primarily by Gram-positive pathogens such as *Staphylococcus*, and *Streptococcus*. However, inulin supplementation exceeding 15% adversely affects enterocytes	Olsen et al. (2001); Ringø et al. (2006)
Sturgeon *Acipenser oxyrinchus desotoi* (Gulf sturgeon)	MOS	Improve overall growth performances, gastrointestinal morphology, and FCR	Pryor et al. (2003)

Abbreviation AXOS = Arabinoxylanoligosaccharide; FCR = feed conversion ratio; FOS = Fructooligosaccharide; GOS = Galactooligosaccharid; MOS = Mannanoligosaccharide; scFOS = Short chain fructooligosaccharide; SGR = Specific growth rate; XOS = Xylooligosaccharide

Survival rates in both fish and shellfish are closely associated with the immunity of GI tract, particularly with gut-associated lymphoid tissue (GALT). GALT can differentiate between pathogenic and probiotic or symbiotic microbes within the gut microbiota. When a pathogen invades, GALT activates cellular and humoral mechanisms that trigger the innate immunity and followed by adaptive immune response (Gómez and Balcázar, 2008). Dietary supplementation with probiotics and/or prebiotics could improve general immunity of the hosts (Burr et al. 2005; Bakke-McKellep et al. 2007; Torrecillas et al. 2007; Sang et al. 2009; Zhang et al. 2011; Mazlum et al. 2011). A

Fig. 3.3 General mechanism of action of prebiotics in finfish and shellfish aquaculture. Prebiotics confer several beneficial effects to hosts, including growth promotion, enhanced survivorship, immunomodulation, disease resistance, modulation of gut microbial ecology, and stimulation of digestive enzyme production

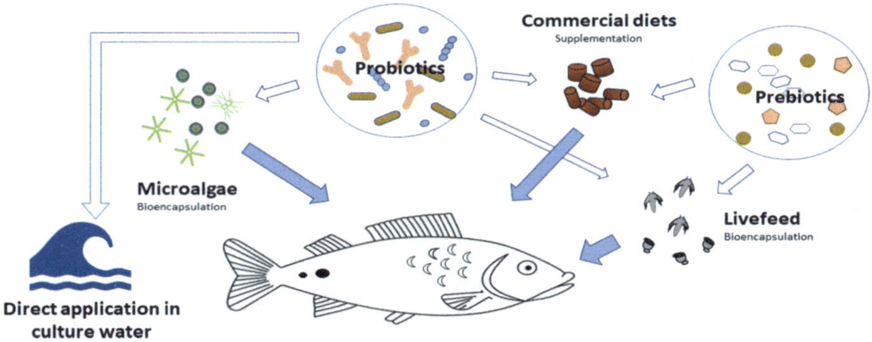

Fig. 3.4 Common methods for administering pro- and prebiotics in finfish and shellfish culture ponds. These products can be supplemented either individually or in combination via live feed (e.g., *Artemia* and rotifers), bioencapsulation, or incorporation into pellet food, and/or directly supplemented to the culture water. The selection of an appropriate administration method is often based on an understanding of the action mechanisms of pro- and prebiotics

healthy gut microflora is important in defending against pathogens through various mechanisms, as discussed in the section on probiotics. However, if the natural balance of the gut microbiota is disrupted by excessive stressors, the gut environment may become favorable for pathogen proliferation. In such cases, administering prebiotics and probiotics is essential to restore and strengthen populations of beneficial bacteria. A consistent supply of prebiotics helps control harmful bacteria; for example, MOS and GroBiotic®-A (contains autolyzed brewer's yeast) have been shown to modify the gut microbiota by establishing well-balanced microbial populations (Dimitroglou et al. 2009; Burr et al. 2008, 2009).

Prebiotics could modulate the gut microflora and intestinal morphology of the GI tracts in fish and shellfish. For instance, one study demonstrated that supplementing rainbow trout with GroBiotic®-A increased the total counts of anaerobic bacteria and LAB in their GI tract (Azari et al. 2011a, b). A similar effect was observed in beluga juveniles (*Huso huso*), where inulin, administered through partial enzymatic hydrolysis at a dosage of 20 g kg^{-1}, significantly increased LAB counts in the GI tract for one week.

Different types of prebiotics and their polymerization can directly influence gut microbiota diversity. For example, the addition of AXOS affected the allochthonous microbiota in Siberian sturgeon (*Acipenser baerii*), leading to the establishment of unique beneficial bacterial species such as *C. colicanis*, *C. baratii*, *L. raffinolactis*, *L. lactis*, *L. aviaries*, and *Eubacterium budayi* exclusively in treated fish (Geraylou et al. 2012). Similarly, prebiotics have been effective in modulating gut microbiota in shellfish. For instance, MOS supplementation in European lobster (*H. gammarus* L.) increased the population count of aerobic-heterotrophic gut microbes at larval stages (Daniels et al. 2010).

Although prebiotics are well-known in modulation of gut microbiota; however, inadequate use of prebiotics could also lead to an increase in pathogenic bacteria. Zhou et al. (2007) found that scFOS could increase *S. faecalis* and *V. parahaemolyticus* in Pacific whiteleg shrimp, *L. vannamei*, raising the risk of endemic infection. Research on the gut microbial community of *L. vannamei* also revealed that using scFOS as a dietary supplement altered the microbial structure, resulting in a significant reduction of potential probiotic species like *Roseobacter* sp., while the abundance of unidentified microbiota increased substantially (Li et al. 2007). Furthermore, inulin supplementation at levels of 1–3% reduced various biological parameters such as weight gain, energy retention, and protein retention in beluga larvae (Akrami et al. 2009). While a study on dietary supplementation with MOS, FOS, and GOS (10 g kg^{-1}) had no observed effect on the growth and digestibility of Atlantic salmon (*Salmo salar*) (Grisdale-Helland et al. 2008).

On a positive side, prebiotics have been shown to enhance the luminal integrity, length of microvilli, and mucosal folds in the microvilli of teleost species (Sweetman et al. 2008; Dimitroglou et al. 2010; Sang et al. 2014). Oligosaccharides like MOS may improve gut development in various species, including cobia larvae (Salze et al. 2008), European lobster (Daniels et al. 2010), rainbow trout (Yilmaz et al. 2007), white sea bream (Dimitroglou et al. 2009), and tiger prawn (Sang et al. 2014). These improvements are primarily attributed to an increased gut surface area, which enhances digestive enzyme production and nutrient absorption, ultimately promoting better feed conversion rates and growth performance. For example, FOS has been shown to support the growth of lactobacilli, bifidobacteria, and *Bacillus* sp., which stimulate or produce digestive enzymes like proteases and amylases for starch and protein digestion (Ben Messaoud et al. 2004; Liu et al. 2009; Kiran et al. 2025; Loh and Ting 2016).

3.4 Synbiotics and Their Applications in Aquaculture

Probiotics and prebiotics exhibit distinct functionalities and provide various beneficial effects to hosts when administered concurrently, a combination referred to as synbiotics. This combinatory treatment aims to enhance the overall number, viability, and proliferation of probiotic bacteria in the gut, with prebiotics often serving as carbon sources for the fermentation of probiotics. An array of commercial synbiotic products is currently available in the market, such as Sanolife PRO-F (INVE Aquaculture), AquaStar (Biomin), and Aquate™ Defender (Alltech). Investigating the efficacy of synbiotics necessitates comprehensive approaches to ensure that the ultimate objectives of benefiting the hosts are met. A key aspect of this investigation involves assessing the impact of synbiotics on host survival and growth parameters, which serve as vital indicators of health and productivity in aquaculture. Additionally, the presence of synbiotics within the microbial ecology of host gastrointestinal tracts and aquatic environments is a subject of increasing interest, offering insights into their influence on the broader ecosystem. The impact of synbiotics on host immunological parameters is also noteworthy, as it underscores their potential to bolster immunity and prevent disease in cultured species (Pardede et al. 2024; Nisha et al. 2025; Rani et al. 2025; Liu et al. 2025).

The incorporation of synbiotics into commercial feed pellets necessitates meticulous attention to preserve the viability of probiotics. While the application of synbiotics in hatcheries or nurseries can be accomplished through live feed bioencapsulation, the use of gnotobiotic cultures and specialized facilities is essential to mitigate the risk of pathogenic contamination prior to the bioencapsulation processes.

Advanced technologies including next-generation sequencing (NGS), fluorescence in situ hybridization (FISH), nanopore sequencing, and terminal restriction fragment length polymorphism (T-RFLP), are now essential tools for scientists investigating probiotic colonization and bacterial adhesion in the gastrointestinal system (Loh and Ting 2017). These advancements in scientific methodologies offer valuable insights that deepen our understanding of microbial ecology and the interactions between these microbes and their host systems. To ensure the biosafety of synbiotics, more in-depth research is crucial to investigate the complex interrelationship between hosts, bacterial community and the surrounding environment.

3.5 Conclusion

In general conclusion, probiotics could enhance the host's immune function through modulating and increasing the activation of immune effectors (cellular immunity) and promoting the secretion of macromolecules like antibodies, complement proteins, and antimicrobial peptides (humoral immunity). Probiotics protect the host from pathogens through different approaches such as nutrient-space-competitive exclusion and the release of inhibitory biocompounds, allowing them to establish a healthy microflora in the GI tract. This colonization strengthens the intestinal barrier against pathogen adhesion while producing substances that inhibit pathogen growth. Additionally, probiotics secrete enzymes that promote food digestion and absorption, energy utilization efficiency and growth. Their ability to break down complex carbohydrates and fibers enables the utilization of prebiotics. Prebiotics, such as non-digestible fiber prebiotics, serve as carbon sources for probiotics, stimulating their growth and activity, which then manipulates the GI microbiota composition in favor of probiotic dominance. Consequently, the health of farmed aquatic animals can be enhanced with pre- and probiotics.

Several commercially available products containing pre- and/or probiotics are used in aquaculture. However, extensive and sustained research is necessary to develop effective pre- and probiotic products. The mechanisms of action for these products must be verified and validated in laboratory settings before clinical trials can be conducted on aquatic farms.

Future Prospects

The application of probiotics and prebiotics, either individually or in combination (synbiotics), is increasingly important in many aquaculture farms, particularly those that have employed various antibiotics for disease management and are now facing more challenging bacterial infections due to the development of antibiotic-resistant pathogens. Unlike terrestrial animal production, where antibiotic treatments are often administered through diets, aquaculture practices present a lower risk of chemical dispersion in the water column. However, the use of antibiotics in aquaculture contributes to the proliferation of AMR genes within both targeted and non-targeted bacterial communities, posing significant concerns for aquaculture operations. The incorporation of probiotics and prebiotics, or synbiotics, may serve as a prophylactic strategy for disease management in aquaculture, enhancing water biosecurity and promoting the host's immune response through the activation of phagocytes such as T and B lymphocytes and immune functional genes in spleen, liver, kidney as well as intestine.

References

Abee T (1995) Pore-forming bacteriocins of gram-positive bacteria and self-protection mechanisms of producer organisms. FEMS Microbiol Lett 129:1–10

Akrami R, Hajimoradloo A, Matinfar A, AbedianKinari A (2009) Effect of dietary prebiotic inulin on growth performance, intestinal microflora, body composition and hematological parameters of Juvenile Beluga, *Huso huso* (Linnaeus, 1758). J World Aquac Soc 40(6):771–779

Al-Dohail MA, Hashim R, Aliyu-Paiko M (2009) Effects of the probiotic, *Lactobacillus acidophilus*, on the growth performance, haematology parameters and immunoglobulin concentration in African Catfish (*Clarias gariepinus*, Burchell 1822) fingerling. Aquac Res 40:1642–1652

Aly SM, Ahmed YAG, Ghareeb AAA, Mohamed MF (2008a) Studies on *Bacillus subtilis* and *Lactobacillus acidophilus*, as potential probiotics, on the immune response and resistance of *Tilapia nilotica* (*Oreochromis niloticus*) to challenge infections. Fish Shellfish Immunol 25:128–136

Aly SM, Mohamed MF, John G (2008b) Effect of probiotics on the survival, growth and challenge infection in *Tilapia nilotica* (*Oreochromis niloticus*). Aquac Res 39:647–656

Amoah K, Huang Y, Cai J, Wang B, Shija VM, Jin X, Anokyewaa MA, Jiang M (2023) Host gut-derived *Bacillus* probiotics supplementation improves growth performance, serum and liver immunity, gut health, and resistive capacity against Vibrio harveyi infection in hybrid grouper (♀ *Epinephelus fuscoguttatus* × ♂ *Epinephelus lanceolatus*). Anim Nutr 14:163–184. https://doi.org/10.1016/j.aninu.2023.05.005

Amoah K, Huang Y, Cai J, Wang B, Shija VM, Jin X, Anokyewaa MA, Jiang M (2024) Identification and characterization of four *Bacillus* species from the intestine of hybrid grouper (*Epinephelus fuscoguttatus* ♀ × *E. lanceolatus* ♂), their antagonistic role on common pathogenic bacteria, and effects on intestinal health. Fish Shellfish Immunol 152:109795. https://doi.org/10.1016/j.fsi.2024.109795

Andrew SR, Sahu NP, Pal AK, Kumar S (2009) Haematological modulation and growth of *Labeo rohita* fingerlings: effect of dietary mannan oligosaccharide, yeast extract, protein hydrolysate and chlorella. Aquac Res 41(1):61–69

Andrino KGS, Apines-Amar MJS, Janeo RL, Corre VL Jr (2014) Effects of dietary mannan oligosaccharide (MOS) and β-glucan on growth, immune response and survival against white spot syndrome virus (WSSV) infection of juvenile tiger shrimp *Penaeus monodon*. AACL Bioflux 7(5):321–332

Anuta JD, Buentello A, Patnaik S, Hume ME, Mustafa A, Gatlin III DM, Lawrence AL (2014) Effects of dietary supplementation of a commercial prebiotic Previda® on survival, growth, immune responses, and gut microbiota of Pacific white shrimp, *Litopenaeus vannamei*. Aquac Nutr. https://doi.org/10.1111/anu.12257

Anuta JD, Buentello A, Patnaik S, Hume ME, Mustafa A, Gatlin DM III, Lawrence AL (2021) Effects of dietary supplementation of a commercial prebiotic Previda® on survival, growth, immune responses, and gut microbiota of Pacific white shrimp *Litopenaeus Vannamei*. Aquac Nutr 27(6):2170–2181

Austin B (2006) The bacterial microflora of fish, revised. Sci World J 6:931–945

Azari AH, Hashim R, Azari Takami G, Farabi SMV, Darvish M, Safari R (2011a) Effect of GroBiotic®-A on the growth performance and intestinal microflora on rainbow trout (*Oncorhynchus mykiss* Walbaum). J Res Biol 1:325–334

Azari AH, Hashim R, Habibi Rezaei M, Sharifzadeh Baei M, Najafpour S, Roohi A, Darvishi M (2011b) The effects of commercial probiotic and prebiotic usage on growth performance, body composition and digestive enzyme activities in juvenile rainbow trout (*Oncorhynchus mykiss*). World Appl Sci J 14:26–35

Bakke-Mckellep AM, Penn MH, Salas PM, Refstie S, Sperstad S, Landsverk T, Ringø E, Krogdahl Å (2007) Effects of dietary soybean meal, inulin and oxytetracycline on gastrointestinal histological characteristics, distal intestinal cell proliferation and intestinal microbiota in Atlantic salmon (*Salmo solar* L.) Br J Nutr 97: 699–713

Balcázar JL, De Blas I, Ruiz-Zarzuela I, Vendrell D, Gironés O, Muzquiz JL (2007) Enhancement of the immune response and protection induced by probiotic lactic acid bacteria against furunculosis in rainbow trout (*Oncorhynchus mykiss*). FEMS Immunol Med Microbiol 51:185–193. https://doi.org/10.1111/j.1574-695X.2007.00294.x

Bandyopadhyay P, Mohapatra PKD (2009) Effect of a probiotic bacterium *Bacillus circulans* PB7 in the formulated diets: on growth, nutritional quality and immunity of *Catla catla* (Ham.) Fish Physiol Biochem 35:467–478

Ben Messaoud E, Ben Ali MB, Elleuch N, Masmoudi NF, Bejar S (2004) Purification and properties of a maltoheptaose- and maltohexaose-forming amylase produced by Bacillus subtilis US116. Enzyme Microb Technol 34:662–666

Berche P (2012) Louis Pasteur, from crystals of life to vaccination. Clin Microbiol Infect 5:1–6

Bevins C, Salzman N (2011) The potter's wheel: the host's role in sculpting its microbiota. Cell Mol Life Sci 68:3675–3685

Bogut I, Milakovic Z, Bukvic Z, Brkicand S, Zimmer R (1998) Influenceof probiotic *Streptococcus faecium* M74 on growth and content of intestinal microfllora in carp *Cyprinus carpio*. Czeh J Anim Sci 43:231–235

Brown M (2011) Modes of action of probiotics: resent developments. J Anim Vet Adv 10(14):1895–1900

Bruno MEC, Montville TJ (1993) Common mechanistic action of bacteriocins from lactic acid bacteria. Appl Environ Microbiol 59:3003–3010

Burr G, Gatlin DM III, Hume M (2009) Effects of the prebiotics GroBiotic® - A and inulin on the intestinal microbiota of red drum, *Sciaenops ocellatus*. J World Aquac Soc 40:440–449

Burr G, Gatlin DM III, Ricke S (2005) Microbial ecology of the gastrointestinal tract of fish and the potential of prebiotics and probiotics in finfish aquaculture. J World Aquac Soc 36:425–436

Burr G, Hume M, Neill WH, Gatlin DM III (2008) Effects of the prebiotics on nutrient digestibility of a soybean-meal-based diet by red drum *Sciaenops ocellatus* (Linnaeus). Aquac Res 39:1680–1686

Cahill M (1990) Bacterial flora of fishes: a review. Microb Ecol 19:21–41

Chiu ST, Tsai RT, Hsu JP, Liu CH, Cheng W (2008) Dietary sodium alginate administration to enhance the nonspecific immune responses, and disease resistance of the juvenile grouper *Epinephelus fuscoguttatus*. Aquaculture 277:66–72

Chong WW, Lim CSY, Lai KS, Loh JY (2019) In vitro antimicrobial assessment on lactic acid bacteria isolated from common freshwater fishes. Asia Pac J Mol Biol Biotechnol 27(2):18–25

Cintas LM, Casaus MP, Herranz C, Nes IF, Hernández PE (2001) Review: bacteriocins of lactic acid bacteria. Food Sci Technol Int 7:281–305

Clark RB, Lister PD, Janda JM (1991) *In vitro* susceptibilities of *Edwardsiella tarda*to 22 antibiotics and antibiotic-beta-lactamase-inhibitor agents. Diagn Microbiol Infect Dis 14:173–175

Coulibaly WH, Kouadio NR, Camara F, Diguta C, Matei F (2023) Functional properties of lactic acid bacteria isolated from Tilapia (*Oreochromis niloticus*) in Ivory Coast. BMC Microbiol 23(1):152. https://doi.org/10.1186/s12866-023-02899-6

Daniels C, Hoseinifar SH (2014) Prebiotic application in shellfish. In: Merrifield D, Ringø E (eds) Aquaculture nutrition: gut health, probiotics and prebiotics, 1st edn. John Wiley & Sons, Ltd, 465 p

Daniels CL, Merrifield DL, Boothroyd DP, Davies SJ, Factor JR, Arnold KE (2010) Effect of dietary *Bacillus* spp. and mannan oligosaccharides (MOS) on European lobster (*Homarus gammarus* L.) larvae growth performance, gut morphology and gut microbiota. Aquaculture 304:49–57

Das S, Lyla PS, Khan SA (2006) Application of *Streptomyces* as a probiotic in the laboratory culture of *Penaeus monodon* (Fabricius). Israeli J Aquac Bamidgeh 58(3):198–204

Delgado GTC, Tamashiro WMDSC, Junior MRM, Moreno YMF, Pastore GM (2011) The putative effects of prebiotics as immunomodulatory agents. Food Res Int 44:3167–3137

Denev S, Staykov Y, Moutafchieva R, Beev G (2009) Microbial ecology of the gastrointestinal tract of fish and the potential application of probiotics and prebiotics in finfish aquaculture. Int Aquat Res 1:1–29

Dimitroglou A, Merrifield DL, Mote R, Davis SJ, Spring P, Sweetman J, Bradley G (2009) Dietary mannan oligosaccharide supplementation modulates intestinal microbiology ecology and improves gut morphology of rainbow trout, *Oncorhynchus mykiss* (Walbaum). J Anim Sci 87:3226–3234

Dimitroglou A, Merrifield DL, Spring P, Sweetman J, Moate R, Davies SJ (2010) Effects of mannan oligosaccharide (MOS) supplementation on growth performance, feed utilization, intestinal histology and gut microbiota of gilthead sea bream (*Sparus aurata*). Aquaculture 300:182–188

Direkbusarakom S, Yoshimizu M, Ezura Y, Ruangpan L, Danayadol Y (1998) *Vibrio* spp. the dominant flora in shrimp hatchery against some fish pathogenic viruses. J Mar Biotechnol 6:266–267

Drider D, Fimland G, Héchard Y, Mcmullen LM, Prévost H (2006) The continuing story of class IIa bacteriocins. Microbiol Mol Biol Rev 70(2):564–582

Ebrahimi G, Ouraji H, Khalesi MK, Sudagar M, Barari A, Zarei Dangesaraki M, Jani Khalili KH (2011) Effects of a prebiotic, Immunogen®, on feed utilization, body composition, immunity and resistance to *Aeromonas hydrophila* infection in the common carp *Cyprinus carpio* (Linnaeus) fingerlings. J Anim Physiol Anim 96:591–599

Fečkaninová A, Koščová J, Mudroňová D, Popelka P, Toropilová J (2017) The use of probiotic bacteria against Aeromonas infections in salmonid aquaculture. Aquaculture 469:1–8

Ganguly S, Dora KC, Sarkar S, Chowdhury S (2013) Supplementation of prebiotics in fish feed: a review. Rev Fish Biol Fisheries 23:195–199

Ganguly S, Paul I, Mukhopadhayay SK (2010) Application and effectiveness of immunostimulants, probiotics, and prebiotics in aquaculture: a review. Israeli J Aquac Bamidgeh 62(3):130–138

Gatesoupe FJ (1999) The use of probiotics in aquaculture. Aquaculture 180:147–165

Gatlin DM III, Li P, Wang X, Burr GS, Castille F, Lawrence AL (2006) Potential application of prebiotics in aquaculture. In: Suarez EC, Marie DR, Salazar MT, Lopez MGN, Cavazos DAV, Cruz ACP, Ortega AG (eds) Avances en Nutricion Acuicola VIII: VIII Simposium International de Nutricion Acuicola. Universidad Autonoma de Nuevo Leon, Monterrey, Nuevo Leon, Mexico, pp 371–376

Genc MA, Yilmaz E, Genc E, Aktas M (2007a). Effects of dietary mannan oligosaccharides (MOS) on growth, body composition, and intestine and liver histology of the hybrid Tilapia (*Oreochromis niloticus* × *O. aureus*). Israeli J Aquac Bamidgeh 59:10–16

Genc MA, Aktas M, Genc E, Yilmaz E (2007b) Effects of dietary manna oligosaccharide on growth, body composition and hepatopancreas histology of *Penaeus semisulcatus* (de Haan 1844). Aquac Nutr 13:156–161

Geraylou Z, Souffreau C, Rurangwa E, D'Hondt S, Callewaert L, Courtin CM, Delcour JA, Buyse J, Ollevier F (2012). Effects of arabinoxylan-oligosaccharides (AXOS) on juvenile Siberian sturgeon (*Acipenser baerii*) performance, immune responses and gastrointestinal microbial community. Fish Shellfish Immunol 33:718–724

Gibson GR, Probert HM, Loo JV, Rastall RA, Roberfroid MB (2004) Dietary modulation of the human colonic microbiota: updating the concept of prebiotics. Nutr Res Rev 17:259–275

Girones R, Jofre JT, Bosch A (1989) Isolation of marine bacteria with antiviral properties. Can J Microbiol 35:1015–1021

Gómez GD, Balcázar JL (2008) A review on the interactions between gut microbiota and innate immunity of fish. Fed Eur Microbiol Soc Immunol Med Microbiol 52:145–154

Grisdale-Helland B, Helland SJ, Gatlin DM III (2008) The effects of dietary supplementation with mannanoligosaccharide, fructooligosaccharide or galactooligosaccharide on the growth and feed utilization of Atlantic salmon (*Salmo salar*). Aquaculture 283:163–176

Gullian M, Thompson F, Rodriguez J (2004) Selection of probiotic bacteria and study of their immunostimulatory effect in *Penaeus vannamei*. Aquaculture 233:1–14

Hai VN, Fotedar R (2009) Comparison of the effects of the probiotics (Bio-Mos® and β-1, 3-D-glucan) and the customized probiotics (*Pseudomonas synxantha* and *P. aeruginosa*) on the culture of juvenile western king prawns (*Penaeus latisulcatus* Kishinouye, 1896). Aquaculture 289:310–316

Hansen GH, Olafsen JA (1999) Bacterial interactions in early life stages of marine cold water fish. Microb Ecol 38:1–26

Harikrishnan R, Balasundaram C, Heo MS (2010) *Lactobacillus sakei* BK19 enriched diet enhances the immunity status and disease resistance to streptococcosis infection in kelp grouper *Epinephelus Bruneus*. Fish Shellfish Immunol 29:1037–1043

He RP, Feng J, Tian XL, Dong SL, Wen B (2017) Effects of dietary supplementation of probiotics on the growth, activities of digestive and non-specific immune enzymes in hybrid grouper (*Epinephelus lanceolatus*♂ × *Epinephelus fuscoguttatus*♀). Aquac Res 48:5782–5790

He S, Xu G, Wu Y, Weng H, Xie H (2003) Effects of IMO and FOS on the growth performance and non-specific immunity in hybrid tilapia. Chin Feed 23:14–15 (in Chinses)

Hernández DR, Santinón JJ, Sánchez D, Domitrovic HA (2012) Effects of prebiotics on growth and survival of catfish (*Rhamdia quelen*) juveniles. Interciencia 37:627–631

Hoseinifer SH, Mirvaghefi A, Merrifield DL, Amiri B, Yelghi S, Bastami K (2011a) The study of some haematological and serum biochemical parameters of juvenile beluga (*Huso huso*) fed oligofructose. Fish Physiol Biochem 37:91–96

Hoseinifer SH, Mirvaghefi A, Mojazi Amiri B, Rostami HK, Merrifield DL (2011b) The effects of oligofructose on growth performance, survival and autochthonous intestinal microbiota of beluga (*Huso huso*) juveniles. Aquac Nutr 17:498–504

Huang JB, Wu YC, Chi SC (2014) Dietary supplementation of *Pediococcus pentosaceus* enhances innate immunity, physiological health and resistance to *Vibrio anguillarum* in orange-spotted grouper (*Epinephelus coioides*). Fish Shellfish Immunol 39:196–205

Ibrahem MD, Fathi M, Mesalhy S, Abd El-Aty AM (2010) Effect of dietary supplementation of inulin and vitamin C on the growth, hematology, innate immunity, and disease resistance of Nile tilapia (*Oreochromis niloticus*). Fish Shellfish Immunol 29:241–246

Iwashita MKP, Nakandakare IB, Terhune JS, Wood T, Ranzani-Paiva MJT (2015) Dietary supplementation with *Bacillus subtilis*, *Saccharomyces cerevisiae* and *Aspergillus oryzae* enhance immunity and disease resistance against *Aeromonas hydrophila* and *Streptococcus iniae* infection in juvenile tilapia *Oreochromis niloticus*. Fish Shellfish Immunol 43:60–66

Jayaprakash NS, Pai SS, Anas A, Preetha R, Philip R, Singh ISB (2005) A marine bacterium, *Micrococcus* MCCB 104, antagonistic to vibrios in prawn larval rearing systems. Dis Aquat Org 68:39–45

Jin Jun L, Bum Jeong J, Huh MD, Chung JK, Choi DI, Lee CH, Jeong HD (2004) Detection of tetracycline-resistance determinants by multiplex polymerase chain reaction in *Edwardsiella tarda* isolated from fish farms in Korea. Aquaculture 240:89–100

Juntunen M, Kirjavainen PV, Ouwehand AC, Salminen SJ, Isolauri E (2001) Adherence of probiotic bacteria to human mucus in healthy infants and during rotavirus infection. Clin Diagn Lab Immunol 8:293–296

Kennedy SB, Tucker JW, Neidig CL, Vermeer GK, Cooper VR, Jarrell JL, Sennett DG (1998) Bacterial management strategies for stock management of warm water marine fish: a case study with common snook (*Centropomus undecimalis*). Bull Mar Sci 62:573–588

Kihara M, Ohba K, Sakata T (1995) Trophic effect of dietary lactosucrose on intestinal tunica muscularis and utilization of this sugar by gut microbes in red seabream, *Pagrus major*, a marine carnivorous teleost, under artificial rearing. Comp Biochem Physiol 112:629–634

Kiran, Kumar N, Sindhu S, Gawaria B, Ansari A, Dhariwal A, Sharma P, Kumar R (2025) GWAS approaches in medicinal plants for enhanced secondary metabolite production. In: Chen JT (ed) Biotechnology, multiple omics, and precision breeding in medicinal plants. CRC Press, pp 119–134. https://doi.org/10.1201/9781003475491-10

Kiron V (2015) Gastrointestinal microorganisms of fish and probiotics. In: Lee CS, Lim C, Gatlin D III, Webster CD (eds) Dietary nutrients, additives and fish health, 1st edn. John Wiley & Sons, Inc, 355 p

Kumar, R. Singh, D., and Yadav, R. (2025). Toxic effects of chlorantraniliprole and fipronil on *Eisenia fetida*: from growth and reproductive parameters to avoidance behavior. *Environmental Engineering and Management Journal,* 24(7), 1499–1508. https://doi.org/10.30638/eemj.2025.117

Kumar R, Gupta R, Dahiya T, Yadav R, Saifi R, Yodha K, Kumari B, Singh P, Shefali (2022a) Surveillance of *Heteropneustes fossilis* (stinging catfish) diseases and health management practices in Haryana. Appl Ecol Environ Sci 10(4):261–267. https://doi.org/10.12691/aees-10-4-12

Kumar R, Gupta RK, Yadav R, Saifi R, Yodha K, Pooja (2022b) *Eisenia fetida* as protein source for growth enhancement of *Heteropneustes fossilis*. Egypt J Aquatic Biol Fish 26(2):577–588. https://doi.org/10.21608/ejabf.2022.234553

Kumar R, Yadav R, Kumar Gupta R, Pooja (2022c) A sustainable way for fish health management by replacement of chemical and drugs by earthworm. In: Environmental degradation in Asia: land degradation, environmental contamination, and human activities. Springer International Publishing, Cham, pp 329–352. https://doi.org/10.1007/978-3-031-12112-8_16

Kumar R, Mukherjee SC, Prasad KP, Pal AK (2006) Evaluation of *Bacillus subtilis* as a probiotic to Indian major carp *Labeo rohita* (Ham). Aquac Res 37:1215–1221

Kumar R, Yadav R, Gupta RK, Pal A, Yodha K, Kumar A (2024) Antioxidant enzyme activities and markers of oxidative stress in the life cycle of different Earthworm species. Indian J Exp Biol 62(07):578–583. https://doi.org/10.56042/ijeb.v62i07.12084

Lara-Flores M, Olvera-Novoa MA, Guzman-Mendez BE, Lopez Madrid W (2003) Use of bacteria *Streptococcus faecium* and *Lactobacillus acidophilus*, and the yeast *Saccharomyces cerevisiae* as growth promoters in the Nile tilapia (*Oreochromis niloticus*). Aquaculture 216:193–201

Lauzon HL, Dimitroglou A, Merrifield DL, Ringø E, Davies SJ (2014) Probiotics and prebiotics: concepts, definitions and history. In: Merrifield D, Ringø E (eds) Aquaculture nutrition: gut health, probiotics and prebiotics, 1st edn. Wiley and Sons Ltd, 465 p

Lazado C, Caipang C, Estante E (2015) Prospects of host-associated microorganisms in fish and penaeids as probiotics with immunomodulatory functions. Fish Shellfish Immunol 45(1):2–12

Li J, Tan B, Mai K (2009) Dietary probiotic *Bacillus* OJ and isomaltooligosaccharides influence the intestine microbial populations, immune responses and resistance to white spot syndrome virus in shrimp *(Litopenaeus vannamei*). Aquaculture 291:35–40

Li JQ, Tan BP, Mai KS, Ai QH, Zhang WB, Liufu ZG, Xu W (2008) Immune responses and resistance against *Vibrio parahaemolyticus* induced by probiotic bacterium Arthrobacter XE-7 in Pacific white shrimp, *Litopenaeus vannamei*. J World Aquac Soc 39:477–489

Li P, Gatlin DM III (2004) Dietary brewer's yeast and the prebiotic GroBiotic™ AE influence growth performance, immune responses and resistance of hybrid striped bass (*Morone chrysops* × *M. saxatilis*) to *Streptococcus iniae* infection. Aquaculture 231:445–456

Li P, Gatlin DM III (2005) Evaluation of the prebiotic Grobiotic®-A and brewer's yeast as dietary supplements for sub-adult hybrid striped sea bass (*Morone chrysops x M. saxatilis*) challenged in situ with *Mycobacterium marinum*. Aquaculture 248:197–205

Li P, Burr GS, Gatlin DM III, Hume ME, Patnaik S, Castille FL, Lawrence AL (2007) Dietary supplementation of short-chain fructooligosaccharides influences gastrointestinal microbiota composition and immunity characteristics of Pacific white shrimp, *Litopenaeus vannamei*, cultured in a recirculating system. J Nutr 137:2763–2768

Liu CH, Chiu CH, Wang SW, Cheng W (2012) Dietary administration of the probiotic, *Bacillus subtilis* E20, enhances the growth, innate immune responses, and disease resistance of the grouper, *Epinephelus coioides*. Fish Shellfish Immunol 33:699–706

Liu CH, Chiu CS, Ho PL, Wang SW (2009) Improvement in the growth performance of white shrimp, *Litopenaeus vinnamei*, by a protease producing probiotic, *Bacillus subtilis* E20, from natto. J Appl Microbiol 107:1031–1041

Liu KF, Chiu CH, Shiu YL, Cheng W, Liu CH (2010) Effects of the probiotic, *Bacillus subtilis* E20, on the survival, development, stress tolerance, and immune status of white shrimp, *Litopenaeus vannamei* larvae. Fish Shellfish Immunol 28(5–6):837–844

Liu Y, Chen QQ, Andin VC, Chor WK, Chong CM, Lee PT, Lim CSY, Loh JY (2025) Effects of immunostimulants on physiological performances, immune gene expression, liver, and intestinal protection against *Vibrio parahaemolyticus* infection in juvenile Asian seabass (*Lates calcarifer*). Aquac Nutr 2025(8855823):15. https://doi.org/10.1155/anu/8855823

Loh JY, Ting ASY (2017) Bioprospecting gastrointestinal microflora of common fishes for disease control in aquaculture. In: Kalia V, Shouche Y, Purohit H, Rahi P (eds) Mining of microbial wealth and metagenomics. Springer, Singapore, pp 161–182

Loh JY, Ting ASY (2016) Effects of potential probiotic *Lactococcus lactis* subsp. *lactis* on digestive enzymatic activities of live feed *Artemia franciscana*. Aquac Int. https://doi.org/10.1007/s10499-016-9991-2.

Loh JY, Chan HK, Yam HC, Lionel In LA, Crystale Lim SY (2020) An overview of the immunomodulatory effects exerted by probiotics and prebiotics in grouper fish. Aquac Int 28:729–750

Loh JY, Kay GL, Ting ASY (2018) Bioencapsulation and colonization characteristics of *Lactococcus lactis* subsp. *lactis* CF4MRS in *Artemia franciscana*: a biological approach for the control of *Edwardsiella tarda* in larviculture. Mar Biotechnol. https://doi.org/10.1007/s10126-018-9813-9

Loh JY, Lim YY, Ting ASY (2017) Bacteriocin-like substances produced by *Lactococcus lactis* subsp. *lactis* CF4MRS isolated from fish intestine: antimicrobial activities and inhibitory properties. Int Food Res J 24(1):394–400

Loh JY, Lim YY, Harmin SA, Ting ASY (2014) *In vitro* assessment on intestinal microflora from commonly farmed fishes for control of the fish pathogen *Edwardsiella tarda*. Turk J Vet Anim Sci 38:257–263

Luna-González A, Almaraz-Salas JC, Fierro-Coronado JA, Flores-Miranda MDC, González-Ocampo HA, Peraza-Gómez V (2012) The prebiotic inulin increases the phenoloxidase activity and reduces the prevalence of WSSV in whiteleg shrimp (*Litopenaeus vannamei*) cultured under laboratory conditions. Aquaculture 362–363:28–32

MacGowan A, Macnaughton E (2017) Antibiotic resistance. Medicine 45(10):622–628

Maeda M, Nogami K, Kanematsu M, Hirayama K (1997) The concept of biological control methods in aquaculture. Hydrobiologia 358:285–290

Mahdhi A, Kamoun F, Messina C, Santulli A, Bakhrouf A (2012) Probiotic properties of *Brevibacillus brevis* and its influence on sea bass (*Dicentrarchus labrax*) larval rearing. Af J Microbiol Res 6:6487–6495

Mahious TS, Gatesoupe FJ, Hervi M, Metailler R, Ollevier F (2006) Effect of dietary inulin and oligosaccharides as prebiotic for weaning turbot, *Psetta maxima* (Linnaeus, C. 1785). Aquacult Int 14:219–229

Manjeet, Airon A, Kumar R, Saifi R (2022) Temporal and spatial impact of lockdown during COVID-19 on air quality index in Haryana, India. Sci Rep 12:20046. https://doi.org/10.1038/s41598-022-20885-2

Mazlum Y, Yílmaz E, Genç M, Guner O (2011) A preliminary study on the use of mannan oligosaccharides (MOS) in freshwater crayfish, *Astacus leptodactylus* Eschscholtz, 1823, juvenile diets. Aquacult Int 19:111–119

Moll G, Hildeng-Hauge H, Nissen-Meyer J, Nes IF, Konings WN, Driessen AJM (1998) Mechanistic properties of the two-component bacteriocin lactococcin G. J Bacteriol 180:96–99

Montville TJ, Winkowski K, Ludescher RD (1995) Models and mechanisms for bacterioncin action and application. Int Dairy J 5(8):797–814

Moriarty DJW (1997) The role of microorganisms in aquaculture ponds. Aquaculture 151:333–349

Moriarty DJW (1998) Control of luminous *Vibrio* species in penaeid aquaculture ponds. Aquaculture 164:351–358

Munita JM, Arias CA (2016) Mechanisms of antibiotic resistance. Microbiol Spectrum. https://doi.org/10.1128/microbiolspec.VMBF-0016-2015

Nayak SK (2010) Role of gastrointestinal microbiota in fish. Aquac Res 41:1553–1573

Nisha, Anshu, Rani S, Kumar S, Kumar R, Kumar A (2025) Recent technologies and strategies for bioremediation of toxic and hazardous waste. In: Dubey KK, Shah MP, Yadav A (eds) Challenges and sustainable solutions in bioremediation. CRC Press, pp 1–16. https://doi.org/10.1201/9781003407317-1

Olsen RE, Myklebust R, Kryvi H, Mayhew TM, Ringø E (2001) Damaging effect of dietary inulin on intestinal enterocytes in Arctic charr (*Salvelinus alpinus* L.). Aquac Res 32:931–934

Ortiz LT, Rebolé A, Velasco S, Rodríguez ML, Treviño J, Tejedor JL, Alzueta C (2012) Effects of inulin and fructooligosaccharides on growth performance, body chemical composition and intestinal microbiota of farmed rainbow trout (*Oncorhynchus mykiss*). Aquac Nutr Early Rev. https://doi.org/10.1111/j.1365-2095.2012.00981.x

Ouwehand AC, Isolauri E, Kirjavainen PV, Tolkko S, Salminen SJ (2000) The mucus binding effect of *Bifidobacterium lactis* BB12 is enhanced in the presence of *Lactobacillus* GG and *Lactobacillus delbrueckii* ssp *Bulgaricus*. Lett Appl Microbiol 30:10–13

Panase A, Thirabunyanon M, Promya J, Chitmanat C (2023) Influences of *Bacillus subtilis* and *fructooligosaccharide* on growth performances, immune responses, and disease resistance of Nile tilapia *Oreochromis Niloticus*. Front Vet Sci 9:1094681. https://doi.org/10.3389/fvets.2022.1094681

Pardede MA, Widanarni, Sukenda, Yuhana M, Loh JY (2024) Effects of dietary synbiotic supplementation on digestive enzyme, total Vibrio count, and hepatopancreas of Pacific white shrimp, *Penaeus vannamei* challenged with *Vibrio parahaemolyticus*. J Appl Aquac. https://doi.org/10.1080/10454438.2024.2436373

Parker RB (1974) Probiotics, the other half of the antibiotics story. Anim Nutr Health 29:4–8

Pirarat N, Kobayashi T, Katagiri T, Maita M, Endo M (2006) Protective effects and mechanisms of a probiotic bacterium *Lactobacillus rhamnosus* against experimental *Edwardsiella tarda* infection in tilapia (*Oreochromis niloticus*). Vet Immunol Immunopathol 113:339–347

Pryor GS, Royes JB, Chapman FA, Miles RD (2003) Mannanoligosaccharides in fish nutrition: effects of dietary supplementation on growth and gastrointestinal villi structure in Gulf of Mexico sturgeon. N Am J Aquac 65:106–111

Ran C, Carrias A, Williams MA, Capps N, Dan BCT, Newton JC, Kloepper JW, Ooi EL, Browdy CL, Terhune JS, Liles MR (2012) Identification of *Bacillus* strains for biological control of catfish pathogens. PLoS ONE 7:9–14

Rani S, Sharma S, Singh D, Sheoran R, Kumar R (2025) Improved production of anti-cancer compounds from medicinal plants through molecular and biotechnological mutagenesis strategies. In: Chen JT (ed) Biotechnology, multiple omics, and precision breeding in medicinal plants. CRC Press, pp 287–297. https://doi.org/10.1201/9781003475491-21

Ray AK, Ghosh K, Ringø E (2012) Enzyme-producing bacteria isolated from fish gut: a review. Aquac Nutr 18:465–492

Refstie S, Bakke-McKellep AM, Penn MH, Sundby A, Shearer KD, Krogdahl Å (2006) Capacity for digestive hydrolysis and amino acid absorption in Atlantic salmon (*Salmo salar*) fed diets with soybean meal or inulin with or without addition of antibiotics. Aquaculture 261:392–406

Rettger LF (1905) The antagonism of bacteria and their products to other bacteria. J Infect Dis 2:562–568

Reyes-Becerril M, Ascencio-Valle F, Gracia-Lopez V, Macias ME, Roa MC, Esteban MÁ (2014) Single or combined effects of *Lactobacillus sakei* and inulin on growth, non-specific immunity and IgM expression in leopard grouper (*Mycteroperca rosacea*). Fish Physiol Biochem 40:1169–1180

Reyes-Becerril M, Tovar-Ramírez D, Ascencio-Valle F, Civera-Cerecedo R, Gracia-López V, Barbosa-Solomieu V, Esteban MÁ (2011) Effects of dietary supplementation with probiotic live yeast *Debaryomyces hansenii* on the immune and antioxidant systems of leopard grouper *Mycteroperca rosacea* infected with *Aeromonas hydrophila*. Aquac Res 42:1676–1686

Rico A, Jacobs R, Van den Brink P, Tello A (2017) A probabilistic approach to assess antibiotic resistance development risks in environmental compartments and its application to an intensive aquaculture production scenario. Environ Pollut 231:918–928

Ringø E, Dimitroglou A, Hoseinifar SH, Davies SJ (2014) Prebiotics in finfish: an update. In: Merrifield D, Ringø E (eds) Aquaculture nutrition: gut health, probiotics and prebiotics, 1st edn. John Wiley and Sons, Ltd, 465 p

Ringø E, Sperstad S, Myklebust R, Mayhew TM, Olsen RE (2006) The effect dietary inulin on aerobic bacteria associated with the hindgut of Arctic charr (*Salvelinus alpines* L.). Aquac Res 37:891–897

Ringø E, Strøm E, Tabacheck J (1995) Intestinal microflora of salmonids: a review. Aquac Res 26:773–789

Ringø EL, Løvmo M, Kristiansen M, Bakken Y, Salinas I, Myklebust R, Olsen RE, Mayhew TM (2010) Lactic acid bacteria versus pathogens in the gastrointestinal tract of fish: a review. Aquac Res 41:451–467

Roberfroid M (2007) Prebiotics: the concept revised. J Nutr 137:830S-837S

Roberts RJ (2012) Fish pathology. Wiley-Blackwell, Sussex, U.K.

Rodrigues-Estrada U, Satoh S, Haga Y, Fushimi H, Sweetman J (2008) Studies of the effects of mannan-oligosaccharides, *Enterococcus faecalis*, and poly hydrobutyric acid as immune stimulant and growth promoting ingredients in rainbow trout diets. In: 5th World fisheries congress, Yokohama, Japan, Abstract, 2d-1-5, 158 p, 20–25 Oct 2008

Romero J, Feijoó CG, Navarrete P (2012) Antibiotics in aquaculture—use, abuse and alternatives. In: Carvalho E (eds) Health and environment in aquaculture. InTech. http://www.intechopen.com/books/health-and-environment-in-aquaculture/antibioticsin-aquaculture-use-abuse-and-alternatives

Rønnestad I, Tonheim SK, Fyhn HJ, Rojas-García CR, Kamisaka Y, Koven W, Finn RN, Terjesen BF, Barr Y, Conceição LEC (2003) The supply of amino acids during early feeding stages of marine fish larvae: a review of recent findings. Aquaculture 227:147–164

Rurangwa E, Delaedt Y, Geraylou Z, Van De Wiele T, Courtin CM, Delcour JA, Ollevier F (2008) Dietary effect of arabinoxylan oligosaccharides on zootechnical performance and hindgut microbial fermentation in Siberian sturgeon and African catfish. In: Aquaculture Europe, Krakow, Poland, pp 569–570, 15–18 Sept 2008

Sado RY, Almeida Bicudo AJD, Cyrino JEP (2008) Feeding dietary mannan oligosaccharides to juvenile Nile tilapia, *Oreochromis niloticus*, has no effect on haematological parameters and showed decreased feed consumption. J World Aquac Soc 39:821–826

Sahoo PK, Mukherjee SC (2002) The effect of dietary immunomodulation upon *Edwardsiella tarda* vaccination in healthy and immunocompromised Indian major carp (*Labeo rohita*). Fish Shellfish Immunol 12:1–16

Sakata T (1990) Microflora in the digestive tract of fish and shellfish. In: Lesel R (ed) Microbiology in poecilotherms. Elsevier, Amsterdam, pp 171–176

Salze G, McLean E, Schwarz MH, Craig SR (2008) Dietary mannan oligosaccharide enhances salinity tolerance and gut development of larval cobia. Aquaculture 274:148–152

Sanchez LM, Wong WR, Riener RM, Schulze CJ, Linington RG (2012) Examining the fish microbiome: vertebrate-derived bacteria as an environmental niche for the discovery of unique marine natural products. PLoS ONE 7:e35398

Sang HM, Fotedar R (2010) Effects of mannan oligosaccharide dietary supplementation of performances of the tropical spiny lobster juvenile (*Panulirus ornatus*, Fabricius 1798). Fish Shellfish Immunol 28:483–489

Sang HM, Kien NT, Thanh Thuy NT (2014) Effects of dietary mannan oligosaccharide on growth, survival, physiological, immunological and gut morphological conditions of black tiger prawn (*Penaeus monodon* Febricius 1798). Aquac Nutr 20:341–348

Sang HM, Ky LT, Fotedar R (2009) Dietary supplementation of mannan oligosaccharide improves the immune responses and survival of marron, *Cherax tenuimanus* (Smith, 1912) when challenged with different stressors. Fish Shellfish Immunol 27:341–348

Seenivasan C, Saravana BP, Radhakrishnan S, Shanthi R (2012) Enrichment of *Artemia nauplii* with *Lactobacillus sporogenes* for enhancing the survival, growth and levels of biochemical constituents in the post-larvae of the freshwater prawn *Macrobrachium rosenbergii*. Turk J Fish Aquat Sci 12:23–31

Sharma MMM, Kapoor D, Rohilla R, Loyal A, Sharma P, Husen A (2024) Role of essential minerals and rhizosphere in plant under changing environmental conditions. In: Husen A (eds) Essential minerals in plant-soil systems. Elsevier, pp 235–252. (ISBN:978-0-443-16082-0). https://doi.org/10.1016/B978-0-443-16082-0.00002-3

Sharma P, Kapoor D, Kumar R, Husen A (2025) Rhizosphere and osmolytes coordination in plant systems under adverse situations. In: Husen A (ed) Roles of osmolytes in changing environment. Academic Press, pp 101–117

Shefali, Kumar R, Sankhla MS, Kumar R, Sonone SS (2021) Impact of pesticide toxicity in aquatic environment. Biointerface Res Appl Chem 11(3):10131–10140. https://doi.org/10.33263/BRIAC113.1013110140

Shelby RA, Lim C, Yildirim-Aksoy M, Welker TL, Klesius PH (2009) Effects of yeast oligosaccharide diet supplementations on growth and disease resistance in juvenile Nile tilapia, *Oreochromis niloticus*. J Appl Aquac 21:61–71

Sivakumar N, Sundararaman M, Selvakumar G (2012) Probiotic effect of *Lactobacillus acidophilus* against vibriosis in juvenile shrimp (*Penaeus monodon*). Afr J Biotech 11(91):15811–15818

Son VM, Chang CC, Wu MC, Guu YK, Chiu CH, Cheng W (2009) Dietary administration of the probiotic, *Lactobacillus plantarum*, enhanced the growth, innate immune responses, and disease resistance of the grouper *Epinephelus coioides*. Fish Shellfish Immunol 26:691–698

Sorroza L, Padilla D, Acosta F, Roman L, Grasso V, Vega J, Real F (2012) Characterization of the probiotic strain *Vagococcus fluvialis* in the protection of European sea bass (*Dicentrarchus labrax*) against vibriosis by *Vibrio anguillarum*. Vet Microbiol 155:369–373

Staykov Y, Denev S, Spring P (2005) Influence of dietary mannan oligosaccharides (Bio-Mos) on growth rate and immune function of common carp (*Cyprinus carpio* L.). In: Howell B, Flos R (eds) Lessons from the past to optimise the future. European Aquaculture Society, Special Publication, vol 35, pp 431–432

Staykov Y, Spring P, Denev S, Sweetman J (2007) Effect of a mannan oligosaccharide on the growth performance and immune status of rainbow trout (*Oncorhynchus mykiss*). Aquacult Int 15:153–161

Sudheesh PS, Al-Ghabshi A, Al-Mazrooei N, Al-Habsi S (2012) Comparative pathogenomics of bacteria causing infectious diseases in fish. Int J Evol Biol 2012:16

Sun YZ, Yang HL, Huang KP, Ye JD, Zhang CX (2013) Application of autochthonous *Bacillus* bioencapsulated in copepod to grouper *Epinephelus coioides* larvae. Aquaculture 392:44–50

Sun YZ, Yang HL, Ma RL, Lin WY (2010) Probiotic applications of two dominant gut *Bacillus* strains with antagonistic activity improved the growth performance and immune responses of grouper *Epinephelus coioides*. Fish Shellfish Immunol 29:803–809

Sun YZ, Yang HL, Ma RL, Song K, Lin WY (2011a) Molecular analysis of autochthonous microbiota along the digestive tract of juvenile grouper *Epinephelus coioides* following probiotic *Bacillus pumilus* administration. J Appl Microbiol 110:1093–1103

Sun YZ, Yang HL, Ma RL, Zhang CX, Lin WY (2011b) Effect of dietary administration of *Psychrobacter* sp. on the growth, feed utilization, digestive enzymes and immune responses of grouper *Epinephelus* coioides. Aquac Nutr 17:733–740

Sun YZ, Yang HL, Ma RL, Song K, Li JS (2012) Effect of *Lactococcus lactis* and *Enterococcus faecium* on growth performance, digestive enzymes and immune response of grouper *Epinephelus coioides*. Aquac Nutr 18:281–289

Swain SM, Singh C, Arul V (2009) Inhibitory activity of probiotics *Streptococcus phocae* PI80 and *Enterococcus faecium* MC13 against vibriosis in shrimp *Penaeus monodon*. World J Microbiol Biotechnol 25:697–703

Sweetman J, Dimitroglou A, Davies S, Torrecillas S (2008) Nutrient uptake: gut morphology a key to efficient nutrition. Int Aquafeed (January–February):26–30

Talpur AD, Munir MB, Mary A, Hashim R (2014) Dietary probiotics and prebiotics improved food acceptability, growth performance, haematology and immunological parameters and disease resistance against *Aeromonas hydrophila* in snakehead (*Channa striata*) fingerlings. Aquaculture 14–20:426–427

Tannock GW (2003) Probiotics: time for a dose of realism. Curr Issues Intest Microbiol 4:33–42

Tinh NTN, Dierckens K, Sorgeloos P, Bossier P (2008) A review of the functionality of probiotics in the larviculture food chain. Mar Biotechnol 10:1–12

Torrecillas S, Makol A, Caballero MJ, Montero D, Robaina L, Real F, Sweetman J (2007) Immune stimulation and improved infection resistance in European seabass (*Decenrarchus labrax*) fed mannan oligosaccharides. Fish Shellfish Immunol 23:969–981

Touraki M, Karamanlidou G, Karavida P, Chrysi K (2012) Evaluation of the probiotics *Bacillus subtilis* and *Lactobacillus plantarum* bioencapsulated in *Artemia* nauplii against vibriosis in European sea bass larvae (*Dicentrarchus labrax*, L.). World J Microbiol Biotechnol 28:2425–2433

Touraki M, Karamanlidou G, Koziotis M, Christidis I (2013) Antibacterial effect of *Lactococcus lactis* subsp. *lactis* on *Artemia franciscana* and *Dicentrarchus labrax* larvae against the fish pathogen *Vibrio anguillarum*. Aquacult Int 21:481–495

Ulkhaq MF, Kenconojati H, Budi DS, Pardede MA, Loh JY (2024) Chloramphenicol residues and bacterial contamination in farmed African catfish (*Clarias gariepinus*) from Banyuwangi traditional markets: a risk assessment. Jurnal Medik Veteriner 7(2):244–254

Urmila, Gupta RK, Sharma S, Kumar R (2025) Earthworms: the environmentally preferred bioengineers for managing heavy metals. In: Dey S, Bhattacharya S (eds) Biotechnological interventions in the removal of emerging pollutants. Springer Nature, Singapore, pp 365–384

Vaseeharan B, Ramasamy P (2003) Control of pathogenic *Vibrio* spp. by *Bacillus subtilis* BT23, a possible probiotic treatment for black tiger shrimp *Penaeus monodon*. Lett Appl Microbiol 36:83–87

Vijaya Baskar P, Kannan S (2009) Marine bacteria as probiotics to control pathogenic *Vibrio* on infected shrimp. SB Acad Rev 1&2:77–85

Villamil L, Reyes C, Martínez-Silva MA (2014) *In vivo* and *in vitro* assessment of *Lactobacillus acidophilus* as probiotic for tilapia (*Oreochromis niloticus*, Perciformes: Cichlidae) culture improvement. Aquac Res 45(7):1116–1125

Wang G, Zhou Y, Huang W, Huang Y, Liu X, Dong S (2010) Effects of xylooligosaccharide on growth, body composition and non-specific immunity in *Litopenaeus vannamei*. Freshw Fish 40:55–58

Wang WS, Wang DH (1997) Enhancement of the resistance of tilapia and grass carp to experimental *Aeromonas hydrophila* and *Edwardsiella tarda* infections by several polysaccharides. Comp Immunol Microbiol Infect Dis 20:261–270

Wang Y, Tian Z, Yao J, Li W (2008) Effect of probiotics, *Enteroccus faecium*, on tilapia (*Oreochromis niloticus*) growth performance and immune response. Aquaculture 277(3–4):203–207

Welker TL, Lim C, Yildirim-Aksoy M, Shelby R, Klesius PH (2007) Immune response and resistance to stress and *Edwardsiella ictaluri* challenge in channel catfish, *Ictalurus punctatus*, fed diets containing commercial whole-cell yeast or yeast subcomponents. J World Aquac Soc 38:24–35

Wibawati PA, Ulkhaq MF, Loh JY (2024) Occurrence of antibiotic-resistant *Proteus mirabilis* and *Proteus vulgaris* in African catfish (*Clarias* sp.) isolates in Banyuwangi, East Java, Indonesia. Asian J Dairy Food Res:1–7. https://doi.org/10.18805/ajdfr.DRF-475

Xu B, Wang Y, Li J, Lin Q (2009) Effects of prebiotic xylooligosaccharides on growth performance and digestive activities of allogynogenetic carp (*Carassius auratus gibelio*). Fish Physiol Biochem 35:351–357

Yadav R, Gupta RK, Kumar R, Kaur T, Moun R (2022) Biochemical response of earthworm, *Eisenia fetida* to heavy metals toxicity. J Appl Nat Sci 14(3):990–998

Yadav R, Kumar R, Gupta RK, Kaur T, Yodha K, Kour A, Kaur S, Rajput A (2023) Heavy metal toxicity in earthworms and its environmental implications: a Review. Environ Adv:100374

Yang HL, Xia HQ, Ye YD, Zou WC, Sun YZ (2014) Probiotic *Bacillus pumilus* SE5 shapes the intestinal microbiota and mucosal immunity in grouper *Epinephelus coioides*. Dis Aquat Org 111:119–127

Yilmaz E, Genc MA, Genc E (2007) Effects of dietary mannan oligosaccharides on growth, body composition, and intestine and liver histology of rainbow trout, *Oncorhynchus mykiss*. Israeli J Aquac Bamidgeh 59:182–188

Yuji-Sado R, Raulino-Domanski F, Freitas PFDF, Baioco-Sales F (2015) Growth, immune status and intestinal morphology of Nile tilapia fed dietary prebiotics (mannan oligosaccharides-MOS). Lat Am J Aquat Res 43(5):944–952

Zhang Q, Tan B, Mai K, Zhang W, Ma H, Ai Q, Wang X, Liufu Z (2011) Dietary administration of *Bacillus* (*B. licheniformis* and *B. subtilis*) and isomaltooligosaccharide influences the intestinal microflora, immunological parameters and resistance against *Vibrio alginolyticus* in shrimp, *Penaeus japonicus* (Decapoda: Penaeidae). Aquac Res 42:943–952

Zhang XJ, Fang H, Chen CZ, Ge MX, Wang XY (2005) Sensitivity of pathogenic *Edwardsiella tarda* isolated from flounder (*Paralichthys olivaceus*) to some antimicrobial agents. Fish Sci 24:15–18

Zheng ZL, Wang KY, Gatlin DM III, Ye JM (2011) Evaluation of the ability of GroBiotic®-A to enhance growth, muscle composition, immune responses, and disease resistance against *Aeromonas hydrophila* in Nile tilapia, *Oreochromis niloticus*. J World Aquac Soc 42:549–557

Zhou XQ, Li YL (2004) The effects of Bio-Mos1 on intestinal microflora and immune function of juvenile Jian carp (*Cyprinus carpio* Var. Jian). In: Nutritional biotechnology in the feed and food industries: Proceedings of Alltech's 20th annual symposium (Suppl. 1-Abstracts of posters presented), Lexington, KY, USA

Zhou Z, Ding Z, Huiyuan LV (2007) Effects of dietary short chain fructoologosaccharides on intestinal microflora, survival, and growth performance of juvenile white shrimp, *Litopenaeus vannamei*. J World Aquac Soc 38:296–301

Zhou XX, Wang YB, Li WF (2009a) Effect of probiotic on larvae shrimp (*Penaeus vannamei*) based on water quality, survival rate and digestive enzyme activities. Aquaculture 287(3–4):349–353

Zhou Z, He S, Liu Y, Shi P, Huang G, Yao B (2009b) The effects of dietary yeast culture or short-chain fructo-oligosaccharides on the intestinal autochthonous bacterial communities in juvenile hybrid tilapia *Oreochromis niloticus* ♀ × *O. aureus* ♂. J World Aquac Soc 40:450–459

Zhou QC, Buentello JA, Gatlin DMIII (2010a) Effects of dietary prebiotics on growth performance, immune response and intestinal morphology of red drum (*Sciaenops ocellatus*). Aquaculture 309:253–257

Zhou XX, Wang YB, Yao JT, Li WF (2010b) Inhibition ability of probiotic, *Lactococcus lactis*, against *A. hydrophila* and study of its immunostimulatory effect in tilapia (*Oreochromis niloticus*). Int J Eng Sci Technol 2(7):73–80

Zhu ZC, Shi XG, Zhang SJ, Jiang GJ, Xing ZB, Zhao YL, Li ZJ, Wu P (2006) The pathogenic bacteria of the ascites in Japanese flounder (*Paralichthys olivaceus*). Fish Sci 7:325–329

Ziaei-Nejad S, Rezaei MH, Takami GA, Lovett DL, Mirvaghefi AR, Shakouri M (2006) The effect of *Bacillus* spp. bacteria used as probiotics on digestive enzyme activity, survival and growth in the Indian white shrimp (*Fenneropenaeus indicus*). Aquaculture 252:516–524

Zokaeifar H, Balcazar JL, Saad CR, Kamarudin MS, Sijam K, Arshad A, Nejat N (2012) Effects of *Bacillus subtilis* on the growth performance, digestive enzymes, immune gene expression and disease resistance of white shrimp *Litopenaeus Vannamei*. Fish Shellfish Immunol 33:683–689

Chapter 4
Revolutionizing Aquatic Production: Harnessing the Power of Biotics for Superior Fish Growth and Health

Priyanka Joshi, Gajender Singh, Baljeet Singh Saharan, Nikita Goyat, Renu Yadav, and Monika Pilania

Abstract The aquaculture industry is showing remarkable growth in recent years, solving the problem of increasing global demand for sea-based food. However, this rapid expansion has also brought about significant challenges, particularly regarding environmental sustainability and resource efficiency. Conventional aquaculture practices often result in the accumulation of waste products, which can lead to water pollution and negatively impact the surrounding ecosystems. In response to these concerns, there is need of innovative solutions to improve the sustainability of aquaculture operations. One of the most promising solutions is biotics, which involves balancing the organic and inorganic levels in the aquaculture system to enhance water quality and produce in-situ proteinaceous feed. The bacterial incorporation in the form of

P. Joshi (✉) · G. Singh · R. Yadav
Department of Zoology, Chaudhary Charan Singh Haryana Agricultural University, Hisar, Haryana, India
e-mail: joshipriyanka212000@gmail.com

G. Singh
e-mail: drgajendersingh1@gmail.com

R. Yadav
e-mail: renuyadav4615@gmail.com

B. S. Saharan · N. Goyat
Department of Microbiology, Chaudhary Charan Singh Haryana Agricultural University, Hisar, Haryana, India
e-mail: baljeetsaharan@hau.ac.in

N. Goyat
e-mail: nikitagoyat6@gmail.com

R. Yadav
Department of Zoology, Desh Bandhu Gupta Government College, Panipat, Haryana, India

M. Pilania
Biotechnology Technician, Health and Life Sciences, Conestoga College, Kitchener, ON, Canada
e-mail: Monika5090@conestogac.on.ca

R. Kumar et al. (eds.), *Sustainability in Aquaculture and Fisheries*,
Springer Water, https://doi.org/10.1007/978-3-032-16605-0_4

various biotics like probiotics, prebiotics, synbiotics and postbiotics has antimicrobial, anti-inflammatory and immunomodulatory effects. These biotics in aquaculture not only improve growth and utilization of feed but also mitigate diseases and reduce the dependency on antibiotics. Microbial consortia in aquaculture offer multifaceted benefits including improved water quality, enhanced nutrient availability, disease prevention and immune system modulation. These advantages contribute to higher productivity, better animal welfare and a more sustainable and environment friendly aquaculture practice. This chapter explores the beneficial effects of beneficial bacterial communities on water quality improvement, disease control and enhanced feed utilization with future challenges that need to be addressed for its widespread adoption.

Keywords Bacterial communities · Environmental sustainability · Conventional practices · Microbial consortia

4.1 Introduction

As the world population is increasing, the competence and demand of food for such a large population is also increasing. The world's population projected to exceed 9.7 billion by 2050, the demand for food is expected to rise by about 70% compared to current levels (Tkemaladze 2025). Alone with plant resources, human food demands cannot be accomplished. So, to achieve this we have to prefer other sources also such as poultry, aquaculture and animal products. Out of all the most prominent is aquaculture. Aquatic culture is further classified on the basis of their habitat and demand as shown in Fig. 4.1. Its demand is increasing day by day and most demanded section in aquaculture is pisciculture (fish culture).

Aquaculture is rapidly transforming the global food landscape. With wild fish populations struggling to keep pace with rising demand, aquaculture is poised to

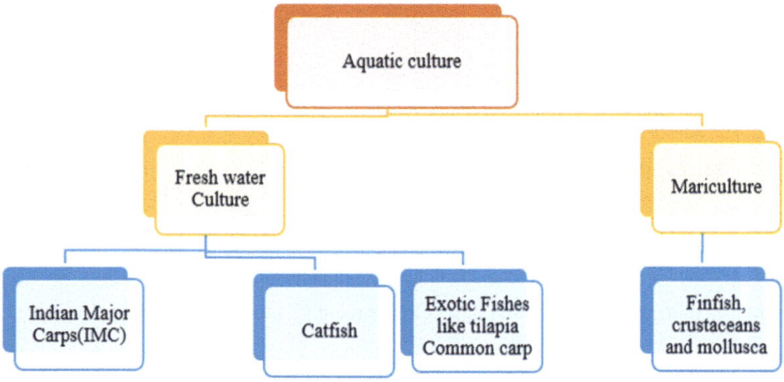

Fig. 4.1 Aquatic culture system overview (*Source* Author)

become the primary source of fish protein. According to the FAO (2022), by 2030, aquaculture will be responsible for 109 million tons of fish production (Patrinou et al. 2022; Kumar et al. 2022a, b). As the human population increases inevitably, the demand for fish as an alternative for protein sources will increase. This industry is significant from both the standpoint of employment and export. An 8.9% average annual growth rate since 1970, aquaculture is the food industry with fastest global growth. By comparison, catch fisheries have had only 1.2% and 2.8% annual growth, respectively. At the national level, Indian major carps (IMCs) account for 70–75% of the country's freshwater fish production. They are the primary source of fish production in India. The exotic species like *Ctenopharyngodon idella, Hypophthalmichthys molitrix* and *Cyprinus carpio* are the second most productive group of fishes, after the Indian main carps (Manjeet et al. 2022; Jayasankar 2018). The fishing sector provides food, nourishment and employment to the surrounding people.

The aquatic sector is at the vanguard of innovation in the ever-changing face of food production working to complete the demand of sustainable proteins sources worldwide. Within aquaculture industry, a paradigm shift is occurring in response to the problems posed by growing populations and environmental concerns (Kumar et al. 2022c).

Aquaculture has historically encountered difficulties like disease outbreaks, environmental damageand inefficient use of resources. However, the incorporation of biotics holds the potential to completely transform this sector by utilizing the inherent strength of biological processes to improve fish health and growth. Microbiome communities consist of many interrelated species, making it difficult for hosts to exercise control over each species (Coyte et al.2015; Nisha et al. 2025; Kiran et al. 2025; Rani et al. 2025). Unprecedented gains in fish growth and health can only be unlocked by utilizing biotics, a novel strategy that combines biology and technology to maximize aquatic productivity.

4.2 Bacterial Dynamics of Aquaculture Environments

The aquaculture industry is increasingly interested in functional feeds that incorporate selected additives to enhance fish growth performance and overall health (Rimoldi et al. 2024). These additives encompass a range of biotics including prebiotics, probiotics, and their related forms such as postbiotics, paraprobiotics, psychobiotics, metabiotics and synbiotics. Understanding the application and impact of these biotics is crucial for developing sustainable and effective aquaculture practices.

Bacteria are extremely common and widespread on Earth. Bacteria are among the most ancient and plentiful forms of life on the planet. Bacteria live in almost every environment, from the deepest ocean trenches to the highest mountains, arctic ice caps to hot springs. They are found in soil, water, air, and even within other organisms as symbionts or parasites. The bacteria are widely distributed due to the following key features.

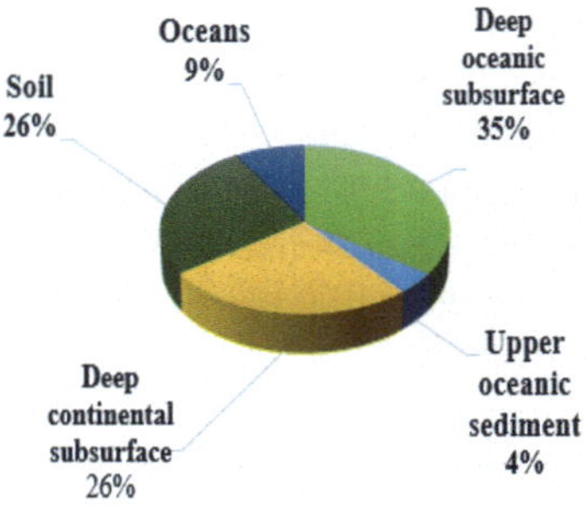

Fig. 4.2 Bacterial abundance in various aquatic habitats

4.2.1 *Abundance*

Bacteria are incredibly abundant. There are trillions of bacteria on Earth outnumbering all other forms of life combined. Most bacteria on Earth are found in the 'big five' habitats shown in Fig. 4.2.

The cell counts in various habitats are estimated as follows: -"deep oceanic subsurface (4×10^{29}), upper oceanic sediment (5×10^{28}), deep continental subsurface (3×10^{29}), soil (3×10^{29}), and oceans (1×10^{29})". Other environments, such as groundwater, atmosphere, ocean surface microlayer, and microbiomes of humans, animals, and plant surfaces (phyllo sphere), contain significantly fewer cells.

Biofilms are predominant across most terrestrial habitats, covering about 80% of bacterial and archaeal cells, with the exception of the open ocean (Flemming and Wuertz 2019). A single gram of soil can contain billions of bacterial cells, and seawater contains millions of bacterial cells per millilitre.

4.2.2 *Diversity*

Bacteria exhibit immense diversity in terms of morphology, physiology and ecological roles. They come in various shapes (e.g., cocci, bacilli, spirilla), sizes and metabolic capabilities. Bacteria can occupy a wide range of ecological niches and play critical roles in nutrient cycling, decomposition, symbiosis and disease.

4.2.3 *Extreme Environments*

Bacteria are known for their ability to thrive in extreme environments where few other organisms can survive. They are found in environments with extreme temperatures (e.g., hot springs, permafrost), pH levels (e.g., acidic springs, alkaline soda lakes), salinity (e.g., salt flats, hypersaline environments) and pressure (e.g., deep-sea hydrothermal vents, deep subsurface environments).

The science, technology and practical applications that are transforming aquaculture in the future along with different ways in which different biotics are bringing an entirely new era in fish farming are discussed in this chapter.

4.3 Major Different Types of Biotics

The major common terms used for the different types of biotics are (Jeon 2022)

- **Antibiotics**: Antibiotics aremedicines that are used for destroying microbes and curing bacterial infections.e.g. Penicillin
- **Probiotics**: Probiotics are "beneficial live bacteria that promote a healthy gut and overall well-being in fish".e.g. *Lactobacillus acidophilus*
- **Prebiotics:** Prebiotics are substances that enhance the growth of beneficial bacteria (probiotics) in the fish's digestive system. eg. Artichokes
- **Para probiotics**: Para probiotics are non-viable or inactivated forms of beneficial bacteria that still confer health benefits to fish.e.g. *Streptococcus thermophilus*
- **Synbiotics**: Synbiotics a "combination of probiotics and prebiotics that work collectively to enhance the health of fish and improve their overall performance".
- **Metabiotic**: Metabiotic word is a combination of two words "metabo-" (referring to metabolism) and "-biotics" (referring to living organisms or substances with a beneficial effect), metabiotic might refer to substances or compounds produced through metabolic processes by living organisms, which have potential beneficial effects on other organisms.
- **Psychobiotics**: Psychobiotics are the probiotics or prebiotics that influence bacteria brain relationships.

We will learn about the major biotics, their history, mode of mechanism, advantages and disadvantages. While prebiotics, probiotics and synbiotics are more established in aquaculture, postbiotics, paraprobiotic, psychobiotics and metabiotic are emerging areas of interest.

4.4 Antibiotics

Antibiotics are substances that kill or hinder the growth of dangerous bacteria, implementing differences between prokaryotes and eukaryotes and making them fairly secure in higher-level organisms. They help fishes combat harmful bacterial illnesses. They assault critical molecular machineries in bacteria, preventing or reducing their operation, by slowing or killing the bacterial cell.

4.4.1 Mode of Action of Antibiotics

Antibiotics work in one of three ways:

1. Disrupts bacterial outer membranes, preventing self-regulation.
2. By preventing DNA or protein production.
3. By inhibiting the function of bacteria-specific enzymes.

Antimicrobials kill bacteria, viruses, fungus, algae and other microorganisms. Bacterial cells (prokaryotes), such as salmonella, are distinct from those of higher-level creatures (eukaryotes), such as fish. Antibiotics are compounds that are intended to kill or hinder the growth of harmful bacteria, while leveraging distinctions between prokaryotes and eukaryotes to render them relatively safe in higher-level organisms.

4.4.2 Classes of Major Antibiotics Based on Their Action

These antimicrobials and antibiotics are classified based on distinct action pattern, each targeting certain components of bacterial physiology.

4.4.2.1 Antibiotics That Inhibit the Synthesis of the Cell Wall

i. **Beta-lactam antibiotics** (e.g., *penicillin, cephalosporins, carbapenems*): These drugs inhibit penicillin-binding proteins (green), which are enzymes responsible for cross-linking peptidoglycan chains.

   Example: Ampicillin (purple).
ii. **Vancomycin:** This antibiotic binds to peptidoglycan building blocks, preventing chain cross-linking and thereby weakening the cell wall.

4.4.2.2 Antibiotics Targeting Protein Synthesis

Aminoglycosides: These drugs bind to the ribosome's small subunit (purple), causing errors in protein assembly, which ultimately leads to cell death.
Example: Paromomycin (red).

4.4.2.3 Antibiotics Inhibiting Bacterial Metabolism

Sulphonamides: These inhibit the enzyme dihydropteroate synthase (blue), essential for synthesizing vitamin B, halting bacterial growth.
Example: Sulfamethoxazole (red).

4.4.2.4 Antibiotics Targeting DNA Synthesis

Quinolones: These drugs inhibit DNA gyrase (orange), an enzyme necessary for unwinding the DNA double helix, which blocks bacterial replication.

Example: Ciprofloxacin.

4.4.3 Role of Antibiotics in Aquaculture

The high incidence of bacterial illnesses in fish necessitates the frequent administration of antibiotics, which leads to their persistence in the aquatic environment, culminating in antibiotic resistance. "Antimicrobial resistance in aquaculture can be transmitted to medicinally critical strains in the natural environment via horizontal gene transfer, affecting the entire ecosystem" (Preena et al. 2020).

Antibiotics have became essential instruments in the complex aquaculture ecosystem for preserving the well-being and production of aquatic life. Antibiotics are vital in treating and preventing bacterial infections that can endanger aquatic species health as a result of excess fish farming techniques. Antibiotic's triple role in disease prevention, control and aquaculture sustainability emphasizes its significance. In aquaculture, antibiotics are used to treat bacteria-related disorders. Antibiotics are occasionally used to treat infections, but they are more usually used to avoid infections by treating the water or fish before they get infected.

4.4.4 Critical Drawbacks of Antibiotic

Antibiotic misuse can result in antibiotic-resistant microorganisms as well as other unexpected consequences. Antibiotic-resistant bacteria can develop on their own when selection pressure produces changes to a bacterium's DNA sequence, allowing it to withstand antibiotic treatment. Antibiotics used to cure fish are additionally effective to treat human diseases. However, pathogenic bacteria that cause infections in humans may develop antibiotic resistance as a result of fish antibiotic therapy.

A systematic analysis of antibiotic resistance in Asian aquaculture and fisheries revealed concerned levels of resistance to medically important antimicrobials in foodborne pathogens (Schar et al. 2021). The study identified "resistance hotspots along major river systems and coastal waters of China and India", emphasizing the need for increased surveillance and targeted interventions in these regions (Schar et al. 2021).

Commercial chicken and aquaculture industries in Bangladesh are significant in the food value chain, but information about antibiotic usage in these systems is inadequate (Chowdhury et al. 2021). Studies have reported that a majority of chicken farm uses antibiotics for disease treatment and prevention (Chowdhury et al. 2021).

The WHO expert meeting on antibiotic usage in aquaculture and antimicrobial resistance was held in Seoul, Republic of Korea, on June 2006, identified two categories of risks:

(a) antimicrobial residues in aquaculture products
(b) selection and spread of Antimicrobial residues (AMR).

According to Bondad-Reantaso et al. (2023), the second possible danger is more substantial than the first due to AMR. In concise, misuse of antibiotics in fish aquaculture (among other agricultural applications) may pose public health risks.

4.4.5 Alternatives to Antibiotics

Antibiotics were historically used as a standard technique for managing fish diseases in order to increase growth and conversion of feed. Because of high incidence of bacterial diseases in fishes, antibiotics are often used and so remain in aquatic environment, leading spread of antibiotic-resistant bacteria. Antimicrobial resistance in aquatic environment can be spread to clinically relevant strains in the natural environment affecting whole ecosystem (Preena et al. 2020). To overcome this, various alternatives are available at the place of antibiotics. Some of the alternatives used in aquaculture are shown in Fig. 4.3.

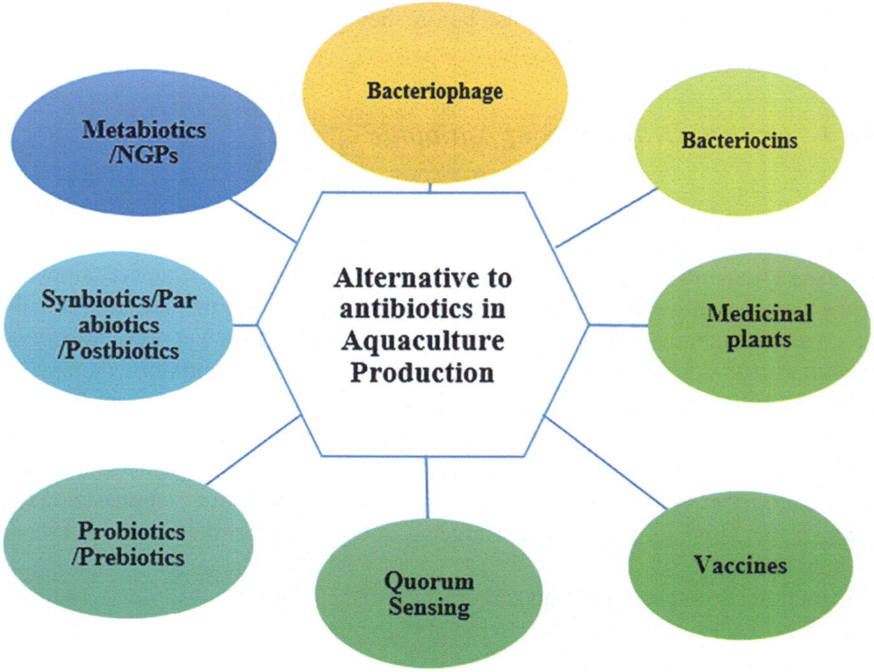

Fig. 4.3 Different alternatives to antibiotics in aquaculture production

4.5 Probiotics

Probiotics have been intensively explored as a substitute to antibiotics (Ushakova et al. 2012). Probiotics are helpful bacteria when supplied in an acceptable number, assist the host by encouraging intestinal-microbial balance, limiting pathogens growth, strengthening immunological response and aiding digestion through enzymatic activities (Balcazar et al. 2006). Probiotics aid in food absorption, promote post-stress adaption, raise microorganism resistance to harmful bacteria and improve digestive system function through increased enzyme production in the digestive tract. There is currently a growing interest in aquaculture production, which has the potential to enhance the supply of nutritious foods accessible to the people and impact of various habitats and feed additives on fish development and well-being were determined. Many writers have investigated the inclusion of probiotics to the diet throughout the years. Recent innovation focuses on the use of microbial communities to enhance fish growth as well as water qualities.

Probiotics are utilized in feed to improve host growth and health by boosting disease resistance (Mocanu et al. 2004). Probiotic bacteria's positive benefits on human, pig, ruminant and poultry nutrition are extensively proven (Bates et al. 2006). Although the use of such probiotics on aquatic animals is a relatively recent notion. The interest in probiotics and changes in gut bacteria arose primarily to promote and preserve the host's health. A Probiotic bacterium that competes with bacterial pathogens for resources and attachment sites, adheres to the gut and inhibits pathogen development is a valuable and valid alternative to antibiotic and biocideprophylaxis. Probiotics included in diets can boost growth rates and illness resistance. They help to improve digestion and reduce permeability stimulated by harmful microorganisms. Encapsulating probiotics with live feed are an effective way to conveying probiotics to aquatic animals. Dosage and duration of time are important elements in achieving the intended results (Hai 2015).

Probiotic bacteria like *Enterococcus, Lactobacillus* and *Saccharomyces* are frequently implemented in aquaculture. Lactic acid forming bacteria from the genus Lactobacillus exist in fish guts as part of the natural microflora, however they are not the dominating species. Lactic Acid Bacteria (LABs) are gram-positive organism having acid-tolerance, nonsporulating, non-respiring, long cylinder -like (bacilli) or circular (coccus) bacteria with comparable metabolic and physiological properties (Akweya et al. 2022; Shefali et al. 2021). LAB bacteria produce lactate as a consequence of their metabolic activities. They exist in fish intestines as part of the regular microflora but are not the prevalent species. LAB bacteria produce lactate as a consequence of their metabolic activities (Ringo et al. 2018). LAB has diverse applications in agriculture, nutrition and therapeutic practice (Kouhi et al.2022). LAB functions as an antagonistic, inhibitory and antibacterial defensive mechanism against infections and spoilage microorganisms. Lactic acid bacteriamay produce organic acids, hydrogen peroxide and bacteriocins, which are highly effective against a variety of illnesses (Haghshenas et al. 2023).

Probiotics are "living bacteria that, when administered in adequate amounts, confer a health benefit on the host" (FAO/WHO 2001). This concept was given by the " World Health Organization (WHO) and the Food and Agriculture Organization (FAO)" in 2002, with subsequent revisions for clarity by International scientific association for probiotics and prebiotics (ISAPP) in 2014. The well-accepted definition of probiotics described by the "Food and Agriculture Organization of the United Nations (FAO)" and the "World Health Organization (WHO)"(2014) is "live microorganisms that, when administered inadequate amounts, confer a health benefit on the host" (Hill et al. 2014). In other words, we can say that "foods or supplements that contain live microorganisms are intended to maintain or improve the "good" bacteria (normal microflora) in the body". Prebiotics are "foods (typically high-fiber foods) that act as food for internal microflora".Probiotics are usually combination of live beneficial bacteria and yeasts.

The probiotic cell wall is made up of cell structural S-layer protein, peptidoglycan, and polysaccharides, which can give amino acids and carbohydrates for starting cultures in fermented goods (Zendeboodi et al. 2020). Ruptured cells of *L. acidophilus*cellwallhavepeptidoglycans, teichoic acids, surface proteins, and anionic-neutral polysaccharides as well as S-layers which can be consumed by yogurt starters during fermentation has been reported (Anjum et al. 2014).

4.5.1 Criteria to Be a Nobel Probiotic

The requirements to become a Nobel Probiotic are high benchmarks that go beyond conventional ideas of microbial supplementation. A Nobel Probiotic should not only be very effective in enhancing host health but also show a deep comprehension of the complex interactions that occur inside the microbiome. To be a noble probiotic it should follow the following standards:

I. It Should be isolated from the same species.
II. It must exert beneficial effect on the host.
III. It must be able to attach to the epithelia of the hostgut.
IV. It should be nonpathogenic and nontoxic.
V. It must be able to inhabit the small intestine and large intestine of the host organism.
VI. It should contain a high number of live cells and maintain viability throughout storage and usage.
VII. It must be able to survive, establish itself and multiply in the gut environment, remaining resilient against gastric juices and bile acids.

4.5.2 Probiotic Strains Used in Aquaculture

It is widely believed that fish have low metabolic rates and can't use carbohydrates efficiently. According to Wilson and colleagues, the plasma glucose clearance rate is sluggish, which makes them seem like they're intolerant of glucose (Wilson 1994). Some authors also noted that fish don't use carbohydrates properly, which means that protein is their main energy source. (Peragon et al. 1999). However, these animals have a great potential for anaerobic metabolic processes, and their muscular framework (40–60% body mass) is primarily made up of weakly vascularized lactic white fibers (75–100% total muscle mass). The effects of commonly used probiotics like *Bacillus sp.*, *Lactobacillus sp.* and *Saccharomyces sp.* have been extensively studied across various aquatic species (Kim et al. 2017). The effectsof dietary *Bacillus* supplementation on growth, immune response, and disease resistance in various fish species like *Catlacatla* (Bandyopadhyay and Mohapatra 2009), *Cirrhinus mrigala* (Bhatnagar and Lamba 2015), *Channa striata* (Talpur et al. 2014) and *Oncorhynchus mykiss* (Irianto and Austin 2002). Most studies have primarily concentrated on using single-species probiotics, while only limited research has explored bacterial consortia—mixtures of multiple probiotic speciesand their effects on fish health (Cerezuela et al. 2012).

Probiotics in aquaculture are usually offered in powder (dry) form. These are dried using spray drying technique resulting in a stable powder. *Bacillus sp.*, *Saccharomyces cereviceae*and *Lactobacillus sp.* are among the most extensively examined probiotic bacteria. Microorganisms are mixed with various carriers such as malt syrup, wheat flour and rice starch. The carrier material's primary goal is to increase storage process survival. Studies have shown that the survival and growth of probiotic bacteria in different carrier materials varies. The contained material supports a wide variety of microbiological development. For the bacteria *B. licheniformis* and *B. subtilis*, the maltodextrin carrier material had the highest number, followed by rice flour for the bacteria *L. brevis* and *L. curvatus* and talc for the bacterium *L. bulgaricus.*

There were some carriers such as tapioca, skim milk and activated carbon, which gave a higher picture of microbial population than that of the standard medium. There was an increase in the number of bacteria after encapsulation because the dry carrier material had already been impacted by the microbial cells. Probiotic microbe counts increased from 40.7–90.1% in activated carbon carriers, 79–94.5% in skimmed milk and 31.7–95.7% in tapioca flour carriers. Such decline in numbers of colonies of probiotic bacteria in suspension formulation may be due to irreversible cell death of bacterial cells during spray drying process of encapsulation (Andriani et al.2022).

The most common method is to incorporate probiotics into the feed, allowing the aquatic animals to ingest them regularly (Rimoldi et al. 2024). Probiotics can also be added directly to the water, where they can help to improve water quality and reduce the levels of harmful bacteria. Probiotics are often used in larval culture to establish a beneficial microbiota in the gut of young aquatic animals, which can improve their survival and growth (Uddin et al. 2015). The introduction of microorganisms requires a specific carrier so their introduction in water is more convenient.

4.5.3 Probiotics Role on Digestive Enzyme of Gastrointestinal Tract

Probiotics are known for their power to enhance the growth of fish by secreting enzymes, which are beneficial for the breakdown of complex compounds, making them easily digestible for the fish. It is also commercially widespread in the market in various forms, either incorporated into the feed or directly used in the medium.

Appropriate probiotic applications improve intestinal microbial balance leading to improved food absorption by improved digestive enzyme activities (Tovar-Ramirez et al. 2004). In a 2015 study, Venkatrayulu and colleagues examined the effects of commercial probiotics on the digestive enzyme activities of the shrimp Penaeus monodon. The study utilized two commercial probiotics, Wunapuo 15 (water probiotic) and Aqualact (feed probiotic), in shrimp culture ponds covering 1 hectare. The water probiotic was applied at a rate of 30 kg/ha every 15 days over a 110-day culture period, while the feed probiotic was added to feed at 5 g/kg every other day, beginning 15 days into the culture until harvest. Enzyme activities for amylase, protease, and lipase were assessed on days 30, 60 and 90 in both control and probiotic-treated ponds during the summer season.

4.5.4 Changes in Biochemical Parameters Due to Probiotics

The variability in biochemical parameters associated with probiotic supplementation in *C. carpio has been observed*. To investigate its effects on *C. carpio,* many laboratory experiments and field studies have been conducted. In controlled laboratory experiments, juvenile or adult *C. carpio* specimens were subjected to different dietary treatments containing varying concentrations of probiotics. Growth performance was assessed by measuring changes in body weight, length, and specific growth rates over particular periods. Metabolic activities were evaluated through biochemical analyses of blood or tissue samples, which included assessments of enzyme activities, metabolite concentrations, and other relevant biochemical parameters.

Dietary supplementation with probiotics led to significant improvements in growth performance, with *C. carpio* specimens exhibiting weight gains and feed conversion efficiencies incompare to control groups (Li et al. 2018). Furthermore, biochemical analyses revealed favourable changes in blood parameters, including increased haemoglobin and haematocrit levels, indicating enhanced metabolic functions and nutrient utilization in probiotic-treated fish. Wang et al. (2016) investigated the effects of probiotics on *C. carpio* metabolism and immune responses. Their findings showed that probiotic supplementation resulted in elevated levels of superoxide dismutase (SOD) and catalase (CAT) in *C. carpio* tissues. These changes indicated improved oxidative stress tolerance and immune system function, enhancing metabolic efficiency and overall health. It suggests that probiotics have the potential to positively influence *C. carpio* growth and metabolism by inducing beneficial changes in biochemical parameters.

4.5.5 Antimicrobial or Antagonistic Properties of Probiotics

Probiotics antimicrobial or antagonistic properties are important, leading to the production of bacteriocins (Kumar et al. 2025). Allosteric exclusion of pathogens acts as an intestinal barrier and enhances resistance to pathogens and the host's immune system to combat pathogens successfully (Sharma et al. 2024; Sharma et al. 2025; Van et al. 2020). *Bacillus* strains of human gut harbour molecular gene biosynthetic locus and show phenotypic antibacterial effects. They also suggested that further intervention studies are required to demonstrate recovery from microbiota dysbiosis, towards eubiosis (Torres-Sanchez et al. 2021).

The antibacterial properties of bacteria and their sensitivity to antibiotics while fighting biofilm-forming Escherichia coli strains. Four fermented food samples, including "Maman pickle, fermented fish, fermented durian, and fermented glutinous rice" were tested for presumptive LAB colonies via API 50 technique. The isolates were identified as *Lactobacillus plantarum, Pediococcus pentosaceus,* and *L. pentosus* by 16S ribosomal DNA sequencing. Many results indicated that all identified. The resistance of LAB to four antibiotics highlights the need for safety and effectiveness testing before commercializing probiotics (Rusli et al. 2023).

4.5.6 Probiotics Used Against Pathogens

Probiotics have also been utilized to treat problems brought on by other pathogenic microorganisms in shrimp and fish farming operations. These bacteria use a competitive mechanism to regulate the proliferation of harmful bacteria in the digestive tract. Parvati et al. (2016) investigated the role of probiotics in combating Epizootic Ulcerative Syndrome (EUS) in mrigal fish (*Cirrhinus mrigala*), a disease caused by the bacterial pathogen *Aeromonas hydrophila* and the fungal pathogen *Aphanomyces invadans*. In vitro tests demonstrated that both probiotic treatments produced inhibition zones against the bacteria and fungi, with Probiotic 2 showing a larger zone of inhibition compared to Probiotic 1. Furthermore, Probiotic 2 significantly decreased the pathogenic load from 8.0×10^{11} to 3.0×10^{5} cfu/mL, surpassing the reduction achieved by Probiotic 1, which was 1.8×10^{5} cfu/mL. This study indicated that probiotic cultures could effectively lower the viable counts of Aeromonas hydrophila in fish. Similarly, studies by Ryan et al. (2009) revealed that live cells of *Lactobacillus salivarius* strains UCC118 and UCC119 were essential to inhibit *Helicobacter pylori* growth.

4.5.7 Probiotics as an Alternative to Antimicrobials

The effects of dietary supplementation with *Lactobacillus plantarum* and *Lactobacillus bulgaricus* on the growth performance and gut microbiota of common carp (*Cyprinus carpio*). Over a 60-day period, fish receiving these probiotics exhibited significantly improved growth indices and a higher proportion of intestinal lactic acid bacteria compared to the control group. Notably, *L. bulgaricus* supplementation led to the highest increase in gut *Lactobacillus* levels. These findings suggest that *L. plantarum* and *L. bulgaricus* can positively influence fish health and development, potentially serving as alternatives to traditional antimicrobials in aquaculture. However, the authors emphasize the need for further large-scale, farm-level trials to fully assess the practical applications and efficacy of these probiotics in diverse aquaculture settings (Alishahi et al. 2018). Additionally, research by Das et al. (2024) highlights that various bacterial isolates can produce antimicrobial compounds effective against pathogens like *Aeromonas* and *Edwardsiella tarda*, underscoring the potential of probiotics in disease prevention within aquaculture. Overall, incorporating probiotics such as *L. plantarum* and *L. bulgaricus* into aquaculture practices holds promise for enhancing fish growth and health, reducing reliance on antibiotics, and promoting sustainable aquaculture.

4.5.8 Probiotics Effecting fish Hemoglobin Level

The assessment of "physiological and hematological parameters in cultured fish species plays a crucial role in advancing aquaculture systems, particularly in differentiating between healthy fish and those that are infected or experiencing stress"(Oneal et al. 2006). A fish's hematological profile can serve as an indicator of its health status, and when used alongside other routine diagnostic methods, it helps identify stressors or diseases that can impact productivity. Thus, monitoring these characteristics is essential for recognizing and managing health conditions that affect production performance in aquaculture (Pavlidis et al. 2007). Mohammadi et al. (2021) showed that the use of probiotics in the diet of Nile tilapia (*Oreochromis niloticus*) significantly improved hemoglobin levels and other hematological parameters. Thus, probiotics led to improved growth performance and hematological indices, including increased hemoglobin content, in various fish species. This study emphasized the role of probiotics in modulating the gut microbiota, which facilitates better digestion and nutrient uptake, ultimately enhancing overall fish health and blood parameters. Moreover, Van Doan et al. (2022) corroborated these findings, showing that probiotics supplementation in diets of *C. carpio* resulted in significant increases in hemoglobin concentration and other blood parameters.

4.5.9 Factors Affecting Probiotic Activity

Both intrinsic and extrinsic factors playrole in determining survival and activity of probiotics in food products. Key elements impacting their viability include physical conditions such as temperature and humidity, which can affect the stability and growth of probiotic cultures. Additionally, chemical factors like oxygen levels, nitrogen presence, and pH also influence probiotic survival, as each of these can either support or hinder the optimal environment needed for probiotic activity. Maintaining the right balance of these conditions is essential to ensure probiotics remain effective in food products.

4.5.10 Effect of Probiotics on Fish Mechanism of Probiotic Activity

I. They can improve fish growth, feed conversion, health, disease resistance, and stress tolerance.
II. by modifying the GIT microbiome.
III. They help in breakdown of chemicals which are not digested by fish host. They assist in carbohydrate digestion and also act as an antinutrient detox.
IV. Example -Probiotic-derived chitinase may enable the adoption of novel feedstuffs like insect meal or krill.
V. Digestion of secondary plant compounds with antinutritional effects by probiotic can increase the rate of fishmeal substitution by using only a specific plant feed ingredient.
VI. Multispecies probiotics have greater probiotic efficacy than single-species applications.
VII. Modulating the gut microbiota through dietary administration of probiotics indicates a potential method to increase microbial metabolite synthesis and boost the immune signals and enhance defensive systems against infections.

4.5.11 Do Probiotics Have Adverse Effects?

Despite their immune-boosting potential, questions persist regarding the safety of traditional probiotics, particularly with repeated use. Studies have shown that the positive effects of live probiotic microbes may be short-lived, intermittent, or may even diminish over extended use. Recent research has also highlighted concerns about the long-term use of probiotics, noting a possible association with opportunistic infections. Probiotics may not always benefit healthy individuals and could even exacerbate inflammation in those with inflammatory bowel disease (IBD). Notably, a well-formulated postbiotic can offer protection against the inflammatory impact of invasive *Salmonella* on healthy tissue and can also help to reduce inflammation in IBD-affected tissue (Tsilingiri et al. 2012). To address these potential drawbacks, a new formulation known as metabiotics, which includes additional nutraceutical benefits, is emerging in food and health sciences as a promising alternative.

4.6 Prebiotic

Growth or beneficial bacteria activity (probiotics) stimulating substances in the fish's digestive system are called prebiotics. These are "non-digestible ingredients of diet that specifically stimulate the growth or activity of beneficial bacteria in the gut, thereby improving host health" (Gupta et al. 2019). Prebiotics are a group of nutrients, primarily fiber, acting as feed for the beneficial microbes in the gut as consumption is expected to promote community stability, provided that such feeding can preferentially weaken interactions between collaborating with species by, for example, providing a substitute carbon source to that that facilitates microbial cross-feeding (Coyte et al. 2015).

Mostly carbohydrates and fiber act as fuel for probiotics. Prebiotics are an alternative to probiotics or their cofactors. This concept was introduced by Gibson and Roberfroid in 1995.Prebiotic usually shows limited breakdown and absorption in the upper gastro-intestinal tract. They lead to selective stimulation of growth and immunostimulation of beneficial bacteria.major properties of prebiotics which are observed in aquaculture are summarized in Fig. 4.4.

In aquaculture, prebiotics modulate the fish gut microbiota and other aquatic animals, leading to improved disease resistance, growth performance and nutrient uptake. Prebiotics can significantly alter the composition and diversity of the gut microbiota in aquatic animals. For instance, alginate oligosaccharides (AlgOS) derived from Laminaria sp. have been shown to shift the intestinal microbiota profile in Atlantic salmon. A low supplementation level of AlgOS facilitated the dominance of Proteobacteria and Spirochaetes in the intestinal content and mucus of the fish (Gupta et al. 2019). Some of these bacteria such as *Aliivibriologei*, *Aquabacterium parvum*, *Brevinemaandersonii*, and *Achromobacter insolitus* possess genes associated with butyrate production, which is beneficial for gut health (Gupta et al. 2019).

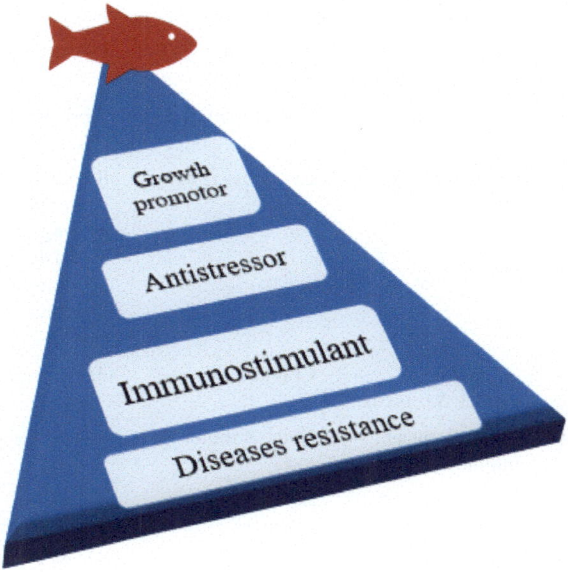

Fig. 4.4 Role of prebiotics in fishes

4.6.1 Types of Prebiotics Used in Aquaculture

Several types of prebiotics have been observed for their potential benefits in aquaculture.

4.6.1.1 Oligosaccharides

These include fructo-oligosaccharides (FOS), galacto-oligosaccharides (GOS), and mannan oligosaccharides (MOS).

4.6.1.2 Polysaccharides

These include alginate oligosaccharides derived from macroalgae (Gupta et al. 2019). Some studies also explore the use of xylo-oligosaccharides (XOS) and inulin as prebiotics in aquaculture feeds. According to Guzman-Villanueva et al. (2014), prebiotic combinations containing β-1,3 + 1,6-glucans have been identified as immunostimulants. Aquaculture farming system species may be under stress as a result of the intensification of the sector, which is primarily intended to provide cost-effective and reasonably priced aquaculture products (Das et al. 2017). Dawood and Koshio (2016) observed that β-glucan alone could reduce stress in red sea bream (*Pagrus major*). FOS by itself has also been shown to lessen common carp (Cyprinus carpio) stress (Hoseinifar et al. 2014).

Table 4.1 Common prebiotics mixture used in aquaculture

Sr. no	Prebiotics mixture	Aquaculture species	Use/benefit	References
1	Beta glucon, MOS	*Oreochromis niloticus*	Work against *Pseudomonas aeruginosa*	El-Nobi et al. (2021)
2	GOS, FOS, Inulin	*Cyprinus carpio*	Regulate Antioxidant-related genes(glutathione S-transferase[GST-α], glutathione reductase and glutathione peroxidase genes	Hoseinifar et al. (2017)
3	Alginate extract, *Sargassum*	Crayfish	*Aeromonas hydrophila*	Amrullah and Wahidah (2019)
4	MOS, β-glucan, inulin	Chinese mitten crab	*Aeromonas hydrophila*	Lu et al. (2019)
5	MOS, β-glucan	Shabout (*Tor grypus*)	Act as an immunostimulant	Mohammadian et al. (2021)
6	β-glucan + MOS + dextrose	Nile tilapia (*Oreochromis niloticus*)	Act as a growth promoter	Ismail et al. (2019)

Beta glucan, inulin, mannan-oligosaccharide (MOS), arabinoxylan oligosaccharide (AXOS), galacto-oligosaccharide (GOS), oligosaccharide, and fructo-oligosaccharides are the main prebiotics utilized in aquaculture. Some of the prebiotics mixture which have been studied in aquaculture are shown in Table 4.1.

4.6.2 Benefits of Prebiotics in Aquatic Culture

The use of prebiotics in aquaculture can lead to several beneficial outcomes. "Prebiotics promote the growth of beneficial bacteria, which can help to maintain a healthy gut environment and prevent the colonization of pathogenic bacteria" (Gupta et al. 2019). "By modulating the gut microbiota, prebiotics can enhance the immune response of aquatic animals, making them more resistant to diseases" (Ke et al. 2021). Some studies have shown that prebiotics can improve growth performance in fish and shrimp by enhancing nutrient absorption and utilization (Rimoldi et al. 2024).With the increasing concern over antibiotic resistance, prebiotics offer a natural alternative to antibiotics for disease prevention and growth promotion in aquaculture (Ke et al. 2021).

4.7 Postbiotics

A postbiotic refers to a "preparation of inactivated microorganisms and/or their components that provides a health benefit to the host". These bioactive substances are produced when probiotic bacteria metabolize prebiotics (fibers). Postbiotics consist of intentionally inactivated microbial cells, which may or may not include metabolites or cell components, yet are associated with proven health benefits. They are "complex mixtures containing metabolic products such as enzymes, secreted proteins, short-chain fatty acids, vitamins, biosurfactants, amino acids, peptides, and organic acids"—typically produced during fermentation and used to enhance health.SCFAs like butyrate are crucial for intestinal health and can influence the gut microbiota (Gupta et al. 2019).

The three main kinds of substances that make up postbiotics are inactivated microbial cells, bacterial cell components, and crude by-products of microbial metabolism. These substances are frequently derived from a wide range of probiotic microbes, including several genera and species (Fijan 2014).

4.7.1 Making Postbiotics in Aquaculture

Heat inactivation, ionizing radiation, UV irradiation, high pressure treatment, and ultrasonication are generally used techniques to turn live probiotics into postbiotics (Li et al. 2023).

The following chemicals can be used as postbiotics:

I. Short-chain fatty acids or lipopolysaccharides
II. Exopolysaccharides or enzymes
III. Fragments of cell wall
IV. Lysates (bacterial components) of bacteria
V. Bacteria and yeast cell-free supernatants
VI. Vitamins and amino acids.

- **Microbial metabolites and vaccines are not considered postbiotics**

Importantly, a postbiotic doesn't need to originate from a probiotic to qualify as such. Postbiotics can work beyond the gut, targeting areas like the oral cavity, stomach, skin, urogenital tract, or nasopharynx, but are not delivered via injections. Their positive health effects must be validated within the specific host species and population intended. Humans, domesticated animals, cattle, and other targets can all serve as hosts.

The advantage of using postbiotics is that they are non-living, making them more stable and easier to handle compared to probiotics. They also do not pose the same risk of antimicrobial resistance transfer as some probiotics (Uddin et al. 2015).

4.8 Para-Probiotics

Para probiotics are also called as ghost, inactivated or non-viable (dead) cells that can lead to health benefits to the host (Siciliano et al. 2021). Para probiotics are "cell fractions or inactivated microbial cells that provide health benefits to the host" (Aguilar-Toala et al. 2018). These might be peptidoglycans, teichoic acids, surface proteins or simple cell extracts in chemical composition. These can include heat-killed probiotics or cell lysates. Para-probiotics can stimulate the immune system, leading to enhanced disease resistance. They can improve the gut barrier function, preventing the leakage of harmful substances into the bloodstream. Some para-probiotics have been shown to have anti-inflammatory effects, which can help to alleviate gut inflammation. Para-probiotics offer the advantage of being safer than live probiotics, as they cannot colonize the gut or transfer antimicrobial resistance genes (Uddin et al. 2015).

4.9 Advantages of Para and Post-Biotics

Certain postbiotics and para-probiotics have been shown to have bioactivities such anti-inflammatory, immunomodulatory, anti-proliferative, antioxidant, and antibacterial properties in both in vitro and in vivo studies (Cuevas-Gonzalez et al. 2020). These provide various benefits over probiotics:

1. Pure availability
2. Simple manufacturing and storage procedures
3. Feasibility of scaling up the production technique for industrial applications.
4. Defined mechanism of action.
5. Enhanced accessibility of microbe-associated molecular patterns (MAMPs) to interact with recognition by pattern recognition receptors (PRRs).
6. Ligand-receptor interactions are more likely to cause tailored reactions.
7. Postbiotics can help regulate and prevent bacteria-related diseases in aquaculture (Quintanilla-Pineda et al. 2023).

4.10 Synbiotics

Probiotics and prebiotics work collectively to enhance the fish health leading to improve in their overall performance. The word "Synbiotic" was firstly used by scientists sir Gibson and Roberfroid in 1995 for this combination.

$$\text{Synbiotics} = \text{Probiotics} + \text{Prebiotics}$$

Synbiotics are combinations of prebiotics and probiotics that work collectively to enhance the health of the host. The prebiotic component selectively stimulates the

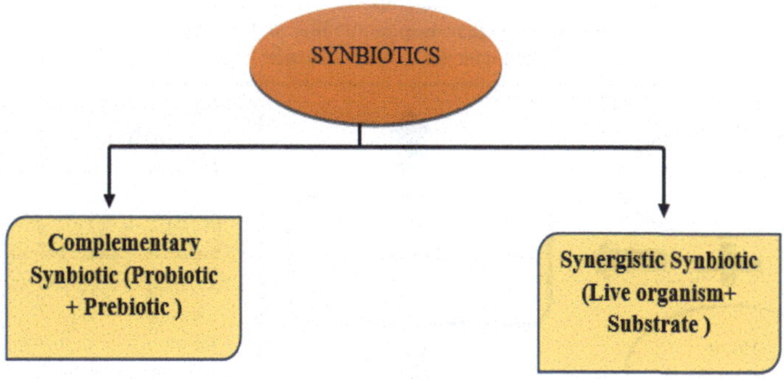

Fig. 4.5 Components of synbiotics

growth and/or activity of the probiotic component, leading to a synergistic effect. The rationale behind using synbiotics is that the prebiotic component can provide a specific substrate for the probiotic bacteria, enhancing their survival and activity in the gut. This can lead to a more pronounced and sustained beneficial effect on the host.

According to mode of action, roles and targets, the synbiotic microbial bioactives can be divided into various categories. Two major classes of synbiotics are complementary synbiotics and synergistic synbiotics (Fig. 4.5).

Synbiotics can have more significant effect on the gut microbiota compared to prebiotics or probiotics alone. In vitro studies have shown that synbiotics can increase the levels of beneficial bacteria, such as Bifidobacterial and Eubacteriumrectale-Clostridium coccoides group, while reducing the levels of harmful bacteria, such as *Escherichia coli* (Nair et al. 2024). Synbiotics can promote a more balanced and diverse gut microbiota, which is essential for maintaining gut health (Nair et al. 2024). Some studies have shown that synbiotics can improve growth performance in fish and shrimp by enhancing nutrient absorption and utilization (Rimoldi et al. 2024).

The production and release of these bioactive low-molecular-weight (LMW) molecules, along with their effects, are carefully controlled. For instance, Gram-negative bacteria use small, diffusible compounds, typically derived from amino acids, to communicate and regulate various responses.

4.11 Metabiotic

"Metabo-"(metabolism) and "-biotics"(living organisms or substances with a beneficial effect), metabiotic might refer to substances or compounds produced through metabolic processes by living organisms, that have potentially beneficial effects on other organisms. Metabiotics are more advanced than traditional probiotics in their chemical structure and have more pronounced positive impacts. Major role and mode

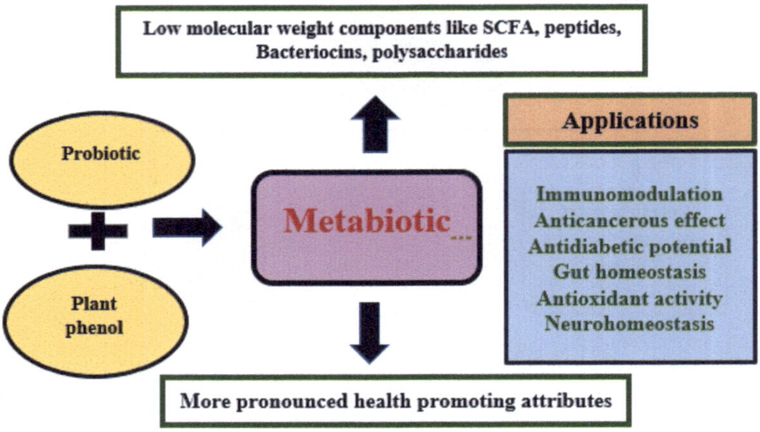

Fig. 4.6 Effect of metabiotics on growth and diseases resistance in aquaculture

of mechanism of metabiotics are summarized in Fig. 4.6.

$$\text{Probiotics} + \text{Plant phenols} = \text{Metabiotic}$$

"Through altering the composition of the gut microbiota, improving intestinal barrier function, and lowering inflammation, microbial metabolites like short-chain fatty acids (SCFAs), vitamins, and amino acids can support gut health" (Speckmann et al. 2024).

4.12 Potential of Bacteria in Aquaculture

Probiotics, prebiotics, metabioticsand synbiotics have great significance in the future of aquaculture. As the demand for sustainable and environmentally friendly food production continues to rise, these approaches offer several benefits:

i. **Improved Sustainability**: Probiotics, prebiotics, and related concepts can contribute to more efficient resource utilization in aquaculture systems. By enhancing nutrient absorption and reducing waste, these approaches can lead to lower feed conversion ratios and decreased environmental impact.
ii. **Disease Management:** The use of beneficial microorganisms can help mitigate disease outbreaks in aquaculture. Probiotics and synbiotics can enhance the immunity of aquatic organisms.
iii. **Water quality Enhancement:** Introducing beneficial bacteria and their metabolites can recuperate water characteristic by aiding breakdown of organic matter and dropping ammonia and nitrate buildup. This can lead to healthier aquatic ecosystems and minimize environmental pollution.

iv. **Enhanced Growth and Production:** Optimizing the gut health of aquatic organisms through prebiotics and probiotics can result in better growth, feed utility and overall results in aquaculture systems.
v. **Reduced Antibiotic Dependency**: The application of probiotics and its associated approaches may help decrease the reliance on antibiotics in aquaculture, addressing concerns about antibiotic resistance and promoting the production of safe and high-quality seafood.
vi. **Customized Nutrition:** Synbiotics and metabiotic can allow for tailored nutrition strategies, optimizing the dietary requirements of different aquatic species for their growth and development.
vii. **Biosecurity and Sustainability:** A balanced microbial ecosystem fostered by these approaches can help create a more resilient and stable aquaculture environment by decreasing the danger of disease outbreaks leading to long-term sustainability.

However, it's essential to consider potential challenges such as strain selection, optimal application methods, and the complex interactions within aquatic ecosystems. The intestinal "microbiome," representing the entire gene pool of organisms within the gut microbiota, is estimated to contain more than 150 times the number of unique genes found in the human genome (Walker and Lawley 2013). Administering live bacterial cells to individuals with weakened immune systems, heightened inflammatory responses, or impaired mucosal barriers may risk transforming typically harmless probiotic bacteria into harmful organisms (Taverniti and Guglielmetti 2011).

Continued research and collaboration between aquaculture experts, microbiologists, and other relevant fields will be crucial to fully harness the potential of these innovative approaches in the future of aquaculture.

4.13 Conclusion

Probiotics are the preferred candidate in aquaculture for better growth rather than antibiotics. Postbiotic supplements are not as widely available yet because they're relatively new compared with prebiotics and probiotics. Though probiotic complements have been used frommanyyears,still some of them also showed pathogenic concerns on longtime use. The lactic acid bacteria are considered one of the good probiotics as they survive through stressful situations.

To counteract growth reduction in fish reared in saline conditions, synbiotics and biofloc meals are being considered as feed additives (Hersi et al. 2023). Paraprobiotics offer a safer alternative to live probiotics, synbiotics combine the benefits of probiotics and prebiotics, and metabiotic harness the power of microbial metabolites (Deshpande et al. 2018).Probiotics, prebiotics, and other biological additives are among the biotic agents that are strategically applied in this revolutionary change to produce a symbiotic environment in aquaculture systems. Through the cultivation of a mutually beneficial association between aquatic creatures and their microbial allies, we can progress towards enhanced fish growth, enhanced disease resistance, and optimized resource utilization. The introduction of probiotic into water also has a significant point from the environmental research concern. Paraprobiotic offer the advantage of being safer than live probiotics, as they cannot colonize the gut or transfer antimicrobial resistance genes (Uddin et al. 2015).

4.14 Future Prospective

In the near future, quantifying low-molecular-weight signaling molecules of probiotic (synbiotic) microbial origin with distinct functional or metabolic roles in various biological samples could reveal links between microbial imbalances and specific health disorders, as well as assess the effectiveness of biotherapy. There is also potential for using LAB and the bacteriocins they produce as natural food preservatives to control food borne pathogens. On the other hand, altering the microbiome could have unfavorable consequences and even open the door for infections. As a result, we must increase our understanding of microbiome regulation. Microbiome regulation. Furthermore, species-specific research is essential prior to probiotic application to a new species. Additionally, numerous Lactobacillus strains have been shown antiretroviral action against murine norovirus (MNV), thus it will be interesting to discover microorganisms that can affect fish-specific viral infectivity.

Acknowledgements The authors are thankful to the faculty of the Department of Zoology and Department of Microbiology, Chaudhary Charan Singh Haryana Agricultural University, Hisar, Haryana, India for their support. Our deepest gratitude goes to our family for their constant motivation and to all researchers whose work laid the foundation for this study.

References

Aguilar-Toala JE, Garcia-Varela R, Garcia HS, Mata-Haro V, Gonzalez-Cordova AF, Vallejo-Cordoba B, Hernandez-Mendoza A (2018) Postbiotics: an evolving term within the functional foods field. Trends Food Sci Technol 75:105–114

Akweya BA, Mghalu JM, Yusuf RUM, Bitange T (2022) Isolation, biochemical characterization and safety screening of potential probiotic lactic acid bacteria from spontaneously fermented cereal products from Western Kenya. Afr J Microbiol Res 16(4):160–166

Amrullah A, Wahidah W (2019) Immune response and growth performance of crayfish *Cheraxquadricarinatus* fed with supplementary diet of synbiotic. Jurnal Aquaculture Indonesia 18(1):33–45

Andriani Y, Pratama RI, Safitri R (2022) Probiotic microbes and their growth in different carriers (a review). South Asian J Res Microbiol 13(1):15–23

Anjum N, Maqsood S, Masud T, Ahmad A, Sohail A, Momin A (2014) *Lactobacillus acidophilus*: characterization of the species and application in food production. Crit Rev Food Sci Nutr 54(9):1241–1251

Balcazar JL, De Blas I, Ruiz-Zarzuela I, Cunningham D, Vendrell D, Muzquiz JL (2006) The role of probiotics in aquaculture. Vet Microbiol 114(3–4):173–186

Bandyopadhyay P, Das Mohapatra PK (2009) Effect of a probiotic bacterium *Bacillus circulans* PB7 in the formulated diets: on growth, nutritional quality and immunity of *Catla catla* (Ham.). Fish Physiol Biochem 35:467–478

Bates JM, Mittge E, Kuhlman J, Baden KN, Cheesman SE, Guillemin K (2006) Distinct signals from the microbiota promote different aspects of zebrafish gut differentiation. Dev Biol 297(2):374–386

Bhatnagar A, Lamba R (2015) Antimicrobial ability and growth promoting effects of feed supplemented with probiotic bacterium isolated from gut microflora of *Cirrhinus mrigala*. J Integr Agric 14(3):583–592

Bondad-Reantaso MG, MacKinnon B, Karunasagar I, Fridman S, Alday-Sanz V, Brun E, Caputo A (2023) Review of alternatives to antibiotic use in aquaculture. Rev Aquac

Cerezuela R, Guardiola FA, Gonzalez P, Meseguer J, Esteban MA (2012) Effects of dietary *Bacillus subtilis*, *Tetraselmis chuii*, and *Phaeodactylum tricornutum*, singularly or in combination, on the immune response and disease resistance of sea bream (*Sparus aurata* L.). Fish Shellfish Immunol 33(2):342–349

Chowdhury S, Ghosh S, Aleem M, Parveen S, Islam MA, Rashid M, Akhtar Z, Chowdhury F (2021) Antibiotic usage and resistance in food animal production: what have we learned from bangladesh? Antibiotics. https://doi.org/10.3390/antibiotics10091032

Coyte KZ, Schluter J, Foster KR (2015) The ecology of the microbiome: networks, competition, and stability. Science 350:663–666

Cuevas-Gonzalez PF, Liceaga AM, Aguilar-Toala JE (2020) Postbiotics and paraprobiotic: from concepts to applications. Food Res Int 136:109502

Das P, Borah P, Bordoloi R, Pegu A, Dutta R, Baruah C (2024) Probiotic bacteria as a healthy alternative for fish and biological control agents in aquaculture. J Appl Nat Sci 16(2):674–689

Das S, Mondal K, Haque S (2017) A review on application of probiotic, prebiotic and synbiotic for sustainable development of aquaculture. Growth 14:15

Dawood MA, Koshio S (2016) Recent advances in the role of probiotics and prebiotics in carp aquaculture: a review. Aquaculture 454:243–251

Deshpande G, Athalye-Jape G, Patole S (2018) Para-probiotics for preterm neonatesthe next frontier. Nutrients

El-Nobi G, Hassanin M, Khalil AA, Mohammed AY, Amer SA, Montaser MM, El-Sharnouby ME (2021) Synbiotic effects of *Saccharomyces cerevisiae*, Mannan oligosaccharides, and β-glucan on innate immunity, antioxidant status, and disease resistance of Nile tilapia, *Oreochromis Niloticus*. Antibiotics 10(5):567

FAO (2022) Sustainability in action. The state of world fisheries and aquaculture. Food and agriculture organization. United Nations, Rome

FAO/WHO (2002) Guidelines for the evaluation of probiotics in food. Report of a Joint FAO/WHO Working Group on Drafting Guidelines for the Evaluation of Probiotics in Food; Food and Agriculture Organization of the United Nations; World Health Organization, London, ON, Canada

FAO/WHO (2001) Health and nutritional properties of probiotics in food including powder milk with live lactic acid bacteria. Food and Agriculture Organization of the United Nations; World Health Organization, Cordoba, Argentina

Fijan S (2014) Microorganisms with claimed probiotic properties: an overview of recent literature. Int J Environ Res Public Health 11(5):4745–4767

Flemming HC, Wuertz S (2019) Bacteria and archaea on Earth and their abundance in biofilms. Nat Rev Microbiol 17(4):247–260

Gupta S, Lokesh J, Abdelhafiz Y, Siriyappagouder P, Pierre R, Srensen M, Fernandes J, Kiron V (2019) Macroalga-derived alginate oligosaccharide alters intestinal bacteria of atlantic salmon. Front Microbiol

Guzman-Villanueva LT, Tovar-Ramirez D, Gisbert E, Cordero H, Guardiola FA, Cuesta A, Esteban MA (2014) Dietary administration of β-1, 3/1, 6-glucan and probiotic strain *Shewanella putrefaciens*, single or combined, on gilthead seabream growth, immune responses and gene expression. Fish Shellfish Immunol 39(1):34–41

Haghshenas B, Kiani A, Mansoori S, Mohammadi-Noori E, Nami Y (2023) Probiotic properties and antimicrobial evaluation of silymarin-enriched *Lactobacillus* bacteria isolated from traditional curd. Sci Rep 13(1):10916

Hai NV (2015) The use of probiotics in aquaculture. J Appl Microbiol 119(4):917–935

Hersi MA, Genc E, Pipilos A, Keskin E (2023) Effects of dietary synbiotics and biofloc meal on the growth, tissue histomorphology, whole-body composition and intestinal microbiota profile of Nile tilapia (*Oreochromis niloticus*) cultured at different salinities. Aquaculture 570:739391

Hill C, Guarner F, Reid G, Gibson GR, Merenstein DJ, Pot B, Morelli L, Canani RB, Flint HJ, Salminen S et al (2014) Expert consensus document. The international scientific association for probiotics and prebiotics consensus statement on the scope and appropriate use of the term probiotic. Nat Rev Gastroenterol Hepatol 11:506–514

Hoseinifar SH, Ahmadi A, Khalili M, Raeisi M, Van Doan H, Caipang CM (2017) The study of antioxidant enzymes and immune-related genes expression in common carp (Cyprinus carpio) fingerlings fed different prebiotics. Aquac Res 48(11):5447–5454

Hoseinifar SH, Soleimani N, Ringø E (2014) Effects of dietary fructo-oligosaccharide supplementation on the growth performance, haemato-immunological parameters, gut microbiota and stress resistance of common carp (Cyprinus carpio) fry. Br J Nutr 112(8):1296–1302

Irianto A, Austin B (2002) Use of probiotics to control furunculosis in rainbow trout, *Oncorhynchus mykiss* (Walbaum). J Fish Dis 25(6):333–342

Ismail M, Wahdan A, Yusuf MS, Metwally E, Mabrok M (2019) Effect of dietary supplementation with a synbiotic (Lacto Forte) on growth performance, haematological and histological profiles, the innate immune response and resistance to bacterial disease in Oreochromis niloticus. Aquac Res 50(9):2545–2562

Jayasankar P (2018) Present status of freshwater aquaculture in India—a review. Indian J Fish 65(4):157–165

Jeon HJ, Ban OH, Bang WY, Yang J, Jung YH (2022) Trends, functionalities, and prospects of probiotics, 465–476

Ke CL, Deng FS, Chuang CY, Lin CH (2021) Antimicrobial actions and applications of chitosan. Polymers 13(6):904

Kim DH, Subramanian D, Heo MS (2017) Dietary effect of probiotic bacteria, *Bacillus amyloliquefaciens*-JFP2 on growth and innate immune response in rock bream *Oplegnathus fasciatus,* challenged with *Streptococcus iniae.* Israeli J Aquac Bamidgeh 69

Kiran, Kumar N, Sindhu S, Gawaria B, Ansari A, Dhariwal A, Sharma P, Kumar R (2025) GWAS approaches in medicinal plants for enhanced secondary metabolite production. In: Chen JT (ed) Biotechnology, multiple omics, and precision breeding in medicinal plants. CRC Press, pp 119–134. https://doi.org/10.1201/9781003475491-10

Kouhi F, Mirzaei H, Nami Y, Khandaghi J, Javadi A (2022) Potential probiotic and safety characterisation of *Enterococcus* bacteria isolated from indigenous fermented Motal cheese. Int Dairy J 126:105247

Kumar R, Gupta R, Dahiya T, Yadav R, Saifi R, Yodha K, Kumari B, Singh P, Shefali (2022a) Surveillance of *Heteropneustes fossilis* (Stinging Catfish) diseases and health management practices in Haryana. Appl Ecol Environ Sci 10(4):261–267. https://doi.org/10.12691/aees-10-4-12

Kumar R, Gupta RK, Yadav R, Saifi R, Yodha K, Pooja (2022b) *Eisenia fetida* as protein source for growth enhancement of *Heteropneustes fossilis*. Egypt J Aquat Biol Fish 26(2):577–588. https://doi.org/10.21608/ejabf.2022.234553

Kumar R, Yadav R, Kumar Gupta R, Pooja (2022c) A sustainable way for fish health management by replacement of chemical and drugs by earthworm. In: Environmental degradation in Asia: land degradation, environmental contamination, and human activities. Springer International Publishing, Cham, pp 329–352. https://doi.org/10.1007/978-3-031-12112-8_16

Kumar R, Singh D, Yadav R (2025) Toxic effects of chlorantraniliprole and fipronil on *Eisenia fetida*: from growth and reproductive parameters to avoidance behavior. Environ Eng Manag J 24(7):1499–1508. https://doi.org/10.30638/eemj.2025.117

Li J, Wei Q, Wu Z, Liu W, Cao Z (2018) Effects of dissolved oxygen and temperature on growth and survival of juvenile Chinese sturgeon (*Acipenser sinensis*). Aquac Res 49(1):97–106

Li S, Yang H, Jin Y, Hao Q, Liu S, Ding Q, Zhou Z (2023) Dietary cultured supernatant mixture of *Cetobacteriumsomerae* and *Lactococcus lactis* improved liver and gut health, and gut microbiota homeostasis of zebrafish fed with high-fat diet. Fish Shellfish Immunol 142:109139

Lu J, Bu X, Xiao S, Lin Z, Wang X, Jia Y, Chen L et al (2019) Effect of single and combined immunostimulants on growth, anti-oxidation activity, non-specific immunity and resistance to Aeromonas hydrophila in Chinese mitten crab (Eriocheir sinensis). Fish Shellfish Immunol 93:732–742

Manjeet, Airon A, Kumar R, Saifi R (2022) Temporal and spatial impact of lockdown during COVID-19 on air quality index in Haryana, India. Sci Rep 12:20046. https://doi.org/10.1038/s41598-022-20885-2

Mocanu EE, Savin V, Popa MD, Dima FM (2022) The effect of probiotics on growth performance, haematological and biochemical profiles in siberian sturgeon (*Acipenser baerii* Brandt, 1869). Fishes 7(5):239

Mohammadi G, Rafiee G, Tavabe KR, Abdel-Latif HM, Dawood MA (2021) The enrichment of diet with beneficial bacteria (single-or multi-strain) in biofloc system enhanced the water quality, growth performance, immune responses, and disease resistance of Nile tilapia (*Oreochromis niloticus*). Aquaculture 539:736640

Mohammadian T, Ghanei-Motlagh R, Molayemraftar T, Mesbah M, Zarea M, Mohtashamipour H, Nejad AJ (2021) Modulation of growth performance, gut microflora, non-specific immunity and gene expression of proinflammatory cytokines in shabout (*Tor grypus*) upon dietary prebiotic supplementation. Fish Shellfish Immunol 112:38–45

Nair R, Paul P, Mahajan S, Maji I, Gupta U, Aalhate M, Singh PK (2024) Emerging era of "biotics": prebiotics, probiotics, and synbiotics. In: Synbiotics in human health: biology to drug delivery. Springer Nature, Singapore, pp 319–348

Nisha, Anshu, Rani S, Kumar S, Kumar R, Kumar A (2025) Recent technologies and strategies for bioremediation of toxic and hazardous waste. In: Dubey KK, Shah MP, Yadav A (eds) Challenges and sustainable solutions in bioremediation. CRC Press, pp 1–16. https://doi.org/10.1201/9781003407317-1

Oneal CC, Hawke JP, Weirich CR (2006) Evaluation of low levels of salinity on hematological parameters and health status of channel catfish reared in multiple-crop ponds. J Aquat Anim Health 18(1):1–10

Patrinou V, Daskalaki A, Kampantais D, Kanakis DC, Economou CN, Bokas D, Tekerlekopoulou AG (2022) Optimization of cultivation conditions for *Tetraselmis striata* and biomass quality evaluation for fish feed production. Water 14(19):3162

Pavlidis M, Futter WC, Katharios P, Divanach P (2007) Blood cell profile of six Mediterranean mariculture fish species. J Appl Ichthyol 23(1):70–73

Peragon J, Barroso JB, Garcıa-Salguero L, de la Higuera M, Lupianez JA (1999) Carbohydrates affect protein-turnover rates, growth, and nucleic acid content in the white muscle of rainbow trout (*Oncorhynchus mykiss*). Aquaculture 179(1–4):425–437

Preena PG, Swaminathan TR, Kumar VJR, Singh ISB (2020) Antimicrobial resistance in aquaculture: a crisis for concern. Biologia 75:1497–1517

Quintanilla-Pineda M, Díaz J, Gutierrez-Falcon A, Ibanez FC, Marzo F (2023) Profiling a new postbiotic product for its application in fish aquaculture. Fishes 8(6):304

Rani S, Sharma S, Singh D, Sheoran R, Kumar R (2025) Improved production of anti-cancer compounds from medicinal plants through molecular and biotechnological mutagenesis strategies. In: Chen JT (ed) Biotechnology, multiple omics, and precision breeding in medicinal plants. CRC Press, pp 287–297. https://doi.org/10.1201/9781003475491-21

Rimoldi S, Di Rosa AR, Armone R, Chiofalo B, Hasan I, Saroglia M, Terova G (2024) The replacement of fish meal with poultry by-product meal and insect exuviae: effects on growth performance, gut health and microbiota of the European seabass, Dicentrarchus Labrax. Microorganisms 12(4):744

Ringo E, Hoseinifar SH, Ghosh K, Doan HV, Beck BR, Song SK (2018) Lactic acid bacteria in finfish—an update. Front Microbiol 9:1818

Rusli NS, Lani MN, Shahabudin S, Azmi NS, Taghavi E, Abdullah WZW, Desa MNM (2023) Antibiotic susceptibility and antimicrobial activity of lactic acid bacteria from Malaysian fermented foods against biofilm-forming *Escherichia coli* strains. J Adv Res Appl Sci Eng Technol 31(1):168–182

Ryan KA, Ohara AM, van Pijkeren JP, Douillard FP, OToole PW (2009) *Lactobacillus salivarius* modulates cytokine induction and virulence factor gene expression in *Helicobacter pylori*. J Med Microbiol 58(8):996–1005

Schar D, Zhao C, Wang Y, Larsson D, Gilbert M, Boeckel TVV (2021) Twenty-year trends in antimicrobial resistance from aquaculture and fisheries in Asia. Nat Commun. https://doi.org/10.1038/s41467-021-25655-

Sharma MMM, Kapoor D, Rohilla R, Loyal A, Sharma P, Husen A (2024) Role of essential minerals and rhizosphere in plant under changing environmental conditions. In: Husen A (eds) Essential minerals in plant-soil systems. Elsevier, pp 235–252. https://doi.org/10.1016/B978-0-443-16082-0.00002-3

Sharma P, Kapoor D, Kumar R, Husen A (2025) Rhizosphere and osmolytes coordination in plant systems under adverse situations. In: Husen A (ed) Roles of osmolytes in changing environment. Academic Press, pp 101–117. https://doi.org/10.1016/B978-0-443-26549-5.00020-5

Shefali, Kumar R, Sankhla MS, Kumar R, Sonone SS (2021) Impact of pesticide toxicity in aquatic environment. Biointerface Res Appl Chem 11(3):10131–10140. https://doi.org/10.33263/BRIAC113.1013110140

Siciliano RA, Reale A, Mazzeo MF, Morandi S, Silvetti T, Brasca M (2021) Paraprobiotics: a new perspective for functional foods and nutraceuticals. Nutrients 13(4):1225

Speckmann B, Ehring E, Hu J, Mateos AR (2024) Exploring substrate microbe interactions: a metabiotic approach toward developing targeted synbiotic compositions. Gut Microbes

Talpur AD, Munir MB, Mary A, Hashim R (2014) Dietary probiotics and prebiotics improved food acceptability, growth performance, haematology and immunological parameters and disease resistance against *Aeromonas hydrophila* in snakehead (*Channa striata*) fingerlings. Aquaculture 426:14–20

Taverniti V, Guglielmetti S (2011) The immunomodulatory properties of probiotic microorganisms beyond their viability (ghost probiotics: proposal of paraprobiotic concept). Genes Nutr 6(3):261–274

Tkemaladze J (2025) Concept of the food security. Longevity Horiz

Torres-Sanchez A, Pardo-Cacho J, Lopez-Moreno A, Ruiz-Moreno A, Cerk K, Aguilera M (2021) Antimicrobial effects of potential probiotics of *Bacillus* spp. isolated from human microbiota: in vitro and in silico methods. Microorganisms 9(8):1615

Tovar-Ramirez D, Infante JZ, Cahu C, Gatesoupe FJ, Vazquez-Juarez R (2004) Influence of dietary live yeast on European sea bass (*Dicentrarchus labrax*) larval development. Aquaculture 234(1–4):415–427

Tsilingiri K, Barbosa T, Penna G, Caprioli F, Sonzogni A, Viale G, Rescigno M (2012) Probiotic and postbiotic activity in health and disease: comparison on a novel polarised ex-vivo organ culture model. Gut 61(7):1007–1015

Uddin GMN, Larsen MH, Christensen H, Aarestrup F, Phu TM, Dalsgaard A (2015) Identification and antimicrobial resistance of bacteria isolated from probiotic products used in shrimp culture. PLoS One

Van Doan H, Soltani M, Leitao A, Shafiei S, Asadi S, Lymbery AJ, Ringo E (2022) Streptococcosis a re-emerging disease in aquaculture: significance and phytotherapy. Animals 12(18):2443

Van Zyl WF, Deane SM, Dicks LM (2020) Molecular insights into probiotic mechanisms of action employed against intestinal pathogenic bacteria. Gut Microbes 12(1):1831339

Venkatrayulu C, Swapna B, Swathi AV, Srinivas D (2015) Influence of commercial probiotics on digestive enzyme activities of black tiger shrimp *Penaeus monodon* (Fabricius) reared in semi-intensive culture ponds. Int J Sci Res 4(10):2195–2200

Walker AW, Lawley TD (2013) Therapeutic modulation of intestinal dysbiosis. Pharmacol Res 69(1):75–86

Wang T, Xu Z, Lu S, Xin M, Kong J (2016) Effects of glutathione on acid stress resistance and symbiosis between *Streptococcus thermophilus* and *Lactobacillus delbrueckii* subsp. *bulgaricus*. Int Dairy J 61:22–28

Zendeboodi F, Khorshidian N, Mortazavian AM, da Cruz AG (2020) Probiotic: conceptualization from a new approach. Curr Opin Food Sci 32:103–123

Chapter 5
Converting Aquaculture By-Products into Biofertilizers Using Microbial and Nanotechnological Approaches: A Way Towards Sustainable Management

Narender Mohan, Pooja Verma, Jyoti Ahlawat, and Mandeep Redhu

Abstract Aquaculture has emerged as a vital component of global food production, providing a significant source of protein to meet increasing demand. This industry, characterized by its rapid growth, contributes substantially to the economy and food security worldwide. However, alongside its benefits, aquaculture generates substantial quantities of organic and inorganic waste. Leftovers from processing plants like, fish waste, and leftover feed, serves as the some of the waste by-products, further, their improper management can result in eutrophication and contamination of the water body. Under these circumstances discussed above management of waste as a sustainable aquaculture has become essential in light of traditional agriculture's problems, such as soil erosion and diminishing fertility. Consequently declining environmental effects and maintain agricultural productivity, has become important to incorporate eco-friendly practices like the production of biofertilizer. Additionally, turning aquaculture by-products into biofertilizers helps to improve soil health and advance sustainable agricultural methods. The present chapter attempts to emphasize on nanotechnological, microbial and other advanced development methods for turning aquaculture by-products into biofertilizers and harnessing the maximum potential capabilities of nanomaterials, optimizing nutrient uptake and utilization

N. Mohan (✉) · P. Verma
Division of Quality and Basic Sciences, ICAR-IIWBR Karnal, Karnal, Haryana, India
e-mail: narendermohan93@gmail.com

P. Verma
e-mail: pooja.verma195@gmail.com

J. Ahlawat
Department of Plant Biotechnology, G.D. Goenka University, Gurugram, Haryana, India
e-mail: jyoti.ahlawat@gdgu.org

M. Redhu
Southern Illinois University, Carbondale, USA
e-mail: mandeep.redhu@siu.edu

R. Kumar et al. (eds.), *Sustainability in Aquaculture and Fisheries*,
Springer Water, https://doi.org/10.1007/978-3-032-16605-0_5

from waste. The ultimate goal is to foster sustainable waste management practices in aquaculture while bolstering agricultural productivity through enhanced soil fertility.

Keywords Aquaculture · Sustainability · Nanotechnology · Biofertilizers

5.1 Introduction

Aquaculture is the cultivation of aquatic creatures such as freshwater or marine fish, molluscs, and crustaceans, and it has emerged as a technique to replace traditional fishing, hence relieving pressure on natural freshwater or marine populations (EL-Saadony et al. 2021). It has emerged as an important factor which provides a significant source of protein to meet the rising global demand (Azra et al. 2021). It has showed the highest average annual growth rate in the last ten years, accounting for 52% of fishery products for human consumption by 2025 (EL-Estado et al. 2022). Every year, this sector created approximately 171 million tons with an estimated value of 36,000 million US dollars, of which 47% came from the aquaculture sector, with Asia being the largest representation at 89% (Ahmed et al. 2019). Aquaculture has numerous advantages, but it also generates a lot of organic and inorganic waste (commonly referred to as "sediment" or "sludge"), which can harm the environment if not managed properly remainder settling and accumulating as a particulate fraction composed primarily of OM, nitrogen (N), and phosphorus (P). The majority of fisheries products used for human consumption were processed by fish processing companies. As a result, it generates considerable amounts of trash and byproducts (Jung and Kim 2016). These byproducts include fish waste, residual feed, fish carcasses, and aquaculture processing facility leftovers (Dauda et al. 2019). Such organic waste has an impact on the environment, negatively affecting aquatic and terrestrial habitats. As aquaculture expands globally, there is a greater demand for effective and sustainable waste management systems to lower pollutant loads preventing eutrophication, hypoxia, and water body pollution resulting efficient waste management. Excess nutrients, particularly nitrogen and phosphorus, can cause algal blooms, some of which are toxic, reducing oxygen levels and killing aquatic creatures (Wurtsbaugh et al. 2019). At the same time, conventional agricultural practices face issues such as diminishing soil health, diminished fertility, and excessive use of chemical fertilizers, all of which contribute to environmental contamination and land degradation (Hossain et al. 2020). The need for environmentally friendly waste management technologies in aquaculture has grown as the world's food supply systems become more interconnected.

In this way, achieving a balance between food production, environmental preservation, and agricultural sustainability is dependent on the sustainable management of aquaculture waste through innovative recycling processes. Adopting integrated approaches that incorporate sustainable agriculture and aquaculture can result in a more resilient and environmentally conscious global food chain (Khandakar et al.

2024; Thomas et al. 2021). Biofertilizer production is one of the innovative and effective strategies in waste management which incorporates green practices to mitigate the negative environmental impacts and enhance agricultural productivity (Onada and Ogunola, 2016). Nutrients with positive effects can be cycled back into the land, enhancing organic content and overall fertility, by converting aquaculture by-products into biofertilizers serving as an effective alternative over synthetic fertilizers (Campanati et al. 2022; Chiquito-Contreras et al. 2022). Biofertilizers obtained from aquaculture can also enhance plant development, enhance soil microbial activity, and enable circular economy that converts waste into a valuable resource (Kumar et al. 2022).

Recent advancement techniques like nanotechnology have provided humans new ways to increase the biofertilizer's efficacy by improved microbial activity, enhance nutrient delivery, and prolong the release of essential nutrients into the soil has been examined (Harish et al. 2022). To minimize nutrient leaching and maximize plant uptake efficiency, biofertilizers can be formulated to supply nutrients in a sustained and controlled manner by encapsulating nutrients or microbial inoculants within nanoscale carriers (Mahapatra et al. 2022; Sharma et al. 2024). Further, another effective technique is the utilization of specific microbial strains for biofertilizer production these microbe might include, nitrogen-fixing bacteria (NFB), plant-growth-promoting rhizobacteria (PGPR), and phosphate-solubilizing bacteria (PSB), which has been found highly responsive in agricultural production and soil quality (Zeng et al. 2022; Wang et al. 2019). In addition, microbial consortia can be tailored to enhance a healthier and sustainable agricultural system by inhibiting plant diseases, speeding up mineralization of nutrients, and enhancing aquatic organic matter decomposition (Vishwakarma et al. 2020; Verma et al. 2023). With emphasis placed on nanotechnological and microbial advancements, this chapter discusses innovative approaches to converting aquaculture waste into biofertilizers (Florez-Jalixto et al. 2021; Sharma et al. 2025).

5.2 Aquaculture By-Products: Challenges and Opportunities

Aquaculture, like any other production system, produces by-products that need to be managed carefully to reduce their negative effects on the environment and to realize their full potential as useful resources. Aquaculture waste differs depending on the species grown, production scale, and farming method (freshwater, marine, etc.). Regarding composition, volume, and environmental impact, every kind of waste poses different difficulties (Dauda et al. 2019). To prevent environmental damage, adhere to legal requirements, and sustainably support the world's food production, aquaculture waste must be managed effectively (Kumar 2014). Aquaculture waste management challenges include public perception of environmental impacts, regulatory complexity, technological limitations in waste treatment and the viability of

Fig. 5.1 Challenges and opportunities associated with aquaculture by-products

waste valorization projects from an economic standpoint (Verdegem et al. 2023). Below is a detailed discussion of the challenges and opportunities associated with aquaculture by-products, specifically fish waste, shellfish shells, uneaten feed, and processing residues. The challenges and opportunities associated with aquaculture by-products have also been shown in Fig. 5.1.

5.2.1 *Challenges Associated with Aquaculture By-Products*

5.2.1.1 Environmental Pollution

Nutrient Overload (Eutrophication) can result from fish waste and leftover feed releasing too much phosphate and nitrogen into aquatic environments. This phenomenon disrupts aquatic life, depletes oxygen, and produces toxic algal blooms (Khan and Mohammad, 2014). Accumulation of organic waste (Hypoxic zones) can develop close to aquaculture plants as a result of decomposing organic materials like carcasses and processing waste (Colombo et al. 2023).

5.2.1.2 Waste Management Issues

Large volumes of by-products are put under stress by the substantial amounts of waste produced by high-intensity aquaculture activities (Campanati et al. 2022). Logistics Challenges: Gathering, shipping, and processing by-products, particularly from far-flung or offshore aquaculture locations can be expensive and difficult (Das and Mandal, 2022).

5.2.1.3 Regulatory and Compliance Barriers

Strict environmental regulations often govern the disposal and treatment of aquaculture waste, requiring investment in waste treatment technologies that small-scale operators may struggle to afford (Jeffery et al. 2022). Improper handling of fish waste and processing residues can lead to the spread of pathogens, potentially affecting both aquatic organisms and human health giving rise to health and biosecurity risks (Adedeji et al. 2012). For instance, recent studies have identified pathogenic and zoonotic *Vibrio* species in the stools of patients with diarrhea, linking these infections to the consumption of undercooked shellfish. These microorganisms are known to cause food poisoning and diarrhea outbreaks in humans (Jamil et al. 2023). Despite their challenges, aquaculture by-products, represent untapped opportunities for sustainable development, circular economy practices, and waste valorization creating opportunities for Sustainable Resource Utilization.

5.2.1.4 Fish Waste

Fish excreta and processing residues can be converted into biogas through anaerobic digestion, providing a renewable energy source giving enough for bioenergy production (Ivanovs et al. 2018). Nutrients such as nitrogen and phosphorus are left behind in the effluent (digestate) following anaerobic digestion, and if properly treated to meet the required standards, they can be used as a bio-fertilizer in crop production (Nges et al. 2012). Nutrient-rich fish waste can be processed into organic fertilizers to enhance agricultural productivity. Fertilizers derived from fish waste are effective in enhancing soil nutrient levels, preventing plant diseases, controlling weeds, and reducing parasite growth which results in managing and producing fertilizers and soil amendments: (Dhar et al. 2024). Moreover, heads, tails, and viscera can be used to produce fish protein hydrolysates and omega-3-rich fish oils, which have applications in nutraceuticals and animal feed (Olaniran et al. 2024). Fisheries also results in production of solid byproducts like heads, shells, tails, fins, viscera, and shellfish waste as well as liquids like wastewater from the processing phase (Otero et al. 2021).

5.2.1.5 Shellfish Shells

Shells from mollusks are rich in calcium carbonate that can be used in construction (e.g., cement), agriculture (e.g., lime for soil pH adjustment), and cosmetics (Topic Popovic et al. 2023). To utilize waste mollusk shells as a source of $CaCO_3$, they must be cleaned first as any remaining organic matter would cause reduction in the purity of the $CaCO_3$ (Murphy and Kerton 2017). In addition, these shells are a primary source of chitin, which is processed into chitosan, a biopolymer with applications in water treatment, pharmaceuticals, and biodegradable packaging (Lakshmi et al. 2024).

5.2.1.6 Uneaten Feed: Recovery and Repurposing

Excess feed can be collected, reprocessed, and used in other aquaculture operations or converted into feed for other livestock (Sandström et al. 2022; Pinotti et al. 2019). Several authors have demonstrated that organic carbon originating from uneaten feed and faecal material produced by cultivated fishes represents a source of available food for shellfish (Mazzola and Sarà 2001).Composting: Uneaten feed combined with other organic materials (animal faeces, decayed plants etc.) can be composted to create high-quality organic fertilizers. By converting organic waste into compost, stakeholders can help reduce waste, mitigate greenhouse gas emissions, and conserve resources, all while fostering community involvement and supporting economic growth (Tanner et al. 1995; Bremaghani, 2024).

5.2.1.7 Processing Residues

Skin and bones of aquatic fishes can be used to extract collagen and gelatin for use in the food, cosmetic, and pharmaceutical industries (Jeevithan et al. 2013). Additional research is needed to clarify the structure of fish gelatin hydrolysates, particularly those that are active as anticancer and antidiabetic agents (Al-Nimry et al. 2021). In addition to the benefits described fish processing by-products can also be utilized as ingredients in foods of pet animals as well as their feeds (Malaweera and Wijesundara 2013). Recent research has highlighted that various bioactive compounds are found in discarded fish and seafood by-products, including the head, skin, viscera, and shells. These ingredients can be used to create pharmaceuticals, functional foods, or nutraceuticals (Özyurt and Özkütük 2019).

5.3 Environmental Impact of Untreated By-Products

A rapidly growing sector, aquaculture, produces substantial amounts of organic and inorganic waste, which, if not managed properly, can lead to environmental degradation can have detrimental effects on aquatic ecosystems, biodiversity, and water quality. Excess nutrients, particularly nitrogen and phosphorus, can cause algal blooms, some of which are toxic, reducing oxygen levels and killing aquatic creatures, which might also results in nutrient leaching, a major issue in aquatic ecosystems (Munguti et al. 2021). Eutrophication, a condition characterized by excessive algal growth that makes hypoxic condition (declining water's dissolved oxygen level), is frequently caused by this uncontrolled nutrient release which further damages aquatic ecosystems, endanger human health and marine life as well (Rathore et al. 2016). Subsequently, the release of untreated byproducts can worsen water pollution by introducing pathogens, antibiotics, and heavy metals into aquatic systems along with nutrient pollution (Akpor et al. 2014). These pollutants degrade overall water quality and alter the well maintained ecological balance necessary for the survival of

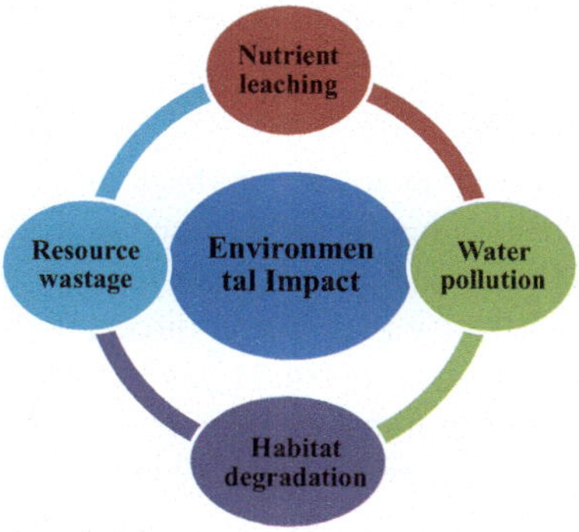

Fig. 5.2 Aquatic waste: By products affecting the environment

several marine species. Furthermore, aquatic habitats' physical structure is changed and biodiversity is decreased when organic sediments from decomposing byproducts build up (Davidson et al. 2018). Anoxic conditions brought on by the accumulation of organic matter in sediments can harm benthic species and hinder the general health of aquatic ecosystems.

These issues are exacerbated by the environmental fatigue caused by untreated byproducts, which regretfully bothers our valuable aquatic ecosystems (Wu et al. 2024). When garbage is not properly managed, a cascade effect occurs: nutrients seep out, water becomes polluted, and critical habitats are damaged (Fig. 5.2). This can ultimately lead to the collapse of local ecosystems and challenge the very idea of sustainable aquaculture (Farrant et al. 2021). It's clear that we urgently need strong initiatives to reduce the negative environmental footprint of aquaculture. Our top priorities should be cutting down on nutrient runoff, preventing harmful substances from being released and building up, and managing organic waste in a way that truly respects and preserves the integrity of our ecosystems (Sharma et al. 2019; Drizo et al. 2022).

5.4 Innovative Approaches and Technologies

Aquaculture provides us with essential protein sources, but also uneaten feed, fish waste, and deceased organisms can pollute our waterways and damage delicate ecosystems creating a significant waste problem. Therefore, sustainable aquaculture practices are crucial to balance our growing need for aquatic food with environmental protection and industry's long-term viability the answer lies in reimagining how we handle aquaculture waste. Let me walk you through some fascinating approaches and technologies that are changing aquaculture waste management.

5.4.1 Trash to Treasure Conversion/Bioconversion

Bioconversion (converting aquaculture waste into biofertilizer) is a promising approach as it turns garbage into high-protein animal feed by utilizing insects like black soldier fly larvae (Ravi et al. 2020; Siddiqui et al. 2024). There has been an old lines found to be true even today "one person's trash is another's treasure" the quote is still very perfect and directs the mindset for waste management in aquaculture. Recently the upcoming and innovative farmers has viewed the waste not a burden rather has transformed it into a valuable resources. Take bioconversion, for instance. Have you ever heard of black soldier fly larvae? These remarkable creatures can devour aquaculture waste and transform it into protein-rich animal feed. I recently learned about a farm that reduced their disposal costs by 40% while creating a new revenue stream through this method. Other approaches include anaerobic digestion, where microbes break down organic matter without oxygen, producing both biogas for energy and nutrient-rich digestate for fertilizing crops. Composting offers another pathway, mixing aquaculture waste with materials like wood chips to create fertile soil amendments. And let's not forget vermiculture wherein the earthworms quietly working to convert waste into premium plant food (Yadav et al. 2022a; b; Yadav et al. 2023).

Advanced Waste Treatment Technologies: Innovations such as biofilters, constructed wetlands, and bioremediation techniques demonstrate promising approaches to treat and utilize aquaculture by-products efficiently. These developments not only lessen aquaculture's negative environmental effects but also pave the stage for long-term, sustainable growth (Tom et al. 2021). Some methods include anaerobic (composting, vermiculture) and anaerobic digestion wherein, the microorganisms break down organic matter in the absence of oxygen, producing biogas and digestate (serving as a biofertilizer) (Saha et al. 2018; Pizzanelli et al. 2023; Adekunle and Okolie 2015). Composting involves mixing aquaculture waste with other organic materials like crop residues or wood chips and allowing it to decompose aerobically. The resulting compost is a nutrient-rich soil amendment that can improve soil fertility and plant growth (Badawi, 2023). Further, vermiculture utilizes earthworms to break down organic waste (Table 5.1). Earthworms consume the waste and produce nutrient-rich castings that can be used as a biofertilizer (Mohite et al. 2024; Kumar et al. 2024).

5.4.2 Integrated Aquaculture Systems/Co-culture

Have you ever wondered why nature rarely has a waste problem? It's because in natural systems, outputs from one organism become inputs for another. This principle inspires integrated aquaculture systems, particularly what experts call Integrated Multi-Trophic Aquaculture (IMTA). For example, shellfish and seaweed can

Table 5.1 Summarizes the microbial and nanotechnological approaches for converting aquaculture by-products into biofertilizers

Approach	Key technologies	Benefits	Challenges	Recent advancements	References
Microbial biotechnology	Controlled composting Microbial inoculants Enzymatic pre-treatment	Enhanced nutrient cycling Improved soil health Pathogen suppression	Process standardization Microbial stability Regulatory approval	Designer microbial consortia Precision fermentation Biofilm-based systems	Tom et al. (2021); Verma et al. (2023)
Nanotechnology	Nano-encapsulation Nutrient nano-formulations Nano-biosensors	Controlled nutrient release Improved nutrient use efficiency Enhanced microbial delivery	Production scaling Environmental fate assessment Cost considerations	Green synthesis methods Multi-functional nanoparticles Biodegradable carriers	Mohan et al. (2023)
Integrated systems	Aquaponics IMTA Smart monitoring systems	Closed nutrient loops Multiple value streams Resource efficiency	System complexity Initial investment Knowledge requirements	IoT integration AI-optimized production Modular scalable systems	Wong and Kerkez (2016)

absorb excess nutrients from fish farming, reducing pollution while creating additional products (Sukhdhane et al. 2018; Nederlof 2022). Moreover, fish waste can be used to fertilize algae or aquatic plants, which can then be used as feed for other fish or shellfish (Naylor et al. 2021). There have been researches implementing these systems and found they not only reduced pollution but also diversified their income streams through multiple harvests.

5.4.3 Water Treatment and Recycling

Advanced water treatment technologies, such as biofiltration and membrane filtration, can remove pollutants from aquaculture wastewater, allowing it to be reused in the system. This reduces water consumption and minimizes waste discharge (Fei et al. 2020; Chen et al. 2015). Researches also showed me a recirculating aquaculture system that reused 98% of its water through such technologies. The long-term savings on water bills and environmental compliance outweighed the sizeable initial investment.

5.4.4 Precision Feeding

Regardless of whether the fish are hungry or not, traditional aquaculture feeding techniques frequently depend on timers and preset amounts. The fish's fluctuating feeding requirements are not taken into consideration by this method, which can result in problems like underfeeding or overfeeding, which can slow the fish's growth and contaminate the aquaculture water (Wu et al. 2015). Besides all the challenges recent feeding technologies has studied the fish behavior deeply and transformed the feeding methods and developed various methods to feed precise amount required at the appropriate time. These innovative and advancements has resulted in improved feed efficiency and prevented unnecessary wastage (Fore et al. 2015; Zhang et al. 2023).

5.4.5 Monitoring and Data Analysis

Monitoring of feed intake, water quality, and so on in real-time can assist in maximizing waste management methods and identifying potential problems beforehand in today's world (Ramesh et al. 2017). Todays advanced and innovative farmers can use the data analytic tool to check and track the amount of feed consumed, they can even monitor water quality traits, and get early warnings by detecting trends and patterns, these methods will allow more sustainable and effective aquaculture practices (Mohanty et al. 2024). Data can be analyzed and remote devices controlled using familiar high-level programming languages such as Python or MATLAB rather than designing firmware code (Wong and Kerkez 2016). Also the users can deal with real-time data transmission, remote device management, and sensor discovery which are all made feasible by IoT platforms.

5.4.6 *Nanotechnological Strategies and Microbial Helpers for Biofertilizer Production/Bioconversion*

New methods to enhance the efficiency and effectiveness of biofertilizer production from aquaculture waste are made possible by nanotechnology. Crop nutrient delivery can be improved, beneficial microbes can be protected, and nutrient recovery can be optimized with the use of nanomaterials. Similarly, a lot of potential is available for sustainable waste management in the microscopic world. Aquaculture waste contains organic materials that certain microbes can break down, recycling nutrients into forms that plants can easily absorb. By employing the metabolic processes of certain microbes for degrading organic matter and recycling nutrients for agriculture, microbial technologies can sustainably transform aquaculture wastes into nutrient-rich biofertilizers. Looking forward for the future food productivity these innovative approaches directed in aquaculture sustainable management can transform waste into a resource and harnessing nature's own mechanisms to nourish increasing populations while safeguarding the environments we all rely upon.

5.5 Biofertilizer Production Using Microbial Strategies

Degradation of fish waste using microbial source solves waste disposal problem and formed hydrolysate could be another efficient technique used to convert aquaculture wastes into biofertilizers providing a renewable means of recycling nutrients, soil enrichment, and waste utilization. This can be effectively used as plant growth enhancer technique as the produced biofertilizers are rich in nitrogen, phosphorous, and potassium nutrients which serves essential for microorganisms such as bacteria, fungi, and algae break down organic materials necessary for plant growth (Itelima et al. 2018). By utilizing processes like fermentation, anaerobic digestion, and composting, microbes assist in converting the complex organic chemicals found in aquaculture waste into plant-consumable forms thus enhancing soil fertility and reducing the need for synthetic fertilizers (Kunjiraman et al. 2024).

5.5.1 Microorganisms

5.5.1.1 Bacteria

Bacillus Species

Bacillus sp. is well known in nature and is considered responsible for fixing atmospheric nitrogen and phosphate solubilization. They can also survive in extreme environments due to the presence of endospores. Moreover, they are widely utilized in biofertilizer production to enhance nutrient availability in soils as well as inhibit the growth of pathogenic fungi (Verma et al. 2023). ***Pseudomonas* species**: These bacteria contribute to organic matter decomposition and nutrient mineralization, improves soil fertility as well as plant's nutrient uptake efficiency. Various microbial species contribute to biofertilizer production from aquaculture by-products (Bargaz et al. 2018). ***Nitrosomonas* and *Nitrobacter***: These bacteria are crucial for nitrification processes, converting ammonia (NH_3) to nitrites (NO_2^-) and then to nitrates (NO_3^-), essential forms of nitrogen for plant growth (Paśmionka et al. 2021).

Phosphobacteria

Genera like *Bacillus* and *Pseudomonas* are referred to as phosphobacteria because of their ability to solubilize insoluble phosphates (e.g., calcium phosphate), releasing phosphate ions (PO_4^{3-}) and making it available for plant uptake (Hui et al. 2011; Rawat et al. 2021). **Lactic Acid Bacteria**: LAB stabilizes the organic matter by producing organic acids (e.g., lactic acid). It also promotes nutrient retention in soils and reduces odors associated with decomposition processes (Raman et al. 2022).

5.5.1.2 Fungi

Fungal species like *Trichoderma* and *Aspergillus* are remarkable for their ability in breaking down complex organic compounds in the environment, releasing nutrients such as nitrogen, phosphorus, and potassium in forms accessible to plants (Mehetre and Mukherjee 2015). This helps in not only recycling nutrients back to soil rather also play a vital part in enhancing health of land and further supporting nutrient cycling in agricultural systems (Nayak et al. 2020).

5.5.1.3 Algae (Tiny Powerhouses for Soil and Climate)

Chlorella and *Spirulina* (filamentous cyanobacterium), are types of green algae, does more than just float in water. These microscopic organisms are experts at absorbing nutrients from their surroundings. By increasing the availability of nutrients in the soil, *Spirulina* helps farmers grow healthier crops while adopting more sustainable practices (Saud et al. 2024). Through their natural exudates, they stimulate soil microbial activity, boost soil fertility, and help support healthy plant growth (Alvarez et al. 2021). *Cyanobacteria* like *Spirulina* captures atmospheric carbon dioxide and packs it in soils and is thus a boon which further play a pivotal role in the fight against climate change. Their impressive ability to sequester carbon helps mitigate global warming by locking away carbon for the long term (Wan et al. 2021).

5.5.2 Transforming Waste to Wealth Utilization of Microbes

5.5.2.1 Composting with Beneficial Microorganisms

Composting is one of the highly effective and commonly used techniques which combine high-carbon-rich materials, such as sawdust or straw, with aquaculture waste, such as fish slime and residual feed (Table 5.1). Beneficial microbes then use enzymes to break down these complex organic compounds into simpler, plant-friendly forms, such as amino acids and sugars (Pal et al. 2021). Some bacteria like *Bacillus* and decomposer fungi like *Trichoderma* are added to accelerate this transformation (Elita et al. 2022). The result is nutrient-rich compost that includes essential elements such as potassium, phosphorus, and nitrogen. Fish hydrolysate is an easy nitrogen source which can serve as an effective biofertilizer utilized in soil amendment resulting in improved texture and soil fertility. Apart from providing nutrition to plants, using this compost to fields enhances soil microbial diversity and overall fertility (Maheshwari et al. 2014).

5.5.2.2 Anaerobic Digestion and Microbial Fermentation

There is another promising technique for aquaculture waste management which offers a sustainable approach and resource recovery recycling organic waste is through anaerobic digestion. This method serves effective as the methanogenic bacteria used in the study degrades waste in oxygen-free environments and results in two important products: a 'nutrient-rich digestate' (serves as a potent biofertilizer), and another one as "biogas" (a renewable source of energy) (Kiyasudeen et al. 2016). Waste produced from aquaculture can also be further fermented using specific important microbes like Lactobacillus, which can help produce liquid biofertilizers further improving soil nourishment and sustainable farming (Amrullah et al. 2021). Therefore it will not be biased to call the technique as a boon as it not only promotes a circular economy by turning garbage into fertilizer and energy, rather, captures methane, which lowers carbon gas emissions, improves soil health by producing digestate that is rich in nutrients and encourages plant development (Lee et al. 2021).

5.5.2.3 Algae and Cyanobacteria-Based Waste Utilization

Aquaculture wastewater is used to develop algae and cyanobacteria, which take up nutrients like phosphorus and nitrogen (Tom et al. 2021). After that, the biomass is collected and turned into biofertilizers, prevents eutrophication by eliminating surplus nutrients from aquaculture effluents, supplies vital micronutrients and organic matter-rich biofertilizers, improves the nutrient cycle and soil moisture retention.

5.5.2.4 Nutrient Cycling

Through species like *Nitrosomonas* and *Nitrobacter*, beneficial microorganisms transform nitrogenous substances (like ammonia) into nitrites and nitrates (nitrification), increasing nitrogen availability for plants (Paśmionka et al. 2021). Furthermore, phosphobacteria increase the availability of phosphorus in soils by solubilizing insoluble phosphates (Rawat et al. 2021). Microbial approaches focus on controlled composting utilizing microbial inoculants along with some enzymatic pre-treatment strategies which could improve soil health and enhance the cycling of nutrients which ultimately enhances the ability of plants to fight pathogens (Tom et al. 2021; Verma et al. 2023) (Table 5.1).

5.5.2.5 Pathogen Suppression

To ensure the safety and effectiveness of biofertilizers, helpful microorganisms, such as Lactic acid bacteria (LAB), produce antimicrobial chemicals and competitively exclude harmful pathogens (Amer et al. 2024). LAB has been proven to be effective as biofertilizers, biocontrol agents, and biostimulants. They can enhance nutrient availability from compost and other organic materials. These bacteria are widely recognized for their ability to control various fungal and bacterial plant pathogens. Furthermore, as biostimulants, they promote plant growth and seed germination, while also helping plants cope with a range of abiotic stress factors (Raman et al. 2022). Additionally, a bioformulation prepared by microbial consortia (different strains of *Bacillus* sp.) also showed inhibition of *Bipolarissorokiniana* and *Fusariumoxysporum* in wheat plants (Verma et al. 2023) (Table 5.1).

5.6 Nanotechnological Strategies for Biofertilizer Production

Nanotechnology introduces innovative solutions to enhance the efficiency and effectiveness of biofertilizer production from aquaculture waste. Nanomaterials can optimize nutrient recovery, protect beneficial microbes, and improve nutrient delivery to crops.

5.6.1 Nano-Adsorbents for Nutrient Recovery

Essential nutrients (such as nitrogen and phosphorus) are extracted from aquaculture wastewater using nanomaterials such as graphene oxide, carbon nanotubes, and zeolite nanoparticles. These substances are perfect for nutrient recovery because of their large surface area and adsorption capacity (Mohan et al. 2023). These particles reduce pollution in the environment by effectively capturing nutrients from waste streams. The nutrients that have been collected are added to biofertilizers to increase their nutrient content. When used on agricultural areas, it further declines nutrient leaching and runoff (Mahapatra et al. 2022). Multi-functional nanoparticles and biodegradable carriers could also help in controlled nutrient release, improved nutrient use efficiency and enhanced microbial delivery (Mohan et al. 2023) (Table5.1).

5.6.2 Nano-Encapsulation for Controlled Nutrient and Microbe Release

Biodegradable nanomaterials, such as chitosan nanoparticles (produced from crab shells), enclose nutrients and helpful bacteria (Ngasotter et al. 2023; Mohan et al. 2023) (Table 5.1). The slow, regulated release of nutrients and microorganisms in the soil is made possible by this nano-encapsulation, which also shields delicate germs, increasing the stability and shelf life of biofertilizers (Hamed 2023). It further minimizes the need for repeated fertilizer applications by delivering a steady flow of nutrients and shielding the beneficial bacteria in the soil from adverse environmental circumstances ensuring their efficacy (Sharma et al. 2023).

5.6.3 Catalytic Nanoparticles to Accelerate Waste Decomposition

Complex organic compounds in aquaculture waste are broken down by metal oxide nanoparticles like as zinc oxide (ZnO) and titanium dioxide (TiO_2), which function as catalysts (Olabisi 2022). By accelerating the oxidation and mineralization of waste, these nanoparticles increase the availability of nutrients for plants. This decreases the amount of time required for waste to break down, improving the production efficiency of biofertilizer which increases the availability of nutrients, encouraging quicker plant development and greater crop yields by reducing the amount of toxic waste that builds up in the environment (Foo et al. 2024).

5.6.4 Integrated Systems for Risk Assessments

Modern approaches like aquaponics, IMTA, smart monitoring systems should be integrated with nanotechnology to reduce overfeeding and minimizing waste. It could also maintain quality of water and help in improving nutrient use efficiency. Challenges for utilizing such systems involve system complexity, initial investment and knowledge requirements (Table 5.1). The recent advancements include IoT integration with AI-optimized production and modular scalable systems (Wong and Kerkez 2016).

5.7 Biofertilizer Production Processes and Optimization

5.7.1 Pre-treatment Technologies

Effective pre-treatment is critical for optimizing aquaculture by-product conversion into biofertilizers. Key approaches include:

- **Physical Pre-treatments**: Mechanical processes such as milling, grinding, and sieving increase surface area and enhance subsequent biological processing (Gao et al. 2020). High-pressure homogenization and ultrasonic treatment have shown particular effectiveness for processing fish waste, reducing particle size while preserving nutrient content. Wang et al. (2022) reported that ultrasonic pre-treatment (20 kHz, 15 min) increased protein solubilization in fish waste by 68%, enhancing subsequent microbial decomposition rates.
- **Chemical Pre-treatments**: Controlled pH adjustment facilitates protein hydrolysis and nutrient solubilization. Mild acid treatment (pH 4–5) can effectively break down recalcitrant materials like chitin in shellfish waste, while alkaline conditions

enhance protein extraction. However, these treatments must be carefully managed to prevent nutrient loss and environmental impacts (Venugopal 2021).

- **Enzymatic Hydrolysis**: Specific enzyme applications target particular waste components, improving digestibility and nutrient availability. Commercial enzyme preparations containing proteases (Alcalase, Neutrase), lipases, and chitinases efficiently break down complex aquaculture waste components. Research by Honradoet et al. (2024) demonstrated that enzymatic hydrolysis of fish processing waste produced peptide-rich hydrolysates with enhanced plant growth-promoting properties and disease suppression capabilities.

5.7.2 Fermentation and Bioconversion Technologies

- **Solid-State Fermentation**: This space-efficient, low-water process is particularly suitable for converting solid aquaculture waste into biofertilizers. Controlled microbial consortia decompose organic matter under aerobic conditions, producing stable, nutrient-rich products (Namasivayam et al. 2024). Modern systems employ environmental monitoring and control systems to maintain optimal temperature, moisture, and aeration conditions.A pioneering study by Kiruba and Saeid (2022) utilized solid-state fermentation of fish waste with *Bacillus subtilis* and *Streptomyces griseus*, producing a biofertilizer with high levels of plant-available nutrients and bioactive compounds that enhanced plant growth and stress resistance.
- **Liquid Fermentation**: Submerged fermentation systems process liquid waste streams and create high-value liquid biofertilizers. These systems often employ specific bacterial or fungal strains to produce metabolites that enhance plant growth and stress resistance (Miranda et al. 2024). Advanced bioreactor designs with precise control of temperature, pH, and dissolved oxygen optimize microbial activity and product quality (Nikita et al. 2023).
- **Vermicomposting**: Earthworms accelerate organic matter decomposition and enhance product quality (Saha et al. 2018; Kumar and Singh 2022). Species like *Eiseniafetida* and *Eudriluseugeniae* efficiently process fish and shellfish waste, increasing nutrient availability and introducing beneficial microorganisms. Field trials by Nurhidayati et al (2017) demonstrated that vermicompost produced from fish waste showed 18–24% higher nutrient content and improved soil physical properties compared to conventional compost.

5.7.3 Quality Control and Standardization

Advanced analytical techniques enable precise nutrient profiling and Nutrient standardization. Modern production facilities employ real-time monitoring and adjustment to maintain consistent product specifications (Saez et al. 2018). Recent developments include rapid spectroscopic methods that provide immediate feedback for process control. Thermal processing, competitive exclusion, and controlled fermentation ensure elimination of the infection (Tucker and Featherstone 2021). Advanced systems monitor critical control points throughout production to verify pathogen reduction. International standards typically require verification of finished biofertilizers that should be free from human pathogens like *Salmonella*, *E. coli*, and helminth egg (Paraguison-Alili et al. 2021). Optimized drying processes, protective packaging, and stabilizing additives extend product viability (Teixidó et al. 2022). Various findings from different groups demonstrates that biofertilizers incorporating specific formulations of humectants, antioxidants, and pH buffers has the ability to sustain viability of microbes for 12–18 months under appropriate storage conditions to enhance stability and shelf life (Sahu and Brahmaprakash 2016).

5.7.4 Field Applications and Performance Assessment

Optimal application strategies maximize biofertilizer effectiveness while minimizing environmental impacts. For example, direct mixing into soil provides immediate nutrient availability and microbial inoculation (Alori et al. 2017). Studies have shown that incorporating fish-derived biofertilizers at a rate of 2–5 tonnes/hectare significantly improves soil organic matter content and microbial activity (Xu and Mou 2017). Whereas, liquid biofertilizers applied directly to plant foliage enable rapid nutrient uptake. Research by Liatile et al. (2022) demonstrated that foliar application of fish hydrolysate at 5–10% concentration significantly enhanced photosynthetic efficiency and stress tolerance in vegetable crops. Integration with irrigation delivers nutrients precisely to the root zone. Modern precision agriculture systems combine soil moisture sensing with automated fertigation, optimizing nutrient delivery based on plant needs and environmental conditions (Xing and Wang 2024). Moreover, different crops respond uniquely to aquaculture-derived biofertilizers, requiring tailored application strategies. High-value crops benefit particularly from the balanced nutrition and bioactive compounds in fish-derived biofertilizers. Field trials conducted by Madende and Hayes (2020) demonstrated yield increases in tomato, pepper, and leafy greens with applications of fish waste-derived biofertilizers compared to conventional organic fertilizers. Staple crops show improved nutrient use efficiency and stress resistance with appropriate biofertilizer applications. Long-term studies indicate that regular application of aquaculture-derived biofertilizers can gradually reduce dependence on synthetic fertilizers while maintaining or improving yields (Goddek et al. 2018).High-value fruits, medicinal plants, and ornamentals benefit from precise nutrient delivery and bioactive compounds. Research indicates that shellfish-derived chitin and chitosan in biofertilizers enhance disease resistance in specialty crops, reducing pesticide requirements (Kumar et al. 2025; Nekvapil et al. 2019).

5.8 Environmental and Economic Benefits Assessment

- **Reduced Environmental Footprint**: Life cycle assessment (LCA) studies demonstrate significant environmental benefits from aquaculture waste valorization (Kiehbadroudinezhad et al. 2023; Vélez-Henao et al. 2021). Research showed that converting fish processing waste into biofertilizers reduced greenhouse gas emissions to several folds compared to landfill disposal while simultaneously reducing eutrophication potential by recycling nutrients (Sohil and Kichloo 2023).
- **Economic Viability Analysis**: Cost–benefit analyses indicate positive economic returns from integrated waste management approaches (Dobraja et al. 2016). Initial investments in processing technology can typically be recovered within 2–4 years through reduced waste management costs and revenue from biofertilizer sales (Yang et al. 2021).
- **Soil Health Improvements**: Long-term studies document improved soil physical, chemical, and biological properties with sustained application of aquaculture-derived biofertilizers. Research by Gómez-Brandón et al (2022) demonstrated significant increases in soil organic carbon, microbial biomass, and enzymatic activity after three years of fish compost application.

5.9 Conclusion

Waste as a resource method has been the center of the study wherein the alteration of aquaculture by-products and transformation into biofertilizers fulfills two important functions: management of aquaculture waste and another most important one and need of an hour i.e. encouraging agricultural sustainability. Utilizing microbial and nanotechnological techniques in development of biofertilizers from waste serves as a usefulresource responsible in soil quality improvement, increase crop yields, and declined environmental effects. The techniques studied and discussed so far, when incorporated into an inclusive waste management program, ensure circular economy models that are beneficial to both aquaculture and agriculture. These methodologies demonstrate that implementing these strategies is not only feasible but also advantageous to not only land systems but humans as well. While nanotechnology improves how the nutrients are delivered to plants, microbial biotechnology supports the effective decomposition of waste and recycling of the nutrients. Considering the futuristic approach, further advancements should focus on refining production processes, deepening our understanding of how plants and microbes interact, and developing even better delivery systems. Precision agriculture and software will also become increasingly important in maximizing the benefits of biofertilizers while minimizing their negative effects on the environment. Integrated solutions that convert waste into useful resources will be more crucial than ever as population growth, resource scarcity, and climate change put increasing strain on the world's food systems. An excellent illustration of this type of integration that points the way toward a more resilient and sustainable food future is the conversion of aquaculture by-products into biofertilizers. Industry participation, supportive legislation, and ongoing research will be necessary to fully realize the potential of these strategies. Aquaculture by-product valorization is an excellent example of the circular economy's tenets, which can help us transition to genuinely sustainable food production systems that benefit both people and the environment.

5.10 Future Prospects

Efficient utilization of microorganisms in biofertilizer production from aquaculture by-products requires optimization of microbial communities, nutrient cycling processes, and application methods. The application of nanotechnology to agricultural processes, such as the manufacturing of biofertilizer, has transformed productivity and resource management. But even with all of its advantages, using nanoparticles presents safety and environmental issues that need to be carefully considered. These worries encompass the prospect of bioaccumulation, leaching, and broader ecological impacts of nanomaterials. There is a need to understand these consequences to ensure that nanotechnology in agriculture is used sustainably and securely. The combination of microbial and nanotechnological innovations in development of biofertilizers is opening up exciting new grounds for sustainable agriculture. Although preliminary primary results are promising, ongoing research is crucial to fully realize the benefits of these technologies. With the distinct character of nanotechnology, the agricultural industry can better control nutrients, enhance yield, and reduce environmental degradation. This innovative approach positions nanotechnology as a key player in the future of sustainable agriculture.

References

Adedeji OB, Okerentugba PO, Innocent-Adiele HC, Okonko IO, Ojeniyi SO, Adejoro SA, Mohamed SA (2012) Benefits, public health hazards and risks associated with fish consumption. New York Sci J 5(9):33–61

Adekunle KF, Okolie JA (2015) A review of biochemical process of anaerobic digestion. Adv Biosci Biotechnol 6(03):205

Akpor OB, Ohiobor GO, Olaolu DT (2014) Heavy metal pollutants in wastewater effluents: sources, effects and remediation. Advances in Biosci Bioengineer 2(4):37–43

Al-Nimry S, Dayah AA, Hasan I, Daghmash R (2021) Cosmetic, biomedical and pharmaceutical applications of fish gelatin/hydrolysates. Mar Drugs 19(3):145
Alori ET, Dare MO, Babalola OO (2017) Microbial inoculants for soil quality and plant health. Sustain Agric Rev 281–307
Alvarez AL, Weyers SL, Goemann HM, Peyton BM, Gardner RD (2021) Microalgae, soil and plants: a critical review of microalgae as renewable resources for agriculture. Algal Res 54:102200
Amer MN, Elmaghraby MM, Abdellatif AA, Elmaghraby IM (2024) Lactic acid bacteria for safe and sustainable agriculture. MetabOmics, Proteomics Gene Ed Approaches Biofertilizer Ind 2:283–297
Amrullah S, Amin M, Ali M (2021) Converting husbandry waste into liquid organic fertilizer using probiotic consortiums (Lactobacillus sp., Rhodopseudomonas sp., Actinomycetes sp., Streptomyces sp.). In: IOP conference series: earth and environmental science. IOP Publishing, vol 679, no 1, p 012001
Anderson JL, Asche F, Garlock T, Chu J (2017) Aquaculture: its role in the future of food. In: World agricultural resources and food security: International food security. Emerald Publishing Limited, pp 159–173
Azra MN, Okomoda VT, Tabatabaei M, Hassan M, Ikhwanuddin M (2021) The contributions of shellfish aquaculture to global food security: assessing its characteristics from a future food perspective. Front Mar Sci 8:654897
Badawi MA (2023) Composting farm waste for production of high quality organic fertilizers and sustainable development. Integr Waste Manag: Circ Econ, UK 196
Bargaz A, Lyamlouli K, Chtouki M, Zeroual Y, Dhiba D (2018) Soil microbial resources for improving fertilizers efficiency in an integrated plant nutrient management system. Front Microbiol 9:1606
Bremaghani A (2024) Utilization of organic waste in compost fertilizer production: implications for sustainable agriculture and nutrient management. Law and Economics 18(2):86–98
Campanati C, Willer D, Schubert J, Aldridge DC (2022) Sustainable intensification of aquaculture through nutrient recycling and circular economies: more fish, less waste, blue growth. Reviews in Fisheries Science Aquaculture 30(2):143–169
Chen S, Yu J, Wang H, Yu H, Quan X (2015) A pilot-scale coupling catalytic ozonation–membrane filtration system for recirculating aquaculture wastewater treatment. Desalination 363:37–43
Chiquito-Contreras RG, Hernandez-Adame L, Alvarado-Castillo G, Martínez-Hernández MDJ, Sánchez-Viveros G, Chiquito-Contreras CJ, Hernandez-Montiel LG (2022) Aquaculture—production system and waste management for agriculture fertilization—a review. Sustainability 14(12):7257
Colombo SM, Roy K, Mraz J, Wan AH, Davies SJ, Tibbetts SM, Turchini GM (2023) Towards achieving circularity and sustainability in feeds for farmed blue foods. Rev Aquac 15(3):1115–1141
Das SK, Mandal A (2022) Diversification in aquaculture resources and practices for smallholder farmers. Agriculture, Livestock Production and Aquaculture: Advances for Smallholder Farming Systems, vol 1. Springer International Publishing, Cham, pp 263–286
Dauda AB, Ajadi A, Tola-Fabunmi AS, Akinwole AO (2019) Waste production in aquaculture: Sources, components and managements in different culture systems. Aquacult Fish 4(3):81–88
Davidson TM, Altieri AH, Ruiz GM, Torchin ME (2018) Bioerosion in a changing world: a conceptual framework. Ecol Lett 21(3):422–438
Dhar M, Jasrotia R, Langer S (2024) Using fish waste and by-products for manufacturing organic fertilizers and manures. In: Fish waste to valuable products. Singapore: Springer Nature Singapore, pp 339–353
Dobraja K, Barisa A, Rosa M (2016) Cost-benefit analysis of integrated approach of waste and energy management. Energy Procedia 95:104–111
Drizo A, Johnston C, Guðmundsson J (2022) An inventory of good management practices for nutrient reduction, recycling and recovery from agricultural runoff in Europe's Northern Periphery and Arctic Region. Water 14(13):2132

Elita N, Illahi AK, Sari DA, Yulensri Y, Maulina F, Karmaita Y, Yanti R (2022) Effect of types of organic materials and microbial enrichment on C/N ratio, nutrition of compost, and microbe population with Trichoderma sp. indigenous activators. Sos Sci J 12(6)
Farrant DN, Frank KL, Larsen AE (2021) Reuse and recycle: Integrating aquaculture and agricultural systems to increase production and reduce nutrient pollution. Sci Total Environ 785:146859
Fei X, Sun S, He S, Huang J, Zhou W (2020) Application of a novel two-stage biofiltration system for simulated brackish aquaculture wastewater treatment. Environ Sci Pollut Res 27:636–646
Florez-Jalixto M, Roldán-Acero D, Omote-Sibina JR, Molleda-Ordoñez A (2021) Biofertilizers and biostimulants for agricultural and aquaculture use: Bioprocesses applied to organic by-products of the fishing industry
Foo WH, Chia WY, Ende S, Chia SR, Chew KW (2024) Nanomaterials in aquaculture disinfection, water quality monitoring, and wastewater remediation. J Environ Chem Eng 113947
Føre M, Frank K, Norton T, Svendsen E, Alfredsen JA, Dempster T, Berckmans D (2018) Precision fish farming: a new framework to improve production in aquaculture. Biosystems Engineer 173:176–193
Gao W, Chen F, Wang X, Meng Q (2020) Recent advances in processing food powders by using superfine grinding techniques: a review. Compr Rev Food Sci Food Saf 19(4):2222–2255
Goddek S, Delaide BP, Joyce A, Wuertz S, Jijakli MH, Gross A, Keesman KJ (2018) Nutrient mineralization and organic matter reduction performance of RAS-based sludge in sequential UASB-EGSB reactors. Aquacult Eng 83:10–19
Gómez-Brandón M, Fornasier F, De Andrade N, Domínguez J (2022) Influence of earthworms on the microbial properties and extracellular enzyme activities during vermicomposting of raw and distilled grape marc. J Environ Manage 319:115654
Hamed RJY (2023) The synthesis of chitosan derivatives as a versatile tool in sustainable agriculture (Doctoral dissertation, faculty of graduate studies the synthesis of chitosan derivatives as a versatile tool in sustainable agriculture by Renad Jalal Yahya Hamed Supervisor Prof. ShehdahJodeh, This Dissertation is Submitted in Partial Fulfillment of the Requirements for the Degree of PhD of Chemistry, Faculty of Graduate Studies, An-Najah National University)
Harish V, Tewari D, Gaur M, Yadav AB, Swaroop S, Bechelany M, Barhoum A (2022) Review on nanoparticles and nanostructured materials: bioimaging, biosensing, drug delivery, tissue engineering, antimicrobial, and agro-food applications. Nanomaterials 12(3):457
Honrado A, Miguel M, Ardila P, Beltrán JA, Calanche JB (2024) From waste to value: fish protein hydrolysates as a technological and functional ingredient in human nutrition. Foods 13(19):3120
Hossain A, Krupnik TJ, Timsina J, Mahboob MG, Chaki AK, Farooq M, Hasanuzzaman M (2020) Agricultural land degradation: processes and problems undermining future food security. Environment, climate, plant and vegetation growth. Springer International Publishing, Cham, pp 17–61
Hui LIU, Xiao-Qin WU, Jia-Hong REN, Jian-Ren YE (2011) Isolation and identification of phosphobacteria in poplar rhizosphere from different regions of China. Pedosphere 21(1):90–97
Itelima JU, Bang WJ, Onyimba IA, Oj E (2018) A review: biofertilizer; a key player in enhancing soil fertility and crop productivity. J MicrobiolBiotechnol Rep 2(1):22–28
Ivanovs K, Spalvins K, Blumberga D (2018) Approach for modelling anaerobic digestion processes of fish waste. Energy Procedia 147:390–396
Jamil M, Abdullah S, Talib F, Bashir R, Ghafoor N, Javed K, Ghafoor A (2023) Vibrionaceae and fish zoonosis. Zoonosis, Unique Scientific Publishers, Faisalabad, Pakistan 4:468–480
Jeevithan E, Qingbo Z, Bao B, Wu W (2013) Biomedical and pharmaceutical application of fish collagen and gelatin: a review. J. Nutr. Ther 2(4):218–227
Jeffery K, MacMillan I, Schiefer P, Muench A, Tew I, Stinton N, Albert R (2022) A scoping study of the technical, regulatory, and economic challenges with respect to the development of the recirculation aquaculture system sector in England
Khan MN, Mohammad F (2014) Eutrophication: challenges and solutions. Eutrophication: causes. ConseqS Control 2:1–15

Khandakar A, Elzein IM, Nahiduzzaman M, Ayari MA, Ashraf AI, Korah L, Badawi A (2024) Smart aquaponics: an innovative machine learning framework for fish farming optimization. Comput Electr Eng 119:109590

Kiehbadroudinezhad M, Hosseinzadeh-Bandbafha H, Varjani S, Wang Y, Peng W, Pan J, Tabatabaei M (2023) Marine shell-based biorefinery: a sustainable solution for aquaculture waste valorization. Renewable Energy 206:623–634

Kiruba NJM, Saeid A (2022) An insight into microbial inoculants for bioconversion of waste biomass into sustainable "bio-organic" fertilizers: a bibliometric analysis and systematic literature review. Int J Mol Sci 23(21):13049

Kiyasudeen SK, Ibrahim MH, Quaik S, Ahmed Ismail SSKK, Ibrahim MH, Ismail SA (2016) An introduction to anaerobic digestion of organic wastes. Prospect Org Waste Manag Significance Earthworms 23–44

Kumar MS (2014) Sustainable management of fisheries and aquaculture for food security and nutrition: policies requirements and actions. Agricultural Research 3:97–103

Kumar R, Singh D (2022) Insecticide- tolerant Bacterial population in Eisenia fetida's Gut and Vermicast exposed to Chlorantraniliprole and Fipronil. Appl Biol Res 24(3):273–279. https://doi.org/10.5958/0974-4517.2022.00033.7

Kumar S, Kumar R, Diksha Kumari A, Panwar A (2022) Astaxanthin: a super antioxidant from microalgae and its therapeutic potential. J Basic Microbiol 62(9):1064–1082

Kumar R, Yadav R, Gupta RK, Pal A, Yodha K, Kumar A (2024) Antioxidant enzyme activities and markers of oxidative stress in the life cycle of different Earthworm species. Indian J Exp Biol 62(07):578–583. https://doi.org/10.56042/ijeb.v62i07.12084

Kumar R, Singh D, Yadav R (2025) Toxic effects of chlorantraniliprole and fipronil on Eisenia fetida: from growth and reproductive parameters to avoidance behavior. Environ Eng Manag J 24(7):1499–1508. https://doi.org/10.30638/eemj.2025.117

Kunjiraman S, Singh IB, Sarasan M, Puthumana J (2024) Immobilized microbial consortia: an eco-friendly and sustainable solution for aquaculture waste management. Microbe 100100

Lakshmi DS, Ravindran L, Mary PM, Rathika G, Sreekala MS, Munisamy S (2024) Valorization of crustacean shells: preparation, characterization, and chemical modification of nano-chitin biopolymer membrane for application in water purification. Waste Biomass Valorization 15(11):6067–6090

Lee ME, Steiman MW, Angelo SKS (2021) Biogas digestate as a renewable fertilizer: Effects of digestate application on crop growth and nutrient composition. Renewable Agric Food Syst 36(2):173–181

Liatile PC, Potgieter G, Moloi MJ (2022) A natural bio-stimulant consisting of a mixture of fish protein hydrolysates and kelp extract enhances the physiological, biochemical and growth responses of spinach under different water levels. Plants 11(23):3374

Madende M, Hayes M (2020) Fish by-product use as biostimulants: an overview of the current state of the art, including relevant legislation and regulations within the EU and USA. Molecules 25(5):1122

Mahapatra DM, Satapathy KC, Panda B (2022) Biofertilizers and nanofertilizers for sustainable agriculture: Phycoprospects and challenges. Sci Total Environ 803:149990

Maheshwari DK, Dheeman S, Agarwal M (2014) Decomposition of organic materials into high value compost for sustainable crop productivity. Compos Sustain Agric 245–267

Malaweera BO, Wijesundara WNM (2013) Use of seafood processing by-products in the animal feed industry. In: Seafood processing by-products: trends and applications. New York, NY: Springer New York, pp 315–339

Mazzola A, Sarà G (2001) The effect of fish farming organic waste on food availability for bivalve molluscs (Gaeta Gulf, Central Tyrrhenian, MED): stable carbon isotopic analysis. Aquaculture 192(2–4):361–379

Mehetre ST, Mukherjee PK (2015) Trichoderma improves nutrient use efficiency in crop plants. Nutr Use Effic: Basics Adv 173–180

Miranda AM, Hernandez-Tenorio F, Villalta F, Vargas GJ, Sáez AA (2024) Advances in the Development of Biofertilizers and Biostimulants from Microalgae. Biology 13(3):199
Mohan N, Ahlawat J, Sharma L, Pal A, Rao P, Redhu M, Yadav S (2023) Engineered nanoparticles a novel approach in alleviating abiotic and biotic stress in millets: a complete study. Plant Stress 10:100223
Mohanty A, Mohanty SK, Mohapatra AG (2024) Real-time monitoring and fault detection in ai-enhanced wastewater treatment systems. In: The AI cleanse: transforming wastewater treatment through artificial intelligence: harnessing data-driven solutions. Cham: Springer Nature Switzerland, pp 165–199
Mohite DD, Chavan SS, Jadhav VS, Kanase T, Kadam MA, Singh AS (2024) Vermicomposting: a holistic approach for sustainable crop production, nutrient-rich bio fertilizer, and environmental restoration. Discover Sustainability 5(1):60
Munguti JM, Kirimi JG, Obiero KO, Ogello EO, Kyule DN, Liti DM, Musalia LM (2021) Aqua-feed wastes: impact on natural systems and practical mitigations—a review
Murphy JN, Kerton FM (2017) Characterization and utilization of waste streams from mollusc aquaculture and fishing industries. Fuels, Chem Mater Ocean Aquat Sources 189–227
Namasivayam SKR, Kumar P, Samrat K, Moovendhan M, Kavisri M, Sivakumar L, Shyamsundar D (2024) Development of high organic-rich low-cost medium derived from microbial consortium decomposed vegetable wastes for the viable inocula production of potential fungal biopesticideMetarhiziumanisopliae. Biomass Convers Biorefinery 14(21):27853–27869
Nayak S, Samanta S, Mukherjee AK (2020) Beneficial role of Aspergillus sp. in agricultural soil and environment. Front Soil Environ Microbiol 17–36
Naylor RL, Hardy RW, Buschmann AH, Bush SR, Cao L, Klinger DH, Troell M (2021) A 20-year retrospective review of global aquaculture. Nature 591(7851):551–563
Nederlof MA (2022) Integrated multi-trophic aquaculture: Nutrient retention efficiency
Nekvapil F, Aluas M, Barbu-Tudoran L, Suciu M, Bortnic RA, Glamuzina B, Pinzaru SC (2019) From blue bioeconomy toward circular economy through high-sensitivity analytical research on waste blue crab shells. ACS Sustain Chem Eng 7(19):16820–16827
Ngasotter S, Xavier KM, Meitei MM, Waikhom D, Pathak J, Singh SK (2023) Crustacean shell waste derived chitin and chitin nanomaterials for application in agriculture, food, and health–a review. Carbohydr Polym Technol Appl 6:100349
Nges IA, Mbatia B, Björnsson L (2012) Improved utilization of fish waste by anaerobic digestion following omega-3 fatty acids extraction. J Environ Manage 110:159–165
Nikita S, Mishra S, Gupta K, Runkana V, Gomes J, Rathore AS (2023) Advances in bioreactor control for production of biotherapeutic products. Biotechnol Bioeng 120(5):1189–1214
Nurhidayati N, Usman A, Murwani I (2017) Chemical composition of vermicompost made from organic wastes through the vermicomposting and composting with the addition of fish meal and egg shells flour. J Pure Appl Chem Res 6(2):127
Olabisi FO (2023) Application of nanotechnology in the treatment of aquaculture effluent
Olaniran AF, Adeoye OE, Oyadeyi OM, Okonkwo CE, Erinle OC, Malomo AA, Faloye OR (2024) Utilization and application of bioactive compounds generated from Fish waste and by product as Functional Food Ingredient: a review. In: IOP conference series: earth and environmental science. IOP Publishing, vol 1342, no 1, p 012014
Onada OA, Ogunola OS (2016) Improving food security in an eco-friendly XE "Eco-friendly" manner through integrated aquaculture. Open Access Libr J 3(3):1–6
Otero P, Carpena M, Fraga-Corral M, Garcia-Oliveira P, Soria-Lopez A, Barba FJ, Prieto MA (2021) Aquaculture and agriculture-by products as sustainable sources of omega-3 fatty acids in the food industry. Efood 2(5):209–233
Özyurt G, Özkütük AS (2019) Advances in discard and by-product processing. In: Innovative technologies in seafood processing. CRC Press, pp 323–350
Pal K, Rakshit S, Mondal KC, Halder SK (2021) Microbial decomposition of crustacean shell for production of bioactive metabolites and study of its fertilizing potential. Environ Sci Pollut Res 1–14

Paraguison-Alili R, Lopez LLMA, Domingo CYJ (2021) Application of molecular tools in detecting zoonotic pathogens in organic fertilizers and liquid supplements. Biochem Biophys Rep 26:101017

Paśmionka IB, Bulski K, Boligłowa E (2021) The participation of microbiota in the transformation of nitrogen compounds in the soil—a review. Agronomy 11(5):977

Pinotti L, Giromini C, Ottoboni M, Tretola M, Marchis D (2019) Insects and former foodstuffs for upgrading food waste biomasses/streams to feed ingredients for farm animals. Animal 13(7):1365–1375

Pizzanelli S, Calucci L, Forte C, Borsacchi S (2023) Studies of organic matter in composting, vermicomposting, and anaerobic digestion by 13C solid-state NMR spectroscopy. Appl Sci 13(5):2900

Raman J, Kim JS, Choi KR, Eun H, Yang D, Ko YJ, Kim SJ (2022) Application of lactic acid bacteria (LAB) in sustainable agriculture: advantages and limitations. Int J Mol Sci 23(14):7784

Ramesh MV, Nibi KV, Kurup A, Mohan R, Aiswarya A, Arsha A, Sarang PR (2017) Water quality monitoring and waste management using IoT. In: 2017 IEEE global humanitarian technology conference (GHTC). IEEE, pp 1–7

Rathore SS, Chandravanshi P, Chandravanshi A, Jaiswal K (2016) Eutrophication: impacts of excess nutrient inputs on aquatic ecosystem. IOSR J Agric Vet Sci 9(10):89–96

Ravi HK, Degrou A, Costil J, Trespeuch C, Chemat F, Vian MA (2020) Larvae mediated valorization of industrial, agriculture and food wastes: Biorefinery concept through bioconversion, processes, procedures, and products. Processes 8(7):857

Rawat P, Das S, Shankhdhar D, Shankhdhar SC (2021) Phosphate-solubilizing microorganisms: mechanism and their role in phosphate solubilization and uptake. J Soil Sci Plant Nutr 21(1):49–68

Saez M, Maturana FP, Barton K, Tilbury DM (2018) Real-time manufacturing machine and system performance monitoring using internet of things. IEEE Trans Autom Sci Eng 15(4):1735–1748

Saha B, Devi C, Khwairakpam M, Kalamdhad AS (2018) Vermicomposting and anaerobic digestion–viable alternative options for terrestrial weed management–a review. Biotechnology Reports 17:70–76

Sahu PK, Brahmaprakash GP (2016) Formulations of biofertilizers–approaches and advances. Microb Inoculants Sustain Agric Prod: Funct Appl 2:179–198

Sandström V, Chrysafi A, Lamminen M, Troell M, Jalava M, Piipponen J, Kummu M (2022) Food system by-products upcycled in livestock and aquaculture feeds can increase global food supply. Nature Food 3(9):729–740

Saud S, Nawaz T, Hassan S, Ur Rahman T, Rasheed MN, Hussain S, Fahad S (2024) Nitrogen-fixing cyanobacteria and soil enrichment for a greener future. In: Environment, climate, plant and vegetation growth. Cham: Springer Nature Switzerland, pp 391–424

Sharma B, Tiwari S, Kumawat KC, Cardinale M (2023) Nano-biofertilizers as bio-emerging strategies for sustainable agriculture development: potentiality and their limitations. Sci Total Environ 860:160476

Sharma B, Vaish B, Monika Singh UK, Singh P, Singh RP (2019) Recycling of organic wastes in agriculture: an environmental perspective. Int J Environ Res 13:409–429

Sharma MMM, Kapoor D, Loyal A, Kumar R, Sharma P, Husen A (2024) Effect of gold nanoparticles on seed germination, plant growth, and plant protection. In Husen A (ed.). Plant Response to Gold Nanoparticles. Springer, pp 155–175. https://doi.org/10.1007/978-981-97-7174-5_8

Sharma MMM, Kapoor D, Loyal A, Kumar R, Sharma P, Husen A (2025) Plant response to silver nanoparticles in terms of growth, development, production, and protection: an overview. In Husen A (ed). Plant response to silver nanoparticles. Springer, Singapore, pp 1–22. https://doi.org/10.1007/978-981-97-7352-7_1

Siddiqui SA, Harahap IA, Osei-Owusu J, Saikia T, Wu YS, Fernando I, Câmara JS (2024) Bioconversion of organic waste by insects–a comprehensive review. Process Saf Environ Prot

Sohil A, Kichloo MA (2023) Sustainable solutions to animal waste: climate change mitigation and bioproduct harvest. In: climate changes mitigation and sustainable bioenergy harvest through

animal waste: sustainable environmental implications of animal waste. Cham: Springer Nature Switzerland, pp 301–332
Sukhdhane KS, Kripa V, Divu D, Vase VK, Mojjada SK (2018) Integrated multi-trophic aquaculture systems: a solution for sustainability. Aquaculture Asia 22(4):26–29
Tanner JC, Holden SJ, Winugroho M, Owen E, Gill M (1995) Feeding livestock for compost production: A strategy for sustainable upland agriculture on Java. Livest Sustain Nutr Cycl Mix Farming Syst Sub-Sahar Afr 2:115–128
Teixidó N, Usall J, Torres R (2022) Insight into a successful development of biocontrol agents: Production, formulation, packaging, and shelf life as key aspects. Horticulturae 8(4):305
Thomas M, Pasquet A, Aubin J, Nahon S, Lecocq T (2021) When more is more: taking advantage of species diversity to move towards sustainable aquaculture. Biol Rev 96(2):767–784
Tom AP, Jayakumar JS, Biju M, Somarajan J, Ibrahim MA (2021) Aquaculture wastewater treatment technologies and their sustainability: a review. Energy Nexus 4:100022
Topić Popović N, Lorencin V, Strunjak-Perović I, Čož-Rakovac R (2023) Shell waste management and utilization: Mitigating organic pollution and enhancing sustainability. Appl Sci 13(1):623
Tucker G, Featherstone S (2021) Essentials of thermal processing. John Wiley and Sons
Vélez-Henao JA, Weinland F, Reintjes N (2021) Life cycle assessment of aquaculture bivalve shellfish production—a critical review of methodological trends. Int J Life Cycle Assess 1–16
Venugopal V (2021) Valorization of seafood processing discards: Bioconversion and bio-refinery approaches. Front Sustain Food Syst 5:611835
Verdegem M, Buschmann AH, Latt UW, Dalsgaard AJ, Lovatelli A (2023) The contribution of aquaculture systems to global aquaculture production. J World Aquaculture Soc 54(2):206–250
Verma P, Chandra P, Rai AK, Kumar A, Prajapat K, Sundha P, Yadav RK (2023) Native rhizobacteria suppresses spot blotch disease, improves growth and yield of wheat under salt–affected soils. Plant Stress 10:100234
Vishwakarma K, Kumar N, Shandilya C, Mohapatra S, Bhayana S, Varma A (2020) Revisiting plant–microbe interactions and microbial consortia application for enhancing sustainable agriculture: a review. Front Microbiol 11:560406
Wan D, Wu Q, Kuča K (2021) Spirulina. In: Nutraceuticals. Academic Press, pp 959–974
Wang C, Jiang C, Gao T, Peng X, Ma S, Sun Q, Zhuang X (2022) Improvement of fish production and water quality in a recirculating aquaculture pond enhanced with bacteria-microalgae association. Aquaculture 547:737420
Wang Z, Chen Z, Xu Z, Fu X (2019) Effects of phosphate-solubilizing bacteria and N2-fixing bacteria on nutrient uptake, plant growth, and bioactive compound accumulation in Cyclocaryapaliurus (Batal.) Iljinskaja. Forests 10(9):772
Wong BP, Kerkez B (2016) Real-time environmental sensor data: an application to water quality using web services. Environ Model Softw 84:505–517
Wu TH, Huang YI, Chen JM (2015) Development of an adaptive neural-based fuzzy inference system for feeding decision-making assessment in silver perch (Bidyanusbidyanus) culture. Aquacult Eng 66:41–51
Wu X, Nawaz S, Li Y, Zhang H (2024) Environmental health hazards of untreated livestock wastewater: Potential risks and future perspectives. Environ Sci Pollut Res 31(17):24745–24767
Wurtsbaugh WA, Paerl HW, Dodds WK (2019) Nutrients, eutrophication and harmful algal blooms along the freshwater to marine continuum. Wiley Interdiscip Rev Water 6(5):e1373
Xing Y, Wang X (2024) Precise application of water and fertilizer to crops: challenges and opportunities. Front Plant Sci 15:1444560
Xu C, Mou B (2017) Drench application of fish-derived protein hydrolysates affects lettuce growth, chlorophyll content, and gas exchange. HortTechnology 27(4):539–543
Yadav R, Gupta RK, Kumar R, Kaur T, Moun R (2022a) "Biochemical response of earthworm, Eisenia fetida to heavy metals toxicity. J Appl Nat Sci 14(3):990–998
Yadav R, Gupta RK, Kumar R, Kaur T (2022b) Assessment of toxicity of lead and nickel on the biochemical and immunological parameters of earthworm. Eudrilus Eugeniae. Agricultural Science Digest 44(5):922–929

Yadav R, Kumar R, Gupta RK, Kaur T, Yodha K, Kour A, Kaur S, Rajput A (2023) Heavy metal toxicity in earthworms and its environmental implications: a review. Environ Adv 100374

Yang L (2021) Bioenergy and biofertilizer recovery from organic fraction of municipal solid waste for environmental protection and sustainable economy in Liaoning province, China (Doctoral dissertation, UTAR)

Zeng Q, Ding X, Wang J, Han X, Iqbal HM, Bilal M (2022) Insight into soil nitrogen and phosphorus availability and agricultural sustainability by plant growth-promoting rhizobacteria. Environ Sci Pollut Res 29(30):45089–45106

Zhang L, Li B, Sun X, Hong Q, Duan Q (2023) Intelligent fish feeding based on machine vision: a review. Biosys Eng 231:133–164

Chapter 6
Antimicrobial Resistance Pollution: A Critical Challenge for Sustainable Aquaculture

Simran Kaur, Prapti Sudan, and Aditya Kumar Upadhyay

Abstract Antimicrobial resistance (AMR) pollution poses multifaceted challenges to fisheries sustainability, aquatic ecosystem health, and human well-being. The careless use of antimicrobials is contributing to emergence of antimicrobial resistant microorganisms. This has led to compromised fish health, jeopardized food safety, and undermined consumer confidence. However, biosecurity measures, improving disease prevention and control strategies, and promoting alternative approaches such as probiotics, phages, and vaccines can help reduce reliance on antimicrobials in aquaculture. Strengthening surveillance and monitoring systems to track resistance patterns in fish population and aquatic environment is crucial for early detection and response to emerging threats. Furthermore, investing in research and innovation to develop sustainable aquaculture practices and novel antimicrobial agents is essential for long-term solutions to AMR pollution in fisheries.

Keywords Antimicrobial resistance · Mechanism · Aquaculture · Antibiotics · One health · Pathogens

6.1 Introduction

With a record-breaking global production of 223.2 million tons, fisheries and aquaculture are essential to feed the expanding population. A projected 165 million tonnes of aquatic animal food were consumed globally in 2022, while 130.9 million tonnes

Present Address:
S. Kaur · P. Sudan · A. K. Upadhyay (✉)
Department of Aquatic Environment Management, College of Fisheries, Guru Angad Dev Veterinary and Animal Sciences University, Ludhiana 141004, Punjab, India
e-mail: adityaupadhyay.nd@gmail.com

S. Kaur
e-mail: simrankaur@gadvasu.in

P. Sudan
e-mail: praptisudan@ymail.com

R. Kumar et al. (eds.), *Sustainability in Aquaculture and Fisheries*, Springer Water, https://doi.org/10.1007/978-3-032-16605-0_6

of aquaculture were produced worldwide, accounting for 59% of all fisheries and aquaculture production. The expected yearly per capita apparent consumption worldwide increased from 9.1 kg in 1961 to 20.7 kg in 2022. High-quality proteins are found in 15% of animal proteins and 6% of all proteins globally in aquatic animal meals. Omega-3 fatty acids, minerals, and vitamins are among the essential nutrients. Prominent global platforms such as the UN Framework Convention on Climate Change and the UN Food Systems Summit are increasingly recognizing the role of aquatic foods in enhancing nutrition, food security, and reducing poverty. In 2022, human consumption is projected to represent 89% of the 185.4 million tonnes of aquatic animals, which equals an expected 20.7 kg per person (Kumar et al. 2022a, b; SOFIA 2024)

Since the early 1990s, the capture fisheries have remained static, while aquaculture production has increased in response to the global demand, representing more than 50% of total production in 2022. This trend is expected to continue through 2030 as demand for aquatic food rises. However, the shift from extensive to intensive farming systems has greatly increased aquaculture productivity, it has also resulted in an increase in disease outbreaks (SOFIA 2024). The development aligns with the growing usage of antimicrobials in the terrestrial food animal industry, where there has been a noticeable increase in antimicrobial use (Boeckel et al. 2015; Boeckel et al. 2017). Notably, antimicrobials consumption rates per kilogram in some fish species, such as catfish, exceeds those observed in terrestrial animals and even humans (Cabello et al. 2023)

Research, over the last few decades, has increased our awareness of the antimicrobial resistance in fisheries and aquaculture (Cabello et al. 2023; Done et al. 2015; Heuer et al. 2009; Kumar and Singh 2022). The antimicrobial consumption in aquaculture, globally, has surged from around 10,259 tonnes in 2017 to an estimated 13,600 tonnes by 2030, marking an increase of 33%. Antibiotics consumption is highest in China 55.9%, followed by India (11.4%), Indonesia (10.1%), and Vietnam (5%) (Gin et al. 2023). In 2017, Africa and Europe shares 2.3% and 1.8% respectively (Schar et al. 2020). Majorly used antibiotics like sulphadiazine, florfenicol, and oxytetracycline accounts for more than 73%, while erythromycin, amoxicillin, sulphadimethoxine, and enrofloxacin are frequently used in around 55% of countries (Ibrahim et al. 2021).

Significant consumers of antibiotics in aquaculture are China and Vietnam, while countries like India, Korea, Thailand, Philippines, and Bangladesh use them less frequently. Farmers use antimicrobials not only for therapeutic purposes but also as a preventive measure. This practice causes rise of antimicrobial resistance (AMR) leading to many treatments ineffective against infections. AMR in aquaculture represents a serious problem. There is evidence that the wrong use of antibiotics results in more resistant bacteria in aquatic habitats, which could harm human health by transferring resistant strains. Ongoing surveillance and regulatory efforts are required to minimize the risks and control the transmission of antibiotic resistance in aquaculture present similar conclusions (Kumar et al. 2022c; Pepi and Focardi 2021, Preena et al. 2020; Schar et al. 2021).

6.2 Current Situation of Antibiotics in Aquaculture

O'Neill (2016) projection shows that bacterial antimicrobial resistance presents a serious and persistent danger worldwide (O'Neill 2016). These findings support the worry, as AMR was attributed to 1.27 million deaths internationally in 2019, and was also connected to 4.95 million deaths resulting from its effects (Tang et al. 2023). Situation in commercial aquaculture is becoming more challenging with each passing year. The practice of keeping aquatic species in small holding areas for intensive farming makes them more vulnerable to a wide variety of diseases. This stress makes them more vulnerable to multiple diseases. Outbreaks of aquatic animal diseases are shaped by interactions between the pathogens, the animals' vulnerability, and environmental conditions (Snieszko et al. 1954).

Best management practices in aquaculture—such as maintaining optimal water quality, proper nutrition, and regulating stocking densities - are commonly employed to minimize disease outbreaks. However, due to production demands and unpredictable environmental factors, antibiotics are now a vital component in disease control. Numerous antibiotics originally developed for human and terrestrial animal have been evaluated for potential application in aquaculture (Pepi and Focardi 2021). Among these, tetracyclines, especially oxytetracyclines have demonstrated strong efficacy against Gram-negative bacterial infections (Snieszko et al. 1954). Despite their widespread use, the estimation of the global scale of antibiotic consumption in aquaculture remains difficult due to inadequate surveillance and monitoring programs and systems. Nevertheless, existing data provides glimpse into the extent of usage. For example, in 2009, 13,000 tons of antibiotics were utilized in various industrial systems in the United States, while Chilean salmon culture utilized 382.5 tons of antibiotics in 2016 alone (Miranda et al. 2018). Excessive use of antibiotics in aquaculture has led to emergence of antibiotic-resistant bacteria in such environments. Antibiotic-resistant strains are often reported from shellfish, in particular, shrimp. Specimens of *Vibrio harveyi* collected from disease outbreaks and mass deaths in *Penaeus monodon* were observed to be broadly resistant to antibiotics indicating that antimicrobial misuse was a concerning issue in open systems (Ferraz 2019). In their work, Ferri and colleagues (2024) revealed that *V. vulnificus* and *V. alginolyticus* are multi-drug resistant in mussels and oysters (Ferri et al. 2024). In Chinese aquaculture, *V. splendidus* and *Pseudoalteromonas* are the main reservoirs of multidrug resistance genes for both abalone and turbot (Dang et al. 2009). Wang et al. (2009) had demonstrated that the strain of *Edwardsiella tarda* causing severe trout disease was chloramphenicol resistant (Wang et al. 2009). Further, the presence of antibiotic-resistant genes in fish feed can find its way into the aquaculture waterways contributing to spreading drug resistant pathogens (Preena et al. 2020).

Overall, antibiotic resistance in aquaculture is a tough problem that intensifies with the use of intensive farming and large amounts of antibiotics. If there are no better ways to monitor or guide antibiotic use in aquaculture, the industry could play a major role global spread of AMR.

6.3 The Rise of Antimicrobial Resistance in Bacterial Species

Historically, antibiotics have been permitted and used in aquaculture with little regard for what happens to them in the environment. While various treatment methods, such as dips, flushes, and injections, are used in aquaculture, the most commonly used approach involves mixing the drug into fish feed (Yasin et al. 2023). However, growing evidence reveals that a significant amount of antibiotics administered to cultured animals, often as feed additives, are neither absorbed nor retained in the body of the animal, instead, they often enter the surrounding aquatic environment via three primary mechanisms listed below.

6.3.1 Loss Through Uneaten Medicated Feed

A portion of uneaten medicated feed settles into the substratum, contributing to environmental contamination. This loss is most likely to occur during disease outbreaks when fish often feed erratically.

6.3.2 Excretion Through Faeces

Even when antibiotics are consumed, they are not fully absorbed through the digestive process. Some portions of the consumed antibiotics are ejected into the environment. Depending on the type of antibiotic administered, 60% to 80% of the active drug may be excreted unchanged in faeces, entering the water system directly (Cravedi et al. 1987).

6.3.3 Excretion Through Urine and Bile

Antibiotics can also be excreted from the body in unmetabolized, microbially active forms via urine and bile. Moreover, antibiotics used in human and veterinary medicine may also enter aquatic habitats. Residues from these frequently and indiscriminately used drugs may enter aquatic ecosystems through waste streams, accumulate in water systems, turning the aquatic habitats into repository for drug residues (Rigos et al. 2004; Watts et al. 2017).

Antibiotics in any environment exert selection pressure on natural microbial communities. Bacteria with metabolic capacity to tolerate antibiotic-mediated selection pressure thrive, whereas antibiotic-sensitive populations may cease to exist. The metabolic capabilities of bacteria to resist antibiotics are mediated by certain genetic elements known as antibiotic resistance genes (ARGs) (Preena et al. 2020). Furthermore, bacteria and fungi naturally manufacture a huge number of antibiotics in their native environments, and these strains have genes that encode antibiotic resistance (Hoskisson 2008). Most of the known antibiotic resistance genes are located on transposons, integrons, or plasmids. These genetic components can be mobilized and transferred to both pathogenic and non-pathogenic bacteria of various species through horizontal gene transfer (HGT) (Michaelis and Grohmann 2023). When human pathogenic bacteria acquire antibiotic-resistance genes, particularly those already possessing highly virulent traits, it often leads to decreased effectiveness of antibiotic treatments, resulting in significant health and economic repercussions. A report from the Organization for Economic Co-operation and Development (OECD) in November 2018 projected that antimicrobial resistance could lead to around 2.4 billion deaths in Europe, North America, and Australia over the next 30 years, with an annual economic impact estimated at $3.5 billion (OECD, 2018). Although data on developing countries is limited, the report indicates that resistance rates in low-income and middle-income nations range from 40 to 60% (Hofer 2019).

6.4 Development of Antimicrobial Resistance

When antibiotics are first introduced, they are typically effective against targeted pathogens. However, with prolonged and repeated exposure, bacteria may develop resistance. This resistance can arise either through chromosomal gene mutations or by acquiring foreign resistance genes through horizontal gene transfer. According to Lee(2019), resistance genes are sometimes transferred through horizontal gene transfer. Usually, three main categories of genes are influenced by these mutations:

(a) Genes encoding the antibiotic's target,
(b) Genes that control the activity of transporters responsible for letting antibiotics into or out of the cell, and
(c) Genes involved in controlling the activity of these transporters, including multi-drug efflux pumps and antibiotic-modifying enzymes, are also regulated by these mutations.

Lee (2019) states that bacteria resistant to antibiotics can acquire genes through intrinsic, acquired, or adaptive mechanisms, thus referred to as natural resistance, acquired resistance and adaptive resistance (Fig. 6.1), which are further elaborated later in this section.

6.4.1 Natural Resistance

Bacteria are said to have natural resistance when they can naturally withstand some antibiotic types. Resistance of this kind is determined by the bacteria's chromosomal

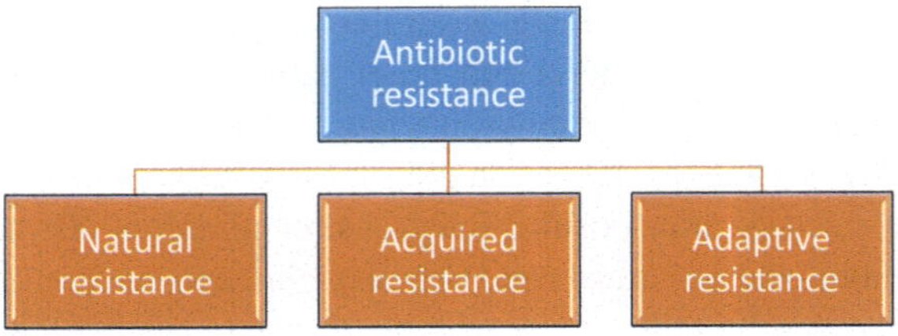

Fig. 6.1 Development of antimicrobial resistance. *Source* Author

genes and is not linked to either gene mutation or gene acquisition. Examples of common mechanisms are the actions of efflux pumps and reduced cell membrane permeability, which together block the entrance of antibiotics (Cox and Wright 2013; Martinez 2014).

6.4.2 Acquired Resistance

When bacteria that were once sensitive to antibiotics become resistant, it is called acquired resistance. Acquired resistance develops either through changes in chromosomal genes or through the acquisition of new genes through horizontal gene transfer, most often by conjugation, transposition, or transformation. The most common method of the movement of resistance genes between bacteria is through conjugation, which regularly involves plasmids. Resistance using this method can be either temporary or permanent subject to genetics and environment (Holmes et al. 2016; Munita and Arias 2016).

6.4.3 Adaptive Resistance

Regardless of an incomplete understanding of the biological aspects of adaptive resistance, scientists have suggested a variety of explanations. These factors are thought to play some part including high mutation rates, gene amplification, efflux pumps, epigenetic inheritance, biofilm development, population structure, and heterogeneity. Both the length of selective pressure and how well the bacteria adapt help decide if adaptive resistance will last a short time or become permanent. Adaptive resistance can develop in humans and livestock when sub-inhibitory antibiotic levels are impacted by environmental things such as stress, growth signals, limited nutrients, pH changes, and ionic conditions (Fernández and Hancock 2012; Salimiyan Rizi et al. 2018).

6.5 Mechanisms of Antimicrobial Resistance Development in Bacteria

Antibiotic activity and microbial adaptation combine intricately to cause the development of antimicrobial resistance in bacteria. Numerous defense mechanisms, such as restricted drug absorption, active drug efflux, target site modification, and drug inactivation, might help bacteria evolve antibiotic resistance (Gauba and Rahman 2023). These mechanisms allow bacteria to survive in antibiotic-rich environments and contribute to the widespread challenge of AMR. Figure 6.2 provides an overview of the general mechanisms by which bacterial cell develops antimicrobial resistance.

6.5.1 Limited Drug Absorption

One of the primary mechanisms of resistance is reduced antibiotic uptake. The lipopolysaccharide layer of Gram-negative bacteria serves as a barrier, providing intrinsic resistance against several major antimicrobial agents. Gram-positive bacteria, unlike Gram-negative bacteria, lack an outer membrane and hence do not

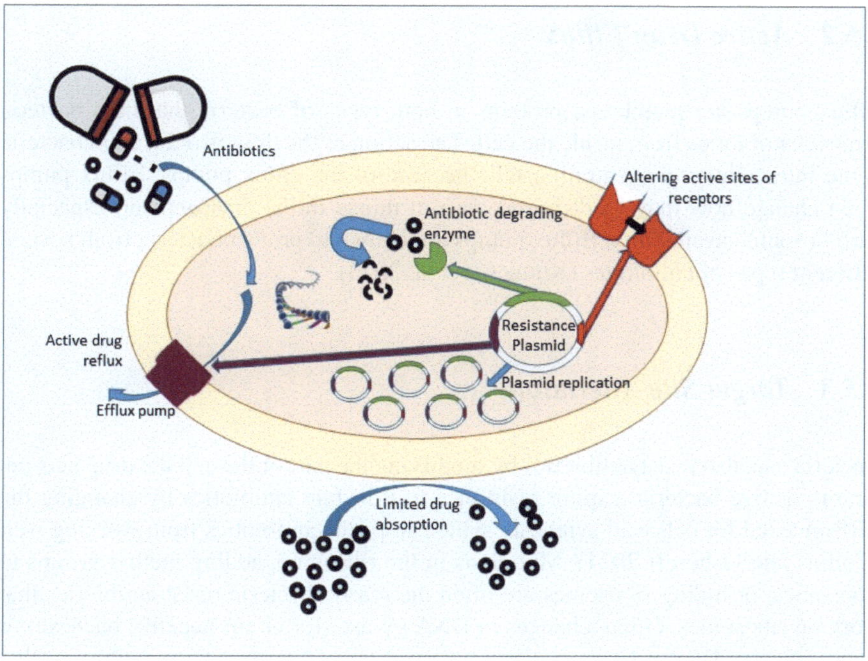

Fig. 6.2 General mechanisms by which bacterial cell develops antimicrobial resistance. *Source* Author

face the same challenges regarding drug access. Mycobacteria, due to the higher lipid content in their outer membranes, allow hydrophobic drugs to enter more easily, while hydrophilic drugs encounter more significant barriers. This characteristic reduces the sensitivity of mycobacteria to antibiotics including tetracyclines, beta-lactams, and some fluoroquinolones (Impey et al. 2020). Additionally, Enterococci exhibit intrinsic resistance to the aminoglycosides because the polar molecules struggle penetrating their thick walls. Porin channels play a characteristic role in material entry in bacteria with large outer membranes. These channels are the main way for Gram-negative bacteria to move hydrophilic compounds. Antibiotic resistance happens when bacteria either lower the number of these channels or change their selectivity, keeping antibiotics out (Winterhalter 2021).

Biofilms form during bacterial colonization, and these are protective communities that are surrounded by a matrix of polysaccharides, DNA, and proteins. Because biofilms protect bacteria from both antimicrobials and the immune response, higher doses of medications are needed for treatment to succeed. Because bacteria in biofilms are often hardly growing and attached, drugs that work on dividing cells have reduced effectiveness. In addition, the closeness of cells in biofilms helps spread resistance genes, making AMR develop more rapidly (Michaelis and Grohmann 2023).

6.5.2 Active Drug Efflux

Efflux pumps are membrane proteins in both types of bacteria that help bacteria remove antibiotics from inside the cell. The action of the drug is reduced in bacteria if the intracellular concentration falls because of the efflux pumps. Efflux pumps often change how they work in response to things in the environment, especially carbon source availability. Efflux pumps are known to protect bacteria from several different types of antibiotics (Kumawat et al. 2023).

6.5.3 Target Site Alteration

Bacteria can also resist antibiotics by modifying the part of the cell the drug acts on. Gram-positive bacteria acquire resistance to β-lactam antibiotics by changing the PBPs needed for cell wall synthesis, which stops the antibiotics from working well (Collins and Osheroff 2024). Mutations in the ribosome, adding methyl groups to ribosomes, or hiding ribosomes are often the reason bacteria resist antibiotics that work on ribosomes. Often, changes in DNA gyrase (for Gram-negative bacteria) or topoisomerase IV (for Gram-positive bacteria) lead to resistance to antibiotics that affect nucleic acid formation, for example fluoroquinolones, by keeping the drug from binding properly (Collins and Osheroff 2024).

Also, enzymes that are part of folate production, including DHPS and DHFR, can make bacteria resistant to drugs that disrupt metabolic functions. Mutations in the enzyme's active site are common, making it possible for bacteria to escape drugs that compete with these enzymes, yet still use their normal substrates (Li et al. 2022).

6.5.4 Drug Inactivation

Antimicrobials can be rendered harmless by bacteria that either degrade them or change their structure. A well-known way bacteria resist antibiotics is by producing hydrolyzing enzymes, beta lactamase and ESBL, that break down the drugs. Beta lactamase and ESBL enzymes from bacteria help antibiotics lose their strength, yet *tetX* gene hydrolysis is what makes bacteria resistant to tetracycline (Wen et al. 2020). For resistance, bacteria produce enzymes including acetyltransferase and others to stop antibiotics from working. Acetylation is the leading method bacteria use to get rid of aminoglycosides, fluoroquinolones, streptogramins, and chloramphenicol in their environment. Tenover explained back in 2006 that phosphorylation and adenylation play the biggest role in how bacteria resist aminoglycosides (Tenover 2006).

6.6 Molecular Mechanism of Antimicrobial Resistance

Learning how resistance grows and spreads at the molecular level helps understand AMR's effects on both water and medicine in fish farming. Wherever AMR is found, the ways that resistance genes spread and remain in bacterial communities are guided by genetic and metabolic processes, without regard to external environmental or human impacts on AMR. Such mechanisms control the outcomes of control practices and influence the ways antibiotics function in aquatic environments.

Molecular support for AMR is diverse, and it may be altered by different bacterial species and their habitats. The methods include target protection, target replacement, ways to get rid of antibiotic activity, and limiting the amount of antibiotic entering the cell as described by Bennett in 2008. Generally, genes for AMR spread and persistence are transmitted mainly by transposable elements, integron systems, and bacterial conjugative plasmids (Fig. 6.3). Bennett (2008) describes how often *tet, qnr, sul, ere*, and other AMR genes join bacterial plasmids through the action of site specific recombination and transposition. Multidrug resistance is largely increased in aquatic habitats by these gene transfer systems, which depend on both antibiotics and a large number of microbes (Bennett 2008).

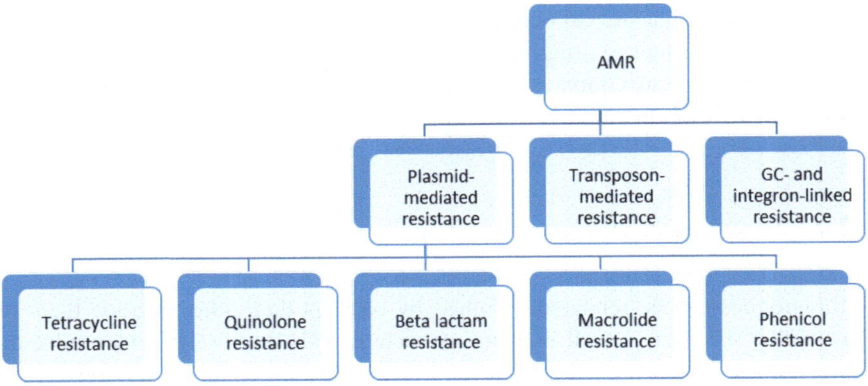

Fig. 6.3 Molecular mechanisms of antimicrobial resistance. *Source* Author

6.6.1 Plasmid Mediated Resistance

Plasmids act as mobile genetic elements and transport genes which give resistance to antibiotics, as well as to heavy metals that are toxic (Yadav et al. 2022a, b; Partridge et al. 2009; Yadav et al. 2023). In aquaculture, these species, including *Vibrio* and *Aeromonas*, acquire plasmid-mediated resistance frequently (Alderman and Hastings 1998; Dixon et al. 1990). *Aeromonas* spp. is a primary source through which MGEs are spread horizontally (Poirel et al. 2012). Resistance to antimicrobials is raised when bacteria have *IncA and IncC* plasmids. Such plasmids are reported in *Escherichia coli, Yersinia pestis,* and *Salmonella enterica* based on work by Harmer and Hall 2015. Verner-Jeffreys et al. (2009), in their research on farmed ornamental fish, showed that incompatible IncA/C plasmids hold resistance genes including *tetA, tetD, tetE,* bla_{TEM21}, *dfr, sul1, qacE2, and floR*, in addition to genes responsible for aminoglycoside resistance (Verner-Jeffreys et al. 2009).

6.6.1.1 Tetracycline Resistance

Bacteria in aquatic environments become resistant to tetracycline because they have genes such as tetA-G, tetM, and tetE, which allow them to avoid tetracycline by both efflux and ribosomal protection mechanisms (Preena et al. 2020).

Part of the reason tetracycline resistance is so common in aquaculture is that genetic transfer systems move *tet34* along with other genes such as *tet39, tetM, and tetE* (Agerso & Guardabassi 2005). Schmidt et al. (2001) reported that tetracycline resistance genes spread broadly because tetracycline-producing microbes are part of many environments (Schmidt et al. 2001).

6.6.1.2 Quinolone Resistance

Both plasmid behaviors and genetic alterations in DNA topoisomerase and gyrase cause quinolone resistance in the *Vibrionaceae* family. Most cases of fluoroquinolone resistance are the result of mutations within the quinolone resistance-determining regions. Many studies show that plasmid-mediated genes for quinolone resistance, including *qnrA, qnrB,* and *qnrS*, are present in *Enterobacteriaceae* and *Aeromonadaceae.* Quinolone resistance plasmids are often tied to integrons and may also give resistance to beta-lactam and aminoglycoside antibiotics (Hooper and Jacoby 2015).

6.6.1.3 Beta Lactam Resistance

The main reasons for beta-lactam resistance are ESBLs and higher expression of the *ampC* beta-lactamase gene found on the chromosome. Antibiotic treatments with third and fourth generation cephalosporins work much worse when ESBLs are present in aquaculture pathogens, according to Verner Jeffreys et al. (2009). Isolates of *Aeromonas* commonly carry genes for ESBLs such as *blaCphA, blaTEM, blaOXA B,* and *blaSHV* (Verner Jeffreys et al. 2009).

6.6.1.4 Macrolide Resistance

Aquatic pathogens become resistant to macrolides by altering ribosomal binding sites, making enzymes that stop macrolides from working, and using active efflux systems (Nguyen et al. 2017).

6.6.1.5 Phenicol Resistance

Plasmids are known to possess chloramphenicol resistance transferases (*cat*) genes that cause resistance to phenol antibiotics. Resistance is also caused by efflux systems, RNA methyl transferase enzymes (*cfr* gene activity), and the action of particular hydrolases (Mendes et al. 2008).

6.6.2 Transposon-Mediated Resistance

Transposons, sometimes called "jumping genes", are DNA parts that can relocate both within the same chromosome and to other chromosomes. These mobile units contribute significantly to spread the antibiotic resistance genes rapidly among bacteria. For instance, *A. salmonicida* possesses the transposon Tn*5393*, which carries multiple antibiotic resistance genes (Babakhani and Oloomi 2018; Bennett 2008; L'abée-Lund and Sørum 2001).

6.6.3 Resistance Linked to Integrons and Gene Cassettes

In aquatic environment, integrons are another major driver of antimicrobial resistance gene disseminating among Gram negative bacteria. They facilitate the acquisition and expression of resistance genes, contributing to multidrug resistance. Integrons are of mainly two primary forms: chromosomal and mobile. Chromosomal integrons, also known as super integrons, can harbour up to 200 gene cassettes (GCs), including proteins with unidentified functions. Resistance genes within chromosomal integrons can transfer to mobile integrons, leading to the evolution and broader dissemination. Mobile integrons, also called multidrug resistance integrons or resistant integrons are found on MGEs and typically contain fewer GCs, predominantly coding antibiotic-resistance determinants. These mobile integrons are particularly significant as they enhance gene transfer (Gao et al. 2012; Rahube and Yost 2010; Stalder et al. 2012).

In addition to resistance associated with gene cassettes, mobile integrons linked to plasmids or transposons also carry additional resistance genes (Li et al. 2022). As a result, the co-selection of resistance through these plasmids, transposons, and mobile integrons containing gene cassettes contributes to the emergence of multidrug-resistant bacterial phenotypes. This phenomenon, often referred to as the "superbug" phenomenon, significantly increases overall resistance levels. Resistance gained through mobile genetic elements is hard to get rid of within the system (Dassanayake et al. 2021; Gopikrishnan et al. 2024).

6.7 Transmission Mechanism of the Antimicrobial Resistance

AMR spread occurs through multiple ecological and biological interactions at once. AMR can spread inside aquatic microbial communities and also move between wider ecosystems like the food web and people. Knowing the routes of AMR transmission is necessary to create useful mitigation methods for aquaculture and public health. Figure 6.4 illustrates the transmission mechanisms of AMR.

6.7.1 Transmission Within Bacterial Communities

Resistance genes move between bacterial communities in water and soil environments thanks to horizontal gene transfer (HGT) processes including conjugation and

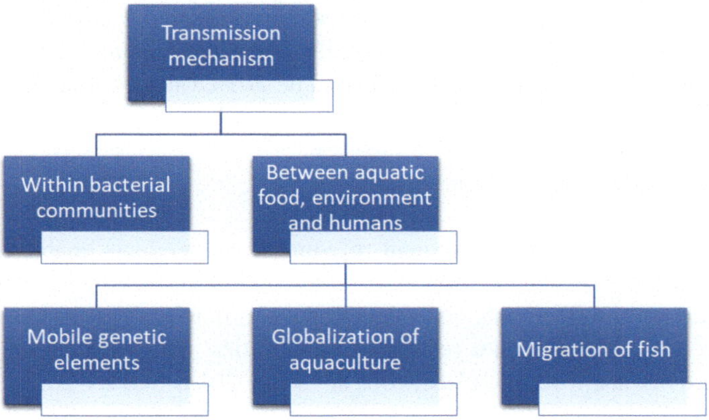

Fig. 6.4 Insights into transmission mechanisms responsible for spread of antimicrobial resistance. *Source* Author

conjugative transposition. Transduction is another important way resistance genes spread, and this happens largely because viruses are so abundant in seawater and marine sediments (Agersø and Guardabassi 2005; Buschmann et al. 2012; Fuhrman 1999). In many developing countries, aquaculture farms are vulnerable to pollution from sewage, farm runoff, and industrial wastes, all of which carry antibiotic-resistant gut bacteria and may contain infectious agents from humans and animals, mixing with both freshwater and seawater (Cabello 2006; Miranda and Zemelman 2001, Sørum 2006). In integrated aquaculture-agriculture systems, the risk grows because contaminants have greater opportunity to spread (Petersen et al. 2002).

6.7.2 *Transmission Between Aquatic Food, Environment and Humans*

The movement of antimicrobial-resistant pathogens and resistance genes from aquatic habitats, mostly through food, is a major public health danger (Cabello 2006; Cabello et al. 2016; Cabello et al. 2013; Ryu et al. 2012; Sørum 2019). This transmission can occur via several interconnected pathways, including mobile genetic elements, the globalization of aquaculture, migratory aquatic species, and the consumption of raw or undercooked seafood (Milijasevic et al. 2024).

6.7.2.1 Mobile Genetic Elements

Mobile genomic elements (MGEs) often contain multiple antibiotic resistance genes (Defoirdt et al. 2011). These MGEs serve as critical vectors for gene exchange not only among aquatic bacteria but also between aquatic and human-associated microbiota. Their presence in both aquaculture systems and the surrounding aquatic environment facilitates the transfer of clinically significant resistance genes to human pathogens (Buschmann et al. 2012; Cabello et al. 2016; Cabello et al. 2013).

6.7.2.2 Globalization of Aquaculture

The international trade and movement of aquaculture products enable the global spread of resistance genes that may have originated in localized environments. By shipping aquacultured animals and their products around the world, the microbes they carry, some of which are resistant to antibiotics, may be released into new environments (Cabello 2006; Cabello et al. 2016, 2013; SOFIA 2024; Thornber et al. 2020).

6.7.2.3 Migration of Fish

Natural movements of fish and aquatic animals between freshwater and marine environments may help spread AMR genes. Migratory animals transfer resistant bacteria between habitats, helping to spread resistance in both nature and aquaculture (Cabello 2006; Goldburg and Naylor 2005; Naylor and Burke 2005).

6.8 Increased Direct AMR Transfer to the Environment via Open Systems

After looking at how AMR spreads within and between aquatic organisms, we must understand how aquaculture system design affects resistance transfer. In particular, open aquaculture systems, which are directly connected to the surrounding natural water bodies, pose a considerable risk for environmental AMR dissemination.

In these open aquaculture systems, antimicrobial agents are commonly administered through medicated feed, and often the dosages are correspondingly larger than those used in terrestrial livestock (On and Resistance 2015). A significant portion of these compounds is not fully metabolized by the fish, leading to persistence of residues in undigested feed, fish faeces, water, products originated from aquaculture, and sediments from aquaculture farms. Depending upon concentrations and biodegradability, the residues may contain the antimicrobials which were not absorbed in the body, antimicrobial metabolites which were excreted from the body. These residues may include unmetabolized antibiotics, partially degraded compounds, and antimicrobial metabolites excreted from the fish (Maldonado et al. 2012).

Fish treated with antibiotics may excrete 70–80% of administered drugs directly into the water (Burridge et al. 2010), which alters the native microbial communities

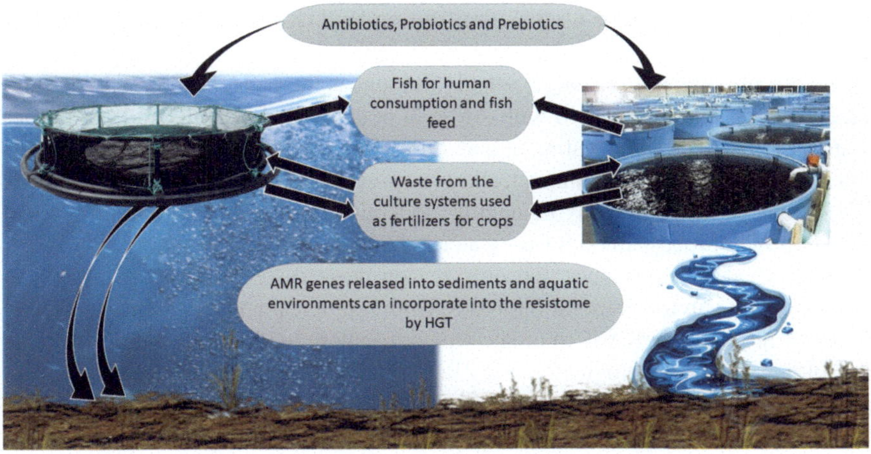

Fig. 6.5 Transfer pathways for antimicrobial resistance (AMR) genes into the water and sediment environmental resistome from closed and open aquaculture systems. *Source* Author

and potentially selects for resistance strains (Watts et al. 2017). Even at low environmental temperatures, selection pressure for AMR bacteria remains effective, causing significant shift in sediment and water microbiota where resistant strains gradually outcompete susceptible ones (Gullberg et al. 2011, Watts et al. 2017).

Moreover, such environmental conditions contribute to an increase in the resistome complexity through enrichment of resistance genes and mobile genetic elements (Gao et al. 2012). These processes are illustrated in Fig. 6.5, which provides a schematic view of AMR transfer dynamics in open and closed aquaculture systems.

6.9 Increased Direct AMR Transfer to the Environment via Closed Systems

Unlike open systems, closed aquaculture systems create a barrier between the growing area and the environment, giving producers more control over elements such as oxygen, temperature, and photoperiods. These systems range from complete near-zero-discharge recirculating systems to simpler flow-through water systems. Due to their ability to manage processes and recycle water, closed systems are often considered a sustainable approach for intensive aquaculture, as they significantly reduce the amount of waste, antimicrobials, and chemical treatments released into the environment. Wastewater generated by closed flow-through systems typically contains high concentrations of phosphorus, microbial loads, and suspended solids. After undergoing several treatment steps, this wastewater can either be directed to the municipal wastewater system, flow into constructed wetlands, or be processed to create sludge that can be used as fertilizer on land (Kumar et al. 2025; Martínez 2008).

The use of aquaculture sludge notably influences the concentration and dissemination of the genes associated with AMR in both the soil systems and food crops (Watts et al. 2017). Recirculating aquaculture systems (RASs), designed to operate with minimal environmental impact, allow the production of species at high densities. This is achieved by efficiently managing, collecting, and treating waste throughout the fish growth process in both freshwater and marine environments.

Under ideal circumstances, the only reason these systems need water replacement is to make up for evaporation losses. Both biological and mechanical filtration processes are used by RASs to create an environment that is effective, productive, physiologically, and biologically safe (Tal et al. 2009). The activities of microbial biofilms that form on tank surfaces, integrated filter media, and the fish themselves are influenced by nutrient availability and biological processes. Under optimal conditions, fish waste—comprising nitrogen from expelled ammonia as well as carbon and nitrogen from uneaten feed and fecal matter—is removed through various interacting physiological processes, including methanogenesis, nitrification, denitrification, and anaerobic ammonium oxidation (anammox) (Schreier et al. 2010).

Since system water is recycled and, ideally, only minimal (if any) water exchange occurs with the external environment, antibiotics added to the feed can accumulate within the system. This accumulation may promote the growth of antimicrobial-resistant (AMR) bacteria, which can thrive in the host, sediment (waste solids), and the biofilter of the RAS. After the medication is discontinued, antibiotic residues would gradually decline to almost undetectable levels (Watts et al. 2017). *TetO*, *qnrA*, and *tetE* are among the antibiotic resistance genes detected in biofilms from RAS mixed bed biofilters. However, biofilms typically withstand antibiotic penetration, making pathogen treatment challenging (Jefferson 2004).

6.10 Impacts of AMR in Aquaculture

As discussed in previous sections, the structure of aquaculture systems. whether open or closed, can greatly influence the dissemination and persistence of AMR. However, beyond the environmental implications, AMR presents significant challenges affecting the environment, human health, animal health, and the aquaculture industry itself. The following subsections highlights the major impacts of AMR in aquaculture and associated sectors.

6.10.1 Limited Treatment Options

The emergence of resistant bacteria significantly compromises the efficacy of the limited antibiotics approved for aquaculture use. As treatment options become increasingly limited, infections become harder to control, leading to prolonged disease outbreaks, higher mortality rates, and significant financial losses for aquafarmers due to lack of effective treatment options (Hanson 2020).

6.10.2 Transmission to Humans

Aquatic environments serve as potential reservoirs and transmission route for resistance bacteria and resistance genes to humans through various pathways, including direct contact with contaminated water or seafood, consumption of contaminated or undercooked seafood, and occupational exposure in aquaculture facilities. Infections caused by resistant pathogens are harder to treat, potentially resulting in increased morbidity, mortality, and medical expenses (Hanson 2020).

6.10.3 Environmental Contamination

During aquaculture practices, antibiotics and antibiotic-resistant bacteria may get into the water, polluting the environment. These residues stay in the sediment, water, and sea life, where they could harm species not meant to be affected and disturb the ecosystem. Antibiotic resistance in natural environments might grow worse when antibiotic-resistant bacteria from aquaculture discharge reach neighbouring systems (Pepi and Focardi 2021).

6.10.4 Risk to Food Safety

Presence of antibiotic-resistant bacteria or antibiotic remnants in aquaculture products poses an important threat to both food and public health safety. Antibiotic residues in seafood are dangerous for your health and might increase your risk of allergies and illnesses that do not respond to antibiotics (Ryu et al. 2012).

6.10.5 Economic Impacts

The economic impacts on aquaculture and related sectors including seafood processing and exporting could be quite large because of AMR. Financial losses can happen to aquaculture producers and exporters if productivity falls, deaths among stocks rise, and trade limitations appear because of antibiotic concerns. In addition, the industry may have to bear higher costs for alternative disease control measures and compliance with new regulations aimed at fighting AMR (Cabello 2006; Henriksson et al. 2018; Watts et al. 2017).

6.10.6 Regulatory and Policy Challenges

Strong and consistent regulation, along with sectoral policies, is needed to prevent AMR in aquaculture. However, effective AMR management in aquaculture can be restricted by regulatory and legislative problems, such as inadequate surveillance systems, weak enforcement, and non-uniform standards (Gentry et al. 2017; Laxminarayan et al. 2016, 2020; Troell et al. 2019).

Solving the issue of AMR in aquaculture and improving sustainability will require a comprehensive strategy that includes all major stakeholders, including aquaculture professionals, government bodies, academic institutions, healthcare providers, and consumers.

6.11 Implementing a One Health Approach to Address the Issue of Antimicrobial Resistance (AMR)

Dealing with AMR in aquaculture needs a One Health method, as aquatic environments, the health of animals, and human health are all related (Thakur and Gray 2019). The "one health" concept places great importance on stopping the spread of AMR-related infections. The approach links many industries so they can understand AMR and work together to set guidelines and reduce associated risks (Dafale et al. 2020).

The following elements should be considered when implementing AMR control using a one healthapproach:

I. Creating and enforcing rules for responsible usage of antibiotics for aquatic animal treatment.
II. Educating the public on how AMR moves through soil, water, and air, making these areas the storage points for resistant bacteria and genes, and also pathways for transfer of AMR from animals to people.
III. Implementing modern technologies in proven methods for measuring AMR.
IV. Implementation of effective communication strategies to educate consumers about the risks of AMR transmission from the food industry and other non-human sources.

6.12 AMR and Zoonoses

One of the most alarming concerns associated with AMR in aquaculture is its link to zoonotic diseases. In addition to prophylactic measures such as maintenance of optimum culture conditions, probiotics, and immunostimulants, antibiotics are often used in farm/hatchery to prevent staggering production losses. The negligent use of antibiotics in aquaculture, frequently without following advised withdrawal times, increases the strain on gut flora to develop resistance, which is why antibiotic-resistant bacteria and antibiotic residues are found in fisheries products (Sudan et al. 2023). Through processes like quorum sensing, vectors, and horizontal gene transfer (HGT), antibiotic-resistant bacteria (ARBs) can spread resistance genes to nearby flora and possibly to people. The gut serves as a bioreactor for their growth. By getting into the food chain, its unmetabolized forms also affect ecological niches (Dafale et al. 2015). *A. hydrophilia*, *V. vulnificus*, *Mycobacterium marinum*, *Photobacterium damselae*, and *Streptococcus iniae* are common infections in cultured fish (Kaur et al. 2024; Rivas-Aravena et al. 2013). The zoonotic pathogens carry ARGs and Extended-spectrum beta-lactamases (ESBL) that propagate throughout the food chain (Kumar and Pal 2018).

AMR epidemiology in terrestrial livestock farming, aquaculture, and human activities is schematically shown in Fig. 6.6. The massive spread of these pathogens and the factors that contribute to their resistance highlight the urgent need for integrated zoonotic surveillance and risk management strategies.

6.13 AMR Control and Mitigation Strategies in Aquaculture

Effective control and mitigation techniques must be put in place immediately due to the significant consequences of AMR in aquaculture, including its involvement in zoonotic transmission, environmental contamination, and concerns to food safety. Later sections focus on cutting down AMR use and finding lasting, preventive solutions, following earlier sections that mainly addressed the widespread misuse of antibiotics as a main AMR driver.

Promoting optimal management practices, boosting immunity with immunizations and immunostimulants, and investigating probiotics, prebiotics, herbal medicines, and bacteriophage therapy are all important components of an integrated strategy for effective AMR mitigation. When combined, these measures can lessen the dependency on antibiotics, promote healthier aquatic life, and stop the spread of infections that are resistant to them.

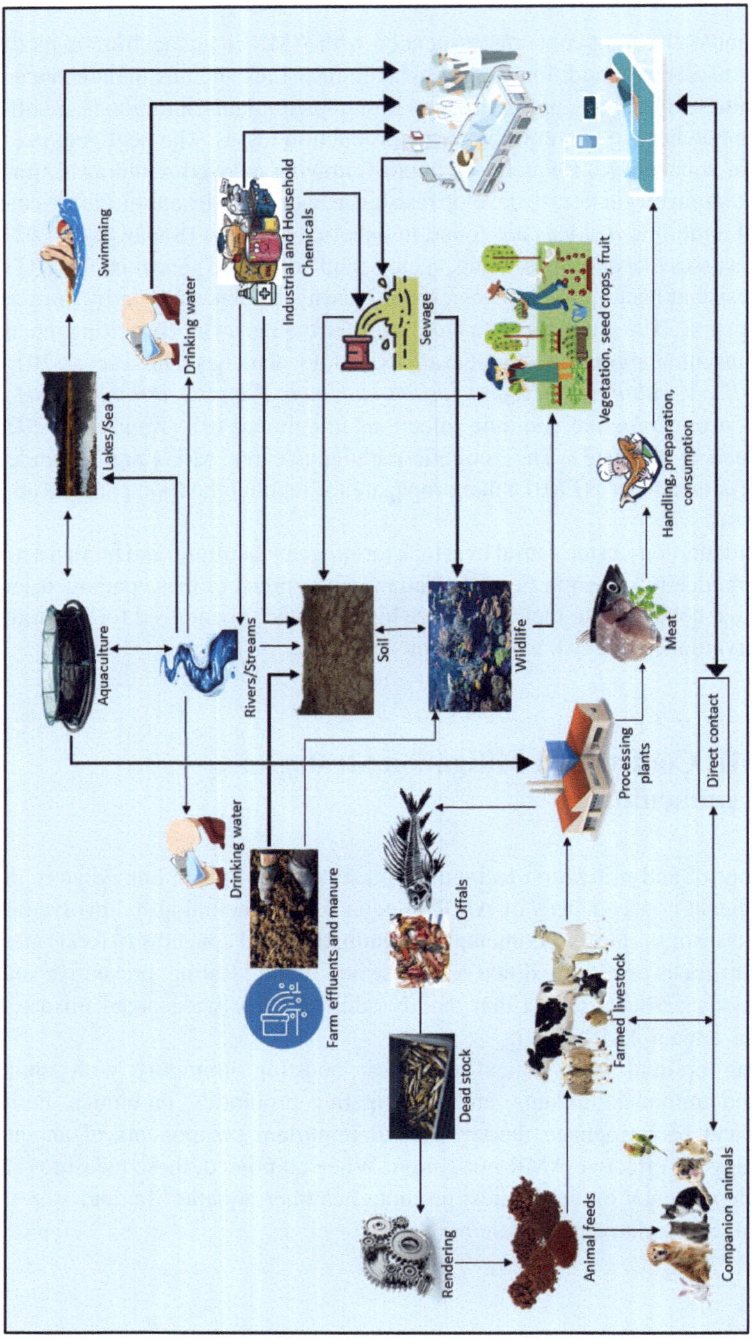

Fig. 6.6 Epidemiology of AMR. *Source* Author

6.13.1 Biosecurity and Best Management Practices (BMPs)

Maintaining ideal culture conditions, regularly applying disinfectants, enforcing stringent quarantine protocols, using disease-free certified seed, managing optimal stocking density, making sure that hands are clean, and disinfecting equipment before and after use as well as in between uses are all important steps that are included in biosecurity measures and farm-level best management practices. It also entails using feed that has been properly stored and disposing off of carcasses in an appropriate manner (Assefa and Abunna 2018). These activities help increase the animals' resilience to disease by reducing stress (Scarfe and Palić 2020).

6.13.2 Vaccine

Vaccination is mostly done to help animals build resistance to certain diseases or disease groups. (Sommerset et al. 2005) reported that most animals receive vaccines while they are well and before illness starts. (Mondal and Thomas 2022) explain that researchers have come up with a lot of fish vaccine types for disease prevention, including DNA vaccines, synthetic peptide vaccines, attenuated vaccines, recombinant vaccines, inactivated or killed vaccines, and mucosal vaccines. Aquaculture professionals have almost 26 vaccines they can use to protect fish from disease (Rai et al. 2023). Immunostimulants lower aquaculture's dependency on antibiotics, which helps protect the environment and keeps antibiotic resistance lower. Administration of vaccines to fish reduces the chance of diseases spreading to wild animals, which helps keep biological diversity safe.

6.13.3 Immunostimulants

Immunostimulants are added to aquatic environments for better disease protection, and these act as boosters that may be manufactured or taken from natural organisms. The immunostimulants used against diseases are chitin, chitosan, peptidoglycan, oligosaccharides, levamisole, lentinan, muramyl dipeptide, schizophyllan, glucan, and blends of vitamins, plus more plant and animal substances. Fish farmers provided with these immunostimulants can more successfully manage viral *hemorrhagic sepicemia, hematopoietic necrosis, yellow head virus, and A. hydrophila, A. salmonicida, V. salmonicida, V. anguillarum, V. vulnificus, Streptococcus spp., and Yersinia ruckeri.* (Mohan et al. 2019).

These immunostimulants prevent diseases in fish, limit the spread of diseases, and minimise the number of deaths in fish hence promoting sustainable aquaculture. These immunostimulants also support sustainable production and contribute to a healthy ecosystem as well. Immunostimulant-treated fish are more resistant to illness due to stress in crowded tanks or water quality heaped up on the fish.

6.13.4 Probiotics

According to (Gatesoupe 1999), probiotics are live microbial supplements that confer health benefits. Aquatic organisms rely heavily on their external environment due to the constant flow of water through their digestive systems. Their gut microbiota is mainly composed of temporary bacteria, with new microorganisms entering from food and water continuously. Thus, probiotics are increasingly being used in aquaculture to support gut health, enhance immune responses, and outcompete pathogenic bacteria. A variety of probiotics are utilized in aquaculture, including *Lactobacillus, Bifidobacterium, Pediococcus, Streptococcus, Carnobacterium, Bacillus, Flavobacterium, Cytophaga, Pseudomonas, Alteromonas, Enterococcus, Nitrosomonas, Nitrobacter,* as well as yeast species such as *Saccharomyces* and *Debaryomyces*. These probiotics work by stimulating the immune system, outcompeting harmful pathogens for nutrients and adhesion sites in an ecologically safe and sustainable manner, as well as enhancing appetite and growth performance. Together, these benefits contribute to improved fish growth and feed efficiency. Once given to fish, they proliferate in the gut, supporting the normal microflora and helping to maintain microbial balance within the host. They combat the pathogen through various mechanisms, including strengthening the epithelial barrier, improving gut mucosal adhesion, inhibiting pathogen adhesion, and competitively excluding harmful microorganisms (Mastan 2015; Vadassery and Pillai 2020).

6.13.5 Bacteriophage

Bacteriophages are viruses that target bacteria directly. They are a promising alternative to antibiotics, especially when it comes to infections that are resistant to many drugs. Phage therapy has focused on a number of harmful bacteria in the aquaculture industry, such as *Vibrio* species, *Aeromonas* species, *Flavobacterium* species, *Pseudomonas* species, and *Edwardsiella* species. These efforts have included, among other things, formulating phage cocktails, isolating and characterizing pathogenic phages, and taking dosage and administration technique into account (Donati et al. 2021).

6.14 Surveillance of AMR in Aquaculture and Fisheries

Educating farmers about rules on antibiotics use and promoting their responsible use in farming and aquaculture is very important. When antibiotics are used incorrectly as growth promoters in aquaculture, antibiotic-resistant bacteria can quickly emerge and become a serious problem, therefore, strict restrictions and regulations are necessary. Tracking antibiotic use and resistance is important for successful management of antimicrobial resistance. The system is open for monitoring and reporting on antimicrobial resistance and use in animal and fishery industries (Schar et al. 2021). India established INFAAR with an aim of helping the country monitor AMR in various animal production industries. The program's goal is to find out how resistant bacteria and ARGs are spread, how resistance grows, and where those bacteria are likely to be found. These results will help experts create strategies to stop AMR in fish and farm animals.

6.14.1 Objectives of INFAAR Include

6.14.1.1 Monitoring of AMR

To know how much antimicrobial resistance exists in India, INFAAR will observe AMR in target microorganisms isolated from fish, shellfish, and healthy farmed animals.

6.14.1.2 To Encourage Farmers and Fish Farmers to Use Antimicrobials Only When Necessary

Helping farmers, veterinarians, fish health experts, and legislators understand antimicrobial resistance is an important goal. Specific education, training, and communication programs with these groups can help reach this goal (Deve et al. 2020).

6.15 Conclusion

Enhanced monitoring and a rise in alternative solutions like vaccines, probiotics, and biosecurity have not stopped the growing problem of antimicrobial use promoting the emergence and spread of resistant bacteria. Earlier sections showed that pollution with AMR negatively impacts controlling aquatic diseases and also distributes resistance genes into the environment and, primarily through seafood consumption and contact with polluted waters, into humans. This cycle is a major public health threat and puts the future sustainability of aquaculture at risk.

However, it is encouraging to see that more and more industry players, regulators, and legislators are becoming aware of the problem. It is becoming clear that antibiotic overuse in aquaculture must be curtailed and that One Health integrated management approaches should be adopted. Reducing the risks from AMR and making aquaculture more sustainable will require group action, good laws, strong research, and education.

6.16 Future Prospects

The ongoing threat of antimicrobial resistance (AMR) to human health and the environment is now making sustainable aquaculture much more difficult. Because seafood consumption is rising worldwide, aquaculture now relies more on antibiotics to protect and heal farmed fish. The misuse of antimicrobial drugs is a major reason resistant diseases can spread from fish to humans and the environment. Aquaculture's pollution problems connected to AMR include releasing untreated or poorly treated effluents that carry antibiotic leftovers, resistant bacteria, and genetic material. These pollutants can move through aquatic environments, changing biodiversity and possibly helping resistance genes move to other places in the environment. Resistant bacteria from infected wild fish can move to humans if we consumption, come near the water, or are in the same environment. We can only promote sustainable aquaculture when we regulate AMR pollution. Better management techniques, strict application of biosecurity, new approaches to disease control through vaccination, probiotics, and plant-based additives should be used to reduce antibiotic utilization in aquaculture. It is also necessary to strengthen regulations on antimicrobial use and better understand how these practices are implemented and monitored. AMR can be reduced in water using the contemporary wastewater treatment concepts. Unless countries work together, crucial AMR in aquaculture could jeopardize the industry and worsen food and health safety. When we feed more people, we have to embrace aquaculture that takes care of ecosystems and keep a tight rein on antimicrobial use.

References

Agersø Y, Guardabassi L (2005) Identification of Tet 39, a novel class of tetracycline resistance determinant in Acinetobacter spp. of environmental and clinical origin. J Antimicrobial Chemotherapy 55(4). https://doi.org/10.1093/jac/dki051

Alderman DJ, Hastings TS (1998) Antibiotic use in aquaculture: development of antibiotic resistance-potential for consumer health risks. Internat J Food Sci Tech 33(2). https://doi.org/10.1046/j.1365-2621.1998.3320139.x

Assefa A, Abunna F (2018) Maintenance of fish health in aquaculture: review of epidemiological approaches for prevention and control of infectious disease of fish. VetY Med Int. https://doi.org/10.1155/2018/5432497

Babakhani S, Oloomi M (2018) Transposons: the agents of antibiotic resistance in bacteria. J Basic Microbiol 58(11). https://doi.org/10.1002/jobm.201800204

Bennett PM (2008) Plasmid encoded antibiotic resistance: acquisition and transfer of antibiotic resistance genes in bacteria. Br J Pharmacol 153 (SUPPL. 1). https://doi.org/10.1038/sj.bjp.0707607

Burridge L, Weis JS, Cabello F, Pizarro J, Bostick K (2010) Chemical use in salmon aquaculture: a review of current practices and possible environmental effects. Aquaculture 306(1–4):7–23. https://doi.org/10.1016/j.aquaculture.2010.05.020

Buschmann AH, Tomova A, López A, Maldonado MA, Henríquez LA, Ivanova L, Moy F, Godfrey HP, Cabello FC (2012) Salmon aquaculture and antimicrobial resistance in the marine environment. PLoS ONE 7(8). https://doi.org/10.1371/journal.pone.0042724

Cabello FC (2006) Heavy use of prophylactic antibiotics in aquaculture: a growing problem for human and animal health and for the environment. Environ Microbiol 8(7):1137–1144. https://doi.org/10.1111/j.1462-2920.2006.01054.x

Cabello FC, Godfrey HP, Tomova A, Ivanova L, Dölz H, Millanao A, Buschmann AH (2013) Antimicrobial use in aquaculture re-examined: its relevance to antimicrobial resistance and to animal and human health. Environ Microbiol 15(7). https://doi.org/10.1111/1462-2920.12134

Cabello FC, Godfrey HP, Buschmann AH, Dölz HJ (2016) Aquaculture as yet another environmental gateway to the development and globalisation of antimicrobial resistance. Lancet Infect Dis 16(7). https://doi.org/10.1016/S1473-3099(16)00100-6

Cabello FC, Millanao AR, Lozano-Muñoz I, Godfrey HP (2023) Misunderstandings and misinterpretations: Antimicrobial use and resistance in salmon aquaculture. Environ Microbiol Rep 15(4):245–253. https://doi.org/10.1111/1758-2229.13147

Collins JA, Osheroff N (2024) Gyrase and topoisomerase IV: recycling old targets for new antibacterials to combat fluoroquinolone resistance. ACS Infect Dis 10(4):1097–1115. https://doi.org/10.1021/acsinfecdis.4c00128

Cox G, Wright GD (2013) Intrinsic antibiotic resistance: Mechanisms, origins, challenges and solutions. In Int J Med Microbiol 303(6–7). https://doi.org/10.1016/j.ijmm.2013.02.009

Cravedi JP, Choubert G, Delous G (1987) Digestibility of chloramphenicol, oxolinic acid and oxytetracycline in rainbow trout and influence of these antibiotics on lipid digestibility. Aquaculture 60(2). https://doi.org/10.1016/0044-8486(87)90305-X

Dafale NA, Semwal UP, Agarwal PK, Sharma P, Singh GN (2015) Development and validation of microbial bioassay for quantification of Levofloxacin in pharmaceutical preparations. J Pharm Anal 5(1):18–26. https://doi.org/10.1016/j.jpha.2014.07.007

Dafale NA, Srivastava S, Purohit HJ (2020) Zoonosis: an emerging link to antibiotic resistance under "one health approach." Indian J Microbiol 60(2):139–152. https://doi.org/10.1007/s12088-020-00860-z

Dang H, Zhao J, Song L, Chen M, Chang Y (2009) Molecular characterizations of chloramphenicol- and oxytetracycline-resistant bacteria and resistance genes in mariculture waters of China. Mar Pollut Bull 58(7). https://doi.org/10.1016/j.marpolbul.2009.02.016

Dassanayake MK, Khoo TJ, An J (2021) Antibiotic resistance modifying ability of phytoextracts in anthrax biological agent *Bacillus anthracis* and emerging superbugs: a review of synergistic mechanisms. Ann Clin Microbiol Antimicrob 20(1). https://doi.org/10.1186/s12941-021-00485-0

Defoirdt T, Sorgeloos P, Bossier P (2011) Alternatives to antibiotics for the control of bacterial disease in aquaculture. Curr Opin Microbiol 14(3). https://doi.org/10.1016/j.mib.2011.03.004

Deve S, Goal SD, Health G (2020) NEWS INFAAR–a research platform for accelerating laboratory-based surveillance of antimicrobial resistance in fisheries and aquaculture in India 119(12)

Dixon BA, Yamashita J, Evelyn F (1990) Antibiotic resistance of aeromonas spp. isolated from tropical fish imported from singapore. J Aquat Anim Health 2(4). https://doi.org/10.1577/1548-8667(1990)002<0295:AROASI>2.3.CO;2

Donati VL, Dalsgaard I, Sundell K, Castillo D, Er-Rafik M, Clark J, Wiklund T, Middelboe M, Madsen L (2021) Phage-mediated control of flavobacterium psychrophilum in aquaculture: in vivo experiments to compare delivery methods. Front Microbiol 12(March):1–16. https://doi.org/10.3389/fmicb.2021.628309

Done HY, Venkatesan AK, Halden RU (2015) Does the recent growth of aquaculture create antibiotic resistance threats different from those associated with land animal production in agriculture? AAPS J 17(3). https://doi.org/10.1208/s12248-015-9722-z

Fernández L, Hancock REW (2012) Adaptive and mutational resistance: role of porins and efflux pumps in drug resistance. Clin Microbiol Rev 25(4). https://doi.org/10.1128/CMR.00043-12

Ferraz LMLS (2019) Antimicrobial residues in aquaculture species for human consumption: analytical determination and assessment of related public health hazards Doctoral dissertation, Universidade de Coimbra Portugal

Ferri G, Olivieri V, Olivastri A, Pennisi L, Vergara A (2024) Multidrug resistant Vibrio spp. identified from mussels farmed for human consumption in Central Italy. J Appl Microbiol 135(4). https://doi.org/10.1093/jambio/lxae098

Fuhrman JA (1999) Marine viruses and their biogeochemical and ecological effects. Nature 399(6736). https://doi.org/10.1038/21119

Gao P, Mao D, Luo Y, Wang L, Xu B, Xu L (2012) Occurrence of sulfonamide and tetracycline-resistant bacteria and resistance genes in aquaculture environment. Water Res 46(7):2355–2364. https://doi.org/10.1016/j.watres.2012.02.004

Gatesoupe FJ (1999) The use of probiotics in aquaculture. Aquac 180(1–2). https://doi.org/10.1016/S0044-8486(99)00187-8

Gauba A, Rahman KM (2023) Evaluation of antibiotic resistance mechanisms in gram-negative bacteria. Antibiotics 12(11). https://doi.org/10.3390/antibiotics12111590

Gentry RR, Froehlich HE, Grimm D, Kareiva P, Parke M, Rust M, Gaines SD, Halpern BS (2017) Mapping the global potential for marine aquaculture. Nat Ecol Evol 1(9). https://doi.org/10.1038/s41559-017-0257-9

Gin KYH, Ng C, Li W, Goh SG, Tong X, Jong MC (2023) Emerging microbial contaminants in the ocean. Ocean Hum Health 315–350

Goldburg R, Naylor R (2005) Future seascapes, fishing, and fish farming. Front Ecol Environ 3(1) SPEC. ISS. https://doi.org/10.2307/3868441

Gopikrishnan M, Haryini S, George Priya Doss C (2024) Emerging strategies and therapeutic innovations for combating drug resistance in Staphylococcus aureus strains: a comprehensive review. J Basic Microbiol 64(5). https://doi.org/10.1002/jobm.202300579

Gullberg E, Cao S, Berg OG, Ilbäck C, Sandegren L, Hughes D, Andersson DI (2011) Selection of resistant bacteria at very low antibiotic concentrations. PLoS Pathog 7(7):1–9. https://doi.org/10.1371/journal.ppat.1002158

Hanson LA (2020) Practical management of bacterial diseases in finfish aquaculture to minimise antimicrobial resistance. Asian Fish Sci 33(S1). https://doi.org/10.33997/j.afs.2020.33.S1.009

Harmer CJ, Hall RM (2015) The A to Z of A/C plasmids. Plasmid 80. https://doi.org/10.1016/j.plasmid.2015.04.003

Henriksson PJG, Rico A, Troell M, Klinger DH, Buschmann AH, Saksida S, Chadag MV, Zhang W (2018) Unpacking factors influencing antimicrobial use in global aquaculture and their implication for management: a review from a systems perspective. Sustainability Sci 13(4). https://doi.org/10.1007/s11625-017-0511-8

Heuer OE, Kruse H, Grave K, Collignon P, Karunasagar I, Angulo FJ (2009) Human health consequences of use of antimicrobial agents in aquaculture. Clin Infect Dis 49(8). https://doi.org/10.1086/605667

Hofer U (2019) The cost of antimicrobial resistance. Nat Rev Microbiol 17(1). https://doi.org/10.1038/s41579-018-0125-x

Holmes AH, Moore LSP, Sundsfjord A, Steinbakk M, Regmi S, Karkey A, Guerin PJ, Piddock LJV (2016) Understanding the mechanisms and drivers of antimicrobial resistance. Lancet 387(10014). https://doi.org/10.1016/S0140-6736(15)00473-0

Hooper DC, Jacoby GA (2015) Mechanisms of drug resistance: Quinolone resistance. Ann N Y Acad Sci 1354(1). https://doi.org/10.1111/nyas.12830

Hoskisson PA (2008) Streptomyces in nature and medicine: the antibiotic makers. Microbe Magazine 3(1). https://doi.org/10.1128/microbe.3.151.1

Ibrahim S, Al-Saryi N, Al-Kadmy IMS, Aziz SN (2021) Multidrug-resistant Acinetobacter baumannii as an emerging concern in hospitals. Mol Biol Rep 48(10):6987–6998. https://doi.org/10.1007/s11033-021-06690-6

Impey RE, Hawkins DA, Sutton JM, Soares da Costa TP (2020) Overcoming intrinsic and acquired resistance mechanisms associated with the cell wall of gram-negative bacteria. Antibiotics 9(9). https://doi.org/10.3390/antibiotics9090623

Jefferson KK (2004) What drives bacteria to produce a biofilm? FEMS Microbiol Lett 236(2):163–173. https://doi.org/10.1016/j.femsle.2004.06.005

Kaur S, Kaur H, Kaur B, Kumar BN, Tyagi A, Singh P, Tanuj Dubey S, Munang'andu HM (2024) Isolating pathogenic multidrug-resistant Aeromonas hydrophila from diseased fish and assessing the effectiveness of a novel lytic Aeromonas veronii bacteriophage (AVP3) for biocontrol. Microb Pathog 196:106914. https://doi.org/10.1016/j.micpath.2024.106914

Kumar R, Gupta R, Dahiya T, Yadav R, Saifi R, Yodha K, Kumari B, Singh P, Shefali (2022a) Surveillance of Heteropneustes fossilis (Stinging Catfish) diseases and health management practices in Haryana. Appl Ecol Environ Sci 10(4):0261–267. https://doi.org/10.12691/aees-10-4-12

Kumar R, Gupta RK, Yadav R, Saifi R, Yodha K, Pooja (2022b) Eisenia fetida as protein source for growth enhancement of Heteropneustes fossilis. Egypt J Aquat Biol Fish 26(2):577–588. https://doi.org/10.21608/ejabf.2022.234553

Kumar R, Yadav R, Kumar Gupta R Pooja (2022c) A sustainable way for fish health management by replacement of chemical and drugs by earthworm. Environ Degrad Asia: Land Degrad, Environ Contam, Hum Act 329–352. https://doi.org/10.1007/978-3-031-12112-8_16

Kumar R, Singh D, Yadav R (2025) Toxic effects of chlorantraniliprole and fipronil on *Eisenia fetida*: from growth and reproductive parameters to avoidance behavior. Environ Eng Manag J 24(7):1499–1508. https://doi.org/10.30638/eemj.2025.117

Kumar A, Pal D (2018) Antibiotic resistance and wastewater: correlation, impact and critical human health challenges. J Environ Chem Eng 6(1):52–58. https://doi.org/10.1016/j.jece.2017.11.059

Kumar R, Singh D (2022) Insecticide- tolerant Bacterial population in *Eisenia fetida*'s Gut and Vermicast exposed to Chlorantraniliprole and Fipronil. Appl Biol Res 24(3):273–279

Kumawat M, Nabi B, Daswani M, Viquar I, Pal N, Sharma P, Tiwari S, Sarma DK, Shubham S, Kumar M (2023) Role of bacterial efflux pump proteins in antibiotic resistance across microbial species. Microb Pathog 181. https://doi.org/10.1016/j.micpath.2023.106182

L'abée-Lund TM, Sørum H (2001) Class 1 integrons mediate antibiotic resistance in the fish pathogen *Aeromonas salmonicida* worldwide. Microb Drug Resist 7(3). https://doi.org/10.1089/10766290152652819

Laxminarayan R, Sridhar D, Blaser M, Wang M, Woolhouse M (2016) Achieving global targets for antimicrobial resistance. Sci 353(6302). https://doi.org/10.1126/science.aaf9286

Laxminarayan R, Van Boeckel T, Frost I, Kariuki S, Khan EA, Limmathurotsakul D, Larsson DGJ, Levy-Hara G, Mendelson M, Outterson K, Peacock SJ, Zhu YG (2020) The lancet infectious diseases commission on antimicrobial resistance: 6 years later. Lancet Infect Dis 20(4). https://doi.org/10.1016/S1473-3099(20)30003-7

Lee JH (2019) Perspectives towards antibiotic resistance: from molecules to population. J Microbiol 57(3). https://doi.org/10.1007/s12275-019-0718-8

Li X, Zhao D, Li W, Sun J, Zhang X (2022) Enzyme inhibitors: The best strategy to tackle superbug ndm-1 and its variants. Int J Mol Sci 23(1). https://doi.org/10.3390/ijms23010197

Maldonado MA, Buschmann AH, Tomova A, Lo A, Ivanova L, Moy F, Godfrey HP, Cabello FC, Henrı LA (2012) Salmon Aquaculture and Antimicrobial Resistance in the Marine Environment. 7(8):26–28. https://doi.org/10.1371/journal.pone.0042724

Martínez JL (2008) Antibiotics and antibiotic resistance genes in natural environments. Sci 321(5887):365–367. https://doi.org/10.1126/science.1159483

Martinez JL (2014) General principles of antibiotic resistance in bacteria. Drug Discov Today: Technol 11(1). https://doi.org/10.1016/j.ddtec.2014.02.001

Mastan SA (2015) Use of Immunostimulants in aquaculture disease management. Int J Fish Aquat Stud (4): 277–280 Use of Immunostimulants in aquaculture disease management. IJFAS 2(4):277–280. www.fisheriesjournal.com

Mendes RE, Deshpande LM, Castanheira M, DiPersio J, Saubolle MA, Jones RN (2008) First report of *cfr*-mediated resistance to linezolid in human staphylococcal clinical isolates recovered in the United States. Antimicrob Agents Chemother 52(6). https://doi.org/10.1128/AAC.00231-08

Michaelis C, Grohmann E (2023) Horizontal gene transfer of antibiotic resistance genes in biofilms. Antibiotics 12(2). https://doi.org/10.3390/antibiotics12020328

Milijasevic M, Veskovic-Moracanin S, Babic Milijasevic J, Petrovic J, Nastasijevic I (2024) Antimicrobial resistance in aquaculture: risk mitigation within the One Health Context. Foods (Basel, Switzerland) 13(15):2448. https://doi.org/10.3390/foods13152448

Miranda CD, Zemelman R (2001) Antibiotic resistant bacteria in fish from the Concepción Bay, Chile. Mar Pollut Bull 42(11). https://doi.org/10.1016/S0025-326X(01)00093-5

Miranda CD, Godoy FA, Lee MR (2018) Current status of the use of antibiotics and the antimicrobial resistance in the chilean salmon farms. Front Microbiol 9:1284. https://doi.org/10.3389/fmicb.2018.01284

Mohan K, Ravichandran S, Muralisankar T, Uthayakumar V, Chandirasekar R, Seedevi P, Abirami RG, Rajan DK (2019) Application of marine-derived polysaccharides as immunostimulants in aquaculture: a review of current knowledge and further perspectives. Fish Shellfish Immunol 86:1177–1193. https://doi.org/10.1016/j.fsi.2018.12.072

Mondal H, Thomas J (2022) A review on the recent advances and application of vaccines against fish pathogens in aquaculture. Aquacult Int 30(4):1971–2000. https://doi.org/10.1007/s10499-022-00884-w

Munita JM, Arias CA (2016) HHS public access mechanisms of antibiotic resistance. Microbiol Spectr 4(2)

Naylor R, Burke M (2005) Aquaculture and ocean resources: raising tigers of the sea. Annu Rev Environ Resour 30. https://doi.org/10.1146/annurev.energy.30.081804.121034

Nguyen TTT, Nguyen HT, Tsai MA, Byadgi O, Wang PC, Yoshida T, Chen SC (2017) Genetic diversity, virulence genes, and antimicrobial resistance of streptococcus dysgalactiae isolates from different aquatic animal sources. Aquaculture 479. https://doi.org/10.1016/j.aquaculture.2017.06.002

Jim O'Niell (2016) Tackling drug-resistant infections globally: final report and recommendations. The review on antimicrobial resistance Chaired by Jim O'neill

OECD (2018) Stemming the superbug tide: just a few dollars more, OECD health policy studies, OECD publishing, Paris, https://doi.org/10.1787/9789264307599-en

On THER, Resistance A (2015) The review on antimicrobial resistance

Partridge SR, Tsafnat G, Coiera E, Iredell JR (2009) Gene cassettes and cassette arrays in mobile resistance integrons: Review article. FEMS Microbiol Rev 33(4). https://doi.org/10.1111/j.1574-6976.2009.00175.x

Pepi M, Focardi S (2021) Antibiotic-resistant bacteria in aquaculture and climate change: a challenge for health in the mediterranean area. Int J Environ Res Public Health 18(11). https://doi.org/10.3390/ijerph18115723

Petersen A, Andersen JS, Kaewmak T, Somsiri T, Dalsgaard A (2002) Impact of integrated fish farming on antimicrobial resistance in a pond environment. Appl Environ Microbiol 68(12). https://doi.org/10.1128/AEM.68.12.6036-6042.2002

Poirel L, Cattoir V, Nordmann P (2012) Plasmid-mediated quinolone resistance; interactions between human, animal, and environmental ecologies. Front Microbiol 3(FEB). https://doi.org/10.3389/fmicb.2012.00024

Preena PG, Swaminathan TR, Kumar VJR, Singh ISB (2020) Antimicrobial resistance in aquaculture: a crisis for concern. Biologia 75:9. https://doi.org/10.2478/s11756-020-00456-4

Rahube TO, Yost CK (2010) Antibiotic resistance plasmids in wastewater treatment plants and their possible dissemination into the environment. Afr J Biotechnol 9(54)

Rai S, Kaur B, Singh P, Singh A, Benjakul S, Vijay Kumar Reddy S, Nagar V, Tyagi A (2023) Perspectives on phage therapy for health management in aquaculture. Aquac Internat (0123456789). https://doi.org/10.1007/s10499-023-01220-6

Rigos G, Nengas I, Alexis M, Troisi GM (2004) Potential drug (oxytetracycline and oxolinic acid) pollution from Mediterranean sparid fish farms. Aquatic Toxicology 69(3). https://doi.org/10.1016/j.aquatox.2004.05.009

Rivas-Aravena A, Sandino AM, Spencer E (2013) Nanoparticles and microparticles of polymers and polysaccharides to administer fish vaccines. Biol Res 46(4):407–419. https://doi.org/10.4067/S0716-97602013000400012

Ryu SH, Park SG, Choi SM, Hwang YO, Ham HJ, Kim SU, Lee YK, Kim MS, Park GY, Kim KS, Chae YZ (2012) Antimicrobial resistance and resistance genes in Escherichia coli strains isolated from commercial fish and seafood. Int J Food Microbiol 152(1–2). https://doi.org/10.1016/j.ijfoodmicro.2011.10.003

SalimiyanRizi K, Ghazvini K, Noghondar M kouhi (2018) Adaptive antibiotic resistance: overview and perspectives. J Infect Dis Ther 06(03). https://doi.org/10.4172/2332-0877.1000363

Scarfe AD, Palić D (2020) Aquaculture biosecurity: practical approach to prevent, control, and eradicate diseases. Aquac Health Manag: Des Oper Approaches. https://doi.org/10.1016/B978-0-12-813359-0.00003-8

Schar D, Klein EY, Laxminarayan R, Gilbert M, Van Boeckel TP (2020) Global trends in antimicrobial use in aquaculture. Scientific Reports 10(1). https://doi.org/10.1038/s41598-020-78849-3

Schar D, Zhao C, Boeckel TP, Van Larsson DGJ, Gilbert M (2021) From aquaculture and fisheries in Asia. Nat Commun 6–15. https://doi.org/10.1038/s41467-021-25655-8

Schmidt AS, Bruun MS, Dalsgaard I, Larsen JL (2001) Incidence, distribution, and spread of tetracycline resistance determinants and integron-associated antibiotic resistance genes among motile aeromonads from a fish farming environment. Appl Environ Microbiol 67(12). https://doi.org/10.1128/AEM.67.12.5675-5682.2001

Schreier HJ, Mirzoyan N, Saito K (2010) Microbial diversity of biological filters in recirculating aquaculture systems. Curr Opin Biotechnol 21(3):318–325. https://doi.org/10.1016/j.copbio.2010.03.011

Snieszko SF (1954) Therapy of bacterial fish diseases. Trans Am Fish Soc 83(1):313–330

Sommerset I, Krossøy B, Biering E, Frost P (2005) Vaccines for fish in aquaculture. Expert Rev Vaccines 4(1):89–101. https://doi.org/10.1586/14760584.4.1.89

Sørum H (2019) Antimicrobial drug resistance in fish pathogens. Antimicrob Resist Bact Anim Orig. https://doi.org/10.1128/9781555817534.ch13

Stalder T, Barraud O, Casellas M, Dagot C, Ploy MC (2012) Integron involvement in environmental spread of antibiotic resistance. Front Microbiol 3(APR). https://doi.org/10.3389/fmicb.2012.00119

Sudan P, Tyagi A, Dar RA, Sharma C, Singh P, Naveen NK, Chandra M, Arora AK (2023) Prevalence and antimicrobial resistance of food safety related Vibrio species in inland saline water shrimp culture farms. Int Microbiol 26(3):591–600. https://doi.org/10.1007/s10123-023-00323-7

Tal Y, Schreier HJ, Sowers KR, Stubblefield JD, Place AR, Zohar Y (2009) Environmentally sustainable land-based marine aquaculture. Aquaculture 286(1–2):28–35. https://doi.org/10.1016/j.aquaculture.2008.08.043

Tang KWK, Millar BC, Moore JE (2023) Antimicrobial Resistance (AMR). Br J Biomed Sci 80. https://doi.org/10.3389/bjbs.2023.11387

Tenover FC (2006) Mechanisms of antimicrobial resistance in bacteria. Am J Med 119(6 SUPPL. 1). https://doi.org/10.1016/j.amjmed.2006.03.011

Thakur S, Gray GC (2019) The mandate for a global "one health" approach to antimicrobial resistance surveillance. Am J Trop Med Hyg 100(2):227–228. https://doi.org/10.4269/ajtmh.18-0973

The State of World Fisheries and Aquaculture (2022) In: the state of world fisheries and aquaculture. https://doi.org/10.4060/cc0461en

The State of World Fisheries and Aquaculture 2022 (2024) In: the State of world fisheries and aquaculture. The State of World Fisheries and Aquaculture (fao.org)

Thornber K, Verner-Jeffreys D, Hinchliffe S, Rahman MM, Bass D, Tyler CR (2020) Evaluating antimicrobial resistance in the global shrimp industry. Rev Aquac 12(2). https://doi.org/10.1111/raq.12367

Troell M, Jonell M, Crona B (2019) The role of seafood for sustainable and healthy diets. The EAT-Lancet Commission report through a blue lens. Stockholm Resilience Centre

Van Boeckel TP, Brower C, Gilbert M, Grenfell BT, Levin SA, Robinson TP, Teillant A, Laxminarayan R (2015) Global trends in antimicrobial use in food animals. In: proceedings of the national academy of sciences of the United States of America, vol 112, pp 18. https://doi.org/10.1073/pnas.1503141112

Van Boeckel TP, Glennon EE, Chen D, Gilbert M, Robinson TP, Grenfell BT, Levin SA, Bonhoeffer S, Laxminarayan R (2017) Reducing antimicrobial use in food animals. Sci 357(6358). https://doi.org/10.1126/science.aao1495

Vadassery DH, Pillai D (2020) Quorum quenching potential of Enterococcus faecium QQ12 isolated from gastrointestinal tract of *Oreochromis niloticus* and its application as a probiotic for the control of *Aeromonas hydrophila* infection in goldfish *Carassius auratus* (Linnaeus 1758). Braz J Microbiol 51(3):1333–1343. https://doi.org/10.1007/s42770-020-00230-3

Verner-Jeffreys DW, Welch TJ, Schwarz T, Pond MJ, Woodward MJ, Haig SJ, Rimmer GSE, Roberts E, Morrison V, Baker-Austin C (2009) High prevalence of multidrug-tolerant bacteria and associated antimicrobial resistance genes isolated from ornamental fish and their carriage water. PLoS ONE 4(12). https://doi.org/10.1371/journal.pone.0008388

Wang Q, Yang M, Xiao J, Wu H, Wang X, Lv Y, Xu L, Zheng H, Wang S, Zhao G, Liu Q, Zhang Y (2009) Genome sequence of the versatile fish pathogen *Edwardsiella tarda* provides insights into its adaptation to broad host ranges and intracellular niches. PLoS ONE 4(10). https://doi.org/10.1371/journal.pone.0007646

Watts JEM, Schreier HJ, Lanska L, Hale MS (2017) The rising tide of antimicrobial resistance in aquaculture: sources, sinks and solutions. Marine Drugs 15(6). https://doi.org/10.3390/md15060158

Wen X, Huang J, Cao J, Xu J, Mi J, Wang Y, Ma B, Zou Y, Liao X, Liang JB, Wu Y (2020) Heterologous expression of the tetracycline resistance gene *tetX* to enhance degradability and safety in doxycycline degradation. Ecotoxicol Environ Saf 191. https://doi.org/10.1016/j.ecoenv.2020.110214

Winterhalter M (2021) Antibiotic uptake through porins located in the outer membrane of Gram-negative bacteria. Expert Opin Drug Deliv 18(4). https://doi.org/10.1080/17425247.2021.1847080

Yadav R, Gupta RK, Kumar R, Kaur T (2022) Assessment of toxicity of lead and nickel on the biochemical and immunological parameters of earthworm. Eudrilus eugeniae. Agricultural Sci Digest 44(5):922–929

Yadav R, Gupta RK, Kumar R, Kaur T, Moun R (2022) "Biochemical response of earthworm, *Eisenia fetida* to heavy metals toxicity. J Appl Nat Sci 14(3):990–998

Yadav R, Kumar R, Gupta RK, Kaur T, Yodha K, Kour A, Kaur S, Rajput A (2023) Heavy metal toxicity in earthworms and its environmental implications: a review. Environ Adv 10037

Yasin ISM, Mohamad A, Azzam-Sayuti M (2023) Control of fish diseases using antibiotics and other antimicrobial agents. Recent Adv Aquac Microb Technol. https://doi.org/10.1016/B978-0-323-90261-8.00010-9

Chapter 7
Viral Pathogenesis to Advanced Preventive Measures in Fisheries

Nisha Beniwal, Rahul Kumar, Reman Mor, and Pankaj Sharma

Abstract Aquaculture is a growing and significant source of animal-based nutrition for human consumption, addressing the challenge of feeding a rapidly growing global population and alleviating poverty. However, there is always a risk of infectious disease-related mortality in live production. Wild fish have a lower vulnerability to bacterial, viral, parasitic, and fungal infections compared to farmed fish due to aquaculture husbandry procedures. The aquaculture industry is constantly at risk from viral illnesses affecting fish, making it crucial to understand the mechanisms within the cell that contribute to the formation of an antiviral state. Single-stranded RNA viruses (ssRNA) are the most common cause of disease in aquaculture habitat, including carp and salmon infectious agents. Rhabdoviruses, particularly Spring Viremia Carp Virus (SVCV), induce interferon expression in zebrafish cells and activate gene transcription, suggesting an IFN-independent stimulation of gene i.e., Mx. Fish reoviruses possess the remarkable capability to induce the synthesis of IFN and IFN-mediated reaction, regardless of the viral particle's deactivation status. In developing nations, aquaculture accounts for 44% of total fish production, producing 74 million tons of seafood worth 160 billion dollars. Disease is the primary issue, threatening the livelihoods of farmers, leading to job losses, wage cuts, and hunger. Improving vaccination rates is crucial for controlling and preventing viral diseases in fish. The implementation of sophisticated preventative measures, including biosecurity protocols and sustainable management practices, has resulted in substantial enhancements in fish populations and ecosystem health. The ECOFISH initiative in

N. Beniwal (✉)
BRIC-National Institute of Immunology, New Delhi, India
e-mail: beniwalnisha7@gmail.com

R. Kumar
Department of Zoology, Chaudhary Charan Singh Haryana Agricultural University, College of Agriculture, Bawal, Rewari, Haryana, India

R. Mor
Department of Veterinary Medicine, Lala Lajpat Rai University of Veterinary & Animal Sciences, Hisar, Haryana, India

P. Sharma
Department of Ophthalmology & Visual Sciences, University of Illinois, Chicago, USA
e-mail: sharmap@uic.edu

R. Kumar et al. (eds.), *Sustainability in Aquaculture and Fisheries*,
Springer Water, https://doi.org/10.1007/978-3-032-16605-0_7

Latin America illustrated the advantages of ecological and socio-economic improvements. Adaptive management employs flexible solutions to guarantee the efficacy of management under varying environmental conditions. Additional research and skill enhancement are essential for sustainable practices, policy formulation, and stakeholder engagement. Future initiatives must focus on enhancing these elements to guarantee the sustainable viability of global fish populations and the welfare of individuals dependent on them.

Keywords Fisheries · Viral Infection · Aquaculture · Pathogenesis · Sustainable management

7.1 Introduction

Aquaculture is a rising and significant animal-based nutrition source for human consumption. A sector of the economy that can address the challenge of feeding a fast-expanding global population and alleviate poverty in various nations (Bank et al. 2013; FAO 2020; Rodrigues et al. 2016). There is always a danger of infectious disease-related mortality in live production (Waagbø 2006; Barber 2007). Wild fish have a lower vulnerability to bacterial, viral, parasitic, and fungal infections compared to farmed fish, because of aquaculture husbandry procedures. Furthermore, the inclination towards more densely populated manufacturing systems, ecological imbalances resulting from industrialization and alterations in the climate, and the anticipated rise in global trade of products from aquaculture and their byproducts have all impacted alterations in the correlation between biological forms, infectious pathogens, and mankind (Kumar et al., 2022a, b). This phenomenon affects the rates at which pathogens replicate and spread, resulting in a wider geographical spread of infectious agents and a surge in the number of species impacted by disease outbreaks (Brugere et al. 2017; Chintagari et al. 2018).

As shown in Fig. 7.1, the interactions that occur in water-based ecosystems among the aquaculture, various environmental elements, and potential potentially harmful pathogens shows numerous problems with respect to the health of aquatic organisms. The health of aquatic organisms has particular challenges due to environmental and zootechnical factors, as well as potential pathogenic agents (Peeler et al. 2011a, b; Hedrick 1996; Oidtmann et al. 2014; Sitjà-Bobadilla and Oidtmann et al. 2017).

Aquaculture industry is constantly at risk from viral illnesses affecting fish. Hence, it is imperative to acquire a more thorough comprehension of the mechanisms within the cell that contribute to the formation of an antiviral state (Kumar et al., 2022c). This state is vital for safeguarding toward the replication of viruses and the ensuing damage, with the ultimate goal of ensuring the sustainability of the aquaculture industry. Viral entities show diverse variability depending upon the site of infection and known as tropism. However, it can be cellular tropism (infection of specific cells), tissues tropism which is the invasion specificity to particular tissue and the

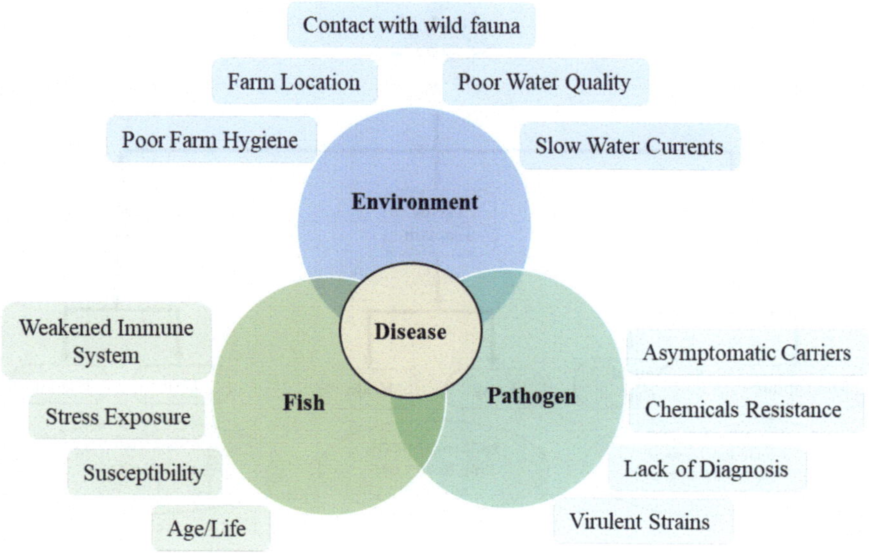

Fig. 7.1 Host-Pathogen interactions and evaluation factors in Fish disease outbreaks in aquaculture

host tropism determines the infectivity to specific host (As shown in Fig. 7.2). An in-depth knowledge of viral processes and resistance mechanisms unraveling the knotty web of viral pathogenesis. Genomic analysis has allowed viruses to be categorized as either ssRNA, dsRNA, or dsDNA. Fish elicit their intrinsic defenses upon viral infection, enabling them to establish an early antiviral defense (Collet 2014; Levraud et al. 2019; Robertsen 2008).

The stimulation of interferon pathway genes that may be the most important ones for fish immune response to viral infection. The most frequently documented viral-induced DNA linked to antiviral action in aquatic creatures (mainly fishes) are: InterferonType I (IFN), are small proteins with a molecular weight of 20–23 KDa. When they attach to a receptor on a membrane of a cell, they trigger various cellular pathways causes transcriptional events that activate genes regulated by interferon known as ISGs. In teleost fish, two commonly present interferons have been identified: INF type I and INF type II (Yu et al. 2010). The myxovirus resistance protein (MX), with a molecular weight ranging from 60 to 70 KDa, are classified as members of the interferon-induced GTPase superfamily. Mxgenes exhibit a significant degree of genetic variation, actually three different Mx variations (Mx1-Mx3) that have been identified in salmonids as well as in perciforms. Additionally, zebrafish have 7 isoforms (MxA to MxG) of Mx (Leong et al. 1998). The interferon, with stimulating gene 15 (ISG15) is a ubiquitin-like protein with a high degree of similarity across species. It has a molecular weight (MW) of 15 KDa and exhibits antiviral activity (Das et al. 2007). Viperin, sometimes referred to as VIG or RSDA2, is a 42 KDa ISG (Interferon-Stimulated Gene). It was initially discovered as a VHSV-triggered genetic material (vig) in the immune cells of rainbow salmon. Viperin

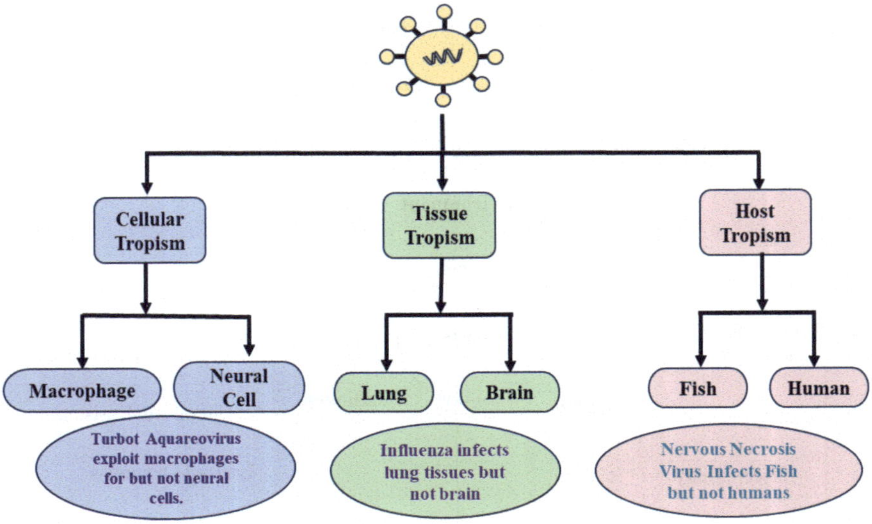

Fig. 7.2 Illustrates the various categories of viral infection tropism

exhibits direct antiviral activity (Wang et al. 2019). Finally, there is a protein called the grass carp reovirus stimulated gene (GIG). Although all aquatic animals possess four vertebrate antiviral genes (Type I Interferon (ifn), Myxovirus Resistance Protein (mx), Interferon-Stimulated Gene 15 (isg15), andViperin (vig or rsad2)), Gig proteins are specific to fish. Their Antimicrobial (specially for viruses) efficacy is beneficial and they can be triggered in a manner that is dependent on or independent of IFN (Zhang and Gui 2004).

7.1.1 Single-Stranded RNA Viruses (ssRNA)

Viruses containing ssRNA, are the most common cause of disease in aquaculture habitat, including carp and salmon infectious agents. As shown in Table 7.1, Rhabdoviruses, particularly Spring Viremia Carp Virus (SVCV), a widespread viral pathogen in cyprinid fish, with zebrafish being an intensively used as paradigm for experimentation to study the innate immune reaction against these viruses (Ashraf et al. 2016). SVCV induces interferon expression in zebrafish cells and activates gene specially Mx transcription as early as 1 h following infection, suggesting an IFN-independent stimulation of gene i.e., Mx. In contrast, infections with Infectious hematopoietic necrosis virus (HNV) and SVCV catalyses interferon manifestation in mature zebrafish, with Mx function remaining Very substantial in zebrafish that survived a Viral hemorrhagic septicemia virus (VHSV) challenge (Ai et al. 2017). Salmon are susceptible to viral infections, such as SsRNA viruses, which induce cell-based IFN activity (Nerbovik et al. 2017). However, gene expression can exhibit temporal variability, with some viruses up-regulating IFN and Mx genes after induction. Salmon infected with a highly pathogenic strain of VHSV exhibit transcription of the Mx gene, while Infectious salmon anemia virus (ISAV) infection results in elevated levels of IFN and Mx afterwards (Campbell et al. 2011). In salmon infected with ISAV, GIG or alike gene regulation has been up-regulated, resembling viral infections of carp, with transcriptional activation seen from two to four weeks following challenge (Bela-Ong et al. 2020).

Many fish species, in particular salmoniod and cyprinid aquatic animals, are vulnerable to infections produced by ssRNA viruses. These infections frequently result in the activation of the IFN cascade. In Pleuronectiformes, a group of flatfish species, expressions of Mx and IFN are significantly upregulated, following infection with rhabdoviruses Hirameno virhabdo virus (HIRRV) and VHSV. This increase can be as high as 16,000 times the typical levels of IFN on the third day post VHSV infection (Diaz-Rosales et al. 2012; Hwang et al. 2020). Similarly, the spleen and kidneys of VHSV-infected perch have shown significant amounts of mx RNA expression.

Table 7.1 List of major fish-infecting viral viruses, their families, illnesses, host species, and transmission methods. Each virus endangers aquaculture and wild fish populations differently

Virus name	Family	Common diseases	Host species	Transmission
Infectious Pancreatic Necrosis Virus (IPNV)	Birnaviridae	Infectious pancreatic necrosis (IPN)	Salmonids (e.g., trout, salmon)	Horizontal transmission through feces
Nervous Necrosis Virus (NNV)	Nodaviridae	Viral encephalopathy and retinopathy	Barramundi, European sea bass, redspotted grouper	Waterborne transmission
Koi Herpesvirus	Alloherpesviridae	Koi herpesvirus disease	Koi, common carp	Direct contact, waterborne
Viral Hemorrhagic Septicemia (VHS)	Rhabdoviridae	Hemorrhagic septicemia	Carp, trout, salmon	Waterborne transmission
Spring Viremia of Carp (SVC)	Rhabdoviridae	Spring viremia of carp	Common carp, koi	Waterborne transmission
Megalocytivirus	Iridoviridae	Megalocytivirus infections	Various freshwater and marine fish	Contaminated water, direct contact
Lymphocystis Disease	Iridoviridae	Lymphocystis disease	Tropical fish (e.g., painted glass fish, angelfish)	Waterborne transmission
Herpesviral Hematopoietic Necrosis	Alloherpesviridae	Hematopoietic necrosis	Various freshwater fish	Direct contact

Nodaviruses present a significant risk to farmed marine fish globally, specifically perciform species. Seabass infected with nervous necrosis virus (NNV) show robust immune responses, specifically in the form of increased expression of ISG15, Mx, and IFN, genes, enters the brain and kidneys within 6–12 h (Moreno et al. 2018; Scapigliati et al. 2009). Studies have shown that Brain cells exhibit elevated Mx transcription rates of barramundi fish that are continuously affected with nodavirus (Wu and Chi 2006). In this context, it has been observed that capsid protein, which of nodaviruses strongly stimulates the transcription of natural immune genes. Both innate immune responses (IFN) and adaptive immune responses (Mx) have been observed in tilapia and larvae of zebrafish infected with tilapia lake virus (TiLV) (Mugimba et al. 2020; Widziolek et al. 2021).

7.1.2 Double Stranded RNA Viruses (dsRNA)

Two significant fish diseases, Infectious Pancreatic Necrosis Virus (IPNV) and Grass Carp Reovirus (GCRV), belong to the families Reoviridae and Birnaviridae and of viruses having genomes as dsRNA (Table 7.1). This RNA is identified by the cell as a molecular pattern that is perceived as unfamiliar. Hence, dsRNA is regarded as a remarkably effective inducer that stimulates the inherent immunity response. More precisely, the artificial version of dsRNA called polyinosinic-poly-cytidylic acidic substances (poly I:C) has been widely used to study how cells react to dsRNA. Its significance in studying the reaction to genuine dsRNA viruses justifies its inclusion in this section. Zebrafish cells infected with GCRV exhibited a significant 100-fold increase in mx transcript levels (Ai et al. 2017). It is noteworthy that fish reoviruses possess the remarkable capability to induce the synthesis of INF as well as the IFN-mediated reaction, irrespective of the deactivation status of the viral particle. The existence of double-stranded RNA alone can stimulate cell responses despite the need for viral multiplication. A study has found that there is a strong correlation between high levels of mx transcription and the formation of effect in cyprinid fish supernatants similar to interferon cells infected with birnavirus (Garcia et al. 2011; Jurado et al. 2013). Upon infection with GCRV, the Mx and ISG15 genes are activated in the kidney and spleen of carp. Greater efficacy has shown among more pathogenic strains of GCRVin inducing this activation (Dai et al. 2017; He et al. 2017). After infecting goldfish with GCRV, the levels of ifn RNA were notably elevated in the dermis (Yan et al. 2016). One well-known example of a dsRNA virus infestation in salmonids is the birnavirus IPNV. Multiple researchers have shown that when salmon were exposed to IPNV, their RNA levels for the ifn, mx, and gig genes increased (Lockhart et al. 2007; Reyes-Lopez et al. 2015). An understudied piscine myocarditis virus (PMCV) infection in salmon is characterized by an increase in the production of the antiviral gene vig/viperin (Timmerhaus et al. 2011). However, vig/viperin, IFN, Mx, and ISG15 gene expression is elevated in salmon infected with piscine reovirus (PRV) (Rosaeg et al. 2017). One virus that can infect red blood cells productively is PRV. Once inside, the virus causes an IFN response to peak at 1 day post infection (dpi) and then mx expression to peak at a later date (Wessel et al. 2015). In vitro and in vivo studies with catfish cell lines encountered withpoly I:C show a quick up-regulation of Mx,IFN, as well asISG15, occurring within 6 h after stimulation (Milev-Milovanovic et al. 2009; Plant et al. 2005). Rock bream and giant yellow croaker show elevated levels of Mx and ISG15 after receiving an intraperitoneal injection of poly I:C (Jung and Jung 2017). IPNV exhibits high concentrations of interferon and Mx in bluegill (Lepomis machrochirus) cells from the earliest stages of infection, following a challenge with the dsRNA virus (Villalba et al. 2017). Beyond cyprinids and salmonids, few research has been conducted on fish. There is a single case report of IPNV infection leading to death in cod, and it involves a stimulation of ISG15 after an IPNV challenge. The transcriptional level of Mx was overexpressed in flounder infected with birnavirus (Pakingking et al. 2004).

7.1.3 Double Stranded DNA Viruses

Similar to ssRNA and dsRNA, dsDNA produced by pathogens can trigger an immune response when detected as "strange" by cell pattern-recognizing receptors (Chang 2020). The families Iridoviridae and Herpesviridae are the two most prominent categories of DNA viruses found in fish. In Senegalese sole, the mx gene expression is activated by the iridovirus LCDV (lymphocystis disease virus) and other cytokines (Carballo et al. 2017). Additionally, turbot infected with RBIV have been found to have elevated amounts of the Mx protein (Zhang et al. 2013). In contrast, sea bream that acquired infection with LCDV failed to produce either raised IFNor Mx mRNA (Cordero et al. 2016). Upon contracting the contagious kidneys and splenic necrosis viruses (ISKNV), mandarin fish similarly show signs of mx transcription activation (Li et al. 2020). There is a strong relationship between viral loads and the interferon response in carp infected with herpesvirus (Xia et al. 2018). Furthermore, it seems that fish exhibit a diverse intrinsic immunological reaction to herpesviruses. Catfish infested with channel catfish virus (CCV) did experience stimulation of Mx transcription, Nevertheless, there were no substantial changes observed in the levels of IFN activity in cyprinid herpesvirus (CyHV)-infected carp (Hwang et al. 2017).

Despite being the most often consumed animals, seafood generally is not afforded the same degree of concern for their welfare as other vertebrates. The scientific investigation of the welfare of fish is still in its early stages, especially when compared to the study of terrestrial animals that are produced for human consumption (Huntingford et al. 2006).

7.2 The Immune System in Fish: A General Overview

Vertebrates are protected from dangerous pathogens by a combined effort of the adaptive as well as the innate immune systems. The body's inherent immune system steps in to combat infections and illnesses when they are in their early stages. The innate immune system is made up of physical barriers, cellular processes, and humoral components (as indicated in Fig. 7.3) (Riera Romo et al. 2016). The adaptive immunological architecture uses memory and specificity to entice persistent pathogens. A long-lasting immunity can be provided by thisimmune system, which is particularly tailored to a specific antigen (Alberts et al. 2002). It is widely believed that the first jawed vertebrates evolved some components belonging to the defensive system, including the major histocompatibility complex (MHC), antibodies (Igs), and T cell receptors (TCRs) (Smith et al. 2019).

Fish and higher animals have comparable immune systems from a physiological standpoint. From their earliest stages of development as embryos, fish rely on nonspecific immune systems to help them survive in the environment, unlike higher vertebrates. Appropriate antigen-specific receptors are the means by which acquired

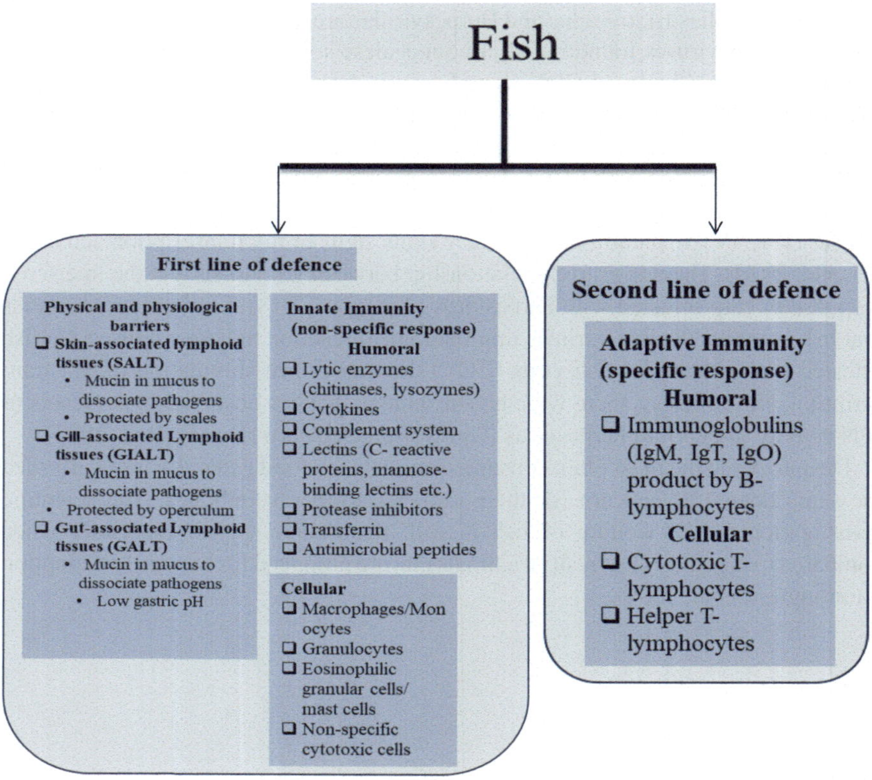

Fig. 7.3 An overview of Immune defense mechanism in Fishes

immunity prevents the spread of infectious diseases, whereas innate immunity detects molecular patterns of harmful microorganisms (Sahoo et al. 2021).

Both the non-specific and specific components make up the structure of the aquatic organism's immune system. To counteract infections, embedded inside the immunological system defenses look for molecular patterns (PAMPs) specific to pathogens, which are not expressed by host cells. The innate immune receptors can identify several PAMPs, including chitin, hemagglutinin, beta-glucan, phosphor-lipomannan, lipoteichoic acid (LTA), and lipopolysaccharides (LPS) (Mogensen 2009; Silva-Gomesa et al. 2015; Taghavi et al. 2017). Major Histocompatibility Complex (MHC), immunoglobulins (Igs), and T cell receptors (TCR) are among the adaptive immunological system's components that are generally accepted to have originated in the first jawed vertebrates (Smith et al. 2019).

7.2.1 Innate Immune System in Fishes

As an effector mechanism of innate immunity, phagocytosis destroys pathogens after detection. In contrast, the adaptive immunity system identifies etiologies by means of highly specific receptors, which trigger the development and multiplication of particular B and T cell clones (Bonilla and Oettgen 2010). While fish do have most of the lymphoid organs that humans do, they are missing several key components, such as lymph nodes and bone marrow. This is a possible limitation and slows down the process of creating fully functional adaptive immune responses (Firdaus-Nawi and Zamri-Saad 2016). Surface or physical, humoral, and cellular components make up the three main areas of fish innate immune system (Magnadóttir 2006; Uribe et al. 2011). Scales, mucus, and the epithelial cells that line the skin, gills, and gastrointestinal canal are all physiological or superficial elements that play an important role in infection resistance (Ellis 2001; Magnadottir 2010). The epidermis is home to homeostasis and a variety of cells that protect the body from harmful invaders, including macrophages and Eosinophil granular cells, and lymphocytes (Ángeles Esteban 2012; Fischer et al. 2006). Maintenance of the osmotic equilibrium and eradication of microbes are critically dependent on the functioning of these cells (Sveen et al. 2020). A variety of soluble chemicals known as humoral immune parameters serve preventative purposes by halting the development of microorganisms and blocking the disease-causing enzymes. Teleost fish include a variety of immune system components, including cytokines, lysozyme, the protein transferrin, lectins and innate immunoglobulins as well as accessory components. and can either kill microbes or slow their growth (Magnadóttir 2006; García-Fernández et al. 2011). Among the several lysins found in fish are lysozyme, complement, and antimicrobial peptides. Bacterial lysis is caused by lysozyme, which targets peptidoglycan, a structural membrane of the cell walls of bacteria (Freedman 1991). Lysozymes can be found in lymphatic tissues and the blood's plasma, mucus, and were generated in the tissue of the liver and other non-hepatic locations. Involved in a vast array of defensive processes, including bacteriolysis, they also have limited antiviral and anticancer effects (Saurabh and Sahoo 2008; Rani et al., 2025). The host's defensive mechanism against microbes and viruses relies heavily on antimicrobial peptides. The gills, liver, and mucus of teleosts are where you can find them (Hellio et al. 2002; Maier et al., 2008). The complement system is crucial for the integration of host defense against pathogenic threats because it links both the immunological responses (Birkemo et al. 2003). The innate immune system of fish is composed of a wide array of cells. When these immune cells detect a pathogen by their PAMP, they switch on and begin attacking the invading microbe. Accordingly, different cell subtypes can cause these cells to react in different ways (Firdaus-Nawi and Zamri-Saad 2016).

7.2.2 Antigen Preparation and Projection

The term "antigen processing and presentation" (App) describes various steps used to convert antigenic proteins into peptides. A subset of MHC-containing proteins is responsible for delivering the peptides to the external surface of the cell (Galluzzi et al. 2017). Researchers have shown that App occurs in teleost fish, specifically the channel catfish (Ictalurus punctatus). They found that, similar to mammals, App is regulated by MHC-like proteins, and that APCs endocytose exogenous antigens, which then induce lymphocyte proliferation (Vallejo et al. 1991, 1992). In addition to transmitting antigens and initiating the adaptive immune response, the spleen is one of the most significant secondary immunological organs in teleosts. Furthermore, the anterior kidney is primarily responsible for regulating melanomacrophage centers, antigen processing, foreign material removal, IgM, and immunological memory development (Page et al. 2013; Stosik et al. 2018).

7.2.3 Distinct Immune Cell Subsets

Natural killer cells, and macrophages, granular leucocytes, clot-forming thrombocytes, dendritic cells, lymphocytes, several types of mast cells, and Eosinophil cells with granules are all components of teleosts' immune systems. Additionally, rodlet cells and melano macrophage centers are present in fish.

7.2.3.1 Anti-Microbial NK Cells

Aside from NK-like cells, teleost fish also have non-specific cytotoxic cells, which are similar to NK cells (NCCs) (Yoshida et al. 1995). The shape of NCCs is small and agranular, similar to that of monocytes in catfish, in contrast to that of NK cells, which are big and bulging (Evans et al. 1987). There is a wide range of morphology in seabream NCCs. Finding NK-like cells in teleosts casts doubt on the similarity between NCCs and NK cells that exist in fish. Research on rainbow trout and catfish has shown certain natural killer-like cells can eliminate both both virally-infected and genetically engineered cells (Meseguer et al. 1994). Different species of fish have cells with different morphologies; for example, seabream have a combination of acidophilic granular cells, monocyte-macrophages, and lymphocytes, whereas catfish have little white cells that resemble monocytes (Evans et al. 1984).

7.2.3.2 Cells Called Macrophages

Macrophages play an essential part in targeted immune responses as they activate lymphocytes and engulf target cells. The immunostimulants enhance the respiratory burst of leukocytes, which in turn produces oxygen species that are reactive with bactericidal activity, because macrophages have particular receptors that can recognize-glucan (Secombes and Fletcher 1992). In fish, the immune cells known as neutrophils and macrophages are responsible for destroying pathogens that have been phagocytosed and for restricting the movement of infectious agents (Neumann et al. 2001). Surface receptors expressed by macrophages include complement and scavenger receptors, as well the cell-line receptors, toll-like receptors, and peptide receptors. The chemokines and cytokines produced by macrophages bridge the gap between both adaptive and innate immunity, allowing for a more effective immune response overall. Furthermore, presenting antigens to T lymphocytes is a critical function of macrophages (Secombes and Wang 2012).

7.2.3.3 Neutrophils

These cells, known as polymorphonuclear cells, are present in the circulatory system, peritoneal fluid and lymphatic organs. They have the ability to phagocytose foreign objects or other cells and generate the bactericidal agent superoxide anions (Secombes 1996). The first granulocytes arrive at the site of damage, followed by macrophages. Both types of cells are directed by chemical signals that are generated by the injured tissue. Neutrophils utilize antimicrobial peptides, proteolytic enzymes, and reactive oxygen compounds (or ROS) at the site of injury to engulf bacteria and eliminate them (Biller-Takahashi and Urbinati 2014). Furthermore, fish neutrophils have the potential to release extracellular filaments known as extracellular neutrophil traps (or NETs), composed of genetic material, histones, and proteins. These compounds have the capability to adhere to, eradicate, and deactivate parasites, fungus, viruses, and bacteria. Furthermore, the myeloperoxidase enzyme obtained from granular leukocytes has the ability to interact with hydrogen peroxide, resulting in the production of hypochlorite. Hypochlorite can be used to produce chloramines, which are powerful oxidative chemicals that specifically attack the membranes of bacteria (Caipang et al. 2012).

7.2.3.4 Eosinophils

Eosinophils are abundant cells found in several locations throughout the body, including connective tissue, the gastrointestinal system, circulation, ovaries, and gills. They release their granules when there are parasite infections (Mokhtar 2019).

7.2.3.5 Basophils

Basophils are a type of big white blood cells with granules that are not commonly observed in species of teleost. They have histamine in their cytoplasmic granules, which is an inflammatory mediator. Both allergic responses and anaphylaxis are influenced by these cells (Hine 1992; Martins et al. 2000). Thrombocytes are cells that are oval-shaped, include a nucleus, and lack granules. They can phagocytize in addition to coagulating blood. They cause cell aggregation at the location of inflammatory processes by exhibiting acid phosphatase activity (Tavares-Dias et al. 1999).

7.2.3.6 Monocytes

Monocytes display phagocytosis as well as non-targeted cell death. During the inflammatory phase, these cells travel to the connected tissue and undergo a transformation into macrophages. They are thought to be transient blood cells. It is possible to deliver specialized phagocytes such as macrophages and granulocytes, along with dendritic- to the area of inflammation (Stuart and Ezekowitz 2005).

One sort of cell that presents antigens is dendritic cells (DCs) found in various tissues. They have the ability to trigger immune responses that are innate as well as acquired (Mokhtar 2017). The distinctive features of dendritic cells (DCs) comprise numerous cytoplasmic dendrites and compressed cell bodies (Sayed et al. 2022). Their presence has been detected in many organs such as the ovaries, skin, gills and digestive tract, many fish species have been shown to possess cell populations that exhibit dendritic cell-like functions and shape (Shao et al. 2015).

7.2.3.7 B and T- Lymphocytes

B lymphocytes mediate humoral immunity by producing antibodies against antigens and serving as specialized antigen-presenting cells (APCs). By undergoing differentiation into memory B cells, they establish immunological memory. Similar to plasma cells found in mammals, B cells secrete antibodies and go by other names like plasmablasts or antibody secreting cells (ASCs). Both T cell-independent and T cell-dependent activation of B cells happen upon interaction with antigens. The outer surface glycoprotein cluster of differentiation 40 ligand (CD40L) is expressed by activated and differentiated T helper cells (Nutt et al. 2015; Kurosaki et al. 2015).

One possible outcome of T cell differentiation is the production of memory B cells, which are involved in the body's secondary antibody response and circulate throughout. Another outcome is the production of plasma cells by plasma blasts. upon early non-specific defenses, T cells provide an immunological response upon antigen recognition by APCs and possess an intrinsic function in regulating immune activity. Surrounding immunological components, like cytokine levels and helper T cells, impact T cell proliferation and regulation. The roles and pathogens that Th1 and Th2 T cells target determine their classification. An efficient T cell biomarker for evaluating expression regions and anatomical patterns is salmonid CD3 (Castro et al. 2011; Koppang et al. 2010).

Important for innate immunity, mast cells are present in teleost fish and other teleost animals. Proteins such as S100, desmin, and CD117 are expressed by these cells, which are situated close to blood arteries in the colon, gills, skin, and ovaries. Teleosts' tissues and organs have an increased amount of mast cells when they are exposed to stress circumstances such as herbicides, inflammation, or parasitic diseases. Interactions between eosinophils and piscidins, host-defense peptides, which are observed in mast cells (Mokhtar 2021; Lauriano et al. 2012).

7.2.3.8 Eosinophilic Granule Cells (EGCs)

Among the several species that frequently possess EGCs are cyprinids and salmonids. The gut mucosa is typically home to EGC followed by gills, fibrous tissue around the spinal cord, and the epidermis of the skin (Zaccone et al. 2017). EGCs are additionally seen in the underlying connective stroma associated with the bile duct (Reite 2005). Their cytoplasm contains spherical granules that are stained reddish by eosin and hematoxylin (HE) due to the presence of basic proteins. Due to the substantial amount of neutrality and sulfated carbohydrates, the granules also show an alcian blue tint and a deep magenta tint when stained with PAS. Because of their metachromatic affinity for toluidine blue and their positive reaction to bromophenol blue, which indicates the existence of protein content. These cells have several activities, such as releasing oxygen radicals and harmful protein molecules on the outermost layer of multicellular organisms in areas of inflammation. Furthermore, acute tissue damage leads to the release of inflammatory mediators through EGC degranulation. However, chronically inflamed tissues often show an elevated number of these cell (Jordanova et al. 2007).

7.2.3.9 Rodlet Cells

Fish cells are unique in that they include cytoplasmic inclusions and thicker capsule-like edges (Siderits and Bielek 2009). They are found in many organs, including the skin, gall bladder, liver, heart, thymus, kidney, spleen, gills, and pancreas. Additionally, they have an actin cytoskeleton (Mokhtar 2021). Secretory defense, transfer of electrolytes or water, regulation of pH, and antibiotic responses are among the many roles played by these cells. Another way to look at them is as immune cells that are involved in immunity but aren't cell specific. Lectin and immunohistochemistry have been used to study the molecular makeup of rodlet cells. Fish with rodlet cells contain complex carbohydrates, mainly N-acetyl-galactosamine glucosidic residues (Rideout et al. 2015).

7.3 Management and Avoidance of Viral Illness in Aquaculture for Better Livelihood

In developing nations, fisheries are an important part of people's daily lives since they provide a steady income and contribute to societal development. In 2014, aquaculture accounted for 44% of the total fish production, producing 74 million tons of seafood worth 160 billion dollars. The vast majority of fish raised in aquariums are intended for human consumption (FAO 2016).

For a variety of factors, aquaculture is experiencing a massive drop in output. Disease is the primary issue since it threatens the livelihoods of farmers, which in turn causes them to lose their jobs, have their wages cut, and end up hungry. Research shows that illnesses, which are more common in developing countries, account for more than 50% of the production loss. The reason behind this is because the developing globe is home to 90% of the aquaculture firm. Disease accounts for as much as $6 billion in yearly revenue loss. Infectious salmon anemia, for example, cost Chile $2 billion and put 20,000 people out of work. Among the world's top aquaculture producers, China loses 15% of its overall fish production to illnesses (Leung and Bates 2013).

In order to compensate for losses caused by infectious diseases in aquaculture, it is critical to tackle these health constraints with locally applicable and scientifically backed methods. The necessity for epidemiological methods in maintaining the health of aquatic animals is fueled by the difficulties posed by climate change, scarcity of water supplies, and the expansion of aquaculture (Peeler et al. 2011a, b).

Optimal strategies for managing infectious illnesses in fish encompass improved animal husbandry and handling practices, restrictions on movement, incorporation of resistant genetic material in the supply, dietary supplements, generalized immunostimulants, vaccination, prebiotics and probiotics, herbal products, water disinfection, biological management, antimicrobial agents, and restriction of movement (Kumar et al. 2016).

The prevention and management of health issues in aquaculture cannot be achieved by relying on a single strategy. It is more effective to use a variety of tactics. It is mandatory to establish a system where farmers and responsible parties may communicate information on a national or regional level. In addition to implementing all of these techniques, having sensitive and specific diagnostic tests and conducting disease surveillance are crucial for ensuring healthy fish (As depicted in Fig. 7.4).

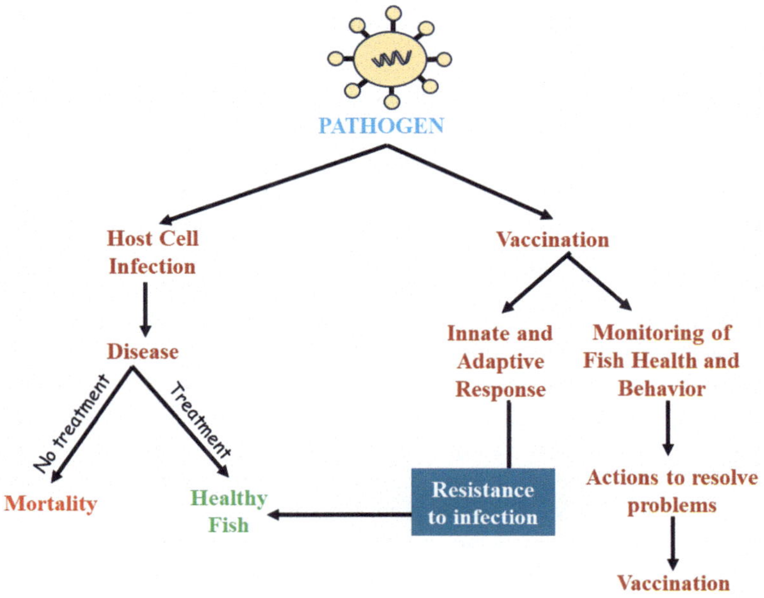

Fig. 7.4 Comprehensive outline of Aquatic animals disease control

When it comes to controlling and preventing viral diseases in fish, improving vaccination rates is among the most significant, if not the most important, strategies. Recently, there have been advancements in the field of fish vaccination. Fewer antibiotics are used in aquaculture, and fish are protected against infectious diseases. Furthermore, the possibility of medication resistance is averted. Compared to antibiotics, the licensing and registration requirements for novel vaccines are significantly lower, and herd immunity allows for protection at the stock level.

Vaccination is a common practice for the majority of animals used for food production. It safeguards fish against contagious infections and lessens the need for antibiotics in aquaculture. Not only that, it keeps drug resistance at bay. It is far easier to license and register new vaccines, which allows for protection at the stock level due to herd immunity, than it is to deal with antibiotics (Plant and LaPatra 2011). When it comes to the administration of vaccinations in fish, there are a few significant factors that need to be kept in the forefront. Points to consider include: Important factors to take into account include the variety of fish that need to be vaccinated, their immune systems' condition, the aquaculture system's life cycle, the production cycle, the ailments that must be prevented in the aquaculture sector, the times when these diseases occur (i.e., sporadic spread of illnesses in the fish tank), methods of farming (coping with and machinery), the surroundings (temperatures range and salt concentration), variables related to stress, dietary habits, and the evaluation of costs and benefits (Gudding and Muiswinkel 2013).

7.4 The Background of Fish Immunization

Vaccinating fish first became widespread in 1942 using the administration of a vaccine against Aeromonas salmonicida infection in Cutthroat fish. At now, the vaccinations being used are of the conventional variety due to limited knowledge about fish immunology. The development of vaccines targeting subcellular bacterial and viral diseases will be a significant undertaking during the next few years (Dadar et al. 2016). Some examples of modern vaccinations include the genome, synthetically produced peptide, recombinant vector, genetically modified, subunit, and lethal vaccines. The benefits of whole organism vaccinations outweighed those of other vaccine kinds. The majority of vaccines, however, do not offer 100% protection against disease (Gudding 2014). Moreover, these sorts of vaccinations may not be easily modifiable to avoid the development of novel disease-causing agents, the occurrence of changes in the antigens that allow infections to evade the immune system, and the proliferation of specific microorganisms that cannot be grown in a laboratory environment. Moreover, the manufacturing of such vaccines is a protracted and labor-intensive procedure, which might occasionally impede the prompt reaction to newly appearing and recurring infections (Pridgeon et al. 2012).

7.4.1 Different Vaccines Used in Fish Virology

Killed vaccinations are traditional forms of vaccines that are created by inactivating the pathogen that causes illness and utilizing it as an immunological response-stimulating vaccination antigen. Most of the immunizations used in aquaculture are potent vaccines. The benefits of these vaccinations can be described as follows: they have a simple design, are stable during preservation, are economical, and do not pose any concerns regarding virulence. Some infections can be prevented by using inactivated vaccines, such as V. salmonicida, contagious hematopoietic necrosis infection, and A. salmonicida (Shoemaker et al. 2009).

Repeated physical and chemical attenuation, as well as laboratory transit, causes the reduction in their pathogenicity of an organism, but not their death, allowing them to be manufactured as attenuated vaccines. The efficacy of live vaccinations in fish has been demonstrated in laboratory trials (Lorenzen et al. 2005).

The antigen itself is not necessary to create a DNA vaccine rather, the gene encoding the antigen can be isolated using molecular techniques and administered as a vaccine (Cho et al. 2017). These vaccinations, when administered intramuscularly, protect farmed salmonids from economically significant illnesses including infectious hemorrhagic septicemia virus (Ballesteros et al. 2015) and contagious hematopoietic necrosis virus (Dhar et al. 2014). For the salmonid rhabdoviruses IHNV and VHSV, DNA vaccinations in fish have been extensively investigated. The aquaculture sector of the salmon industry has discovered that these vaccines effectively lessen the effects of these viruses. Furthermore, both DNA vaccines have been tested against other fish viral pathogens, such as hiramerhab-dovirus and spring viraemia caused by carp virus (Vallejos-Vidal et al. 2014).

Recombinant vector-based vaccinations are produced through the use of biotechnology. This involves reproducing only the immune-stimulating parts of a pathogen in a different host organism. The organism's immunogenic components are presented to carriers. A significant quantity of proteins is produced outside of a living organism and then purified for the purpose of being used in vaccines. When selecting vectors, important factors to consider are the ability to produce large quantities and the ability to express antigenic proteins effectively. In order to safeguard salmon, vaccinations have been developed by expressing viral anemia in salmon as well as hematopoietic necrosis infection pathogens in vectors (Gudding 2014).

While subunit immunizations are thought to be safe to administer, their immunogenicity is far lower than that of inactivated, whole organism injections. Adjuvants are essential for augmenting the immunogenicity of a substance. When the cultivation of the organism is difficult, these vaccines are advantageous as they remove the immunological component and use it as a vaccine (Gudding 2014).

The method of laboratory passaging of organisms leads to an accumulation of genome mutations, which ultimately causes the organism to become weakened. Microorganisms can be genetically modified using molecular techniques to remove the genes that make them pathogenic, therefore reducing their ability to cause disease. A live attenuated vaccination is designed to replicate at a reduced concentration and can efficiently stimulate both the production of antibodies and the activation of immune cells (Gudding 2014).

Immunizations against disease using synthetic peptides. These vaccines are made from synthetic antigens that are short chains of amino acids (Shoemaker et al. 2009). For the purpose of preventing infectious diseases such as nodavirus, viral hemorrhagic septicaemia, rhabdovirus, and birnavirus, these vaccines have been administered (Gudding 2014).

7.4.2 Routes of Vaccine Administration

Along with choosing which illnesses to vaccinate against, it is important to have a firm grasp on when and how to administer booster doses. To maximize protection, it is recommended to have the vaccine well in advance of potential exposure to the virus. This will allow the immune system enough time to mature. The size of the fish is the most essential element in controlling the development of immunological competence, although water temperature is also a potential consideration when choosing whether to vaccinate (Harikrishnan et al. 2011).

Oral immunization is a stress-free method that is straightforward to provide in the activity of fishing. The vaccination is included into the food itself, while it is being made, or it can be covered in pellets or put inside a capsule (Plumb and Hanson 2011).

The fish receive topical vaccinations throughout the immersion procedure. The skin, gills, and lateral line all absorb the antigen. For a short period of time, the fish can be submerged in a highly concentrated vaccine solution. Another way to administer the vaccine solution is to spray it on the fish. Since small fish and young fingerlings are difficult to manage for injection, immersion immunization is particularly beneficial for them. Additionally, this method minimizes stress in fish and allows for the reuse of the vaccine solution. The drawbacks of this approach are its reliance on manual work and the high cost associated with purchasing tanks and specialized equipment (Gudding 2014).

For adjuvant vaccines, the only acceptable route of administration is injection, which also happens to be the most protective form of vaccination overall (Rogers and Basurco 2009). The benefits of injecting a vaccination include achieving good protection with a relatively minimal dose, quick and economical dosage calculation, and the ability to deliver a multivalent vaccine. This is especially true for larger fish. Here are some drawbacks of this administration: incompatible with small fish, can cause adhesions, stress, and reduced feed intake; injection injury can result in multiple fish fatalities; and very young stock should not be administered this because their immune systems may not be developed enough (Plumb and Hanson 2011).

Antimicrobial dosing in aquaculture is distinct from that in terrestrial animal husbandry. Disruption of the microbiota in the environment is a consequence of this variation, in particular the addition of medications to water, whether with or without feeding (Defoirdt et al. 2011). The frequent utilization of antimicrobial agents in aquatic environments poses a significant risk to human health. Aquaculturists should try to avoid using antibiotics whenever possible; nevertheless, if antibiotics are absolutely necessary, they should only give their fish medications that have been approved for use with fish (Maqsood et al. 2011).

Research on ways to improve fish natural immunity is an exciting new area with many potential benefits (Sharifuzzaman and Austin 2017). When it comes to managing infectious illnesses in aquaculture, one of the most successful strategies is to use biological containment measures (Mastan 2015).

7.4.3 Probiotics

It is a collection of non-pathogenic bacterial strains that are specifically cultured for fish (De et al. 2014). Another meaning of probiotics is the introduction of living microorganisms into hosts to establish a defensive immunological state. Upon administration to fish, these organisms undergo multiplication within the fish's gut, aiding the usual microflora and promoting microbial equilibrium in the hosts (Hermosillo et al. 2012). One option for the feed in the reservoir or tank is to add probiotics. Several research have been carried out to propose the most effective method of delivering, ideal dosage, and necessary technical measures, particularly to maintain the viability of probiotics in dehydrated pellets (Kumar et al. 2016).

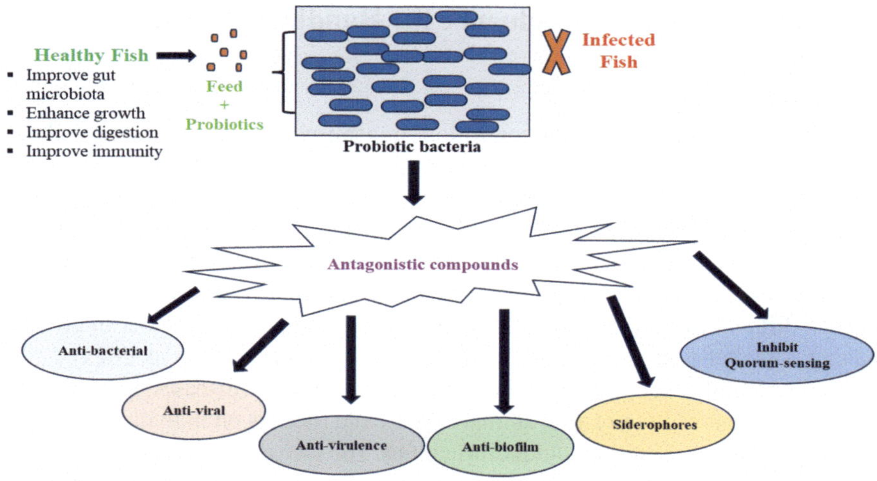

Fig. 7.5 Application of probiotics and immunostimulantto enhance the immunityin fish

7.4.3.1 Direct Incorporation in Feed

An essential and highly versatile application of probiotics is their direct incorporation into feed pellets. (Fig. 7.5). The main type of probiotics used in food is primarily into the form of of spores. It is crucial to constantly evaluate the viability of probiotics when introducing them to ensure the fish's enhanced immune response. Frozen cultures can be added and mixed with lipids to be used as top coatings in animal feed (Kumar et al. 2016).

Another often employed method for administering probiotics is through a process known as encapsulation. During this process, cells of the microorganism are surrounded by a gel-like substance composed of alginate, chitosan, a substance known as, or pectin. This serves to offer consequently physical as well as chemical safeguarding to the microorganisms, particularly whenever they are present in a dense population (Peeler and Taylor 2011a, b). The application of encapsulating microbes of Shewanella putrefaciens in calcium alginate has been successfully used in aquaculture. This has shown that the encapsulated probiotic cells can survive the journey through the alimentary canal of sole (Solea senegalensis) (Savoca et al. 2020).

The method of catching and constraining cells which were discharged into a gel matrix composed of alginates surrounding the core element is known as immobilization wall. Probiotic-based immobilization is a modern method extensively used in the livestock and pharmaceutical industries, particularly for the utilization of lactic acid bacteria (LAB). Cell immobilization offers significant advantages for the generation of feedstock and metabolites, especially as compared with liberated cell systems (Kumar et al. 2016).

7.5 Components That Influence the Handling of Fish Diseases

In contrast, it is true to say that controlling transmissible illnesses in aquaculture is more complex than it is for land animals. The dissemination of infectious agents among fish is impacted by both the environmental conditions in which they reside and the inherent qualities of the fish. Due to the nature of their habitat, fish are not easily subject to close observation, which poses a challenge in monitoring their health. Additionally, the aquatic environment can facilitate the rapid spread of diseases. Fish are also difficult to capture without causing stress, and they often gather in groups, making it challenging to detect and characterize diseases (Frazen et al. 2022).

Another significant problem lies in the identification of fish diseases. In contrast to identifying diseases in terrestrial animals, where the focus is on individual animals, fish disease diagnosis presents unique difficulties. The scenario in aquaculture disease inquiry differs due to the unique characteristics of the aquatic environment in which fish reside. When it comes to healthy livestock, the whole system may be the source of an infectious ailment that spreads quickly. For this case, the focus of examination and diagnosis is not limited to an individual fish, but rather encompasses the entire tank. It is necessary to collect samples not only from fishes, but also from water in order to determine vital characteristics including viscosity, pH levels, and soil bottom characteristics. All of these things add complexity and difficulty to the process of aquatic animal diagnosis (Commonwealth Fisheries Harvest Strategy Policy and Guidelines, Department of Agriculture, Fisheries and Forestry, https://www.agriculture.gov.au/agriculture-land/fisheries/domestic/harvest_strategy_policy, Last Accessed: 01/09/2024).

7.6 Case Studies and Success Stories

Illustrations of Effective Execution of Proactive Measures in Fisheries. Effective Bycatch Management in Latin America and the Caribbean: The Global Environment Facility (GEF) and FAO have launched a project with the objective of mitigating bycatch in bottom trawling fisheries in many countries, such as Brazil and Costa Rica. The initiative successfully reduced bycatch by 25–50% by making simple technical modifications, such as altering net mesh sizes. This effort also promoted the development of management structures that encourage active participation, therefore strengthening the relationship of trust between governments and local players in the fishing industry. Additionally, it empowered marginalized populations, including women, to actively participate in the decision-making processes (Savoca et al. 2020).

The Ecosystems Improved for Sustainable Fisheries (ECOFISH) Project is being executed in the Philippines. Commenced in 2012, this initiative included collaboration between national and local governments and USAID to advance the Ecosystem

Approach to Fisheries Management (EAFM). The initiative yielded a 24% augmentation in fishing biomass and a 12% enhancement in employment throughout eight Marine Key Biodiversity Areas. ECOFISH successfully enhanced coastal and marine resource management by using a participatory and decentralized strategy, resulting in the improvement of over 1.8 million hectares of municipal marine waterways (Frazen et al. 2022).

Australia introduced a harvest strategy policy in 2007 with the goal of preserving important commercial fish populations at environmentally sustainable levels while maximizing economic profits. This strategy has effectively implemented management frameworks that provide openness and predictability in fisheries management, which are essential for maintaining industry stability and sustainability (https://www.agriculture.gov.au/agricultureland/fisheries/domestic/harvest_strategy_policy).

Indonesia's Implementation of Community-Based Fisheries Management: The people of Menarbu, a fishing hamlet, established community-driven management strategies, which included the establishment of marine protected zones and the regulation of fishing activities. This endeavor resulted in enhanced fish populations and biodiversity, demonstrating the efficacy of localized government in the management of fisheries(https://doi.org/https://doi.org/10.1016/j.ocecoaman.2023.106766).

To the Blue Ports Initiative of the Food and Agriculture Organization (FAO): The objective of this endeavor is to promote the long-term viability of fishing ports by including practices that support ecological preservation, community well-being, and financial sustainability. The effort has successfully engaged a network of 26 nations to implement best practices in port management. This has resulted in a decrease in the environmental impact of fishing operations and an enhancement in adherence to sustainability standards (https://www.fao.org/in-action/blue-ports-initiative/en).

7.7 Effects of Enhanced Preventive Measures

The adoption of sophisticated preventative measures, such as biosecurity procedures and sustainable management methods, has resulted in significant enhancements in fish populations and the general health of the ecosystem. An example of this is the ECOFISH project, which not only raised the number of fish in the area but also improved job prospects for local residents. This initiative effectively showcased the advantages of both ecological and socio-economic enhancements. In Latin America, the effective reduction of bycatch has not only preserved marine biodiversity but also improved the economic conditions of local fishing communities. Adaptive management involves the use of flexible and adaptable strategies in fisheries management, which may be modified based on continuous monitoring and assessment. This strategy guarantees that management techniques maintain their effectiveness in the face of evolving environmental circumstances and rising challenges (Frazen et al. 2022).

7.8 Investigation and Development of Skills

It is crucial to continue researching the dynamics of fisheries, the health of ecosystems, and the socio-economic effects of management strategies. Enhancing the implementation of sustainable practices may be achieved by increasing the capacity of local communities and government authorities. Policy development is essential for attaining sustainability in fisheries management. It involves creating strong policies that combine best practices, scientific data, and input from stakeholders. Policymakers should give utmost importance to the implementation of legislation and the creation of frameworks that encourage and support responsible fishing practices.

Preventative approaches in fisheries management are vital for achieving long-term sustainability, aligning with the United Nations' Sustainable Development Goals (SDGs), particularly SDG 14 ("Life Below Water"). These approaches emphasize ecological, economic, and social sustainability to ensure the health of fish populations and the well-being of communities dependent on them.

Engaging stakeholders, including fishers, policymakers, scientists, and local communities, fosters inclusive decision-making and ensures that management measures are context-specific and widely accepted. Participatory processes improve compliance and accountability while integrating local knowledge into management strategies. Strengthen collaboration through transparent mechanisms like the Fisheries Transparency Initiative (FiTI) to build trust and enhance data-sharing.

Ecosystem-based fisheries management (EBFM) has emerged as a cornerstone for sustainable practices. This approach accounts for ecological complexities, climate change impacts, and socio-economic factors. Techniques such as harvest control rules, marine protected areas (MPAs), and seasonal closures allow adaptive responses to changing conditions. Expand the use of innovative technologies (e.g., real-time monitoring systems) and data-driven tools to refine adaptive measures. Tailor these techniques to specific fisheries while ensuring they remain precautionary and science-based. Effective fisheries management requires robust legal frameworks that provide a legitimate mandate for enforcement. Regulations must balance ecological conservation with equitable resource distribution to ensure sustainable use while reducing inequalities. Strengthen enforcement capacity through international cooperation, particularly in areas beyond national jurisdiction. Promote equitable resource-sharing agreements under frameworks like the Biodiversity Beyond National Jurisdiction treaty. Sustainable fisheries contribute to climate change mitigation by enhancing blue carbon ecosystems (e.g., mangroves, seagrasses) and reducing greenhouse gas emissions through low-carbon fishing practices. Integrate climate resilience into fisheries management plans by restoring coastal ecosystems and adopting renewable energy technologies in fishing operations. Reliable data on fish stocks, habitats, and socio-economic factors are critical for informed decision-making. Data gaps hinder responsive action and effective implementation of sustainable practices. Invest in capacity-building for data collection, analysis, and interpretation. Encourage community participation in monitoring efforts to enhance data accuracy and inclusivity.

7.9 Conclusions

To guarantee the long-term viability of global fish populations and the well-being of dependent communities, future endeavors must prioritize these components. By fostering collaboration among stakeholders, implementing adaptable techniques, strengthening regulatory frameworks, addressing climate impacts, and improving data systems, fisheries management can achieve its sustainability goals. These efforts will not only protect marine biodiversity but also secure livelihoods for millions worldwide while contributing to broader global objectives like poverty reduction and food security.

7.10 Future Prospects

Aquaculture's growing importance in food security and poverty reduction necessitates addressing infectious diseases, notably viral infections. As single-stranded RNA viruses continue to cause most aquaculture disease outbreaks, especially in economically important species like carp and salmon, future research must focus on fish antiviral immunity's cellular and molecular mechanisms. IFN-independent mechanisms such Mx gene activation and new immune modulators generated by rhabdoviruses and reoviruses are being studied.

Increasing discovery of innate antiviral responses, even without live virus particles, implies untapped promise for next-generation vaccines and immunostimulants. Nanotechnology, mucosal vaccination, and reverse vaccinology can greatly improve aquaculture vaccination coverage by improving fish vaccine immunogenicity and delivery.

Using systems biology models can speed up biomarker identification for early illness detection, immune state monitoring, and vaccination responsiveness. Diagnostic infrastructure and real-time surveillance are needed for targeted interventions. Expand biosecurity frameworks and adapt management tactics to local environmental conditions. The success of ECOFISH shows the importance of integrating ecological, socio-economic, and technical interventions. Scaling such ideas globally requires improving public–private partnerships, community-based management, and international coordination.

Finally, training, stakeholder participation, and policy formation are essential to applying scientific findings. A multi-pronged strategy based on innovation, sustainability, and inclusivity is needed to protect aquaculture systems and millions of aquatic food resource users.

References

Ai K, Li Y, Hu W, Gao W, Fang L, Tian G, Ruan G, Xu Q (2017) Expression pattern analysis of IRF4 and its related genes revealed the functional differentiation of IRF4 paralogues in teleost. Fish Shellfish Immunol 60:59–64

Alberts B, Johnson A, Lewis J, Raff M, Roberts K, Walter P (2002) 'Innate immunity', in Molecular Biology of the Cell, 4th edn. Garland Science, New York

Ángeles Esteban M (2012) An overview of the immunological defenses in fish skin. Int Sch Res Not 2012:29

Ashraf U, Lu Y, Lin L, Yuan J, Wang M, Liu X (2016) Spring viremia of carp virus: Recent advances. J Gen Virol 97:1037–1051

Ballesteros NA, Alonso M, Saint-Jean SR, Perez Prieto SI (2015) An oral DNA vaccine against infectious haematopoietic necrosis virus (IHNV) encapsulated in alginate microspheres induces dose-dependent immune responses and significant protection in rainbow trout (Oncorrhynchus mykiss). Fish Shellfish Immunol 45(2):877–888

Bank TW (2013) Fish to 2030—prospects for fisheries and aquaculture. International Bank for Reconstruction and Development/International Development Association (The World Bank), Washington, DC, p 77

Barber I (2007) Parasites, behaviour and welfare in fish. Appl Anim Behav Sci 104:251–264

Bela-Ong DB, Greiner-Tollersrud L, van der Wal YA, Jensen I, Seternes OM, Jorgensen JB (2020) Infection and microbial molecular motifs modulate transcription of the interferon-inducible gene ifit5 in a teleost fish. Dev Comp Immunol 111:103746

Biller-Takahashi JD, Urbinati EC (2014) Fish immunology. The modification and manipulation of the innate immune system: Brazilian studies. Anais da Academia Brasileira de Ciências, 86:1484–1506

Birkemo GA, Lüders T, Andersen Ø, Nes IF, Nissen-Meyer J (2003) Hipposin, a histone-derived antimicrobial peptide in Atlantic halibut (Hippoglossus hippoglossus L.). Biochim Biophys Acta (BBA)-Proteins Proteomics 1646:207–215

Bonilla FA, Oettgen HC (2010) Adaptive immunity. J Allergy Clin Immunol 125:S33–S40

Brugere C, Onuigbo DM, Morgan KL (2017) People matter in animal disease surveillance: Challenges and opportunities for the aquaculture sector. Aquaculture 467:158–169

Caipang CMA, Lazado CC, Brinchmann MF, Kiron V (2012) Transcription of selected immune-related genes in spleen cells of cod (Gadus morhua) following incubation with alginic acid and β-glucan. J Exp Mar Biol Ecol 416:202–207

Campbell S, McBeath A, Secombes CJ, Snow M, Collet B (2011) Interferon response following infection with genetically similar isolates of viral haemorrhagic septicaemia virus (VHSV) exhibiting contrasting virulence in rainbow trout. Fish Shellfish Immunol 30:287–294

Carballo C, Castro D, Borrego JJ, Manchado M (2017) Gene expression profiles associated with lymphocystis disease virus (LCDV) in experimentally infected Senegalese sole (Solea senegalensis). Fish Shellfish Immunol 66:129–139

Castro R, Bernard D, Lefranc M, Six A, Benmansour A, Boudinot P (2011) T cell diversity and TcR repertoires in teleost fish. Fish Shellfish Immunol 31:644–654

Chang CJ (2020) Immune sensing of DNA and strategies for fish DNA vaccine development. Fish Shellfish Immunol 101:252–260

Chintagari S, Hazard N, Edwards G, Jadeja R, Janes M (2018) Risks associated with fish and seafood. Preharvest Food Saf:123–142

Cho SY, Kim HJ, Lan NT et al (2017) Oral vaccination through voluntary consumption of the convict grouper Epinephelus septemfasciatus with yeast producing the capsid protein of red spotted grouper nervous necrosis virus. Vet Microbiol 204:159–164

Collet B (2014) Innate immune responses of salmonid fish to viral infections. Dev Comp Immunol 43:169–173

Commonwealth Fisheries Harvest Strategy Policy and Guidelines (n.d.) Department of Agriculture, fisheries and forestry. https://www.agriculture.gov.au/agriculture-land/fisheries/domestic/harvest_strategy_policy (Accessed: 1 Sep 2024)
Cordero H, Cuesta A, Meseguer J, Esteban MA (2016) Characterization of the gilthead (Sparus aurata L.) immune response under a natural lymphocystis disease virus outbreak. J Fish Dis 39:1476
Dadar M, Dhama K, Vakharia VN et al (2016) Advances in aquaculture vaccines against fish pathogens: Global status and current trends. Rev Fish Sci & Aquac 25(3):184–217
Dai Z, Li J, Hu C, Wang F, Wang B, Shi X, Hou Q, Huang W, Lin G (2017) Transcriptome data analysis of grass carp (Ctenopharyngodon idella) infected by reoviruses provides insights into two immune-related genes. Fish Shellfish Immunol 64:68–77
Das BK, Collet B, Snow M, Ellis AE (2007) Expression kinetics of ISG15 and viral major capsid protein (VP2) in Atlantic cod (Gadus morhua L.) fry following infection with infectious pancreatic necrosis virus (IPNV). Fish Shellfish Immunol 23:825–830
De BC, Meena DK, Behera BK, Das P, Das Mohapatra PK, Sharma AP (2014) Probiotics in fish and shellfish culture: Immunomodulatory and ecophysiological responses. Fish Physiol Biochem 40(3):921–971
Defoirdt T, Sorgeloos P, Bossier P (2011) Alternatives to antibiotics for the control of bacterial disease in aquaculture. Curr Opin Microbiol 14(3):251–258
Dhar AK, Manna SK, Allnutt FCT (2014) Viral vaccines for farmed finfish. Virus Disease 25(1):1–17
Diaz-Rosales P, Romero A, Balseiro P, Dios S, Novoa B, Figueras A (2012) Microarray-based identification of differentially expressed genes in families of turbot (Scophthalmus maximus) after infection with viral haemorrhagic septicaemia virus (VHSV). Mar Biotechnol 14:515–519
Ellis A (2001) Innate host defense mechanisms of fish against viruses and bacteria. Dev Comp Immunol 25:827–839
Evans DL, Hogan KT, Graves SS, Carlson RL Jr, Floyd E, Dawe DL (1984) Nonspecific cytotoxic cells in fish (Ictalurus punctatus). III. Biophysical and biochemical properties affecting cytolysis. Dev Comp Immunol 8:599–610
Evans DL, Smith EE Jr, Brown FE (1987) Nonspecific cytotoxic cells in fish (Ictalurus punctatus). VI. Flow cytometric analysis. Dev Comp Immunol 11:95–104
FAO (2016) The State of World Fisheries and Aquaculture: Contributing to food security and nutrition for all. Food and Agriculture Organization of the United Nations, Rome
FAO (2020) The state of world fisheries and aquaculture 2020: Sustainability in action. FAO, Rome
FAO (n.d.) Blue ports initiative. https://www.fao.org/in-action/blue-ports-initiative/en (Accessed: 7 October 2025)
Firdaus-Nawi M, Zamri-Saad M (2016) Major components of fish immunity: A review. Pertanika J Trop Agric Sci 39:393–420
Fischer U, Utke K, Somamoto T, Köllner B, Ototake M, Nakanishi T (2006) Cytotoxic activities of fish leucocytes. Fish Shellfish Immunol 20:209–226
Australian Government Department of Agriculture, Fisheries and Forestry (n.d.) Harvest strategy policy. https://www.agriculture.gov.au/agricultureland/fisheries/domestic/harvest_strategy_policy (Accessed: 7 Oct 2025)
Frazen Tolentino-Zondervan NA, Zondervan N (2022) Sustainable fishery management trends in Philippine fisheries. Ocean Coast Manag 223:106149
Freedman SJ (1991) The role of alpha 2-macroglobulin in furunculosis: A comparison of rainbow trout and brook trout. Comp Biochem Physiol, B: Comp Biochem 98:549–553
Galluzzi L, Baehrecke EH, Ballabio A, Boya P, Bravo-San Pedro JM, Cecconi F, Choi AM, Chu CT, Codogno P, Colombo MI (2017) Molecular definitions of autophagy and related processes. EMBO J 36:1811–1836
Garcia I, Galiana A, Falco A, Estepa A, Perez L (2011) Characterization of an infectious pancreatic necrosis (IPN) virus carrier cell culture with resistance to superinfection with heterologous viruses. Vet Microbiol 149:48–55

García-Fernández C, Sánchez JA, Blanco G (2011) Characterization of the gilthead seabream (Sparus aurata L.) transferrin gene: Genomic structure, constitutive expression and SNP variation. Fish Shellfish Immunol 31:548–556

Gudding R (2014) Vaccination as a preventive measure. Fish Vaccination, 1st edn. John Wiley & Sons, Oxford, pp 12–21

Gudding R, Van Muiswinkel WB (2013) A history of fish vaccination: Science-based disease prevention in aquaculture. Fish Shellfish Immunol 35(6):1683–1688

Harikrishnan R, Balasundaram C, Heo M-S (2011) Fish health aspects in grouper aquaculture. Aquaculture 320(1–2):1–21

He L, Zhang A, Pei Y, Chu P, Li Y, Huang R, Liao L, Zhu Z, Wang Y (2017) Differences in responses of grass carp to different types of grass carp reovirus (GCRV) and the mechanism of hemorrhage revealed by transcriptome sequencing. BMC Genomics 18:452

Hedrick RP (1996) Movement of pathogens with the international trade of live fish: Problems and solutions. Rev Sci Tech 15:523–531

Hellio C, Pons AM, Beaupoil C, Bourgougnon N, Le Gal Y (2002) Antibacterial, antifungal and cytotoxic activities of extracts from fish epidermis and epidermal mucus. Int J Antimicrob Agents 20:214–219

Hermosillo OAM, Mart P, Ib AL, Ram HC (2012) Use of probiotics in aquaculture. Int Sch Res Not, Article ID 916845, 13

Hine P (1992) The granulocytes of fish. Fish Shellfish Immunol 2:79–98

Huntingford FA, Adams C, Braithwaite VA, Kadri S, Pottinger TG, Sandøe P, Turnbull JF (2006) Current issues in fish welfare. J Fish Biol 68:332–372

Hwang JA, Kim JE, Kim HS, Lee JH (2017) Immune response to koi herpesvirus (KHV) of koi and koi red common carp (Cyprinus carpio). Development & Reproduction 2017:361–370

Hwang JY, Lee UH, Heo MJ, Jeong JM, Kwon MG, Jee BY, Park C, Park JW (2020) RNA-seq transcriptome analysis in flounder cells to compare innate immune responses to low- and high-virulence viral hemorrhagic septicemia virus. Adv Virol 166:191–206

Jordanova M, Miteva N, Rocha E (2007) A quantitative study of the hepatic eosinophilic granule cells and rodlet cells during the breeding cycle of Ohrid trout (Salmo letnica Kar.). Fish Shellfish Immunol 23:473–478

Jung M-H, Jung S (2017) Innate immune response against rock bream iridovirus (RBIV) infection in rock bream (Oplegnatus fasciatus) following poly (I:C) administration. Fish Shellfish Immunol 71:171–176

Jurado MT, Garcia-Valtanen P, Estepa A, Perez L (2013) Antiviral activity produced by an IPNV-carrier EPC cell culture confers resistance to VHSV infection. Vet Microbiol 166:412–418

Koppang EO, Fischer U, Moore L, Tranulis MA, Dijkstra JM, Köllner B, Aune L, Jirillo E, Hordvik I (2010) Salmonid T cells assemble in the thymus, spleen and in novel interbranchial lymphoid tissue. J Anat 217:728–739

Kumar R, Gupta R, Dahiya T, Yadav R, Saifi R, Yodha K, Kumari B, Singh P, Shefali (2022a) Surveillance of Heteropneustes fossilis (Stinging Catfish) diseases and health management practices in Haryana. Appl Ecol Environ Sci 10(4):261–267. https://doi.org/10.12691/aees-10-4-12

Kumar R, Gupta RK, Yadav R, Saifi R, Yodha K, Pooja (2022b) Eisenia fetida as protein source for growth enhancement of Heteropneustes fossilis. Egypt J Aquat Biol Fish 26(2): 577–588. https://doi.org/10.21608/ejabf.2022.234553

Kumar R, Yadav R, Kumar Gupta R, Pooja (2022c) A sustainable way for fish health management by replacement of chemical and drugs by earthworm. In: Environmental degradation in Asia: Land Degradation, environmental contamination, and human activities, pp 329–352. Springer International Publishing, Cham. https://doi.org/10.1007/978-3-031-12112-8_16

Kumar V, Roy S, Meena DK, Sarkar UK (2016) Application of probiotics in shrimp aquaculture: Importance, mechanisms of action, and methods of administration. Rev Fish Sci & Aquac 24(4):342–368

Kurosaki T, Kometani K, Ise W (2015) Memory B cells. Nat Rev Immunol 15:149–159

Lauriano ER, Calò M, Silvestri G, Zaccone D, Pergolizzi S, Cascio PL (2012) Mast cells in the intestine and gills of the sea bream, Sparus aurata, exposed to a polychlorinated biphenyl, PCB 126. Acta Histochem 114:166–171

Leong JC, Trobridge G, Kim CH, Johnson M, Simon B (1998) Interferon-inducible Mx proteins in fish. Immunol Rev 166:349–363

Leung TLF, Bates AE (2013) More rapid and severe disease outbreaks for aquaculture at the tropics: Implications for food security. J Appl Ecol 50(1):215–222

Levraud JP, Jouneau L, Briolat V, Laghi V, Boudinot P (2019) IFN-stimulated genes in zebrafish and humans define an ancient arsenal of antiviral immunity. J Immunol 203:3361–3373

Li L, Chen SN, Laghari ZA, Hou HJ, Hou J, Huang L, Nan L, Nie P (2020) 'Myxovirus resistance (Mx) gene and its differential expression regulated by three type I and two type II IFNs in mandarin fish. Siniperca Chuatsi', Dev Comp Immunol 105:103604

Lockhart K, McBeath AJA, Collet B, Snow M, Ellis AE (2007) Expression of Mx mRNA following infection with IPNV is greater in IPN-susceptible Atlantic salmon post-smolts than in IPN-resistant Atlantic salmon parr. Fish Shellfish Immunol 22:151–156

Lorenzen N, Lapatra S (2005) Vacunas de ADN para peces de vivero. Revue Scientifique Et Technique De l'OIE 24(1):201–213

Magnadóttir B (2006) Innate immunity of fish (overview). Fish Shellfish Immunol 20:137–151

Magnadottir B (2010) Immunological control of fish diseases. Mar Biotechnol 12:361–379

Maier VH, Dorn KV, Gudmundsdottir BK, Gudmundsson GH (2008) Characterisation of cathelicidin gene family members in divergent fish species. Mol Immunol 45:3723–3730

Maqsood S, Singh P, Samoon MH, Munir K (2011) Emerging role of immunostimulants in combating the disease outbreak in aquaculture. Int Aquat Res 3:147–163

Martins ML, de Moraes FR, de Moraes JRE, Malheiros EB (2000) 'Falha na resposta do cortisol ao estresse por captura e por carragenina em Piaractus mesopotamicus Holmberg, 1887 (Osteichthyes: Characidae)', Acta Scientiarum. Biological Sciences 22:545–552

Mastan SA (2015) Use of immunostimulants in aquaculture disease management. Int J Fish Aquat Stud 2(4):277–280

Meseguer J, Esteban MA, Lopez-Ruiz A, Bielek E (1994) Ultrastructure of nonspecific cytotoxic cells in teleosts. I. Effector-target cell binding in a marine and a freshwater species (Seabream: Sparus aurata L., and Carp: Cyprinus carpio L.). Anat Rec 239:468–474

Milev-Milovanovic I, Majji S, Thodima V, Deng Y, Hanson L, Arnizaut A, Waldbieser G, Chinchar G (2009) Identification and expression analyses of poly (I:C)-stimulated genes in channel catfish (Ictalurus punctatus). Fish Shellfish Immunol 26:811–820

Mogensen TH (2009) Pathogen recognition and inflammatory signaling in innate immune defenses. Clin Microbiol Rev 22:240–273

Mokhtar DM (2017) Fish histology: From cells to organs. Apple Academic Press, Palm Bay, FL, USA

Mokhtar DM (2019) Characterization of the fish ovarian stroma during the spawning season: Cytochemical, immunohistochemical and ultrastructural studies. Fish Shellfish Immunol 94:566–579

Mokhtar DM (2021) Fish histology: From cells to organs, 2nd edn. Apple Academic Press, New York, NY, USA

Moreno P, Lopez-Jimena B, Randelli E, Scapigliati G, Buonocore F, Garcia-Rosado E, Borrego JJ, Alonso MC (2018) Immuno-related gene transcription and antibody response in nodavirus (RGNNV and SJNNV)-infected European sea bass (Dicentrarchus labrax L.). Fish Shellfish Immunol 78:270–278

Mugimba KK, Lamkhannat M, Dubey S, Mutoloki S, Munang'andu HM, Evensen O (2020) Tilapia lake virus downplays innate immune responses during early stage of infection in Nile tilapia (Oreochromis niloticus). Scientific Rep 10:20364

Nerbovik IKG, Solheim MA, Eggestol HO, Ronneseth A, Jakobsen RA, Wergeland HI, Haugland GT (2017) 'Molecular cloning of MDA5, phylogenetic analysis of RIG-I-like receptors

(RLRs) and differential gene expression of RLRs, interferons and proinflammatory cytokines after in vitro challenge with IPNV. ISAV SAV Salmon Cell Line TO', J Fish Dis 40:1529–1544

Neumann NF, Stafford JL, Barreda D, Ainsworth AJ, Belosevic M (2001) Antimicrobial mechanisms of fish phagocytes and their role in host defense. Dev Comp Immunol 25:807–825

Nutt SL, Hodgkin PD, Tarlinton DM, Corcoran LM (2015) The generation of antibody-secreting plasma cells. Nat Rev Immunol 15:160–171

Oidtmann BC, Peeler EJ, Thrush MA, Cameron AR, Reese RA, Pearce FM, Dunn P, Lyngstad TM, Tavornpanich S, Brun E et al (2014) Expert consultation on risk factors for introduction of infectious pathogens into fish farms. Prev Vet Med 115:238–254

Page DM, Wittamer V, Bertrand JY, Lewis KL, Pratt DN, Delgado N, Schale SE, McGue C, Jacobsen BH, Doty A (2013) An evolutionarily conserved program of B-cell development and activation in zebrafish. Blood J Am Soc Hematol 122:e1–e11

Pakingking R, Okinaka Y, Mori K, Arimoto M, Muroga K, Nakai T (2004) In vivo and in vitro analysis of the resistance against viral haemorrhagic septicaemia virus in Japanese flounder (Paralichtys olivaceus) precedingly infected with aquabirnavirus. Fish Shellfish Immunol 17:1–11

Peeler EJ, Taylor NG (2011a) The application of epidemiology in aquatic animal health—opportunities and challenges. Veterinary Res 42(1), article 94

Peeler EJ, Taylor NGH (2011b) The application of epidemiology in aquatic animal health—opportunities and challenges. Vet Res 42:94

Plant KP, Harbottle H, Thune RL (2005) Poly I:C induces an antiviral state against Ictalurid Herpesvirus 1 and Mx1 transcription in the channel catfish (Ictalurus punctatus). Dev Comp Immunol 29:627–635

Plant KP, LaPatra SE (2011) Advances in fish vaccine delivery. Dev Comp Immunol 35(12):1256–1262

Plumb JA, Hanson LA (2011) Health maintenance and principal microbial diseases of cultured fishes. Blackwell Publishing Ltd., Iowa, USA

Pridgeon JW, Klesius PH (2012) Major bacterial diseases in aquaculture and their vaccine development. CAB Rev: Perspect Agric, Vet Sci, Nutr Nat Resour 7, article 48:1–16

Rani S, Sharma S, Singh D, Sheoran R, Kumar R (2025) Improved production of anti-cancer compounds from medicinal plants through molecular and biotechnological mutagenesis strategies. In: Chen JT (ed) Biotechnology, multiple omics, and precision breeding in medicinal plants, pp 287–297. CRC Press. https://doi.org/10.1201/9781003475491-21

Reite OB (2005) The rodlet cells of teleostean fish: Their potential role in host defence in relation to the role of mast cells/eosinophilic granule cells. Fish Shellfish Immunol 19:253–267

Re-thinking Indonesian marine fisheries quota-based policy (2023) Ocean Coast Manag 243:106766. https://doi.org/10.1016/j.ocecoaman.2023.106766

Reyes-Lopez F, Romero JS, Vallejos-Vidal E, Reyes-Cerpa S, Sandino AM, Tort L, MacKenzie S, Imarai M (2015) Differential immune gene expression profiles in susceptible and resistant full-sibling families of Atlantic salmon (Salmo salar) challenged with infectious pancreatic necrosis virus (IPNV). Dev Comp Immunol 53:210–221

Rideout R, Smith S, Morgan M (2015) High-density aggregations of rodlet cells in the gonads of Greenland halibut Reinhardtius hippoglossoides, a deep-water marine flatfish. J Fish Biol 86:1630–1637

Riera Romo M, Pérez-Martínez D, Castillo Ferrer C (2016) Innate immunity in vertebrates: An overview. Immunology 148:125–139

Robertsen B (2008) Expression of interferon and interferon-induced genes in salmonids in response to virus infection, interferon inducing compounds and vaccination. Fish Shellfish Immunol 25:351–357

Rodrigues PM, Schrama D, Campos A, Osório H, Freitas M (2016) Applications of proteomics in aquaculture. In: Salekdeh GH (ed) Agricultural Proteomics, Volume 1: Crops, horticulture, farm animals, food, insect and microorganisms. Springer International Publishing, Cham, pp 165–199

Rogers C, Basurco B (2009) Antimicrobial agents in aquaculture: Practice, needs and issues. CIHEAM 59:41–59
Rosaeg MV, Lund M, Nyman IB, Markussen T, Asperhaug V, Sindre H, Dahle MK, Rimstad E (2017) Immunological interactions between Piscine orthoreovirus and Salmonid alphavirus infections in Atlantic salmon. Fish Shellfish Immunol 64:308–319
Sahoo S, Banu H, Prakash A, Tripathi G (2021) Immune system of fish: An evolutionary perspective. In: del Mar Ortega-Villaizan M, Chico V (eds) Antimicrobial immune response. IntechOpen, London, p 1. https://www.intechopen.com/chapters/78026 (Accessed 17 Dec 2022)
Saurabh S, Sahoo P (2008) Lysozyme: An important defence molecule of fish innate immune system. Aquac Res 39:223–239
Savoca MS, Brodie S, Welch H et al (2020) Comprehensive bycatch assessment in US fisheries for prioritizing management. Nature Sustainability 3:472–480. https://doi.org/10.1038/s41893-020-0506-9
Sayed RK, Zaccone G, Capillo G, Albano M, Mokhtar DM (2022) Structural and functional aspects of the spleen in Molly fish Poecilia sphenops (Valenciennes, 1846): Synergistic interactions of stem cells, neurons, and immune cells. Biology 11:779
Scapigliati G, Buonocore F, Randelli E, Casani D, Meloni S, Zarletti G, Tiberi M, Pietretti D, Boschi I, Manchado M et al (2009) Cellular and molecular immune responses of the sea bass (Dicentrarchus labrax) experimentally infected with betanodavirus. Fish Shellfish Immunol 28:303–311
Secombes C (1996) The nonspecific immune system: Cellular defenses. Fish Immune Syst: Org, Pathog Environ 15:63–103
Secombes C, Fletcher T (1992) The role of phagocytes in the protective mechanisms of fish. Annu Rev Fish Dis 2:53–71
Secombes C, Wang T (2012) The innate and adaptive immune system of fish. Infectious disease in aquaculture. Elsevier, Amsterdam, pp 3–68
Shao T, Zhu L-Y, Nie L, Shi W, Dong W-R, Xiang L-X, Shao J-Z (2015) Characterization of surface phenotypic molecules of teleost dendritic cells. Dev Comp Immunol 49:38–43
Sharifuzzaman SM, Austin B (2017) Probiotics for disease control in aquaculture shellfish. In: Austin B, Newaj-Fyzul A (eds) Diagnosis and control of diseases of fish and shellfish, 1st edn. Wiley & Sons Ltd., Oxford, pp 189–222
Shoemaker CA, Klesius PH, Evans JJ, Arias CR (2009) Use of modified live vaccines in aquaculture. J World Aquaculture Soc 40(5):573–585
Siderits D, Bielek E (2009) Rodlet cells in the thymus of the zebrafish Danio rerio (Hamilton, 1822). Fish Shellfish Immunol 27:539–548
Silva-Gomesa S, Decouta A, Nigoua J (2015) Pathogen-associated molecular patterns (PAMPs). In: Parnham ME (ed) Encyclopedia of Inflammatory Diseases. Springer, Basel, pp 1–16
Sitjà-Bobadilla A, Oidtmann B (2017) Integrated pathogen management strategies in fish farming. In: Jeney G (ed) Fish Diseases. Academic Press, Cambridge, MA, pp 119–144
Smith NC, Rise ML, Christian SL (2019) A comparison of the innate and adaptive immune systems in cartilaginous fish, ray-finned fish, and lobe-finned fish. Front Immunol 10:2292
Stosik MP, Tokarz-Deptuła B, Deptuła W (2018) Specific humoral immunity in Osteichthyes. Cent Eur J Immunol 43:335–340
Stuart LM, Ezekowitz RAB (2005) Phagocytosis: Elegant complexity. Immunity 22:539–550
Sveen L, Karlsen C, Ytteborg E (2020) Mechanical induced wounds in fish—A review on models and healing mechanisms. Rev Aquac 12:2446–2465
Taghavi M, Khosravi A, Mortaz E, Nikaein D, Athari SS (2017) Role of pathogen-associated molecular patterns (PAMPs) in immune responses to fungal infections. Eur J Pharmacol 808:8–13
Tavares-Dias M, Schalch SHC, Martins ML, Silva ÉD, Moraes FR, Perecin D (1999) Hematologia de teleósteos brasileiros com infecção parasitária. I. Variáveis do Leporinus macrocephalus Garavelo e Britski, 1988 (Anostomidae) e Piaractus mesopotamicus Holmberg, 1887 (Characidae). Acta Scientiarum Biol Sci 21:337–342

Timmerhaus G, Krasnov A, Nilsen P, Alarcon M, Afanasyev S, Rode M, Takle H, Jorgensen SV (2011) Transcriptome profiling of immune responses to cardiomyopathy syndrome (CMS) in Atlantic salmon. BMC Genomics 12:459

Uribe C, Folch H, Enríquez R, Moran G (2011) Innate and adaptive immunity in teleost fish: A review. Veterinární Medicína 56:486–503

Vallejo A, Miller N, Clem L (1991) Phylogeny of immune recognition: Role of alloantigens in antigen presentation in channel catfish immune responses. Immunology 74:165–172

Vallejo AN, Miller NW, Harvey NE, Cuchens MA, Warr GW, Clem LW (1992) Cellular pathway(s) of antigen processing and presentation in fish APC: Endosomal involvement and cell-free antigen presentation. Dev Immunol 3:51–65

Vallejos-Vidal E, Reyes-López F, MacKenzie S (2014) Immuno-stimulant diets and oral vaccination in fish. In: Austin B, Fyzul AN (eds) Diagnosis and control of diseases of fish and shellfish. John Wiley & Sons, Oxford, pp 77–89

Villalba M, Pérez V, Herrera L, Stepke C, Maldonado N, Freddericksen F, Yáñez A, Olavarría VH (2017) Infectious pancreatic necrosis virus infection of fish cell lines: Preliminary analysis of gene expressions related to extracellular matrix remodeling and immunity. Vet Immunol Immunopathol 193–194:10–17

Waagbø R (2006) Feeding and disease resistance in fish. In: Mosenthin R, Zentek J, Zebrowska T (eds) Biology of growing animals, vol 4. Elsevier, Amsterdam, pp 387–415

Wang F, Jiao H, Liu W, Chen B, Wang Y, Chen B, Lu Y, Su J, Zhang Y, Liu X (2019) The antiviral mechanism of viperin and its splice variant in spring viremia of carp virus infected fathead minnow cells. Fish Shellfish Immunol 86:805–813

Wessel O, Olsen CM, Rimstad E, Dahle MK (2015) Piscine orthoreovirus (PRV) replicates in Atlantic salmon (Salmo salar L.) erythrocytes ex vivo. Veterinary Res 46:26

Widziolek M, Janik K, Mojzesz M, Pooranachandran N, Adamek M, Pecio A, Surachetpong W, Levraud JP, Boudinot P, Chadzinska M (2021) Type I interferon-dependent response of zebrafish larvae during tilapia lake virus (TiLV) infection. Dev Comp Immunol 116:103936

Wu YC, Chi SC (2006) Persistence of betanodavirus in Barramundi brain (BB) cell line involves the induction of interferon response. Fish Shellfish Immunol 21:540–547

Xia S, Wang H, Hong X, Lu J, Xu D, Jiang Y, Lu L (2018) Identification and characterization of a type I interferon induced by cyprinid herpesvirus 2 infection in crucian carp (Carassius auratus gibelio). Fish Shellfish Immunol 76:35–40

Yan J, Peng L, Li Y, Fan H, Tian Y, Liu S, Feng H (2016) IFNα of triploid hybrid of goldfish and allotetraploid is an antiviral cytokine against SVCV and GCRV. Fish Shellfish Immunol 54:529–536

Yoshida SH, Stuge TB, Miller NW, Clem LW (1995) Phylogeny of lymphocyte heterogeneity: Cytotoxic activity of channel catfish peripheral blood leukocytes directed against allogeneic targets. Dev Comp Immunol 19:71–77

Yu F-F, Zhang Y-B, Liu T-K, Liu Y, Sun F, Jiang J, Gui J-F (2010) Fish virus-induced interferon exerts antiviral function through Stat1 pathway. Mol Immunol 47:2330–2341

Zaccone D, Icardo JM, Kuciel M, Alesci A, Pergolizzi S, Satora L, Lauriano ER, Zaccone G (2017) Polymorphous granular cells in the lung of the primitive fish, the bichir Polypterus senegalus. Acta Zoologica 98:13–19

Zhang J, Hu Y, Xiao Z, Sun L (2013) Megalocytivirus-induced proteins of turbot (Scophthalmus maximus): Identification and antiviral potential. J Proteomics 91:430–443

Zhang YB, Gui JF (2004) Identification of two novel interferon-stimulated genes from cultured CAB cells induced by UV-inactivated grass carp hemorrhage virus. Dis Aquat Org 60:1–9

Chapter 8
Sustainable Practices for Shrimp Disease Management

Urfi Siddiqui, Runit Isaac, and Sheersh Massey

Abstract Shrimp farming is a traditional practice that plays a vital role in global seafood production and economic growth. However, disease outbreaks caused by bacterial, viral, and parasitic infections pose significant threats to shrimp farming, leading to substantial economic losses. Traditional disease management practices have relied heavily on antibiotics and chemicals, which have resulted in antimicrobial resistance, environmental damage, and public health challenges. As a result, the development of sustainable shrimp disease management strategies is becoming increasingly important. These strategies include biosecurity measures, improved agricultural practices, the use of probiotics and phytobiotics, and vaccination to promote shrimp health and reduce disease incidence. Essential biosecurity practices such as water quality monitoring, quarantine protocols, and disinfection are crucial for preventing pathogen outbreaks. Probiotics and phytobiotics, which serve as alternatives to antibiotics, can enhance shrimp gut health and immune response, thereby lowering disease risks and protecting the environment. Additionally, targeting pathogens such as the White Spot Syndrome Virus (WSSV) and various Vibrio species through vaccination and immunostimulants can bolster disease resistance. Implementing polyculture systems and Recirculating Aquaculture Systems (RAS) can enhance productivity while minimizing environmental impact. These practices contribute to improved water quality, reduced waste, and decreased pathogen levels. Long-term reductions in disease susceptibility can be achieved through the genetic selection and breeding of disease-resistant shrimp strains. In conclusion, it is essential to focus on sustainable methods for shrimp disease management to ensure long-term benefits with minimal adverse effects. By reducing disease outbreaks, conserving

U. Siddiqui
Department of Biosciences, Jamia Millia Islamia, New Delhi, India

R. Isaac
Department of Chemistry, Sam Higginbottom University of Agriculture Technology and Sciences (SHUATS), Prayagraj, Uttar Pradesh, India

S. Massey (✉)
Human Genetics Laboratory, Department of Biosciences, Jamia Millia Islamia, New Delhi, India
e-mail: masseysheersh@gmail.com

R. Kumar et al. (eds.), *Sustainability in Aquaculture and Fisheries*,
Springer Water, https://doi.org/10.1007/978-3-032-16605-0_8

the environment, and minimizing chemical use while improving shrimp resistance, these strategies support the sustainability of shrimp farming.

Keywords Immunostimulants · Shrimp farming · Diseases · Phytobiotics · Polyculture

8.1 Introduction

Shrimp farming is the practice of raising or harvesting shrimp species in artificial environments. It is the fastest-growing sector in food production and offers an important and sustainable source of animal protein. However, to meet the increasing global demand, shrimp production must increase by 20% by the year 2050 (Report: Global Seafood Production Could Rise 20% to 2050—The Food Institute, n.d.; Mandal and Singh 2025). The crustacean industry has seen rapid development due to the high global demand for prawns and shrimp. Prawns and shrimp are significant in terms of market value and revenue, leading many aquaculturists, particularly in tropical countries, to focus on their cultivation as a primary protein source (Bondad-Reantaso et al. 2012). In 2023, commercial shrimp production was mainly dominated by two species of penaeid shrimp, with 95% being either *Litopenaeus vannamei* (white leg shrimp) or *Penaeus monodon* (black tiger shrimp) (Mandal and Singh 2025). In India, Andhra Pradesh is one of the leading producers of global shrimp farming (Singh et al. 2021). In 2021–2024, the production rate has increased to over 500,000 metric tons. The most dominant species for substantial growth of shrimp farming is Whiteleg shrimp (*L. vannamei*). The growing demand for aquaculture increases the economic production rate (Liao and Chien 2011). Proper management of the feed is yet another crucial factor in sustainability. Shrimp feeds have traditionally used fishmeal and fish oil as ingredients, with the consequence being overfishing and depletion of marine resources. Today, because of desire to reduce their dependency on wild fisheries, sustainable farms are exploring alternative sources of proteins like plant components, microbial protein, and insect meals. Also, advancements in feed formulation and nutrition technology have yielded more efficient feed conversion ratios, which reduce the cumulative impact on marine ecosystems (Kumar et al. 2022a). One of the key elements of sustainable shrimp farming is disease management. Shrimp farms have already suffered from viral disease outbreaks like Early Mortality Syndrome (EMS) and White Spot Syndrome Virus (WSSV) (Roy et al. 2020). The use of probiotics, selective selection for disease-resistant strains, biosecurity practices, and minimizing antibiotic misuse to avoid the development of antimicrobial resistance are all critical elements of sustainable practices (Kumar and Singh 2022; Butt et al. 2021).

Social responsibility has a big influence on sustainability. Certification programs such as Global G.A.P., Best Aquaculture Practices (BAP), and Aquaculture Stewardship Council (ASC) promote ethical work practices, transparency, traceability, and environmentally responsible business practices. Apart from assisting customers to make informed choices, these certifications also inspire producers to adhere to rigorous environmental standards. Eco-friendly shrimp farming is currently a necessity to the survival of the industry and not an option. Shrimp aquaculture can thrive in the future without putting the ecological health of the planet at risk by employing holistic practices that respect the environment and serve local people.

8.2 Shrimp Diseases

In an artificially created aquaculture environment, shrimp face a variety of adverse effects that are not present in the natural environment. These effects include physical and chemical factors such as water pollution, variation in water pH and temperature, low dissolved oxygen content, and poor quality or contaminated food (Páez-Osuna 2001). These factors can weaken the immunity of shrimps, induce stress, and lead to increased mortality and poor growth. As a result, the sustainability and profitability of shrimp production are significantly impacted. The intensive culture of shrimp also increases the likelihood of disease outbreaks, which can have a significant negative impact on shrimp farming (Kautsky et al. 2000; Páez-Osuna 2001). In 2020, reports highlighted that viruses and some bacterial and fungal infections were major contributors to the adverse effects on shrimp farming. It is worth noting that while approximately twenty different shrimp viruses are known, one particular virus has had a devastating impact on shrimp populations (Roy et al. 2020).

8.2.1 White Spot Syndrome Disease (WSSD)

WSSD is a DNA virus disease caused by White Spot Syndrome Virus (WSSV) and is highly contagious and deadly virus that affects approximately half of the shrimp population. It can cause 80–100% mortality in shrimps within the time span of 10 days (Roy et al. 2020; Sánchez-Martínez et al. 2007). It can thereby lead to a significant decline in shrimp farming production, causing substantial economic losses of up to $15 million USD. WSSV belongs to the virus family Nimaviridae, and it is the only member of the genus Whispovirus. This double-stranded circular DNA virus significantly impacts aquaculture production. About 100 species of shrimp can act as suitable hosts for WSSV, contributing to its rapid spread within shrimp farming (Desrina et al. 2022; Kumar et al. 2022b). It has also been shown that the WSSV disease can spread horizontally through the waterborne route or by feeding infected organisms such as rotifer eggs, microalgae, and polychaete worms (Kim et al. 2023). The symptoms of white spot disease (WSSD) include a 1–2 mm diameter white spot, enlargement and yellowish discolouration of the hepatopancreas, swelling of branchiostegites, loosening of the cuticle, and abnormal behavior (decreased swimming ability, disorientation, and swimming on one side) (Hossain et al. 2015). Immunological, molecular detection methods such as PCR, and morphological identification are among the diagnostic approaches used for WSSD (Talukder et al. 2021).

8.2.2 Infectious Hypodermal and Hematopoietic Necrosis Virus (IHHNV)

Infectious Hypodermal and Hematopoietic Necrosis Virus (IHHNV) is listed as a reportable crustacean disease pathogen by the World Organization for Animal Health (OIE) over the past 25 years. It is a deadly pathogen and can cause more than 50% economic loss in shrimp aquaculture. The prevalence of IHHNV infection is high in America, Australia, and Asian countries with huge mortality rates in *Penaeus stylirostris* (Blue shrimp) (Yu et al. 2021). The transmission of IHHNV can occur by cannibalism, contaminated water, or by infected eggs. The infection of IHHNV causes retarded growth known as runt-deformity syndrome in *Penaeus monodon* (Black Tiger Shrimp) and *Penaeus vannamei* (Pacific white shrimp) (Jagadeesan et al. 2019). Other clinical symptoms of IHHNV infection in shrimp include loss of appetite, morphological and behavioral changes. The IHHNV infection can be diagnosed by monitoring clinical signs, histopathology, in situ-hybridizations using a virus-specific gene probe, and PCR (Pazir et al. 2011).

8.2.3 White Tail Disease (WTD)

White-tailed disease often occurs in freshwater shrimp like *M. rosenbergii*, and exhibits high mortality rates leading to considerable economic losses. It is caused by *Macrobrachium rosenbergii* that is RNA virus and is accompanied with another virus XSV (extra small virus) (Arcier et al. 1999). It was first discovered in Guadeloupe (French West Indies) in the late 20th century and later in Australia, Martinique, and South Asia (Arcier et al. 1999; Lee et al. 2022). Lethargy, telson degeneration and opaqueness of the abdominal muscle, within 4 days of start, are among the clinical symptoms of WTD-infected shrimp (Sahul Hameed and Bonami 2012). Moreover, WTD infection starts in parts of the tail, spreads to the muscles in the abdomen, and in its advanced stages causes whitish coloring in every muscle (Sahul Hameed and Bonami 2012; Murwantoko et al. 2016). The virus can spread horizontally through formulations of M. rosenbergii infected with WTD, natural hosts of surrounding ecosystems and culture systems, dead tissue, live carriers, and free virions. RT-PCR and histopathology are the most sensitive diagnostic methods for the detection of WTD (Woodburyet al. 2008; Sahul Hameed and Bonami 2012).

8.2.4 Yellow Head Virus Genotype 1

The YHV genotype 1 is one of the most virulent viruses that were first discovered in *P. monodon* in Thailand, in 1990. It exhibits significantly high mortality rates and causes economic losses to the shrimp industry (Flegel 1997). It was first reported in *P. monodon* in 1990 and was taken under the category of notifiable disease by the OIE in 1995. YHV genotype 1 infection presents typical disease symptoms with yellow coloration of the cephalothorax and gills, but YHV-1 infection can exist for long periods without any signs of disease, such as with the WSSV outbreaks (Flegel 1997; Wijegoonawardane, Cowley and Walker 2008). YHV-1 can spread horizontally through the formula of an infected shrimp individual or by direct contact with the YHV-1 virus released into the water. Reverse transcriptase-polymerase chain reaction (RT-PCR), nested RT-PCR, loop mediated isothermal amplification (RT-LAMP), in situ hybridization, and real-time RT-LAMP are among the methods currently in use for YHV detection (Yang et al. 2016).

8.2.5 Taura Syndrome Virus (TSV)

TSV was first identified in prawn farms close to the mouth of the Taura River in the Gulf of Guayaquil, Ecuador, in the middle of 1992. Based on its virion structure, TSV has been categorized as a picornavirus (Griffith 2024). The cumulative mortality rates of the disease ranged between 60 to 90% of impacted pond-cultured juvenile *P. vannamei,* and it resulted in devastating losses. Disoriented shrimp, Reddish coloration of the body (especially the tail fan), black spots on the body parts, and softened shell are some of the prominent symptoms in shrimp infected with TSV (Nielsen et al. 2005; Griffith 2024). The transmission of TSV in shrimps can potentially occur through contaminated water or cannibalism. TSV infection in shrimp can be diagnosed histopathologically and based on the mentioned symptoms (Lightner 1999).

8.3 Medicinal Plants as Remedy

Medicinal plants have been widely used in shrimp farming to protect shrimp from deadly diseases. Using plant extracts for disease management provides an effective and beneficial treatment for pathogen infections. There are several strategies used to control and prevent bacterial and viral diseases in shrimp farming, including farming specific pathogen-free and resistant shrimp, maintaining limited stocking density, practicing polyculture, and using recombinant vaccines. To prevent disease outbreaks, the government should take preventative measures such as minimizing waste discharge into the water, implementing proper disinfection practices, and introducing pathogen-free shrimp species for stocking and breeding. Diseases in aquaculture have severe effects on human populations through the food chain. Therefore, it is important to adopt a strategy that is safe and reliable for aquaculture, humans, and the environment. Antibiotic and antiviral medicines have been used in aquaculture, but excessive use can lead to the development of resistant strains and harmful compounds (Holmström et al. 2003; Citarasuet al. 2022; Okeke et al. 2022). This has resulted in the discontinued use of these remedies in commercial production. The disadvantage of antibiotic treatment is that it is ineffective against viruses and antibiotic-resistant bacteria. In contrast, bioactive plants have several advantages. They pose a reduced risk of side effects, are harmless, do not cause water contamination, and do not produce resistant effects on pathogens (Okeke et al. 2022). Many aquatic and terrestrial plants have been tested against viral and bacterial infections in shrimps. These plants produce compounds that inhibit infections. The use of medicinal plants is an eco-friendly and safe alternative approach to fighting viral and bacterial contaminations. Medicinal plants possess antimicrobial and antiviral properties that can block the synthesis of bacterial walls and viral mRNA (Citarasu et al. 2022). They also contain secondary metabolites that are bioactive and immune-modulating, which can stimulate the shrimp's immune system to fight against infections. Additionally, the plants provide antioxidants that act as a shield to protect the shrimp from oxidative stress (Balasubramanian et al. 2007). A wide variety of medicinal plants are used as home remedies to treat certain types of infections caused by pathogens. For example, plant extracts provide a valuable remedy to treat the white spot syndrome (Balasubramanian et al. 2007). The use of beneficial plant extracts in shrimp feed can increase the survival rate by up to 60%. These plant extracts can be administered in various forms such as powder, crude extract, and aqueous solutions of bioactive compounds, and can be given to shrimp orally as feed supplements or via intra-muscular injections (Lee et al. 2024b).

8.3.1 Cynodon Dactylon

Cynodon dactylon is a widely cultivated herbal plant known for its benefits in treating diseases such as epilepsy. It is traditionally used to purify blood and treat conditions such as anuria, gonorrhea, and diarrhea (Sharma and Bafna 2012). A well-known herbal extract called RAVs or Sanjivani possesses antiviral properties that can combat white spot syndrome in *P. monodon* and *L. vannamei* (Balasubramanian et al. 2007). Although a recombinant DNA vaccine against WSSV has not been very successful due to high production costs and management of side effects, treating shrimp with natural methods has led to increases in both shrimp lifespan and production rate (Das et al. 2021). The herbal extract, obtained through aqueous and ethanol extraction methods, contains bioactive compounds such as catechin, syringic acid, quercitin, and epicatechin in ideal concentrations (Das et al. 2021). Lab tests have confirmed the efficacy of these extracts, showing their antiviral properties against WSSV infections.

C. dactylon extract is used as a therapeutic agent and feed additive to protect against diseases (Kaleeswaran et al. 2011).

8.3.2 *Xylocarpus granatum*

Xylocarpus granatum is a type of mangrove plant that has pharmaceutical properties, including antimicrobial, antibiotic, anti-diarrheal, and anti-cytotoxic activity in its leaf extract. This plant has been traditionally used by coastal communities to treat infections (Hadrich et al. 2014). Mangrove plants contain secondary metabolites such as flavonoids, phenolic compounds, and steroids, which possess antibacterial, antifungal, and antioxidant properties (Sulaiman et al. 2022). The active ingredients in the leaf extract of *X. granatum* have immune-stimulatory effects, increasing the activity of phagocytic cells, which helps in treating *V. harveyi* infections in shrimp (*Penaeus monodon*). When tiger prawns (*P. monodon*) consume these leaf extracts in water for three days, it enhances their immune granulocytes and hyaline cells without any side effects (Saptiani et al. 2020).

8.3.3 *Eleutherine bulbosa*

Eleutherine bulbosa acts as an immunomodulator, protecting and providing resistance to the white shrimp Litopenaeus vannamei against bacterial intestinal infection from Vibrio parahaemolyticus (Munaeni et al. 2020a). The extract of *E. bulbosa* contains bioactive components, including flavonoids, tannins, steroids, quinones, and phenolic compounds. These metabolites were found effective against pathogenic organisms such as *Aeromonas hydrophila*, *Vibrio parahaemolyticus*, and *V. harveyi*. *E. bulbosa* contains oligosaccharides inulin and raffinose, which act as a probiotic for increasing the gut microflora (Munaeni et al. 2019, 2020b). *E. bulbosa* is an effective antioxidant that removes the DPPH (2.2-diphenyl-1-picrylhydrazyl) free radical and protects the cell membrane from damage. Supplementation of shrimp feed with *E. bulbosa* powder significantly increases the cell-mediated and humoral immune responses to infections (Cortez-Mago et al. 2025). The addition of *E. bulbosa* extract or powder to the feed improves the gastric parameters and growth and survival of shrimp. Optimal dosage and usage of the extract are significantly important for enhancing the performance and survival rate of shrimp species (Munaeni et al. 2020b).

8.3.4 *Aloe baradensis (Aloe Vera)*

Aloe baradensis (Aloe vera) is a perennial plant belongs to the Alaeaceae family. It is a wound healing, immunostimulants and antibacterial agent consist mainly water, hemicellulose and mannose derivatives. It also contains immunostimulant compounds Aloe emodin which possess antiviral activity (Boudreau and Beland 2006). It increases the survival rate of *L. vannamei* infected with both virus (WSSV) and bacteria (*Vibrio parahaemolyticus*) (Trejo-Flores et al. 2016). Addition of aloe vera extract in shrimp feed increase the growth, appetite phagocytic activity and immune cell response. Aloe vera has enhanced immunity by increasing the expression Hsp70, and SOD (Superoxidase dismutase) as defense mechanism for protection against the pathogenic organism and antioxidant stress (Ramírez et al. 2012). Aloe vera also stimulates the absorption of nutrients by shrimp. It is a safe and natural process for the protection of shrimp farms from viral and bacterial infection (Trejo-Flores et al. 2016).

8.3.5 *Centella asiatica*

Centella asiatica is a subtropical plant belongs to the Apiaceae family. It has great traditional medicinal value used in the treatment of leprosy, ulcers and vein diseases. Extract of *C. asiatica* used in animal feed to treat the fungal and parasitic infections (Gohil et al. 2010). The extract contains bioactive ingredients such as centelosides, saponine, madecassic acid, brahmoside and centoic acid (Gohil et al. 2010). *Macrobrachium rosenbergii* fed with diet containing the medicinal *Centella asiatica* plant extract, improves gut health, growth performance and protects the shrimp from the pathogenic organism-*Vibrio harveyi and Aeromonas hydrophila* (Barad et al. 2024) (Table 8.1).

8.4 Sustainable Approach

Shrimp farming has been an important industry for food production, providing employment and improves economic growth in developed countries. Major challenges are environmental degradations, loss of habitat, water contamination, destruction of forest ecosystem, untreatable discharge of waste into water bodies, and overuse of antibiotics produces the negative effects on the growth and survival rate of shrimp species, which leads to loss of diversity of some economical important species (Ponce-Palafox et al. 1997; Furtado et al. 2011). To address these challenges, aquaculture sector produces the sustainable approach to conserve the shrimp and improve their growth performance.

Table 8.1 Medicinal plants with its bioactive compounds used against the shrimp diseases

Sl. No.	Medicinal plant	Active ingredient	Against pathogen	Shrimp	Beneficial effects	References
1	*Olea europaea* (Olive)	Oleuropei, verbascoside, rutin, apigenin, luteolin	*Campylobacter jejuni*	*Penaeussemisulcatus*	Antibacterial, antioxidant	Ahmed et al. (2014)
2	*Phyllanthus emblica*	Sesquiterpene, tannins, cineole	*Vibrio parahaemolyticus*	*Penaeus monodon*	Immunomodulant	Ghosh et al. (2024)
3	*Zingiber officinalis*	Artemia	Photobacterium damselae	*P. monodon*, *Litopenaeus vannamei*	Improve growth performance, feeding efficiency	Shahraki et al. (2021)
4	*Citrus limon*	Citric acid, ascorbic acid, limonoids, essential oils	Bacillus, vibrio	*P. vannamei*, *Litopenaeus vannamei*	Antimicrobial, antioxidant, Antibacterial, anticancer, wound healing	Lee et al. (2024a)
5	*Allium sativum*	Allicin, sulfur compounds	Gram+ bacteria—Vibrio, E. coli, Micrococcus	*P. monodon*	Antimicrobial, immunostimulants,	Chirawithayaboon et al. (2020)
6	*Moringa oleifera*	Quercetin, chlorogenic acid, isothiocyanates	*V. alginolyticus*	*P. vannamei*	Enhances immune response, antifungal, antiviral, antibacterial, improve growth	Abidin et al. (2021)
7	*Curcuma longa*	Curcumin	Vibrio species	*P. monodon*, *Litopenaeus vannamei*	Improve survival, growth and feeding efficiency	Lawhavinit et al. (2010)
8	*Mentha spicata*	Naringin, quercitin	Bacteria, yeast, fungi	*P. monodon*	Improve gut health, immunomodulant	Naidu et al. (2014)
9	*Eucalyptus globulus*	Eucalyptol, cineole, rutin, saponin, tannin	*V. parahaemolyticus*	*P. monodon*	Improve gut health, survival rate	Sivaraj et al. (2017)
10	*Psidium guajava*		*V. cholerae*	*P. vannamei*	Disease resistance, improve nutrient absorption	Dewi et al. (2021)

(continued)

Table 8.1 (continued)

Sl. No.	Medicinal plant	Active ingredient	Against pathogen	Shrimp	Beneficial effects	References
11	*Azadirachta indica*	Azadirachtin, nimbin, salannin	Enterocytozoon hepatopanaei	*L. vannamei*, *P. monodon*	Hepatopancreative, immunomodulant	Madesh et al. (2024)
12	*Ocimum basilicium*	Rosmarinic, caffeic acid	*V. parahaemolyticus*	*P. indicus*	Provide resistance against pathogen, innate immunity, tissue repair	Abdel-Tawwab et al. (2021)

8.4.1 Integrated Multi Trophic Aquaculture (IMTA)

Integrated multi-trophic aquaculture system helps in maintains the ecological stability or balance by integrated farming with the introduction of other species such as algae, mollusk and fishes (Sukhdhane 2018). These species occupy their niche or trophic levels to avoid the competition with other species. Shrimp produce the waste in the water which is utilized by filter feeding organism and nutrient rich water supports the growth of algae. This eco-friendly approach improves the water quality, reduce the accumulation of toxins in the water bodies, and produces the high productivity which supports economic growth (Chang et al. 2020). This system is most practiced in northern and central coast of China. ITMA promotes the recycling of nutrients, which supports the growth shrimp farming (Chang et al. 2020).

8.4.2 Closed Recirculating Aquaculture System (CRAS)

Recirculating aquaculture system (RAS) is the innovative environmental friendly approach to reduce the negative impact on shrimp culture. Unlike the open system, the closed RAS system prevents the exchange of water with the surrounding environment, and reduces the risk of outbreaks of diseases and pollution. This approach is significantly important because less water consumption allows the same water to be used repeatedly, and reduces the disease and contaminants to enter the system (Suantika et al. 2018). This system consists of mechanical filter used to capture solid waste, such as shrimp feces and biological filter removes the toxic ammonia effluents. Water is pretreated with UV lights; before it is recirculate back into the shrimp tanks. Proper aeration or oxygenation is provided to ensure the optimal growth of shrimp in oxygen rich environment. These systems include the temperature and pH regulator to maintain the optimum condition for shrimp. RAS is highly sustainable footstep, prevents the habitat destruction, water pollution, and eutrophication (Nugraha et al. 2023). Water is pretreated and recirculates, reduces the chances of introduction of bacterial, fungal and viral infections to the shrimp. Unlike the traditional farming, RAS can be practiced in both rural and urban area, do not require large land for farming (Badiola et al. 2012).

It is done in small tanks with controlled conditions, also promotes the integration system where multiple tanks are stacked leads to the optimize the land use. This system can be more sustainable for long run, by utilizing the solar energy for power requirements. Certain challenges are faced by RAS system, includes high cost set up, technical expertise, energy consumptions. To overcome the challenges, RAS improved their efficiency and affordability, so that easily accessible to the farmers (Badiola et al. 2012).

8.4.3 Eco-Certification and Best Management Practices

As people become more aware about the importance of aquaculture, eco-certification programme such as Aquaculture stewardship council (ASC) and Global aquaculture alliance (GAA) provide information to the customers about the product. GAA focuses on issues food security and environmental sustainability (Rector et al. 2023; Hammarlund et al. 2024).

8.4.4 Feed Innovation

With shrimp feed comprised of components and formulations provided sustainably, sustainable shrimp farming relies on choosing shrimp feed which contains reduced fishmeal as well as fewer fish oils. This is a deliberate choice when using other protein sources rather than the traditional use of fishmeal and fish oil from wild fisheries, such as fishmeal and fish oil, in shrimp feed. Large-scale implementations of such traditional diets for marine shrimps have increased ecosystem disturbance and exploitation (Darodes de Tailly et al. 2021).

In order to reduce dependency on limited marine resources, sustainable shrimp farming relies heavily on responsible feeding techniques. This strategy promotes the use of ecologically friendly shrimp feed materials, guaranteeing that the nutritional requirements of shrimp are effectively satisfied. In addition to reducing the environmental impact of shrimp farming, the thoughtful selection of shrimp feed that includes alternative protein sources helps to promote a more sustainable and balanced ecology (Vanichkul et al. 2010). This thoughtful approach emphasizes the need of making responsible decisions when formulating shrimp feed and is consistent with the overall objective of sustainable farming (Al Eissa et al. 2022). Such a strategy helps in dilution of pressure on the wild fish stock as there is much less reliance on fishmeal and fish oil. Alternative protein sources such as insect-based meals can be produced more sustainably, leaving behind a smaller carbon footprint than the fish meal (Obaldo et al. 2002).

8.5 Mangrove Conservation and Restoration

Mangroves are peculiar coastal ecosystems that play a crucial role in biodiversity, shoreline protection, and the very important lives of many communities across the coast. Shrimp farming is believed to be among the main causes of mangrove destruction on large scales, especially in Southeast Asia and Latin America (Ashton 2008). The removal of mangroves for shrimp ponds depresses the local ecosystems, reduces carbon sequestration, and ruins the capabilities of coastal protection that make them prone to storms and erosion. An issue that can be brought into solution by simply implementing a sustainable approach to shrimp farming, involving mangrove conservation and restoration, can be addressed. Shrimp farming can be balanced with economic growth while preserving the environment through measures such as restoring degraded areas, reducing the destruction and impact of shrimp farming on mangroves, and following environmentally friendly aquaculture practices (Alam et al. 2022).

In order to minimize the impacts that shrimp farming creates on mangroves, there should be an embrace of responsible practices which encourage mangrove-friendly aquaculture. These include.

8.5.1 Integrated Mangrove-Shrimp Farming (Silvo-Aquaculture)

This is the integrated shrimp farming within the mangrove by putting shrimp ponds within mangroves without clearing the entire forest. Mangrove and shrimp are made to coexist, through benefiting off each other's existence. This promotes the several benefits:

(a) Natural Balance within an Ecosystem: Mangrove forests naturally filter the water, absorb nutrients, and are a habitat for shrimp and other marine living organisms thus reducing any chemical needed and enhancing the quality of the water (Bosma et al. 2016).
(b) Protecting against the Coastal Erosion: Mangroves help stabilize the soil, and this makes the coast end not erode easily thus protecting the shrimp farms from storm damage.
(c) Conservation of Biodiversity: This approach conserves both marine and terrestrial biodiversity since it conserves the natural mangrove ecosystem.

8.5.2 Mangrove Restoration Project

(a) Replantation and rehabilitation: Degraded mangrove areas can be rehabilitated using indigenous species seeded within the area. This helps reconstruct the natural habitat and ecosystem services.
(b) Social Participation: Involving local communities in mangrove restoration makes living sustainable. Educating and training local farmers on the importance of the mangrove builds upon better means of conserving and rehabilitating these critical ecosystems (Datta et al. 2012).
(c) Mangrove rehabilitation: Supports the efforts on a global platform to counter climate change because it enhances the capacity of coastal regions to seize carbon. Program, for instance, Blue Carbon promotes the protection and restoration of coastal ecosystem as a measure to combat climate change (Song et al. 2023).

8.5.3 *Zero Exchange System (ZES)*

The zero-exchange system also referred to as the biofloc system or closed-loop system is a new and sustainable shrimp-farming technology with the potential for countering some of the greatest environmental concerns that have long been linked to aquaculture (Nguyen 2008). Conventional shrimp farming relies heavily on water exchange to manage water quality but results in large-scale water consumption, pollution, and disease breakouts. On the contrary, however, the zero-exchange system minimizes or eliminates the exchange of water by recycling it and reusing the water in the same farm, reducing environmental degradation while at the same time improving shrimp production efficiency (Burford et al. 2004).

8.6 Challenges Faced by Traditional Shrimp Farming

Traditional shrimp farming has been associated with several environmental issues mainly owing to the large amounts of water used and constantly discharged. It can consume large amounts of water in areas where freshwater is scarce, while its discharge of nutrient-rich wastewater into natural water bodies could also produce various consequences such as eutrophication, harmful algal blooms, and degradation of aquatic ecosystems due to water pollution. In addition, shrimp farms frequently apply antibiotics and chemicals to prevent diseases, but such treatments may lead to antibiotic-resistant bacteria as well as contamination of the surrounding environment (Paul and Vogl 2011). These issues need a more sustainable and responsible shrimp farm system, and the Zero Exchange System addresses this issue.

8.6.1 *Process of Zero Exchange System*

The Zero Exchange System works on the principle of creating a self-sustaining ecosystem in a shrimp farm whereby water constantly gets recycled and reused without regular exchange. This becomes possible because of biofloc technology, which incorporates helpful microorganisms. These microorganisms' cultivation maintains the water quality and recycles nutrients. Biofloc is a collection of bacteria, algae, and other microorganisms that aggregate in the water in what is called "flocs." Such bioflocs are very important in the Zero Exchange System because they degrade organic wastes, including shrimp excreta and uneaten feed, into microbial biomass. The microbial biomass can then be provided to the shrimp as a natural, nutrient feed supplement that reduces reliance on external feed inputs (Suita et al. 2015).

In addition to biofloc technology, the system utilizes mechanical filtration, aeration, and correct pond management for the maintenance of a healthy, balanced aquatic environment. The aeration systems are believed to suspend the bioflocs in water and thus ensure an ample supply of oxygen to both the shrimp and the microorganisms (Wasielesky et al. 2006). As the system controls water quality using both biological and mechanical means, it reduces or eliminates the need to exchange the water used in the cultivation, saving water and preventing the detrimental effects of polluted water discharge into the environment around it.

8.6.2 Benefits of Zero Exchange System

One of the greatest merits of the Zero Exchange System is low water consumption. Owing to repeated recycling and reuse of water, ZES could manage with a fraction of the water consumed by the traditional shrimp farms. Hence, it is highly attractive in areas where water is scarce or costly. The other advantage of the system is its closed-loop nature that does not allow the nutrient-rich wastewater to find their way into local water bodies and thus negates the possibility of environmental pollutions and destruction of the neighboring ecosystem like mangrove and estuaries.

In addition, with ZES, another major good benefit is the improved health and productivity of shrimps. The presence of beneficial microorganisms in the biofloc suppresses any harmful pathogens and hence reduces the disease occurrence. This would result in shrimp that will be raised without the use of antibiotics and other chemicals, thus maintaining a health status and providing safer products to consumers (Wasielesky et al. 2006; Zhao et al. 2012). Additionally, bioflocs add supplement protein and other essential nutrients, giving more accelerated growth rates and better feed conversion ratios. Translation into higher yields and improved profitability can be achieved with shrimp cultivation.

8.7 Biofloc Technology

This system appears to be one of the core principles that define biofloc technology: a bio controlled, self-sustaining ecosystem inside the shrimp pond or tank, whereby the shrimp's uneaten feed, excreta, and other wastes are utilized by beneficial microorganisms like bacteria, algae, and protozoa to create bioflocs. In this process, organic waste, such as nitrogen compounds like ammonia, is converted into microbial biomass. These biofloc aggregates then become a natural feed for the shrimp, providing it with additional nutrition and lessening the quantity of exogenous feeds to be added (Zhao et al. 2012).

In maintaining an efficient and healthy biofloc, the correct proportions of carbon, nitrogen, and other nutrient balances are necessary. The water, according to shrimp farming, is supplemented with carbonaceous sources from molasses or rice bran, which fosters the growth of favorable bacteria that feed on the nitrogenous wastes (El-Sayed 2021; Yu et al. 2024). Sufficient aeration ensures optimal conditions for the suspensions of biofloc within the water column and adequate oxygen levels for both shrimp and microorganisms.

8.7.1 Environmental Benefits

Probably one of the biggest advantages of biofloc technology is its environmental-friendly nature. In traditional shrimp culture, water exchange at high intensity is a prerequisite for maintaining water quality and, eventually, it leads to the discharge of nutrient-rich wastewater into nearby water bodies, causing pollution and inducing eutrophication. In a biofloc system, there is usually minimal or zero water exchange as the waste products is recycled within the same system. This will help in conserving water and at the same time not dispose of pollutants into the environment that are protecting natural ecosystem, such as rivers, lakes, and coastal areas (Zhao et al. 2012).

However, the use of biofloc technology also avails such benefits that the exchange of water is reduced, and if the region is in an area where water is scarce or even with regions where the use of water is expensive, then applying biofloc technology will be useful. Water reuse within the system leads to a substantial saving of water, thus making shrimp farming more sustainable and environmentally responsible (Yu et al. 2024).

8.7.2 Economic Benefits

Biofloc technology has the following economic benefits: cutting down feed expense, as biofloc can be consumed by shrimp as a natural and nutritious feed that supplements their diet, thus allowing less commercial feed to be used. Decreased production costs and higher profit margins for the farmer are some of the outputs from this scenario. Additionally, biofloc has a lot of protein and other nutrient factors that will facilitate a potential increase in growing rates for shrimp and efficiencies of feed conversion.

The ability to improve health and survival for shrimp also leads to economic benefits (Kasan et al. 2018). This is because of the microorganisms in the biofloc contributes to disease control because they can outcompete pathogenic organisms for any available resources, thus reducing the likelihood of disease outbreaks. This, therefore, means that the farmers reduce or eliminate the use of antibiotics and chemicals (Zhao et al. 2012). Healthy shrimp results when there is reduced use of antibiotics, as well as safer products without antibiotics for consumers. Reduction of antibiotics decreases costs but will still address concerns about antibiotic resistance and food safety.

8.8 Polyculture System

This technique is an innovative and sustainable system of shrimp culture, in which a number of species share the same aquaculture environment. It is not like the traditional monoculture systems, where all the energy goes into the raising of single species. In this method, shrimp are integrated with other compatible aquatic species such as fish, seaweeds, or mollusks so that by natural species interactions, a balanced ecosystem is formed that will be self-sustaining, providing environmental, economic, and social benefits. The higher the interest for polyculture as an environmentally more acceptable alternative to the conventional shrimp farming. It deals with a host of issues that include resource use efficiency, waste management, and disease control (Martínez-Porchas et al. 2010).

In the polyculture system, shrimp are farmed in partnership with other species that complement each other in terms of habitat and feeding habits. The species is selected accordingly but depends on particular environmental conditions and local biodiversity. General examples include shrimp with tilapia, shrimp with different types of seaweeds, or shrimp with bivalves such as mussels or oysters. Each species in these groups would have a unique role in the ecosystem while maintaining water quality and reducing waste. For instance, excreta and leftovers from the shrimp can become aquatic pollutants if not managed properly (Tian et al. 2001). The shrimp reared in such polyculture systems, such as tilapia, consume the said organic wastes in a bid to reduce pollution and maintain water purity. Excess availability of nutrients, such as nitrogen and phosphorus, is absorbed by seaweeds or aquatic plants for additional water purification purposes. It limits the amount of waste material produced and makes the environment healthier for all these species (Junior et al. 2012).

8.8.1 Benefits of Polyculture to the Environment

One of the most notable benefits of this polyculture system is environmental sustainability. Conventional forms of shrimp farming often lead to water pollution, especially because of organic waste buildup and nutrient overload, which can be destructive to the local ecosystem. Waste products are fed by one species as a resource in polyculture, hence drastically minimizing the overall negative effects on the environment. This recycling of nutrients almost stimulates natural ecosystem and hence makes the farming process more efficient and sustainable (Martínez-Porchas et al. 2010).

The most significant advantage of water conservation is that, through polyculture, several species are brought into aquaculture together. Thus, there will be less replenishment of water. This process is highly useful in dry regions where the availability of water resources is inadequate. Apart from this, due to natural biological processes, polyculture systems maintain the quality of water by preventing eutrophication as well as other types of water pollution, conserving neighboring aquatic ecosystem and maintaining biodiversity (Martínez-Porchas et al. 2010; Junior et al. 2012).

8.8.2 Economic Benefits

With polyculture systems comes environmental friendliness, so there is also economic value. One of the major economic gains is a high production. For instance, from the same area yields are harvested at various times or in multifarious ways, hence there are income streams that are more diversified and also reduce risks faced by many farms. Instead of selling shrimp, a farmer can sell both shrimp and fish or seaweed. Income from other species could cushion loss of income due to shrimp production problems brought about by diseases or market fluctuations (Citarasu et al. 2022).

Feed costs are also lower and therefore more economically viable for the polyculture systems. In a monoculture system, feed supply is only on shrimp, which in many cases may be very costly. In a polyculture system, some species, like tilapia or carp, may feed on natural feed found in a system, such as algae and plankton or even organic wastes produced in the system (Tian et al. 2001; Junior et al. 2012). This would decrease the dependency on commercial feeds, save costs, and increase profitability. Other species, such as seaweeds or shellfish, require less feeding and can be associated with shrimp to tap into the existing nutrients within the water.

8.9 Reducing Antibiotic Use

Shrimp farming has gained a lot in the last decades and is currently one of the most profitable sectors in world aquaculture. However, the methods applied in intensive shrimp farming often suffer from major problems such as huge disease outbreaks. In response, many shrimp farmers have relied on antibiotics to control disease and production. But antibiotics have a lot of side effects such as antibiotic resistance, pollution of the environment, and inhumane health hazards against human beings. As concerns with these issues grow, reducing antibiotic use is now increasingly recognized as a sustainable option for shrimp aquaculture-both for the industry and the environment (Luu et al. 2021). Reducing the use of antibiotics in shrimp farming is no way tantamount to compromising on shrimp health and production (Suzuki et al. 2023). The erstwhile situation has only led to alternative a method for the use of which has proven to be effective as well as sustainable. A few of them include.

8.9.1 Probiotics and Prebiotics

Probiotics are best known as beneficial bacteria that can be introduced into shrimp ponds to advance a healthy gut microbiome, with improved shrimp immunity. Probiotics reduce the use of antibiotics since they build the defensive structures within the shrimp. Prebiotics are undigested compounds, and good bacteria often use it as food during digestion, thereby growing stronger. Thus probiotic combined with prebiotic can provide an environment that balances itself and can be applied in shrimp ponds without causing diseases within them (Butt et al. 2021).

8.9.2 Vaccine

Vaccines for shrimp that are presently under research and development. It might one day be used as a prophylactic prevention against bacterial infections. It will, one day, yield long-term immunity to specific pathogens, curtailing the use of antibiotics. There is much research and testing needed before this will become a viable form of disease management in shrimp farming (Johnson et al. 2008).

8.9.3 Best Management Practices

Sometimes simple but effective management practices in the farms may help in reducing disease incidence in shrimp farms. For instance, monitoring of some of the water quality parameters, such as pH, temperature, dissolved oxygen, and ammonia, would avoid stress that may induce susceptibility to infections in the shrimp. Proper stocking densities, removal of dead shrimp, and good hygiene and biosecurity measures would prevet diseases and minimize the use of antibiotics.

8.9.4 Plant-Based Immunostimulants

Plant-based immunostimulants, such as herb or spice extracts, can stimulate the immune system of shrimp. Such natural products would strengthen the disease resistance of shrimp and even improve their overall health status. Since they are organic and biodegradable, they do not pollute the environment and are therefore considered an ideal substitute for antibiotics (Kumar et al. 2022a, 2025; Rani et al. 2025).

8.10 Case Study of Shrimp Management in Andhra Pradesh

Andhra Pradesh, located on the southeast coast of India, has emerged as a prominent hub for shrimp aquaculture and a significant contributor to the country's exports of seafood. The Pacific white shrimp or Litopenaeus vannamei is the species that the state primarily cultivates due to its fast growth rate and greater market value. Groundwater salinization, epidemic outbreaks, antibiotic overuse, and biodiversity decline are just a few among the environmental and socioeconomic problems generated by the uncontrolled expansion of the shrimp culture industry in this region over the last 20 years. The problems prompted an action towards more sustainable shrimp farming practices, and that encouraged starting various programs intended to promote responsible aquaculture practices (Muralidhar et al. 2021).

The cluster farming approach, promoted by the Andhra Pradesh Shrimp Quality Improvement Program (APSQIP), is one of the strongest approaches employed in the region. The idea groups small farmers into clusters based on shared assets and proximity. Ten to twenty farms typically form each cluster and have the same protocols for biosecurity, disease management, and water use. Besides ensuring cost-effectiveness, this joint strategy helps to maintain consistency in production standards (Umesh et al. 2010; Koteswari et al. 2014). The risk of contamination and disease spread between farms is also reduced by common infrastructure like settling ponds, reservoir tanks, and water filtering systems. Effective water and waste management is key to successful shrimp production. To manage water flow and quality, most farmers in Andhra Pradesh employed improved water management methods, including the utilization of input and exit canals. To maintain a healthy microbial population and reduce dependence on pesticides and antibiotics, probiotics are often incorporated in ponds (Yadav et al. 2022a, b, 2023; Muralidhar et al. 2021). Adequate oxygenation to enable the growth and survival of shrimp as optimally as possible is ensured using aerators. Sediment tanks and sludge removal methods are also employed to reduce the organic load as well as improve water clarity, which improves the overall health of the shrimp.

Implementation of traceability and certification systems is another significant trend in the field. Today, many farms are accredited under internationally recognized programs such as Best Aquaculture Practices (BAP) and the Aquaculture Stewardship Council (ASC). Such certifications require compliance with high environmental, social, and food safety standards in addition to giving entry to premium export markets. To enhance openness and trustworthiness, traceability systems based on QR codes have been implemented to allow consumers and buyers to track the shrimp supply chain from start to finish (Murray 2025). At the state level, a disease surveillance and response program has been developed to manage disease problems. Biosecurity measures, quarantine procedures, and early disease detection are all instructed to farmers. Only authorized hatcheries can supply shrimp seed, and regular monitoring helps in early detection and management of diseases like White Spot Syndrome Virus (WSSV). In addition, a shift towards plant-based or alternative protein sources has been encouraged, reducing the pressure on wild fish supplies used in traditional feed (Lee et al. 2024a).

8.11 Conclusions

Sustainable shrimp farming has become one of the most important issues for the aquaculture industry, striking a balance between economic productivity and both environmental stewardship and social responsibility. Present day traditional shrimp farming practices characterized by high water exchange rates, heavy use of antibiotics, and significant waste discharge often result in environmental degradation, prevalence of diseases, and economic instability (Lightner 1999). However, now the industry realizes that the adoption of more sustainable practices is the only way this industry will ensure long-term profitability while trying to minimize ecological impacts and community welfare. Since sustainable approaches in shrimp farming are actually very much needed for the practice of dealing with ecological, economic, and social burdens haunting the industry for several decades, this study overviewed the technologies, such as biofloc, polyculture, and zero-exchange systems, and minimization of antimicrobial usage in its practices and the protection of mangrove ecosystems, in itself paving the way to a responsible and resilient shrimp farming industry. These methods offer viable solutions that allow farmers to be productive with the usage of available natural resources and minimal environmental impact. Additionally, sustainable practice would raise possibilities to attract new customers, build consumer confidence, and enhance the contribution of shrimp farming towards food security.

The future of shrimp farming will depend upon the capacity of the industry to innovate toward sustainability (Nguyen 2008; Junior et al. 2012; Sukhdhane 2018; Rector et al. 2023). This is, for example, how the aquaculture industry can create green, robust shrimp farming systems that cater to the ever-growing demand for seafood by promoting responsible practices and investment in education, research, and technology. Sustainable shrimp farming isn't just a necessity, but also an opportunity for a thriving industry that supports communities, protects ecosystems, and delivers on the promise of safe, nutritious, and high-quality food to consumers worldwide.

8.12 Future Perspective

Future sustainable shrimp farming focus on ongoing advancement and the practices of environmentally friendly techniques that sustains environmental integrity as well as increases productivity. Integrated multi-trophic aquaculture, Recirculating aquaculture systems and biofloc for examples are vital to lowering waste and disease outbursts. Artificial intelligence based management policies and real-time monitoring are two crucial farming methods that will further enhance resource usage and decrease environmental effects. Growing studies of alternate protein sources for prawn feed, like insect or plant based diets, can diminish dependence on fishmeal and increase the sustainability of production. Adopting the sustainable approach is a crucial effort that will benefit the shrimp farming sector flourish in the coming decades.

Acknowledgements The authors extend their heartfelt gratitude to Sam Higginbottom University of Agriculture, Technology and Sciences (SHUATS) for their invaluable support and resources that made this chapter possible. Special thanks to Dr. Shaziya Haseeb Siddiqui and the faculty and research staff at SHUATS for their guidance, access to state-of-the-art facilities, and collaborative spirit in advancing sustainable aquaculture practices.

References

Abdel-Tawwab M et al (2021) Effects of dietary sweet basil (Ocimum basilicum) oil on the performance, antioxidants and immunity welfare, and resistance of Indian shrimp (Penaeus indicus) against Vibrio parahaemolyticus infection. Aquac Nutr 27(4):1244–1254. https://doi.org/10.1111/ANU.13265

Abidin Z et al (2021) 'Moringa oleifera leaves' extract enhances nonspecific immune responses, resistance against Vibrio alginolyticus, and growth in whiteleg shrimp (Penaeus vannamei). Animals 12(1):42. https://doi.org/10.3390/ANI12010042

Ahmed AM et al (2014) Antibacterial effect of olive (Olea europaea L.) leaves extract in raw peeled undeveined shrimp (Penaeus semisulcatus). Int J Vet Sci Med 2(1):53–56. https://doi.org/10.1016/J.IJVSM.2014.04.002

Al Eissa A et al (2022) Effects of feed formula and farming system on the environmental performance of shrimp production chain from a life cycle perspective. J Ind Ecol 26(6):2006–2019. https://doi.org/10.1111/JIEC.13370

Alam MI et al (2022) Mangrove forest conservation vs shrimp production: uncovering a sustainable co-management model and policy solution for mangrove greenbelt development in coastal Bangladesh. For Policy Econ 144:102824. https://doi.org/10.1016/J.FORPOL.2022.102824

Arcier JM et al (1999) A viral disease associated with mortalities in hatchery-reared postlarvae of the giant freshwater prawn Macrobrachium rosenbergii. Dis Aquat Org 38(3):177–181. https://doi.org/10.3354/DAO038177

Ashton EC (2008) The impact of shrimp farming on mangrove ecosystems. CAB Rev: Perspect Agric, Vet Sci, Nutr Nat Resour 3. https://doi.org/10.1079/PAVSNNR20083003

Badiola M, Mendiola D, Bostock J (2012) Recirculating aquaculture systems (RAS) analysis: main issues on management and future challenges. Aquac Eng 51:26–35. https://doi.org/10.1016/J.AQUAENG.2012.07.004

Balasubramanian G et al (2007) Screening the antiviral activity of Indian medicinal plants against white spot syndrome virus in shrimp. Aquaculture 263(1–4):15–19. https://doi.org/10.1016/J.AQUACULTURE.2006.09.037

Barad RR et al (2024) Herbal feed additives: natural boost for aquatic health and growth, pp 405–431. https://doi.org/10.1007/978-981-97-4279-0_18

Bondad-Reantaso MG et al (2012) The role of crustacean fisheries and aquaculture in global food security: past, present and future. J Invertebr Pathol 110(2):158–165. https://doi.org/10.1016/J.JIP.2012.03.010

Bosma RH et al (2016) Shrimp-based livelihoods in mangrove silvo-aquaculture farming systems. Rev Aquac 8(1):43–60. https://doi.org/10.1111/RAQ.12072

Boudreau MD, Beland FA (2006) An evaluation of the biological and toxicological properties of Aloe barbadensis (Miller), aloe vera. J Environ Sci Health C 24(1):103–154. https://doi.org/10.1080/10590500600614303

Burford MA et al (2004) The contribution of flocculated material to shrimp (Litopenaeus vannamei) nutrition in a high-intensity, zero-exchange system. Aquaculture 232(1–4):525–537. https://doi.org/10.1016/S0044-8486(03)00541-6

Butt UD et al (2021) Overview of the latest developments in the role of probiotics, prebiotics and synbiotics in shrimp aquaculture. Fish Shellfish Immunol 114:263–281. https://doi.org/10.1016/J.FSI.2021.05.003

Chang ZQ et al (2020) Development and current state of seawater shrimp farming, with an emphasis on integrated multi-trophic pond aquaculture farms, in China—a review. Rev Aquac 12(4):2544–2558. https://doi.org/10.1111/RAQ.12457

Chirawithayaboon P, Areechon N, Meunpol O (2020) Hepatopancreatic antioxidant enzyme activities and disease resistance of Pacific white shrimp (Litopenaeus vannamei) fed diet supplemented with garlic (Allium sativum) extract. Agric Nat Resour 54(4):377–386. https://li01.tci-thaijo.org/index.php/anres/article/view/247954. Accessed 6 Apr 2025

Citarasu T, Babu MM, Yilmaz E (2022) Alternative medications in shrimp health management for improved production. Aquaculture 561:738695. https://doi.org/10.1016/J.AQUACULTURE.2022.738695

Cortez-Mago R, Borges L, Wasielesky W (2025) Medicinal plants and their applications in shrimp culture. Lat Am J Aquat Res 53(1):22–38. https://doi.org/10.3856/VOL53-ISSUE1-FULLTEXT-3251

Darodes de Tailly JB et al (2021) Monitoring methods of feeding behaviour to answer key questions in penaeid shrimp feeding. Rev Aquac 13(4):1828–1843. https://doi.org/10.1111/RAQ.12546

Das S et al (2021) A review of the pharmacological and nutraceutical properties of Cynodon dactylon. Pharmacogn Res 13(3):104–112. https://doi.org/10.5530/pres.13.3.1

Datta D, Chattopadhyay RN, Guha P (2012) Community based mangrove management: a review on status and sustainability. J Environ Manag 107:84–95. https://doi.org/10.1016/J.JENVMAN.2012.04.013

Desrina et al (2022) White spot syndrome virus host range and impact on transmission. Rev Aquac 14(4):1843. https://doi.org/10.1111/RAQ.12676

Dewi NR et al (2021) Guava (Psidium guajava) leaf extract enhances immunity, growth, and resistance against Vibrio parahaemolyticus in white shrimp Penaeus vannamei. Fish Shellfish Immunol 118:1–10. https://doi.org/10.1016/J.FSI.2021.08.017

El-Sayed AFM (2021) Use of biofloc technology in shrimp aquaculture: a comprehensive review, with emphasis on the last decade. Rev Aquac 13(1):676–705. https://doi.org/10.1111/RAQ.12494

Flegel TW (1997) Major viral diseases of the black tiger prawn (Penaeus monodon) in Thailand. World J Microbiol Biotechnol 13(4):433–442. https://doi.org/10.1023/A:1018580301578/METRICS

Furtado PS, Poersch LH, Wasielesky W (2011) Effect of calcium hydroxide, carbonate and sodium bicarbonate on water quality and zootechnical performance of shrimp Litopenaeus vannamei reared in bio-flocs technology (BFT) systems. Aquaculture 321(1–2):130–135. https://doi.org/10.1016/j.aquaculture.2011.08.034

Ghosh AK et al (2024) Effects of Piper betle and Phyllanthus emblica leaf extracts on the growth and resistance of black tiger shrimp, Penaeus monodon, against pathogenic Vibrio parahaemolyticus. Aquac Int 32(3):3689–3708. https://doi.org/10.1007/S10499-023-01345-8/FIGURES/7

Gohil KJ, Patel JA, Gajjar AK (2010) Pharmacological review on Centella asiatica: a potential herbal cure-all. Indian J Pharm Sci 72(5):546. https://doi.org/10.4103/0250-474X.78519

Griffith DRW (2024) White spot and taura syndrome virus: disease as drivers in the shrimp farming industry. In: The shrimp book, pp 429–441. https://doi.org/10.1079/9781800629349.0015/ASSET/C1465E54-B058-4DF6-9AD8-ED1448F61349/ASSETS/9781800629349.0015.FP.PNG

Hadrich F et al (2014) Antioxidant and lipase inhibitory activities and essential oil composition of pomegranate peel extracts. J Oleo Sci 63(5):515–525. https://doi.org/10.5650/JOS.ESS13163

Hammarlund C et al (2024) Eco-certification in aquaculture—economic incentives and effects. Rev Fish Sci Aquac [Preprint]. https://doi.org/10.1080/23308249.2024.2440712

Holmström K et al (2003) Antibiotic use in shrimp farming and implications for environmental impacts and human health. Int J Food Sci Technol 38(3):255–266. https://doi.org/10.1046/J.1365-2621.2003.00671.X

Hossain A et al (2015) Prevalence and distribution of white spot syndrome virus in cultured shrimp. Lett Appl Microbiol 60(2):128–134. https://doi.org/10.1111/LAM.12353

Jagadeesan V et al (2019) Classical runt deformity syndrome cases in farmed Penaeus vannamei along the East Coast of India. J Coast Res 86(SI):107–111. https://doi.org/10.2112/SI86-016.1

Johnson KN, van Hulten MCW, Barnes AC (2008) "Vaccination" of shrimp against viral pathogens: phenomenology and underlying mechanisms. Vaccine 26(38):4885–4892. https://doi.org/10.1016/J.VACCINE.2008.07.019

Junior APB et al (2012) Polyculture of Nile tilapia and shrimp at different stocking densities. Revista Brasileira de Zootecnia 41(7):1561–1569. https://doi.org/10.1590/S1516-35982012000700002

Kaleeswaran B, Ilavenil S, Ravikumar S (2011) Dietary supplementation with Cynodon dactylon (L.) enhances innate immunity and disease resistance of Indian major carp, Catla catla (Ham.). Fish Shellfish Immunol 31(6):953–962. https://doi.org/10.1016/J.FSI.2011.08.013

Kasan NA, Dagang AN, Abdullah MI (2018) Application of biofloc technology (BFT) in shrimp aquaculture industry. IOP Conf Ser: Earth Environ Sci 196(1):012043. https://doi.org/10.1088/1755-1315/196/1/012043

Kautsky N et al (2000) Ecosystem perspectives on management of disease in shrimp pond farming. Aquaculture 191(1–3):145–161. https://doi.org/10.1016/S0044-8486(00)00424-5

Kim MJ et al (2023) Evaluation of the horizontal transmission of white spot syndrome virus for Whiteleg shrimp (Litopenaeus vannamei) based on the disease severity grade and viral shedding rate. Animals 13(10):1676. https://doi.org/10.3390/ANI13101676/S1

Kiran, Kumar N, Sindhu S, Gawaria B, Ansari A, Dhariwal A, Sharma P, Kumar R (2025) GWAS approaches in medicinal plants for enhanced secondary metabolite production. In: Chen JT (ed) Biotechnology, multiple omics, and precision breeding in medicinal plants. CRC Press, pp 119–134

Koteswari N et al (2014) Impact of aqua societies on shrimp farming in Andhra Pradesh, India

Kumar R, Yadav R, Kumar Gupta R, Pooja (2022a) A sustainable way for fish health management by replacement of chemical and drugs by earthworm. In: Environmental degradation in Asia: land degradation, environmental contamination, and human activities. Springer International Publishing, Cham, pp 329–352

Kumar S et al (2022b) Immunostimulants for shrimp aquaculture: paving pathway towards shrimp sustainability. Environ Sci Pollut Res 30(10):25325–25343. https://doi.org/10.1007/S11356-021-18433-Y

Kumar R, Singh D, Yadav R (2025) Toxic effects of chlorantraniliprole and fipronil on Eisenia fetida: from growth and reproductive parameters to avoidance behavior. Environ Eng Manag J 24(7):1499–1508. https://doi.org/10.30638/eemj.2025.117

Kumar R, Singh D (2022) Insecticide-tolerant bacterial population in Eisenia fetida's gut and vermicast exposed to chlorantraniliprole and fipronil. Appl Biol Res 24(3):273–279. https://doi.org/10.5958/0974-4517.2022.00033.7

Lawhavinit O, Kongkathip N, Kongkathip B (2010) Antimicrobial activity of curcuminoids from Curcuma longa L. on Pathogenic bacteria of shrimp and chicken. Agric Nat Resour 44(3):364–371. https://li01.tci-thaijo.org/index.php/anres/article/view/244941. Accessed 6 Apr 2025

Lee D et al (2022) Viral shrimp diseases listed by the OIE: a review. Viruses 14(3):585. https://doi.org/10.3390/V14030585

Lee YC et al (2024a) Effect of fermented lemon peel as a functional feed additive on growth, non-specific immune responses, and Vibrio alginolyticus resistance in Whiteleg shrimp, Litopenaeus Vannamei. Aquac Rep 34:101918. https://doi.org/10.1016/J.AQREP.2024.101918

Lee YS et al (2024b) Functional and mechanistic studies of a phytogenic formulation, shrimp best, in growth performance and vibriosis in whiteleg shrimp. Sci Rep 14(1):1–18. https://doi.org/10.1038/s41598-024-62436-x

Liao IC, Chien Y-H (2011) The Pacific white shrimp, Litopenaeus vannamei, in Asia: the world's most widely cultured alien crustacean. In: The wrong place—alien marine crustaceans: distribution, biology and impacts, pp 489–519. https://doi.org/10.1007/978-94-007-0591-3_17

Lightner DV (1999) The penaeid shrimp viruses TSV, IHHNV, WSSV, and YHV. J Appl Aquac 9(2):27–52. https://doi.org/10.1300/J028V09N02_03

Luu QH et al (2021) Antibiotics use in fish and shrimp farms in Vietnam. Aquac Rep 20:100711. https://doi.org/10.1016/J.AQREP.2021.100711

Madesh S et al (2024) Exploring neem aqueous extracts as an eco-friendly strategy to enhance shrimp health and combat EHP in aquaculture. Aquac Int 32(3):3357–3377. https://doi.org/10.1007/S10499-023-01326-X/FIGURES/5

Mandal A, Singh P (2025) Global scenario of shrimp industry: present status and future prospects. In: Shrimp culture technology, pp 1–23. https://doi.org/10.1007/978-981-97-8549-0_1

Martínez-Porchas M et al (2010) Shrimp polyculture: a potentially profitable, sustainable, but uncommon aquacultural practice. Rev Aquac 2(2):73–85. https://doi.org/10.1111/J.1753-5131.2010.01023.X

Munaeni W et al (2019) Phytochemical analysis and antibacterial activities of Eleutherine bulbosa (Mill.) Urb. extract against Vibrio parahaemolyticus. Asian Pac J Trop Biomed 9(9):397–404. https://doi.org/10.4103/2221-1691.267669

Munaeni W et al (2020a) Effect in white shrimp Litopenaeus vannamei of Eleutherine bulbosa (Mill.) Urb. powder on immune genes expression and resistance against Vibrio parahaemolyticus infection. Fish Shellfish Immunol 102:218–227. https://doi.org/10.1016/J.FSI.2020.03.066

Munaeni W et al (2020b) Impact of dietary supplementation with Eleutherine bulbosa (Mill.) Urb. on intestinal microbiota diversity and growth of white shrimp, Litopenaeus vannamei. Aquaculture 528:735466. https://doi.org/10.1016/J.AQUACULTURE.2020.735466

Muralidhar M et al (2021) Impacts of climate change and adaptations in shrimp aquaculture: a study in coastal Andhra Pradesh, India. Aquat Ecosyst Health Manag 24(3):28–38. https://doi.org/10.14321/AEHM.024.03.05

Murray F (2025) Certification in tilapia health management. In: Tilapia: aquaculture, biology and health management, pp 779–845. https://doi.org/10.1079/9781800629455.0017/ASSET/F98CC069-06C3-4F0C-86F9-CEF8BF646B9A/ASSETS/9781800629455.0017.FP.PNG

Murwantoko M et al (2016) Macrobrachium rosenbergii nodavirus infection in a giant freshwater prawn hatchery in Indonesia. Springerplus 5(1):1–8. https://doi.org/10.1186/S40064-016-3127-Z/FIGURES/6

Naidu JR, Ismail R, Sasidharan S (2014) Acute oral toxicity and brine shrimp lethality of methanol extract of Mentha spicata L. (Lamiaceae). Trop J Pharm Res 13(1):101–107. https://doi.org/10.4314/TJPR.V13I1.15

Nguyen D et al (2008) Lab-scale zero exchange shrimp aquaculture. http://deepblue.lib.umich.edu/handle/2027.42/58676. Accessed 6 Apr 2025

Nielsen L et al (2005) Taura syndrome virus (TSV) in Thailand and its relationship to TSV in China and the Americas. Dis Aquat Org 63(2–3):101–106. https://doi.org/10.3354/DAO063101

Nugraha MAR et al (2023) Recirculating aquaculture system (RAS) towards emerging Whiteleg shrimp (Penaeus vannamei) aquaculture. Int Aquat Res 15(1):1–14. https://doi.org/10.22034/IAR.2023.1973316.1361

Obaldo LG, Divakaran S, Tacon AG (2002) Method for determining the physical stability of shrimp feeds in water. Aquac Res 33(5):369–377. https://doi.org/10.1046/J.1365-2109.2002.00681.X

Okeke ES et al (2022) Antibiotic resistance in aquaculture and aquatic organisms: a review of current nanotechnology applications for sustainable management. Environ Sci Pollut Res 29(46):69241–69274. https://doi.org/10.1007/S11356-022-22319-Y

Páez-Osuna F (2001) The environmental impact of shrimp aquaculture: causes, effects, and mitigating alternatives. Environ Manag 28(1):131–140. https://doi.org/10.1007/S002670010212/METRICS

Paul BG, Vogl CR (2011) Impacts of shrimp farming in Bangladesh: challenges and alternatives. Ocean Coast Manag 54(3):201–211. https://doi.org/10.1016/J.OCECOAMAN.2010.12.001

Pazir MK et al (2011) Detection and identification of white spot syndrome virus (WSSV) and infectious hypodermal and hematopoietic necrosis virus (IHHNV) of Litopenaus vannamei from Bushehr and Sistan and Baloochestan provinces (Iran), during 2009–2010. Iran J Fish Sci 10(4):708–726. https://doi.org/10.18869/ACADPUB.IJFS

Ponce-Palafox J, Martinez-Palacios CA, Ross LG (1997) The effects of salinity and temperature on the growth and survival rates of juvenile white shrimp, Penaeus vannamei, Boone, 1931. Aquaculture 157(1–2):107–115. https://doi.org/10.1016/S0044-8486(97)00148-8

Ramírez I et al (2012) Superoxide dismutase is a critical enzyme to alleviate oxidative stress in Aloe vera (L.) Burm. plants subjected to water deficit. Plant Ecol Divers 5(2):183–195. https://doi.org/10.1080/17550874.2011.615033

Rani S, Sharma S, Singh D, Sheoran R, Kumar R (2025) Improved production of anti-cancer compounds from medicinal plants through molecular and biotechnological mutagenesis strategies. In: Chen JT (ed) Biotechnology, multiple omics, and precision breeding in medicinal plants. CRC Press, pp 287–297

Rector ME et al (2023) Sustainability outcomes of aquaculture eco-certification: challenges and opportunities. Rev Aquac 15(2):840–852. https://doi.org/10.1111/RAQ.12763

Roy S et al (2020) Trained immunity and perspectives for shrimp aquaculture. Rev Aquac 12(4):2351–2370. https://doi.org/10.1111/RAQ.12438

Sahul Hameed AS, Bonami JR (2012) White tail disease of freshwater prawn, macrobrachium rosenbergii. Indian J Virol 23(2):134–140. https://doi.org/10.1007/S13337-012-0087-Y/FIGURES/4

Sánchez-Martínez JG, Aguirre-Guzmán G, Mejía-Ruíz H (2007) White spot syndrome virus in cultured shrimp: a review. Aquac Res 38(13):1339–1354. https://doi.org/10.1111/J.1365-2109.2007.01827.X

Saptiani G et al (2020) Response of hemocytes profile in the black tiger shrimp (Penaeus monodon) against Vibrio harveyi induced by Xylocarpus granatum leaves extract. Vet World 13(4):751. https://doi.org/10.14202/VETWORLD.2020.751-757

Shahraki N et al (2021) Dietary administration of aqueous Zingiber officinale extract on growth performance, antioxidant activity and resistance of shrimp Litopenaeus vannamei against Photobacterium damselae. Iran J Fish Sci 20(1):32–44. https://doi.org/10.18869/ACADPUB.IJFS

Sharma N, Bafna P (2012) Effect of Cynodon dactylon on rotenone induced Parkinson's disease. Orient Pharm Exp Med 12(3):167–175. https://doi.org/10.1007/S13596-012-0075-1/TABLES/1

Singh ND et al (2021) Performance and perspective analysis of Indian shrimp exports. Agric Res 10(3):457–467. https://doi.org/10.1007/S40003-020-00513-Z/FIGURES/5

Sivaraj A et al (2017) Antibacterial activity of Eucalyptus oblique aqueous leaf extract on selective Vibrio species of Penaeus monodon culture hatchery. World J Pharm Sci 36–39. https://wjpsonline.com/index.php/wjps/article/view/antibacterial-activity-eucalyptus-oblique-vibrio-species. Accessed 6 Apr 2025

Song S et al (2023) Mangrove reforestation provides greater blue carbon benefit than afforestation for mitigating global climate change. Nat Commun 14(1):1–11. https://doi.org/10.1038/s41467-023-36477-1

Suantika G et al (2018) Development of a zero water discharge (ZWD)—recirculating aquaculture system (RAS) hybrid system for super intensive white shrimp (Litopenaeus vannamei) culture under low salinity conditions and its industrial trial in commercial shrimp urban farming in Gresik, East Java, Indonesia. Aquac Eng 82:12–24. https://doi.org/10.1016/J.AQUAENG.2018.04.002

Suita SM et al (2015) Dextrose as carbon source in the culture of Litopenaeus vannamei (Boone, 1931) in a zero exchange system. Lat Am J Aquat Res 43(3):526–533. https://doi.org/10.3856/VOL43-ISSUE3-FULLTEXT-13

Sukhdhane KS et al (2018) Integrated multi-trophic aquaculture systems: a solution for sustainability. https://enaca.org/?id=1012. Accessed 6 Apr 2025

Sulaiman M et al (2022) Antimicrobial secondary metabolites from the mangrove plants of Asia and the Pacific. Mar Drugs 20(10):643. https://doi.org/10.3390/MD20100643/S1

Suzuki A, Nam VH, Lee G (2023) Reducing antibiotics use among smallholders: experimental evidence from the shrimp aquaculture sector in Vietnam. Aquaculture 572:739478. https://doi.org/10.1016/J.AQUACULTURE.2023.739478

Talukder AS et al (2021) Molecular identification of white spot syndrome virus (WSSV) and associated risk factors for white spot disease (WSD) prevalence in shrimp (Penaeus monodon) aquaculture in Bangladesh. J Invertebr Pathol 179:107535. https://doi.org/10.1016/J.JIP.2021.107535

The Food Institute (n.d.) Report: global seafood production could rise 20% to 2050. The Food Institute. https://foodinstitute.com/focus/report-global-seafood-production-could-rise-20-to-2050/. Accessed 5 Apr 2025

Tian X et al (2001) An experimental study on closed-polyculture of penaeid shrimp with tilapia and constricted tagelus. Aquaculture 202(1–2):57–71. https://doi.org/10.1016/S0044-8486(01)00559-2

Trejo-Flores JV et al (2016) Protective effect of aloe vera in Litopenaeus vannamei challenged with Vibrio parahaemolyticus and white spot syndrome virus. Aquaculture 465:60–64. https://doi.org/10.1016/J.AQUACULTURE.2016.08.033

Umesh NR et al (2010) Shrimp farmers in India: empowering small-scale farmers through a cluster-based approach. In: Success stories in Asian aquaculture, pp 41–66. https://doi.org/10.1007/978-90-481-3087-0_3

Vanichkul K et al (2010) Immunological and bactericidal effects of turmeric (Curcuma longa Linn.) extract in Pacific white shrimps (Litopenaeus vannamei Boone). Agric Nat Resour 44(5):850–858. https://li01.tci-thaijo.org/index.php/anres/article/view/244998. Accessed 9 Sept 2024

Wasielesky W et al (2006) Effect of natural production in a zero exchange suspended microbial floc based super-intensive culture system for white shrimp Litopenaeus vannamei. Aquaculture 258(1–4):396–403. https://doi.org/10.1016/J.AQUACULTURE.2006.04.030

Wijegoonawardane PKM, Cowley JA, Walker PJ (2008) Consensus RT-nested PCR detection of yellow head complex genotypes in penaeid shrimp. J Virol Methods 153(2):168–175. https://doi.org/10.1016/J.JVIROMET.2008.07.009

Woodbury MR, Chirino-Trejo M, Mihajlovic B (2008) Diagnostic detection methods for Mycobacterium avium subsp. paratuberculosis in white-tailed deer. Can Vet J 49(7):683. https://pmc.ncbi.nlm.nih.gov/articles/PMC2430399/. Accessed 5 Apr 2025

Yadav R, Gupta RK, Kumar R, Kaur T, Moun R (2022a) Biochemical response of earthworm, Eisenia fetida to heavy metals toxicity. J Appl Nat Sci 14(3):990–998

Yadav R, Gupta RK, Kumar R, Kaur T (2022b) Assessment of toxicity of lead and nickel on the biochemical and immunological parameters of earthworm, Eudrilus eugeniae. Agric Sci Dig 44(5):922–929

Yadav R, Kumar R, Gupta RK, Kaur T, Yodha K, Kour A, Kaur S, Rajput A (2023) Heavy metal toxicity in earthworms and its environmental implications: a review. Environ Adv 100374

Yang HL et al (2016) A novel method of real-time reverse-transcription loop-mediated isothermal amplification developed for rapid and quantitative detection of a new genotype (YHV-8) of yellow head virus. Lett Appl Microbiol 63(2):103–110. https://doi.org/10.1111/LAM.12591

Yu JY et al (2021) Research progress on hosts and carriers, prevalence, virulence of infectious hypodermal and hematopoietic necrosis virus (IHHNV). J Invertebr Pathol 183:107556. https://doi.org/10.1016/J.JIP.2021.107556

Yu YB et al (2024) The use, application and efficacy of biofloc technology (BFT) in shrimp aquaculture industry: a review. Environ Technol Innov 33:103345. https://doi.org/10.1016/J.ETI.2023.103345

Zhao P et al (2012) The application of bioflocs technology in high-intensive, zero exchange farming systems of Marsupenaeus japonicus. Aquaculture 354–355:97–106. https://doi.org/10.1016/J.AQUACULTURE.2012.03.034

Chapter 9
Advancement in Integrated Fish Farming for Prevention of Fish Diseases

Mostafa A. M. Soliman

Abstract Similar to capture fisheries, the aquaculture sector provides fish products for human consumption. Aquaculture has been characterized by great development and expansion, both horizontally and vertically. Consequently, this caused a notable rise in the amount of production per unit area used and the amount of water available, through increased fish density during the production phase. Notwithstanding the significant advancements in the field of fish farming, certain challenges still exist that have an impact on output. One of the most significant of these challenges is the presence of diseases that affect fish during the production stage, that greatly influences the amount as well as fish produced. This can be avoided or managed by integrating fish farms and keeping water quality standards within the ideal ranges for fish, which will keep them healthy and able to withstand these types of diseases. Integration systems in fish farming have advanced significantly in various forms, whether they use bacteria (such as aquamimicry and biofloc) or plants (such as the aquaponics system) to increase yield and lower pathogens. Therefore, the process of integration in fish farming is not only an opportunity to increase production but also to prevent diseases and maintain the health and vitality of fish.

Keywords Fish farming · Integration · Aquaponics · Biofloc · Aquamimicry

9.1 Introduction

A few of the biggest food-producing sectors in the world is fish farms, and both the intensity of aquaculture and the consumption of aquaculture products have steadily grown over the last few decades (Kümmerer 2009; Sun et al. 2020a, b). Globally, finfish aquaculture is steadily growing in both freshwater and marine settings (FAO 2020), and there is ample evidence that the marine environment could see a significant increase in the future (Gentry et al. 2017). Fish aquaculture is currently dominated

M. A. M. Soliman (✉)
Animal System Production Research Department, Animal Production Research Institute (APRI), Agriculture Research Center, Ministry of Agriculture, Dokky-Giza, Egypt
e-mail: mostafa.soliman242@gmail.com

R. Kumar et al. (eds.), *Sustainability in Aquaculture and Fisheries*,
Springer Water, https://doi.org/10.1007/978-3-032-16605-0_9

by freshwater, especially in China and Indonesia, but it is possible to expand this industry in Europe and the Americas by using recirculation technology, which is being given more attention and funding. Similar to capture fisheries, the aquaculture sector provides fish products for human consumption, and the sector is currently choosing new fish species for domestication (FAO 2020). Aquaculture processes are typically optimized to be economically viable, and intensive organism administration can result in malady spread and financial losses (Abdel-Latif et al. 2020). Disease issues may occur in a fish production system regardless of whether stocking is done with free of illness fish, fortunately the farm is subjected to infectious agents (water intake from exterior sources). Aquaculture fish may become infected and spread phases of parasites found in communities of fish from nature (environmental spillover). As the severity of ailments rises in restricted conditions of aquaculture, some of these organisms may cause disease because of their notable capacity to spread in aquaculture settings.

The economy of aquaculture businesses as well as the wellbeing and good health of the aquatic animals will be put to the test (Shinn et al. 2015). Additional, environmental elements known as abiotic stressors have an impact on fish survival, growth, and health, which lowers output, profit, and product quality. Aquaculture needs to pay particular attention to abiotic stress (Shoko et al. 2014; Mateus et al. 2017; Ahmed et al. 2020; Kumar et al. 2022a). Whenever any abiotic element, like temperature, salinity, or The level of oxygen, is not nearly ideal, fish cells, tissues, and organs suffer (Cossins and Crawford 2005; Chowdhury and Saikia 2020). Diseases account for up to 50% of aquaculture production losses (Assefa and Abunna 2018). Inadequate water quality and high stocking densities provide ideal conditions for parasite infestation and reproduction, which swiftly escalates to pathogenic levels. Additionally, the inevitable transportation of fish and tools contributes to the spread of microbes that are transmissible (Subasinghe et al. 2001). Nonetheless, a number of methods, including vaccinations, antibiotics, and chemotherapeutics, have been used to prevent and control illnesses (Ramesh and Souissi 2018). Aquaculture infectious agents can be prevented and controlled by using immunostimulants as nutrition additives to maintain fish well-being and improve their development. These natural alternatives are becoming more widely accepted (Heuer et al. 2009; Kumar et al. 2022b). To reduce the effects of using materials that resist fish diseases, whether chemical or feed additives, whether manufactured or natural, so as not to affect the surrounding environment, we must turn to new ideas to protect fish from diseases.

9.2 Side Effects of Solo Aquaculture

Aquaculture's successful conversion of existing feed into aquatic food has led to a rise in fish and seafood output over the past several years (EEA 2016). Around 49% of all fisheries production came from aquaculture worldwide in 2022, with 87.5 million tonnes of aquaculture produced overall (FAO 2024). In contrast, the output from aquaculture was only 20.8 million tonnes in 2000; it represents a significant increase

(Ende et al. 2024). The global increase in fish farming has resulted in expanding environmental issues due to the discharge of residue from ponds and excess diet. Lost fish meals and waste from aquaculture are rich in phosphorus (P) and nitrogen. This can damage the aquatic environment by causing an excess of nitrogen and phosphorus in adjacent water sources like lakes, rivers, and oceans. As a result, increased the production of aquaculture facilities has sparked ecological worries worldwide (Crab et al. 2007). The possibility of waste production exists in any process that involves production, including aquaculture. Dumping is divided into two distinct groups: solid wastes and dissolving wastes. Waste can come from sources of ignored inputs or leftovers, such as feed (Gao et al. 2020; Olusegun et al. 2016), chemicals (Hesni et al. 2020; Sun et al. 2020a, b), or pathogens (Naughton et al. 2020; Rosa et al. 2020). Typically, each one kilogram of fish biomass requires one to three kilograms of drought-prone food that the fish do not fully eat or digest and instead expel into the water as stiff and dissolving waste (Crab et al. 2007). In addition to species and diet composition, other factors like temperature, the degree of nutrient retention is significantly influenced by fish dimensions, consuming level, and administration. For instance, marine fish species have nutrient retention rates of 13–43% for N, 18–36% for P, and 14–38% for C (Nederlof et al. 2022). In a typical aquaculture system, it is estimated that 83% of the P and 79% of the N utilized to produce aquatic animals end up going into the waters that surround (Moura et al. 2016). Aquaculture discharges released directly into water bodies increase the levels of COD, N, and P, causing excessive richness of nutrients. This has numerous negative consequences for the ecosystem and the financial system (Moura et al. 2016). Initially the output of aquaculture improves, so will waste, causing more environmental damage (Abdel-Shafy and Mansour 2018; Khiari et al. 2019). Aquaculture inputs include water provide, fish offspring, nutrition, and fertilizer to promote fish growth with the goal of harvesting as an output. Wastewater, chemical residue, and microorganisms are among the expected wastes generated throughout the process of culture (Ahmad et al. 2021), are illustrated in Fig. 9.1. Consequently, the need to find aquaculture systems that could regulate inputs and outputs and minimize production-related waste surfaced. Insufficient productivity, water quality degradation, and excessive usage of antibiotics are all challenges to the sustainability of conventional aquaculture systems. Alternatively, RAS-based fish production is ecologically propitious and able to be upheld, resulting in increased fish goodness and eco-safety (Ahmed and Turchini 2021).

9.3 Recirculating Aquaculture Systems (RAS)

Circulate water again systems for aquaculture (RAS) are fully controlled cultivation of fish conditions that reuse and recycling water. RAS provides a number of benefits over traditional flow-through aquaculture systems, such as increased productivity, lower water consumption, improved biosecurity, and waste recycling opportunities. However, in order to be widely adopted, RAS must overcome its economic

Fig. 9.1 Basic inputs and outputs for the aquaculture sector operation. *Source* Author

and environmental obstacles (Ahmed and Turchini 2021). Consequently, RAS was previously proposed as a method to reduce nutrient discharge. Utilization of RAS is rapidly expanding in Europe; in 2014, RAS produced 62,779 million tonnes in total. The Netherlands (49%) and Denmark (26%) are the countries that use RAS the most, then came Germany as well as the United Kingdom (both at 7%), according to Report AS-1201 (2024). Furthermore, RAS systems may obtain nourishment from waste and convert it to useful production by incorporating micro-algae growing. This ability is critical to ensuring environmental sustainability and system effectiveness. RAS's conventional circulation framework consists of multiple separate units, such as waste solids elimination, culture containers, drained system, mechanical/granular filters, settling basins and tanks, solid destruction, biological filtration, oxygenation, CO_2 deletion, and monitoring/control (Ende et al. 2024) as shown in Fig. 9.2. A regular RAS requires an ideal water conversion rate of 0.1–3.0 m^3 kg^{-1} diet (Bregnballe 2022), with 50–70% of feed nitrogen distributed as trash (Schneider et al. 2005).

Fortunately, due to greater water purification expenses compared to traditional aquaculture systems, this framework is still not widely used worldwide (Xifan et al. 2020). Furthermore, the dumping of collected sewage (primarily drum filter leftovers) and the need for oxygen consumption create obstacles in achieving a cost-effective, ecologically sound mechanism. In addition, massive quantities of N and P continue to escape into the environment through RAS effluent and sewage (Ende et al. 2024). On the basis of Ebeling et al. (2006), providing 1 kg of feed, involving 32% crude protein, is fed into a 1 m^3 RAS, causes the expulsion of 30 g of ammonia-N is expelled into the surrounding water. Ammonia-N ought to be limited to 1 mg L^{-1} in RAS water because it can damage fish (Labatut et al. 2007). The breathing of fish and the decomposition of bacteria in RAS continuously produce dissolved CO_2. A degassing procedure must be used to regulate CO_2 levels, and if the pH and dissolved CO_2 are not adequately managed, the fish may be more susceptible to TAN toxicity (Ende et al. 2024). Continuous exposure to these concentrations of carbon dioxide

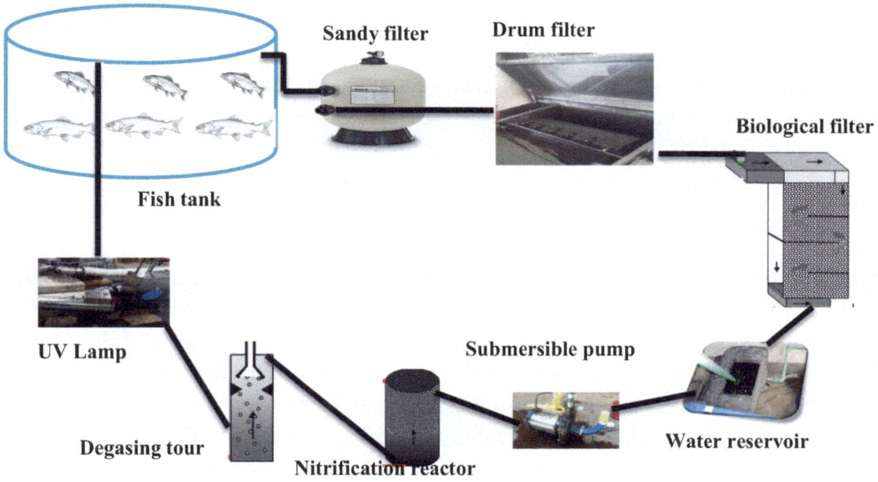

Fig. 9.2 Diagrammatic representation of the essential components of a recirculating aquaculture system (RAS). *Source* Author

and ammonia weakens fish immunity and exposes them to diseases, reducing growth rates and productivity. Consequently, integrated systems are thought to be a highly valuable solution to this issue in order to prevent fish diseases.

9.4 Integrated Aquaculture Systems

Aquaculture is currently not receiving many favorable reviews from the public for a number of reasons, including its effects on water quality (turbidity, pH, eutrophication) (Murray et al. 2021), changes in bacterial development (Moriarty 1986), and sediment (alterations in chemistry, biology, and physical properties) (Vezzulli et al. 2002; Carroll et al. 2003), and effects on landscape (Domínguez and Martín 2004), and biodiversity conservation (i.e., relying on apprehend to produce fodder) (Crab et al. 2012). Therefore, it necessary to identify, support, and employ more sustainable aquaculture production models, like aquaponics, biofloc technology, Integrated Multi-Trophic Aquaculture (IMTA), and Recirculating Aquaculture System (RAS), which could improve the food supply (increased species and trophic level diversification) and lessen the strain on natural resources while also adhering to the circular financial system's fundamentals (Lothmann and Sewilam 2023). Every part of RAS units collaborates to guarantee proper Water purification and secure surroundings for the aquatic animals. Regardless of the benefits, there remain opportunities to enhance financial viability as well as minimize their environmental impact. That may be achieved by putting policies in place like integrated multi-trophic aquaculture, which reuses all nutrients (Nederlof et al. 2022) and utilizing Using sources of sustainable energy to lower the release of carbon dioxide while lowering operational expenses (Ahmed and Turchini 2021). Therefore, creative integrated strategies for affordable RAS in conjunction with value-added compound production may aid in overcoming the obstacle to global industrialization (Ende et al. 2024). Aquaculture requires the establishment of different organic technologies and the production of aquatic species with environmental concerns (Kumar et al. 2018).

a. **Aquamimicry technology**

Aquamimicry is a creative approach for effectively presenting a sustainable challenge to the shrimp production sector. Aquaculture mimicry, as well referred to as aquamimicry, is an innovative approach which includes mimicking naturally coastal environments by increasing zooplankton diversity (primarily copepods) to provide additional nutrition to cultured shrimp while also promoting beneficial bacteria in culture water (Puja et al. 2018). Aquamimicry uses sources of carbon like rice bran and probiotics to abundance microscopic plant and crustaceans massiveness, and the resulting explodes mimic natural pond conditions while also providing additional nourishment to farmed fish and shrimp and safeguarding the state of the water (Vijayan and Panigrahi 2019), was clarify in Fig. 9.3. The best feed supplements are fermented carbohydrate sources, like fermented rice bran, due to their beneficial effects as probiotics and prebiotics (Haryati et al. 2015; Yang et al. 2021). By absorbing nitrogen compounds and producing in situ microbial protein, these microorganisms aid in the improvement of a system's water quality by eliminating inorganic nitrogen from wastewater (Emerenciano et al. 2017; Nisar et al. 2021). Secondly, it helps pond zooplankton as well as gives shrimp additional nourishment (Khanjani et al. 2023). Furthermore, The carbon source provides heterotrophic organisms with an adequate carbon–nitrogen (C:N) proportion for development and operation; as a result, harmful nitrogenous metabolites as ammonia are transformed into bacterial proteins (Avnimelech 2009). This technique significantly reduces cultures that require a lot of stocking and water exchange (Nisar et al. 2021). By reusing organic nutrients, it offers high-quality outputs and affordable, environmentally friendly aquaculture production (Romano et al. 2018; Cho and Yiğit 2022). Furthermore, by promoting the production of natural food organisms and the gradual breakdown of minerals, aquamimicry technology promotes the natural remediation of ecosystems and offers a culture environment free of disease (Romano and Kumar 2017).

Through altering the gut microbiota, probiotics—via diet or water, alive microorganisms are brought into the aquaculture surroundings to improve the host's health and growth performance (Merrifield et al. 2010; Dawood et al. 2018; Vidal et al. 2018). Appropriate substitute for medications in the administration of infectious expansions of diseases, and they can help to enhance the quality of water (Jahangiri and Esteban 2018; Biswas et al. 2019; Nathanailides et al. 2021). With the lowering the pH of the gut, probiotics create enzymes that inhibit the reproduction of damaging microbes and make it more difficult for them to survive at low pH (Martínez-Cruz et al. 2012).

Fig. 9.3 **a** Fermenting rice bran for production of aquamimicry, and **b** distributing the mixture into the fish ponds. *Source* Author

b. **Biofloc technology**

Among aquaculture innovations, biofloc technology (BFT) has received an excessive amount of interest due to its capacity to promote environmentally friendly fish farming aims (Khanjani et al. 2022). When properly applied, BFT can improve farm biological safety while drastically lowering water consumption and adverse environmental effects (Avnimelech 2012). Biofloc microbes (BFOs) serve three primary functions in BFT: (i) absorbing nitrogen-containing substances to maintain the state of the water; (ii) supplying nourishment for species under cultivation (Emerenciano et al. 2012; Wang et al. 2016); and (iii) competitors for infectious agents as well the development of probiotic qualities (Ferreira et al. 2015), which are frequently performed by heterotrophic organisms (e.g., Bacillus spp. bacteria). BFOs contribute to water quality maintenance via metabolizing feces and unconsumed food, which reduces unsafe nitrogenous substances like nitrite and ammonia (Suita et al. 2015; Wang et al. 2016). Environmental TAN regulate is linked to BFO development (Khanjani and Sharifinia 2022). Bacteria, microalgae, yeast, rotifers, ciliates, protozoans, nematodes, copepods, and crustaceans are among the organisms found in bioflocs (Collazos-Lasso and Arias-Castellanos 2015; Khanjani et al. 2022). A wider range of BFOs can improve flocs' ability to mineralize waste (Martínez-Córdova et al. 2015), Protein consumption (Avnimelech 2007); also inhibition of pathogens (Emerenciano et al. 2013) (Fig. 9.4).

BFT systems, which maintain high stocking density while eliminating or minimizing water exchange, have new aquaculture advancement. When other forms of carbon are added to such systems after the C/N ratio approaches 15:1, the organisms that are actually in the water are prompted to make utilize of the nitrogen-rich waste that has accumulated (as unconsumed feed or excrement). This accelerates their growth and causes flocs, or agglomerations, to form (Khanjani and Sharifinia 2022; Khanjani et al. 2022). According to Cienfuegos-Martínez et al. (2020), these communities are dwelling to a variety of species that engage in various interactions with one another and the environment. Fish and crustacean farming have used these microscopic communities as feed and environmental enhancers owing to BFT (Khanjani et al. 2020, 2021; Emerenciano et al. 2022).

Fig. 9.4 **a** Polyethylene lined ponds used in a biofloc system for cultivating Nile tilapia; **b** biofloc sampling through the growth season; **c** measurements of the floc volume to maintain the suitable range. *Source* Author

To promote aquatic farming being sustainable both economically and environmentally, especially under intensive conditions, optimal water quality is a crucial prerequisite (Emerenciano et al. 2017). The BFT is an intricate nutritional system in which the processes (photosynthesis, nitrification, assimilation) and MOs (photoautotrophs, chemoautotrophs, and heterotrophs) are essential for preserving and improving water quality. The "core" of BFT is thought to be heterotrophic bacteria that regulate growth of algae and determine the well-being of the ecosystem. By using organic matter that is dissolved, re-incorporating as much as 50% of the carbon distributed by microalgae, accelerating up mineralization, and making substances (like carbon) accessible to greater trophic level MOs, heterotrophs are placed at the bottom of the dietary chain in accordance with the "microbial loop concept" (Azam et al. 1983), protozoans then eat the bacteria, and more substantial microbes in the BFO food chain eat the protozoans (Nevejan et al. 2018). Given the high nutrient input loads and decreased water exchange in BFT, this recycling process is particularly crucial (Samocha et al. 2007; Asaduzzaman et al. 2010; Emerenciano et al. 2022).

BFOs' probiotic qualities improve the cultivated animals' growth, digestion, metabolism, and resistance to disease (Daniel and Nageswari 2017; Hostins et al. 2019). Additionally, BFOs can be useful in pathogen command. BFT doesn't make the pathogenic species more resistant because its effects stem from competition between beneficial and harmful microbes, in contrast to traditional approaches to pathogen control. By giving possible probiotic bacteria an appropriate material, Biofloc has probiotic qualities (Cienfuegos-Martínez et al. 2020).

c. **Integrated microalgae cultivation**

The traditional RAS technologies were created to deliver dependable efficiency, as well as the algae's ability to improve the relevant technology is illustrated in

Table 9.1. The term "uncoupled" refers to a microalgae cultivation system that is set up separately from the RAS system in the framework of an incorporated RAS and microalgae method. An alternative is to create a coupled system in which the fish in the RAS system are directly cultivated alongside microalgae. Depending on particular operational objectives, resource usage, and overall system efficiency considerations, one can choose between uncoupled and coupled configurations (Ende et al. 2024). The two most crucial phases in creating biomass from microalgae are the cultivation procedure and the dewatering phase, and they both need to be improved further to lower the whole price tag of production (Kroumov et al. 2017). Two of the popular widely used systems for microalgae cultivation are closed photobioreactors (PBRs) and open raceway ponds. Closed PBRs are more expensive than raceway reservoirs, but they have a number of benefits, such as quicker rate of development, lower pollution, and more effectively cultivation state control. The main limitation of continuous cultivation in PBR is natural biomass wash-out that is the depletion of microalgal biomass caused by an elevated dissolution rate or insufficient residence duration (Discart et al. 2014). Thus, this problem might be resolved by decoupling the dilution rate from the storage period for microalgal biomass (MRT) (Soroosh et al. 2022). It was recommended to install a membrane cleaning section regarding the cultivation reservoir to operate the PBR in a membrane photobioreactor (MPBR) method (Discart et al. 2014; Senatore et al. 2021), as present in Fig. 9.5. Even though green biosynthesis of these products has advantages for the environment, it is crucial to find creative ways to reduce production costs. To increase the profitability of the fish industry and improve the revenue of farmers, a merged biological refinery of microalgae-assisted RAS is crucial. Relying upon the micro-algal kind and waste category that are grown, the desired product can change. For instance, employing aqua farming out-flowing as a developmental substance significantly raised the green microalga Nannochloris maculate's lipid yield (Khatoon et al. 2016). Moreover, when cultivated in wastewater medium, *N. maculate* and *Tetraselmis chuii* displayed higher protein contents of 10% and 8%, respectively. As a business supplier of chemicals with significant value like β-carotene, astaxanthin, phycobilins, as well as vital unsaturated fatty acids, microalgae have also been discussed (Ende et al. 2024).

Furthermore, algal biomass extracted from significant amounts substances will be utilized as fish food or as feedstock for the manufacture of several different biofuels, resulting in a waste-free strategy. The granular fertilizer created from Aq-S is another valuable outcome which is capable of being generated from RAS waste streams. Nonetheless, safety measures ought to be implemented to prevent the introduction of unwanted substances, like antibiotics and heavy metals, into the agricultural system or the transmission of sickness (Yadav et al. 2022a, b, 2023; Ende et al. 2024). The optimization of the purity of water, boosting energy conservation, controlling stock densities, utilizing sophisticated feed techniques, and putting stringent biosecurity measures in place to lower emissions are all ways to enhance RAS performance. These techniques require larger beginning investments and more complex administration, but they additionally boost efficiency in manufacturing, reduce general expenses, and enhance the availability of the system. The integration of such approaches alongside novel innovations, for example enhanced microalgae

Table 9.1 Clarifying the role of algae in improving the efficiency of recirculating aquaculture systems

Technique	Refining by mechanical methods	Filtration by biological processes	Chemical filtration	Techniques for oxygenation	Control of temperature	Management of solid waste	Exchange of water	References
Justification	To purify the water of particulate matter (solids), including debris, fish waste, and food remnants	Using beneficial bacteria, harmful ammonia from fish waste is converted into nitrites, which are then converted into less harmful nitrates	The application of energize carbon or another chemical-based mediums to eliminate or counterbalance disintegrate tailings, toxins, or antibiotics from liquid	Making sure that the water has a sufficient amount of diffuse oxygen, this is essential to the survival and well-being of aquatic life	To control the water's temperature and maintain it within the ideal range for the particular species being cultivated	To successfully evacuate solid wastes from the system using methods such as swirl separators, drum filters, and settling basins	For proper control of nutrient levels and dilute pollutants, a portion of the system water should be replaced on a regular basis with fresh, treated water	Ende et al. (2024)
Superiority	Helps maintain the water unambiguous that is necessary for the fish's well-being, and improves condition monitoring	Essential for avoiding ammonia toxicity in RAS systems' dense populations	Efficient in managing a range of water quality problems those biological and mechanical filters are unable to resolve	Encourages raised stocking densities by making up for bacterial activity and the oxygen that fish consume	Maintaining a suitable temperature is critical for controlling metabolic rates, optimizing growth, and preventing disease	Lessens the strain on filters by eliminating a major origin of ammonia prior to it arrives at biological liquidation phase	A quick and easy method to cut down on nitrate accumulation and other substances that cannot be filtered	Ende et al. (2024)

(continued)

Table 9.1 (continued)

Technique	Refining by mechanical methods	Filtration by biological processes	Chemical filtration	Techniques for oxygenation	Control of temperature	Management of solid waste	Exchange of water	References
Restriction	Dissolved wastes cannot be eliminated by mechanical filters	It takes time to set up a new system, is responsive to shifts in the water quality and may be disturbed when not maintained appropriately	Overuse of synthetic media may eliminate useful substances from water, so it must be replaced on a regular basis	Vital and energy-intensive to maintain; if it fails, fish life may quickly be lost	Heating and cooling needed energy, which can be expensive; breakdowns in devices can cause stress or stock loss	Necessitate more waste product handling and disposal; effective removal system designing and handling are critical	Water replacement is expensive and, if done carelessly, can cause system disruptions. In comparison to fully recirculated systems, it also leads to less water conservation	Ende et al. (2024)
Algal role	Dissolved wastes can be removed with the aid of algae integration	The mutually beneficial interactions among algae and bacteria can improve the effectiveness of cleanup	Algae are highly effective at removing these toxins and chemicals	The oxygenation process is improved by microalgae's ability to produce oxygen and sequester CO_2	Not applicable	Anaerobic digestion of solid waste or algae integration with hydrothermal liquefaction (HTL) guarantees a zero-waste path	Using microalgae to treat wastewater lowers the RAS water exchange rate	Ende et al. (2024)

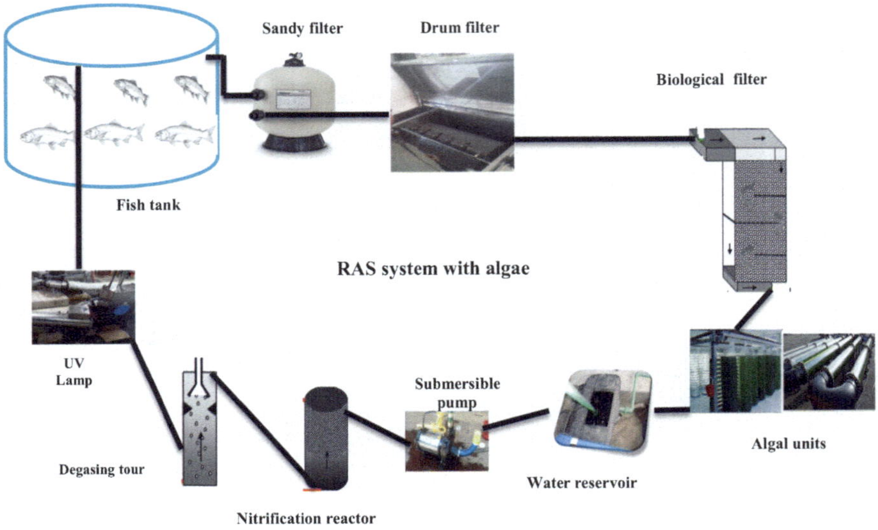

Fig. 9.5 Emphasize algae culture to improve the efficiency of the recycled aquaculture system and minimize the impact of environmental factors on fish. *Source* Author

production, is vital for aquaculture's future development, ecological consciousness, and financial stability (Ende et al. 2024).

d. **Aquaponics system**

Aquaponics is an innovative method of farming that takes advantage of the mutually beneficial interaction between aquatic microorganisms and terrestrial vegetation in a specially designed recirculating system. Benefiting microorganisms transform fish waste into a nutrient-dense solution in this closed-loop system, which uses a nutrient-rich water circulation network to support the development of numerous crops, such as vegetables and other horticultural products (Jha et al. 2024). Therefore, in the current climate, soilless cultivation is becoming more and more significant. Soil-less culture plants are perfect for urban environments because of their lightweight characteristics and resource efficiency (Li et al. 2018). Food production using soilless-based farming techniques is growing in popularity, according to Abusin and Mandikiana (2020). More effective methods of producing food through soilless culture have demonstrated some promising outcomes globally (Sardare and Admane 2013). The most basic definition of aquaponics is the amalgamation of hydroponics and aquaculture, whereby plants utilize nourishment released by growing fish to grow (Yep and Zheng 2019), as shown in Fig. 9.6.

To optimize the utilization of water and natural assets, aquaponics system designs differ depending on the climate. Because they effectively recycle and re-mineralize aquatic animal waste using naturally occurring microbes, recirculating aquaculture systems (RAS) are preferred over non-recirculating systems. This makes nutrients available for plant uptake (Yep and Zheng 2019). As a water-efficient substitute for

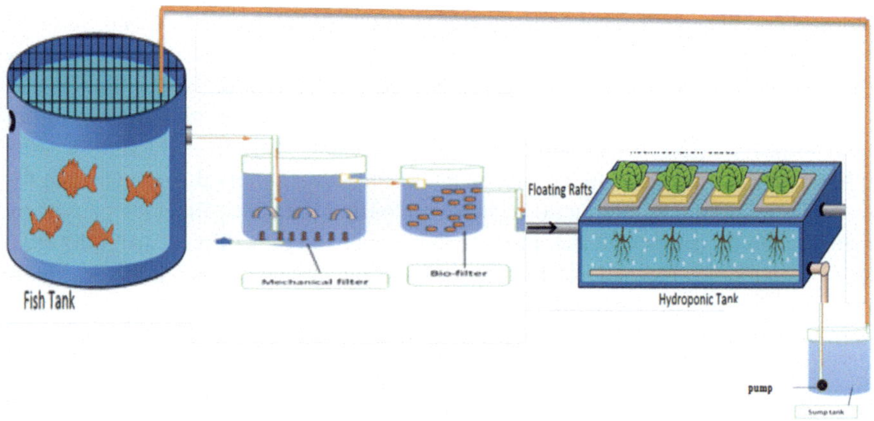

Fig. 9.6 Traditional aquaponic systems as one-loop or coupling systems. *Source* Author

conventional aquaculture, RAS has emerged as a useful production system in regions where climatic and environmental factors present difficulties (Ahmed and Turchini 2021). When compared to traditional aquaculture systems, it can save 90–99% of water use (Timmons and Ebling 2010). Decoupled systems offer more flexibility in various climates by separating hydroponic and aquaculture units. Decoupled systems make it possible to optimize conditions in every component, which improves nutrient utilization and boosts output. Compared to coupled designs, these systems allow for more efficient pH and fertilizer management, which results in higher yields (Monsees et al. 2017). By modifying the hydroponic area according to plant transpiration rates, these systems can be tailored for various climates (Goddek and Körner 2019).

Bacteria are essential for nutrient cycling and system stability, according to recent research on microbial communities in aquaponic systems (Pablos et al. 2024). These dynamic microbial communities play an important part in nitrogen-related processes and change in response to environmental changes. An important stage in the nitrogen cycling within RAS is the transformation of ammonium into nitrite, which is facilitated by ammonia-oxidizing bacteria (AOB) such as *Nitrosomonas* communist (Revsbech et al. 2006; Van Rijn et al. 2006). Water recirculation (Calone et al. 2019; Yang and Kim 2020), food safety (Pantanella et al. 2012; Moriarty et al. 2018), fish-plant pathogens (Mori and Smith 2019; Stouvenakers et al. 2019), beneficial microorganisms (Schmautz et al. 2017; Joyce et al. 2019), stocking density (Andriani et al. 2017; Baßmann et al. 2020), functional setup (Kyaw and Ng 2017; Lee and Wang 2020), sustainability (Forchino et al. 2017; Danner et al. 2019), and cultivation media (Zou et al. 2016; Fang et al. 2017) have been the main topics of aquaponics research up to this point.

Aquaponic systems can also be based on the biosecurity and fish welfare principles of responsible aquaculture. Fish, plants, bacteria in biological filters, and co-cultured organisms in aquaponics must all be considered in all of their applications. To create particular feeds for aquaponic conditions, more research is required. The well-being of the fish in aquaponic systems is among of the majority crucial elements of aquaculture's long-term viability. However, through the instance of aquaponics, optimizing water parameters while taking into account The features of collaboration generation of fish and plants are an important topic over aquaponics' sustainability and achievement (Yavuzcan Yildiz et al. 2017). Controlling these factors benefits both welfare and production because many of the same factors that are Critical for welfare is also vital to production, such as avoiding parasites, illnesses caused by bacteria and viruses, as well as injuries to the leather and gills (Yavuzcan Yildiz et al. 2017). Aquaponics is considered to be among of the greatest significant ecosystems through which diseases that affect fish or plants can be prevented, due to the great control over all environmental factors.

9.5 Conclusion

As aquaculture has advanced significantly, particularly in terms of increasing production by increasing the number of fish per unit area, this has had a substantial impact on the environment in which the fish live. As a result, diseases have emerged in the fish, and a significant portion of the fish production has been lost. Therefore, one of the main objectives of maintaining high fish productivity is to preserve the surrounding environment to ensure disease control and prevent its occurrence during the stages of aquatic life production.

Integration in aquaculture is considered one of the good methods for controlling diseases by maintaining water quality and eliminating factors that cause stress on aquatic life, thus preserving the vital immunity of fish to resist these diseases and reaching the final stage with full production capacity per unit area. There are many forms of integration in aquaculture, either with phytoplankton, zooplankton, and bacteria as in aquamimicry and biofloc systems, or with separate phytoplankton, or with plants as in the aquaponics system.

9.6 Future Prospects

As a result of significant developments in intensive aquaculture systems, which aim to increase production per unit area. This is accomplished by increasing the live mass of the fish, exposing them to significant stress and resulting in the loss of some production. In order to sustain fish production, modern aquaculture systems aim to drastically alter traditional systems and enhance their functionality. Furthermore, aquaculture systems can preserve water as a fundamental resource in a number of ways, such as by growing microorganisms (as in biofloc and mimicry systems), growing plants and algae (as in aquaponics), or integrating algae with water recycling systems.

References

Abdel-Latif HMR, Dawood MAO, Menanteau-Ledouble S, El-Matbouli M (2020) The nature and consequences of co-infections in tilapia: a review. J Fish Dis 43:651–664. https://doi.org/10.1111/jfd.13164

Abdel-Shafy HI, Mansour MSM (2018) Solid waste issue: sources, composition, disposal, recycling, and valorization. Egypt J Petrol 27:1275–1290. https://doi.org/10.1016/j.ejpe.2018.07.003

Abusin SA, Mandikiana BW (2020) Towards sustainable food production systems in Qatar: assessment of the viability of aquaponics. Glob Food Sec 25:100349. https://doi.org/10.1016/j.gfs.2020.100349

Ahmad A, Abdullah SRS, Hasan HA, Othman AR, Ismail NI (2021) Aquaculture industry: supply and demand, best practices, effluent and its current issues and treatment technology. J Environ Manag 287:112271. https://doi.org/10.1016/j.jenvman.2021.112271

Ahmed N, Turchini GM (2021) Recirculating aquaculture systems (RAS): environmental solution and climate change adaptation. J Clean Prod 297:126604

Ahmed I, Reshi QM, Fazio F (2020) The influence of the endogenous and exogenous factors on hematological parameters in different fish species: a review. Aquac Int 28:869–899

Andriani Y, Dhahiyat Y, Zahidah Z, Zidni I (2017) The effect of stocking density ratio of fish on water plant productivity in aquaponics culture system. Nusant. Biosci. 9:31–35

Asaduzzaman M, Rahman MM, Azim ME, Islam MA, Wahab MA, Verdegem MCJ, Verreth JAJ (2010) Effects of C/N ratio and substrate addition on natural food communities in freshwater prawn monoculture ponds. Aquaculture 306(1–4):127–136

Assefa A, Abunna F (2018) Maintenance of fish health in aquaculture: review of epidemiological approaches for prevention and control of infectious disease of fish. Vet Med Int 2018:1–10

Avnimelech Y (2007) Feeding with microbial flocs by tilapia in minimal discharge bioflocs technology ponds. Aquaculture 264:140–147

Avnimelech Y (2009) Biofloc technology. A practical guide books. The World Aquaculture Society, Baton Rouge

Avnimelech Y (2012) Biofloc technology. A practical guide book. The World Aquaculture Society, Baton Rouge, p 272

Azam F, Fenchel T, Field JG, Gray JS, Meyer-Reil LA, Thingstad F (1983) The ecological role of water-column microbes in the sea. Mar Ecol Prog Ser 10:257–263

Baßmann B, Harbach H, Weißbach S, Palm HW (2020) Effect of plant density in coupled aquaponics on the welfare status of African catfish, Clarias gariepinus. J World Aquac Soc 51:183–199

Biswas PC, Sultana S, Kabiraj M, Sm SH (2019) Role of probiotics in aquaculture practice of Satkhira region of Bangladesh. J Fish Aquat Stud 7:174–181

Bregnballe J (2022) A guide to recirculation aquaculture: an introduction to the new environmentally friendly and highly productive closed fish farming systems. Food and Agriculture Org.

Calone R, Pennisi G, Morgenstern R, Sanyé-Mengual E, Lorleberg W, Dapprich P, Winkler P, Orsini F, Gianquinto G (2019) Improving water management in European catfish recirculating aquaculture systems through catfish-lettuce aquaponics. Sci Total Environ 687:759–767

Carroll M, Cochrane S, Fieler R, Velvin R, White P (2003) Organic enrichment of sediments from salmon farming in Norway: environmental factors, management practices, and monitoring techniques. Aquaculture 226:165–180. https://doi.org/10.1016/S0044-8486(03)00475-7

Cho Y, Yiğit ÜMÜT (2022). Biomass gain, feed efficiency and survival rates in Whiteleg shrimp (*Litopenaeus vannamei*) cultured in Aquamimicry concept and conventional methods with water exchange and settling chamber. Mar Rep 1(2) https://doi.org/10.5281/zenodo.7393853

Chowdhury S, Saikia SK (2020) Oxidative stress in fish: a review. J Sci Res 12:145–160

Cienfuegos-Martínez K, Monroy-Dosta MC, Hamdan-Partida A, Hernández-Vergara MP, Becerril-Cortés D, López-García E (2020) A review of the use of probiotics in freshwater prawn (Macrobrachium sp.) culture in biofloc systems. Lat Am J Aquat Res 48(4):518–528

Collazos-Lasso LF, Arias-Castellanos JA (2015) Fundamentals of bioflocs technology (BFT). An alternative for fish farming in Colombia: a review. Orinoquia 19(1):77–86

Cossins AR, Crawford DL (2005) Fish as models for environmental genomics. Nat Rev Genet 6:324–333
Crab R, Avnimelech Y, Defoirdt T, Bossier P, Verstraete W (2007) Nitrogen removal techniques in aquaculture for a sustainable production. Aquaculture 270:1–14
Crab R, Defoirdt T, Bossier P, Verstraete W (2012) Biofloc technology in aquaculture: beneficial effects and future challenges. Aquaculture 356–357:351–356. https://doi.org/10.1016/j.aquaculture.2012.04.046
Daniel N, Nageswari P (2017) Exogenous probiotics on biofloc based aquaculture: a review. Curr Agric Res J 5(1):88–107
Danner RI, Mankasingh U, Anamthawat-Jonsson K, Thorarinsdottir RI (2019) Designing aquaponic production systems towards integration into greenhouse farming. Water 11:2123
Dawood MAO, Koshio S, Esteban MA (2018) Beneficial roles of feed additives as immunostimulants in aquaculture: a review. Rev Aquac 10:950–974
Discart V, Bilad MR, Marbelia L, Vankelecom IFJ (2014) Impact of changes in broth composition on Chlorella vulgaris cultivation in a membrane photobioreactor (MPBR) with permeate recycle. Bioresour Technol 152:321–328
Domínguez LM, Martín JMV (2004) Aquaculture environmental impact assessment. In: Popov V, Itoh H, Brebbia CA, Kungolos S (eds) Waste management and the environment II. WIT Press, pp 321–333. https://doi.org/10.2495/WM040321
Ebeling JM, Timmons MB, Bisogni JJ (2006) Engineering analysis of the stoichiometry of photoautotrophic, autotrophic, and heterotrophic removal of ammonia-nitrogen in aquaculture systems. Aquaculture 257:346–358
EEA (2016) Seafood in Europe. A food system approach for sustainability. European Environment Agency, Copenhagen
Emerenciano M, Ballester EL, Cavalli RO, Wasielesky W (2012) Biofloc technology application as a food source in a limited water exchange nursery system for pink shrimp *Farfantepenaeus brasiliensis* (Latreille, 1817). Aquac Res 43(3):447–457
Emerenciano MGC, Rombenso AN, Vieira FdN, Martins MA, Coman GJ, Truong HH, Noble TH, Simon CJ (2022) Intensification of penaeid shrimp culture: an applied review of advances in production systems, nutrition and breeding. Animals 12:236
Emerenciano MGC, Gaxiola G, Cuzon G (2013) Biofloc technology (BFT): a review for aquaculture application and animal food industry. In: Matovic MD (ed) Biomass now: cultivation and utilization. InTech, Belfast, Canada, pp 301–308
Emerenciano MGC, Martínez-Córdova LR, Martínez-Porchas M, Miranda-Baeza A (2017) Biofloc technology (BFT): a tool for water quality management in aquaculture. In: Water quality. InTechopen. https://doi.org/10.5772/66416
Ende S, Henjes J, Spiller M, Elshobary M, Hanelt D, Abomohra A (2024) Recent advances in recirculating aquaculture systems and role of microalgae to close system loop. Bioresour Technol 131107
Fang Y, Hu Z, Zou Y, Fan J, Wang Q, Zhu Z (2017) Increasing economic and environmental benefits of media-based aquaponics through optimizing aeration pattern. J Clean Prod 162:1111–1117
FAO (2020) The State of World Fisheries and Aquaculture (SOFIA). United Nations Organisation for Food and Agriculture, Rome, Italy. ISBN: 978-92-5-136364-5
FAO (2024) The state of world fisheries and aquaculture 2022. Food and Agriculture Organization (FAO), Rome, Italy. https://doi.org/10.4060/CA9231EN
Ferreira GS, Bolívar NC, Pereira SA, Guertler C, Vieira FN, Mouriño JLP, Seiffert WQ (2015) Microbial biofloc as source of probiotic bacteria for the culture of *Litopenaeus vannamei*. Aquaculture 448:273–279
Forchino AA, Lourguioui H, Brigolin D, Pastres R (2017) Aquaponics and sustainability: the comparison of two different aquaponic techniques using the life cycle assessment (LCA). Aquac Eng 77:80–88
Gao Y, Guo L, Shao M, Hu F, Wang G, Zhao Y, Gao M, Jin C, She Z (2020) Denitrification performance evaluation and kinetics analysis with mariculture solid wastes (MSW) derived carbon

source in marine recirculating aquaculture systems (RAS). Bioresour Technol 313. https://doi.org/10.1016/j.biortech.2020.123649
Gentry RR, Froehlich HE, Grimm D, Kareiva P, Parke M, Rust M, Gaines SD, Halpern BS (2017) Mapping the global potential for marine aquaculture. Nat Ecol Evol 1:8
Goddek S, Körner O (2019) A fully integrated simulation model of multi-loop aquaponics: a case study for system sizing in different environments. Agric Syst 171:143–154. https://doi.org/10.1016/J.AGSY.2019.01.010
Haryati T, Susanti T, Susana IWR (2015) Nutritional value of rice bran fermented by Bacillus amyloliquefaciens and humic substances and its utilization as a feed ingredient for broiler chickens. Asian-Australas J Anim Sci 28(2):231. https://doi.org/10.5713/ajas.14.0039
Hesni MA, Hedayati A, Qadermarzi A, Pouladi M, Zangiabadi S, Naqshbandi N (2020) Using Chlorella vulgaris and iron oxide nanoparticles in a designed bioreactor for aquaculture effluents purification. Aquacult Eng 90. https://doi.org/10.1016/j.aquaeng.2020.102069
Heuer OE, Kruse H, Grave K, Collignon P, Karunasagar I, Angulo FJ (2009) Human health consequences of use of antimicrobial agents in aquaculture. Clin Infect Dis 49:1248–1253. https://doi.org/10.1086/605667
Hostins B, Wasielesky W, Decamp O, Bossier P, De Schryver P (2019) Managing input C/N ratio to reduce the risk of acute hepatopancreatic necrosis disease (AHPND) outbreaks in biofloc systems—a laboratory study. Aquaculture 508:60–65
Jahangiri L, Esteban MA (2018) Administration of probiotics in the water in fin fish aquaculture systems: a review. Fishes 3:33
Jha S, Batra A, Dewangan AK (2024) Integrated aquaponics through smart sensors. J Harbin Eng Univ 45(11):321–329
Joyce A, Timmons M, Goddek S, Pentz T (2019) Bacterial relationships in aquaponics: new research directions. In: Aquaponics food production systems, Springer, Cham, Switzerland, pp 145–161
Khanjani MH, Sharifinia M (2022) Biofloc as a food source for Banana shrimp (*Fenneropenaeus merguiensis*) postlarvae. N Am J Aquac. https://doi.org/10.1002/naaq.10261
Khanjani MH, Sharifinia M, Hajirezaee S (2020) Effects of different salinity levels on water quality, growth performance and body composition of Pacific white shrimp (Litopenaeus vannamei Boone, 1931) cultured in a zero water exchange heterotrophic system. Ann Anim Sci 20(4):1471–1486
Khanjani MH, Alizadeh M, Sharifinia M (2021) Effects of different carbon sources on water quality, biofloc quality, and growth performance of Nile tilapia (Oreochromis niloticus) fingerlings in a heterotrophic culture system. Aquac Int 29:307–321
Khanjani MH, Sharifinia M, Hajirezaee S (2022) Recent progress towards the application of biofloc technology for tilapia farming. Aquaculture 552:738021
Khanjani MH, da Silva LOB, Fóes GK, Vieira FdN, Poli MA, Santos M, Emerenciano MGC (2023) Synbiotics and aquamimicry as alternative microbial-based approaches in intensive shrimp farming and biofloc: novel disruptive techniques or complementary management tools? A scientific-based overview. Aquaculture 567. https://doi.org/10.1016/j.aquaculture.2023.739273
Khatoon H, Banerjee S, Syakir Syahiran M, Mat Noordin NB, Bolong MA, Endut A (2016) Re-use of aquaculture wastewater in cultivating microalgae as live feed for aquaculture organisms. Desalin Water Treat 57:29295–29302
Khiari Z, Kaluthota S, Savidov N (2019) Aerobic bioconversion of aquaculture solid waste into liquid fertilizer: effects of bioprocess parameters on kinetics of nitrogen mineralization. Aquaculture 500:492–499. https://doi.org/10.1016/j.aquaculture.2018.10.059
Kroumov AD, Scheufele FB, Trigueros DEG, Modenes AN, Zaharieva M, Najdenski H (2017) Modeling and technoeconomic analysis of algae for bioenergy and coproducts. Algal Green Chem Recent Prog Biotechnol 201–241 https://doi.org/10.1016/B978-0-444-63784-0.00011-4
Kumar G, Engle C, Tucker C (2018) Factors driving aquaculture technology adoption. J World Aquac Soc 49(3):447–476
Kumar R, Gupta RK, Yadav R, Saifi R, Yodha K, Pooja (2022a) *Eisenia fetida* as protein source for growth enhancement of *Heteropneustes fossilis*. Egypt J Aquat Biol Fish 26(2):577–588

Kumar R, Yadav R, Kumar Gupta R, Pooja (2022b) A sustainable way for fish health management by replacement of chemical and drugs by earthworm. In: Environmental degradation in Asia: land degradation, environmental contamination, and human activities. Springer International Publishing, Cham, pp 329–352

Kümmerer K (2009) Antibiotics in the aquatic environment—a review—part II. Chemosphere 75(4):435–441. https://doi.org/10.1016/j.chemosphere.2008.12.006

Kyaw TY, Ng AK (2017) Smart aquaponics system for urban farming. Energy Procedia 143:342–347

Labatut RA, Ebeling JM, Bhaskaran R, Timmons MB (2007) Hydrodynamics of a large-scale mixed-cell raceway (MCR): experimental studies. Aquac Eng 37:132–143. https://doi.org/10.1016/j.aquaeng.2007.04.001

Lee C, Wang Y-J (2020) Development of a cloud-based IoT monitoring system for fish metabolism and activity in aquaponics. Aquac Eng 90:102067

Li C, Lee CT, Gao Y, Hashim H, Zhang X, Wu WM, Zhang Z (2018) Prospect of aquaponics for the sustainable development of food production in urban. Chem Eng Trans 63:475–480

Lothmann R, Sewilam H (2023) Potential of innovative marine aquaculture techniques to close nutrient cycles. Rev Aquac 15(3):947–964. https://doi.org/10.1111/raq.12781

Martínez Cruz P, Ibáñez AL, Monroy Hermosillo OA, Ramírez Saad HC (2012) Use of probiotics in aquaculture. Int Sch Res Not 2012(1):916845

Martínez-Córdova LR, Emerenciano M, Miranda-Baeza A, Martínez-Porchas M (2015) Microbial-based systems for aquaculture of fish and shrimp: an updated review. Rev Aquac 7:131–148

Mateus AP, Power DM, Canário AVM (2017) Stress and disease in fish. Elsevier, Amsterdam, The Netherlands. ISBN 9780128045640

Merrifield DL, Dimitroglou A, Foey A, Davies SJ, Baker RT, Bogwald J, Castex M, Ringo E (2010) The current status and future focus of probiotic and prebiotic applications. Aquaculture 302:1–18

Monsees H, Kloas W, Wuertz S (2017) Decoupled systems on trial: eliminating bottlenecks to improve aquaponic processes. PLoS ONE 12(9):e0183056

Mori J, Smith R (2019) Transmission of waterborne fish and plant pathogens in aquaponics and their control with physical disinfection and filtration: a systematized review. Aquaculture 504:380–395

Moriarty MJ, Semmens K, Bissonnette GK, Jaczynski J (2018) Inactivation with UV-radiation and internalization assessment of coliforms and Escherichia coli in aquaponically grown lettuce. LWT 89:624–630

Moriarty DJW (1986) Bacterial productivity in ponds used for culture of penaeid prawns. Microb Ecol 12(3):259–269. http://www.jstor.org/stable/4250886

Moura RST, Valenti WC, Henry-Silva GG (2016) Sustainability of Nile tilapia net-cage culture in a reservoir in a semi-arid region. Ecol Indic 66:574–582

Murray A, Ives S, Smith R, Moriarty M (2021) A preliminary assessment of indirect impacts on aquaculture species health and welfare in Scotland during COVID-19 lockdown. Vet Anim Sci 11:100167. https://doi.org/10.1016/j.vas.2021.100167

Nathanailides C, Kolygas M, Choremi K, Mavraganis T, Gouva E, Vidalis K, Athanassopoulou F (2021) Probiotics have the potential to significantly mitigate the environmental impact of freshwater fish farms. Fishes 6:76

Naughton S, Kavanagh S, Lynch M, Rowan NJ (2020) Synchronizing use of sophisticated wet-laboratory and in-field handheld technologies for real-time monitoring of key microalgae, bacteria and physicochemical parameters influencing efficacy of water quality in a freshwater aquaculture recirculation system. Aquaculture 526. https://doi.org/10.1016/j.aquaculture.2020.735377

Nederlof M, Verdegem MCJ, Smaal AC, Jansen HM (2022) Nutrient retention efficiencies in integrated multi-trophic aquaculture. Rev Aquac 14:1194–1212

Nevejan N, De Schryver P, Wille M, Dierckens K, Baruah K, Van Stappen G (2018) Bacteria as food in aquaculture: do they make a difference? Rev Aquac 10:180–212

Nisar U, Peng D, Mu Y, Sun Y (2021) A solution for sustainable utilization of aquaculture waste: a comprehensive review of biofloc technology and aquamimicry. Front Nutr 8. https://doi.org/10.3389/fnut.2021.791738

Olusegun A, Babatunde D, Abiodun O (2016) Haematological response of *Clarias gariepinus* juveniles reared in treated wastewater after waste solids removal using alum or Moringa oleifera seed powder. Int J Aquac. https://doi.org/10.5376/ija.2016.06.0011

Pablos C, Molina MD, Vaz-Moreira I, Silva MIL, Nunes O, Manaia C, Marugán J (2024) Study of bacterial microbiota over the course of tilapia growth cycle in an aquaponics system. Aqua 2024:1

Pantanella E, Cardarelli M, Colla G, Rea E, Marcucci A (2012) Aquaponics vs. hydroponics: production and quality of lettuce crop. Acta Hortic 927:887–893

Puja C, Abhijith M, Muralidhar PAA, Karthiredyy S (2018) Aquamimicry: an innovative method for shrimp farming. Aquac Int 26:48–50

Ramesh D, Souissi S (2018) Effects of potential probiotic Bacillus subtilis KADR1 and its subcellular components on immune responses and disease resistance in Labeorohita. Aquac Res 49:367–377. https://doi.org/10.1111/are.13467

Report AS-1201 (2024) Recirculating aquaculture system market global industry research analysis and forecast 2022 to 2030

Revsbech NP, Risgaard-Petersen N, Schramm A, Nielsen LP (2006) Nitrogen transformations in stratified aquatic microbial ecosystems. Antonie Van Leeuwenhoek 90:361–375

Romano N, Kumar V (2017) Vegetarian shrimp: pellet-free shrimp farming. World Aquac 37:36–38

Romano N, Dauda AB, Ikhsan N, Karim M, Kamarudin MS (2018) Fermenting rice bran as a carbon source for biofloc technology improved the water quality, growth, feeding efficiencies, and biochemical composition of African catfish Clarias gariepinus juveniles. Aquac Res 49(12):3691–3701. https://doi.org/10.1111/are.13837

Rosa J, Lemos MFL, Crespo D, Nunes M, Freitas A, Ramos F, Pardal MÂ, Leston S (2020) Integrated multitrophic aquaculture systems—potential risks for food safety. Trends Food Sci Technol 96:79–90. https://doi.org/10.1016/j.tifs.2019.12.008

Samocha TM, Patnaik S, Speed M, Ali AM, Burger JM, Almeida RV, Ayub Z, Harisanto M, Horowitz A, Brock DL (2007) Use of molasses as carbon source in limited discharge nursery and growout systems for Litopenaeus vannamei. Aquac Eng 36(2):184–191

Sardare MD, Admane SV (2013) A review on plant without soil—hydroponic. Int J Res Eng Technol 2(3):299–304

Schmautz Z, Graber A, Jaenicke S, Goesmann A, Junge R, Smits THM (2017) Microbial diversity in different compartments of an aquaponics system. Arch Microbiol 199:613–620

Schneider O, Sereti V, Eding EH, Verreth JAJ (2005) Analysis of nutrient flows in integrated intensive aquaculture systems. Aquac Eng 32:379–401

Senatore V, Buonerba A, Zarra T, Oliva G, Belgiorno V, Boguniewicz-Zablocka J, Naddeo V (2021) Innovative membrane photobioreactor for sustainable CO_2 capture and utilization. Chemosphere 273:129682

Shinn AP, Pratoomyot J, Bron JE, Paladini G, Brooker EE, Brooker AJ (2015) Economic costs of protistan and metazoan parasites to global mariculture. Parasitology 142:196–270

Shoko AP, Limbu SM, Mrosso HDJ, Mgaya YD (2014) A comparison of diurnal dynamics of water quality parameters in Nile tilapia (*Oreochromis niloticus*, Linnaeus, 1758) monoculture and polyculture with African sharp tooth catfish (Clarias gariepinus, Burchell, 1822) in earthen ponds. Int Aquat Res 6:56

Soroosh H, Otterpohl R, Hanelt D (2022) Influence of hydraulic retention time on municipal wastewater treatment using microalgae-bacteria flocs in sequencing batch reactors. Bioresour Technol Rep 17:100884

Stouvenakers G, Dapprich P, Massart S, Jijakli MH (2019) Plant pathogens and control strategies in aquaponics. In: Aquaponics food production systems. Springer, Berlin/Heidelberg, Germany, pp 353–378

Subasinghe RP, Bondad-Reantaso MG, McGladdery SE (2001) Aquaculture development, health and wealth. Food and Agriculture Organization of the United Nations

Suita SM, Ballester ELC, Abreu PCOV, Wasielesky W (2015) Dextrose as carbon source in the culture of Litopenaeus vannamei (Boone, 1931) in a zero exchange system. Lat Am J Aquat Res 45(3):526–533

Sun R, Chen J, Pan C, Sun Y, Mai B, Li QX (2020a) Antibiotics and food safety in aquaculture. J Agric Food Chem 68(43):11908. https://doi.org/10.1021/acs.jafc.0c03996

Sun H, He Z, Zhang M, Yen L, Cao Y, Hu Z, Peng Y, Lee SY (2020b) Spatial variation of soil properties impacted by aquaculture effluent in a small-scale mangrove. Mar Pollut Bull 160. https://doi.org/10.1016/j.marpolbul.2020.111511

Timmons M, Ebling J (2010) Recirculating aquaculture, 2nd edn. Ithaca NY

Van Rijn J, Tal Y, Schreier HJ (2006) Denitrification in recirculating systems: theory and applications. Aquac Eng 34(3):364–376

Vezzulli L, Chelossi E, Riccardi G, Fabiano M (2002) Bacterial community structure and activity in fish farm sediments of the Ligurian Sea (Western Mediterranean). Aquac Int 10:123–141. https://doi.org/10.1023/A:1021365829687

Vidal JMA, Pessoa MNDC, Santos FLD, Mendes PDP, Mendes MS (2018) Probiotic potential of Bacillus cereus against Vibrio spp. in post-larvae shrimps. Revista Caatinga 31:495–503

Vijayan KK, Panigrahi A (2019) Biofloc technology for nursery and grow out aquaculture. World Aquaculture Society, Baton Rouge, LA

Wang C, Pan L, Zhang K, Xu W, Zhao D, Mei L (2016) Effects of different carbon sources addition on nutrition composition and extracellular enzymes activity of bioflocs, and digestive enzymes activity and growth performance of *Litopenaeus vannamei* in zero-exchange culture tanks. Aquac Res 47(10):3307–3318

Xifan N, Mubashar M, Zhang S, Qin Y, Zhang X (2020) Current progress, challenges and perspectives in microalgae-based nutrient removal for aquaculture waste: a comprehensive review. J Clean Prod 277:124209

Yadav R, Gupta RK, Kumar R, Kaur T, Moun R (2022a) Biochemical response of earthworm, *Eisenia fetida* to heavy metals toxicity. J Appl Nat Sci 14(3):990–998

Yadav R, Gupta RK, Kumar R, Kaur T (2022b) Assessment of toxicity of lead and nickel on the biochemical and immunological parameters of earthworm, *Eudrilus eugeniae*. Agric Sci Dig 44(5):922–929

Yadav R, Kumar R, Gupta RK, Kaur T, Yodha K, Kour A, Kaur S, Rajput A (2023) Heavy metal toxicity in earthworms and its environmental implications: a review. Environ Adv 100374

Yang T, Kim H-J (2020) Effects of hydraulic loading rate on spatial and temporal water quality characteristics and crop growth and yield in aquaponic systems. Horticulturae 6:9

Yang L, Zeng X, Qiao S (2021) Advances in research on solid-state fermented feed and its utilization: the pioneer of private customization for intestinal microorganisms. Anim Nutr 7(4):905–916. https://doi.org/10.1016/j.aninu.2021.06.002

Yavuzcan Yildiz H, Robaina L, Pirhonen J, Mente E, Domínguez D, Parisi G (2017) Fish welfare in aquaponic systems: its relation to water quality with an emphasis on feed and faeces—a review. Water 9(1):13

Yep B, Zheng Y (2019) Aquaponic trends and challenges—a review. J Clean Prod 228:1586–1599

Zou Y, Hu Z, Zhang J, Xie H, Guimbaud C, Fang Y (2016) Effects of pH on nitrogen transformations in media-based aquaponics. Bioresour Technol 210:81–87

Chapter 10
Biosafety and Biosecurity in Fish Farms: Necessity for Sustainable Aquaculture

Monita Dhiman

Abstract Biosafety and biosecurity are key components of assuring long-term aquaculture, especially in fish farming, where disease outbreaks and environmental concerns represent considerable obstacles. This chapter delves into the importance of biosafety and biosecurity measures in maintaining fish health, reducing environmental consequences, and promoting economic sustainability. It analyzes the conceptual framework of biosafety and biosecurity, distinguishing their functions in avoiding disease spread and alleviating biohazards. The chapter focuses on major issues that jeopardize biosecurity in fish farms, including resource depletion, climate change, and a lack of tech adoption. Strategies for establishing biosecure facility designs, effective disease management protocols, and water quality control measures are outlined, highlighting their significance in attaining the United Nations Development Goals (UNDGs). Moreover, the incorporation of technical advancements, including as biosensors and automation, is examined as a revolutionary strategy for enhancing biosecurity systems. This chapter analyzes the economic and environmental ramifications of biosecurity measures, promoting legislative and regulatory frameworks that endorse sustainable practices. This document also emphasizes the critical importance of biosafety and biosecurity in promoting sustainable aquaculture practices that conform to global sustainability objectives through actionable insights and case examples.

Keywords Biosafety · Biosecurity · Fish Farms · Sustainable aquaculture

10.1 Introduction

Aquaculture, the farming of aquatic organisms such as fish, crustaceans, and molluscs, has emerged as a cornerstone of global food security. The global population's dependence on aquaculture is undeniable. According to the Food and Agriculture Organization (Kumar et al. 2022a, 2022b; FAO 2022), aquaculture now supplies

M. Dhiman (✉)
Department of Zoology, Khalsa College for Women, Civil Lines, Ludhiana, Punjab, India
e-mail: monid.84@gmail.com

R. Kumar et al. (eds.), *Sustainability in Aquaculture and Fisheries*,
Springer Water, https://doi.org/10.1007/978-3-032-16605-0_10

over 50% of all fish consumed by humans, a figure projected to rise as wild fisheries stagnate and global demand for protein surges. By 2030, aquaculture is expected to supply nearly 60% of edible fish, driven by increasing per capita consumption (from 20.5 kg in 2018 to 21.5 kg by 2030) and declining wild catches (FAO 2022). Developing nations, particularly in Asia, dominate production: China, India, and Indonesia collectively account for 80% of global aquaculture output (Ybanez and Gonzales 2023). However, the sector's rapid expansion (growing at an average annual rate of 5.3% since 2000) has introduced significant risks that threaten its long-term viability (Bartley 2022). These include escalating disease outbreaks, environmental degradation, and the overuse of antimicrobials, all of which undermine the principles of sustainability. Additionally, intensive farming practices, such as high-density stocking and monoculture systems, create ideal conditions for pathogen proliferation and environmental stress (Kumar et al. 2022c; Pant and Kumar 2025).

The global demand for fish and seafood has increased significantly due to population growth, rising incomes, and an awareness of the health benefits of fish consumption. As wild fish stocks face overexploitation and environmental degradation, aquaculture has emerged as a vital solution for meeting this demand. However, with the expansion of fish farming comes the increased risk of disease outbreaks, environmental degradation, and biosecurity threats. The practice of meeting current fish production needs without compromising the ability of future generations to meet theirs, while minimizing ecological harm and ensuring animal welfare leads to the Sustainable aquaculture (Bohnes et al. 2022). To achieve this balance the necessary challenges to taken care are: biosecurity (in prevention of disease spread and introduction), biosafety (minimizing the unpredictable danger to ecosystem), and socioeconomic equity (equal opportunities for small-scale farmers). Biosafety in aquaculture refers to the collection of procedures designed to safeguard farming operations from the accidental introduction or transmission of harmful biological agents, such as infectious pathogens or invasive species. In contrast, biosecurity involves a set of proactive and strategic interventions aimed at preventing, managing, and mitigating biological threats. These measures minimize the harm to the ecosystem as well as protect the health of aquatic. Furthermore, promotion of socioeconomic equity in the aquaculture sector provide fair distribution of benefits and resources, particularly among small-scale producers. Small scale aquaculture makes more than 85% of global aquaculture and mainly from Asia and Africa continent and also play a vital role in enhancing food security and supporting rural economies (FAO 2024).

Biosafety and biosecurity mainly manage the economic damage caused by disease outbreaks, protect public health by reducing the risk of zoonotic transmission, and promote environmentally sustainable aquaculture practices. White Spot Syndrome Virus (WSSV) in shrimp or *Streptococcus* infections in fish farms are the notable examples and emphasize the urgent need for stringent protective protocols. Perera et al. (2005) emphasized that the biosecurity is essential to safeguarding aquatic life, human health, and environmental quality from emerging biological threats. It has become necessary to adopt a strong on-farm biosecurity framework to develop the aquaculture industry, increase global trade and minimize the emergence of novel diseases.

Biosecurity measures also helps in mitigate fish mortality, which directly influences production efficiency and the economic sustainability of aquaculture avenues (Adah et al. 2022). However, due to lack of awareness or resources, many fish farmers not able to avail these measures, which further highlights the dire need for targeted education and implementation support (Adah et al. 2022). Improving biosecurity has become a cornerstone for the long-term viability and resilience of the aquaculture sector (Hine et al. 2010).

In this chapter, topics like the fundamental principles of biosafety and biosecurity in aquaculture, their essential role in controlling disease spread and contaminant exposure, and their contribution to sustainable fish farming are explored. It also deals with global aquaculture trends and their intersection with the United Nations Sustainable Development Goals (SDGs). As eliminating aquatic diseases remains a persistent challenge, preventive strategies at the farm level are crucial for maintaining healthy, productive, and sustainable aquaculture systems.

10.2 Sustainable Aquaculture

10.2.1 Sustainable Aquaculture and the United Nations Development Goals (UNDGs)

Incorporating biosafety and biosecurity measures into aquaculture is not just a technical requirement, it is a vital strategy for advancing the United Nations Sustainable Development Goals (SDGs) to attain the blue transformation. Out of 17 SDGs at least six key SDGs play a supportive role in attaining the sustainable aquaculture. These are Zero Hunger (SDG 2), Good Health and Well-Being (SDG 3), Decent Work and Economic Growth (SDG 8), Responsible Consumption and Production (SDG 12), Climate Action (SDG 13), and Life Below Water (SDG 14). This part of the chapter explores how the establishment of strong biosafety and biosecurity systems contributes to these global objectives to achieve the sustainable aquaculture (Fig. 10.1).

10.2.1.1 Contribution to SDG 2: Zero Hunger

Aquaculture is pivotal to eradicating hunger, providing affordable protein to over 3 billion people (FAO 2022). The FAO's *Blue Transformation* initiative emphasizes biosecurity to boost aquaculture productivity while ensuring equitable food access, directly supporting SDG 2 targets (Ligtvoet et al. 2019). However, disease outbreaks and poor biosecurity undermine this goal by destabilizing food supplies. For example: The 2021 outbreak of *Tilapia Lake Virus (TiLV)* in Egypt and Thailand reduced tilapia yields by 40%, threatening nutrition security in low-income households (Debnath et al. 2023). Implementing quarantine protocols and vaccines

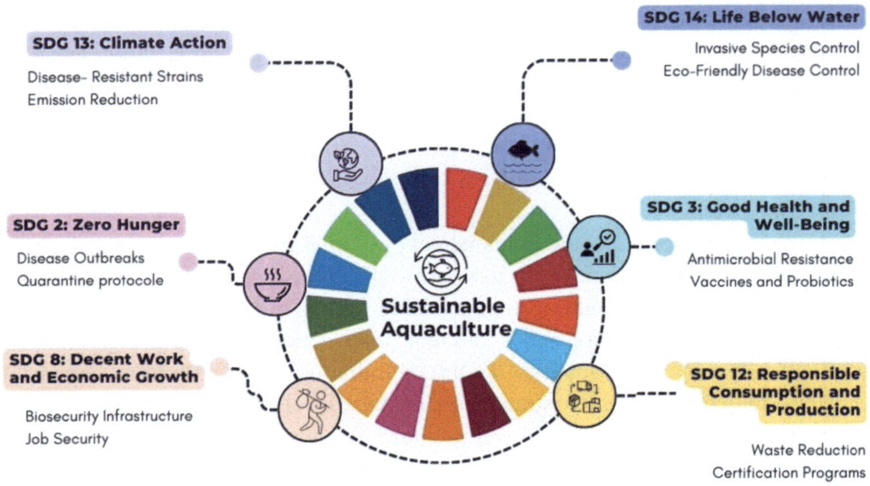

Fig. 10.1 Biosafety and Biosecurity in Sustainable Aquaculture

in Indian carp farms reduced mortality rates by 60%, securing livelihoods for 15 million small-scale farmers (Othman et al. 2024).

10.2.1.2 Link to SDG 3: Good Health and Well-Being

Antimicrobial resistance (AMR), exacerbated by irresponsible antibiotic use in aquaculture, poses a global health crisis. Approximately 75% of aquaculture antibiotics are classified as "critically important" for human medicine (WHO 2019). Key connections include: Resistant *E. coli* strains from Vietnamese shrimp farms have been linked to human urinary tract infections (Erickson et al. 2024) and Norway's shift to vaccines and probiotics cut antibiotic use in salmon farming by 99% between 1987 and 2017, reducing AMR risks (Sommerset et al. 2005). In Bangladesh, training programs on biosecure pond management reduced antibiotic use in carp farms by 70%, lowering AMR gene prevalence in local water systems (Mohsin et al. 2020).

10.2.1.3 Supporting SDG 8: Decent Work and Economic Growth

Aquaculture employs over 20 million people globally, but disease outbreaks and environmental crises jeopardize jobs. Investing in biosecurity infrastructure (e.g., diagnostic labs, certified hatcheries) creates skilled jobs while protecting existing ones, aligning with SDG 8's focus on inclusive economic growth. For instance: Chile's ISA outbreak (2007–2009) cost $2 billion and displaced 26,000 workers (Asche et al. 2022). **Biosecurity as Job Security**: Ecuador's adoption of farm zoning and pathogen surveillance in shrimp farming stabilized exports, now valued at $5.6 billion annually (Wurmann 2022).

10.2.1.4 Alignment with SDG 12: Responsible Consumption and Production

Unsustainable practices, such as excessive feed use and chemical reliance, violate SDG 12's call for resource efficiency. Thai shrimp farms adopting ASC standards reduced nitrogen waste by 50% and increased market access to premium EU markets (Bush et al. 2013). Biosecurity innovations address this by: **Reducing Waste**: Recirculating Aquaculture Systems (RAS) recycle 95% of water and minimize effluent discharge (Bostock et al. 2018) and **Certification Programs**: The Aquaculture Stewardship Council (ASC) mandates biosecurity compliance, ensuring eco-labeling for responsibly farmed products.

10.2.1.5 Role in SDG 13: Climate Action

Climate change amplifies aquaculture risks, as warmer waters accelerate pathogen growth and stress fish immunity. The Paris Agreement's adaptation funds could subsidize biosecure infrastructure in climate-vulnerable regions like the Pacific Islands. Biosecurity mitigates these impacts through: **Disease-Resistant Strains**: Selective breeding of heat-tolerant tilapia in Ghana reduced climate-linked mortality by 30% (Ragasa et al. 2022). **Emission Reductions:** RAS systems lower carbon footprints by reducing transportation needs for wild-caught fish (Hallström et al. 2019).

10.2.1.6 Advancing SDG 14: Life Below Water

Aquaculture can degrade marine ecosystems through pollution and invasive species, contradicting SDG 14. Robust biosecurity reverses this by: **Containing Invasives**: Sterile triploid oysters in U.S. aquaculture prevent genetic contamination of wild populations (Nichols 2001). **Eco-Friendly Disease Control**: Sea lice mitigation using cleaner fish (e.g., lumpfish) in Scottish salmon farms reduced pesticide use by 80% (Brooker et al. 2018). Such as Mangrove restoration in Indonesia's shrimp farms enhanced coastal biodiversity while acting as natural biofilters for farm effluents (Alam et al. 2022).

Despite these synergies, barriers persist: such as **financial constraints**: Small-scale farmers lack funds for biosecurity technologies, **policy fragmentation**: Only 30% of nations have enforceable aquaculture biosecurity laws (Bartley 2022) and **knowledge gaps**: Limited awareness of SDG linkages among stakeholders. **To overcome these policy makers can subsidize biosecurity**: Allocate SDG-aligned grants for RAS and diagnostic tools, **strengthen regulations**: Enact laws mandating pathogen surveillance and effluent treatment, **promote certification**: Incentivize

ASC/BAP certification through tax breaks and enhancing **global partnerships**: Expand the FAO-OIE-WHO Tripartite Alliance to address AMR in aquaculture. Biosafety and biosecurity are linchpins for achieving the UNDGs through sustainable aquaculture. By curbing diseases, reducing pollution, and safeguarding livelihoods, these practices ensure aquaculture's role as a driver of equitable development. Multi-lateral collaboration, guided by SDG frameworks, is essential to scale these solutions globally. By adopting biosafety and biosecurity measures, aquaculture can address environmental concerns, ensure economic viability, and support global sustainability goals, creating a balance between food production and ecological preservation.

10.3 Conceptual Framework of Biosafety and Biosecurity

To achieve the sustainable aquaculture, biosafety and biosecurity serve as the backbone, which are interrelated but distinct components for the aquatic ecosystem. With these two components a farmer can ensure the prevention of biological risks and the promotion of safe and secure aquaculture practices. In this section we comprehensively explore of these concepts, their distinctions, and their critical roles in disease management and prevention within fish farms.

- Biosafety involves the application of precautionary strategies aimed at safeguarding aquatic species, farm personnel, and nearby ecosystems from unintended contact with hazardous biological agents. Its primary goal is to reduce risks by promoting safe handling and containment of pathogens (Shahid et al. 2022; Chandhana et al. 2025).
- Biosecurity encompasses a broader range of strategies aimed at preventing the introduction, spread, and amplification of infectious diseases and invasive species in aquaculture systems. Quarantine protocols, pathogen surveillance, vaccination, and site disinfection are the main measures (Shahid et al. 2022; Nagothu et al. 2025).

Biosecurity and biosafety play pivotal roles in aqua farming, managing safe practices to get the increasing demand for seafood with the same time these are also helping in minimizing environmental and public health risks (Stentiford 2022). Although biosafety and biosecurity share overlapping goals, they differ in scope and application given in the Table 10.1.

Understanding these differences is critical for designing effective management protocols tailored to specific aquaculture setting (Dadar et al. 2018).

Table 10.1 Aspects of Biosafety and Biosecurity in Aquaculture

Aspect	Biosafety	Biosecurity	References
Focus	Preventing accidental exposure to biological risks	Preventing intentional or unintentional introduction and spread of pathogens	Shahid et al. (2022), Chandhana et al. (2025)
Scope	Risk containment within operational systems	Broader, encompassing external and internal risk mitigation	Shahid et al. (2022), Nagothu et al. (2025)
Example	Safe handling of diseased fish, controlled lab practices	Quarantine, pathogen screening, vaccination programs	Shahid et al. (2022), Chandhana et al. (2025), Nagothu et al. (2025)

10.4 Role of Biosafety and Biosecurity in Disease Management and Prevention

Aquaculture systems are highly vulnerable to biological risks, including disease outbreaks that can devastate fish populations, disrupt supply chains, and cause economic losses. Biosafety and biosecurity play crucial roles in mitigating these risks. It refers to measures that prevent the introduction of infectious agents into defined geographical locations or facilities. It also includes bio containment referring to control of diseases already present in an area to prevent transmission to other areas. To ensure biosafety in aquaculture, biosecurity measures should be implemented at all stages of the aquaculture production cycle in:

a. **Preventing Disease Introduction**

Biosecurity measures such as quarantine and health certification of broodstock prevent the entry of pathogens into farming systems. Climate change and human migration directly influence microbial diversity and pathogen transmission, making the traceable detection of infections crucial. Biological nanoscale technology and biosensor technology have arisen at an opportune time. Ongoing advancements in biological detection and recognition, along with the swift, sensitive, and precise identification of pathogens, have facilitated prompt preliminary screening and early warning (Zhou et al. 2019). For instance, the implementation of stringent quarantine protocols in shrimp aquaculture has been effective in minimizing the spread of White Spot Syndrome Virus (WSSV) (Islam et al. 2023).

b. **Reducing Disease Transmission**

Biosafety protocols like proper water quality management and the use of disinfectants minimize the spread of diseases within aquaculture systems. Recent advancements in water treatment technologies, including ultraviolet sterilization, further enhance biosafety (FAO 2024). Rapid disease identification is essential for the sustainability of aquaculture, and significant advancements in biotechnology over the past decade have facilitated the invention and enhancement of several immunodiagnostic and molecular approaches, and reagents and kits have become increasingly accessible.

Immunodiagnostic methods currently used, such as immunohistochemistry (IHC), the fluorescence antibody test (FAT) and indirect fluorescence antibody test (IFAT) enable rapid specific detection of pathogens in tissue samples without the need to first isolate the pathogen. Fifteen fish infections have been identified via microarray technology, and multiple teams are engaged in assay development. The process involves hybridizing DNA fragment samples (amplicons), amplified by PCR, onto particular DNA detector fragments affixed to a solid substrate. A multitude of DNA spots from various diseases can be incorporated onto a single plate, facilitating multiplexing for diverse infections. The method has great sensitivity, specificity, substantial throughput capacity, cost reduction, and expedited diagnosis; yet, it remains nascent in aquaculture (Kostić-Vuković et al. 2021).

c. **Early Detection and Response**

Aquaculture pathogens are traditionally identified using conventional PCR techniques or antibody-based assays, which are largely restricted to laboratory settings. However, recent advancements in assay technologies and biosensors present promising opportunities for developing highly sensitive, specific, and portable diagnostic tools suitable for field use. For the detection of pathogens and their identification various innovations in DNA/RNA analysis, immunoassays, and nanotechnology have played significant role by improving response time, sensitivity, and the ability to distinguish closely related species. Moreover, advanced research in electrochemical and optical signal transduction have enabled the integration of these assays into compact, practical devices (Pires et al. 2021). Furthermore, electrochemical signal transduction also offers a cost-effective pathway for developing biosensors. These biosensors can be effectively utilized in aquaculture environments. Continuous surveillance and monitoring are also essential for primary detection of disease spread. The biosensors is a recent innovation contributing to proactive disease management for real-time pathogen detection (Subasinghe et al. 2022). To achieve the sustainable and secure aquaculture system to need to adopt these innovative tools.

c. **Promoting Resilience Against Emerging Diseases**

Climate change leads to emergence of various diseases, pose a growing threat to aquaculture systems and demanding adaptive strategies that are both proactive and sustainable. Resilience can be promoted by strengthening ecosystem health through integrated and precautionary management approaches that reduce stress on aquatic species and environments. Diversifying markets, livelihoods, and products can reduce dependency on single species or systems, thereby lowering exposure to disease-related shocks. Insurance schemes for fish stocks and infrastructure, coupled with early warning systems and robust post-outbreak recovery protocols, further help mitigate losses. Infrastructure improvements, such as better water quality systems, protected hatcheries, and biosecure landing sites, are essential to reduce pathogen transmission (De Young et al. 2012). At the community level, education, capacity-building, and support for local governance institutions enhance awareness and enable rapid response to emerging biosecurity threats. However, measures like increasing fishing effort or expanding production without regulation may intensify stressors and promote disease proliferation. Thus, a balanced, ecosystem-based approach integrating technological innovation, economic diversification, and institutional support is crucial for building resilience to emerging diseases in aquaculture under changing climate conditions.

Biosafety and biosecurity measures provide a framework for addressing these evolving risks, ensuring adaptability and sustainability (Stentiford 2022).

10.5 Challenges in Maintaining Biosafety and Biosecurity in Fish Farms

The last three to four decades have seen a rise in the farming of aquatic animals global, it is further enhanced due to incorporation of biosecurity and biosafety aspects in aquaculture. Additionally the international movement of juvenile animals and broodstock, with the little knowledge regarding the health status and source of the animal leads to the frequent emergence of new diseases. Despite the critical importance of biosafety and biosecurity in ensuring sustainable aquaculture, several challenges impede their effective implementation. These challenges stem from biological, environmental, and technological factors that demand urgent attention to safeguard fish farms and promote sustainable practices.

10.5.1 Biological Challenges

Aquaculture systems are susceptible to a range of threats that compromise biosafety and biosecurity. The common threats are disease outbreaks, pathogen spread and contamination.

High stocking densities commonly found in fish farming operations often create favorable environments for the rapid transmission of infectious diseases. Pathogen outbreaks, including those caused by *Vibrio* species and White Spot Syndrome Virus (WSSV), have led to significant economic setbacks on a global scale (Dadar et al. 2018). Additionally, the expansion of aquaculture has increased the trade and movement of fish, raising the risk of introducing stocks with unknown or unverified health statuses (Opiyo et al. 2018). The international transport of live fish, particularly broodstock, contributes to the trans boundary spread of aquatic diseases (Bondad-Reantaso 2005). The presence of endemic, emerging, and re-emerging diseases in aquaculture poses serious threats to both livelihoods and public health, especially in cases involving zoonotic pathogens, which can potentially compromise future market stability. Studies indicate that nearly half of the production losses in aquaculture are attributed to disease outbreaks and parasitic infestations (Assefa and Abunna 2018). Further, the pathogens can enter aquaculture systems through contaminated water, infected broodstock, or vectors like birds and insects. Inadequate quarantine measures exacerbate this issue, increasing the likelihood of pathogen amplification (Shefali et al. 2021; Boyd et al. 2022). Additionally, the use of chemicals for disease control, such as antibiotics, can lead to antimicrobial resistance, while environmental contaminants like pesticides can stress aquatic organisms, making them more susceptible to diseases (Yadav et al. 2022a, 2022b, 2023; Stentiford 2022).

10.5.2 Environmental Challenges

Climate change and global warming have introduced additional complexities to maintaining biosafety and biosecurity in aquaculture. Rise in water temperatures leading to more frequent and severe disease outbreaks by accelerating the metabolism of fish and pathogens. For example, the prevalence of harmful algal blooms has increased in warmer climates, affecting water quality and fish health (FAO 2024). Extreme Weather Events such as floods, storms, and droughts disrupt aquaculture systems, leads to the spread of pathogens across regions. Such events also damage biosecurity infrastructure, making farms more vulnerable to biological threats (Subasinghe et al. 2022). Changing environmental conditions force the relocation of aquaculture activities and habitat shift for the fish, potentially exposing farms to new pathogens and ecological challenges (Stentiford 2022).

10.5.3 Technological Limitations

Real-time biosensors, although effective but remain underutilized in most regions (Dadar et al. 2018). As many fish farms lack the resources for continuous monitoring of pathogens and water quality. Additionally, there is a gap in knowledge of advanced technology among farmers due to lack of training in implementing biosecurity protocols effectively, which increase vulnerability to disease outbreaks (Boyd et al. 2022). Additionally, the lack of global collaboration limits the sharing of best practices and innovations (Subasinghe et al. 2022).

Weak regulatory policy frameworks and poor enforcement hinder the adoption of standardized biosafety and biosecurity measures. Furthermore, in small-scale fish farms there is a financial constraints in adopting advanced technologies and infrastructure for biosecurity (FAO 2022). This economic disparity leaves many farms exposed to avoidable risks.

10.6 Strategies for Implementing Biosafety and Biosecurity Protocols

To ensure the long-term sustainability of aquaculture, implementing robust biosafety and biosecurity protocols is essential. These strategies involve a holistic approach combining facility design, operational practices, and disease management systems to minimize risks and promote healthy aquaculture systems at all stages of the aquaculture production cycle, which include broodstock-stations, individual farms, national and international organizations responsible for aquaculture exports and imports.

10.6.1 Biosecure Facility Design and Construction

The design and construction of aquaculture facilities play a pivotal role in maintaining biosecurity. First of all the physical barrier, the farms should be designed with fencing, bird nestin, sealed ponds to prevent the entry of disease vectors which can contaminant the aquaculture production cycle at any stage (Boyd et al. 2022). Further there should be efficient waste management systems to ensure that the effluent water does not contaminate surrounding water bodies, this also reduce the risk of pathogen spread (FAO 2022). Separate quarantine zones should be established for new or sick stock to prevent potential pathogen transmission to healthy populations (Subasinghe et al. 2022).

10.6.2 Best Practices in Hatchery Management and Water Quality Control

Hatchery management and water quality control are critical components of biosafety. Using certified pathogen-free broodstock is essential to prevent the introduction of diseases into aquaculture systems (Stentiford 2022). Similarly, maintaining optimal water parameters (e.g., pH, temperature, dissolved oxygen) minimizes stress on fish, reducing their susceptibility to diseases. Advanced water filtration systems, such as biofilters and UV sterilizers, are highly effective in maintaining water quality (Boyd et al. 2022). This can also be enhanced by regular cleaning and disinfection of tanks, equipment, and surrounding areas reduce the buildup of harmful microorganisms. Rotational fallowing of ponds is another effective measure to disrupt pathogen life cycles (FAO 2022). Keeping the nutritional management in the hatchery by providing a balanced and nutrient-rich diet enhances fish immunity, making them more resilient against diseases (Dadar et al. 2018; Nisha et al. 2025).

10.6.3 Integrated Disease Management Systems

Integrated Disease Management (IDM) is a holistic approach combining preventive, diagnostic, and therapeutic measures. Continuous monitoring and early detection of diseases enable timely interventions, reducing mortality rates. Recent advances in molecular diagnostic tools, such as PCR and biosensors, provide rapid and accurate pathogen identification (Subasinghe et al. 2022). Stentiford (2022) noticed that the vaccination is a cost-effective and environmentally friendly method to prevent disease outbreaks. For example, vaccines against bacterial pathogens like *Aeromonas hydrophila* have shown promising results. Similarly, the use of probiotics and immunostimulants enhances gut health and immunity in fish, serving as an alternative to antibiotics in disease control (Boyd et al. 2022).

Developing and implementing comprehensive biosecurity plans tailored to specific farms ensures the systematic management of risks. These plans should include contingency measures for outbreak response.

10.7 Economical and Environmental Ramification of Biosecurity Measures

The adoption of robust biosecurity measures in aquaculture has profound economic and environmental consequences. While initial implementation may involve significant investments, the long-term benefits, including reduced disease outbreaks, improved farm productivity, and minimized environmental impact, outweigh the costs. These measures align with sustainable development goals (SDGs) by promoting efficient resource use and environmental conservation. Implementing biosecurity and biosafety measures, incures upfront costs. However, studies revealed that these investments reduces economic losses from disease outbreaks by over 50% (Pa et al. 2021). Small-scale aquaculture can face challenges in adopting technology as these are expensive but the subsidies, cooperative funding models and low cost innovations resolved the issue and making the biosecurity accessible to all (Subasinghe and Bondad–Reantaso 2005). Effective biosecurity measures prevent disease outbreaks and reducing the environmental contamination (Dadar et al. 2018). Similarly, Improved biosecurity decreases the reliance on antibiotics and chemical treatments, mitigating the risks of antimicrobial resistance and chemical pollution in water bodies (Boyd et al. 2022). Biosecure practices prevent the escape of farmed species and pathogens into natural ecosystems, safeguarding local biodiversity and maintaining ecological balance.

10.8 Conclusion

The future of biosafety and biosecurity in aquaculture lies in embracing technological advancements, fostering education, and promoting global collaboration. As the aquaculture industry continues to expand, managing emerging threats and promoting proactive measures is crucial to ensure sustainable practices. This whole system ensure the sustainability of aquaculture and its contribution to global food security. This chapter explored key aspects of biosafety and biosecurity, including their role in disease prevention, technological innovations, economic implications, emphasizing their significance for sustainable aquaculture.

10.9 Future Prospective

The future perspective of biosafety and biosecurity in aquaculture is crucial and pivotal in ensuring sustainable, resilient and productive fish farming system. As the demand for aquatic products increasing globally, the fish farms faces significant challenges such as disease management, environmental sustainability and food security. Biosecurity measures helping to mitigate these challenges and ensuring the health of aquatic animals. To maintain the biosafety and biosecurity in future, the following strategies can be considered:

- Integration of smart technologies (real-time sensors, automated monitoring systems, AI models etc.) in the aquaculture. This will lead to early detection of pathogen and quick response and reduces the risk of disease outbreaks.
- Government and international bodies should introduce strict regulatory frameworks as guidelines and bio-certifications for fish farms.
- Eco-Friendly disinfection and biological strategies will be the priority by considering the growing concerns using chemical and antibiotics.
- Genetic and genomic advancements will be helpful to produce disease-resistant fish species, reducing biosecurity risks and improving overall farm health.
- Training programmes and capacity building for farmers about the biosecurity protocols will be key to effective implementation.
- Strong surveillance systems will be essential for the proactive management of biosecurity threats across aquaculture zones.

References

Adah DA, Saidu L, Oniye SJ, Adah AS (2022) An assessment of the impacts of biosecurity measures on mortality of fish from fish farms. Aquaculture Studies 23(5). https://doi.org/10.4194/AQUAST1060

Agriculture Organization of the United Nations. Fisheries Department (2018) The state of world fisheries and aquaculture. Food and Agriculture Organization of the United Nations

Alam MI, Rahman MS, Ahmed MU, Debrot AO, Ahsan MN, Verdegem MCJ (2022) Mangrove forest conservation vs shrimp production: uncovering a sustainable co-management model and policy solution for mangrove greenbelt development in coastal Bangladesh. Forest Policy Econ 144:102824. https://doi.org/10.1016/j.forpol.2022.102824

Asche F, Eggert H, Oglend A, Roheim CA, Smith MD (2022) Aquaculture: externalities and policy options. Rev Environ Econ Policy 16(2):282–305. https://doi.org/10.1086/721055

Assefa A, Abunna F (2018) Maintenance of fish health in aquaculture: review of epidemiological approaches for prevention and control of infectious disease of fish. Veterinary Med International 2018(1):5432497. https://doi.org/10.1155/2018/5432497

Bartley DM (2022) World aquaculture 2020–a brief overview

Bohnes FA, Hauschild MZ, Schlundt J, Nielsen M, Laurent A (2022) Environmental sustainability of future aquaculture production: analysis of Singaporean and Norwegian policies. Aquaculture 549:737717. https://doi.org/10.1016/j.aquaculture.2021.737717

Bondad-Reantaso MG, Subasinghe RP, Arthur JR, Ogawa K, Chinabut S, Adlard R, Shariff M (2005) Disease and health management in Asian aquaculture. Vet Parasitol 132(3–4):249–272. https://doi.org/10.1016/j.vetpar.2005.07.005

Bostock J, Fletcher D, Badiola M, Murray F (2018) An update on the 2014 report: Review of recirculation aquaculture system technologies and their commercial application

Boyd CE, McNevin AA, Davis RP (2022) The contribution of fisheries and aquaculture to the global protein supply. Food Security 14(3):805–827. https://doi.org/10.1007/s12571-021-01246-9

Brooker AJ, Papadopoulou A, Gutierrez C, Rey S, Davie A, Migaud H (2018) Sustainable production and use of cleaner fish for the biological control of sea lice: recent advances and current challenges. Veterinary Record 183(12):383–383. https://doi.org/10.1136/vr.104966

Bush SR, Belton B, Hall D, Vandergeest P, Murray FJ, Ponte S, Kusumawati R (2013) Certify sustainable aquaculture? Science 341(6150):1067–1068. https://doi.org/10.1126/science.1237314

Chandhana BL, Vijayan A, Rejish Kumar VJ (2025) Biosecurity in aquaculture. In: Antimicrobial resistance in aquaculture and aquatic environments, pp 335–358. Singapore: Springer Nature Singapore. https://doi.org/10.1007/978-981-97-7320-6_14

Dadar M, Chakraborty S, Dhama K, Prasad M, Khandia R, Hassan S, Chaicumpa W (2018) Advances in designing and developing vaccines, drugs and therapeutic approaches to counter human papilloma virus. Front Immunol 9:2478. https://doi.org/10.3389/fimmu.2018.02478

De Young C, Soto D, Bahri T, Brown D (2012) Building resilience for adaptation to climate change in the fisheries and aquaculture sector. Building Resilience Adaptation Climate Change Agric Sector 23:103. https://doi.org/10.1007/s10499-023-01117-4

Debnath SC, McMurtrie J, Temperton B, Delamare-Deboutteville J, Mohan CV, Tyler CR (2023) Tilapia aquaculture, emerging diseases, and the roles of the skin microbiomes in health and disease. Aquacult Int 31(5):2945–2976. https://doi.org/10.1007/s10499-023-01117-4

Erickson VI, Dung TT, Hounmanou YMG, Phu TM, Dalsgaard A (2024) Genomic Insights into *Edwardsiella ictaluri*: Molecular Epidemiology and Antimicrobial Resistance in Striped Catfish (*Pangasianodon hypophthalmus*) Aquaculture in Vietnam. Microorganisms 12(6):1182. https://doi.org/10.3390/microorganisms12061182

FAO (2022) The state of world fisheries and aquaculture 2022. Towards blue transformation. Rome, FAO. https://doi.org/10.4060/cc0461en

FAO. (2024) the state of world fisheries and aquaculture 2024—blue transformation in action. Rome. https://doi.org/10.4060/cd0683en

Hallström E, Bergman K, Mifflin K, Parker R, Tyedmers P, Troell M, Ziegler F (2019) Combined climate and nutritional performance of seafoods. J Clean Prod 230:402–411. https://doi.org/10.1016/j.jclepro.2019.04.229

Hine M, Adams S, Arthur JR, Bartley D, Bondad-Reantaso MG, Chávez C, Wardle R (2010) Improving biosecurity: a necessity for aquaculture sustainability. In: Global conference on aquaculture 2010

Islam SI, Mou MJ, Sanjida S, Mahfuj S (2023) A review on molecular detection techniques of white spot syndrome virus: perspectives of problems and solutions in shrimp farming. Veterinary Med Sci 9(2):778–801. https://doi.org/10.1002/vms3.979

Kostić-Vuković J, Kolarević S, Kračun-Kolarević M, Višnjić-Jeftić Ž, Rašković B, Poleksić V, Vuković-Gačić B (2021) Temporal variation of biomarkers in common bream Abramis brama (L., 1758) exposed to untreated municipal wastewater in the Danube River in Belgrade. Serbia. Environ Monitoring Assess 193:1–18. https://doi.org/10.1007/s10661-021-09232-6

Kumar R, Gupta R, Dahiya T, Yadav R, Saifi R, Yodha K, Kumari B, Singh P, Shefali (2022a) Surveillance of *Heteropneustes fossilis* (Stinging Catfish) diseases and health management practices in Haryana Appl Ecol Environ Sci 10(4):261–267. https://doi.org/10.12691/aees-10-4-12

Kumar R, Gupta RK, Yadav R, Saifi R, Yodha K, Pooja (2022b) *Eisenia fetida* as protein source for growth enhancement of *Heteropneustes fossilis*. Egypt J Aquatic Biol Fisheries 26(2):577–588. https://doi.org/10.21608/ejabf.2022.234553

Kumar R, Yadav R, Kumar Gupta R, Pooja (2022c) A sustainable way for fish health management by replacement of chemical and drugs by earthworm. In: Environmental degradation in Asia: land degradation, environmental contamination, and human activities, pp 329–352. Cham: Springer International Publishing. https://doi.org/10.1007/978-3-031-12112-8_16

Ligtvoet A, Maier F, Sijtsma L, Van Den Broek LAM, Doranova A, Eaton D, Zhechkov R (2019) Blue Bioeconomy forum: Roadmap for the blue bioeconomy. European commission.

Mohsin M, Shad AA, Ali J (2020) Antimicrobial resistance, food systems and climate change. Sustainable agriculture reviews 46: mitigation of antimicrobial resistance Vol 1 Tools and Targets, 59–81. https://doi.org/10.1007/978-3-030-53024-2_3

Nagothu SK, Bindu Sri P, Anitha G, Vincent S, Kumar OP (2025) Advancing aquaculture: fuzzy logic-based water quality monitoring and maintenance system for precision aquaculture. Aquacult Int 33(1):32. https://doi.org/10.1007/s10499-024-01701-2

Nichols D (2001) Implications of the introduction and transfer of non-indigenous marine species with particular reference to Canadian marine aquaculture (Doctoral dissertation, Memorial University of Newfoundland)

Nisha A, Rani S, Kumar S, Kumar R, Kumar A (2025) Recent technologies and strategies for bioremediation of toxic and hazardous waste. In: Dubey KK, Shah MP, Yadav A (Eds), Challenges and sustainable solutions in bioremediation, pp 1–16. CRC Press. https://doi.org/10.1201/978 1003407317-1

Opiyo MA, Marijani E, Muendo P, Odede R, Leschen W, Charo-Karisa H (2018) A review of aquaculture production and health management practices of farmed fish in Kenya. Int J Veterinary Sci Med 6(2):141–148. https://doi.org/10.1016/j.ijvsm.2018.07.001

Othman R, Elabd H, Pratiwy FM, Barzkar N, Lim LS, Rodrigues KF (2024) Biosecurity in aquaculture: nurturing health and ensuring sustainability. In: Essentials of aquaculture practices, pp 139–182. Singapore: Springer Nature Singapore. https://doi.org/10.1007/978-981-97-6699-4_8

Pa A, Ca E, Mc E (2021) The role of technology on aquaculture. Academia Lett. https://doi.org/10.20935/AL2304

Pant DR, Kumar J (2025) Review on fish polyculture practices: sustainability, innovations and challenges. Environ Conserv J 26(1):257–263. https://doi.org/10.36953/ECJ.29052881

Perera RP, Jones BRIAN, Beers PETER, Kleeman S, McGladdery S (2005) Maintaining biosecurity in aquaculture systems: a constraint or a challenge. In: 6th Symposium on diseases in Asian aquaculture

Pires NM, Dong T, Yang Z, da Silva LF (2021) Recent methods and biosensors for foodborne pathogen detection in fish: Progress and future prospects to sustainable aquaculture systems. Crit Rev Food Sci Nutr 61(11):1852–1876. https://doi.org/10.1080/10408398.2020.1767032

Ragasa C, Agyakwah SK, Asmah R, Mensah ETD, Amewu S, Oyih M (2022) Accelerating pond aquaculture development and resilience beyond COVID: ensuring food and jobs in Ghana. Aquaculture 547:737476. https://doi.org/10.1016/j.aquaculture.2021.737476

Shahid F, Iftikhar I, Rehman HM, Javaid S, Fatima M, Rehman I, Babar M (2022) Biosecurity and biosafety concerns of research and diagnostic laboratory under international guidelines. Advancements Life Sci 9(2):151–156

Shefali, Kumar R, Sankhla MS, Kumar R, Swaroop SS (2021) Impact of pesticide toxicity in aquatic environment. Biointerface Res Appl Chem 11(3):10131–10140. https://doi.org/10.33263/BRIAC113.1013110140

Sommerset I, Krossøy B, Biering E, Frost P (2005) Vaccines for fish in aquaculture. Expert Rev Vaccines 4(1):89–101. https://doi.org/10.1586/14760584.4.1.89

Stentiford GD (2022) One Health Aquaculture–a personal perspective. Bull Europ Assoc Fish Pathologists 41(5):188–191. https://doi.org/10.48045/001c.35858

Subasinghe RP (2022) An introduction to global aquaculture. In: Aquaculture pathophysiology, pp 45–48. Academic Press. https://doi.org/10.48045/001c.35858

Subasinghe RP, Bondad–Reantaso MG (2005) Biosecurity in aquaculture: International agreements and instruments, their compliance, prospects, and challenges for developing countries. Aquaculture biosecurity: prevention, control, and eradication of aquatic animal disease, 9–16. https://doi.org/10.1002/9780470376850.ch2

World Health Organization (2019) Measles vaccines: WHO position paper, April 2017–Recommendations. Vaccine 37(2):219–222. https://doi.org/10.1016/j.vaccine.2017.07.066

Wurmann CS (2022) Regional review on status and trends in aquaculture development in Latin America and the Caribbean–2020

Yadav R, Gupta RK, Kumar R, Kaur T, Moun R (2022a) "Biochemical response of earthworm, *Eisenia fetida* to heavy metals toxicity. J Appl Nat Sci 14(3):990–998

Yadav R, Gupta RK, Kumar R, Kaur T (2022b) Assessment of Toxicity of Lead and Nickel on the Biochemical and Immunological Parameters of Earthworm. Eudrilus Eugeniae. Agricultural Science Digest 44(5):922–929

Yadav R, Kumar R, Gupta RK, Kaur T, Yodha K, Kour A, Kaur S, Rajput A (2023) Heavy metal toxicity in earthworms and its environmental implications: a review. Environ Adv, 100374

Ybanez Jr CO, Gonzales RC (2023) Challenges and progress of grouper aquaculture in Asia: a review. Davao Res J 14(2):6–29. https://doi.org/10.59120/drj.v14i2.109

Zhou D, Song H, Wang J, Li Z, Xu S, Ji X, Xu J (2019) Biosafety and biosecurity. J Biosafety Biosecur 1(1):15–18. https://doi.org/10.1016/j.jobb.2019.01.001

Chapter 11
Eco-Friendly and Sustainable Products from Shrimp Waste

Khushbu Sharma, Rachna Gulati, Sushma Singh, Nisha Beniwal, and Pankaj Sharma

Abstract Shells and other byproducts are a major source of waste for the quickly growing global shrimp farming industry. Conventionally disposed off in landfills or the ocean, this waste represents serious environmental hazards, including eutrophication, habitat destruction, and the emission of sulfurous fumes. However, the valuable elements found in shrimp waste—such as chitin, proteins, lipids, and minerals—have both ecological and economic value. This study investigates several environmental friendly ways to turn shrimp waste into useful products. Important substances like chitosan, astaxanthin, and bioactive proteins can be extracted by using bioconversion methods like microbial fermentation. These substances are used in a wide range of products, including bioplastics, biofertilizers, and medications. The study also explores the possibility of using shrimp waste as packaging materials, soil amendments, and animal feed. This strategy promotes the growth of a circular economy in addition to lessening environmental deterioration. The study aims to incorporate these solutions into sustainable practices for the shrimp farming industry by highlighting the potential for shrimp waste valorization through a thorough analysis of these green technologies.

Keywords Eco-friendly · Chitosan · Sustainable · Shrimp waste · Valorization

K. Sharma (✉) · R. Gulati · S. Singh
Department of Zoology, Chaudhary Charan Singh Haryana Agriculture University, Hisar, Haryana, India
e-mail: khushbu181997@gmail.com

N. Beniwal
BRIC- National Institute of Immunology, New Delhi, India

P. Sharma
Department of Ophthalmology and Visual Sciences, University of Illinois Chicago, Chicago, USA

R. Kumar et al. (eds.), *Sustainability in Aquaculture and Fisheries*,
Springer Water, https://doi.org/10.1007/978-3-032-16605-0_11

11.1 Introduction

Seafood is essential for food security and sustainability, as well as for human health and economic development. Global fish production reached 178.5 million tons in 2018 Venugopal (2022). If demand rises by 60% as predicted, there may be 9.8 billion people on the planet by 2050 (Islam et al. 2019; Manjeet et al. 2022). Fish like tuna, pollock, whiting, herring, and mackerel as well as crustaceans like prawns, shrimp, crab, mollusks, and lobster can be found in marine habitats (Sharma et al. 2023a, b). The head and body shell make up around 40% of the shrimp that are thrown away as garbage. A portion of the shrimp waste is utilized for aquaculture and animal husbandry, while the majority is discarded in landfills (Khushbu et al. 2022), burned, or dumped into the ocean. Shrimp waste disposal sites may turn into sulfurous fumes, gas, and dust hotspots (Susetyaningsih et al. 2020). Nutrient-rich shrimp waste (SW) can be extracted with a variety of chemical and biological techniques (Ambigaipalan and Shahidi 2017). Proteins, chitin, lipids, and pigments are among the essential elements found in this trash (Ahmadkelayeh et al. 2020). All these nutrients have a considerable market value (Kandra et al. 2012). The second most common biopolymer on the Earth is chitin, after cellulose. In addition to shrimp, it is present in the exoskeletons of fungus and insects (Forsberg et al. 2014). Because chitin is highly crystalline and its chains are joined by hydrogen bonds, chitin has a strong, rigid structure (Shahidi et al. 2005). Chitin is utilized extensively in the pharmaceutical industry and is rich in nitrogen. Another useful compound, calcium carbonate, has uses in agriculture, medicine, and in the paper industry (Trushina et al. 2022). Despite the potential for contamination from heavy metals and microplastics, calcium carbonate can also be made from marble and limestone (Shefali et al. 2021; Pattanaik et al. 2020). Its direct use for human purposes is therefore restricted. A shell biorefinery's concept is to turn waste shell materials, like shrimp and crab shells, into useful products by using biotechnological techniques. This approach aims to convert shell waste into a variety of biofuels, biopolymers, and nutraceuticals, such as chitin, proteins, and minerals. Using shell is a profitable and sustainable way to manage shell waste and reduce dependency on fossil fuels and raw materials (Vicente et al. 2024).This can also result in the establishment of new businesses and jobs in coastal areas where the seafood industry produces a lot of shell trash. The proteins in shellfish can be used to make fish and animal feed, and chalk made from shells is a useful ingredient in tablets (Pattnaik et al. 2021). Conventional methods of processing SW produce a lot of effluent during demineralization and deproteination. This effluent contains notable amounts of calcium and proteins. However, it is often wasted due to its high salinity, corrosiveness, and high concentrations of sodium and chloride ions (Djumanto et al. 2018).

11.2 Shrimp Waste a Great Challenge in Shrimp Farming

Shrimp farming sector is the rapidly expanding industry worldwide (Khushbu et al. 2022). The waste from shrimp, crab, and lobster is estimated to be 10 million tons annually, of which 2.5 million come from South Asia alone (Nguyen et al. 2017). Out of all processed seafood, shrimp comprise 45%. The substantial amount of waste produced by the quick rise in shrimp production is cause for concern (Iber and Kasan 2021). Given its chemical makeup and ease of breakdown, this waste poses a risk to the environment. Ten percent of the world's shrimp come from Latin American nations like Ecuador, Brazil, and Mexico, and the remaining twenty-five percent come from Asian nations like China, Thailand, and Bangladesh (Action and Organization 2020). The output of shrimp has increased dramatically by major producers between 2000 and 2021. Regrettably, the ecosystem is severely harmed by all varieties of shrimp farms. Thirty-five percent of mangrove forests were destroyed by shrimp farming in the early 1990s. As a result of this ongoing tendency, these forests are currently suffering additional harm from shrimp aquaculture. Although intensive shrimp farming has caused erosion and flooding, mangroves help stabilize coasts and trap sediment (Kumar et al. 2022a, b; Phan et al. 2022). The natural cycles of coastal wetlands are upset and nitrogen levels are raised when salt marshes are converted to shrimp ponds, which can have a negative impact on the ecosystem. A byproduct of shrimp ponds made up mostly of organic materials is produced by the waste of the shrimp and leftover feed. The aquaculture environment and adjacent waters may become contaminated by this waste. Due to the drainage and supply canals connecting shrimp farms, large volumes of effluent end up in coastal seas (Hargan et al. 2020). Approximately 43 billion tons of effluent from shrimp enters coastal seas just in China. Up to 40% of the feed is left uneaten by shrimp producers because they frequently use more feed than is required to optimum growth. This nutrient-rich, uneaten feed soon decomposes, resulting in both financial losses and environmental damage (Herbeck et al. 2013). The feed's high phosphorus and nitrogen content can produce eutrophication, a condition in which an overabundance of nutrients promotes the fast growth of algae and phytoplankton. This growth damages aquatic life by reducing the amount of oxygen in the water and obstructing light from reaching underwater plants. Fish that are kept in low oxygen environments may get smaller, have malformed eggs, have respiratory issues, and become lethargic. Serious ecological imbalances are caused by the majority of the disintegrating feed that ends up in the environment. Antibiotic resistance is a serious global concern that is increased when antibiotics, which are present in many shrimp feeds, enter the food chain (Alderman et al. 1998). If shrimp waste is not properly disposed of, it may draw pests such as cockroaches, flies, rats, and mosquitoes that can spread disease. For this reason, controlling shrimp waste is essential to safeguarding both the environment and public health (Latt 2002). Considerable environmental issues can arise from shrimp farming, particularly because of Vietnam's high saline levels. To keep shrimp ponds at the proper salinity, farmers frequently use large amounts of groundwater pumps. This may cause saltwater to leak into the land, endangering the food supply (Moustafa, 2018; Kumar et al., 2022c; Kumar and Singh 2022). Many creatures cannot withstand high salinity levels, which can cause issues for the surrounding ecology. For instance, variations in salinity have an impact on specific genes in white leg shrimp (*L. vannamei*) (Rahi et al. 2021). To maintain the health of shrimp, farmers also utilize a variety of pesticides and pharmaceuticals, which might wind up in public rivers and endanger public health. Shrimp provide a further risk to human and marine life because they can transport heavy metals, infectious agents, and microplastics (Unuofin and Igwaran 2023). Shrimp aquaculture generates large volumes of waste, with Bushehr province in Iran producing the most of it. Shrimp output has been rising globally, with notable increase in Thailand and Brazil. However, a lot of debris, including heads and shells, is left over after shrimp production (Patmasiriwat et al. 1998). Even though this waste is high in oil and protein, it is frequently wasted. Farmers also use various chemicals and medications to keep shrimp healthy, but these substances can end up in public waterways, posing health risks to humans. Shrimp can carry microplastics, heavy metals, and infectious agents, further threatening marine and human life (Grossi 2024). The waste produced by shrimp farming is substantial, with significant amounts generated in places like Bushehr province, Iran. Global shrimp production has been increasing, with significant growth in countries like Brazil and Thailand. However, shrimp farming leaves behind a lot of waste, such as heads and shells (Sheikhi et al. 2024). In many cases, this waste is not utilized, despite being rich in protein and oil. Waste material can decompose quickly, giving off an unpleasant odor and endangering the surrounding area (Rossi et al. 2024).

11.3 Strategies for Shrimp Waste Management

Shrimp farming has grown a lot due to more intensive methods and larger farms, which has increased its environmental impact. Problems like habitat loss, disease outbreaks, and water pollution are concerning due to the discharge of waste from shrimp farms (Kumar et al. 2022a, b, c; Rathod et al. 2024). Shrimp shells contain useful substances like calcium carbonate, protein, and chitin, but we often don't use them effectively (Saravanan et al. 2023). Scientists are trying different ways to extract valuable chemicals from shrimp waste for things like animal feed, cleaning up pollutants, and other uses. However, some current methods use chemicals that can harm the environment. Finding better, sustainable ways to process shrimp waste remains a big challenge, despite of various efforts in different countries to develop new methods like using bacteria or fermentation to break down waste.The key to solving these problems is figuring out how to keep shrimp waste from decomposing too quickly and turn it into useful product by eco-friendly methods Fig 11.1.

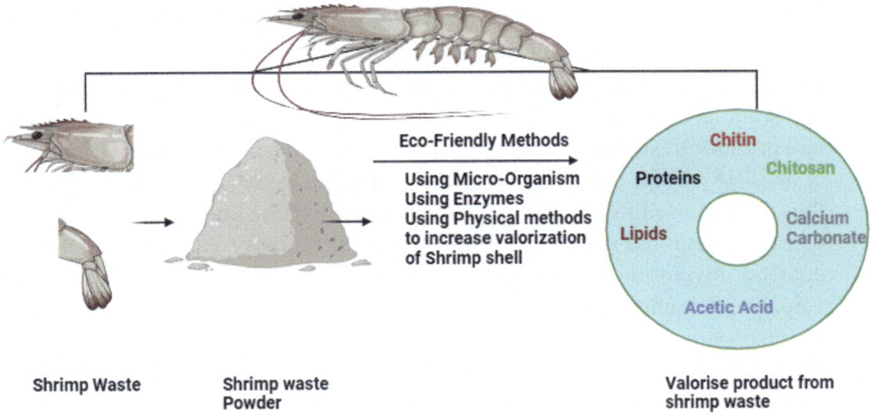

Fig. 11.1 Eco friendly methods to extract valorise product from shrimp waste (*Source* Author)

11.3.1 Bioconversion of Shrimp Waste

Bioconversion processes facilitate fermentation, demineralization, and deproteinization. These methods are safe, cost-effective, and environmentally benign. Two additional green technologies are methanogenesis and the use of microbes that produce oil to improve the quality of extracted SW products (Deng et al., 2020). Because of their versatility and abundance, microorganisms are efficient agents for SW bioconversion (Wani et al. 2024). Anaerobic, aerobic, facultative mycelium, and microalgae can all perform fermentation (Nagappan et al. 2021). *Streptococcus thermophilus*, *Lactobacillus bulgaricus*, and *Lactobacillus acidophilus* are common microorganisms used in SW fermentation (Duan 2011). Microorganisms are beneficial agents for SW bioconversion due to their abundance and adaptability. Common microbes used in the fermentation of SW include *Streptococcus thermophilus*, *Lactobacillus bulgaricus*, and *Lactobacillus acidophilus* (Duan 2011). This study examines a number of environmentally, human, and animal health-safe techniques for valuing waste from shrimp aquaculture. A variety of subjects are covered in the study, such as how to recover valuable chemicals from shrimp waste, turn it into energy, and use it as a sustainable source of fertilizer and animal feed. This study's main objective is to list the most practical ways to make money off of the waste from shrimp farming while also pointing out areas that require more investigation and development.

(A) **Valorization**: Waste valorization refers to the process of transforming waste materials into valuable resources such as raw materials, fuels, chemicals, or energy through methods like recycling, repurposing, or composting (Abdel-Shafy and Mansour 2018). When seafood debris, such as shrimp shells, is not properly handled, it can produce toxic gasses like methane and ammonia. By repurposing seafood waste rather than simply discarding it, value-adding it has positive effects on the economy and the environment. Chemical treatments, drying, and grinding are steps in the process used to recover valuable materials such as chitin from seafood trash in order to extract minerals and proteins (Singh et al. 2022). Conventional techniques involve the use of chemicals such as HCl or NaOH to extract chitin from shells and acids such as sulfuric acid or formic acid to remove calcium carbonate. These substances can be harsh and produce trash that is acidic, which can harm the environment (Hahladakis et al. 2018). Safer substitutes for these compounds are being discovered, and they will be used in the processes of demineralization and deproteination. To extract chitin more sustainably, for instance, enzymatic treatments and natural eutectic solvents are employed (Table 11.1). These techniques seek to recover valuable elements from shrimp and seafood trash while minimizing harm to the environment (Roy et al. 2023).

(B) **Fermentation**: Seafood waste (SW) is processed primarily in two ways. One method involves the use of chemicals like sodium hydroxide and hydrochloric acid to extract proteins, lipids, and minerals from the waste (Arvanitoyannis and Kassaveti 2008). Chitin is the principal substance extracted by this process;

Table 11.1 Different methods for valorisinng shrimp waste

S. no	Product	Methods	References
1	Chitin	Extraction using alkali-acid treatment, biocomposite materials	Boussetta et al. (2022)
2	Shrimp Shell Waste	Fermentation with molasses for silage, hydrolysis with enzymes	Chamsart et al. (2022)
3	Chitin	Conversion to bioactive compounds using ionic liquids, enzymatic demineralization	Kumari et al. (2023)
4	Chitosan	Development of composite films with natural fibers, use in food packaging	Elhussieny et al. (2020)
5	Astaxanthin	Extraction for aquafeed and supplements, antioxidant properties	Abun et al. (2022)
6	Bioactive Compounds	Isolation for antimicrobial and antioxidant applications	Ngo et al. (2014)
7	Protein-rich Extracts	Utilization in animal feed, nutrient supplements	Liu et al. (2023)
8	Enzymes (e.g., chitinase)	Production via microbial fermentation, biotechnological applications	Singh et al. (2021)

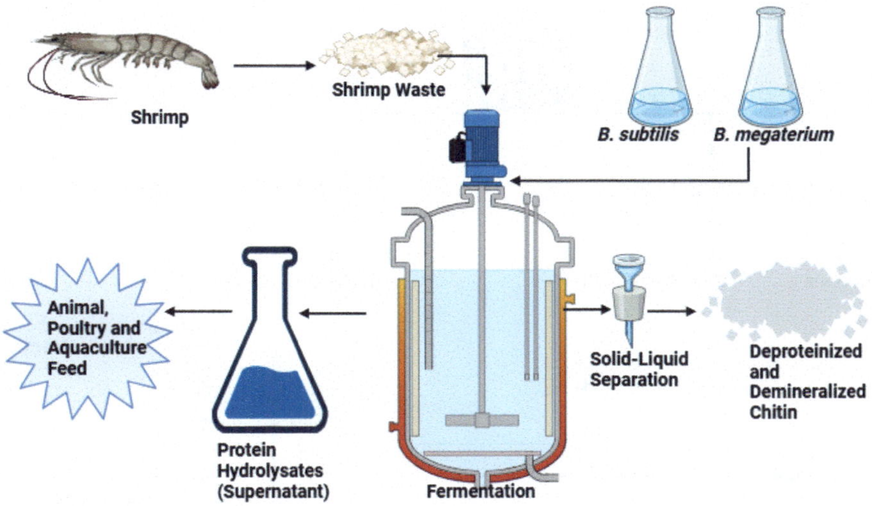

Fig. 11.2 Methods using fermentation to break down shrimp waste (*Source* Author)

other materials are difficult to recover. Utilizing exogenous proteinases or autolysis to recover proteins is a further strategy. But this could result in a protein hydrolysate that tastes unpleasant and smells fishy. Lactic acid bacteria are used in fermentation to break down SW, providing a more ecologically friendly alternative. Chitin, proteins, and carotenoids can all be recovered with this method without polluting the environment (Gortari and Hours 2013). Numerous investigations have improved the output of useful compounds from shrimp waste by employing bacteria such as *Lactobacillus* to optimize fermentation processes (Fig 11.2).

(C) **Dimenralization**: It is the process of separating the organic elements, such as chitin, from the minerals, mainly calcium carbonate, found in shrimp shells (Fig 11.3). This is usually accomplished by agitating the ground shrimp shells at room temperature while treating them with diluted acids like citric acid or hydrochloric acid (HCl). The acid dissolves the mineral content, which is then separated by filtration, followed by washing until neutral pH. The final product is free of minerals and prepared for additional processing. To lessen the impact on the environment while preserving efficiency, alternative environmentally friendly acids and methods like citric acid and deep eutectic solvents have been employed. Under ideal circumstances, typical demineralization rates surpass 95%–98% (Al Shaqsi et al. 2020).

(D) **Deproteinization**: Proteins attached to the chitin matrix of the shrimp shell are removed through a process known as deproteination (Fig 11.3). After demineralization, this step is frequently carried out with alkaline solutions, such as sodium hydroxide (NaOH), at high temperatures (about 80–95°C) while stirring constantly. Proteins are broken down and rendered soluble by the alkali, allowing for their eventual removal through washing. To purify the chitin and get

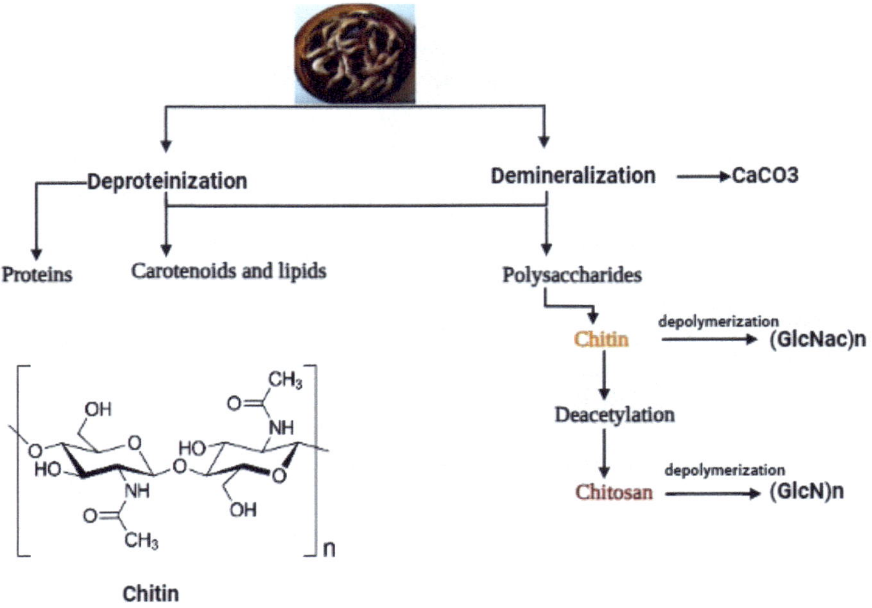

Fig 11.3 Deproteinzation and demineralization of shrimp waste (*Source* Author)

rid of any remaining nitrogenous materials, this step is essential. Efficiency can be increased by combining physical techniques like microwave assistance with mild alkaline treatments. Although they are generally slower, microbial and enzymatic approaches are alternatives. With chemical methods, deproteination efficiencies of 85–95% are typical (Al Shaqsi et al. 2020).

(E) **Chitin hydrolysis**: Chitin hydrolysis is the process of dissolving chitin using chemical, physical, or biological techniques to release its constituent N-acetylglucosamine (GlcNAc) units (Fig 11.3). Strong acids, such as hydrochloric acid, or bases, such as sodium hydroxide, are used in chemical processes at high pressure and temperature, and the GlcNAc monomers are then purified (Cao et al. 2022). Physical methods such as ultrasonication or high-pressure homogenization mechanically degrade chitin with minimal energy use and environmental impact. Enzymes such as chitosanases and chitinases are used in biological hydrolysis to break down chitin into smaller pieces. Enzymatic hydrolysis works best at pH values of 5–6, and it is environmentally benign when it operates at these temperatures. In the fields of medicine, cosmetics, and agriculture, hydrolyzed chitin has a variety of uses. It can be an antibiotic, fertilizer, and feed supplement for animals. Additionally, it serves as a precursor for the production of chitosan, which has uses in both medicine and a variety of industries (Iber et al. 2022).

11.4 Application of Shrimp Waste

Shrimp waste is a valuable resource for sustainable value creation because it can be used in a variety of industries. Its main function is to extract chitin and chitosan, which are biodegradable materials used in medicine, cosmetics, agriculture, and environmentally friendly packaging, such as feed packaging with antimicrobial qualities. Furthermore, shrimp waste is turned into high-protein additives for animal feed, which enhance nutrition and support sustainable livestock management. The recovery of bioactive substances like antioxidants and their transformation into fertilizers high in calcium or renewable energy sources are additional applications. By transforming shrimp byproducts into useful products, these applications support waste reduction, environmental sustainability, and the circular economy.Different valuable products extracted from shrimp waste are summarised in Table 11.2.

11.4.1 Industrial Application of Shrimp Waste

The extremely nutritious bioactive substances found in seafood waste (SW) include lipids, carotenoids, and amino acids. It is possible to extract elements like chitin, proteins, colors, and minerals from up to 30% of SW. For many years, Japan has

Table 11.2 List of Shrimp waste and their application for extracting valuable products

S. no	Waste	Function	References
1	Shrimp shell	Chitin extraction, biocomposite materials, bioadsorbents	Akhbarizadeh et al. (2020)
2	Astaxanthin	Aquafeed, antioxidant supplements	Abun et al. (2022)
3	Chitin	Biomedical applications, bioplastics, antimicrobial agents	Yarnpakdee et al. (2022)
4	Chitosan	Food packaging, composite films, pharmaceutical applications	Kumar et al. (2024)
5	Bioactive Compounds	Antimicrobial, antioxidant, anti-cancer properties	Abd El-Hack et al. (2020)
6	Shrimp shell waste	Biofertilizers, soil amendments, nutrient source in hydroponics	Goud and Lanka (2020)
7	Protein-rich extracts	Animal feed, nutrient supplements	Liu et al. (2023)
8	Carotenoids (e.g., beta-carotene)	Food coloring, antioxidant properties	Rossi et al. (2024)
9	Enzymes (e.g., chitinase)	Biotechnological applications, enzyme production	Singh et al. (2021)

dominated the chitin/chitosan producing industry. The chitosan market is forecast to expand at a compound yearly growth rate of 24.7% by 2027 from its estimated valuation of $476.6 million in 2016 to $1088.0 million by 2022 (Shefali et al. 2020; Wani et al. 2024). Even with the significant increase in the value of SW and other seafood for bioprospecting, more research and strategy optimization are still required to make the most of this nutrient-rich, ecologically benign waste

11.4.2 Shrimp Waste as Agrochemicals and Soil Conditioners

Although agrochemicals, such as pesticides and fertilizers, have greatly increased agricultural output, they also present major environmental dangers, including air, water, and soil contamination. Through the release of greenhouse gases like nitrous oxide, they contribute to climate change (Sharma et al. 2024; Sharma et al. 2025; Rkhaila et al. 2021). It's important to look for alternatives, and bio-organic fertilizers made from natural sources like shrimp waste (SW) present a viable one. In comparison to unfertilized or manure-fertilized plants, SW extracts have been shown to raise glucosinolate levels and increase crop yield when applied to broccoli. Rich in calcium, phosphorus, and nitrogen, SW chitin serves as a beneficial biofertilizer to enhance the development and yield of wheat and potatoes (Table 11.3). Fermenting SW with bacteria that produce chitinase and protease is an eco-friendly method of lowering the risks connected with chemical extraction (Kaur and Dhillon 2015). It has been demonstrated that using this fermented hydrolysate as a biofertilizer enhances soil quality and promotes the development of advantageous bacteria in plant roots. Furthermore, compounds derived from SW exhibit potential as natural fungicides that inhibit the growth of various plant diseases and lessen the severity of fruit rot (Biswas et al. 2021). Furthermore, SW is an inexpensive culture medium for a range of microorganisms that produce chitinase and protease, two enzymes essential for a number of commercial applications, such as agriculture, food processing, and environmental control (Vivekanandhan et al. 2022). Shrimp waste (SW) is a versatile resource that can enhance farming methods in a variety of ways. It can be composted and used as a soil amendment due to its high organic matter content and abundance of nutrients like calcium, phosphorus, and nitrogen. Adding SW encourages plant growth and productivity by improving the soil's fertility, structure, and water-holding capacity (Li et al. 2018).

In hydroponic systems, SW serves as a source of nutrients, providing essential minerals such as calcium, magnesium, phosphorus, potassium, and nitrogen that are required for plant growth. Vermicomposting SW with worms produces nutrient-rich vermicompost that is ideal for enhancing soil fertility and plant health (Passioura and Angus 2010). Chitin and its derivative, chitosan, found in SW, aid plants in surviving stress by promoting defense mechanisms. Specifically, plants that are exposed to chitosan develop systemic acquired resistance (SAR), which protects them from infections and increases their tolerance to abiotic stressors like dehydration and heavy

Table 11.3 List of Shrimp waste product, their mode of application and functions in plant growth

Shrimp waste product	Mode of application	Functions in plant	References
Chitosan	Foliar	Resistance against drought stress	Emami Bistgani and Sefidkon (2023)
Green chitosan Nanoparticles	Foliar	Source of nano nitrogen	Saad et al. (2023)
Chitin microfibers	Foliar	Resistance against *Colletotrichum Fructicola* and *Alternaria brassicicola*	Parada et al. (2018)
Nanochitin whisker	Suspension culture	Resistance against fungal infections	Zhou et al. (2020)
Pennisetum glaucum	Calcium-rich biochar Tuber immersion	Improving antioxidant defense	Abo-Elyousr et al. (2022)
Chitin oligosaccharide	Leaf infiltration	Resistance against *Candidatus liberibacter*	Shi et al. (2019)
Chitosan-based microparticles	Foliar	Improvement defense responses	Martin-Saldaña et al. (2018)
0.01 and 0.015% chitosan	Shoot spraying	Accumulation of defense enzymes	Fooladi Vanda et al. (2019)
1% chitosan	Foliar	*Phytophthora capsica* resistance	Esyanti et al. (2019)
Chitosan	Endogenous	Drought resistance	Li et al. (2020)

metal toxicity. Because it promotes root development, improves photosynthesis, and enhances nutrient uptake, it increases plant biomass and yield (Parewa et al. 2014).

11.4.3 Shrimp Waste as Biosorbents for Wastewater Treatment

Shrimp waste (SW) and its derivatives, such as chitin and chitosan, are utilized to remove harmful substances through the use of adsorption techniques (Santos et al. 2020). The surface chemistry of these materials promotes interactions with pollutants in aqueous solutions, including adsorption, surface complexation, electrostatic attraction, ion exchange, and chelation, making them efficient bio-adsorbents. Chitosan's abundance of amino and hydroxyl groups allows it to form complexes with heavy metals through amine groups, which improves metal removal efficiency. The adsorption process is greatly impacted by the pH of the solution, which also affects the ionization state of the contaminants and the availability of adsorption sites (Delle Site and Data 2001).

For example, the removal efficiency of contaminants like colors, manganese, barium, vanadium, arsenic, and chromium is affected by the pH levels of chitin and chitosan-based polymers obtained from SW. To improve the efficiency of pollution removal, the ideal pH must be attained. This demonstrates the versatility and potential environmental benefits of bio-adsorbents derived from SW (Sudha et al. 2014).

11.4.4 Shrimp Waste in Animal Feed

Shrimp waste (SW), which contains approximately 35% tissue protein, calcium carbonate, and chitin, can be a rich source of silage and animal feed if it is properly preserved. Research has examined the possibility of fermenting shrimp and crab waste with molasses to enhance its nutritional value and ease of digestion. Studies have shown that SW shells contain essential amino acids that are similar to soybean meal, even though current processing methods often fail to fully recover these proteins. Converting SW into high-protein animal feed becomes increasingly important as demand rises (Yadav et al. 2022a; 2022b; Jain et al. 2024).

Chitin and chitosan extracted from SW have proven to be both environmentally safe and economically feasible, particularly for applications such as crustacean canning and carotenoid recovery. Astaxanthin, which is frequently used in fish diets and to improve the quality of eggs, is one significant material that is eliminated during these procedures. During safe testing as a component of livestock feeds, chitosan has demonstrated high rates of digestion in a range of animals.SW's immune-stimulating properties, attributed to compounds like beta-carotene and astaxanthin, enhance animal health and productivity (Yadav et al. 2023; Uyanga et al. 2023). Chitin-rich SW feeds can improve the quality of the meat and eggs that birds produce and boost the amount of milk that cattle produce. Additionally, SW-derived chitin's antibacterial, antioxidant, and possibly anti-cancer properties support its biocompatibility and sustainable development applications (He et al. 2024).

In order to extract important components from SW as much as possible, both chemical and biological approaches are used. Alkali-acid treatments are used in chemical processes to produce a large number of products, although there may be adverse effects (Sasidharan et al. 2011; Rani et al. 2025; Nisha et al. 2025; Kiran et al. 2025). Because prebiotic bacteria have a lesser metabolic activity than other approaches, biological methods that use them for demineralization offer sustainable options with lower yields. In conclusion, improving animal nutrition, health, and sustainable agriculture practices depend on refining processing techniques to fully utilize the many nutritional and functional qualities of SW-derived chitin and chitosan.

11.4.5 Shrimp Waste as Packaging Material

The use of bio-composite materials is continuously promoted because of its low cost, biodegradability, non-petroleum origin, and low carbon emissions (Mohanty et al., 2020). Derived from chitin, chitosan has many benefits for the manufacturing, pharmaceutical, and agricultural sectors (de Oliveira et al. 2021). Chitosan is especially useful as a substitute for synthetic polymers in food packaging because of its antibacterial and biodegradable qualities.

11.5 Challenges in Waste Processing and Utilization

The current techniques for making shrimp waste valuable are pricy, harmful, and unprofitable. To produce items of superior quality, shrimp waste is subjected to many chemical processing cycles. A significant investment in the use of cutting-edge technology is required to decrease waste, maintain product quality, enhance performance, and lengthen product shelf life in order to maximize operational efficiency. However, the expanded application of shrimp waste bioconversion is hindered by some restrictions. Compared to processes mediated by chemicals, the processing pace is noticeably slower. Another issue is the foul smell that is produced when shrimp waste is bioprocessed. Additionally, when using microorganisms to digest shrimp waste, there is a greater chance of unintended microbial development that might result in contamination, which has detrimental negative effect.

11.6 Conclusion

The article explains the growing concerns about the spread of underutilized SW in the world's leading shrimp production nations. A serious cause for concern is the extensive loss of nutrient-rich SW. There are substantial, significant ecological drawbacks to the chemical procedures employed to value SW. Microorganisms have been used during the past ten years to mediate the demineralization and deproteination processes for SW bioprospecting. Fermentation and enzymatic processes are the preferred methods for breaking down marine waste, particularly seafood waste (SW). These techniques enable innovative applications across the food, pharmaceutical, cosmetic, and agricultural industries, advancing efforts toward a green economy. By harnessing bioactive compounds from SW, environmental sustainability can be supported. Therefore, it is crucial to standardize and improve SW-processing methods to produce high-quality products while minimizing environmental impact.

11.7 Future Prospective

In order to properly fractionate seafood waste's key components while minimizing waste and avoiding dangerous chemicals, creative green chemistry methodologies are necessary to establish a profitable and sustainable industry from SW. Although methods such as ball milling and steam explosion have proven effective in refining woody biomass, there is still much to learn about how these techniques might be used to treat marine wastes like shrimp shells. Ball milling without the use of solvents has potential for cell fractionation, and acids and steam pressure can release important components of solar water. Subsequent investigations have to concentrate on identifying innovative pathways for chitin conversion, enhancing productivity, and streamlining product separation, maybe utilizing ionic liquids for effective dissolution of carbohydrate polymers. Accelerating microbial fermentation processes through genetic engineering could enhance productivity and yield valuable molecules from SW. Collaborations between seafood producers, researchers, and policymakers are essential to commercialize new products like nutraceuticals, cosmetics, and bioplastics derived from SW, integrating seafood waste management into circular economy models to maximize resource recovery and minimize environmental impact. Genetic engineering can enhance microbial fermentation processes, speeding up productivity and enabling the production of valuable chemicals from seafood waste (SW). Successful commercialization of products such as nutraceuticals, cosmetics, and bioplastics derived from SW requires strong collaboration between researchers, policymakers, and seafood producers. Integrating seafood waste management into circular economy models is essential to maximize resource recovery and reduce environmental impact.

References

Abd El-Hack ME, El-Saadony MT, Shafi ME, Zabermawi NM, Arif M, Batiha GE, Khafaga AF, Abd El-Hakim YM, Al-Sagheer AA (2020) Antimicrobial and antioxidant properties of chitosan and its derivatives and their applications: A review. 164:2726–2744

Abdel-Shafy HI, Mansour MS (2018) Solid waste issue: Sources, composition, disposal, recycling, and valorization. 27(4):1275–1290

Abo-Elyousr KA, Mousa MA, Ibrahim OH, Alshareef NO, Eissa MA (2022) Calcium-rich biochar stimulates salt resistance in pearl millet (Pennisetum glaucum L.) plants by improving soil quality and enhancing the antioxidant defense. 11(10):1301

Abun A, Haetami K, Widjastuti T (2022) The effect of feed containing fermented shrimp waste on the conversion of protein, carcass weight, and abdominal fat of native chicken. 9(1):046–053

Action SJF, Organization A (2020) World fisheries and aquaculture. 2020:1–244

Ahmadkelayeh S, Hawboldt K (2020) Extraction of lipids and astaxanthin from crustacean by-products: A review on supercritical CO2 extraction. 103:94–108

Akhbarizadeh R, Moore F, Monteiro C, Fernandes JO, Cunha SC (2020) Occurrence, trophic transfer, and health risk assessment of bisphenol analogues in seafood from the Persian Gulf. 154:111036

Alderman DJ, Hastings TS (1998) Antibiotic use in aquaculture: development of antibiotic resistance–potential for consumer health risks. 33(2):139–155
Ambigaipalan P, Shahidi F (2017) Bioactive peptides from shrimp shell processing discards: Antioxidant and biological activities. 34:7–17
Arvanitoyannis IS, Kassaveti AJIS (2008) Fish waste management: treatment methods and potential uses of treated waste. 861
Biswas I, Mitra D, Senapati A, Mitra D, Chattaraj S, Ali M, Basak G (2021) Valorization of vermicompost with bacterial fermented chicken feather hydrolysate for the yield improvement of tomato plant: A novel organic combination. 10(1):29–42
Boussetta A, Ait Benhamou A, Ihammi A, Ablouh EH, Barba FJ, Boussetta N, Grimi N, Moubarik A (2022) Shrimp waste protein for bio-composite manufacturing: Formulation of protein-cornstarch-mimosa-tannin wood adhesives. 187:115323
Cao S, Liu Y, Shi L, Zhu W, Wang HJGC (2022) N-Acetylglucosamine as a platform chemical produced from renewable resources: opportunity, challenge, and future prospects. 24(2):493–509
Chamsart S, Rattanpradit K, Kuakaew J (2022) Nutritive values of protein hydrolysate products from shrimp waste utilization by fermentation with Lactobacillus plantarum
Delle Site A, Data CR (2001) Factors affecting sorption of organic compounds in natural sorbent/water systems and sorption coefficients for selected pollutants. A review. 30(1):187–439
Deng JJ, Mao HH, Fang W, Li ZQ, Shi D, Li ZW, Zhou T, Luo XC (2020) Enzymatic conversion and recovery of protein, chitin, and astaxanthin from shrimp shell waste. 271:122655
Djumanto D, Rustadi R, Triyatmo BJAI (2018) 1907 Utilization of Wastewater from Vannamei shrimp pond for rearing milkfish in Keburuhan Coast Purworejo Sub-District, pp 38–46
Duan J-S (2011) Convenient analytic recurrence algorithms for the Adomian polynomials. 217(13):6337–6348
Elhussieny A, Faisal M, D'Angelo G, Aboulkhair NT, Everitt NM, Fahim IS (2020) Valorisation of shrimp and rice straw waste into food packaging applications. 11(4):1219–1226
Emami Bistgani Z, Sefidkon F (2023) Optimizing planting density of satureja spicigera in dryland conditions under organic sources of nutrients to enhance some biochemical characteristics. Crop Prod, Antioxid Enzym Act
Esyanti RR, Dwivany FM, Mahani S, Nugrahapraja H, Meitha K (2019) Foliar application of chitosan enhances growth and modulates expression of defense genes in chilli pepper ('Capsicum annuum'L.). 13(1):55–60
Fooladi Vanda G, Shabani L, Razavizadeh RJB (2019) Chitosan enhances rosmarinic acid production in shoot cultures of Melissa officinalis L. through the induction of methyl jasmonate. 60:1–10
Forsberg KJ, Patel S, Gibson MK, Lauber CL, Knight R, Fierer N, Dantas GJN (2014) Bacterial phylogeny structures soil resistomes across habitats. 509(7502):612–616
Gortari MC, Hours RA (2013) Biotechnological processes for chitin recovery out of crustacean waste: A mini-review. 16(3):14–14
Goud S, Lanka S (2020) Chitosan from Crustacean shell waste in heavy metal removal from industrial effluents
Grossi G (2024) Corso di Perfezionamento "La valorizzazione dei co-prodotti della pesca e delle industrie ittiche: il fondamento scientifico alla base del progetto EcoeFishent"
Hahladakis JN, Velis CA, Weber R, Iacovidou E, Purnell P (2018) An overview of chemical additives present in plastics: Migration, release, fate and environmental impact during their use, disposal and recycling. 344:179–199
Hargan KE, Williams B, Nuangsaeng B, Siriwong S, Tassawad P, Chaiharn C, McAdoo B, Los Huertos M (2020) Understanding the fate of shrimp aquaculture effluent in a mangrove ecosystem: Aiding management for coastal conservation. 57(4):754–765
He Z, Lin H, Sui J, Wang K, Wang H, Cao L (2024) Seafood waste derived carbon nanomaterials for removal and detection of food safety hazards.172332

Herbeck LS, Unger D, Wu Y, Jennerjahn TC (2013) Effluent, nutrient and organic matter export from shrimp and fish ponds causing eutrophication in coastal and back-reef waters of NE Hainan, tropical China. 57:92–104

Iber BT, Kasan NAJH (2021) Recent advances in Shrimp aquaculture wastewater management. 7(11)

Iber BT, Kasan NA, Torsabo D, Omuwa JW (2022) A review of various sources of chitin and chitosan in nature. 10(4):1097

Islam SM, Karim Z (2019) World's demand for food and water: The consequences of climate change

Jain I, Kaur R, Kumar A, Paul M, Singh N (2024) Emerging protein sources and novel extraction techniques: a systematic review on sustainable approaches

Kandra P, Challa MM, Kalangi Padma Jyothi H (2012) Efficient use of shrimp waste: present and future trends. 93:17–29

Kaur S, Dhillon GS (2015) Recent trends in biological extraction of chitin from marine shell wastes: a review. 35(1):44–61

Khushbu, Gulati R, Sushma, Pankaj, Kour A (2022) Ecological impact of heavy metals on aquatic environment with reference to fish and human health. 14(4):1471–1484

Kiran Kumar N, Sindhu S, Gawaria B, Ansari A, Dhariwal A, Sharma P, Kumar R (2025) GWAS approaches in medicinal plants for enhanced secondary metabolite production. In: Chen JT (ed) Biotechnology, multiple omics, and precision breeding in medicinal plants, pp 119–134. CRC Press. https://doi.org/10.1201/9781003475491-10

Kumar R, Gupta R, Dahiya T, Yadav R, Saifi R, Yodha K, Kumari B, Singh P, Shefali (2022a) Surveillance of Heteropneustes fossilis (Stinging Catfish) diseases and health management practices in Haryana. Appl Ecol Environ Sci 10(4):261–267. https://doi.org/10.12691/aees-10-4-12

Kumar R, Gupta RK, Yadav R, Saifi R, Yodha K, Pooja (2022b) Eisenia fetida as protein source for growth enhancement of Heteropneustes fossilis. Egypt J Aquat Biol Fish 26(2):577–588. https://doi.org/10.21608/ejabf.2022.234553

Kumar R, Yadav R, Kumar Gupta R, Pooja (2022c) A sustainable way for fish health management by replacement of chemical and drugs by earthworm. In: Environmental degradation in Asia: Land degradation, environmental contamination, and human activities, pp 329–352. Springer International Publishing, Cham. https://doi.org/10.1007/978-3-031-12112-8_16

Kumar JA, Sathish S, Prabu D, Giri J, Makki E, Jayaprabakar J, Ziyayeva GK, Baigenzhenov O, Sathish T, Praveenkumar TJAEJ (2024) Waste shrimp shell mediated Chitosan-Magnesium Oxide nanocomposite: Synthesis, characterization and exploitation towards acenaphthene removal from aqueous solution. 104:124–135

Kumar R, Singh D (2022) Insecticide- tolerant Bacterial population in *Eisenia fetida*'s Gut and Vermicast exposed to Chlorantraniliprole and Fipronil. Appl Biol Res 24(3):273–279

Kumari R, Kumar M, Vivekanand V, Pareek NJR, Reviews SE (2023) Chitin biorefinery: A narrative and prophecy of crustacean shell waste sustainable transformation into bioactives and renewable energy. 184:113595

Latt UW (2002) Shrimp pond waste management. 7(3):11–48

Li Z, Schneider RL, Morreale SJ, Xie Y, Li C, Li JJG (2018) Woody organic amendments for retaining soil water, improving soil properties and enhancing plant growth in desertified soils of Ningxia, China. 310:143–152

Li K, Xing R, Liu S, Li P (2020) Chitin and chitosan fragments responsible for plant elicitor and growth stimulator. 68(44):12203–12211

Liu Z, de Souza TS, Holland B, Dunshea F, Barrow C, Suleria HAJP (2023) Valorization of food waste to produce value-added products based on its bioactive compounds. 11(3):840

Manjeet, Airon A, Kumar R, Saifi R (2022) Temporal and spatial impact of lockdown during COVID-19 on air quality index in Haryana, India. Scientific Rep 12:20046. https://doi.org/10.1038/s41598-022-20885-2

Martin-Saldaña S, Chevalier MT, Iglesias MJ, Colman SL, Casalongué CA, Álvarez VA, Chevalier AA (2018) Salicylic acid loaded chitosan microparticles applied to lettuce seedlings: recycling shrimp fishing industry waste. 200:321–331

Mohanty B, Majumdar RK, Mohanty U, Mehta NK (2020) Partial purification and characterization of proteases from the visceral waste of Indian major carp, Labeo rohita (Hamilton, 1822) 9(2):1326–1342
Moustafa K (2018) Seawater for sustainable agriculture. Emerging technologies for sustainable desalination handbook. Elsevier, pp 493–514
Nagappan S, Das P, AbdulQuadir M, Thaher M, Khan S, Mahata C, Al-Jabri H, Vatland AK, Kumar G (2021) Potential of microalgae as a sustainable feed ingredient for aquaculture. 341:1–20
Ngo DH, Kim SK (2014) Antioxidant effects of chitin, chitosan, and their derivatives. 73:15–31
Nguyen TT, Barber AR, Corbin K, Zhang W (2017) Lobster processing by-products as valuable bioresource of marine functional ingredients, nutraceuticals, and pharmaceuticals. Bioresour Bioprocessing 4:1–19
Nisha, Anshu, Rani S, Kumar S, Kumar R, Kumar A (2025) Recent technologies and strategies for bioremediation of toxic and hazardous waste. In: Dubey KK, Shah MP, Yadav A (eds) Challenges and sustainable solutions in bioremediation, pp 1-16. CRC Press. https://doi.org/10.1201/978 1003407317-1
de Oliveira AL, Cavalcante FT, da Silva Moreira K, Lima PJ, de Castro Monteiro RR, Pinheiro BB, dos Santos KP, dos Santos JC (2021) Chitosan nanoparticle: alternative for sustainable agriculture. Nanomater Nanotechnol: Biomed, Environ, Ind Appl, 95–132
Parada RY, Egusa M, Aklog YF, Miura C, Ifuku S, Kaminaka H (2018) Optimization of nanofibrillation degree of chitin for induction of plant disease resistance: Elicitor activity and systemic resistance induced by chitin nanofiber in cabbage and strawberry. 118:2185–2192
Parewa HP, Yadav J, Rakshit A, Meena VS, Karthikeyan N (2014) Plant growth promoting rhizobacteria enhance growth and nutrient uptake of crops. 2(2):101–116
Passioura JB, Angus JF (2010) Improving productivity of crops in water-limited environments. 106:37–75
Patmasiriwat D, Kuik O, Pednekar S (1998) The shrimp aquaculture sector in Thailand: A review of economic, environmental and trade issues. CREED
Pattanaik SS, Sawant PB, Xavier KM, Dube K, Srivastava PP, Dhanabalan V, Chadha NJA (2020) Characterization of carotenoprotein from different shrimp shell waste for possible use as supplementary nutritive feed ingredient in animal diets. 515:734594
Pattnaik F, Tripathi S, Patra BR, Nanda S, Kumar V, Dalai AK, Naik S (2021) Catalytic conversion of lignocellulosic polysaccharides to commodity biochemicals: a review. 19:4119–4136
Phan MH, Stive MJJO, Management C (2022) Managing mangroves and coastal land cover in the Mekong Delta. 219:106013
Rahi ML, Azad KN, Tabassum M, Irin HH, Hossain KS, Aziz D, Moshtaghi A, Hurwood DA (2021) Effects of salinity on physiological, biochemical and gene expression parameters of black tiger shrimp (Penaeus monodon): Potential for farming in low-salinity environments. 10(12):1220
Rani S, Sharma S, Singh D, Sheoran R, Kumar R (2025) Improved production of anti-cancer compounds from medicinal plants through molecular and biotechnological mutagenesis strategies. In: Chen JT (ed) Biotechnology, multiple omics, and precision breeding in medicinal plants. CRC Press, pp 287–297
Rathod SV, Saras P, Gondaliya SM (2024) Environmental pollution: Threats and challenges for management, eco-restoration of polluted environment. CRC Press, pp 1–34
Rkhaila A, Chtouki T, Erguig H, El Haloui N, Ounine KJM (2021) Chemical proprieties of biopolymers (chitin/chitosan) and their synergic effects with endophytic Bacillus species: Unlimited applications in agriculture. Molecules 26(4):1117
Rossi N, Grosso C, Delerue-Matos CJMD (2024) Shrimp waste upcycling: Unveiling the potential of polysaccharides. Proteins, Carotenoids, Fat Acids Emphas Extr Tech Bioact Properties 22(4):153
Roy VC, Islam MR, Sadia S, Yeasmin M, Park JS, Lee HJ, Chun BS (2023) Trash to treasure: An up-to-date understanding of the valorization of seafood by-products, targeting the major bioactive compounds. 21(9):485

Saad GK, Hassanein AM, El-Deep BA, Salem JM, Farghal ZE (2023) Effects of chitosan on growth and physiological changes associated adaptation of in vitro obtained banana plantlets to ex vitro conditions. J Environ Studies 31(1):16–27. https://doi.org/10.21608/jesj.2023.237831.1059

Santos VP, Marques NS, Maia PC, Lima MA, Franco LD, Campos-Takaki GM (2020) Seafood waste as attractive source of chitin and chitosan production and their applications. 21(12):4290

Saravanan A, Kumar PS, Yuvaraj D, Jeevanantham S, Aishwaria P, Gnanasri P, Gopinath M, Rangasamy GJER (2023) A review on extraction of polysaccharides from crustacean wastes and their environmental applications. 221:115306

Sasidharan S, Chen Y, Saravanan D, Sundram KM, Latha LY (2011) Extraction, isolation and characterization of bioactive compounds from plants' extracts. Afr J Tradit, Complement Altern Med 8(1)

Shahidi F, Abuzaytoun R (2005) Chitin, chitosan, and co-products: chemistry, production, applications, and health effects. 49(4):93–137

Sharma K, Gulati R, Singh S, Kumari A, Sharma PJJ (2023a) Potentiality of natural live food organisms in shrimp culture: A review. J Appl & Nat Science 15(4):1373–1385

Sharma K, Gulati R, Poonam SS, Rani AJASA (2023b) Mineral fortified inland low saline water for shrimp culture. 7(5)

Sharma MMM, Kapoor D, Rohilla R, Loyal A, Sharma P, Husen A (2024) Role of essential minerals and rhizosphere in plant under changing environmental conditions. In: Husen A (eds) Essential minerals in plant-soil systems, pp 235–252. Elsevier. https://doi.org/10.1016/B978-0-443-16082-0.00002-3

Sharma P, Kapoor D, Kumar R, Husen A (2025) Rhizosphere and osmolytes coordination in plant systems under adverse situations. In: Husen A (ed) Roles of osmolytes in changing environment, pp 101–117. Academic Press. https://doi.org/10.1016/B978-0-443-26549-5.00020-5

Kumar R, Sankhla MS, Kumar R, Sonone SS (2021) Impact of pesticide toxicity in aquatic environment. Biointerface Res Appl Chem, 11(3):10131–10140

Sheikhi H, Mirzaei RJM (2024) Occurrence and abundance of macro, meso and microplastics in the coasts of the Persian Gulf (Case Study: Bushehr Province Coast)

Shi Q, George J, Krystel J, Zhang S, Lapointe SL, Stelinski LL, Stover E (2019) Hexaacetyl-chitohexaose, a chitin-derived oligosaccharide, transiently activates citrus defenses and alters the feeding behavior of Asian citrus psyllid. 6

Singh R, Upadhyay SK, Singh M, Sharma I, Sharma P, Kamboj P, Saini A, Voraha R, Sharma AK, Upadhyay TK, Khan F (2021) Chitin, chitinases and chitin derivatives in biopharmaceutical, agricultural and environmental perspective. 11(3):9985–10005

Singh S, Negi T, Sagar NA, Kumar Y, Tarafdar A, Sirohi R, Sindhu R, Pandey A (2022) Sustainable processes for treatment and management of seafood solid waste. 817:152951

Sudha PN, Gomathi T, Vinodhini PA, Nasreen KJA (2014) Marine carbohydrates of wastewater treatment. 73:103–143

Susetyaningsih R, Suntoro S, Gunawan T, Budiastuti MTS (2020) Impact of shrimp pond waste on water quality (case study of trisik lagoon in yogyakarta). In: AIP conference proceedings

Trushina DB, Borodina TN, Belyakov S, Antipina MNJMTA (2022) Calcium carbonate vaterite particles for drug delivery: Advances and challenges. 14:100214

Unuofin JO, Igwaran AJJOSR (2023) Microplastics in seafood: Implications for food security, safety, and human health. 194:102410

Uyanga VA, Ejeromedoghene O, Lambo MT, Alowakennu M, Alli YA, Ere-Richard AA, Min L, Zhao J, Wang X, Jiao HJJOFF (2023) Chitosan and chitosan-based composites as beneficial compounds for animal health: Impact on gastrointestinal functions and biocarrier application. 104:105520

Venugopal V (2022) Green processing of seafood waste biomass towards blue economy. Curr Res Environ Sustain 4:100164. https://doi.org/10.1016/j.crsust.2022.100164

Vicente FA, Hren R, Novak U, Čuček L, Likozar B, Vujanović AJR, Reviews SE (2024) Energy demand distribution and environmental impact assessment of chitosan production from shrimp shells. 192:114204

Vivekanandhan P, Swathy K, Chordia MA (2022) Pseudomonas species-derived chitinase mass multiplication, production cost analysis, and marketing: As a biocontrol agent for crop protection, Industrial microbiology based entrepreneurship: Making money from microbes. Springer, pp 179–192

Wani AK, Akhtar N, Mir TUG, Rahayu F, Suhara C, Anjli A, Chopra C, Singh R, Prakash A, El Messaoudi NJES, Research P (2024) Eco-friendly and safe alternatives for the valorization of shrimp farming waste. 31(27):38960–38989

Yadav R, Kumar R, Gupta RK, Kaur T, Yodha K, Kour A, Kaur S, Rajput A (2023) Heavy metal toxicity in earthworms and its environmental implications: A Review. Environ Adv:100374

Yadav R, Gupta RK, Kumar R, Kaur T (2022) Assessment of toxicity of lead and nickel on the biochemical and immunological parameters of earthworm. Eudrilus Eugeniae Agric Sci Dig 44(5):922–929

Yadav R, Gupta RK, Kumar R, Kaur T, Moun R (2022) "Biochemical response of earthworm, *Eisenia fetida* to heavy metals toxicity. J Appl Nat Sci 14(3):990–998

Yarnpakdee S, Kaewprachu P, Jaisan C, Senphan T, Nagarajan M, Wangtueai SJP (2022) Extraction and physico–chemical characterization of chitosan from mantis shrimp (Oratosquilla nepa) shell and the development of bio-composite film with agarose. 14(19):3983

Zhou Y, Jing M, Levy A, Wang H, Jiang S, Dou D (2020) Molecular mechanism of nanochitin whisker elicits plant resistance against Phytophthora and the receptors in plants. 165:2660–2667

Chapter 12
Sustainable and Eco-Friendly Management Practices of Fish Waste

Anshu, Jyoti, Nisha, Rahul Kumar, and Mayur Mukut Murlidhar Sharma

Abstract Rapid population growth, resource depletion, and environmental pollution are some of the major issues that the world's population is currently confronting. Food scarcity, especially as a result of erratic climate change, is one of the major problems resulting from these difficulties. In order to resolve the world's food problem, the aquaculture sector is becoming more and more significant. Aquaculture contributes a major portion to global food production. China and India are the two largest producers of fish through aquacultural practices. In commercial aquaculture and fishing industries, bulk production of fish and fishery products leads to a vast production of waste. More than half of the total fish production is estimated to be wasted, which causes economic and environmental problems. This chapter explores the diverse and sustainable methods of utilizing fish waste within the aquaculture industry, demonstrating how these practices can contribute to a circular economy, improve resource efficiency, and support environmental sustainability. Valuable components found in fish waste i.e., gelatin, collagen, hydrolysate, oils, enzymes and protein, are discussed along with their sustainable uses. Fish silage, fish oil (containing essential omega-3 fatty acid), fish waste composting, nanoparticle synthesis, antimicrobial peptides extraction from mucus and aquaponics are major waste mitigation strategies discussed in this chapter. Further, this chapter highlights future directions, including the need for technological innovations, regulatory frameworks and cost-effective alternatives.

Anshu
Department of Zoology and Environmental Sciences, Punjabi University, Patiala, India
e-mail: jyotibishnoi930@gmail.com

Jyoti
Government College Mangali, Hisar, Haryana, India

Nisha
Department of Biochemistry, Kurukshetra University, Kurukshetra, Haryana, India

R. Kumar
Department of Zoology, Chaudhary Charan Singh Haryana Agricultural University, College of Agriculture, Bawal, Rewari, Haryana, India

M. M. M. Sharma
Department of Agriculture and Life Industry, Kangwon National University, Chuncheon, South Korea

R. Kumar et al. (eds.), *Sustainability in Aquaculture and Fisheries*,
Springer Water, https://doi.org/10.1007/978-3-032-16605-0_12

Keywords Aquaculture industry · Environmental sustainability · Waste mitigation · Fish meal and compost

12.1 Introduction

In recent times, the global population is facing significant challenges, including rapid population growth, depletion of resources, and environmental pollution. One of the critical issues arising from these challenges is the scarcity of food, particularly due to unpredictable climate changes. The aquaculture industry is increasingly important in addressing global food shortages. It is anticipated that the share of fish consumed by humans from aquaculture will rise from 52% (the average for 2016–2018) to 58% by 2028. However, traditional aquaculture systems have a major negative impact, particularly in terms of water quality and secondary pollution, which can severely harm aquatic ecosystems (Manjeet et al. 2022; Tom et al. 2021). Aquaculture plays a major role in global food production, growing at an average rate of 3.2% each year. In 2014, the consumption of farmed fish surpassed that of wild fish (FAO 2016). China was the world's top producer and exporter of aquaculture goods, accounting for 60.5% of global production (FAO 2016). India ranks as the second-largest producer of fish through aquaculture, contributing 1.07% to the country's GDP. The Indian fishery sector has reached a value of 334.41 billion, driven by the success of the Blue Revolution.

Sustainable aquaculture plays an essential role in meeting the global demand for seafood while reducing the environmental footprint of food production. However, the waste generated from fish farming comprising offal, carcasses, bones, scales, uneaten feed, and more has long posed challenges in terms of disposal and environmental management. Instead of viewing fish waste as a by-product to be discarded or processed in costly ways, innovative solutions are emerging that convert this waste into valuable resources (Frankic and Hershner 2003; Kumar et al. 2022a, b). This chapter explores the diverse and sustainable methods of utilizing fish waste within the aquaculture industry, demonstrating how these practices can contribute to a circular economy, improve resource efficiency, and support environmental sustainability.

12.2 Origin of Fish Waste

In commercial aquaculture and fishing industries waste is generated during fish processing, as several parts are not used and enter the waste stream. Boneless waste is known as "offal". Filleting (removal of bones from fish flesh) can account for 40–60% of total processed waste (Ramadhani et al. 2018). Fish waste can be generated due to the selective use of fish parts such as harvesting of "roe" and flesh only. Fish parts such as viscera, scales, bones, tails and fins fall under fish waste. Fish waste poses significant environmental hazards due to its contribution to high chemical and biological oxygen demands when disposed of in oceans, as well as its content of pathogenic microbe category (Chowdhury et al. 2010; Kumar et al. 2022c; Arvanitoyannis et al. 2008).

a. **Reason of discarding Fishcatches**:

- Catch surpasses the specific quota limit set by regulation.
- Fishes are below the minimum prescribed size.
- If fish are of poor quality and are not of commercial use (Guillen et al. 2018).

12.3 Types of Fish Waste

In aquaculture systems, fish waste is generated as a byproduct of the fish farming process. The waste can be categorized into several types based on its form, composition, and source. Understanding these types of fish waste is crucial for developing effective management and utilization strategies (Boyd and Tucker, 2012).

12.3.1 Solid Waste

Murugan (2024) explained the type of fish waste as follows:

(A) **Fecal Matter**

Fish excrete solid waste in the form of feces, which contains undigested feed particles, metabolic by-products, and microbial matter. The composition of fish feces varies depending on the species, the feed they consume, and their digestive processes (Tú 2019). This waste is typically organic and rich in nitrogen and phosphorus, which, if not properly managed, can lead to environmental pollution, particularly in enclosed aquaculture systems. The main components of fish feces include inorganic minerals and organic materials like proteins, lipids, and carbohydrates. Fecal buildup in high-density aquaculture systems can worsen water quality, lower oxygen levels, and raise the possibility of disease outbreaks.

(B) **Feed Residue**

A portion of the fish feed that is not consumed sinks to the bottom of ponds or aquaculture tanks. This leftover feed may add to the water's excessive nutritional load, especially in the form of nitrogen and phosphorus. One of the main sources of organic matter in aquaculture wastewater is frequently feed residue. Generally speaking, feed residue is made up of grains, protein sources (such as fishmeal), lipids, vitamins, and undigested or extra feed pellets (Tangendjaja 2022). If feed residue is not adequately managed, it can cause eutrophication and problems with water quality by adding organic sludge in tanks or ponds.

12.3.2 Organic Matter and Inorganic Matter

Fish farming operations generate dissolved organic matter (DOM) and nutrients in addition to solid waste, which are discharged into the water as a result of feed digestion, waste excretion, and metabolic processes. DOM is made up of a variety of substances, such as urea, amino acids, and fatty acids. The overall health of the aquaculture system as well as the quality of the water, may be significantly impacted by these dissolved materials.

(A) **Dissolved Organic Matter (DOM)**

Fish excretion, the microbial breakdown of solid waste, and organic debris discharged from uneaten feed are the usual causes of DOM. Amino acids, fatty acids, and other dissolved organic molecules are among the substances it contains. Fish may become stressed or die as a result of oxygen depletion brought on by elevated DOM levels, which can also raise the water's biochemical oxygen demand (BOD). Additionally, it may encourage the development of toxic algal blooms, which would lead to eutrophication (Shefali et al. 2020; Lv et al. 2024).

(B) **Inorganic Matter**

Fish waste is rich in nitrogen (mainly as ammonia) and phosphorus (mainly as phosphate). Ammonia excreted by fish is toxic in high concentrations but can be converted into less harmful forms (nitrate) by nitrifying bacteria in well-managed systems (Parvathy et al. 2023). Phosphorus is often present in the form of phosphate; excess phosphorus can fuel algal growth and result in eutrophication, where excessive nutrients lead to oxygen depletion and fish kills (Purohit 2018). These nutrients are essential for aquatic plant growth but can lead to environmental problems if present in excess.

12.3.3 Chemical and Biological By-Products

Aquaculture operations generate various chemical and biological by-products described below.

(A) **Chemical By products**

(a) **Ammonia**

Fish protein metabolism produces ammonia, a hazardous byproduct. It is excreted into the water as either ammonium (NH_4^+) or ammonia gas (NH_3), depending on the pH of the water. Ammonia concentrations must be carefully monitored and controlled in aquaculture systems to prevent toxicity, especially in intensive farming operations where waste builds up quickly Parvathy et al. 2023). High ammonia levels can impair fish gill function, decrease growth rates, and increase mortality. Ammonia is also a key contributor to eutrophication.

(B) **Carbon Dioxide (CO_2)**

As fish respire, they release carbon dioxide into the water. In intensive systems, elevated CO_2 concentrations can lead to a decrease in pH, making the water more acidic and potentially harmful to fish. High CO_2 levels can reduce fish growth, impair immune responses, and lead to acidification of the water, affecting the entire aquaculture ecosystem.

(C) **Pharmaceuticaland Chemical Residues**

Aquaculture involves the use of antibiotics, antiparasitics, and other chemicals to prevent or treat diseases (Dinesh et al. 2022). These chemicals can be excreted by the fish into the water and, if not carefully managed, can accumulate and pose risks to both the aquaculture environment and the surrounding ecosystems. Antimicrobial resistance, aquatic ecosystem disruption, and environmental contamination can result from the buildup of pharmaceutical residues in aquaculture wastewater.

(B) **Biological By-products**

Fish are hosts to various microorganisms, including bacteria, viruses, and parasites. As part of their metabolism, fish excrete waste that contains these microorganisms, which can contribute to the spread of disease within the aquaculture system if not managed properly. Unmanaged biological waste can lead to outbreaks of diseases such as vibriosis, furunculosis, and other infections, which can impact fish health and productivity (Suresh et al. 2022).

In aquaculture, fish waste can take many different forms, including as solid waste (feces and feed residue), nutrients, dissolved organic matter, and a variety of chemical and biological byproducts. Maintaining the sustainability of the aquaculture sector, reducing its negative effects on the environment, and preserving the health of aquaculture systems all depend on the efficient management and use of these wastes. (Ahmad et al. 2022). Proper waste treatment, resource recovery, and waste-to-value strategies can help mitigate the negative effects of fish waste while unlocking opportunities for sustainable aquaculture practices.

Fish parts are rich in nutrients, and free from toxic metals and inner contaminants (Isangedighi and David 2019). Offal as well as whole fish waste are abundant in proteins, lipids, phosphate, calcium, and nitrogen and thus can be used as fertilizer. Fish tissue is easily decomposed by exoenzymes and autolytic enzymes. The intestine and muscles of fish contain hydrolytic enzymes that can degrade protein, scale, and skin (Martin and Patel 1991).

12.4 Valuable Components Found in Fish Waste

12.4.1 Gelatin

Fish bones, scales, and cartilage contain gelatin, which exhibits antioxidative properties (Bruno et al. 2019). A study was conducted by Pang et al. (2017) on Tilapia skin gelatin to replace bovine gelatin from yogurt preparation. Compared to bovine gelatin, tilapia gelatin had a lower melting point and gelling temperature as well as a greater capacity to store water. Due to its low antigenicity and strong biocompatibility, fish gelatin can also be utilized in tissue engineering applications and wound closure (Lv et al. 2019).

12.4.2 Collagen

It is the most abundant protein found in fish that can be used with gelatin to improve cell adhesion and production of cartilage (Li et al. 2020). It is used for wound healing dressings in burn wounds.

12.4.3 Hydrolysate and Peptide

Hydrolysis of fish byproducts generate 2-20 amino-acid peptides that have antihypertensive and ACE- inhibitory activity. Many peptides are shown to have antimicrobial activity (Caruso et al. 2020).

12.4.4 Oil and Fatty Acids

Fatty acids are generally stored in subcutaneous tissue, and their lower impurity levels and low oxidation potential make them an excellent source of oil (Thirukumaran et al. 2022). Fish oil have anti-inflammatory, anti-hypertensive and anti-diabetic properties (Mutalipassi et al. 2021).

12.4.5 Enzymes

Fish waste is a good source of enzymes, majorly proteases which can be used in the fermentation industry (Ghaly et al. 2013).

12.4.6 *Chitin/Chitosan*

Chitin is primarily obtained from fish scales. Chitin is a biopolymer known to have nutraceutical and pharmaceutical applications. Deacetylation of chitin leads to the production of chitosan (Joseph et al. 2021).

12.4.7 *Minerals*

Bones of fish are great sources of minerals such as calcium, phosphate, magnesium, iodine, selenium, and hydroxyapatite (Caruso et al. 2020). Hydroxyapatites and Calcium oxide have potential applications in dentistry as these are used in manufacturing teeth and bone implants (Marchmeyer et al. 2020). Fish bones can be converted into biofertilizers using phosphorus-solubilizing microbes that have agricultural applications (Alfio et al. 2021). Table 12.1, presented below, shows various waste components and their probable uses.

Fish waste management includes steps such as input of raw material, processing of material and final conversion into valuable product as shown in Fig. 12.1.

12.5 Fish Waste Management Strategies

12.5.1 *Fish Meal Production*

Bones, guts, heads, and tails are among the portions of fish waste that are inappropriate for human consumption but can be processed into fishmeal. Half of the fish that are caught are reportedly not eaten (Kristinsson and Rasco 2000). One of the most well-known uses for fish waste is the production of fish oil and fish meal, two crucial ingredients in aquaculture feed. Fish meal is a great source of protein, and fish oil offers vital omega-3 fatty acids that are important for fish health and growth. Usually made from low-value fish or fish processing waste, these materials are cooked, pressed, and dried to produce a nutrient-rich feed additive. Low-trophic level fish may be able to use fishmeal derived from fish waste since it has a lower crude protein content (58%) than high-quality fish meal (60–70%) (Esteban et al. 2007). For gilthead bream, smoked trout intestines may be a great source of fatty acids. Fishmeal derived from fish waste may be a viable source for the culture of Chinese carp and tilapia, as these species require very little fishmeal (Kotzamanis et al. 2001).

Table 12.1 Showing various fish waste components and its uses

Sr. no	Fish waste derivatives	Sources	Applications	References
1.	Collagen Gelatin	Bone, skin, scales and fins Partial hydrolysis and heat denaturation of collagen	• Greater bioavailability to collagen from pigs and cows • Functional compounds used in the food, biomaterial, pharmaceutical, cosmetic, and nutraceutical sectors • Cell and cartilage regeneration	Sripriya et al. (2015), León-López et al. (2019) Xu et al. (2020)
2.	Bioactive Peptides Hydrolysed fish collagen	Full body fish waste or Specific body part waste, including frame, head, bone, scale, gonads and viscera	• Antimicrobial, antioxidant and antihypertensive neuroprotective activities • Cause bone marrow mesenchymal stromal cells (BMSCs) to differentiate into osteoclastin, • Upregulating the expression of anti-inflammatory mediators such as inteleukin-6 (IL-6), prostaglandin E2 (PGE2), and TGF-β1 • Hydrolyzed collagen lowers the expression of pro-inflammatory mediators such tumor necrosis factor alpha (TNF-α) and interleukin 1 beta (IL-1β)	Urakova et al. (2012) Liu and Sun (2019)
3.	Chitin and chitosan (deacetylation of chitin)	Scales	They chelate metals, and their optical structural properties make them desirable compounds for use in a variety of industries, including nutraceuticals, bioremediation, gene therapy, the pharmaceutical, food, and cosmetics sectors	Joseph et al. (2021)

(continued)

Table 12.1 (continued)

Sr. no	Fish waste derivatives	Sources	Applications	References
4.	Oil	Fish intestines, flesh, skin, head, frames, fin, and tail	• PUFA are mostly utilized in the manufacturing of biofuel and as feed and food supplements • Omega-3 fatty acids: they prevent atherosclerosis, arrhythmias, lower blood pressure, help diabetics, protect against manic-depressive illness, lessen asthma symptoms, prevent chronic obstructive pulmonary diseases, lessen cystic fibrosis symptoms, increase cancer patient survival, lower cardiovascular disease, and enhance learning capacity • Manufacturing of sustainable fuels, particularly biodiesel The use of fat liquoring agents in the leather industry	Kumar (2010) Zampolli et al. (2006), Jung et al. (2006) Saranya et al. (2020)
	Antimicrobial peptides	Mucus from skin, gills and serum	• Antibacterial • Antifungal • Antiviral • Antiparasitic activity Immune modulation	Hellio et al. (2002)

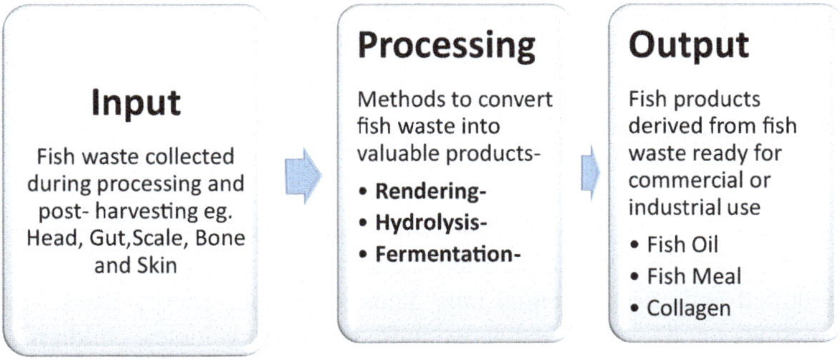

Fig. 12.1 Steps of fish waste management

12.5.2 Fish Silage

Fish silage is a liquid made from the entire fish or portions of it that have been treated with acids, enzymes, or bacteria that create lactic acid. The fish's enzymes cause the bulk to liquefy (FAO 2003).

Preparation of Silage:

The raw material is broken up into tiny bits in order to prepare the chemical silage. Acid is then added to enable its operation till liquefaction occurs. Because it permits the intended biochemical changes, the procedure can run at room temperature. Once the untreated portions of the material have begun to decompose, the mixture needs to be agitated to bring the raw material into touch with the acid. The silage process automatically starts after the first combination, but periodic stirring aids in achieving the required homogeneity (Oetterer 2002).

The most plentiful source of animal protein used to make meals for household animals is fish meal. The global market has consistently searched for a viable substitute for fish meal (Nogueira Junior et al. 1997). Fish silage is a desirable substitute for fish meal.

The following are the benefits of silage production over fish meal

- This procedure is essentially scale-independent.
- Technology is very simple.
- Even in large-scale production, the investment is less.
- A decrease in odour issues and effluents.
- In tropical regions, the silage process proceeds quickly, and the final product is suitable for local usage.

The product's substantial nature, if consumed in its pasty condition, suggests an additional drying cost, which is a drawback (Beerli et al. 2004).

12.5.3 Fish Oil

About 30% essential omega-3 fatty acids, including roughly 18% EPA (Eicosapentaenoic Acid) and 12% DHA (Docosahexaenoic Acid) are typically found in fish oil. The animal body uses essential fatty acids to produce energy. These are used in many nations in pharmaceutical active ingredients and vehicles, children's nutrition, parenteral nutrition, medicinal and functional foods, and nutritional supplements (Arruda et al. 2007). For thousands of years, omega-3 fatty acids have been a component of the human diet. According to estimates, early humans' diets contained a 1:1 ratio of omega-6 to omega-3 fatty acids. Due to a combination of decreased consumption of omega-3 fatty acids and the extensive usage of vegetable oils high in linoleic acid, the ratio has increased to 10:1 in the United States (Blasbalg et al. 2011; Kiran et al. 2025).

Consumption of fish oil can be beneficial for coronary heart disease, blood pressure regulation and stroke management (Kris-Etherton et al. 2002). By converting fish waste into valuable products like meal and oil, aquaculture can reduce the need for wild-caught fish in feed production, lessening the pressure on marine ecosystems and promoting the sustainable use of marine resources.

12.5.4 Antimicrobial Agents From Fish Mucus

Mucus contains high molecular weight glycoproteins known as mucins which belong to the O-linked glycoprotein category. Mucus is responsible for pathogen entrapment and physical protection of the skin. Mucus exhibits action by inhibiting the adhesion of pathogens to prevent their colonization because of its continuous production and routine shedding (Arasu et al. 2013) and it contains AMPs (anti-microbial peptides), CRPs, lysozymes, lectins, proteases, immunoglobulins (Charroux et al. 2018). Fish mucus exhibits strong antimicrobial, antiviral, and antiparasitic activity. This can be explored as an alternative to conventional antibiotics to which bacteria show resistance. Fish mucus and gills can be used for the exploration of peptides that possess broad-spectrum antibacterial, antiviral, antifungal, and antiparasitic activity (Hellio et al. 2002).

12.5.5 Fish Compost and Fertilizer

When fish are processed, a lot of liquid and solid waste is produced, including fish heads, viscera, skin, and carcasses. Leaving these trash neglected endangers both the ecosystem and the people who live nearby (Jaies et al. 2024) An effective and environmentally friendly way to manage fish waste is by repurposing it into manure and fertilizer. This approach not only ensures scientific disposal but also reduces environmental impact. Fish waste serves as an excellent source of essential nutrients like nitrogen, phosphorus, and potassium, which are crucial for crop growth.

12.5.5.1 Nutrient Composition of Fish Body Parts

(i) **Fish Bones**-Fish bones are rich in calcium and phosphorus—especially those of species like cod and salmon—are frequently used as fertilizer or as supplements. According to the study, they are a valuable source of minerals because of their bioavailability.

(ii) **Fish Heads**: They have notable concentrations of trace metals like iron and zinc, as well as calcium and magnesium. These nutrients are essential for both enzymatic and bone health. Fish heads are good for both agricultural and human consumption due to their high nutritional density.

(iii) **Fish Viscera**-Fish viscera's nutrient makes them perfect for organic fertilizer because they are high in nitrogen and potassium. This portion of fish is convert into compost and other soil improving materials.
(iv) **Fish Muscle (Flesh)**-It contain modest amount of magnesium and phosphorus, other proteins and amino acids that provide nourishment to both plants and humans (Jensen et al. 2024).

Fish waste can be composted to form stable fertilizer or it can be combined with other waste such as cattle manure to improve compost quality (Ahuja et al. 2020). According to Kalbani et al. (2016) Agro-fish pellet improved fruit output, fruit quality and stem diameter of tomato plant. Gagnon and Berrouard (1994) also checked effect of fish scale meal in combination with peat compost on tomato plant and found improved stem growth. The global management of fish residues presents a substantial challenge, as improper handling can lead to environmental disturbances and health risks (Lopes et al. 2015).

12.5.6 Nanoparticle Synthesis

Fish waste contains valuable organic components such as proteins and lipids, along with inorganic substances like hydroxyapatite. Bulk production of fish and fisheries products leads to vast production of waste. More than half of total fish production is estimated to be wasted which causes economic and environmental problems. Use of fish waste for manufacturing nanostructures of high value is increasing day by day. Fish waste hold significant potential when integrated with nanotechnology as it can provide cost-effective and sustainable raw material for nanoparticle synthesis (Yesilsu et al. 2024).

12.5.7 Sustainable Aquaponics

Food production is severely hampered by agriculture's need on good soil, since estimates now indicate that 38% of agricultural land worldwide is degraded. Soil degradation refers to the natural decline in the soil's ability to support productive use, often leading to reduced yields caused by insufficient nutrients and limited water availability (Surnar et al. 2015). Aquaponics is a combination of both aquaculture (fish cultivation) and hydroponics (soil-free plant farming) (Jung et al. 2017), as it provides a sustainable solution for the main issues:

- In aquaculture, nutrient-rich fish waste must be filtered or disposed away.
- To give plants grown hydroponically all the vital nutrients and minerals they need, nutrient-rich water must be used as fertilizer.

Therefore, combining these two technologies removes a waste product that is frequently disposed of as wastewater while providing a natural fertilizer solution for plant growth (Surnar et al. 2015). Because it can completely recycle waste and nutrients, it is a completely environmentally beneficial production system. The working of aquaponics system shown in Fig. 12.2.

The fish tank, grow beds, and a little pump that moves and filters water between them are the main parts of an aquaponics system. Both the plant and fish tanks receive air through tubes, which supplies oxygen to the fish as well as the biofilm that forms on the tank surfaces and the underside of the Styrofoam sheets (Marklin et al. 2013). Fish feed, a small quantity of replacement water, and electricity for heating are examples of external inputs required by the system. As microorganisms or nitrifying bacteria are important in metabolizing the ammonia from waste of fish, to maintain the water quality hydroponic beds are serve as biofilters, for the production of fish by eliminating ammonia, nitrites, and nitrates (Mishra et al. 2020). To maintain a suitable habitat for fish, the fish tank's water supplies organic nutrients to the plants in the growth beds, which absorb the waste and organically filter the water (Silva et al. 2015; Goddek et al. 2016; Andriani et al. 2017). By reusing aquaculture effluent, this water cycle avoids the environmental contamination that is usually connected to traditional fisheries (Endut et al. 2011; Hao et al. 2020).

12.5.7.1 Aquaponics Systems are of These Types:

a. **Floating Raft or Deepwater Culture**

Plants are cultivated on floating rafts in this approach. Plants are planted in net pots in tiny holes carved into the rafts. In the water where nutrient intake takes place, the roots dangle freely (Surnar et al. 2015).

b. **Flood and Drain**

The plants are cultivated in a grow bed that is filled halfway. The substrate media serves as a mechanical and biological filter, assisting in the better upkeep of the plants that are grown (Kulkarni et al. 2019).

c. **Nutrient Film (NFT)**

Water is continuously pushed from the tank to the filter while a tiny amount of water flows via pipes to grow plants (Graber and Junge 2009).

12.5.8 Fish Scale as Biosorbent

Tons of fish scales are removed every day, making them one of the main waste products of the fishing industry. As a special kind of skin protection, fish scales are mineralized plates that cover the dermis layer of the epidermis. Fish scales are

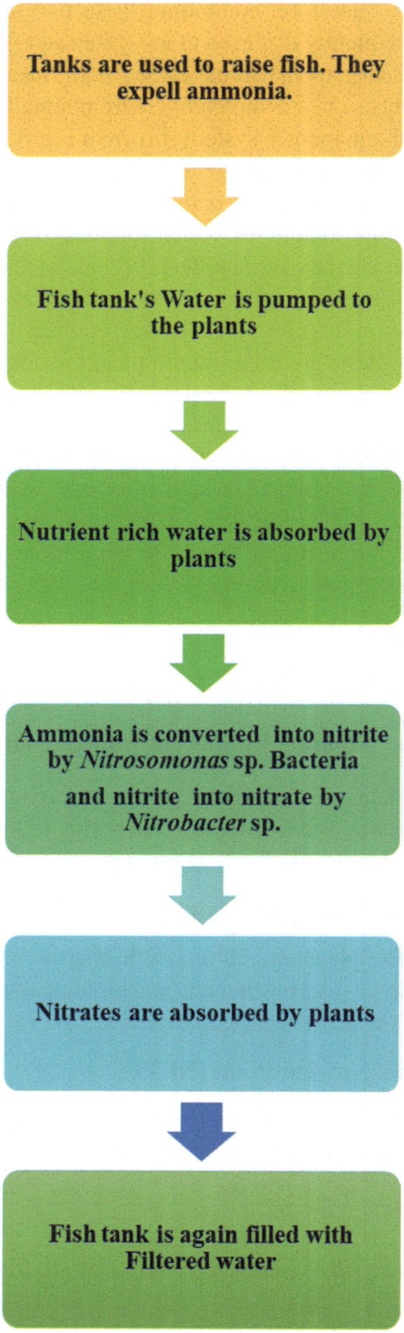

Fig. 12.2 Illustration of the working of aquaponics system

Table 12.2 Showing metal absorption by scales of different fish species

Fish scales	Metal absorbed	References
Rutilus kutum	Cu^{2+} ion	Ahmadifar and Dadvand Koohi (2018)
Cyprinus carpio	As ions	Veličković et al. (2018)
Labeo rohita	Cadmium (Cd) and lead (Pb) ions	Pal and Maiti (2020)
Tilapia	Cr ions	Li et al. (2016)
Oreochromis niloticus	Zinc ions	Ribeiro et al. (2019)

composed of approximately 41–45% organic components (collagen, fat, lecithin, sclerotin, vitamins, etc.), 38–46% inorganic components (calcium-deficient hydroxyapatite, calcium phosphate, etc.), and trace elements like calcium, magnesium, zinc, and iron (Chinh et al. 2019). Because scales cannot be eaten, they are categorized as fish waste. In several investigations, researchers have employed fish scales as a biosorbent material. A living material capacity to absorb metals from its environment is known as biosorption, and it has been demonstrated that fish scales are effective biosorbents (Ahalya et al. 2003; Nisha et al. 2025).

Mustafiz et al. (2003) suggested that scales of *Gadus morhua* fish can be used to remove lead, arsenic and chromium from water. An experiment was performed to evaluate biosorbent property of scales of *Tilapia niloticus* fish for the removal of copper from water. Industrial effluent is a major source of copper in waterbodies. Its level in water should be within permissible limits because excess copper can damage human liver and kidneys. Results of study showed that larger amount of scales added and longer contact time has increased the sorption percentage of copper from water.

There are two possible ways by which scale can remove copper:

- Direct adsorption of copper on surface of scale
- Microbes present on the scale helped in the removal of copper (Huang and Huang 2007).

Scales can absorb different ions from the surrounding water as stated in Table 12.2. This confirms the biosorbent property of scales.

Because fish scales offer an environmentally beneficial way to remove heavy metals from water, they can be a powerful substitute. If we can create effective methods to use fish scales as biosorbents, we can manage fish scale trash in a sustainable manner.

12.5.9 Scales as Ornament for Designing Industry

Fish scales are frequently used as a handicraft material such as decoration and jewellery in some countries. Fish scales are a common solid waste item that ends up in landfills, but they may also find use in the clothing business. Studies have indicated

Table 12.3 Fish scales as biosorbents for industrial dye removal

Fish scales	Dyes absorbed	References
Oreochromis niloticus	Azo dye (AB113 dye)	Vieira et al. (2012)
Tilapia	Ponceau 4R	Zhu et al. (2013)
Leporinus elongatus	Anionic Remazol	Ooi et al. (2017)

that the scales of certain fish species are capable of absorbing aqueous dyes (Kabir 2018). When the potential of black drum *(Pogonias cromis)* scales for the absorption of acid blue 45 dye from textile effluent was examined, promising findings were found and. Shows scales from different fishes and dyes adsorbed. Wastewater dye that harms the water system was absorbed from fish scales that would have otherwise ended up in the landfill. A new design was then inspired by these dyed scales. Future design studies could look into additional fish scale application techniques, such as machine sewing and fusing (Stannard and Kabir 2018) (Table 12.3).

12.5.10 Biogas and Biomethane

Fish waste can act as substrate for bioenergy production, particularly through the production of biogas and biomethane via anaerobic digestion. In this process, microorganisms in an oxygen-free environment break down the organic matter in fish waste to produce biogas, which is mainly made up of methane and can be converted to biomethane for use as a sustainable energy source. Fish waste used for bioenergy helps in reducing environmental impacts by decreasing waste accumulation, reducing greenhouse gas emissions, and providing a sustainable substitute to fossil fuels. The process also benefit economicallyby generating electricity, fabricating renewable natural gas, and creating a nutrient-rich mixture that can be utilized as fertilizer (Pedziwiatr 2017; Yadav et al. 2022; Yadav et al. 2023). Challenges like management of high nitrogen content in fish waste, control of odor, and the primary investment in biogas systems need to be recognised for worldwide adoption. Among these barriers, fish waste-based bioenergy offers a most promising solution for enhancing the sustainability of aquaculture operations (Cadavid-Rodríguez 2019).

12.6 Conclusion

Fish waste can cause immense pollution and harm to the environment. In order to prevent this, various waste management strategies can be deployed, such as composting, silage, fish meal, and oil production, etc. These will reduce the physical quantity of fish waste and also prevent chemical pollutant actions produced by waste piles. Scientific management of fish waste not only reduces its quantity but it will also helps in its conversion into marketable goods. Waste from the aquaculture industry can serve as raw material for other industries. Conversion of fish waste into valuable products will support the security of food and reduce the environmental impacts and marking aquaculture as more sustainable and efficient.

12.7 Future Directions

The increased progression in waste processing technologies will continue to increase the efficiency of fish waste utilization and promote sustainable aquaculture practices, as well as reduce environmental impacts. Establishing standards of quality and regulations will ensure safe and effective waste removal to support public health and environmental sustainability. Sustainable management of fish waste can significantly decrease greenhouse gas emissions and reduce pollution, promoting healthier aquatic ecosystems. Incorporating fish waste into aqua feeds can lower reliance on expensive fishmeal, thus reducing operational costs for aquaculture producers.

Acknowledgements The authors are highly thankful to the parents' university for the valuable guidance.

References

Ahalya N, Ramachandra TV, Kanamadi RD (2003) Biosorption of heavy metals. Res J Chem Environ 7(4):71–79

Ahmad AL, Chin JY, Harun MHZM, Low SC (2022) Environmental impacts and imperative technologies towards sustainable treatment of aquaculture wastewater: A review. J Water Process Eng 46:102553

Ahmadifar Z, Dadvand Koohi A (2018) Characterization, preparation, and uses of nanomagnetic Fe_3O_4 impregnated onto fish scale as more efficient adsorbent for $Cu2^+$ ion adsorption. Environ Sci Pollut Res 25(19):19687–19700

Ahuja I, Dauksas E, Remme JF, Richardsen R, Løes AK (2020) Fish and fish waste-based fertilizers in organic farming – with status in Norway: A review. Waste Manage 115:95–112

Alfio VG, Manzo C, Micillo R (2021) From fish waste to value: An overview of the sustainable recovery of omega-3 for food supplements. Molecules 26(4):1002

Andriani Y, Dhahiyat Y, Zahidah Z, Zidni I (2017) The effect of stocking density ratio of fish on water plant productivity in aquaponics culture system. Nusantara Bioscience 9(1):31–35

Arasu A, Kumaresan V, Sathyamoorthi A, Palanisamy R, Prabha N, Bhatt P, Roy A, Thirumalai MK, Gnanam AJ, Pasupuleti M, Marimuthu K (2013) Fish lily type lectin-1 contains β-prism architecture: Immunological characterization. Mol Immunol 56(4):497–506

Arruda LFD, Borghesi R, Oetterer M (2007) Use of fish waste as silage: A review. Braz Arch Biol Technol 50:879–886

Arvanitoyannis IS, Kassaveti A (2008) Fish industry waste: Treatments, environmental impacts, current and potential uses. Int J Food Sci Technol 43(4):726–745

Beerli EL, Beerli KM, Logato PVR (2004) Silagem ácida de resíduos de truta (Oncorhynchus mykiss) com a utilização de ácido muriático. Ciência Agrotecnológica 28:195–198

Blasbalg TL, Hibbeln JR, Ramsden CE, Majchrzak SF, Rawlings RR (2011) Changes in consumption of omega-3 and omega-6 fatty acids in the United States during the 20th century. Am J Clin Nutr 93(5):950–962

Boyd CE, Tucker CS (2012) Pond aquaculture water quality management. Springer Science & Business Media

Bruno SF, Ekorong FJAA, Karkal SS, Cathrine MSB, Kudre TG (2019) Green and innovative techniques for recovery of valuable compounds from seafood by-products and discards: A review. Trends Food Sci Technol 85:10–22

Cadavid-Rodríguez LS, Vargas-Muñoz MA, Plácido J (2019) Biomethane from fish waste as a source of renewable energy for artisanal fishing communities. Sustainable Energy Technol Assess 34:110–115

Caruso G, Floris R, Serangeli C, Di Paola L (2020) Fishery wastes as a yet undiscovered treasure from the sea: Biomolecules sources, extraction methods and valorization. Mar Drugs 18(12):622

Charroux B, Capo F, Kurz CL, Peslier S, Chaduli D, Viallat-Lieutaud A, Royet J (2018) Cytosolic and secreted peptidoglycan degrading enzymes in Drosophila respectively control local and systemic immune responses to microbiota. Cell Host Microbe 23(2):215–228

Chinh NT, Manh VQ, Trung VQ, Lam TD, Huynh MD, Tung NQ, Trinh ND, Hoang T (2017) Characterization of collagen derived from tropical freshwater carp fish scale wastes and its amino acid sequence. Nat Prod Commun 14(7):1934578–1986628

Chinh NT, Manh VQ, Trung VQ, Lam TD, Huynh MD, Tung NQ, Trinh ND, Hoang, T. (2019). Characterization of collagen derived from tropical freshwater carp fish scale wastes and its amino acid sequence. Natural Prod Commun 14(7):1934578X19866288

Chowdhury P, Viraraghavan T, Srinivasan A (2010) Biological treatment processes for fish processing wastewater—a review. Biores Technol 101(2):439–449

Dinesh R, Anand C, John KR, George MR, Bharathi S, Kumar JSS (2022) An overview of chemicals and drugs in aquaculture disease management. Indian J Anim Health 62(1):1–20

Endut A, Jusoh A, Ali N, Nik WBW (2011) Nutrient removal from aquaculture wastewater by vegetable production in aquaponics recirculation system. Desalin Water Treat 32:422–430

Esteban MB, García AJ, Ramos P, Márquez MC (2007) Evaluation of fruit–vegetable and fish wastes as alternative feedstuffs in pig diets. Waste Manage 27(2):193–200

FAO (2003) Animal feed resources information system. Retrieved from http://www.fao.org

FAO (2016) The state of world fisheries and aquaculture: Contributing to food security and nutrition for all. FAO, Rome

Frankic A, Hershner C (2003) Sustainable aquaculture: developing the promise of aquaculture. Aquacult Intern 11(6):517–530

Gagnon B, Berrouard S (1994) Effects of several organic fertilizers on growth of greenhouse tomato transplants. Can J Plant Sci 74(1):167–168

Ganjeh AM, Saraiva JA, Pinto CA, Casal S, Silva AM (2023) Emergent technologies to improve protein extraction from fish and seafood by-products: An overview. Appl Food Res. https://doi.org/10.1016/j.afres.2023.100339

Ghaly AE, Ramakrishnan VV, Brooks MS, Budge SM, Dave D (2013) Fish processing wastes as a potential source of proteins, amino acids and oils: A critical review. Journal of Microbial & Biochemical Technology 5(4):107–129

Goddek S, Schmautz Z, Scott B, Delaide B, Keesman KJ, Wuertz S, Junge R (2016) The effect of anaerobic and aerobic fish sludge supernatant on hydroponic lettuce. Agronomy 6(2):37

Graber A, Junge R (2009) Aquaponic systems: Nutrient recycling from fish wastewater by vegetable production. Desalination 246(1–3):147–156

Guillen J, Holmes SJ, Carvalho N, Casey J, Dörner H, Gibin M, Mannini A, Vasilakopoulos P, Zanzi A (2018) A review of the European Union landing obligation focusing on its implications for fisheries and the environment. Sustainability 10(4):900

Hao Y, Ding K, Xu Y, Tang Y, Liu D, Li G (2020) States, trends, and future of aquaponics research. Sustainability 12(18):7783

Hellio C, Pons AM, Beaupoil C, Bourgougnon N, Le Gal Y (2002) Antibacterial, antifungal and cytotoxic activities of extracts from fish epidermis and epidermal mucus. Int J Antimicrob Agents 20(3):214–219

Huang E, Huang CP (2007) Use of fish scales as biosorbent for the removal of copper in water. Water Res 30:1985–1990

Isangedighi IA, David GS (2019) Heavy metals contamination in fish: Effects on human health. J Aquat Sci Mar Biol 2(4):7–12

Jensen MB, Jakobsen J, Jacobsen C, Sloth JJ, Ibarruri J, Bald C, Iñarra B, Bøknæs N, Sørensen ADM (2024) Content and bioaccessibility of minerals and proteins in fish-bone containing side-streams from seafood industries. Marine Drugs 22(4):162

Joseph SM, Krishnamoorthy S, Paranthaman R, Moses JA, Anandharamakrishnan C (2021) A review on source-specific chemistry, functionality, and applications of chitin and chitosan. Carbohydr Polym Technol Appl 2:100036

Jung WK, Karawita R, Heo SJ, Lee BJ, Kim SK, Jeon YJ (2006) Recovery of a novel Ca-binding peptide from Alaska Pollack (Theragra chalcogramma) backbone by pepsinolytic hydrolysis. Process Biochem 41(9):2097–2100

Junge R, König B, Villarroel M, Komives T, Jijakli MH (2017) Strategic points in aquaponics. Water 9(3):182

Kabir SMF (2018) Waste management by waste: Removal of acid dyes from wastewaters of textile coloration using fish scales. (Unpublished thesis), Louisiana State University, Baton Rouge, Louisiana, USA

Kalbani FOSA, Salem MA, Cheruth AJ, Kurup SS, Senthilkumar A (2016) Effect of some organic fertilizers on growth, yield and quality of tomato (Solanum lycopersicum). Int Lett Nat Sci 53:1–9

Kiran, Kumar N, Sindhu S, Gawaria B, Ansari A, Dhariwal A, Sharma P, Kumar R (2025) GWAS approaches in medicinal plants for enhanced secondary metabolite production. In: Chen JT (ed) Biotechnology, multiple omics, and precision breeding in medicinal plants, pp 119–134. CRC Press

Kotzamanis YP, Alexis MN, Andriopoulou A, Castritsi-Cathariou I, Fotis G (2001) Utilization of waste material resulting from trout processing in gilthead bream (Sparus aurata L.) diets. Aquac Res 32:288–295

Kris-Etherton PM, Harris WS, Appel LJ (2002) Fish consumption, fish oil, omega-3 fatty acids, and cardiovascular disease. Circulation 106(21):2747–2757

Kristinsson HG, Rasco BA (2000) Fish protein hydrolysates: Production, biochemical, and functional properties. Crit Rev Food Sci Nutr 40:43–81

Kulkarni AA, Dhanush P, Chethan BS, Thamme CS, Shrivastava PK (2019) A brief study on aquaponics: An innovative farming technology. Indian J Sci Technol 12(48):1–5

Kumar R, Gupta RK, Yadav R, Saifi R, Yodha K, Pooja (2022b) Eisenia fetida as protein source for growth enhancement of Heteropneustes fossilis. Egypt J Aquat Biol Fish 26(2):577–588. https://doi.org/10.21608/ejabf.2022.234553

Kumar R, Yadav R, Kumar Gupta R, Pooja (2022c) A sustainable way for fish health management by replacement of chemical and drugs by earthworm. In: Environmental degradation in Asia: Land degradation, environmental contamination, and human activities, pp 329–352. Springer International Publishing, Cham. https://doi.org/10.1007/978-3-031-12112-8_16

Kumar R, Gupta R, Dahiya T, Yadav R, Saifi R, Yodha K, Kumari B, Singh P, Shefali (2022a) Surveillance of heteropneustes fossilis (Stinging Catfish) diseases and health management practices in Haryana. Appl Ecol Environ Sci 10(4):261–267. https://doi.org/10.12691/aees-10-4-12

Kumar MNR (2000) A review of chitin and chitosan applications. React Funct Polym 46(1):1–27

Kurtovic I, Marshall SN, Zhao X, Simpson BK (2009) Lipases from mammals and fishes. Rev Fish Sci 17(1):18–40

Lands WE (1992) Biochemistry and physiology of n-3 fatty acids. Fed Am Soc Exp Biol 6:2530–2536

León-López A, Morales-Peñaloza A, Martínez-Juárez VM, Vargas-Torres A, Zeugolis DI, Aguirre-Álvarez G (2019) Hydrolyzed collagen—sources and applications. Molecules 24(22):4031

Li C, Huang B, Li C, Chen X, Huang Y (2016) In situ formation of nanoscale zero-valent iron on fish-scale-based porous carbon for Cr(VI) adsorption. Water Sci Technol 73(9):2237–2243

Li H, Chen R, Jia Z, Wang C, Xu Y, Li C, Xia H, Meng D (2020) Porous fish collagen for cartilage tissue engineering. Am J Transl Res 12(10):6107–6120

Li Q, Yi L, Marek P, Iverson BL (2013) Commercial proteases: Present and future. FEBS Lett 587(8):1155–1163

Liu C, Sun J (2019) Osteogenically differentiated mesenchymal stem cells induced by hydrolyzed fish collagen maintain their immunomodulatory effects. Life Sci 238:116970

Lopes C, Antelo LT, Franco-Uría A, Alonso AA, Pérez-Martín R (2015) Valorisation of fish by-products against waste management treatments—comparison of environmental impacts. Waste Manage 46:103–112

Lv LC, Huang QY, Ding W, Xiao XH, Zhang HY, Xiong LX (2019) Fish gelatin: The novel potential applications. J Funct Foods 63:103581

Lv Z, Ran X, Liu J, Feng Y, Zhong X, Jiao N (2024) Effectiveness of chemical oxygen demand as an indicator of organic pollution in aquatic environments. Ocean-Land-Atmos Res 3:0050

Manjeet, Airon A, Kumar R, Saifi R (2022) Temporal and spatial impact of lockdown during COVID-19 on air quality index in Haryana, India. Scientific Rep 12:20046. https://doi.org/10.1038/s41598-022-20885-2

Marklin Jr RW, Mathison M, Mayer B, Nagurka ML, Cariapa V, Schabelski J (2013) Aquaponics: A sustainable food production system that provides research projects for undergraduate engineering students

Martin AM, Patel TR (1991) Bioconversion of wastes from marine organisms. In: Martin AM (ed) Bioconversion of waste materials to industrial products. Elsevier Applied Science, London, pp 417–440

Maschmeyer T, Luque R, Selva M (2020) Upgrading of marine (fish and crustaceans) biowaste for high added-value molecules and bio (nano)-materials. Chem Soc Rev 49(13):4527–4563

Mishra RP, Das SK, Kumar N, Mishra J (2020) Sustainable aquaponics system and its challenges: A review. Int J Curr Microbiol App Sci 9(5):577–588

Murugan G, Ahilan K, Prakasam VPA, Malreddy J, Benjakul S, Nagarsaajan M (2024) Fish waste composition and classification. In: Fish waste to valuable products, pp 1–26. Springer Nature Singapore, Singapore

Mustafiz S, Rahaman MS, Kelly D, Tango M, Islam MR (2003) The application of fish scales in removing heavy metals from energy-produced waste streams: The role of microbes. Energy Sources 25(9):905–916

Mutalipassi M, Esposito R, Ruocco N, Viel T, Costantini M, Zupo V (2021) Bioactive compounds of nutraceutical value from fishery and aquaculture discards. Foods 10(7):1495

Nisha A, Rani S, Kumar S, Kumar R, Kumar A (2025) Recent technologies and strategies for bioremediation of toxic and hazardous waste. In: Dubey KK, Shah MP, Yadav A (eds) Challenges and sustainable solutions in bioremediation, pp 1–16. CRC Press. https://doi.org/10.1201/9781003407317-1

Nogueira Junior S, Negri Neto A, Tsunechiro A, Okano C, Sato GS, Giulietti N, Diehl SRL (1997) Alimentação animal: Realidade e perspectivas (Coleção cadeias de produção da aqüicultura). Secretaria da Agricultura e Abastecimento, São Paulo

Oetterer M (2002) Industrialização do pescado cultivado Guaíba: Agropecuária, 200p

Ooi J, Lee LY, Hiew BYZ, Thangalazhy-Gopakumar S, Lim SS, Gan S (2017) Assessment of fish scales waste as a low-cost and eco-friendly adsorbent for removal of an azo dye: Equilibrium, kinetic and thermodynamic studies. Biores Technol 245:656–664

Pal D, Maiti SK (2020) An approach to counter sediment toxicity by immobilization of heavy metals using waste fish scale-derived biosorbent. Ecotoxicol Environ Saf 187:109833

Pang Z, Deeth H, Yang H, Prakash S, Bansal N (2017) Evaluation of tilapia skin gelatin as a mammalian gelatin replacer in acid milk gels and low-fat stirred yogurt. J Dairy Sci 100(5):3436–3447

Parvathy AJ, Das BC, Jifiriya MJ, Varghese T, Pillai D, Rejish Kumar VJ (2023) Ammonia induced toxico-physiological responses in fish and management interventions. Rev Aquac 15(2):452–479

Pędziwiatr P (2017) Aquaculture waste management. Acta Innov (22):20–29

Purohit R (2018) Eutrophication: Harmful algal blooms. Int J Sci Res Multidiscip Stud 1:267–274

Ramadhani S, Iswanto B, Purwaningrum P (2018) Waste utilization of red snapper (Lutjanus sp.) fish bone to improve phosphorus contents in compost. In: IOP conference series: Earth and environmental science, vol 106, no 1, p 012091

Ribeiro C, Scheufele FB, Alves HJ, Kroumov AD, Espinoza-Quiñones FR, Módenes AN, Borba CE (2019) Evaluation of hybrid neutralization/biosorption process for zinc ions removal from automotive battery effluent by dolomite and fish scales. Environ Tech

Sahar RA (2024) Pemanfaatan limbah ikan menjadi pakan bernutrisi tinggi: Solusi inovatif dalam sektor perikanan di Kabupaten Kepulauan Selayar. Jurnal Ilmiah Wahana Laut Lestari (JIWaLL) 2(1):1–7

Sampathkumar K, Yu H, Loo SCJ (2023) Valorisation of industrial food waste into sustainable aquaculture feeds. Future Foods 7:100240

Saranya R, Tamil Selvi A, Jayapriya J, Aravindhan R (2020) Synthesis of fat liquor through fish waste valorization, characterization and applications in tannery industry. Waste Biomass Valorization 11:6637–6647

Shefali, Kumar R, Sankhla MS, Kumar R, Sonone SS (2021) Impact of pesticide toxicity in aquatic environment. Biointerface Res Appl Chem 11(3):10131–10140. https://doi.org/10.33263/BRIAC113.1013110140

Silva L, Gasca-Leyva E, Escalante E, Fitzsimmons KM, Lozano DV (2015) Evaluation of biomass yield and water treatment in two aquaponic systems using the dynamic root floating technique (DRF). Sustainability 7(11):15384–15399

Sripriya R, Kumar R (2015) A novel enzymatic method for preparation and characterization of collagen film from swim bladder of fish Rohu (Labeo rohita). Food Nutr Sci 6(15):1468–1478

Stannard CR, Kabir F (2018) Fishy finery. In: International textile and apparel association annual conference proceedings (75:1). Iowa State University Digital Press

Suresh JI, Janani MS, Sowndharya R (2022) Bacterial diseases in fish with relation to pollution and their consequences—A global scenario. In: Bacterial fish diseases, pp 113–131. Academic Press

Surnar SR, Sharma OP, Saini VP (2015) Aquaponics: Innovative farming. Int J Fish Aquat Stud 2(4):261–263

Tangendjaja B (2022) Quality control of feed ingredients for aquaculture. In: Feed and feeding practices in aquaculture, pp 165–194

Thirukumaran R, Priya VKA, Krishnamoorthy S, Ramakrishnan P, Moses JA, Anandharamakrishnan C (2022) Resource recovery from fish waste: Prospects and the usage of intensified extraction technologies. Chemosphere 299:134361

Tom AP, Jayakumar JS, Biju M, Somarajan J, Ibrahim MA (2021) Aquaculture wastewater treatment technologies and their sustainability: A review. Energy Nexus 4:100022

Tú TLC (2019) Factors affecting the quantity and quality of faecal waste in striped catfish. Doctoral dissertation, Wageningen University and Research

Urakova IN, Pozharitskaya ON, Demchenko DV, Shikov AN, Makarov VG (2012) The biological activities of fish peptides and methods of their isolation. Russ J Mar Biol 38(6):417–422

Veličković ZS, Karkalić R, Bajić ZJ, Marinković AD, Nikolić A, Otřísal P, Florus S (2018) Cerium supported on high porous carbon from fish scales carp, as a novel low cost adsorbent to remove As(V) ions from water. Mater Methods Technol 12:110–122

Vieira EFS, Cestari AR, Carvalho WA, Oliveira RA (2012) The use of freshwater fish scale of the species Leporinus elongatus as adsorbent for anionic dyes. J Therm Anal Calorim 109(3):1407–1412

Xu Y, Duan L, Li Y, She Y, Zhu J, Zhou G, Jiang G, Yang Y (2020) Nanofibrillar decellularized Wharton's jelly matrix for segmental tracheal repair. Adv Func Mater 30(14):1910067

Yadav R, Kumar R, Gupta RK, Kaur T, Yodha K, Kour A, Kaur S, Rajput A (2023) Heavy metal toxicity in earthworms and its environmental implications: A review. Environ Adv 100374

Yadav R, Gupta RK, Kumar R, Kaur T, Moun R (2022) Biochemical response of earthworm, Eisenia fetida to heavy metals toxicity. J Appl Nat Sci 14(3):990–998

Yesilsu AF, Alp-Erbay E, Kahyaoğlu LN, Aydin İ (2024) Fish waste to valuable products: nanotechnological approach. Bio Nano Sci 14(5):4733–4751

Zampolli A, Bysted A, Leth T, Mortensen A, De Caterina R, Falk E (2006) Contrasting effect of fish oil supplementation on the development of atherosclerosis in murine models. Atherosclerosis 184(1):78–85

Zhu K, Gong X, He D, Li B, Ji D, Li P, Peng Z, Luo Y (2013) Adsorption of Ponceau 4R from aqueous solutions using alkali boiled Tilapia fish scales. RSC Adv 3(47):25221–25230

Chapter 13
Agriculture Waste Utilization for Aqua Feeds: Way Forward to Sustainability

Krina D. Patel, Rajkumar M. Dabhi, and Kailash H. Vadher

Abstract The demand for sustainable and functional feed components in aquaculture has sparked a growing curiosity in utilizing agricultural waste, which is high in bioactive chemicals. These substances, including antioxidants, peptides, polysaccharides, and essential fatty acids, can provide significant health benefits for aquatic species when integrated into aqua feeds. Agricultural waste, an underutilized resource, presents an exciting opportunity to address both environmental and economic concerns in the aquaculture business. This chapter investigates the possibilities of agricultural waste, including as crop residues, oilseed cakes, and fruit processing byproducts, as sustainable aqua feed components. The nutritional makeup of these wastes, which includes proteins, fibers, and critical micronutrients, is evaluated to determine their viability as substitutes for traditional feed components. Furthermore, the chapter investigates the possible advantages of introducing these fermented bioactive-rich feeds into aquaculture diets, such as increased fish growth, immunological response, and disease resistance. Scaling up these technologies presents both obstacles and opportunities, as do regulatory issues and market acceptance. Finally, the chapter emphasizes the future prospects of incorporating agricultural waste-based bioactive chemicals into sustainable aquaculture systems, which promote both animal health and environmental sustainability.

Keywords Aquaculture · Sustainable · Bioactive compounds · Fermentation · Agriculture waste

K. D. Patel (✉)
Department of Aquaculture, College of Fisheries Science, Kamdhenu University, Veraval, Gujarat, India
e-mail: kinu44165@gmail.com

Present Address:
R. M. Dabhi · K. H. Vadher
Department of Aquaculture, College of Fisheries Science, Kamdhenu University, Navsari, Gujarat, India
e-mail: dabhiraj1996@gmail.com; dabhiraj@kamdhenuuni.edu.in

K. H. Vadher
e-mail: khvadher@kamdhenuuni.edu.in

R. Kumar et al. (eds.), *Sustainability in Aquaculture and Fisheries*, Springer Water, https://doi.org/10.1007/978-3-032-16605-0_13

13.1 Introduction

Sustainability needs to be addressed in all sectors in order to reduce resource depletion, establish ecological balance and enhance utilization of resources. Society's dependence on the environment is deteriorating at an accelerated rate, resulting in various worldwide environmental, social, and economic issues emerging. The worldwide population has constantly increased and consumer demand for food exceeded what traditional farming techniques are capable of producing. Global consumption of food is predicted to increase between 35 and 56% by 2050 (Zlaugotne et al. 2022; Sampathkumar et al. 2023).

Seafood, especially fish, is crucial for nutrition and food security worldwide, particularly in underdeveloped countries. The world's supply of fish comes from two connected sectors: caught in the wild fisheries and Aquaculture (the farming of fish). For decades, the aquaculture sector has served as the world's fastest extending food production business. Capture fisheries have reached a plateau, requiring constant output increases to maintain or enhance consumption per person (Kumar et al. 2022a, 2022b; Bjorndal et al. 2024).

The UN Food and Agriculture Organization reports that aquaculture is increasing faster than any other animal food production industry (FAO 2022). Over the last 20–30 years, commercial and industrial aquaculture's contribution to world output has steadily increased. Small and medium-sized aquaculture organizations have also profited from the aquaculture sectors expansion. The sector expanded and commercialized, increasing its contribution to food security, revenue and trade. Despite these financial benefits, aquaculture development is always related with its emissions of nutrients and toxins into aquatic environment, relationships with regional biodiversity, utilization of energy and reliance solely on capture-based feed manufacturing only (Bohnes and Lauraent 2021; Manjeet et al. 2022; Verdegem et al. 2023).

The quality and composition of feeds are necessity in aquaculture for significant development in terms of growth and body texture of fishes. Furthermore, dependency of aquaculture sector on wild fish stocks for feed production leads to ocean over exploitation and leads to rapid collapse in fish stocks due to unstainable pressure (Fantatto et al. 2024). Therefore, the primary challenge for aquaculture is finding novel techniques that are more sustainable for the environment. Enhancing feed efficiency and growth performance is the most effective strategy to reduce aquaculture's environmental impact. However, as feed significantly contributes to the ecological effects of the aquaculture sector, new feed formulation techniques must be implemented which replace fish meal and fish oil with eco-friendly alternatives (Wilfart et al. 2023).

Thus, aquaculture feed industry has gone through many years for sustainable alternatives which can be utilized for completion of dietary requirement of warm, cool and cold-blooded aquatic species. These kinds of agriculture-based alternatives will effective in terms of their growth with minimum environment impact and produce high quality meat for human consumption in cost effective manner (Shefali et al. 2021; Gatlin et al. 2007).

Managing agricultural waste is a key focus in worldwide waste management strategies. Since the 1990s, there has been a growing concern about the environmental impact of agriculture. Today's agricultural techniques have been reviewed for their environmental, human and animal health impacts which lead to more stringent legislation (Kumar et al. 2022c; Al Seadi and Holm-Nielsen 2004).

India ranks among the top producers globally for various agricultural products producing a large amount of agricultural waste that is both harmful to environment but rich in nutrition. It can be cost effective and utilization for aquaculture feed manufacturing is way more sustainable and eco-friendlier. This chapter offers an overview of how agricultural waste may be converted into nutrient-rich meal for aquaculture feeds, aiming to replace unsustainable resources.

13.2 Agriculture Waste: An Overview

Agriculture accounts for approximately 37% of the earth's land surface and consumes a substantial number of resources, such as water, which accounts for 70% of global abstractions. The agricultural industry is regarded as a major driver of the industrial sector in various countries throughout the world, offering thousands of opportunities for employment and functioning as our preliminary source of feed, food, fiber and fuel. Due to the rising demand for agricultural products, massive amounts of agricultural waste are predicted to be generated (Yadav et al. 2023; Green 2019; El-Ramady et al. 2020).

The amounts of waste and negative consequences on the ecosystem are significant. This poses a significant challenge for humans to manage trash in an integrated manner to promote environmental sustainability. Many of the garbage produced can be recycled and turned into new items. This can be an effective method of trash management (Ajila et al. 2012; Bonciu et al. 2021).

The residue left over after generating and preparing raw agricultural and related products, such as vegetables, meat, produce, poultry, dairy, fish and crops are known as agricultural waste. Although they may contain material that can benefit people, these non-product outcomes of agricultural product manufacturing and utilization have economic worth below the expense of gathering, transferring and processing for safe and useful use. Agro-waste include agricultural waste (maize stalks, bagasse from sugarcane, drops, culls from vegetables, fruits and prunings) waste from animals (manure, carcasses), food processing waste (80% of maize is wasted) and harmful agricultural waste (pesticides, insecticides, herbicides, etc.). Increased agricultural production has led to more waste, including livestock waste, crop residues and industrial byproducts. Intensive farming techniques in developing nations are expected to lead to a considerable increase in global agricultural waste. The annual production of agricultural waste is over 998 million tons (Kumar et al. 2024; Agamuthu 2009). Approximately 15% of the total waste produced in every single country is made up of agricultural waste (Hsing et al. 2004).

13.2.1 Types of Agriculture Waste

There are numerous types of agricultural wastes that fall into various categories, including crop residues, byproducts from the processing of fruits and vegetables, sugar, starch, and confections, byproducts from the milling process of grains and lentils, byproducts from the manufacturing and oil fields and waste products from alcohol distilleries and breweries. Different types of agriculture waste details are given in Fig. 13.1.

Agriculture waste can be classified into two primary groups or categories: field crop residue and waste generated from agro-based industry and which holds significant potential for sustainable utilization. Another type of class comprises agriculture livestock waste such as cow dung, poultry dung, fish processing waste and slaughterhouse by-products including bones and feathers, which can be used as feed additives or organic fertilizers (Kumar and Singh 2022; Kumar et al. 2025; Yadav et al. 2022; Bharagava 2020).

Agriculture waste is often utilized to make essential products like fertilizer, gas for heating and cooking and as a low-cost option for treating wastewater polluted with heavy metals. Low-cost waste includes coconut husk, sawdust, sugarcane bagasse, neem bark and rice husk. Rice husk is an affordable catalyst for biodiesel production. Bamboo leaves and bagasse from sugarcane are additional waste materials that contain silica. Carbon nanostructures and waste-create bricks are two more sustainable building materials that may be made from agricultural waste (Raut et al. 2023).

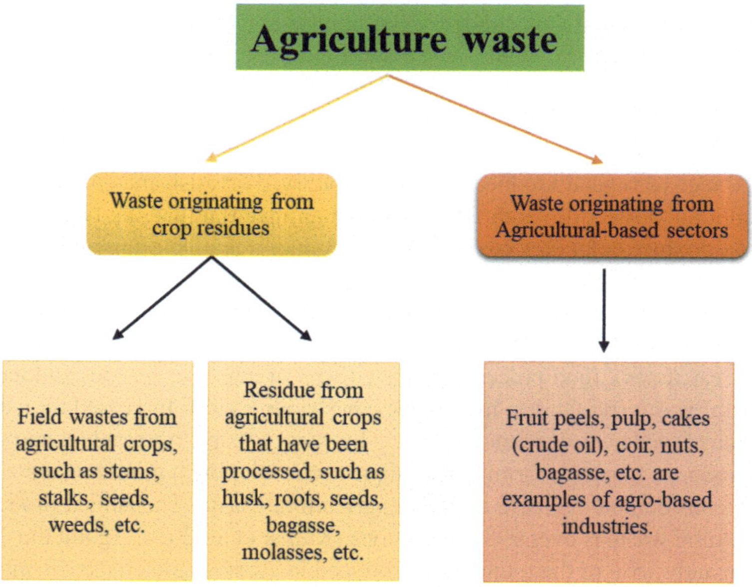

Fig. 13.1 Agriculture waste

13.2.1.1 Waste from Crop Residues

According to Owen and Jayasuriya (1989), the term "crop residues" has been the go-to term in tropical research and development circles for characterizing the fiber leftovers of grains, sugarcane, tubers, roots, oilseeds, pulses, oil plants, vegetables and fruits. After crops are harvested, the leftovers are known as crop residues and can be used as animal feed. They are distinct from agricultural waste products (such as brans and oil cakes) that are created during the processing of crops. Typically, residues are categorized according to the type of crop: cereals, grain legumes, stems, stover, straws, sea pods, tubers, roots, leaves and tubers and bulbs (Renard 1997; Akram and Fırıncıoglu 2019).

These crops have low protein content, but they include other nutrients that we can include in feed for various animals based on their nutritional needs by mixing them with additional supplements. Because of their availability and important nutrients, many crop residue can be used to prepare livestock feed, however there are certain challenges, as shown in Table 13.1.

13.2.1.2 Waste from Agro-Based Industries

Agro-industrial wastes are classified into three types: recyclable and compostable or naturally occurring agricultural and industrial wastes, non-recyclable and non-compostable agricultural and industrial wastes, and hazardous agricultural and industrial wastes (Ogbu and Okey 2023). Recyclable and compostable naturally available agro-food byproducts will be utilized as natural resources that are renewable since they contain polysaccharides, proteins, carbohydrates, and other substances. In addition to that, these byproducts are completely environmentally friendly, easily accessible, economical, and sustainable, making them ecological resources (Shrivastava and Singh 2022).

There is a tendency toward discovering novel sources of beneficial substances such as plant food byproducts, which were previously underestimated. The word "by-product" indicates that plant food wastes may be useful and have their own market. About 60% of cultivated plants may be made up of these byproducts. Due to sustainability issues in several different sectors, these wastes are exceptionally perishable goods that are challenging to manage (Freitas et al. 2021).

Agro-industrial wastes that have been closely examined include skins, seeds, pits, the pulp, pressed cakes, and leaves. Phenolic molecules, which are the most prevalent family of bioactive compounds with antioxidant effects found in fruit, are among the many secondary plant metabolites that originate from these residues. According to several studies, these ingredients have anti-inflammatory properties, anti-diabetic properties, anticancer, antioxidant, antimicrobial, and antiproliferative qualities, among additional beneficial health effects (Kodagoda and Marapana 2017). The trash generated by the fruit and vegetable sector often consists of both fresh fruit and vegetable waste and waste from highly nutritious processed goods. The findings

Table 13.1 Nutritional composition and utilization of agricultural residues in livestock

Crop	Residue	Nutritional highlights	Applications	Challenges/Notes	References
Rice	Straw, bran	OM 777 g/kg, CP 56 g/kg, Celluloid 293 g/kg, better digestibility in stems than leaves	Cattle feed, compost, bioenergy	Requires supplementation to improve protein content	Avadhanam et al. (2020)
Wheat	Straw, bran	CP 4.2%, NDF 77.5%, ADF 50%	Silage, roughage feed	Often combined with legumes to balance nutrients	Rosenfelder et al. (2013)
Barley	Straw	CP 5.5%, CF 38.2%	Feed for ruminants, bedding material	Limited digestibility without treatment	Bhandari (2019)
Maize (Corn)	Stover, cobs, bran	Stover CP 3.7–6.9%, NDF 82.4%, husks have high digestibility	Silage, biofuel	High fiber may limit intake in non-ruminants	Bhandari (2019)
Millet	Straw	CP 5%, CF 35.9%	Feed supplement, dry fodder	Better suited for regions with water scarcity	Hassan et al. (2021)
Sorghum	Straw, stover	CP 3.9%, CF 35.6%	Mixed with other feeds	Should not be fed fresh after drought due to prussic acid toxicity	Bhandari (2019)
Sugarcane	Bagasse, tops, molasses	Molasses: digestible carbs; low protein	Energy source, silage, mulch	Molasses used as binder or energy enhancer	Bhandari (2019)
Pulses	Hay, meal	High protein content, better than cereals	Silage, dried hay	Difficult recovery of leaves and stems	Sherasia et al. (2017)
Sunflower	Cakes, straw, pruning	CP 5.7–13%, moderate feed quality	Silage, mixed rations	Palatability issues if not treated	Bhandari (2019)
Roots & Tubers	Waste	Sweet potato vines: CP 8.4%, Ca 1.47%, good carbohydrate sources	Feed for small ruminants	Risk of spoilage; peeling waste varies in quality	Obafemi et al. (2022)
Vegetables	Tops, peelings	Includes carrot tops, cabbage leaves; rich in fiber	Compost, fresh feed	Seasonal availability, quick spoilage	Wadhwa and Bakshi (2013)

(continued)

Table 13.1 (continued)

Crop	Residue	Nutritional highlights	Applications	Challenges/Notes	References
Soybean	Straw, meal	Straw: CP 5.1%, NDF 80.8%; Meal: CP 53.5%, high nutritive value	Protein-rich feed, silage	Straw requires treatment for better digestibility	Bhandari (2019)
Groundnut	Tops	Dried hay from haulms; rich in protein	Valuable in mixed rations	Risk of aflatoxin contamination if improperly stored	Bhandari (2019)
Miscellaneous	Tree pruning, fruit residues	Olive leaves, banana pseudo-stems	Compost, low-quality feed	Collection and transportation logistics; seasonal/sporadic availability	Bhandari (2019)

Where: OM—Organic Matter; CP—Crude Protein; CF—Crude Fiber; NDF—Neutral Detergent Fiber; ADF—Acid Detergent Fiber; Ca—Calcium

in the scientific literature revealed that vegetable and fruit byproducts have great nutritional value which is given in Tables 13.2 and 13.3.

13.3 Bioactive Compounds in Agriculture Waste

Plants as feed additives have good impacts on fish health because of the enormous amount and they include a range of secondary bioactive metabolites, or phytochemicals, which are compounds, produced by the secondary metabolism of plants and provide them with their characteristics and activities.

Fruit peels and vegetable peels are rich in nutrition which has potential bioactive compounds we can replace with commercial expensive ingredients for making fish feeds. Bioactive substances derived from many plants have been used to enhance the health of aquatic animals (Table 13.4). Many bioactive active compounds are used to improve the health of aquatic animals in various ways, including their antioxidant, antimicrobial, immunostimulant, growth-promoting, and anti-inflammatory properties. Another innovative method is the incorporation of bioactive substances into aquatic feed, which is a reliable and restoring way to improve the feed quality (Vijayaram et al. 2022).

Terpenes, polyphenols, and alkaloids are the most significant phenolic compounds, which also include glycosides, saponins, anthraquinones, steroids, and coumarins. The main constituents of essential oils are terpenes (mono-, sesqui-, di-, tri-, and tetra-terpenes [carotenoids]) and terpenoids (alcohols, aldehydes, ketones, or esters). Flavonoids, anthocyanins, tannins, and lignins are the different types of polyphenols. Flavonoids are further categorized as flavones, flavonols (anthocyanins,

Table 13.2 The approximate dry weight composition of the fruit peels

Waste/ By-product	Moisture content (%)	DM (%)	OM (%)	CP (%)	CF (%)	Carbohydrate (%)	References
Orange peels	9.68	90.3	85.4	11.0	13.99	54.17	Egbuonu and Osuji (2016)
Papaya peels	9.82	90.2	84.2	11.7	32.51	47.33	Martial-Didier et al. (2017)
Pineapple peels	5.10	94.9	91.1	5.78	4.10	75.67	Owoeye et al. (2022)
Watermelon peels	14.75	85.2	72.2	6.02	17.10	65.41	Ho et al. (2016)
Banana peels	9.83	90.1	80.6	3.23	12.67	63.82	Tsadoet al. (2021)
Apple pomace	9.4	90.6	97.5	1.3	16.4	43.2	Masoodi and Chauhan (1998)
Banana leaves	8.3	78.8	–	8.5	12.5	–	Fernandeset al.(2013)
Cashew nut shell	–	–	98.6	5	–	–	Manju et al. (2015)
Citrus pulp	64.3	37.8	96.0	8.0	6.1	10	Pascual and Carmona (1980)
Cull dates	8.6	91.4	99.0	6.44	14.2	70.2	Attia et al. (2021)
Date palm pulp	2.20	97.8	96.5	2.98	1.78	42.02	Ibrahim et al. (2019)
Date seeds/ pits	8.64	91.36	99.04	5.84	7.92	12.56	Habib and Ibrahim (2009) and Ibrahim (2019)
Grape pomace	5.94	94.06	86.73	11.6	15	53.65	Rennaet al. (2024)
Dragon fruit peel	6.66	–	98.3	4.35	21.24	–	Cacatian and Guittap (2018)

Dry matter (DM), organic matter (OM), crude protein (CP), and crude fiber (CF)

condensed tannins, or proanthocyanidins), and isoflavonoids. Phenylpropanoids are polyphenols that manifest as aldehydes, alcohols, or acids (phenolic acids) (Beltrán and Esteban 2022).

Phenolic compounds are the most common secondary metabolic byproducts in plants and are quite common in nature. The main source of naturally occurring antimicrobial compounds that can inhibit bacteria development is plants. To increase their effectiveness and stop the emergence of germ resistance, they can be administered in conjunction with earlier antibiotics. Alkaloids, sulfur-containing compounds, terpenoids, and phenolic compounds are among the plant chemicals that are frequently employed for antimicrobial applications. Over 30,000 antimicrobial

Table 13.3 Proximate composition of the vegetable waste on dry weight basis

Vegetable Waste	DM	Ash	OM	CP	EE	HC	CEL	References
Baby corn husk	–	7.62	92.38	10.16	0.9	24.35	35.0	Bakshi et al. (2017)
Bottle gourd pulp	12.3	9.3	90.7	24.3	2.4	10.4	10.5	Bakshi et al. (2016)
Carrot	10.1	7.97	91.8	8.01	1.87	–	–	Agiriga et al. (2015)
Carrot pulp	93.72	6.95	93.05	7.34	1.36	–	–	Cozma et al. (2024)
Cucumber	87.1	11.6	75.45	15.9	0.13	–	–	Niyi et al. (2019)
Cucumber peels	78.3	7.85	70.45	26.5	1.44	–	–	Niyi et al. (2019)
Grocery by-product waste	17.5	7.3	92.7	14.6	11.3	3.9	–	Bakshi et al. (v)
Jackfruit waste	86.93	3.01	83.92	10.06	1.49	7.74	–	Amadi et al. (2018)
Jackfruit seeds	71.92	3.92	68.00	10.09	4.29	7.89	–	Amadi et al. (2018)
Jackfruit leaves	85.33	4.91	80.42	1.19	0.73	5.33	–	Amadi et al. (2018)
Pea pods	14.1	8.04	92	19.8	1	12.7	24	Bakshi et al. (2016)
Pea vines ensiled	25	9	91	13.1	3.3	10	34	Bakshi et al. (2016)
Potato	12	4.8	95.2	9.5	0.4	–	–	Bakshi et al. (2016)
Liquid potato feed	11.7	7.8	88.9	14.0	0.5	14.2	5.8	Bakshi et al. (2016)
Pumpkin waste	98.75	1.25	97.50	1.30	1.43	6.36	10.8	Khanam et al. (2019)
Pumpkin oil meal	92.77	7.23	85.54	–	–	–	–	Bakshi et al. (2016)

Dry matter (DM), organic matter (OM), crude protein (CP), ether extract (EE), HC- hemi cellulose, CEL- cellulose

chemicals have been extracted from plants, and over 1,350 plants having antibacterial qualities have been found. Utilizing plant components that would otherwise be thrown away as trash is a safe and sustainable way to find new antimicrobial agents, and plants are a great source of natural antimicrobials (Shirahigue and Ceccato-Antonini 2020).

Natural or plant-based immunostimulants are increasingly being used to improve animal defense mechanisms and protect them from bacterial and viral infections. Plant-based substances such as berberine, gelselegine, piperine (alkaloids), Zligustilide, tetramethylpyrazine, luteolin, apigenin (flavones), phloroglucinol, quinones, and others have been used to strengthen the immune system in aquaculture as well as enhance fish health. Plant-based products have been shown to improve phagocytic cell function, boost the ability to kill bacteria, and activate Antibody response, lysozyme activity, complement, and natural killer cells in fish and shellfish. Plant-based immunostimulants have a crucial role in shellfish aquaculture, where the innate immune system is primarily responsible for combating infections (Sharma et al. 2024, 2025; Rani et al. 2025; Kumar and Bossier 2018).

Plant waste contains bioactive compounds that act as immunostimulants and enhance gut microbiota, improving fish health. These compounds, including omega-3 fatty acids (like ALA), bioactive peptides, antioxidants, vitamins, and minerals,

Table 13.4 Bioactive Compounds from Agro-Industrial Waste: Sources, examples and properties

Bioactive Compound	Source	Examples	Properties	References
Phenolic Compounds	Plant residues (husks, seeds, etc.)	Flavonoids, lignins, stilbenes	Antioxidants, antibacterial, antiparasitic, antiviral, anti-inflammatory, anticancer, antihypertensive	Oleszek et al. (2023)
Terpenes	Essential oils, carotenoids	Thymol, carvacrol	Antioxidants, antimicrobial	Graßmann (2005)
Dietary Fiber (β-glucans)	Cell walls of plants, algae, fungi	Glycans with β-1,3 and β-1,6	Immunostimulatory, prebiotic, resistant to digestion	Leyva-Lopez et al. (2020)
Glucosinolates	Brassica family (broccoli, mustard)	Isothiocyanates	Antioxidants, antibacterial, anticancer	Leyva-Lopez et al. (2020)
Saponins	Legumes, agave	Triterpene saponins, steroidal saponins	Immunostimulatory, antimicrobial, microbiota modulating	da Silva et al. (2024)

regulate metabolism, support immune function, and promote fish growth and reproduction. Additionally, plant-based bioactive compounds influence enzymatic activity, hormonal balance, and metabolic efficiency, ensuring optimal nutrient absorption in fish. They also contribute to cellular membrane integrity, muscle development, and neurological functions. In aquaculture, these compounds improve fish health, feed efficiency, and resilience against environmental stressors (Nisha et al. 2025; Kiran et al. 2025; Leyva-Lopez et al. 2020).

13.4 Utilizing Agricultural Waste as Aquaculture Feed Additives

Agriculture waste may be used as feed additives to improve development, immunological response, and disease resistance in aquatic species since it contains bioactive components such phenols, tannins, and saponins. As feed additives some agriculture co-products can be utilized by drying them and incorporating them as powder form or in extract form in diet of commercially important fishes.

A wide range of feed additives made from agricultural waste, including as leaves, stems, roots, seeds, tubers, and shrubs, are referred to as phytogenics. In addition to extracts and essential oils, they come in powdered, solid, and dried forms.

Phytogenics in aquatic farm animals enhance appetite, support healthy gut flora, and provide analgesic, antibacterial, anti-carcinogenic, antioxidant, and antiparasitic qualities (Onomu and Okuthe 2024). Plant waste use as feed additives in diet of commercially important fishes which is given in Table 13.5.

13.4.1 Processing of Agriculture Waste as Aquaculture Feeds

Due to their accessibility and affordability, upcycling agricultural waste and co-products into animal feed can be a profitable endeavor. It can be challenging to use some of these by-products in animal feeds, though, because of their high moisture content, which makes them extremely susceptible to spoilage. Other biomass materials have low feeding value because of high concentrations of lignin and indigestible fiber, lacking protein content and anti-nutritional compounds.

Therefore, there is a good chance that the bioprocessing process these feedstuffs will increase their feeding value and, therefore, their use as a significant portion of complete feeds to guarantee optimum productivity while lowering the environmental effect of farmed livestock production (Sun et al. 2024).

13.4.1.1 Fermentation of Agriculture Waste

Byproducts like vegetables and fruit peels are rich in carbohydrates which are ideal for fermentation and that fermented ingredients have more nutrients as compared to raw form. The various contents of agricultural wastes promote the growth of microbes, which ultimately produces a variety of useful enzymes through fermentation. They are utilized as raw materials for making fermented feed products or produce enzymes. Fermented waste can be used as feed supplements for aquatic organisms (Sadh et al. 2018). Fermentation involves using enzymes from microbial secondary metabolites to simplify complicated chemical molecules. Fermenting feed materials improves quality, digestibility, and creates fish-friendly aromas and flavorings. Fermenting agricultural and food waste improves feed quality through increasing the amount of protein, decreasing the amount of fiber and improving the digestibility (Cahya et al. 2022).

Here is common process of fermentation which is given in Fig. 13.2. (Dai et al. 2020a, b). There are some reviews of fermented fruits and vegetable waste and their use as feed which are given in Table 13.6.

There are major two types of fermentation processes which are given below:

I. **Solid-State Fermentation (SSF):** Fermentation of agricultural waste by using microorganisms like yeast, such as *Saccharomyces cerevisiae*. SSF enhances nutritional value of the fish feed by increasing nutrition and reducing anti-nutritional factors. Additionally, SSF enhance immune response and improves nutrient digestibility in fish (Das et al. 2021).

Table 13.5 Agricultural waste feed additives for enhancing aquaculture health and performance

Feed additives (Source and Dosage)	Species	Biological properties and effects	References
Guava Leaf Compounds (Guava leaf, 0.025–0.4%)	*Penaeus monodon*	Growth enhancer, immunostimulant; improved growth, prophenoloxidase, nitric oxide synthase, lysozyme, and antioxidant enzyme activities	Yin et al. (2014)
Papaya Leaf Extract (Papaya leaf, 1–2%)	*Oreochromis niloticus*	Growth enhancer, feed efficiency; increased final weight, improved weight gain, specific growth rate, and feed efficiency	Hamid et al. (2022)
Giloy leaf powder, (0.25–1%)	*Catla catla*	Growth enhancer, immunostimulant; improved gross conversion efficiency, feed conversion ratio, survival	Upreti and Chauhan (2018)
Sugarcane Bagasse (10–80 g/kg)	*Oreochromis niloticus*	Antioxidant, immunostimulant; higher immune-related gene expressions (IL1, IL8, GPX), with growth improvement at 20–40% levels	Lumsangkul et al. (2021)
Passionfruit Peel Powder (10–80 g/kg)	*Oreochromis niloticus*	Antioxidant, immunostimulant; elevated immune gene expressions (IL1, IL8, GPX), with slight growth improvement	Outama et al. (2022a, b)
Cocoa bean shells and others	*Oreochromis niloticus*	Cost-effective growth enhancer; improved weight gain, feed conversion ratio, and reduced feeding cost	Bamba et al. (2014)
Fruit waste (Apple, Grapes and Orange) powder	*Macrobrachium rosenbergii*	Digestive and growth enhancer; enhanced survival, growth, digestive enzymes, antioxidants, and biochemical constituents	Saravana Bhavan et al. (2013)
Pineapples waste (10-30%)	*Oreochromis niloticus*	Growth enhancement improved survival but no effect on fillet quality	Sukri et al. (2022)
Pomegranate Peel (PP) (1–20%)	*Oreochromis niloticus*	Growth enhancer, feed efficiency; 3–5% PP improved survival, growth, feed utilization, villi length, and goblet cell count; 20% PP reduced body composition quality	Abozahra et al. (2023)
Yuzu Peel Paste (10–100 g/kg diet)	(*Seriola quinqueradiata*) Yellowtail	Antioxidant, flesh freshness enhancer; reduced muscle discoloration, higher fecal fat levels, and lower digestibility at 100 g/kg	Fukada et al. (2014)
Pea Meal (15–25%)	*Oreochromis niloticus*	Alternative protein source; 15% replacement enhanced growth and feed efficiency	Desouky et al. (2023)

(continued)

Table 13.5 (continued)

Feed additives (Source and Dosage)	Species	Biological properties and effects	References
Mango Peel Powder (MGPP) (6.25–50 g/kg diet)	*Oreochromis niloticus*	Immunostimulant; 25 g/kg improved immunity (lysozyme, peroxidase, IL-1, IL-8) but no significant enhancement in growth or antioxidant activity	Outama et al. (2022a, b)
Mandarin Peel Powder (MP) (10–20%)	*Oreochromis niloticus*	Antioxidant, anti-inflammatory; 20% MP improved improved gut health, decreased pro-inflammatory cytokines, and enhanced antioxidant activity	Reda et al. (2024)
Pineapple Peel Powder (PAPP) (5–40 g/kg diet)	*Oreochromis niloticus*	Immunostimulant; 10 g/kg PAPP enhanced immunity (lysozyme, peroxidase, complement, phagocytosis), immune gene expressions, and survival rate	Van Doan et al. (2021)
Banana Peel Powder (2–8% inclusion)	*Labeo rohita*	Improved proximate composition; enhanced digestibility; minimum FCR at 4% inclusion level	Sanvriya et al. (2022)

II. **Liquid fermentation:** On the contrary liquid fermentation needed liquid medium which has various bacteria and yeast. This method effectively breaks down complex carbohydrates and proteins, making them more accessible for fish consumption (Siddik et al. 2024).

13.4.1.2 Production of Pectin from Agriculture waste

Agriculture waste also contains pectin, plant cell walls contain the majority of non-starch polysaccharides (NSPs), which function as anti-nutritional agents. NSPs are categorized into two types: insoluble and soluble. NSPs are typically regarded as fish food ingredients that are indigestible because they lack enzymes matching (Sagar et al. 2018).

Pectin is a common non-starch polysaccharides primarily found in the cell walls of plants and layers of intracellular cells, as well as in considerable concentrations in the peels of grapefruit, oranges, lemons and apples. Pectin has recently been proved to be effective as a binder for aqua feed or addition for feeds. According to earlier studies, apple peel pectin effectively strengthened the immune system with other potential to resist pathogenic germs. Similarly, pectin produced from citrus showed antibacterial activity against dangerous microorganisms (Zhou et al. 2024). The detailed process of making pectin from fruit residue is given in (Gama et al. 2015) Fig. 13.3.

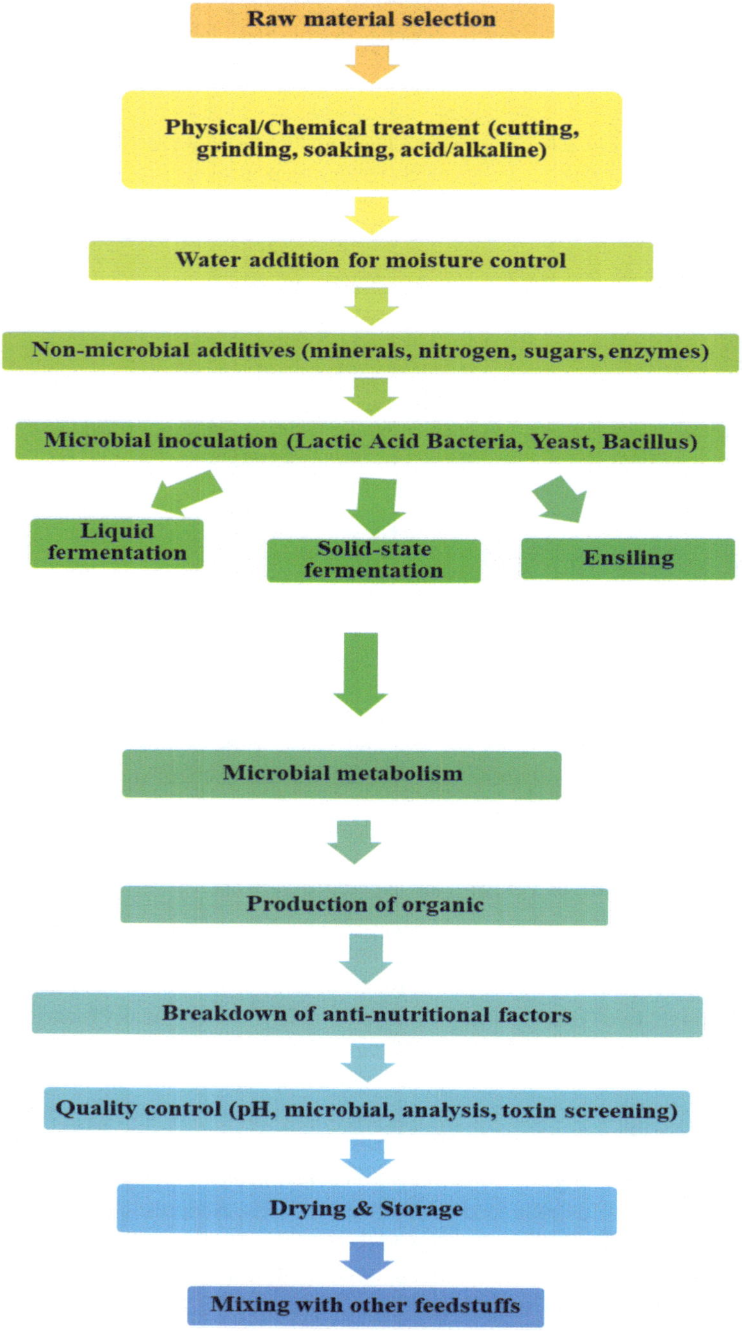

Fig. 13.2 Fermentation process of agriculture waste

Table 13.6 Fermentation of peels and their uses as feedstuff in aquaculture

Study parameter	Findings/Results	Applications/ Conclusions	References
Fermented banana peel flour (MFBP) in tilapia feed	Substitution up to 50% MFBP showed comparable growth, feed utilization, and survival rates to the control diet; above 50%, performance decreased	MFBP can replace up to 50% fishmeal in tilapia feed without compromising growth. Reduces feed cost and utilizes agricultural waste	Intharathat et al. (2024)
Fermented lemon peel in grouper diet	Supplementation of 1–3% improved weight gain, feed efficiency, lysozyme activity, and disease resistance against *P. damselae*	As a functional ingredient, fermented lemon peel improves immunity and growth. Recommended inclusion: 1–3%	Zhuo et al. (2021)
Fermented flour with coffee peel in tilapia feed	Increased HDL, DHA, EPA deposition, reduced LDL, cholesterol, and triglycerides in fish fillets with supplementation of up to 24%	Fermented coffee-peel flour enhances the nutritional quality of fish fillets, promoting healthier fish production. Optimal dose: 24%	Fitria et al. (2020)
Cassava peel meal in *C. gariepinus* Diet	Replacement up to 50% of maize with cassava peel meal improved growth and nutrient utilization. Fermentation reduced phytate and cyanide levels significantly	Fermented cassava peel is suitable for replacing up to 50% maize in catfish feed, reducing dependency on conventional feedstuffs	Adewumi (2018)
Citrus by-products in olive flounder diet	Supplementation with Bacillus-fermented citrus by-products enhanced disease resistance and innate immunity without affecting growth performance	Fermentation with Bacillus subtilis enhances bioactivity of citrus by-products. Promising for immunity and resistance to infectionin flounder	Lee et al. (2013)

13.4.2 Classification of Pectins

Pectins are classified based on their degree of esterification (DE). If more than 50% of carboxyl groups are methylated, they are high-methoxy pectins (HMP); if less, they are low-methoxy pectins (LMP). LMP can also be processed into amidated (AM) pectins using ammonia (Sundarraj and Ranganathan 2017).

1. **High-Methoxyl Pectins (HMP):** Require high solid content (>55%) to gel, usually with sucrose, making them suitable for sweetened products.

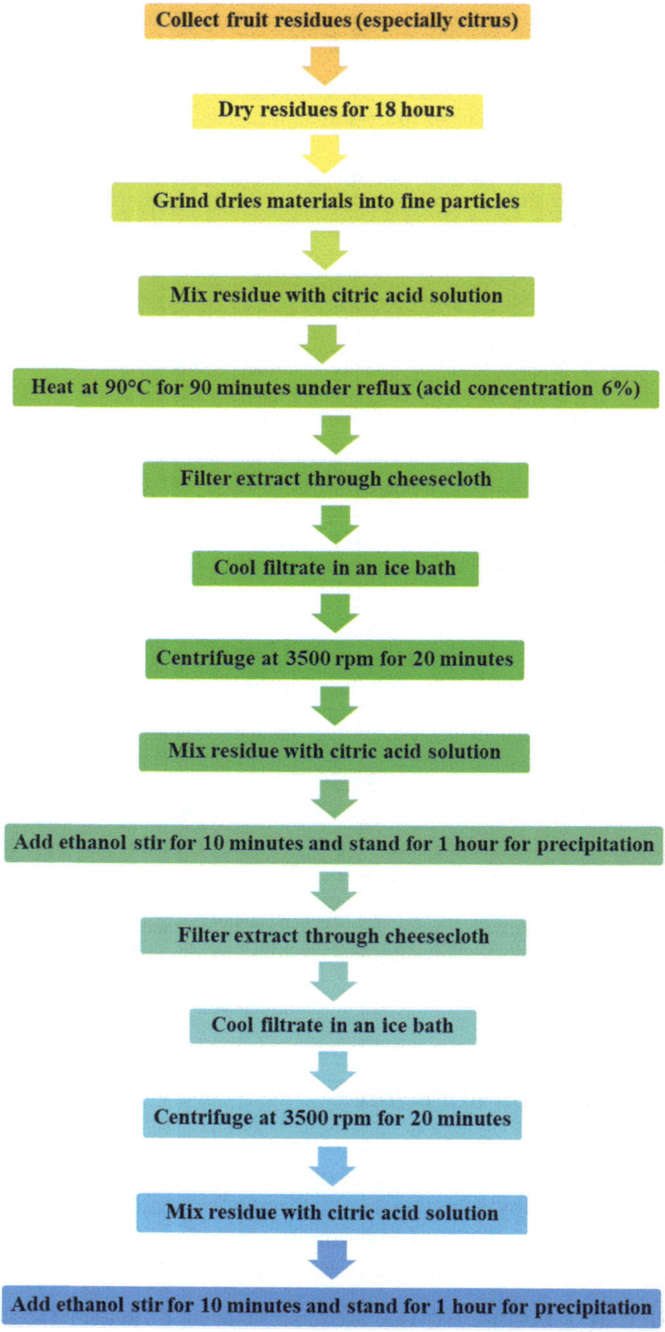

Fig. 13.3 Production of pectin from fruit residues

2. **Low-Methoxyl Pectins (LMP):** Gel in the presence of calcium through an 'egg box' mechanism. They work within a pH range of 2–6 and don't need high solid content, but excessive calcium may cause a bitter taste.

In aquaculture feed production Low-Methoxyl Pectins (LMP) are commonly used. They form a gel in presence of calcium which improve feed stability as well as control nutrient release in water (Chandel et al. 2022).

Pectin has ability to work without high sugar content and across a wide pH range makes them ideal for aqua feed formulations. They also support digestion, promote gut health, and improve feed efficiency in aquatic species (Liu et al. 2023). Here are detailed reviews about agriculture waste derived pectin used as aquaculture feed which is given in Table 13.7.

Table 13.7 Waste derived pectin and their uses as feedstuff in aquaculture

Study parameter	Findings/Results	Applications/ Conclusions	References
Red Yeast + Pectin from Passion Fruit Peel in Tilapia Feed	Combined supplementation significantly improved intestinal anatomy, growth, immunity, and the intestinal microbiota composition	Red yeast cell walls and pectin function well as prebiotics. Combined use enhances tilapia health and growth	Lubis et al. (2024)
Pectin from Apple Peel in the Carp Diet	Supplementation up to 2% improved growth, feed conversion ratio, and serum antioxidant activity (SOD, CAT). Enhanced immune responses like lysozyme and total immunoglobulin levels	Apple peel pectin improves carp growth and immune function. Optimal inclusion: up to 2%	Hoseinifar et al. (2021)
Fruit Peels for Thraustochytrids Production	Fermented fruit peels enriched with *Aurantiochytrium sp.* optimized lipid production, yielding 8.37% fatty acids	Utilization of agricultural waste for essential fatty acid production supports a circular bioeconomy for aquaculture	Saikia et al. (2024)
Orange Peel Pectin + *L. plantarum* in Tilapia Diet	Combination significantly increased growth, Disease resistance, serum immunity, and skin mucus immunity to *Streptococcus agalactiae*	Combination: pectin and *L. plantarum* is an effective additive for tilapia under biofloc systems, enhancing growth and disease resistance	Van Doan et al. (2019)

13.5 Sustainable Utilization of Agriculture Waste as Aquaculture Feed

Utilization of agriculture waste as feed for fish food production is important for long term sustainability. Incorporating this byproduct into aquaculture feed supports sustainable practice by reducing waste. By incorporating agriculture waste into fish feed the sector can move towards cost-effective and eco-friendly solutions which maintain long term sustainability. This book chapter describes detailed view about agriculture waste types, their origin, use, processing and their importance. Here is flowchart given in Fig. 13.4 to describe about the process of agriculture waste utilization which is given by (Sun et al. 2024; Urmila et al. 2025).

13.6 Conclusion

Agricultural wastes are abundant, cost-effective, nutrient-rich, and a valuable source of bioactive compounds. These active compounds will enhance growth and immunity of different fish species. Furthermore, process like fermentation and enzyme extraction like pectin also improve the quality of feeds which have more potential. Additional researches are needed for more prominent conclusion because waste have also antinutritional factors and pesticides residues so appropriate processing treatment are needed to understand their nutritive value and their suitability for aquaculture diets.

13.7 Future Prospects and Challenges

Implementing agriculture waste as aquaculture feed faces several challenges that need to be considered. Challenges like less efficient processing technologies and limited capacity to store and handle wide range of agriculture wastes must be considered. Further collaborative research needed to overcome these problems and enhance efficiency. Feed preparation involves high cost of production, so more efficient and viable technologies needed that have less production cost.

Furthermore, research is needed to close the knowledge gaps in environmental impact assessment, nutritional quality feed preparation and scalable integration. Sustainable solutions like waste characterization, feed material selection and proper optimization will improve efficiency and support widespread adaptations.

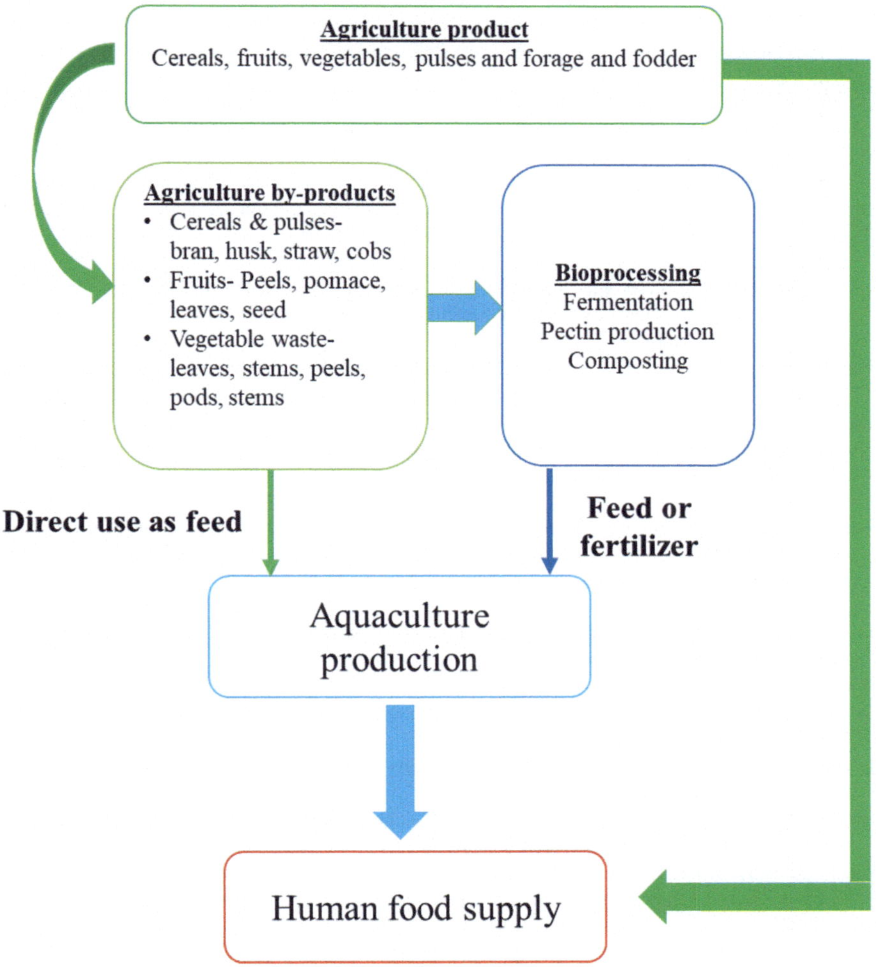

Fig. 13.4 Agriculture byproduct utilization

References

Abozahra NA, Albalawi A, Althobaiti N, Mohamed R, Diab A (2023) Effects of pomegranate (*Punica granatum*) Peel Methanolic extract dietary supplementation on *Oreochromis Niloticus* performance, blood health, Intestine Morphometry and immunity. Egypt J Veterinary Sci 54(7):221–236

Adewumi AA (2018) Evaluation of fermented cassava (*Manihot esculenta*) peel meal on the growth of *Clarias gariepinus*. J Biosci Biotechnol Discov 3(5):90–98

Agamuthu P (2009) Challenges and opportunities in agro-waste management: an Asian perspective. In: Inaugural meeting of first regional 3R forum in Asia (Vol 11, p 12). University of Malasiya

Agiriga AN, Iwe MO, Etoamaihe UJ, Olaoye OA (2015) Impact of different blanching treatments on the nutritional and sensory properties of oven dried carrot slices. Sky J Food Sci 4(7):102–107

Ajila CM., Brar SK, Verma M, Prasada Rao UJS (2012) Sustainable solutions for agro processing waste management: an overview. Environmental protection strategies for sustainable development, 65–109

Akram MZ, Fırıncıoğlu SY (2019) The use of agricultural crop residues as alternatives to conventional feedstuffs for ruminants: a review. Eurasian J Agric Res 3(2):58–66

Al Seadi T, Holm-Nielsen JB (2004) Agricultural wastes. In Waste Management Series. Elsevier 4:207–215

Amadi JA, Ihemeje A, Afam-Anene OC (2018) Nutrient and phytochemical composition of jackfruit (*Artocarpus heterophyllus*) pulp, seeds and leaves. Int J Innov Food Nutrition Sustain Agric 6(3):27–32

Attia AI, Reda FM, Patra AK, Elnesr SS, Attia YA, Alagawany M (2021) Date (*Phoenix dactylifera* L.) by-products: chemical composition, nutritive value and applications in poultry nutrition, an updating review. Animals 11(4):1133

Avadhanam S, Polumat A, Ellamraju S (2020) Usage of crop residue for making livestock feed. ILRI project report. Nairobi, Kenya: International livestock research institute (ILRI)

Bakshi MPS, Wadhwa M, Kumar B (2017) Nutritional evaluation of baby corn fodder and conventional maize fodder in buffaloes. Development 29(7)

Bakshi MPS, Wadhwa M, Makkar HP (2016) Waste to worth: vegetable wastes as animal feed. CABI Rev, 1–26

Bamba Y, Ouattara N, Soro Y, Ouattara A, Yao K, Gourène G (2014) Evaluation of production efficiency of Nile tilapia (*Oreochromis niloticus* L.) fed diets containing crop residues in combination with cocoa bean shell and coconut oil cake in Côte d'Ivoire. Livest Res Rural Dev 26(10):8

Beltrán JMG, Esteban MÁ (2022) Nature-identical compounds as feed additives in aquaculture. Fish Shellfish Immunol 123:409–416

Bhandari B (2019) Crop residue as animal feed. Paper Rev, 1–18

Bharagava RN (ed) (2020) Emerging eco-friendly green technologies for wastewater treatment. Springer, Berlin, Heidelberg, Germany

Bjorndal T, Dey M, Tusvik A (2024) Economic analysis of the contributions of aquaculture to future food security. Aquaculture 578:740071

Bohnes FA, Laurent A (2021) Environmental impacts of existing and future aquaculture production: comparison of technologies and feed options in Singapore. Aquaculture 532:736001

Bonciu E, Paunescu RA, Rosculete E, Paunescu G (2021) Waste management in aquaculture. Scientific papers series management, economic engineering in agriculture & rural development 21(3)

Cacatian SB, Guittap LJV (2018) Production, proximate analysis and functional properties of dragon fruit peel powder. IAMURE Int J Ecol Conserv 25(1):1–1

Cahya MD, Andriani Y, Haetami K, Risdiana R (2022) Application of fermented product feed to growth performance of fish: a review. Depik 11(3):333–340

Chandel V, Biswas D, Roy S, Vaidya D, Verma A, Gupta A (2022) Current advancements in pectin: extraction, properties and multifunctional applications. Foods 11(17):2683

Cozma A, Mihut C, Velciov A, Chis C, Mircov V, POopescu S, Rada M (2024) Determination of the nutritional composition of carrot pomace. Scientific papers. Series B. Horticulture 68(1)

Da Silva RMG, Kacew S, Granero FO (2024) Saponins: a class of bioactive natural products with wide applications in human health. Stud Nat Prod Chem 82:185–233

Dai Z, Cui L, Li J, Wang B, Guo L, Wu Z, Wu G (2020) Fermentation techniques in feed production. Animal agriculture, 407–429

Dai Z, Cui L, Li J, Wang B, Guo L, Wu Z, Wu G (2020) Fermentation techniques in feed production. In: Animal agriculture, pp 407–429. Academic Press

Das KC, Mohanty S, Sahoo PK, Das R, Sahoo L, Swain P (2021) Effect of solid-state fermented aquafeed on growth performance, digestive enzymes and innate immunity of Rohu, *Labeo rohita*. Agric Sci Digest 41(3)

Desouky AM, Hwihy HM, Shaban WM, Azab AM (2023) Evaluating of pea peels meal as a fishmeal alternative in formulated diet ingredients of *Oreochromis niloticus*. Egypt J Aquatic Biol Fisheries 27(6)

Egbuonu ACC, Osuji CA (2016) Proximate compositions and antibacterial activity of *Citrus sinensis* (sweet orange) peel and seed extracts. Europ J Med Plants 12(3):1–7

El-Ramady H, El-Henawy A, Amer M, Omara AED, Elsakhawy T, Elbasiouny H, El-Mahrouk M (2020) Agricultural waste and its nano-management: mini review. Egypt J Soil Sci 60(4):349–364

Fantatto RR, Mota J, Ligeiro C, Vieira I, Guilgur LG, Santos M, Murta D (2024) Exploring sustainable alternatives in aquaculture feeding: the role of insects. Aquac Reports 37:102228

FAO (2022). The state of world fisheries and aquaculture 2022. Towards blue transformation.https://doi.org/10.4060/cc0461en

Fernandes ERK, Marangoni C, Souza O, Sellin N (2013) Thermochemical characterization of banana leaves as a potential energy source. Energy Convers Manage 75:603–608

Fitria PD, Amin M, Lokapirnasari WP, Lamid M (2020) Supplementation of fermented coffee-peel flour to increase high-density lipoprotein (HDL) cholesterol, docosahexaenoic acids (DHA) and eicosapentaenoic acids (EPA) deposition in tilapia fillet. Biocatal Agric Biotechnol 24:101502

Freitas LC, Barbosa JR, da Costa ALC, Bezerra FWF, Pinto RHH, de Carvalho Junior RN (2021) From waste to sustainable industry: How can agro-industrial wastes help in the development of new products? Resour Conserv Recycl 169:105466

Fukada H, Furutani T, Shimizu R, Masumoto T (2014) Effects of yuzu (Citrus junos) peel from waste as an aquaculture feed supplement on growth, environmental load, and dark muscle discoloration in yellowtail *Seriola quinqueradiata*. J Aquat Food Prod Technol 23(5):511–521

Gama B, Silva CDF, Da Silva LO, Abud ADS (2015) Extraction and characterization of pectin from citric waste. Chem Eng Trans 44:259–264

Gatlin DM III, Barrows FT, Brown P, Dabrowski K, Gaylord TG, Hardy RW, Wurtele E (2007) Expanding the utilization of sustainable plant products in aquafeeds: a review. Aquac Res 38(6):551–579

Graßmann J (2005) Terpenoids as plant antioxidants. Vitam Horm 72:505–535

Green A (2019) Agricultural waste and pollution. Academic Press, In Waste, pp 531–551

Habib HM, Ibrahim WH (2009) Nutritional quality evaluation of eighteen date pit varieties. Int J Food Sci Nutr 60(1):99–111

Hamid NKA, Somdare PO, Harashid KAM, Othman NA, Kari ZA, Wei LS, Dawood MA (2022) Effect of papaya (*Carica papaya*) leaf extract as dietary growth promoter supplement in red hybrid tilapia (*Oreochromis mossambicus* × *Oreochromis niloticus*) diet. Saudi J Biol Sci 29(5):3911–3917

Hassan ZM, Sebola NA, Mabelebele M (2021) The nutritional use of millet grain for food and feed: a review. Agric Food Secur 10:1–14

Ho LH, Suhaimi MA, Ismail I, Mustafa KA (2016) Effect of different drying conditions on proximate compositions of red-and yellow-fleshed watermelon rind powders. J Agrobiotechnol 7:1–12

Hoseinifar SH, Jahazi MA, Mohseni R, Yousefi M, Bayani M, Mazandarani M, El-Haroun ER (2021) Dietary apple peel-derived pectin improved growth performance, antioxidant enzymes and immune response in common carp, *Cyprinus carpio* (Linnaeus, 1758). Aquaculture 535:736311

Hsing HJ, Wang FK, Chiang PC, Yang WF (2004) Hazardous wastes transboundary movement management: a case study in Taiwan. Resour Conserv Recycl 40(4):329–342

Ibrahim S, Sadau SB, Musa ZB, Namadina MM (2019) Comparative proximate analysis of some date palm cultivars (*Phoenix dactylifera* l.). FUDMA J Sci 3(4):522–529

Intharathat B, Ponza P, Karaket T (2024) Feasibility of molasses-fermented banana peel as a protein source in practical diet for hybrid tilapia (*Oreochromis spp.*): effect on growth and feed efficiency. *Acta Scientiarum.* Animal Sci 46:e68154

Khanam T, Malik S, Gururani P, Chaudhary V (2019) Nutritionally enriched cake using vegetable and fruit waste: a review. J Pharmacognosy Phytochemistry 8(3):722–725

Kiran, Kumar N, Sindhu S, Gawaria B, Ansari A, Dhariwal A, Sharma P, Kumar R (2025) GWAS approaches in medicinal plants for enhanced secondary metabolite production. In: Chen JT (Ed), Biotechnology, multiple omics, and precision breeding in medicinal plants, pp 119–134. CRC Press. https://doi.org/10.1201/9781003475491-10

Kodagoda KHGK, Marapana RAUJ (2017) Utilization of fruit processing by-products for industrial applications: a review

Kumar R, Singh D, Yadav R (2025) Toxic effects of chlorantraniliprole and fipronil on *Eisenia fetida*: from growth and reproductive parameters to avoidance behavior. Environ Eng Manag J 24(7):1499–1508. https://doi.org/10.30638/eemj.2025.117

Kumar R, Singh D (2022) Insecticide-tolerant bacterial population in *Eisenia fetida*'s Gut and Vermicast exposed to Chlorantraniliprole and Fipronil. Appl Biol Res 24(3):273–279. https://doi.org/10.5958/0974-4517.2022.00033.7

Kumar R, Gupta R, Dahiya T, Yadav R, Saifi R, Yodha K, Kumari B, Singh P, Shefali (2022a) Surveillance of *Heteropneustes fossilis* (Stinging Catfish) Diseases and Health Management Practices in Haryana. Appl Ecol Environ Sci 10(4):261–267. https://doi.org/10.12691/aees-10-4-12

Kumar R, Gupta RK, Yadav R, Saifi R, Yodha K, Pooja (2022b) *Eisenia fetida* as protein source for growth enhancement of *Heteropneustes fossilis*. Egypt J Aquatic Biol Fisheries 26(2):577–588. https://doi.org/10.21608/ejabf.2022.234553

Kumar R, Yadav R, Gupta RK, Pal A, Yodha K, Kumar A (2024) Antioxidant enzyme activities and markers of oxidative stress in the life cycle of different Earthworm species. Indian J Exp Biol 62(07):578–583

Kumar R, Yadav R, Kumar Gupta R, Pooja (2022c) A sustainable way for fish health management by replacement of chemical and drugs by earthworm. In: Environmental degradation in asia: land degradation, environmental contamination, and human activities, pp 329–352. Cham: Springer International Publishing. https://doi.org/10.1007/978-3-031-12112-8_16

Kumar V, Bossier P (2018) Importance of plant-derived compounds and/or natural products in aquaculture. Aquafeed 10:28–31

Lee BJ, Kim SS, Song JW, Oh DH, Cha JH, Jeong JB, Lee KJ (2013) Effects of dietary supplementation of citrus by-products fermented with a probiotic microbe on growth performance, innate immunity and disease resistance against *Edwardsiella tarda* in juvenile olive flounder, *Paralichthys olivaceus*. J Fish Dis 36(7):617–628

Leyva-Lopez N, Lizarraga-Velazquez CE, Hernandez C, Sanchez-Gutierrez EY (2020) Exploitation of agro-industrial waste as potential source of bioactive compounds for aquaculture. Foods 9(7):843

Liu Y, Zhou H, Fan J, Huang H, Deng J, Tan B (2023) Potential mechanisms of different methylation degrees of pectin driving intestinal microbiota and their metabolites to modulate intestinal health of *Micropterus salmoides*. Int J Biol Macromol 2(51):126297

Lubis AR, Linh NV, Srinual O, Fontana CM, Tayyamath K, Wannavijit S, Van Doan HV (2024) Effects of passion fruit peel (*Passiflora edulis*) pectin and red yeast (*Sporodiobolus pararoseus*) cells on growth, immunity, intestinal morphology, gene expression, and gut microbiota in Nile tilapia (*Oreochromis niloticus*). Sci Rep 14(1):22704

Lumsangkul C, Tapingkae W, Sringarm K, Jaturasitha S, Le Xuan C, Wannavijit S, Van Doan H (2021) Effect of dietary sugarcane bagasse supplementation on growth performance, immune response, and immune and antioxidant-related gene expressions of Nile tilapia (*Oreochromis niloticus*) cultured under Biofloc system. Animals 11(7):2035

Manjeet, Airon A, Kumar R, Saifi R, (2022) Temporal and spatial impact of lockdown during COVID-19 on air quality index in Haryana, India. Scientific Reports 12:20046. https://doi.org/10.1038/s41598-022-20885-2.

Manju Wadhwa, M. W., Bakshi, M. P., & Makkar, H. P. (2015). Waste to worth: fruit wastes and by-products as animal feed. *CABI Reviews*, 1–26.

Martial-Didier AK, Hubert KK, Parfait KEJ, Kablan T (2017) Phytochemical properties and proximate composition of papaya (*Carica papaya*) peels. Turkish J Agric Food Sci Technol 5(6):676–680

Masoodi FA, Chauhan GS (1998) Use of apple pomace as a source of dietary fiber in wheat bread. J Food Process Preserv 22(4):255–263

Nisha A, Rani S, Kumar S, Kumar R, Kumar A (2025) Recent technologies and strategies for bioremediation of toxic and hazardous waste. In: Dubey KK, Shah MP, Yadav A (Eds), Challenges and sustainable solutions in bioremediation, pp 1–16. CRC Press. https://doi.org/10.1201/978 1003407317-1

Niyi OH, Jonathan AA, Ibukun AO (2019) Comparative Assessment of the proximate, mineral composition and mineral safety index of peel, pulp and seeds of Cucumber (*Cucumis sativus*). Open J Appl Sci 9(09):691

Obafemi AI, Ajayi MO, Fagbenro RB, Komolafe PL, Olarinde AI, Samba BA, Babayemi OJ (2022) Effect of root and tuber peels on livestock and poultry production in Nigeria (A Review). Niger J Anim Prod 49(4):39–57

Ogbu CC, Okey SN (2023) Agro-industrial waste management: the circular and bioeconomic perspective. Agricultural waste-new insights

Oleszek M, Kowalska I, Bertuzzi T, Oleszek W (2023) Phytochemicals derived from agricultural residues and their valuable properties and applications. Molecules 28(1):342

Onomu AJ, Okuthe GE (2024) The role of functional feed additives in enhancing aquaculture sustainability. Fishes 9(5):167

Outama P, Le Xuan C, Wannavijit S, Lumsangkul C, Linh NV, Montha N, ... Van Doan H (2022) Modulation of growth, immune response, and immune-antioxidant related gene expression of Nile tilapia (*Oreochromis niloticus*) reared under biofloc system using mango peel powder. Fish Shellfish Immunol 131:1136–1143

Outama P, Linh NV, Xuan CL, Wannavijit S, Tongsiri S, Chitmanat C, ... Van Doan H (2022) Passionfruit (*Passiflora edulis*) peel powder stimulates the immune and antioxidant defense system in nile tilapia, *Oreochromis niloticus*, cultivated in a biofloc system. Fishes 7(5):233

Owen EMCN, Jayasuriya MCN (1989) Use of crop residues as animal feeds in developing countries. Res Dev Agric 6(3):129–138

Owoeye TF, Akinlabu DK, Ajayi OO, Afolalu SA, Popoola J O, Ajani OO (2022) Phytochemical constituents and proximate analysis of dry pineapple peels. In: IOP conference series: earth and environmental science (Vol 993, No 1, p 012027). IOP Publishing

Pascual JM, Carmona JF (1980) Composition of citrus pulp. Anim Feed Sci Technol 5(1):1–10

Rani S, Sharma S, Singh D, Sheoran R, Kumar R (2025) Improved production of anti-cancer compounds from medicinal plants through molecular and biotechnological mutagenesis strategies. In: Chen JT (Ed.), Biotechnology, multiple omics, and precision breeding in medicinal plants, pp 287–297. CRC Press. https://doi.org/10.1201/9781003475491-21

Raut NA, Kokare DM, Randive KR, Bhanvase BA, Dhoble SJ (2023) Introduction: fundamentals of waste removal technologies. 360-Degree Waste Manag 1:1–16

Reda RM, El-Rahim MIA, Elkerdawy DA, Metwally MM, Said N (2024) Effects of mandarin peel powder on growth, biochemical, immune, and intestinal health in *Oreochromis niloticus* at suboptimal temperatures. BMC Vet Res 20(1):446

Renard C (1997) Crop residues in sustainable mixed crop/livestock farming systems

Renna M, Gasco L, Livorsi L, Mele M, Conte G, Meneguz M, Lussiana C (2024) Growth performance, proximate composition and fatty acid profile of black soldier fly larvae reared on two grape pomace varieties. Animal 18(8):101240

Rosenfelder P, Eklund M, Mosenthin R (2013) Nutritive value of wheat and wheat by-products in pig nutrition: a review. Anim Feed Sci Technol 185(3–4):107–125

Sadh PK, Duhan S, Duhan JS (2018) Agro-industrial wastes and their utilization using solid state fermentation: a review. Bioresour Bioprocess 5(1):1–15

Sagar NA, Pareek S, Sharma S, Yahia EM, Lobo MG (2018) Fruit and vegetable waste: bioactive compounds, their extraction, and possible utilization. Compreh Rev Food Sci Food Safety 17(3):512–531

Saikia DK, Chikkaputtaiah C, Velmurugan N (2024) Nutritional enrichment of fruit peel wastes using lipid accumulating Aurantiochytrium strain as feed for aquaculture in the North-East Region of India. Environ Technol 45(6):1215–1233

Sampathkumar K, Yu H, Loo SCJ (2023) Valorisation of industrial food waste into sustainable aquaculture feeds. Future Foods 7:100240

Sanvriya H, Ojha ML, Sharma SK, Jain HK, Gupta L (2022) Efficacy of dietary supplementation of banana peel powder on proximate composition and digestibility of *Labeo rohita* (Hamilton, 1822) fingerlings. J Exp Zool India 25(1)

Saravana Bhavan P, Kirubhanandhini V, Muralisankar T, Manickam N, Srinivasan V (2013) Effect of fruits wastes (apple, grape and orange) incorporations on the growth of freshwater prawn (*Macrobrachium rosenbergii*). Asian J Sci Technol 4(10):075–081

Sharma MMM, Kapoor D, Rohilla R, Loyal A, Sharma P, Husen A (2024) Role of essential minerals and rhizosphere in plant under changing environmental conditions. In: Husen A (Eds), Essential minerals in plant-soil systems, 235–252. Elsevier. https://doi.org/10.1016/B978-0-443-16082-0.00002-3

Sharma P, Kapoor D, Kumar R, Husen A (2025) Rhizosphere and osmolytes coordination in plant systems under adverse situations. In: Husen A (Ed), Roles of osmolytes in changing environment, pp 101–117. Academic Press. https://doi.org/10.1016/B978-0-443-26549-5.00020-5

Shefali, Kumar R, Sankhla MS, Kumar R, Swaroop S, Sonone (2021) Impact of pesticide toxicity in aquatic environment. Biointerface Res Appl Chem 11(3):10131–10140. https://doi.org/10.33263/BRIAC113.1013110140

Sherasia PL, Garg MR, Bhanderi BM (2017) Pulses and their by-products as animal feed. Food and agriculture organization, 222 pp. https://openknowledge.fao.org/handle/20.500.14283/i7779en

Shirahigue LD, Ceccato-Antonini SR (2020) Agro-industrial wastes as sources of bioactive compounds for food and fermentation industries. Ciência Rural 50:e20190857

Shrivastava R, Singh NK (2022) Agro-wastes sustainable materials for wastewater treatment: review of current scenario and approaches for India. Mater Today: Proc 60:552–558

Siddik MA, Julien BB, Islam SM, Francis DS (2024) Fermentation in aquafeed processing: achieving sustainability in feeds for global aquaculture production. Rev Aquac 16(3):1244–1265

Sukri SAM, Andu Y, Harith ZT, Sarijan S, Pauzi MNF, Wei LS, Kari ZA (2022) Effect of feeding pineapple waste on growth performance, texture quality and flesh colour of nile tilapia (*Oreochromis niloticus*) fingerlings. Saudi J Biol Sci 29(4):2514–2519

Sun X, Dou Z, Shurson GC, Hu B (2024) Bioprocessing to upcycle agro-industrial and food wastes into high-nutritional value animal feed for sustainable food and agriculture systems. Resour Conserv Recycl 201:107325

Sundarraj AA, Ranganathan TV (2017) A review-Pectin from Agro and industrial waste. Int J Applied Environ Sci 12(10):1777–1801

Tsado AN, Okoli NR, Jiya AG, Gana D, Saidu B, Zubairu R, Salihu IZ (2021) Proximate, minerals, and amino acid compositions of banana and plantain peels. BIOMED Nat Appl Sci 1(01):032–042

Upreti P, Chauhan RS (2018) Effect of leaf powder of giloy (*Tinospora cordifolia*) in fish feed on survival and growth of post larvae of *Catla catla*. J Appl Nat Sci 10(1):144–148

Urmila, Gupta RK, Sharma S, Kumar R* (2025) Earthworms: the environmentally preferred bioengineers for managing heavy metals. In: Dey S, Bhattacharya S (Eds), Biotechnological interventions in the removal of emerging pollutants, pp 365–384. Singapore: Springer Nature Singapore

Van Doan H, Hoseinifar SH, Harikrishnan R, Khamlor T, Punyatong M, Tapingkae W, El-Haroun E (2021) Impacts of pineapple peel powder on growth performance, innate immunity, disease resistance, and relative immune gene expression of Nile tilapia, *Oreochromis niloticus*. Fish Shellfish Immunol 114:311–319

Van Doan H, Hoseinifar SH, Naraballobh W, Jaturasitha S, Tongsiri S, Chitmanat C, Ringo E (2019) Dietary inclusion of orange peels derived pectin and *Lactobacillus plantarum* for Nile tilapia (*Oreochromis niloticus*) cultured under indoor biofloc systems. Aquaculture 508:98–105

Verdegem M, Buschmann AH, Latt UW, Dalsgaard AJ, Lovatelli A (2023) The contribution of aquaculture systems to global aquaculture production. J World Aquaculture Soc 54(2):206–250

Vijayaram S, Sun YZ, Zuorro A, Ghafarifarsani H, Van Doan H, Hoseinifar SH (2022) Bioactive immunostimulants as health-promoting feed additives in aquaculture: a review. Fish Shellfish Immunol 130:294–308

Wadhwa M, Bakshi MPS (2013) Utilization of fruit and vegetable wastes as livestock feed and as substrates for generation of other value-added products. RAP Publ 4(2013):67

Wilfart A, Garcia-Launay F, Terrier F, Soude E, Aguirre P, Skiba-Cassy S (2023) A step towards sustainable aquaculture: Multiobjective feed formulation reduces environmental impacts at feed and farm levels for rainbow trout. Aquaculture 562:738826

Yadav R, Gupta RK, Kumar R, Kaur T (2022) Assessment of toxicity of lead and nickel on the biochemical and immunological parameters of earthworm, *Eudrilus eugeniae*. Agric Sci Digest 44(5):922–929

Yadav R, Kumar R, Gupta RK, Kaur T, Yodha K, Kour A, Kaur S, Rajput A (2023) Heavy metal toxicity in earthworms and its environmental implications: a review. Environ Adv, 100374

Yin XL, Li ZJ, Yang K, Lin HZ, Guo ZX (2014) Effect of guava leaves on growth and the non-specific immune response of *Penaeus monodon*. Fish Shellfish Immunol 40(1):190–196

Zhou H, Liu Y, Meng X, Lin X, Zhang L, Mi H, Deng J (2024) Effects of dietary pectin on the growth performance, intestinal barrier, and antioxidant status of juvenile rainbow trout (*Oncorhynchus mykiss*). Aquaculture Reports 36:102055

Zhuo LC, Chen CF, Lin YH (2021) Dietary supplementation of fermented lemon peel enhances lysozyme activity and susceptibility to *Photobacterium damselae* for orange-spotted grouper, *Epinephelus coioides*. Fish Shellfish Immunol 117:248–252

Zlaugotne B, Pubule J, Blumberga D (2022) Advantages and disadvantages of using more sustainable ingredients in fish feed. Heliyon 8(9)

Chapter 14
Bycatch Utilization: A Way Towards Sustainability

Nehal A. Tandel, Bhavika R. Tandel, Deepak T. Vaghela, Trushti R. Tandel, and Priya Singh

Abstract Unselective fishing practices lead to significant bycatch, which refers to non-target organisms that are caught and often discarded, raising critical concerns for ocean conservation and resource management. It has gained attention in recent years as an area needing effective management as it discard practices contribute to the unnecessary waste of valuable marine resources, exacerbating the depletion of marine populations. As target fish stocks approach collapse and profits decline, bycatch has inadvertently acted as a buffer for sustaining fisheries. It is increasingly recognized that management strategies must focus on the sustainability of both bycatch and target species to ensure long-term viability. A balanced approach that incorporates economic and ecological considerations is essential for maximizing the benefits of fisheries. Trawling, in particular, has been identified as a destructive fishing method, that causes a substantial amount of bycatch, needs critical evaluation from a sustainability standpoint. Reducing waste and managing bycatch effectively are vital for fostering resilient ecosystems and safeguarding the sustainability of fisheries, thereby promoting global food security and alleviating poverty. This chapter

N. A. Tandel (✉)
Department of Fisheries Resource Management, College of Fisheries Science, Kamdhenu University, Veraval, Gujarat, India
e-mail: tandelsonu8997@gmail.com

B. R. Tandel
Department of Fisheries Resource Management College of Fisheries Science, Kamdhenu University, Navsari, Gujarat, India
e-mail: bhavikatandel@kamdhenuuni.edu.in

D. T. Vaghela
Department of Aquatic Environment Management, College of Fisheries Science, Kamdhenu University, Veraval, Gujarat, India
e-mail: dtvaghela@kamdhenuuni.edu.in

T. R. Tandel
Department of Aquaculture, College of Fisheries Science, Kamdhenu University, Veraval, Gujarat, India

P. Singh
Department of Aquaculture, College of Fisheries Science, Chaudhary Charan Singh Haryana Agricultural University, Hisar, Haryana, India

R. Kumar et al. (eds.), *Sustainability in Aquaculture and Fisheries*,
Springer Water, https://doi.org/10.1007/978-3-032-16605-0_14

discuss the potential of transforming bycatch, particularly from low-value species, into edible resources, addressing both waste and ethical concerns surrounding the ocean's loss of billion tons of protein.

Keywords Discards · Target · Non-target varieties · Socio-ecological consideration · Technology · Novel products · Fisheries Management

14.1 Introduction

Bycatch has long been a challenge in fisheries, but it has emerged as a more prominent issue in fisheries management in recent decades. This issue has evolved over time, with its roots extending back to the early days of human fishing practices in aquatic environments for food (Horsten and Kirkegaard 2002). Scientists began investigating a problem of bycatch and discards only over these past 40–50 years and this concern often attributed to either an overexploitation of certain populations and underutilization of another (Manjeet et al. 2022; Sultana et al. 2014; Kumar and Singh 2022). It quickly became a significant concern in fisheries research as it become clear that incidental capture was affecting various marine species, potentially endangering non-targeted resources like mammals, marine birds, turtles and sometimes a variety of fishes (Alverson 1999; Kumar et al. 2022a, 2022b). Development in modern fishing gears tremendously increases amount of by-catch landings in last 50 years efficiently catching target species with additional non target varieties (Sultana et al. 2014). Consequently, bycatch was recognized as a critical fisheries issue. However, heightened awareness of bycatch has grown significantly due to the expansion of global fisheries and increased efforts in conservation and environmental advocacy aimed at protecting marine turtles, birds, and mammals suffering from intensive fishing practices (Alverson and Hughes 1996).

Fish is a highly nutritious form of affordable, excellent high-quality protein with a biological value of 15% to 23% (Susheela 2022). Rich in omega-3 polyunsaturated fatty acids (PUFAs), particularly eicosapentaenoic acid (EPA) and docosahexaenoic acid (DHA), it also supplies essential amino acids, notably lysine, which is often deficient in grains (Susheela 2022). Its low sodium and high potassium, calcium, vitamin, and mineral content contribute to a diverse and cost-effective diet (Shefali et al. 2021; Susheela 2022). Utilizing bycatch can enhance nutritional security and serve as a valuable supplement in daily diets, particularly in fish-consuming nations such as India, Thailand, Vietnam, Indonesia, the Philippines, Bangladesh, and Sri Lanka. The issue of fisheries bycatch in our complicated and evolving environment cannot be solved easily by using only one method (Kumar et al. 2022c; Komoroske and Lewison 2015). Instead, investing in a wide range of flexible tools and strategies tailored to specific economical, sociocultural, and ecological factors provides the most effective means to tackle global bycatch and promote sustainable fisheries for the foreseeable future (Komoroske and Lewison 2015). Striking a balance between minimizing bycatch and implementing utilization strategies is crucial, particularly in developing countries.

The term "bycatch" is interpreted and defined differently by different scientists and fisheries managers.

14.2 By-Catch

"By-catch" generally defines as an unintended capture of non-target species in fisheries that aim to target specific animals, such as fish or shrimp or a group of species (as shown in Fig. 14.1) (Lobo 2012). The term 'discards' (implies a portion of the catch thrown back into the sea) is often used interchangeably with bycatch in some regions (Lobo 2012). It is either not utilized or not properly managed (Davies et al. 2009).

Alverson et al. (1994) established definitions that will be utilized, as their approach has frequently served as the benchmark for evaluating discards and bycatch. This definition includes:

- Landed catch = Incidental catch + Target catch
- Bycatch = Incidental catch + Discarded catch
- Overall catch = Discarded catch + Landed catch

Fig. 14.1 Non-target varieties (such as sole fishes, grunter, *eels, other sp.)* with target species (Mackerel) (*Source* Author)

- Retained bycatch (Incidental catch)

It is essential to clearly defining the criteria which is used as basis to assess bycatch. Hall (1994) proposed a valuable classification system that categorizes bycatch based on its level and impact as follows:

- **Critical bycatch**—unintentional capture of endangered species as bycatch
- **Non-sustainable bycatch**—the capture of species that are threatened, whose populations might decrease if current bycatch rates persist
- **Sustainable bycatch**—bycatch levels that don't cause a reduction in population numbers
- **Non-biologically significant bycatch**—bycatch that is so minimal it has insignificant impact on the affected populations
- **Bycatch of unknown levels**—situations where abundance or mortality data is insufficient to assess if bycatch is critical or sustainable
- **Ecosystem-level impacts**—when bycatch affects a broad range of species within an ecosystem
- **Charismatic bycatch**—acknowledges that different groups assign varying values to species, and some species are seen as special or important, regardless of their conservation status or the impact of bycatch on them.

In fisheries, the term "**bycatch**" is used in at least four different ways (FAO 2009a, b).

- Catch that is not the fishery's target species but is kept and sold (Sultana et al. 2014).
- Fish species, sizes and sexes that fishermen throw out to sea (Karp et al. 2011).
- Non-target fish species, either they kept for trade or thrown out (OECD 1997).
- Undesirable species of invertebrate, like non-commercial crustaceans, and echinoderms, as well as other threatened species categories, consists of elasmobranchs, sea turtles, seabirds, and marine mammals (Sultana et al. 2014).
- A 2002 European Commission (EC) report (European Commission 2002) describes discards: commercially targeted varieties that are kept by fishing gear that were hauled on-board a vessel and discarded at sea, excluding non-commercial varieties (Kelleher 2005).

Horsten and Kirkegaard (2002) suggest a model to evaluate bycatch and apply subsequent management measures, as shown in the below diagram (Fig. 14.2), in accordance with Hall's (1994) framework for classifying bycatch.

Bycatch was extremely variable across years. In 2015 and 2016, bycatch represented > 50% of the overall catch, but in subsequent years, it has typically been 25% or less of the overall catch (SIOFA Secretariat 2024). According to an estimate mainly based on data from trawl fisheries, it accounts for roughly 40.4% of the total global marine catch, which is estimated at 95 million tonnes on an annual basis (Davies et al. 2009). Over the past two decades, estimates of global discards have dropped significantly—from 27 million tonnes (Alverson et al. 1994), to 7.3 million tonnes (Kelleher 2005), and then to 9.1 million tonnes (Pérez Roda et al. 2019). According

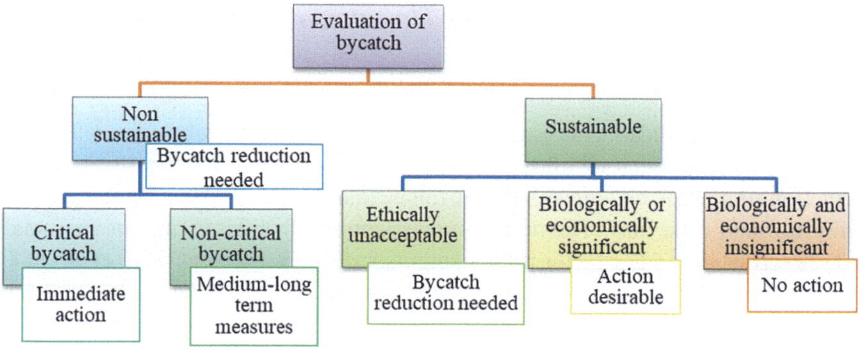

Fig. 14.2 A model to evaluate bycatch and apply subsequent management measure based on different categories of bycatch (*Source* Author)

to FAO, the gear type which contributed the most to global fisheries discards and discard rate per annum was bottom trawls (4.2 million t) followed by boat seine (0.48 million t), dredges (0.2 million t), gill net mainly bottom gillnets (0.80 million t), bottom longlines (0.36 million t), purse seines (1.02 million t) and mid water trawls (0.9 million t) (Pérez Roda et al. 2019).

Fish that were commonly discarded in the past have steadily gained significance as bycatch that is now retained and landed (Pérez Roda et al. 2019). FAO estimates indicated the volume of fish discards has declined gradually over the years (Kelleher 2005; Zeller and Pauly 2005). A key contributor in this reduction is the rising utilization of formerly discarded non-commercial fish bycatch, which is now being processed for aquaculture and livestock feed, used as fertilizer on farms, and utilize for human consumption driven by technological advancement in processing industries and growing market demand (Simon et al. 2005; Nunoo et al. 2009).

Detecting bycatch is challenging unless fishing vessels have scientific observers onboard. Fishing operations generate a substantial amount of bycatch, whether done commercially or on a small artisanal or large industrial scale basis. According to Sultana et al. (2014), the primary causes for bycatch landings and discards include inadequate management methods, modern fishing gear practices, and a lack of enforcement of fisheries policies. Addressing bycatch and creating effective mitigation plans is an intricate ecological issue that involves technological, socio-cultural, and socio-economic factors (Komoroske and Lewison 2015). Recently, research on bycatch has expanded significantly, leading to a deeper understanding of the species impacted, the scale of the problem, and its geographical extent.

14.2.1 Bycatch and Discards of Threatened, Endangered and Protected Species

Bycatch poses significant concerns threatening marine biodiversity, especially when it involves species with long lifespans and slow reproductive rates, mostly at a risk of extinction or already endangered that involves marine megafauna such as seabirds, sea turtles and cetaceans. The unintentional capture of seabirds, marine mammals and sea turtles is a major factor influencing fisheries management (Hall et al. 2000; Lewison et al. 2004). Incidental capture of various megafauna under the categories of nearly threatened or endangered species such as devil ray, tigersharks, sword fishes, mammal, and bowmouth guitarfish landed as bycatch can be seen in Fig. 14.3. Each year, 20 million protected, endangered, and vulnerable species—such as sharks (10 million), sea turtles (8.5 million), seabirds (1 million), sea snakes (225,000), and mammals (650,000)—are caught and dumped worldwide as part of global capture fisheries (Gray and Kennelly 2018).

An abundance of several marine mammal species has drastically decreased as a result of fishing bycatch mainly due to entanglement in fishing gear (Wade et al. 2021) and this threat is largely or entirely responsible for the Critically Endangered status and near extinction of the vaquita (*Phocoena sinus*) (Rojas-Bracho and Taylor

Fig. 14.3 Incidental capture of devil ray, tiger sharks, sword fishes, mammal, and guitarfish landed as bycatch (*Source* Author)

2017). The danger of different types of fishing gear to marine mammals varies. Each variety (trawl net, trammel, anchored, drift, set gill net, purse seine) pose significant risk for all population of marine mammals. Large whales (e.g., right, humpback), bottlenose dolphins and pinnipeds are seriously threatened by buoy lines of pot/traps and longlines (Wade et al. 2021). Dredges, handlines, jigs, pound nets, cast and ring nets, hook and line, and demersal seines are other low-risk gear examples, but some reports are there of interactions of sea lions, dolphins, harbor porpoises, and seals (Wade et al. 2021) that resulted in entanglement and trapping.

Bycatch of long-lived marine megafauna has significant immediate and secondary environmental impacts (Lewison et al. 2004), including a decline in megafauna stocks (Heppell et al. 2000) and alterations in ocean trophic cascades (Estes et al. 2011; McCauley et al. 2015). This issue, encompassing elasmobranchs, seabirds, marine turtles and marine mammals poses an international resource management concern. As these species traverse vast oceanic regions, it requires coordinated international efforts and policy measures to effectively tackle large scale bycatch (Komoroske and Lewison 2015).

Bycatch of regulated or protected species typically prohibited to be brought ashore or traded. However, the National Bycatch Reduction Strategy in 2016 promotes "increased utilization" of fishes which are legally caught thrown out due to its low market price. Identifying strategies to utilize this legal provision can reduce overall bycatch and benefits fishing industry financially (NOAA 2016). Increased utilization focuses on building markets and offering subsidies in order to decrease economic discards and enhance the amount of capture that is landed and traded. Lowest discard rates are observed in the western central Pacific, eastern Indian and southeast Pacific Oceans (Pérez Roda et al. 2019). The International Guidelines on Bycatch Management and Reduction of Discards (FAO 2011) advised member countries to assess and minimize the mortality effects of ghost fishing and pre-catch losses, as well as to enhance the longevity rates of discarded species (Pérez Roda et al. 2019).

14.3 Bycatch: Indian Coast

According to Kelleher's (2004) estimation that the bycatch discards in Indian fisheries were 58,000 tons, accounting for approximately 2% of the overall landings. Kumar and Deepthi (2006) examined the issue of bycatch in India and explored various strategies for mitigating this While Najmudeen and Sathiadhas (2008) estimated that juvenile fishing along the Indian coast results in an annual economic loss of approximately US$ 19.445 billion. Bycatch is a normative issue in tropical nations like India with diverse fisheries and social groups, prompting concerns regarding its effects for equity and justice in fisheries (Lobo 2012). In early decades, unreported catches made up 45% of total catches, declining to 25% by late 2010s. Discarding reached a peak 11 lakh tonnes in 1993 and declined to approximately 600,000 tonnes annually by the late 2010s, with the industrial sector accounting for 94% of total discards (Zeller et al. 2023). Dineshbabu et al. (2022) reported 80% of trawl catches from

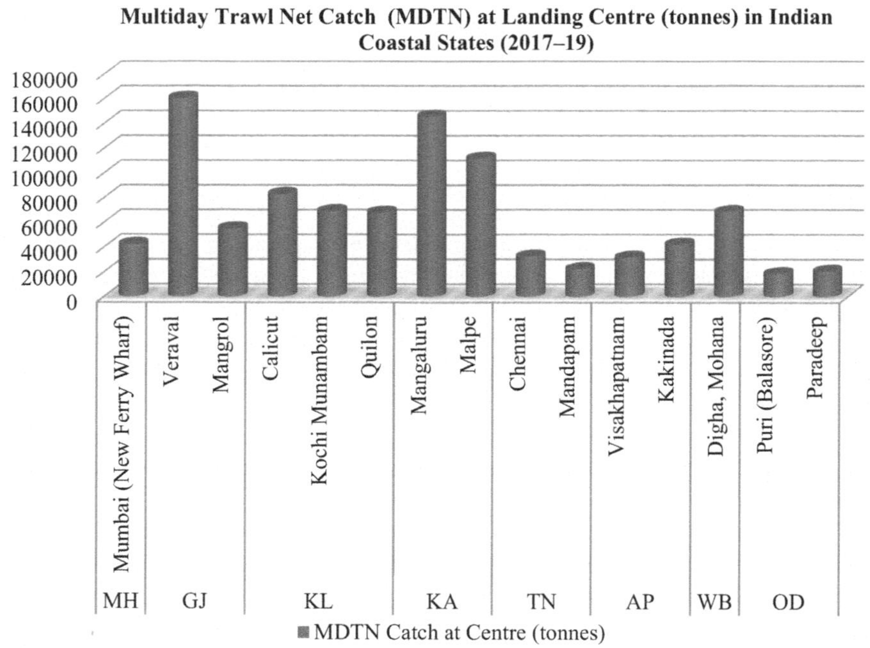

Fig. 14.4 Multiday Trawl Net Catch (MDTN) in Indian Coastal States (2017–19) (*Source* Dineshbabu et al. 2022)

multiday trawl net catch (MDTN) along 30–60% low Value Bycatch (LVB) from the different landing centres of Indian coastal states during 2017 to 2019 demonstrated in Figs. 14.4 and 14.5 (Dineshbabu et al. 2022). Findings from the previous studies of bycatch and trash fishes caught with target varieties from different trawlers in Indian waters were presented in Table 14.1.

Bycatch species composition in India consists various varieties such as Sciaenids, Bombay duck, Ribbon fish, Anchovy, Yellow goatfish, Croaker, Clupeidae, Carangid, Gobiidae, Russell scad, Soles, Shad, Drift fish, Puffer fish, Squilla, Squid, Cuttle fish, Crab, Shrimp, Paste shrimp, Red prawn, Lizardfish, Silver belly, Threadfin bream, Pink perch, Acetes, Solenocera (Sultana et al. 2014; Abdul Azeez et al. 2021). Figure 14.6 shows the sorting of various bycatch species caught along with the target species from Indian waters. Also, previous reports of bycatch species composition of Indian coastal states were depicted in Table 14.2.

When bycatch results in discards or the waste of a resource, it is deemed undesirable, or if it causes mortalities of endangered or threatened species, it heightens the risk to the stock or the ecosystem (Horsten and Kirkegaard 2002).

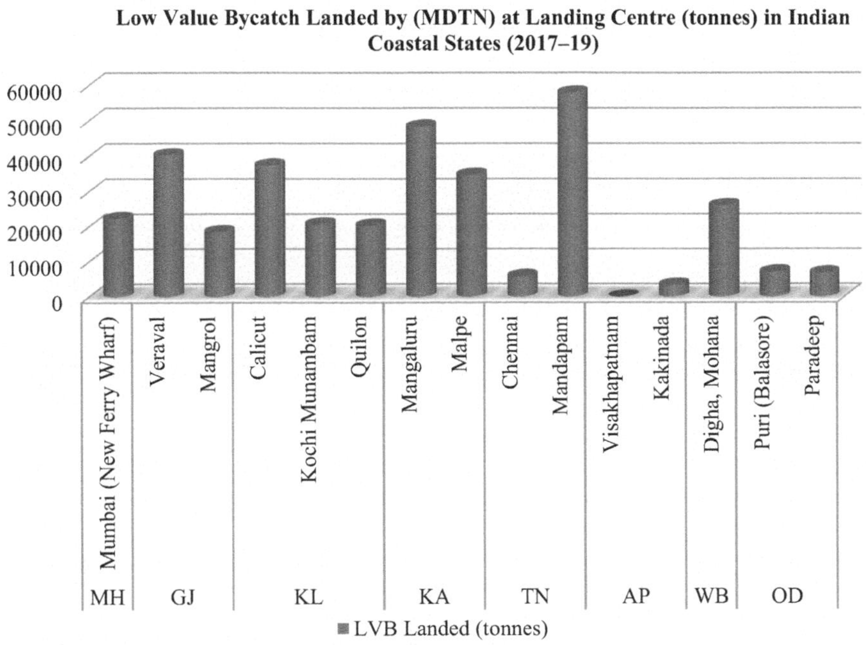

Fig. 14.5 Low-Value Bycatch (LVB) in Indian Coastal States (2017–19) (Dineshbabu et al. 2022)

14.4 Can Bycatch Ever Be Beneficial?

Bycatch isn't always a negative aspect; in many cases, the incidental capture of valuable commercial species can actually be beneficial for fishers, turning otherwise unprofitable trips into successful ones (Clucas and Teutsher 1999). In economically developing nations, bycatch from shrimp trawlers and commercial fishing boats can be transferred to smaller local boats, where they subsequently sell it domestically, providing important protein and enhancing local job opportunities (Clucas and James 1997). Additionally, discards can serve as a crucial food source for species such as seabirds, and reducing discards might adversely affect these populations (Gislason 1994). The main challenge is not in the bycatch itself but in effectively managing and controlling it to ensure sustainable harvesting while protecting sensitive species.

The ecological importance of low-value or so-called trash fish is often underestimated due to their minimal economic worth. They play an important role in marine ecosystems as they serve as meal to various fish and marine animals (Simon et al. 2005). Large-scale removal of them from the environment disrupts the food chain, potentially causing declines in larger fish species beyond their own. Moreover, using demersal fishing gear that harms habitats exacerbates the overall ecological impact. Thus, their conservation is vital not only for their own species but also for maintaining the balance and health of marine environments (Simon et al. 2005).

Table 14.1 Key findings on bycatch in various trawl fishing studies in Indian waters

Bycatch findings	Study and Year
Observed that the target to bycatch ratios in the south-west and south-east parts of India ranged from 1: 4.6 and 1: 1.3	Menon et al. (2000)
Revealed that in 2003, trawlers operating along the Indian coast contributed 10–20% from trash fish of the total landed catch	Jayaraman (2004)
Reported bycatch produced via shrimp trawling across the coast of central Kerala	Boopendranath et al. (2008)
Estimated a total of 283 species as a component of trawl bycatch near central Kerala	Gibinkumar (2008)
Showed that the quantity of juveniles exploited by trawl varies: 20–60% for seer fishes and groupers, and 12% for squids in Gujarat	Mohamed et al. (2009)
Documented 1.2 million tonnes discards from mechanised trawlers (Indian EEZ). Also, bycatch constitute of the total catch 56.3% from shrimp trawlers	Pramod (2010)
Revealed that low-value bycatch (LVB) landed in trawl fisheries rise from 14% in 2008 to 25% in 2011	Dineshbabu et al. (2013)
Found that 89% (174 sp.) were bycatch, with 36% of the species being discarded and 53% being landed based on their size with respect to 11% (22 sp.) of target varieties off Goa	Velip and Rivonker (2015)
Noted a 21% of low value bycatch landed by multiday trawlers and 38.2% by single day trawler in trawl fisheries of Karnataka Coast in 2012–14	Mahesh et al. (2017)
Recorded bycatch primarily composed of sciaenids (35%), subsequently sharks and rays (10%), anchovies (10%), shrimps (8%), Bombay duck (6%) from a single-day shrimp trawling with overall monthly bycatch ranging from 11.82 to 20.65 kg/h of Mumbai coast in 2015–16	Samanta et al. (2018)
Found that juveniles of 20 species of Andhra Pradesh (Visakhapatnam) and Odisha coasts trawl fisheries constituted 12,757.16 tonnes and 286.86 million individuals annually, contributing 55.30% by weight and 57.03% by numbers to total landings in trawl bycatch	Behera et al. (2021)
Calculated that bycatch accounted for approximately 53.36% from mid-water trawlers of total catch operating off Gujarat coast	Abdul Azeez et al. (2021)
Studied the effect of trawling and its bycatch and reported trash fish landings constituted 20.5% of overall trawl catches in 2010–12, with an average ratio of 1:4.07	Panda et al. (2022)
Reported that 30–60% of trawl landing in India was comprised by LVB during 2017–19, mainly used for fishmeal preparation	Dineshbabu et al. (2022)
Reported a total of 184 species landed as bycatch from the trawlers operated off Thoothukudi Coast, Gulf of Mannar	Santhoshkumar et al. (2023)

Fig. 14.6 Sorting of various bycatch species caught along target species (*Source* Author)

There are different ways to reduce loss and waste (bycatch) (FAO 2025);

I. **Policy**

In 2011, the Committee on Fisheries approved the International Guidelines on Bycatch Management and Reduction of Discards, which are voluntary guidelines that help States and regional fisheries management organizations implement the ecosystem approach to fisheries. They can be used as a guide to develop appropriate bycatch management and discard reduction strategies in all fisheries and regions.

II. **Appropriate Technology**

Technology associated with discards, bycatch, and reducing food loss and waste (FLW) comprise data collection, bycatch usage, and selective fishing gear. There are two ways to add value to fish bycatch: mass transformation and sorting.

(a) **Mass transformation** is the process of converting all discards into a single good like hydrolysates, fertilizer, fish oil, and fish meal.
(b) **Sorting** is the process of separating various species or waste materials (such as bones, guts, and fins) for sale or the creation of exclusive products. For example, omega 3, protein, sports foods and beverages, gelatin, liver oil, calcium, cosmetics, biotechnological uses, and medications.

Table 14.2 Bycatch species composition of Indian costal states

State	Species composition in the bycatch	Study and year
Gujarat	Bullseye, engraulids, lizardfishes, mackerels, ribbon fish, threadfin breams, sardines, sciaenids, sharks, penaeid and non-penaeid prawns, cuttlefish, squids, Octopus, crabs, lobster (*Panuliruspolyphagus*), jelly fish, squilla	Zynudheen et al. (2004) Abdul Azeez et al. (2024)
Karnataka	*Decapterusrusselli, Dussumieriaacuta, Encrasicholinadevisi, Epinephelusdiacanthus, Lagocephalusinermis, Mene maculate, Muraenesox sp., Nemipterusrandalli, Platycephalusindicus, Rastrelligerkanagurta, Sauridatumbil, Sardinellalongiceps, Trichiuruslepturus, Uranoscopus spp.*	Mahesh et al. (2019)
Maharashtra	Anchovies, bombay ducks, catfishes, eels, flat fishes, goby, pomfrets, puffer fish, ribbon fish, rays, sharks, sciaenids, shrimps and prawns, lobster, crabs, hermit crab, mantis shrimps, molluscsshells, jellyfish, cephalopods	Samanta et al. (2018) Wanjari and Ramteke (2024)
Andhra Pradesh	Apogonidae, carangidae, dasyatidae, engraulidae, fetraodontidae, leiognathidae, loliginidae, mullidae, nemipteridae, penaeidae, portunidae, priacanthidae, sciaenidae, scombridae, sepiidae, squillidae, synodontidae *Ariomma indica, Johniuscarutta, Lepturacanthussavala, Nemipterusrandalli, Nemipterus japonicas, Nibea maculate, Pennahiaanea, Platycephalusindicus, Priacanthushamrur, Rastrelligerkanaguta, Sauridatumbil, Sauridaundosquamis, Trichiuruslepturus, Upeneusmoluccensis, Upeneus sulphurous, Upeneusvittatus, Sepia aculeata, Sepia pharaonis, Sepiellainermis,, Uroteuthisduvauceli*	Behera et al. (2017) Behera et al. (2021)
Kerala	Fin fishes, soles, juvenile shrimps, crabs, stomatopods, cephalopods, gastropods, jellyfishes, eggs, echinoderms, snakes	Kurup et al. (2003)
Goa	Catfishes, carangids, clupeoids, eels, elasmobranchs, flatheads, gobies, leiognaths, nemipterids, pleuronectiforms, puffer fishes, sauries, sciaenids, shrimps, crabs, stomatopods, gastropods, jellyfish, echinoderms, cnidarian and reptiles	Velip and Rivonker (2015)

(continued)

Table 14.2 (continued)

State	Species composition in the bycatch	Study and year
Tamilnadu (Chennai, Cuddalore and Parangipettai coastal waters)	*Equuliteslineolatus, Gazzaachlamys, Karalla dussumieri, Leiognathusequulus, Nibeamaculata&Otolithesruber* Electric rays, stingrays and butterflyrays, sardines, shads and sprats, herrings, anchovies and whitebaits, conger eels, sea cat fishes, lizard fishes, flying fishes, Halfbeaks, cornet fishes, pipe fishes and seahorses, scorpion fishes, flatheads, sea basses groupers, grunters, jacks and pampanos, pony fishes, snappers, threadfin breams, drummer and croakers, bigeyes or Catalufas, goat fishes, wrasses, scats, barracudas, surgeon fishes, burrowing gobies, rabbit fishes, ribbonfishes, mackerels and tunas, butter fishes, mullets, eyes sinistral, flounders, soles, tongue soles, filefishes, boxfishes, puffers, porcupinefishes	Kodeeswaran et al. (2020) Sambandamoorthyet al. (2015)
West Bengal	*Brachygobiusnunus, Butishumeralis, Coiliadussumieri, Coiliareynaldi (Anchovy), Gobiopterusbrachypterus, Harpadonnehereus, Liza macrolepis (Large-scale Mullets), Megalopscyprinoides (Indo—Pacific Terapon), Oryziusdancena, Parapocryptesserperaste, Plotosuscanius (Gray eel Cat Fish), Pseudapocrypteselongatus (Pointed tailed Goby), Teraponjarbua (Terapon), TetraodonfluviatilisScartelaoshistophorus (Goby), Stigmatogobiussadanundio, Stolephorusdubiosus, Zenarchopterus sp.,* Sciaenids, clupeids, elasmobranchs, angler fish, puffer fish, tripod fish, juveniles of fish, crabs, shrimp, stomatopods, mollucsc, gastropods, bivalves and octopus	Chakraborty et al. (2021) Jana et al. (2024)
Odisha	*Coiliadussumieri, Cynoglossusarel, Megalaspiscordyla, Opisthopetrustardoore, Stolephoruscommersonii, Trichiuruslepturus, Thryssa spp.*	Dineshbabu et al. (2022)

III. **Skills and Knowledge**

Reducing food loss and waste requires knowledge and skills including being aware of the reasons of discards, fisheries management, bycatch control methods, legal compliance, food safety maintenance, hygiene and sanitation, waste storage and disposal.

IV. **Services and Infrastructure**

Fishing gear manufacturers, suppliers, producer enterprises, fisheries management agencies, inspection facilities, fish meal production companies, and research and development activities are among the infrastructure and services for reducing discard, using bycatch, and eliminating food loss and waste.

V. **Regulatory Environment**

Decisions on quotas (Total allowable catch) and other aspects of fisheries management might affect food waste and loss. In order to control bycatch and minimize discards, states and regional fisheries management organizations should implement measures including input/output controls (e.g., licence limitation, minimum mesh size, and minimum legal size), improved fishing gear design, spatial measures, bycatch limits or quotas, discard bans, and incentives for fishermen to follow these regulations.

VI. **Social and Gender Equity**

Low-value bycatch and discard markets are directly related to the socio-economics of the fish trade, particularly in developing nations, offering crucial trade and processing opportunities for women.

VII. **Markets**

Fishing industry can increase value by establishing markets for edible species, minimizing discards, and increasing interest to the bycatch utilization. EU regulations oblige fishers to record their catches and avoid discarding, while buyers must source responsibly and ensure fish is not illegal, unreported, or unregulated.

Bycatch markets are growing due to poverty, population growth, and growing pressure on target species, making discard reduction financially viable. Because of seasonal changes and declining profitability, shrimp trawlers are interested in creating bycatch marketplaces. Many species in the Asia-Pacific region have gelling properties, which is used to produce surimi and other items include fish noodles,

flakes, pickles, satay, jelly, and fish mince e.g., Through the Fishing for the Markets initiative, the UK's Department for Environment, Food and Rural Affairs (DEFRA) is working with the retail industry to promote underutilised species.

14.5 Challenges in Bycatch Utilization

Bycatch holds significant importance for developing value-added products for non-edible applications and human consumption through efficient utilization (Aswathy et al. 2012). Bycatch and discards pose significant threats to sustainable fisheries by causing avoidable deaths. According to FAO's 1995Code of Conduct for Responsible Fisheries, states should increase their utilization of bycatch as far as it aligns with responsible fisheries management. Bycatch utilization encounters different legal and regulatory frameworks with various rules and regulation & legislation for different region or countries as shown in Table 14.3.

Table 14.3 Bycatch legislation (examples)

Nation	Rule or regulation	Major strategies	References
USA	Magnuson Act (the Fisheries Conservation and Management Act of 1976, Public Law 94–265, 90 Stat. 331, 16 US Code 1801–1882)	Prioritizes bycatch minimization as a top priority for conservation and management strategies (Sect. 106.9)	Pascoe (1997)
European Union (EU)	European Union (EU) Common Fisheries Policy (CFP) (Article 15 of REGULATION (EU) No 1380/2013)	Ban on discarding/no-discard regimes	Pérez Roda et al. (2019)
Norway	A policy of 'No discards'	Prohibited discards by law and fishermen are obliged to bring all their catch ashore	Olsen (1995) Isaksen (1997)
Canada	Canadian Code of Conduct for Responsible Fishing Operations, Consensus Code 1998)	Own versions of a Code of Conduct	DFO (1998)
Thailand	Thai Fisheries Act B.E. 2490 (1947), Sect. 32 (Trawl fisheries & bycatch management)	No direct action but Department Of Fisheries has lowers number of push net fishing boats for couple of years (1997–2003) result in fewer trash fish	Supongpan and Boonchuwong (2010)
Australia and New Zealand	–	Quota management systems	Baulch and Pascoe (1992)
	FAO International Guidelines on Bycatch Management and Reduction of Discards	Assist states and regional fisheries management organizations and arrangements RFMO/As in managing bycatch and discards in fisheries around the world by enforcing the proper measures	FAO (2011)
	FAO Code of Conduct for responsible fisheries	Aims to minimize discards, simplify fish trade laws, effective implementation of regulations and administrative processes for fish trade; promoting responsible fisheries management practices and minimizing waste and post-harvest losses	FAO (1995)
	Kyoto Convention International Conference on the sustainability contribution of fisheries to Food security (Dec 1995)	Aims to reduce waste and discards by developing selective, environmentally safe, and cost-effective fishing gear and techniques; information sharing about technologies and procedures to minimize waste and discards	Clucas (1997)

(continued)

Table 14.3 (continued)

Nation	Rule or regulation	Major strategies	References
	FAO International Plan of Action for Reducing Incidental Catch of Seabirds in Longline Fisheries	–	FAO (1999)
	FAO International Plan of Action for the Conservation and Management of Sharks	–	FAO (1999)
	FAO Guidelines to Reduce Sea Turtle Mortality in Fishing Operations	–	FAO (2009a, b)

14.5.1 Legal and Regulatory Frameworks

14.5.2 Socioeconomic Factors

Bycatch carries ethical implications because it involves the unintended discarding of significant amounts of valuable fish protein (Kelleher 2005).

Challenges encountered during the marketing and utilization of bycatch (Aswathy et al. 2012);

- Fish's perishable nature and inadequate supply chain mechanisms: Insufficient infrastructure and basic amenities, Monopoly Impact on Producers and Consumers.
- Limited consumer demand and fluctuating prices: Lower and middle income groups, demand fluctuate with seasonality, availability, and high cost of valuable varieties.
- Interstate traders competition leads to forced sales in retail market
- Post-harvest losses: inadequate ice for preservation and poor hygienic in handling
- Inadequate techniques to prevent post-harvest spoilage: Preservation of low-value fish is often carried out in unhygienic conditions leads to reduced preference of consumer
- Absence of value added products: Bycatch varies in quantity and species mix, leads to inconsistent texture and taste in products. Also marketing novel products and winning over customers poses challenges for producing goods with value-addition.
- High costs are incurred for storing and transporting bycatch to the coast in a healthy state through multiday fishing operations.
- Prices fluctuation in non-edible applications: based on necessities of fish meal in livestock and fish culture and by the affordability of raw ingredients for fertilizer orfish feed.

14.6 Innovative Approaches to Bycatch Utilization

Bycatch management and discard reduction strategies include a range of approaches, such as bycatch limits, effort restriction, improvements in fishing equipment or techniques, implementing spatial and temporal device limitations, enforcing discard ban (landing requirement) (Pérez Roda et al. 2019). Furthermore, discard reduction may also achieved through better fleet communication, raising awareness, training, efficient utilization of bycatch, and providing financial benefits (Pérez Roda et al. 2019). Here we are going to discuss about the one of the approach bycatch utilization.

Nowadays, bycatch is widely used in America, Africa, and Asia, minimising discards (Pérez Roda et al. 2019). In terms of quantity, Asia leads the world in both the aquaculture and capture fishing sectors. Bycatch makes up 13–90% of total landings in Asia, based on current estimates determined by region, fishing gear, and fishing season (Sultana et al. 2014). Asian bycatch is known for its abundant biodiversity, comprising a wide variety of fish species ranging from 30 to over 100, as well as shrimp, crab, squid, cuttlefish, paste shrimp, etc. Although bycatch has a high value, losses has been kept to a minimum by using it as much as possible for animal feed and human consumption (Sultana et al. 2014). Table 14.4 represents the bycatch utilisation in Asia.

South East Asia has been most successful in using bycatch and discard, and due to the market-driven demand former bycatch has now become the target species (Ordoñez 1985; Suwanrangsi 1986; James 1998).

Manufacturing and marketing value-added and pharmaceutical-grade marine goods from low-cost fish can provide a promising opportunity to earn premium prices in both domestic and international markets (Aswathy et al. 2012; Kumar et al. 2023). Asian nations produce a variety of fish products from bycatch, ranging from traditional to high-value products. These products are processed using a range of techniques, from low-tech techniques employed at home or in small businesses to

Table 14.4 Bycatch utilization in Asia (*Source* Sultana et al. 2014)

East Asia	South East Asia	South Asia
Bycatch utilized as		
Feed for livestock and aquaculture	Innovative and Value-added goods	Animal feed, conventional products, and fresh fish (depending on bycatch sp. availability) • Small pelagic fish, such as croakers, sardines, and scads, are dried, salted, and sold as an inexpensive protein source • Low-gel strength minced fish meat is utilised in local products, typically as a binder in rolls, deep-fried fish cakes and other dishes

Fig. 14.7 **a** Non target catches utilised for surimi production (ribbon fish), dried or salted, fish mince or (*Source* Author), **b** Fish meal production (Bulle's eye, lizard fish, groupers, eels, puffer fish, etc.) (*Source* Author)

high-tech techniques used for mass production (Sultana et al. 2014). Various non-target catches such as small juveniles of ribbon fishes [Fig. 14.7a] utilised for surimi production and Bulle's eye, lizard fish, groupers, eels, puffer fish, etc. utilised dried or salted, as fish mince or fish meal production shown in Fig. 14.7b. The products made in many nations are significantly influenced by their ancestral, religious, societal, and cultural beliefs. These products fall into the following general categories: (i) hydrolysed, (ii) minced, (iii) dehydrated, (iv) cured, (v) fermented foods high in protein, (vi) canned and (vii) other items (Sultana et al. 2014).

14.6.1 Utilization as Food/quality Protein/animal and aquaculture Feeds

Bycatch could be used for food and increase food security (Clucas 1997). It is an excellent raw ingredient for making minced fish that may be exported and turned into a variety of novel products such as fish flakes hydrolysates, fish soup powder, moulded products etc. (Gopakumar 1998). Its utilization minimizes waste; reduce environmental impact and maximizing the economy of local markets and supports livelihood of fishermen. Previous works of manufacturing of novel products by various authors from non-target species in India were presented in Table 14.5.

14.6.2 Increased Bycatch Retention and Utilization

Nowadays, a number of fish species that were once considered as deemed bycatch are being targeted more frequently. However, it remains uncertain how much of the

Table 14.5 Novel products manufactured from Non-target fish species at India

Products	Varity of Fish Species	References
Fresh fish (human consumption)	Commercially profitable fish and shellfish in the bycatch (Bigeyes, catfish, carangids, eels, flatheads, goatfish, lizardfish, sardines, sharks, sciaenids, silverbellies, sole threadfin breams, wolf-herrings)	Zynudheen et al. (2004) Doddamani et al. (2022)
Dried fish & fish powder (sun-dried or salt-dried)	Anchovy *(Stolephorous spp.)* carangids, cluepeids, croakers, Bombay-duck, driftfish, gobids, Lactarius, lizard fishes, mackerels, ribbonfish, sardines, soles	Salagrama (1998), Susheela (2022), Bijukumar and Deepthi (2006)
Cutlets, wafers, spirals, soup powder	Fresh *Acetes indicus*	Zynudheen et al. (1998)
Protein powder (flavoring agent)	*Acetes* sp.	Garg et al. (1977)
Fish balls, crackers and burgers	Sciaenids, engraulids and carangids	Yu (1993)
Fish meal (Grade A, B & C)	low-value bycatch (LVB)/ Trash fishes (*Acetes* sp., carangids, clupeids, driftfish, pufferfish, Sardine, Silverbellies, small crabs, *Solenocera* sp., soles)	George et al. (1981), Salagrama (1998), Simon et al. (2005), Zynudheen et al. (2004), Chandrapal (2005) Dineshbabu et al. (2022)
Fish oil	Low value fishes	Simon et al. (2005), Hulkoti et al. (2024), Dineshbabu et al. (2013), Ponnusamy et al. (2012)
Fish Manure/fertilizer	Trash fishes, Non-penaeid shrimps, squilla and non-edible crabs	George et al. (1981), Zynudheen et al. (2004), Dineshbabu et al. (2013),
Fish paste, fish sausages, fish pappads, fish wafers, fish spirals, fish save, fish diamondcuts, fish jam, fish noodles and canned fish paste fish silage (poultry and animal feed	Cheaper bycatch fish species (Bombay duck Lizard fish Threadfin breams Goat fishes, silverbellies, flat fishes, stomatopods)	Bijukumar and Deepthi (2006), Aswathy et al. (2012), Dineshbabu et al. (2022)
Bacteriological peptone	Threadfin bream	George et al. (1981)

(continued)

Table 14.5 (continued)

Products	Varity of Fish Species	References
Fish protein hydrolysates	Anchovies, Barracuda, flat head, Jew fish, Lizard Fish, Ribbon fish, Threadfin bream, soles	Gopakumar (1998)
Surimi	Low value fish Threadfin bream, triggerfishes (*Odonus niger*)	Hulkoti et al. (2024), Dineshbabu and Radhakrishnan (2009) Dineshbabu et al. (2022)

increment in international marine landings is attributable to the rising in catches of these formerly discarded species (Kelleher 2005).

- Fisheries in Southeast Asia generally have minimal discard rates, with rising utilization driven by higher demand for aqua farm feed and advances in development of novel products.
- Anti-discard rules and better implementation of regulations, and growing bycatch utilization for human intake (generally processed or preserved) or as animal feed, as a result of advanced processing technology and growing market demands (Madhu 2022).
- In African industrial trawl fishery, there is a growing trend of marketing previously discarded demersal finfish, especially in urban African markets.
- Additionally, factory vessels are enhancing at-sea processing, manufacturing surimi and related goods, though this could lead to rising disposal of carcasses and processing waste that aren't categorized as discards.

Numerous variables are associated with the growing utilization of bycatch (Kelleher 2005):

- Rising populations and incomes are increasing demand and prices of fish products, especially across developing nations.
- Trash fish or discards is increasingly being used as animal feed and for aquaculture especially in South and Southeast Asia.
- Advances in technology are enabling the use of small-sized fish from various varieties to create novel products like surimi.
- Emerging market channels for previously discarded or unfamiliar varieties, like deep-water sharks, alongside lower supply and higher costs of desired varieties, are shifting demand.
- Decline in catches of target varieties, potentially resulting from overexploitation, are leading to more capture of non-target varieties or trash fish.
- Smaller fishing voyages aimed at improving the quality of fish are creating additional hold space suitable for bycatch use.
- Harvesting of bycatch at sea is increasing, especially among tropical shrimp trawl operations in Central and South America and Africa.

- Modifications in management practices are encouraging or requiring the landing or at-sea harvesting of bycatch.
- Regulatory advances, such as lowering minimum landing sizes to match trawl mesh sizes and allowing the trading of quotas between vessels or fishers are impacting bycatch management.
- Economic incentives are driving efforts to maximize returns from catches.

14.7 Technological Advancements for Bycatch Utilization

Technology has not only contributed to the emergence of various issues but also serves as a significant source of potential solutions (Prado 1997; Alverson 1998).

There is hardly any technological advancement for utilizing bycatch. An Asian country primarily relies on basic, traditional, and economical methodologies with the exception of Japan, where technological advancements have allowed for the creation of novel fishery products using high-value fish and advanced equipment's even though the amount of bycatch is relatively small (Sultana et al. 2014; Rani et al. 2025; Nisha et al. 2025).

14.7.1 Innovations in Processing and Utilization

An advancement of technologies in processing industries and developing market potential for trash fish or bycatch varieties has grown the utilization of bycatch as animal food and for human consumption. A "no-discards" strategy is a crucial component of fisheries management, and the necessity of striking a harmony among bycatch utilization and minimisation has significantly decreased fishery losses and, consequently, the losses to fishing communities (Sultana et al. 2014).

Recent declination in bycatch is primarily responsible for the introduction of more selective fishing gear, implementation of bycatch and discard legislation, and enhanced compliance of laws and policies (Sultana et al. 2014; Kiran et al. 2025). Development of new technologies and look into ways to modify fishing methods developed to reduce bycatch of fish (such as deep-sea corals, shallow tropical corals and sponges) and endangered species (like sea turtles, seabirds, and marine mammals), and also reduce bycatch mortality and injury (that involves mortality and post-release injury) (Sultana et al. 2014). This approach supports sustainable food supply over a longer period and helps reduce poverty, particularly in fishing communities and small island developing states where fish are a primary food source, fisheries provide essential employment, and fishing is integral to their way of life by utilising effective management efforts crucial for sustainable fisheries (Pérez Roda et al. 2019).

Recent advancements that utilize telemetry and satellite data to address dynamic circumstances and bycatch risks show potential for reducing bycatch in the context of sustainable fisheries (e.g., Turtle Watch) (Howell et al. 2008; Howell et al. 2015). Social platforms and community-orientated organizations or authorities have proven to be highly effective in mitigating bycatch (Hall et al. 2007; Peckham and Maldonado-Diaz 2012). Establishing these organizational structures can enhance collaborations and knowledge and expertise sharing among fishermen, managers, and scientists, fostering a collaborative environment where every interested party is involved in the development, assessment, remarks, and execution of mitigation techniques. This approach improves fishermen's comprehension of the reasons behind fishery laws and policies and integrates regional insights and community necessities into sustainable management actions (Campbell and Cornwell 2008; Alfaro-Shigueto et al. 2010; Peckham and Maldonado-Diaz 2012; Piovano et al. 2012).

14.8 Conclusions

Nowadays, bycatch is one of the biggest concerns influencing the fisheries management. The sustainable utilisation of bycatch presents a crucial incentive to strengthening fishery' environmental and economic viability. We may decrease waste, alleviate environmental effect, and promote responsible fishing practices by converting previously deemed bycatch and discards into useful resources such as animal feed, fertilisers, novel products, and even pharmaceuticals also provides nutritional security. Integrating innovation and regulatory frameworks that encourage bycatch use not only conserve marine biodiversity but also promotes the livelihoods of coastal communities that rely on fisheries. Moving forward, stakeholder engagement, research into new technologies and policy enforcement will be crucial for realising the full potential of bycatch as a sustainable resource.

14.9 Future Prospects

Bycatch is an inherent aspect of multi-species fishing affecting numerous unintentionally captured species in addition to targeted ones. One effective strategy to mitigate the negative impacts of bycatch is to enhance its utilization. The growing acceptance of bycatch as edible fish can alleviate pressure on excessively harvested target catch such as fish and shrimp species. Promoting data sharing and technology transfer across regional, national, and international levels is vital for advancing product development and improving post-harvest techniques related to bycatch utilization. This highlights the need for increased research to protect fishery resources from commercial extinction while supporting biodiversity conservation. Utilizing bycatch responsibly, in line with sustainable fisheries management practices, can also aid in enhancing fish food security. A comprehensive approach that considers both economic and ecological factors is necessary to ensure that countries can benefit holistically. Continued research is crucial for developing bycatch management policies in India that focus on reducing juvenile catch, assessing the value of low-value bycatch (LVB) species, and addressing ecological concerns related to the unregulated removal of prey species from ecosystems. Establishing guidelines that align with sustainability goals, such as the Code of Conduct for Responsible Fisheries (CCRF), alongside case studies focusing on specific fisheries, can provide valuable insights for addressing discard-related challenges.

Acknowledgements Authors would like to express their sincere gratitude to the Vice chancellor of Kamdhenu University, Gandhinagar, Gujarat, the Dean & other college authorities of the college of Fisheries Science, Veraval for providing valuable guidance, academic support and necessary facilities throughout the study.

References

Abdul Azeez P, Rohit P, Shenoy L, Jaiswar AK, Koya M, Vase VK, Divu D (2024) Analysis of bycatches from mid-water trawl fishery targeting ribbonfish Trichiurus lepturus on the northwest coast of India. Indian J Fish 71(1):95–104

Abdul Azeez P, Rohit P, Shenoy L, Jaiswar AK, Raman M, Koya KM, Vase VK, Damodaran D (2021) Species composition and spatio temporal variation of bycatch from mid-water trawlers operating in the Arabian Sea along north-west coast of India. Reg Stud Mar Sci 43:101692. https://doi.org/10.1016/j.rsma.2021.101692

Alfaro-Shigueto J, Mangela JC, Pajuelo M et al (2010) Where small can have a large impact: structure and characterization of small-scale fisheries in Peru. Fish Res 106:8–17

Alverson DL (1998) Discarding practices and unobserved fishing mortality in marine fisheries: an update. Washington Sea Grant, pp 98–06

Alverson DL (1999) Some observations on the science of bycatch. marine technology society. Mar TechnolSoc J 33(2):6

Alverson DL, Freeberg MH, Murawski SA et al (1994) A global assessment of fisheries bycatch and discards. FAO Fisheries Technical Paper 339, 235 pp

Alverson DL, Hughes SE (1996) Bycatch: from emotion to effective natural resource management. In: University of Alaska Fairbanks (eds) Solving Bycatch: Considerations for today and tomorrow. Proceedings of the solving bycatch workshop, University of Alaska Sea Grant College Progress Report 96–03, pp 13–28

Aswathy N, Sathiadhas R, Narayanakumar R et al (2012) Marketing and utilization of marine by catch: Problems and prospects. J Fish Eco Dev 12(2):1–8

Baulch K, Pascoe S (1992) Bycatch management options in the south east fishery, ABARE Research Report 92.18, ABARE, Canberra

Behera PR, Ghosh S, Muktha M et al (2017) Species composition and temporal variation of trawl by-catch in fishing grounds off northern Andhra Pradesh, western Bay of Bengal. India. J Mar Sci 46(10):2037–2045

Behera PR, Ghosh S, Ramulu KS et al (2021) The cost of fishing on juveniles of finfish and shellfish: assessment of economic impacts of trawl juvenile bycatch along the coast of North Andhra Pradesh, India. Thalassas:An Int. J Mar Sci 37:409–426

Biju Kumar A, Deepthi GR (2006) Trawling and by-catch: implications on marine ecosystem. Curr Sci 90:922–931

Boopendranath MR, Pravin P, Gibinkumar TR et al (2008) Bycatch reduction devices for selective shrimp trawling. Final report on ICAR Adhoc Project (Code No.0644003), Central Institute of Fisheries Technology, Cochin, 220pp

Chakraborty S, Sarkar K, Sadukhan S et al (2021) Evaluation of bycatch discarded fishes during tiger prawn (*Penaeusmonodon*) seed collection in indian sundarbans: implication for sustainable management. Int J Aquatic Sci 12(2):4711–4727

Chandrapal GD (2005) Status of trash fish utilization and fish feed requirements in aquaculture-India. In: Regional workshop on low value and trash fish in the Asia-Pacific region. Hanoi, Vietnam 7–9 June 2005, pp 1–7

Clucas IJ (1997) A study of the options for utilization of bycatch and discards from marine capture fisheries. FAO Fisheries Circular No. 928, FAO, Rome, 59 pp

Clucas IJ, James DG (1997) Reduction of wastage in fisheries. In: Technical consultation (eds) on Tokyo. FAO Fisheries Report No. 547 (Suppl.), FAO Rome

Clucas IJ, Teutsher F (1999) Bycatch Utilization in tropical fisheries. In: Report and Proceedings of FAO/DFID expert consultation (eds), Natural Resources Institute, Beijing, China, Chatham, UK

Davies RWD, Cripps SJ, Nickson A et al (2009) Defining and estimating global marine fisheries bycatch. Mar Policy 33:661–672

DFO (1998) Canadian code of conduct for responsible fishing operations [online]. Fisheries and Oceans Canada. Ottawa, Ontario, Canada. [Cited April 09 2025] https://www.dfo-mpo.gc.ca/about-notre-sujet/publications/fisheries-peche-eng.html

Dineshbabu AP, Radhakrishnan EV (2009) Trawl fishery of juvenile fishes along Mangalore-Malpe Coast of Karnataka and its impact on fish stock. Asian Fish Sci 22(2):491–500

Dineshbabu AP, Radhakrishnan EV, Thomas S (2013) Appraisal of trawl fisheriesof India with special reference on the changing trends in bycatch utilization. J Mar Biol Ass India 55(2):69–78

Dineshbabu AP, Thomas S, Josileen J et al (2022) Bycatch in Indian trawl fisheries and some suggestions for trawl bycatch mitigation. Current Sci 123(11):1372–1380

Doddamani PL, Benakappa S, Naveenkumar BT et al (2022) By-catch and its utilization, socio-economic impact of bycatch reduction off Mangalore Coast. J Exp Zool India 25(1):155–165

Estes JA, Terborgh J, Brashares JS et al (2011) Trophic downgrading of planet earth. Sci 333:301–306

FAO (1995) Code of conduct for responsible fisheries. FAO, Rome, p 41

FAO (1999) International plan of action for reducing incidental catch of seabirds in longline fisheries. International plan of action for the conservation and management of sharks. International plan of action for the management of fishing capacity. FAO, Rome, 26pp

FAO (2009) Technical guidelines for responsible fisheries. Number 1: fishing operations. Supplement 2: best practices to reduce incidental catch of seabirds in capture fisheries. FAO, Rome

FAO (2009a) Guidelines to reduce sea turtle mortality in fishing operations. FAO, Rome, p 128

FAO (2011) International guidelines on bycatch management and reduction of discards. Food and Agricultural Organization of the United Nations, Rome

FAO (2025) Discards and bycatch. Capture fisheries. In: Food Loss and Waste in Fish Value Chains [online]. Rome. [Cited April 02 2025] https://www.fao.org/flw-in-fish-value-chains/value-chain/capture-fisheries/discards-and-bycatch/en/

Funge-Smith S, Lindebo E, Staples D (2005) Asian Fisheries today-The production and use of low value trash fish from marine fisheries of the Asia Pacific region, Asia-Pacific fisher commission. In: Food and Agriculture Organization of the United Nations, RAP Publication 16, 38p

George MJ, Suseelan C, Balan K (1981) By-catch of shrimp fishery in India. Mar Fish Inf Ser t&e Ser 28:1–13

Gibinkumar TR (2008) Studies on hard bycatch reduction devices. PhD Dissertation. Cochin Univerisity of Science and Technology, Cochin

Gislason H (1994) Ecosystem effects of fishing activities in the North Sea. Mar Pollution Bulletin 29:6–12

Gopakumar K (1998) Utilization of trawl bycatches in Southeast Asia. In: Proceedings of the APFIC symposium, RAP Publication, Beijing, pp 29–48.

Gray C, Kennelly S (2018) Bycatches of endangered, threatened and protected species in marine fsheries. Rev Fish Biol Fish 28:521–541

Hall M, Nakano H, Clarke S et al (2007) Working with fishers to reduce bycatches. In: Kennelly S (ed) Bycatch Reduction in the World's Fisheries. Springer-Verlag, Dordrecht, pp 235–288

Hall MA (1994) A classification of bycatch problems and some approaches to their solutions. Workshop on bycatches. In: Pritcher TJ, Chuenpagdee R (eds), Fisheries and their impact on the ecosystem.University of British Columbia. Fisheries Centre Research Reports, 2(1): 65–74

Hall MA, Alverson DL, Metuzals KI (2000) Bycatch: problems and solutions. Mar Pollution Bulletin 41(1–6):204–219

Heppell SS, Caswell H, Crowder LB (2000) Life histories and elasticity patterns: perturbation analysis for species with minimal demographic data. Eco 81:654–665

Horsten MB, Kirkegaard E (2002) Bycatch from a perspective of sustainable use. IUCN SSC European sustainable use specialist group: fisheries working group comment on the European commission Gren paper on the common fisheries policy, 16 pp

Hulkoti SH, Somashekara SR, Sheethal KU et al (2024) A study on trawl bycatch and its utilization along Udupi coast of Karnataka. J Exp Zool India 27(1):745

Isaksen B (1997) The Norwegian approach to reduce bycatch and avoid discards. In: Clucas, James D (eds) Technical consultation of reduction of wastage in fisheries, FAO, Fisheries Report No. 547 (Suppl.), FAO, Rome, pp 89–93

James DG (1998) Fish utilization in Asia Pacific. In: Proceedings of the APFIC symposium. RAP Publication (eds), Beijing, 323pp

Jana S, Talagunda NS, Samanta R et al (2024) Quantitative assessment and temporal fluctuation of a multiday Shrimp Trawl Bycatch off Digha Coast, West Bengal, India: assessment of bycatch from a shrimp trawler off Digha coast. Fish Technol 61(3):2582–2632

Jayaraman R (2004) Overview of status and trend of 'trash fish' from marine fisheries and their utilization, with special reference to aquaculture: India. Mar Fish Inf Ser T&E Ser 149, 6 pp

Karp WA, Desfosse LL, Brooke SG (2011) U.S. national bycatch report. National Marine Fisheries Service, NOAA Technical Memo NMFS-F/SPO-117E

Kelleher K (2004) Discards in the world's marine fisheries. An update. FAO Fisheries Technical paper. No. 470. Food and Agriculture Organization of the United Nations. Rome

Kelleher K (2005) Discards in the world's marine fisheries. An update. Food and Agriculture Organization of the United Nations, Rome, 131pp

Kiran, Kumar N, Sindhu S, Gawaria B, Ansari A, Dhariwal A, Sharma P, Kumar R (2025) GWAS approaches in medicinal plants for enhanced secondary metabolite production. In: Chen JT (Ed), Biotechnology, multiple omics, and precision breeding in medicinal plants, pp 119–134. CRC Press

Kodeeswaran P, Jayakumar N, Ranjith L (2020) Assessing the ichthyofaunal diversity and trophic level from trawl bycatch of Chennai Fishing Harbour, Southeast Coast of India. Reg Stud Mar Sci 40:101530

Komoroske LM, Lewison RL (2015) Addressing fisheries bycatch in a changing world. Front Mar Sci 2:83

Kumar AB, Deepthi GR (2006) Trawling and bycatch: implications on marine ecosystem. Current Sci 90(7):922–931

Kumar R, Gupta R, Dahiya T, Yadav R, Saifi R, Yodha K, Kumari B, Singh P, Shefali (2022a) Surveillance of *Heteropneustes fossilis* (Stinging Catfish) diseases and health management practices in Haryana. Appl Ecol Environ Sci 10(4):261–267. https://doi.org/10.12691/aees-10-4-12

Kumar R, Gupta RK, Yadav R, Saifi R, Yodha K, Pooja (2022b) *Eisenia fetida* as protein source for growth enhancement of *Heteropneustes fossilis*. Egypt J Aquatic Biol Fisheries 26(2):577–588. https://doi.org/10.21608/ejabf.2022.234553

Kumar R, Singh D (2022) Insecticide- tolerant Bacterial population in *Eisenia fetida*'s Gut and Vermicast exposed to Chlorantraniliprole and Fipronil. Appl Biol Res 24(3):273–279

Kumar R, Yadav R, Gupta RK, Kiran, Kataria SK, Kadyan P, Pooja, Kaur S (2023) The earthworms: Charles Darwin's ecosystem engineer. In: Organic fertilizers (Eds. Khalid Rehman Hakeem), 1–7, Intech Open. https://doi.org/10.5772/intechopen.1001339

Kumar R, Yadav R, Kumar Gupta R, Pooja (2022c) A sustainable way for fish health management by replacement of chemical and drugs by earthworm. In: Environmental degradation in Asia: land degradation, environmental contamination, and human activities, pp 329–352. Cham: Springer International Publishing. https://doi.org/10.1007/978-3-031-12112-8_16

Kurup BM, Premlal P, Thomas JV et al (2003) Bottom trawl discards along Kerala coast: a case study. J Mar Biol Ass India 45(1):99–107

Lewison RL, Crowder LB, Read AJ et al (2004) Understanding impacts of fisheries bycatch on marine megafauna. Trends Ecol Evol 19:598–604

Lobo AS (2012) Managing fisheries in an ocean of bycatch. Position Paper for CBD-COP 11. Dakshin foundation, Bengaluru and foundation Coor ecological security, Anand

Madhu VR (2022) Overview on bycatch from harvesting systems. In: ICAR-sponsored short course on bycatch reduction in fisheries: recent advances. CIFT, Kochi, pp 1–3

Mahesh V, Benakappa S, Dineshbabu AP et al (2017) Occurrence of low value bycatch in trawl fisheries off Karnataka, India. Fish Technol 54:227–236

Mahesh V, Dineshbabu AP, Naik ASK, Anjanayappa HN, Vijaykumar ME, Khavi M (2019) Characterization of low value by-catch in trawl fisheries off Karnataka coast, India and its impact juveniles of commercially important fish species. Indian J GeoMar Sci 48(12):1733–1742

Manjeet, Airon A, Kumar R, Saifi R (2022) Temporal and spatial impact of lockdown during COVID-19 on air quality index in Haryana, India. Scientific Reports 12:20046. https://doi.org/10.1038/s41598-022-20885-2.

McCauley DJ, Pinsky ML, Palumbi SR et al (2015) Marine defaunation: animal loss in the global ocean. Sci 347:247

Menon NG, Nammalwar P, Zachariah PU, Jagadis I (2000) Investigations on the impact of coastal bottom trawling on demersal fishes and macrobenthos. In: Central Marine Fisheries Research Institute, Annual Report 1999–2000, Cochin, pp 55–57

Mohamed KS, Pravin P, Asokan PK et al (2009) Demonstration of responsible fishing for trawl fisheries of Gujarat—Project Final report MPEDA: CMFRI/CIFT, 37pp

Najmudeen TM, Sathiadhas R (2008) Economic impact of juvenile fishing in a tropical multigear multi-species fishery. Fish Res 92:322–332

Nisha, Anshu, Rani S, Kumar S, Kumar R, Kumar A (2025) Recent technologies and strategies for bioremediation of toxic and hazardous waste. In: Dubey KK, Shah MP, Yadav A (Eds.), Challenges and sustainable solutions in bioremediation, pp 1–16. CRC Press. https://doi.org/10.1201/9781003407317-1

NOAA (2016) National bycatch reduction strategy. https://repository.library.noaa.gov/view/noaa/17062

Nunoo FKE, Boateng JO, Ahulu AM et al (2009) When trash fish is treasure: the case of Ghana in West Africa. Fish Res 96:167–172

OECD (1997) Towards sustainable fisheries: economic aspects of the management of living marine resources. OECD, Paris

Olsen VJ (1995) Ways in which Norway is solving the bycatch problem. In: Solving bycatch: considerations for today and tomorrow, Alaska sea grant college program, Alaska Sea Grant College Program Report No. 96–03, University of Alaska, Fairbanks, pp 289–291

Ordoñez J (1985) A study of trash fish caught by otter trawl in the Visayan Sea. Philipp J Fish 18:1–76

Panda K, Benakappa S, Doddamani PL et al (2022) Effect of trawling and its by-catch on marine ichthyofauna off south west coast of India. Pharma Innov 11(11):349–355

Pascoe S (1997) Bycatch management and the economics of discarding. FAO Fisheries Technical Paper. No. 370. FAO, Rome, p 137

Pérez Roda MA, Gilman E, Huntington T et al (2019) A third assessment of global marine fisheries discards. FAO Fisheries and Aquaculture Technical Paper No. 633, FAO, Rome, 78pp

Ponnusamy K, Ambasankar K, Ponniah AG (2012) Production and marketing of fish meal in India—a study. Indian J Fish 59(1):147–149

Prado J (1997) Technical measures for bycatch reduction. In: Clucas IJ, James DG (eds) Technical consultation on reduction of wastage in fisheries, Tokyo, Japan, pp 25–44

Pramod G (2010) Illegal, Unreported and unregulated marine fish catches in the Indian exclusive economic zone. In: Pitcher TJ (ed) Policy and Ecosystem Restoration in Fisheries. University of British Columbia, BC, Vancouver, Canada, Fisheries Centre, p 29

Rani S, Sharma S, Singh D, Sheoran R, Kumar R* (2025) Improved production of anti-cancer compounds from medicinal plants through molecular and biotechnological mutagenesis strategies. In: Chen JT (Ed), Biotechnology, multiple omics, and precision breeding in medicinal plants, pp 287–297. CRC Press

Rojas-Bracho L, Taylor BL (2017) *Phocoena sinus*, Vaquita. *The IUCN Red List of Threatened Species, e. T17028A5*

Salagrama V (1998) Bycatch utilisation in Indian fisheries: an overview. In: Report and proceedings of FAO/DFID expert consultation on bycatch utilisation in tropical fisheries, Beijing, China, 21–23 pp

Samanta R, Chakraborty SK, Shenoy L et al (2018) Bycatch characterization and relationship between trawl catch and lunar cycle in single day Shrimp Trawls from Mumbai Coast of India. Reg Stud Mar Sci 17:47–58

Sambandamoorthy P, Nagamuthu J, Arumugam S et al (2015) Diversity and trophic level of ichthyofauna associated with the trawl bycatches of Cuddalore and Parangipettai, south-east coast of India. Mar Biodivers Rec 8:1–8

Santhoshkumar S, Jawahar P, Srinivasan A et al (2023) Analysis of temporal Patterns of Ichthyofaunal Bycatch Landed Along the Thoothukudi Coast, Gulf of Mannar, India. Thalassas 39:513–526

Shefali, Kumar R, Sankhla MS, Kumar R, Swaroop S Sonone (2021) Impact of pesticide toxicity in aquatic environment. Biointerface Res Appl Chem 11(3):10131–10140. https://doi.org/10.33263/BRIAC113.1013110140

SIOFA Secretariat (2024) Overview of SIOFA Fisheries 2024. Southern Indian Ocean Fisheries Agreement (SIOFA), 70 pp

Sultana R, Jamil K, Khan SI (2014) Bycatch utilization in Asia. In: Kim SK (ed) Seafood Processing By-Products: Trends and Applications. Springer, New York, pp 243–284

Supongpan M, Boonchuwong P (2010) National report bycatch management in trawl fisheries in the Gulf of Thailand. Thailand, 28 May 2010, pp 1–108

Susheela M (2022) Recent advances in the utilization of bycatch as quality protein. In: ICAR-sponsored short course on bycatch reduction in fisheries: recent advances. CIFT, Kochi, pp 1–3

Suwanrangsi S (1986) Improved bycatch utilization in Thailand. In: Maclean JL, Dizon LB, Hosillos LV (eds) Proceedings of the first Asian fisheries forum. Asia Fisheries Society, Manila, pp 467–469

Velip DT, Rivonker CU (2015) Trends and composition of trawl bycatch and its implications on tropical fishing grounds off Goa. India. Reg Stud Mar Sci 2:65–75

Wade PR, Long KJ, Francis TB et al (2021) Best Practices for Assessing and Managing Bycatch of Marine Mammals. Front Mar Sci 8:757330

Wanjari RN, Ramteke KK (2024) Exploring the diversity of trawl catch composition in Mumbai coastal waters through experimental trawling. Indian J Fish 71(4):6–19

Yu SY (1993) Utilization of *Decapterusrusselli* in fish cakes and surimi. Trop Sci 33:330–337

Zeller D, Ansell M, Andreoli V et al (2023) Trends in Indian Ocean marine fisheries since 1950: synthesis of reconstructed catch and effort data. Mar Freshw Res 74(4):301–319

Zeller D, Pauly D (2005) Good news, bad news: global fisheries discards are declining but so are total catches. Fish Fish 6:156–159

Zynudheen AA, Ninan G, SenA et al (2004) Utilization of by-catch in Gujarat (India). NAGA, World Fish Center Quart 27(3&4):20–23

Zynudheen AA, Sen A, Ravishankar CN, Kumar KA, Badonia R, Solanki KK (1998) Utilization of Jawla (Acetes spp.). In: Balachandran KK, Iyer TSG, Madhavan P, Joseph J, Perigreen PA, Raghunath MR, Varghese MD (eds) Proceedings of the symposium on advances and priorities in fisheries technology, SOFT (I), Cochin, 11–13 Feb 1998, pp 264–266

Chapter 15
Innovative Nanotechnology Approaches for Preventing and Treating Fish Diseases

Swati Rani, Randeep Singh, Bhupnesh, Shweta, and Sindhu Sheoran

Abstract Aquaculture has contributed significantly to food security in recent decades by meeting the enormous demand for animal protein. However, disease prevalence and environmental contamination are crucial issues facing the industry. In this sense, new technological strategies have successfully been developed to address such difficulties. Nanotechnology—a new and creative instrument—has many applications and enormous promise in aquaculture and seafood preservation. It can guarantee the civilized defence of farmed fish against pathogens that cause disease by offering new technologies for drug management and vaccine release. An overview of nanotechnology and its uses in aquaculture is given in this article. It also provides a quick overview of fish diseases and traditional methods of pathogen control. However, this review clarifies nanotechnology as a potentially innovative tool that could improve disease prevalence management and control. Thus, it has also been emphasised how crucial this technology is to advancing sustainable aquaculture.

S. Rani
Department of Biotechnology, Guru Jambheshwar University of Science and Technology, Hisa, Haryana, India
e-mail: swatijangra297@gmail.com

R. Singh (✉)
Department of Vegetable Science, Maharana Pratap Horticultural University, Karnal, Haryana, India
e-mail: randeepsuthar123@gmail.com

Bhupnesh
Department of Botany and Plant Physiology, Chaudhary Charan Singh Haryana Agriculture University, Hisar, Haryana, India
e-mail: bhupneshsharma186@gmail.com

Shweta
Department of Physiology, AIIMS, Delhi, India
e-mail: shwetasaraswat@hau.ac.in

S. Sheoran
Senior Research Program Coordinator, Hawkesbury, Western Sydney University, Sydney, Australia
e-mail: S.Sheoran@westernsydney.edu.au

R. Kumar et al. (eds.), *Sustainability in Aquaculture and Fisheries*,
Springer Water, https://doi.org/10.1007/978-3-032-16605-0_15

Keywords Aquaculture · Viruses · Sensors · Nanotechnology · Infectious

15.1 Introduction

Aquaculture involves cultivating aquatic organisms like fish, molluscs, crustaceans, and aquatic plants in various water environments, including lakes, ponds, and oceans. Aquaculture practice is not new; it was mentioned in China at least 4000 years ago. However, even though aquaculture has been around since antiquity, its use has increased since the middle of the twentieth century (Beveridge and Little 2002). Currently, aquaculture is an evolving industry that demands innovation and research. In 2014, aquaculture produced 70 million tonnes (MT), worth US$160.2 billion. The Food and Agriculture Organisation (FAO) of the United Nations estimates that the expansion has been taking place at about 8% per year. This expansion helps to produce wholesome food that the world's growing population can easily access. Aquaculture has grown all over the world as an important part of food security, which offers an important supply of high-quality protein. But as industry has risen very fast, disease outbreaks have increased as well, harming fish health, financial stability, and environmental sustainability. Infectious diseases caused by bacteria, viruses, fungi, and parasites cause a big loss in aquaculture and endanger wild fish populations. Antimicrobial resistance (AMR) and environmental degradation have made traditional disease prevention and treatment strategies, like antibiotics and chemical treatments are becoming increasingly ineffective. To overcome these problems, innovative sustainable solutions are immediately required. Fish health management has become more difficult due to aquaculture's explosive growth, especially when it comes to preventing infectious diseases brought on by bacteria, viruses, and parasites. Conventional methods, such as chemical treatments and antibiotics, pose important risks due to environmental contamination and the rise of antimicrobial resistance. The difficulties thus call for innovative and sustainable solutions. Nanotechnology has considerably transformative potential in aquaculture. Through this, there is specific prevention, diagnostics, and treatment for diseases in aquaculture. Nanoparticles and nanodelivery systems have special physicochemical properties that allow for improving the bioavailability and targeted delivery of therapeutic agents and enhancing immune responses through nano-vaccines. More advanced technologies are being integrated to allow aquaculture to break dependency on conventional treatments and increase fecundity while improving sustainability.

15.1.1 Overview of Fish Diseases in Aquaculture

A major difficulty in aquaculture is diseases, which lower fish production in both natural habitats and farming systems. The sustainability of the industry is threatened by disease outbreaks, which raise costs because of stock losses, treatment costs, and declining yield quality. Because of their high death rates and quick spread, infectious diseases brought on by pathogens—such as bacteria, viruses, fungi, and parasites—are especially worrying. High stocking densities, inadequate nutrition, and poor water quality are frequently the causes of non-infectious diseases. Small-scale farmers in developing nations are disproportionately affected by these issues, and their lack of resources and expertise in disease management makes matters vulnerable (Kumar et al. 2022a, b; Mishra et al. 2017a; Alfred et al. 2020).

Prevention is given first over treatment in effective fish health management. The establishment of diagnostic labs and seed certification systems, adequate nutrition, biosecurity protocols, and ideal water quality are important preventive measures. Controlling infectious diseases mainly requires vaccination strategies that use oral, injection, or immersion administration of killed, attenuated, DNA, recombinant, or synthetic peptide vaccines. Probiotics and nanotechnology are two new strategies that commit improving fish health and disease resistance (Kumar et al. 2022c; Faruk and Anka 2017; Assefa and Abunna 2018). By integrating immune prophylaxis, biosecurity, and innovative technologies, the industry can overcome the effects of disease, boost productivity, and guarantee long-term growth and resilience.

15.1.1.1 Bacterial Diseases

Bacteria are most commonly found in aquatic environments can turn harmful due to impact of various factors like low water quality, a poor diet, and overcrowding. One of the main causes of death for both wild and farmed fish is bacterial infections, which frequently leads aquaculture to suffer large financial losses (Mukherjee 2002; Sandeep et al. 2016). These infections can be systemic illnesses that impact internal organs (Mishra et al. 2017b) or surface illnesses like fin and gill infections. *Aeromonas hydrophila, Flavobacterium columnare, Edwardsiella ictaluri, Renibacterium salmoninarum*, and *Vibrio* species are examples of prevalent bacterial pathogens (Alfred et al. 2020). One of the most widespread and destructive illnesses in freshwater fish culture is motile aeromonassepticaemia (MAS), which is brought on by *A. hydrophila* and related species. Bacillary necrosis of Pangasius *(E. ictaluri*), bacterial kidney disease (*R. salmoninarum*), and columnaris disease (*F. columnare*) are other noteworthy bacterial diseases. The latter is responsible for large losses in Pangasius farming (Faruk et al. 2004; Faruk and Anka 2017). When fish are under stress, bacterial diseases repeatedly start as opportunistic infections and can expeditiously progress for systemic conditions with some necrotic damage to organs such as the liver, kidney, and spleen. Antibiotics are generally used for many treatments, but their huge use has resulted in antimicrobial resistance, which is making management of diseases even more difficult. Attenuating bacterial disease outbreaks in aquaculture systems requires preventative measures, such as preserving ideal water quality, lowering stressors, and putting biosecurity approaches in place.

15.1.1.2 Viral Diseases

Since viruses are tiny, obligatory intracellular pathogens that can replicate only inside host cells, they are very hard to find which can be problematic for fish aquaculture. While adults may act as carriers, viral infections frequently cause high mortality rates in fish larvae. Latent viral infections can be brought on by environmental stressors like handling, crowding, and low water quality. Koi herpesvirus disease (KHVD), which is extremely contagious and can kill up to 100% of *Cyprinus carpio*, infectious pancreatic necrosis (*Aquabirnavirus*), viral hemorrhagic septicaemia (*Novirhabdovirus*), and lymphocystis (*Lymphocystivirus*) are notable viral diseases in aquaculture (Mishra et al. 2017a; Wolf 2019).

Viral epidemics have no substantial impact on freshwater aquaculture in India, because Indian Major Carps (IMCs) are resilient and the environment is less favourable for viral growth. However, there have been reports of viral diseases like Cyprinid herpesvirus 2 (CyHV-2), koi ranavirus (KIRV), and carp oedema virus (CEV) in ornamental fish culture. Since there are currently no effective treatments for viral outbreaks, preventive measures like biosecurity, hygiene, and quarantine are essential. In Indian aquaculture, seasonal fluctuations and parasitic infestations present greater obstacles than viral diseases, highlighting the necessity of focused preventive measures (Ganguly 2017).

15.1.1.3 Fungal Diseases

Fungal diseases are less common than bacterial or parasitic infections, but these diseases pose a serious threat to aquaculture, especially when fish are stressed not living in ideal conditions. Oomycetes like Saprolegnia, Achlya, and Aphanomyces are the majority of fungal pathogens in aquaculture. They cause diseases like saprolegniasis, which shows that as white to grey cotton-like growths on the skin, fins, and eggs. These fungi are opportunistic pathogens that regularlycolonise under injured tissues after being infected by bacteria or parasites. Fish are more sensitive to these infections due to poor water quality, high organic loads, and overcrowding (Yanong 2003; Panchai et al. 2007).

Saprolegniasis, which is well known disease in aquaculture system which primarily affects fish eggs, hatcheries, and overwintering ponds, decreasing hatchability and fry survival. Many Clinical symptoms include mycelial growth on skin and eggs. A revival of Saprolegnia-related problems has resulted from the global ban of malachite green, which was once used to manage the disease. This underscores the necessity for alternative control measures (Patel et al. 2018). Epizootic Ulcerative Syndrome (EUS) is primarily caused by Aphanomyces invadans, another economically significant fungal disease. A number of freshwater fish species are afflicted by EUS, which indicates ulcers and hemorrhagic lesions that can develop into systemic infections. EUS risk is increased by many elements such as water from rivers or from rice fields, mainly in Southeast Asia, which includes Bangladesh and India (Das and Mishra, 2014). The management of fungal diseases places a strong significance on preventive actions like preserving water quality, lowering stress levels, minimising crowding, and making sure people are eating a healthy food. In hatcheries, proactive approaches like cleaning and removing dead eggs are essential. EUS and related bacterial infections have been effectively managed with CIFAX, a formulation which is created by ICAR-CIFA. However, the need for long-term and efficient antifungal strategies is highlighted by the growing significance of fungal infections in aquaculture (Mishra et al. 2017a, b).

15.1.1.4 Parasitic Diseases

Parasitic diseases have threat for the survival and growth of fingerlings and fry in many fish nurseries. Protozoan and metazoan infections are critical problem in aquaculture systems, constantly resulting in large-scale carp fry and fingerling mortality.

Protozoan Diseases

(A). **Ichthyophthiriasis (White Spot Disease)**

The ciliated protozoan *Ichthyophthirius multifiliis*, commonly known as "Ich," is mainly responsible for white spot disease, or Ichthyophthiriasis. This condition is one of the most detrimental protozoan infections affecting freshwater fish hatcheries and farms that raise fry and fingerlings (Mohan 2007). The disease has been observed in the fingerlings and fry of fish species such as *Catla catla*, *Labeo rohita*, and *Cirrhinus mrigala* in nurseries and rearing ponds (Chandra 2004).

The clinical hallmark of the disease includes tiny white spots or nodules on the skin, fins, and gills of fish. Infected fish tend to be irritable, move erratically, and rub against objects in their environment. In the early stages of the infection, you may notice scratching and a decreased appetite. In severe cases, cysts can merge, creating asymmetrical patches on the fish's body. The parasite is characterised by a spherical body, uniformly spaced cilia, and a horseshoe-shaped nucleus, which rotates distinctly when the organism is alive.

(B). **Trichodiniasis**

This disease is caused by peritrichous ciliates of the *Trichodina* genus, which can severely impact hosts, particularly fry, fingerlings, and adult fishes such as *C. catla*, *L. rohita*, *C. cirrhosus*, and *Hypophthalmichthys molitrix* (Chandra 2004). The parasite attaches to the skin and gills, leading to excessive mucus secretion, darkened skin, and a greyish-blue coating over the body.

Infections often correlate with poor water quality and high stocking densities. Effective treatments include formalin dip baths (250 ppm for 3–5 min), sodium chloride (2–3% for 5 min), or potassium permanganate dips (0.1% for 30–45 s).

Prolonged treatment with formalin (25–30 ppm) has also proven effective (Tonguthai et al. 1997).

(C). **Ichthyobodosis**

The flagellated ectoparasite *Ichthyobodo necator* causes Ichthyobodosis, a condition prevalent in freshwater hatcheries. The parasite's free-living stage exists in water, while the parasitic stage attaches to dorsal fins and secondary gill lamellae. Symptoms include small grayish-white spots on the skin, excessive mucus production, and reddened, hemorrhagic patches. Affected fish may exhibit damaged fins, pale gills covered with mucus, anorexia, and lethargy. Severely infected fish may die after becoming moribund.

(D). **Cryptobiosis**

Caused by the flagellated protozoan *Cryptobia* spp. this disease primarily affects the stomach but may infect other organs. Fish become lethargic, emaciated, and develop dark skin pigmentation. Nutritional management has been found to play a critical role in controlling this infection.

Metazoan Parasites and Their Impact

Infestations by monogenean trematodes, such as *Gyrodactylus* spp. (skin flukes) and *Dactylogyrus* spp. (gill flukes), cause significant morbidity and economic losses in freshwater aquaculture. These parasites disrupt host metabolism, impair respiratory functions, and lead to tissue destruction. Additionally, ectoparasitic crustaceans like *Argulus* spp. (fish lice) and *Lernaea* spp. Contribute to dermal ulceration, osmotic imbalance, and immunosuppression.

The prevalence of *Argulosis* in major Indian carps is notably higher than that of Chinese and European carps, with infestation intensity influenced by seasonal changes in host physiology and ecology (Chandra 2004). The economic impact of *Argulus* infestations has been estimated at INR 30,000 (US$ 615) per hectare per year in carp culture systems in India, primarily due to reduced growth rates and additional chemical treatment costs (Menezes et al. 1990; Shinn et al. 2015).

15.2 Economic and Ecological Impacts of Fish Diseases

Fish diseases significantly impact fisheries resources, causing substantial economic losses due to mortality and morbidity of broodstock, reduced market value of infected fish, and production declines. Studies conducted in areas such as Rajshahi, Naogaon, Natore, and Bogra revealed average production losses in ponds ranging from 29.12% to 38.0%, attributed to diseases caused by bacteria, fungi, protozoa, monogeneans, crustaceans, malnutrition, oxygen deficiency, and water pollution (Shefali et al. 2021;

Chowdhury et al. 2011). Fingerlings and adult fish mortality averaged 38.62%, highlighting the vulnerability of young fish with underdeveloped immune systems. Indiscriminate pesticide use to combat parasites like *Argulus* sp. further exacerbates the problem by deteriorating pond ecosystems and harming non-target species. Rural farmers, often lacking resources and knowledge, face significant financial hardships due to inadequate fish health management and limited support from the government and non-government sectors. Practical solutions include improved water quality monitoring, disease diagnosis facilities, farmer training on health management practices, and stricter regulations on chemotherapeutic agents. Promoting sustainable practices, including proper stocking density, feeding regimens, and research facilities, can mitigate fish health challenges, prevent environmental hazards, and enhance fish production, ultimately safeguarding the livelihoods of fish farmers and the nation's fisheries resources.

The costs of controlling infectious agents on aquaculture farms represent the marine diseases'most visible economic impact. Additionally, infectious diseases are unpredictable and can potentially disrupt markets and deter investment rapidly (Begum et al. 2002). Another less apparent and contentious consequence is exporting infectious agents from aquaculture to wild stocks, which could negatively affect wild fisheries and threaten biodiversity. However, wild fisheries, which are already struggling with issues like declining catch rates and competition from aquaculture, might also experience indirect benefits from the impact of infectious diseases on farms (Lafferty et al. 2015). Consumers are particularly affected by marine illnesses, as these issues result in a decreased availability of seafood and higher prices. Given its increasing role in meeting consumer demand for fishery products, aquaculture must adhere to safety and quality standards. From an economic and health perspective, infectious fish diseases pose significant challenges in aquaculture production, alongside concerns such as residues from veterinary medications, contamination from organic and inorganic heavy metals, and agrochemical residues. These diseases lead to increased costs due to the loss of dead fish, the expense of treatment, and the slower growth rates of sick and recovering fish, ultimately reducing profits. Fish often encounter pathogens, but the relationship between these pathogens and host fish in fish diseases is not always clear. The stressful environmental conditions in intensive fish farming systems make fish more susceptible to various pathogens. Implementing effective management techniques in fish farming can help reduce the spread of infectious diseases and improve product safety in aquaculture, thus minimising potential risks to production. Fish health management can be more successful by focusing on disease prevention rather than treatment through integrated measures at the production level. This global approach will produce safer fish products (Blanco et al. 2000; Moran and Fofana 2007).

15.3 Current Methods of Fish Disease Prevention and Treatment and Their Limitations

For aquaculture production to be sustainable, fish health management must be done well. Historically, a variety of tactics have been used to prevent and treat fish diseases, such as vaccination, water quality control, probiotics, antibiotics, and herbal remedies. Despite the fact that these techniques have greatly decreased disease outbreaks and mortality, they have a number of drawbacks that compromise their sustainability and long-term effectiveness. An outline of the main techniques and their limitations is provided below in Table 15.1.

Rapid evolution of pathogens frequently results in their resistance to current therapies. At the same time, overuse of chemicals and antibiotics damages beneficial microbiota, upsetting aquatic ecosystems and endangering fish health in general. The high expense of cutting-edge therapies and technologies further restricts their availability in low-income areas. Furthermore, it is difficult to apply universal disease management strategies due to the variety of aquaculture practices and growing production scales. Traditional fish disease prevention and treatment techniques have been vital to aquaculture, but their drawbacks underscore the need for creative, long-lasting, and accurate solutions. A promising substitute is nanotechnology, which offers more effective delivery methods, reduced environmental impact, and targeted delivery systems. The next chapter delves into how nanotechnology-based interventions can transform fish disease management, addressing the shortcomings of conventional approaches.

15.4 Basics of Nanotechnology

Nanotechnology, a field of transforming matter at the nanoscale (1–100 nanometers), has gained recognition as having enormous potential in all industries, such as aquaculture. Aquaculture will benefit from innovative solutions provided by nanotechnology regarding fish health improvement, feed efficiency, and environmental sustainability. The following section explains the introduction of nanotechnology and its possible applications in aquaculture, focusing on several nanotechnology techniques. Nanotechnology integrated into aquaculture practices can benefit various challenges that arise, including the management of diseases, water quality monitoring, and optimisation of feeds. Nanomaterials are solely important to these innovations because of their unique physicochemical characteristics. Nano-enhanced feed additives, an application of nanotechnology, aim to enhance fish's nutrient assimilation and absorption rate. For example, encapsulating essential nutrients within nanoparticles may protect them from degradation and promote targeted release, thus increasing growth rates and fish health, emphasising their role in nutrient absorption and feed efficiency. To maximise treatment effectiveness while minimising potential environmental impacts, nanoparticles serve as effective carriers for delivering

15.1 Contains overview of fish diseases in aquaculture, environment and economic impacts with effective need of control, prevention and treatment

Sr. No.	Current methods of fish disease prevention and treatment	Approach	Limitations	References
1.	Water quality management	Maintaining optimal water parameters (temperature, pH, dissolved oxygen) to prevent stress-induced disease susceptibility	-Requires continuous monitoring and advanced infrastructure -Susceptible to external environmental changes, such as climate variability -Does not address pathogen load directly	Mohapatra et al. (2013)
2.	Vaccination	Administration of specific vaccines to prevent infections caused by bacterial or viral pathogens.	-Limited availability of vaccines for many pathogens -High production and storage costs -Ineffectiveness against rapid pathogen mutations or secondary infections -Requires skilled personnel for administration, especially in small-scale operations	Mohamad et al. (2021)

(continued)

15.1 (continued)

Sr. No.	Current methods of fish disease prevention and treatment	Approach	Limitations	References
3.	Antibiotic therapy	Use of antibiotics to treat bacterial infections	Development of antimicrobial resistance (AMR) due to overuse and misuse Residual antibiotics in aquatic ecosystems lead to environmental contamination Concerns over human health risks from the consumption of antibiotic-laden fish Non-target effects on beneficial microbiota	Salma et al. (2022)
4.	Probiotics	Use of beneficial microorganisms to enhance immune response and outcompete pathogens	Variable efficacy depending on fish species and environmental conditions Short shelf-life and storage constraints Difficulty in standardising dosing protocols for mass application	Zorriehzahra et al. (2016)
5.	Herbal remedies	Use of plant extracts with antimicrobial or immunostimulatory properties	Limited scientific validation and consistency in efficacy Requires large-scale extraction and standardisation. Potential toxicity at high concentrations	Harikrishnan et al. (2011)

(continued)

15.1 (continued)

Sr. No.	Current methods of fish disease prevention and treatment	Approach	Limitations	References
6.	Chemical treatments	Disinfectionants (e.g. formalin, potassium permanganate) can be applied to control pathogen load in water or on fish surfaces	Potential toxicity to fish and the environment Residual chemicals in water bodies disrupt aquatic ecosystems Over-reliance can lead to pathogen resistance	Assefa and Abunna, (2018)
7.	Genetic approaches	Selective breeding or gene editing to develop disease-resistant fish strains	Time-consuming and resource-intensive Ethical concerns and regulatory hurdles, especially for genetically modified organisms Resistance traits may not be effective against all pathogens	Ferdous et al. (2022)
8.	Immunostimulants	Use of dietary supplements or bioactive compounds to boost innate immunity in fish	Inconsistent results due to species-specific responses. Limited protection against virulent pathogens	Farooqi and Qureshi, (2018)
9.	Quarantine and biosecurity measures	Isolation of infected fish and implementation of biosecurity protocols to prevent disease spread	High operational costs and labour-intensive processes Ineffective against pathogens with asymptomatic carriers Requires strict compliance, which is challenging in small-scale aquaculture	Palic et al. (2015)

therapeutic agents, vaccines, or immune stimulants directly to the fish. Targeting the release of active compounds at specific locations reduces the amount of the medicine required and thus possible side effects. The same study reports using nanoparticles in fish medicine to manage diseases effectively through controlled drug delivery systems.

15.5 Nanotechnologies Used for Prevention of Disease

15.5.1 Nanocapsulation

Aquaculture stands to gain significantly from the advancements in nanotechnology, which have a variety of applications. It can provide new genetic, biotechnology, fish reproduction, and aquatic health resources. Techniques in nanotechnology include innovative drug delivery systems like DNA nano-vaccines, as well as the use of nanomaterials and nanosensors. Water treatment, pond sterilisation, infection detection and control, and efficient drug and nutrient delivery are some of the current applications of nanotechnology in aquaculture. For example, DNA nanovaccines have been used to strengthen fish's immune systems, while iron nanoparticles can promote fish growth. One method involves encasing a solid nanoparticle in a fatty acid shell to protect the medication. When mild inflammatory agents are combined with nanoparticle carriers such as poly-lactide-co-glycolic acid (PLGA) and chitosan, fish and shellfish gain significant protection against bacterial and certain viral diseases that could otherwise lead to side effects from vaccination.

Additionally, pharmaceuticals can be delivered effectively using porous nanoparticles as a matrix. Mesoporous silica particles also play a role in this delivery system. For example, the controlled release of pharmaceuticals can benefit from oral nanodelivery systems that utilise nanoparticles (Sharma et al. 2024a; Sharma et al. 2025a; Mahboub et al. 2022). Some documented applications of these systems include: improved control over drug release, targeted tissue delivery, increased bioavailability for drugs with low absorption rates, stabilisation of drugs through prolonged retention in the gut, and enhanced absorption capabilities resulting from a higher molecular dispersion rate.

The aquaculture sector has begun utilising DNA nanovaccines, tiny strands of DNA enclosed in nanocapsules, to boost fish's immune systems. This approach integrates timed-release drug delivery with iron nanoparticles and has been shown to accelerate fish development. For instance, Asian carp have been effectively protected against *Listonella anguillarum* infections using nanoencapsulated vaccines. Additionally, supplementing fish feed with selenium nanoparticles can significantly enhance growth and improve the antioxidant defense system. Nanomaterials can be created from a variety of chemical components. However, the most commonly used types include metal oxides, carbon nanotubes, carbon spheres, nano-metals, and composites such as nanoshells, quantum dots, and nanoceramics (Ahmed et al. 2023).

15.5.1.1 Nanocapsulation and its Approaches

Factors such as size, distribution, shape, solubility, surface area, and the mechanisms of particle release play essential roles and are influenced by the delivery system and encapsulation technique (Dar et al. 2020; Sharma et al. 2024b; Sharma et al. 2025b). The necessary size, physicochemical characteristics, type of wall and core material, and application in a specific field should all be considered when choosing an encapsulation technique. Only specific top-down methods can be scaled down and applied to nanoencapsulation instead of microencapsulation. However, nano-systems' uniqueness is often based on bottom-up creation, which is generally simple and confers unique features (Ahmed et al. 2023). When a microcapsule's width falls between 0.2 and 5000 microns, it is called an encapsulated particle; when it exceeds 5000 microns, it is called a microcapsule; when it drops below 1.0 microns, it is called a nanocapsule. Nanocapsules offer several intriguing advantages over their micro-sized counterparts, such as increased surface area, enhanced solubility, significantly higher bioaccessibility, and the ability to release substances under controlled conditions at the intended site of action. These differences from traditional molecular solvents and the potential for natural tuning through careful selection of constituents contribute to several desirable properties for applications such as chemical extraction, catalysis, reaction media, novel lubricants, materials chemistry, and electrochemistry.

An emulsion is a mixture of two or more liquids that do not easily blend. A nanoemulsion is a specific emulsion characterised by dispersed droplet diameters of 500 nanometers or less. Their unique characteristics pointed to a point of attraction because of their small size and enormous kinetic energy. The study on the stabilisation of freeze-dried oil and the encapsulation of fish oil focused on beta-cyclodextrin, which is known for its strong fragrances, rapid oxidation processes, and health benefits. Research revealed that water-insoluble polymers were more effective at preserving fish oil due to their water-insoluble properties. This conclusion was reached after analysing the rate of fish oil release under various storage conditions, including different humidity and temperature levels. The findings are particularly relevant for small fish or prawns weighing 1–5 grams or less, for which bioencapsulation may be used in an oral distribution system. The researcher may concentrate on the various methods of incorporating the vaccine into artificial feed, including (1) using adhesives like edible oil or gelatin, (2) spraying the vaccine in liquid form onto the finished feed, and (3) combining the antigen with the feed during manufacturing. However, there are certain disadvantages, such as inconsistent feed distribution and the possibility that the antigen will be revealed to harsh stomach conditions right after consumption, which could lead to antigen degradation. As a result, many encapsulation techniques have been formed and analysed to shield the embedded antigens from the harsh stomach environment (Sharma et al. 2024c; Ahmed et al. 2023).

15.5.2 Nanoliposomes

It is known that by adding some essential oils to surfactant nanometric micelles results in improved and strong antibacterial activity. Nano-liposomes based on essential oils can release their contents constantly, which enhances the film's antibacterial properties. Wu et al. (2015), for instance, created nano-liposomes with cinnamon essential oil integrated into gelatin films. This method improved the oil's antibacterial efficiency while also slowing its release. Furthermore, it has been demonstrated that carvacrol and eugenol encapsulated in nano-metric surfactant micelles have very strong antibacterial activity (Gaysinsky et al. 2005).

Moreover, soy lecithin and rapeseed were used to encapsulate orange essential oil, producing immensely stable nano-liposomes that were then added to caseinate starch films (Valencia-Sullca et al. 2016). This shows that lecithin-based nano-liposomes have adequate encapsulating capacity. Moreover, lecithin nano-liposomes that contain eugenol can hold 40–50% of the eugenol, while those that add it by direct emulsification only manage to hold 1–2% (Dar et al. 2020).

15.5.3 Nanobiosensors

In order to encourage corrective action, nanosensors have been developed to detect contaminants and monitor water parameters in real-time. Aquatic organisms can live in a healthy environment because these sensors can identify pollutants at very low concentrations. Nanotechnology has been used to create nano-coatings that prolong the shelf life of fish products and stop biofouling on aquaculture equipment. Because of many antimicrobial qualities, these coatings lower the probability of contamination and the spread of disease. They offer a feasible way to monitor and control environments, nanobiosensors have become innovative instrument in the advancement of sustainable aquaculture. These devices united nanomaterials and biological recognition elements to detect analytes with high sensitivity and specificity. Therefore, in addition to helping with disease detection, applications of nanobiosensors include optimized feeding and water quality monitoring, which enhance fish health and environmental sustainability.

'Nano biosensors were designed for the early and fast detection of disease-causing pathogens, so suitable intervention may occur before these conditions become problematic, and dependency on antibiotics will also decrease. A system that focus on detecting specific insecticides using nanosensors applied in aquaculture to monitor harmful substances to the health of the fish. Water quality monitoring is another field where nanobiosensors play a crucial role. Optimal water conditions are essential for aquaculture productivity. Nanobiosensors can detect various environmental pollutants, including heavy metals and pesticides, to ensure a safe habitat for aquatic organisms. A detailed analysis emphasised the effectiveness of biosensors and nanobiosensors for ecological monitoring, focusing on their role in the eco-friendly detection of agro-based pesticides and potential public health risks. Nano biosensors contribute to feed optimisation through real-time monitoring of nutrient levels and metabolic byproducts. The information helps to adjust feeding regimes to optimise growth rates while minimising waste and increasing the sustainability of aquaculture.

15.6 Conclusion

In summary, nanotechnology is a very important tool for revolution and management of aquaculture which offers numerous chances to boost the productivity and sustainability in these crucial industries. Researchers and practitioners can effectively approach long-standing problems such as affecting feed supplementation, disease control, environmental monitoring, and water quality management in aquaculture and fisheries. Though the creative use of nanomaterials and nanodevicesin aquaculture lead to revolutionary advancements. The accuracy and sensitivity of sensors and nanomaterial-based devices enable real-time monitoring and investigation of aquatic environments. However, vaccines and antimicrobial coatings based on nanoparticles show remarkable efficacy for managing disease outbreaks, maintaining fish health, and minimising losses. Furthermore, nanotechnology enables the formation of implemented nanoparticles, encapsulation techniques, and nanoscale nutrient delivery systems, all of them enhance fish health overall, feed efficiency, and nutrient assimilation. These advancements make many improvements in immunological response, growth performance, and sustainability in aquaculture. Moving forward, achieving the full potential of nanotechnology in aquaculture and fisheries will require sustained research efforts, interdisciplinary partnerships, and the development of strong regulatory frameworks. We can use the revolutionary potential of nanotechnology to advance aquaculture and fisheries toward a more promising and resilient future by encouraging innovation, guaranteeing sustainability, and protecting the environment.

15.7 Future Prospects

'There are promising opportunities to improve sustainability and productivity through nanotechnology in aquaculture and fishery management. To realise the potential of nanomaterials in aquatic industries, several issues and research methods must be resolved. The two major concerns surrounding their use in aquaculture are the potential environmental impact of nanomaterials and assuring their safety for marine organisms and ecosystems. Even though nanomaterials have unique properties and benefits, their unexpected release into the environment could also endanger human health and aquatic ecosystems. Therefore, future investigation efforts should focus on comprehensive risk assessments and toxicity studies to evaluate the environmental fate, bioaccumulation, and ecotoxicological effects of nanomaterials used in aquaculture applications. Using nanotechnology in aquaculture, it represents several challenges, including understanding the regulatory environment and ensuring the current laws governing the use of nanomaterials in food production and environmental management. Regulatory agencies worldwide are continuous working on development of frameworks for assessing the safety and efficacy of nanomaterials, including their use in aquaculture practices and products.

Therefore, it is important to work with many regulatory agencies, industry stakeholders, and scientific communities to develop precise guidelines and standards that are responsible for application of nanotechnology in aquaculture and fishery management. The scalability and affordability of nanotechnology-enabled solutions substantially hamper the widespread adoption in aquaculture operations, specifically for small-scale farmers and resource-constrained environments. Because nanomaterials are very expensive to produce, formulate, and use in aquaculture practitioners may find it challenging to obtain them, especially in developing countries. Therefore, future research should mainly focus on development of cost-effective manufacturing processes, scalable production methods, and economical nanomaterial formulations that meets the need of diverse aquaculture systems and their stakeholders. Approaching the complex opportunities and issues associated with nanotechnology in aquaculture requires collaboration and knowledge integration. To develop comprehensive solutions that consider environmental, economic, and social factors, researchers, practitioners, policy makers from various disciplines—including nanotechnology, aquaculture, environmental science, and public health—must be collaborated. Research funding organisations, academic institutions, and business partners should prioritise supporting interdisciplinary research projects, training programs, and capacity-building initiatives focused on aquaculture nanotechnology applications. By facilitating collaboration, knowledge exchange, and skill development, can foster innovation. Nanotechnology offers many opportunities to enhance sustainable aquaculture and fishery management. Together, we can use nanotechnology as a transformative potential tool for advance the resilience, sustainability of the world's fisheries and aquaculture industries.

References

Ahmed J, Vasagam KK, Ramalingam K (2023) Nanoencapsulated aquafeeds and current uses in fisheries/shrimps: a review. Appl Biochem Biotechnol 195(11):7110–7131

Alfred O, Shaahu A, Orban DA, Egwenomhe M (2020) An overview on understanding the basic concept of fish diseases in aquaculture. IRE J 4:83–91

Assefa A, Abunna F (2018) Maintenance of fish health in aquaculture: review of epidemiological approaches for preventing and controlling infectious disease of fish. VetY Med Int 2018(1):5432497

Begum A, Alam SN (2002) Social and economic impacts of shrimp disease among small-scale, coastal farmers and communities in Bangladesh. FAO Fish Tech Pap 191–200

Beveridge MC, Little DC (2002) The history of aquaculture in traditional societies. Ecological aquaculture. The evolution of the blue. Revolution 3–29
Blanco MM, Gibello A, Fernández-Garayzábal JF (2000) Influence of fish health management: bases, procedures and economic implications. Cah Options Méditerranéennes 51:45–49
Chandra KJ (2004) Fish parasitology. Published by K Ray Choudhury
Dar AH, Rashid N, Majid I, Hussain S, Dar MA (2020) Nanotechnology interventions in aquaculture and seafood preservation. Crit Rev Food Sci Nutr 60(11):1912–1921
Das BK, Mishra SS (2014) Diseases in freshwater aquaculture. Training manual on model training course on preventive health management practices in freshwater aquaculture. ICAR-Central Institute of Freshwater Aquaculture, Bhubaneswar, Odisha, India
Farooqi FS, Qureshi WUH (2018) Immunostimulants for aquaculture health management. J Pharmacogn Phytochem 7(6):1441–1447
Faruk MAR, Alam MJ, Sarker MMR, Kabir MB (2004) Status of fish disease and health management practices in rural freshwater aquaculture of Bangladesh. Pakistan J Biol Sci 7(12):2092–2098
Faruk MAR, Anka IZ (2017) An overview of diseases in fish hatcheries and nurseries. Fundam Appl Agric 2(3):311–316
Ferdous MA, Islam SI, Habib N, Almehmadi M, Allahyani M, Alsaiari AA, Shafie A (2022) CRISPR-Cas genome editing technique for fish disease management: current study and future perspective. Microorganisms 10(10):2012
Ganguly S (2017) Viral diseases infecting finfishes and ornamental fishes: a review of relevance to sustainable aquaculture. Int J Pure App Biosci 5(1):282–284
Gaysinsky S, Davidson PM, Bruce BD, Weiss J (2005) Growth inhibition of Escherichia coli O157: H7 and listeria monocytogenes by carvacrol and eugenol encapsulated in surfactant micelles. J Food Prot 68(12):2559–2566
Harikrishnan R, Kim MC, Kim JS, Balasundaram C, Heo MS (2011) Probiotics and herbal mixtures enhance the growth, blood constituents, and nonspecific immune response in Paralichthysolivaceus against streptococcus parauberis. Fish Shellfish Immunol 31(2):310–317
Hossain MK, Islam KT, Hossain MD, Rahman MH (2011) Environmental impact assessment of fish diseases on fish production. J Sci Found 9(1–2):125–131
Kumar R, Gupta R, Dahiya T, Yadav R, Saifi R, Yodha K, Kumari B, Singh P, Shefali. (2022a) Surveillance of *Heteropneustes fossilis* (stinging catfish) diseases and health management practices in Haryana. Appl Ecol Environ Sci 10(4):261–267. https://doi.org/10.12691/aees-10-4-12
Kumar R, Gupta RK, Yadav R, Saifi R, Yodha K, Pooja. (2022b) *Eisenia fetida* as protein source for growth enhancement of *Heteropneustes fossilis*. Egypt J Aquat Biol Fish 26(2):577–588. https://doi.org/10.21608/ejabf.2022.234553
Kumar R, Yadav R, Kumar Gupta R, Pooja (2022c) A sustainable way for fish health management by replacement of chemical and drugs by earthworm. In: Environmental degradation in Asia: land degradation, environmental contamination, and human activities. Springer International Publishing, Cham, pp 329–352. https://doi.org/10.1007/978-3-031-12112-8_16
Lafferty KD, Harvell CD, Conrad JM, Friedman CS, Kent ML, Kuris AM et al (2015) Infectious diseases affect marine fisheries and aquaculture economics. Annu Rev Mar Sci 7(1):471–496
Mahboub HH, Shahin K, Mahmoud SM, Altohamy DE, Husseiny WA, Mansour DA et al (2022) Silica nanoparticles are novel aqueous additives mitigating heavy metals toxicity and improving the health of African catfish. Clarias Gariepinus Aquat Toxicol 249:106238
Menezes J, Ramos MA, Pereira TG, da Silva AM (1990) Rainbow trout culture failure in a small lake due to massive parasitosis related to careless fish introductions. Aquaculture 89(2):123–126
Mishra SS, Das R, Das BK, Choudhary P, Rathod R, Giri BS, Swain P (2017a) A survey report shows the status of aquamedicine, drugs, and chemicals used in India. J Aquac Fisheries 1(004):1–15
Mishra SS, Rakesh D, Dhiman M, Choudhary P, Debbarma J, Sahoo SN, Mishra CK (2017b) Present status of fish disease management in freshwater aquaculture in India: state-of-the-art review. J Aquac Fish 1(003):14

Mohamad A, Zamri-Saad M, Amal MNA, Al-Saari N, Monir MS, Chin YK, Md Yasin IS (2021) Vaccine efficacy of a newly developed feed-based whole-cell polyvalent vaccine against vibriosis, streptococcosis and motile aeromonad septicemia in asian seabass. Lates Calcarifer Vaccines 9(4):368

Mohan CV (2007) 8.1 seed quality in freshwater fish production. Assess Freshw Fish Seed Resour Sustain Aquac 499

Mohapatra S, Chakraborty T, Kumar V, DeBoeck G, Mohanta KN (2013) Aquaculture and stress management: a review of probiotic intervention. J Anim Physiol Nutr 97(3):405–430

Moran D, Fofana A (2007) An economic evaluation of controlling three notifiable fish diseases in the United Kingdom. Prev Vet Med 80(2–3):193–208

Mukherjee, S C (2002) Fish diseases in India, their causes and control measures- winter school on recent advances in diagnosing and managing diseases in mariculture, 7th to 27th November 2002, Course Manual

Palić D, Scarfe AD, Walster CI (2015) Standardised approach for meeting national and international aquaculture biosecurity requirements for preventing, controlling, and eradicating infectious diseases. J Appl Aquac 27(3):185–219

Panchai, K, Hanjavanij, C, &Kitancharoen, N (2007) Characteristics of Achlyabisexualis isolated from eggs of Nile tilapia (Oreochromis niloticus Linn)

Patel AS, Patel SJ, Bariya AR, Pata BA, Ghodasara SN (2018) Fungal diseases of fish: a review. J Vet Sci Res 3(3):000–164

Salma U, Shafiujjaman M, Al Zahid M, Faruque MH, Habibullah-Al-Mamun M, Hossain A (2022) Widespread use of antibiotics, pesticides, and other aqua-chemicals in finfish aquaculture in the Rajshahi District of Bangladesh. sustainability 14(24):17038

Sandeep P, Chamundeswari Devi B, Kumar KP (2016) Present status of parasitic and bacterial diseases in fresh water fish seed farms in east Godavari District, Andhra Pradesh. Int J Appl Pure Sci Agric 2:117–121

Sharma MMM, Kapoor D, Loyal A, Kumar R, Sharma P, Husen A (2024a) Mitigation of drought stress in plants using gold nanoparticles. In: Husen A (ed) Plant response to gold nanoparticles. Springer, pp 193–209. https://doi.org/10.1007/978-981-97-7174-5_10

Sharma, M.M.M., Kapoor, D., Loyal, A., Kumar, R., Sharma, P. and Husen, A., (2024c). Environmental toxicity of engineered carbon nanoparticles. In: A.J. Bachheti, R.K. Bachheti, A. Husen, (eds) Carbon-based Nanomaterials: synthesis, agricultural, biomedical, and environmental interventions, 337–353.Springer https://doi.org/10.1007/978-981-97-0240-4_16

Sharma MMM, Kapoor D, Loyal A, Kumar R, Sharma P, Husen A (2025a) Plant response to silver nanoparticles in terms of growth, development, production, and protection: an overview. In: Husen A (ed) Plant response to silver nanoparticles. Springer, Singapore, pp 1–22. https://doi.org/10.1007/978-981-97-7352-7_1

Sharma MMM, Kapoor D, Rohilla R, Loyal A, Sharma P, Husen A (2024b) Role of essential minerals and rhizosphere in plant under changing environmental conditions. In: Husen A (ed) Essential minerals in plant-soil systems. Elsevier, pp 235–252. https://doi.org/10.1016/B978-0-443-16082-0.00002-3

Sharma P, Kapoor D, Kumar R, Husen A (2025b) Rhizosphere and osmolytes coordination in plant systems under adverse situations. In: Husen A (ed) Roles of Osmolytes in changing environment. Academic, pp 101–117. https://doi.org/10.1016/B978-0-443-26549-5.00020-5

Shefali, Kumar R, Sankhla MS, Sonone SS (2021) Impact of pesticide toxicity in aquatic environment. Biointerface Res Appl Chem 11(3):10131–10140. https://doi.org/10.33263/BRIAC113.1013110140

Shinn AP, Pratoomyot J, Bron JE, Paladini G, Brooker EE, Brooker AJ (2015) Economic costs of protistan and metazoan parasites to global mariculture. Parasitology 142(1):196–270

Tonguthai K (1997) Control of freshwater fish parasites: a Southeast Asian perspective. Int J Parasitol 27(10):1185–1191

Valencia-Sullca C, Jiménez M, Jiménez A, Atarés L, Vargas M, Chiralt A (2016) Influence of liposome-encapsulated essential oils on properties of chitosan films. Polym Int 65(8):979–987

Wolf K (2019) Fish viruses and fish viral diseases. Cornell University Press
Wu J, Liu H, Ge S, Wang S, Qin Z, Chen L, Zhang Q (2015) The preparation, characterisation, antimicrobial stability and in vitro release evaluation of fish gelatin films incorporated with cinnamon essential oil nanoliposomes. Food Hydrocoll 43:427–435
Yanong RP (2003) Fungal diseases of fish. VetY Clin Exot Anim Pract 6(2):377–400
Zorriehzahra MJ, Delshad ST, Adel M, Tiwari R, Karthik K, Dhama K, Lazado CC (2016) Probiotics as beneficial microbes in aquaculture: an update on their multiple modes of action: a review. Vet Q 36(4):228–241

Chapter 16
Antibacterial Activity of Green Synthesized Silver Nanoparticles against Fish Pathogens

Rishu Yadav, Rajender Kumar Gupta, Sunidhi, Nikki, and Pankaj Sharma

Abstract Nanotechnology has revolutionised the world by its effective use in research including drug delivery, medical science, therapeutic use of nanomaterials in different fields which has helped in diversifying the structures and devices. Nanoparticles can be prepared utilising different methods out of which the biological practice has proved to be eco- friendly and cost effective. Different metals like silver, zinc, copper, selenium, gold etc. can be utilized to synthesize nanoparticle out of which silver nanoparticles are much effective against the pathogenic bacterial as well as fungal strains. Excessive use of antibiotics has made fishes resistant to them so silver nanoparticles have a great efficacy against different strains of bacteria including *Aeromonas, Providencia, Psychrobacter, Moraxella, Pseudomonas fluorescens, Acinetobacter johnsonii*. Many diseases causing pathogens have been found to infect the fishes and are known to cause serious damage to different organs of fishes. The antimicrobial qualities silver and other nano metals are utilised in an extensive range of medical, industrial, agricultural, packaging, cosmetic and health-related fields. Size, shape, stability and various other factors affect the efficacy of silver nanoparticles as smaller is the size higher is the effectiveness of silver nanoparticles.

Keywords Antibacterial · Bactericidal · Fishes · Nano plates · Nanoparticles · Pharmaceutical

---

Rishu Yadav (✉) · R. K. Gupta · Sunidhi · Nikki
Department of Zoology, CCS Haryana Agricultural University, Hisar, Haryana, India
e-mail: rishuydv871@gmail.com

P. Sharma
Department of Ophthalmology & Visual Sciences, University of Illinois at Chicago, Chicago, USA
e-mail: sharmap@uic.edu

R. Kumar et al. (eds.), *Sustainability in Aquaculture and Fisheries*,
Springer Water, https://doi.org/10.1007/978-3-032-16605-0_16

16.1 Introduction

Nanotechnology is a swiftly distending field of modern research which entangles designing, synthesizing, characterizing, forming different kinds of structures and implementing them into systems by manipulating size and structure at the nanoscale. Furthermore, a recent development in nanotechnology is called bio-nanotechnology, which adds on the concepts of biology along with various physical and chemical processes to produce nanoparticles accompanied by particular purposes (Keat et al. 2015). The particles with minimum one dimension between 1 and 100 nm are known as nanoparticles (NPs). Their tiny size outcomes in distinctive physicochemical characteristics lead to enhanced features related to mechanical, optical, magnetic and structural in contrast to the initial bulk material (Beer et al. 2012; Sharma et al. 2024a).

On broader scale nano metals can be generated via different methods including biological as well as chemical methods. The biological method has gained a tremendous attention popularly called as green synthesis which is environment- friendly and cheaper. On the other hand, chemical method is costly, deteriorates the environment, includes use of costly and lethal chemicals. Owing to their unusual characteristics, nanoparticles have found a comprehensive usage in industries and their products such as antibacterial agents, consumer goods, in coating of various medical equipments, optical sensors, beauty products; medical and food industries; diagnostics, orthopaedics, delivery of different drugs and eventually, they have ameliorated the tumour supressing and killing capabilities of the medicines that have been used in treating cancer. AgNPs have been widely utilized in wound dressings, devices used in medical field, keyboards and many textiles recently. As nanoparticles have effective surface-to-volume ratio so because of their uniqueness and high capacity to absolutely alter physico- chemical as well as the biological features, there is multipurpose cast-off of metallic nanoparticles. (Zhang et al. 2016; Sharma et al. 2024b). The morphological characteristics of metallic nanoparticles make them interesting for a variety of requisitions together with sensing, optics, antibacterial activity and data storage.

The antibacterial features of various particles made from nano metals like colloids of silver, seems to be indirectly proportional to their size; – i.e., smaller the silver nuclei, greater the antibacterial activity. Furthermore, the morphological features and chemical-physical environment all affect their catalytic activity. Hence, it is of great importance to generate nanoparticles of a defined range. Usually, altering the synthesis techniques, agents which assist in reducing as well as in stabilising allows for precise control over their size distribution and shape (Nour et al., 2010). Various metals like silver, gold, zinc, selenium and many more have been utilised in developing nano materials; however, nano material prepared from silver have been found to be most efficient having the robust antimicrobial activity against bacteria, viruses and various microbes. Now a day's findings have reported that nanoparticles made from silver and metallic silver as well can be utilized in medicine for different purposes, together with treating burns, coating stainless steel materials, treating water, treating textiles, treating burns, and making sunscreen lotions. They also exhibit low volatility, high thermal stability, and minimal harm to human cells. (Rai et al. 2009).

16.2 Green Synthesis of Silver Nanoparticles

Fresh neem leaves were garnered from neem plant which were rigrously cleansed thrice with demineralised water so as to get rid of the dust particles and were finely crushed using motor and pestle. 10 gm of finely crushed leaves were put into 100 ml of water and boiled. This temperature minimizes the phytochemical degradation in the neem extract while simultaneously promoting the reduction reaction. Increased temperatures could speed up the reduction process, but they could also cause the bioactive compounds to degrade thermally, which could reduce the effectiveness of the biological method. On the other hand, when silver ions are transformed or converted into nanoparticles low temperature could hinder the reduction kinetics and cause incomplete or prolonged synthesis times (Ramya et al. 2024). After boiling, the solution was purified using filter paper. To prepare the solution, 1 mM of $AgNO_3$ was dissolved in 90 ml of demineralized water and then 10 ml of neem leaf extract was added to it as shown in Fig. 16.1. This mixture was stirred vigorously at an optimum temperature. During this period the conical flask was enclosed with aluminium foil to avoid photo activation of silver nanoparticles. After vigorous shaking the colour of solution turned dark brown from transparent which confirmed the formation of silver nanoparticles demonstrated in Figs. 16.1 and 16.2. Instead of vigorous shaking this whole solution can be kept overnight in dark room which also shows the similar results i.e. formation of silver nanoparticles.

The neem extract constitutes of various phytochemicals which helps in reduction process putting an ease in production of silver nanoparticles. Bioactive substances with reducing and stabilizing qualities such as terpenoids, phenolic compounds and flavonoids are abundant in the extract of neem plant. The silver ions (Ag^+) and chemicals present in the neem extract react and function as strong reducing as well as stabilising agents, which helps Ag^+ ions get converted into silver nanoparticles(Ag). The Ag^+ ions are reduced as there is transfer of electrons from the phytochemicals to the ions results in the production of silver nanoparticles. Furthermore, neem extract has plenty of phytochemicals that are essential for maintaining the stability of the generated nanoparticles, stopping their aggregation and guaranteeing their even dispersion (Ramya et al. 2024).

Silver nanoparticles can be characterized using various techniques like PSA (Potential size analyser), XRD, SEM, TEM, UV Visible, FTIR. The average particle size can be measured using PSA and stability of nanoparticles can be analysed using zeta potential.

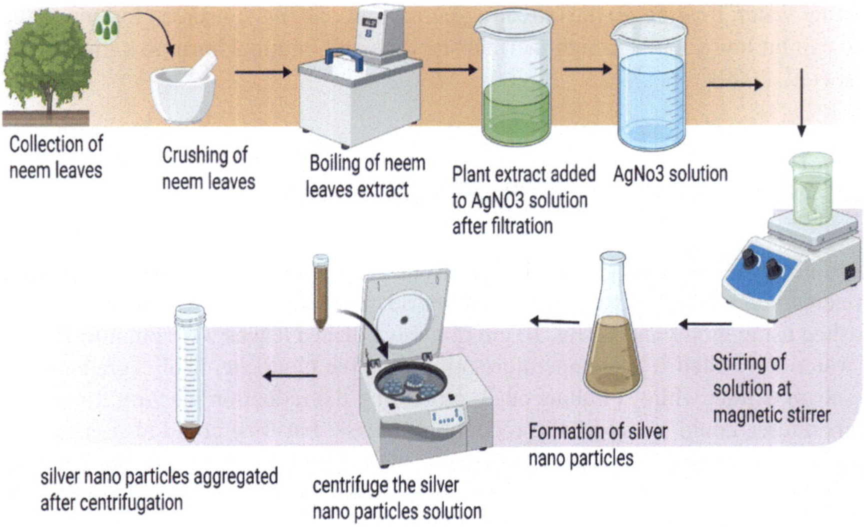

Fig. 16.1 Process of formation of silver nanoparticles

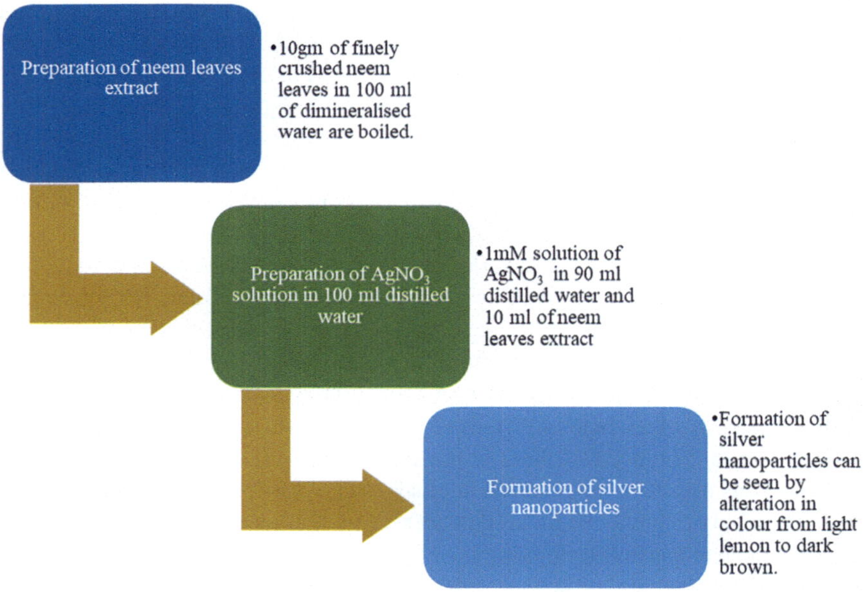

Fig. 16.2 Formation of silver nanoparticles

16.3 Antibacterial Property of Silver Nanoparticles

Microorganisms, including bacteria, molds, yeasts and viruses can infect humans in their daily environments. There has been a lot of research done on antibacterial materials made of different inorganic and natural substances. A lot of research has been done on nanoparticles which have been made from metals, having a high specific surface area as well as surface atoms. Their specific physical and chemical properties, which include catalytic activity, optical, electronic, and magnetic qualities have made them part of keen observation. Among all the Me-NPs silver nanoparticles have been found to have bactericidal and inhibitory effects (Shahverdi et al. 2007). The pathogens responsible for disease are becoming increasingly resistant to new antibiotic treatments these days. Researchers have reported using metallic nanoparticles, such as nanoparticles made from silver (AgNPs), act as efficient antimicrobial agent in a lot of studies to address this problem. AgNPs are primarily used in cure of burns and other abrasions in order to prevent infectious pathogens and wound pathogen contamination. AgNPs' advantageous physicochemical characteristics play a major role in research and medicine. Antifungal, antibacterial, anti-angiogenesis anti-inflammatory, antiviral, and antiplatelet characteristics are present in AgNPs (Kumar et al. 2021). Particularly in comparison to bulk silver metal, it is expected that Ag-NPs' strong antimicrobial action is because of high fraction of surface atoms and surface area. As of late, it has been observed that there is resistance to antibiotics by the pathogenic bacteria, which poses a serious threat to health of people and other organisms. In this investigation, Ag-NPs were assessed for their potential to boost up antimicrobial efficacy of various antibiotics against *Escherichia coli* and *Staphylococcus aureus.*

Since ancient times, the antimicrobial property of silver metal salts are known to have great impact and attention that's why even today, Ag is employed in a range of applications, such as burn wounds etc. Indeed, it is renowned that silver ions and compounds based on silver have a strong biocidal effect on up to 12 different species of bacteria, including *E. coli*, and are extremely toxic to microorganisms. Recently, it was demonstrated that hybrids of amphiphilic hyper branched macromolecules and Ag nanoparticles act as antimicrobial surface coating agents having a great specificity and efficacy (Kim et al. 2007).

16.3.1 Silver Nanoparticles and Their Antimicrobial Activity

Since the beginning of time, metallic silver and its compounds have been known to have antibacterial qualities. Silver is harmless for human cells at low quantities but deadly for most germs and viruses, which is why they are frequently employed in daily life to disinfect food and water and in medicine to prevent infections (Krutyakov et al. 2008). Widely used antiseptics based on silver have a lower potential to make bacteria resistant to them. The excellent binding efficiency and responsiveness with cell disruption and death are caused by proteins that alter the structure of the bacterial cell membrane and nuclear membrane (Das et al. 2020). The ability of microbes to evolve resistance to silver through mutations appears unlikely, given silver ions target a wide range of different proteins within a cell because there are now more pathogenic bacterial species that are resistant to narrow spectrum antibiotics and pose a major threat to human life and health, this priceless characteristic has become even more crucial (Wright 2005).Silver nanoparticles are also efficient to kill the viral particles as glycoproteins in the viral membrane are bound by silver nanoparticles. Silver nanoparticles readily attach to the thiol groups found in cysteine-rich residues seen in TCRV glycoproteins. Three phenomena result from this interaction:

a. It can prevent the internalization of the viral particle by interfering with receptor binding.
b. It can internalize silver nanoparticles that prevent the viral replication of TCRV L protein; and bind to the viral glycoprotein to stop the viral coating in the endosome.
c. Ag-NPs inactivate the virus prior to its entry into the cell since they have no impact on viral replication in cells treated with them (Khandelwal et al. 2014).

16.3.2 Defence Mechanism of Silver Nanoparticles

Silver is well recognized for its antimicrobial properties and uses in combating infections. Because of its high binding force to sulphur, it strongly binds to the thiol (-SH) groups of vital compounds in living organisms and neutralizes them. Moreover, it modifies the dynamics of the cellular membrane, resulting in reduced membrane stability and energy levels. Changes in membrane dynamics result in the death of microorganisms. Silver nanoparticles (AgNPs) operate through the same mechanism as silver ions, but they also possess an extra benefit of dimension that enables their straightforward absorption into the cells (Dube 2024). Silver ions (Ag^+) exhibit a high affinity for the peptidoglycan matrix of bacterial cell walls, thereby obstructing oxygen diffusion into the cytoplasm, ultimately leading to bacterial immobilization. Additionally, silver exhibits reactivity toward phosphorus and sulphur containing biomolecules—key constituents of the plasma membrane, proteins, and nucleic acids—which facilitates its accumulation on the cell envelope. This interaction induces cytoplasmic contraction, membrane dissociation, pore formation, and eventual membrane rupture, culminating in cellular lysis. Intracellularly, Ag^+ ions interact with nucleotides and cytochrome complexes, impairing phosphate-associated energy pathways and inhibiting DNA replication and mitotic processes. Silver nanoparticles (AgNPs), typically ranging from 1 to 100 nm in diameter and characterized by elevated surface energy, enable the sustained release of Ag^+ ions into the surrounding milieu, thereby extending their antimicrobial potency over time (Huang et al. 2015; Shaalan et al. 2018; Tang and Zheng, 2018; Sharma et al. 2025a). A notable advantage of AgNPs over traditional antimicrobial agents is their multifaceted mechanism of

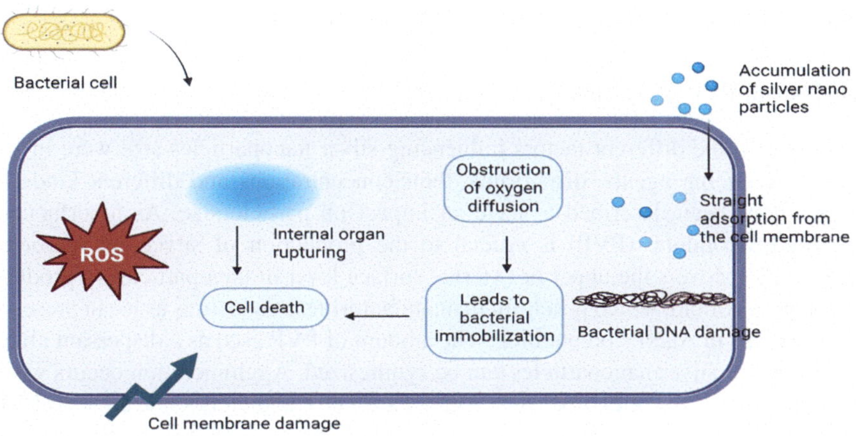

Fig. 16.3 Cellular damage done by silver nanoparticles

action, targeting several essential bacterial pathways simultaneously, which significantly reduces the probability of resistance development. Cell death may ensue from the internal organelles rupturing due to cytoplasmic membrane denaturation. Additionally, oxidative stress in the bacterium was caused by blocking channels that transport electrons, which disrupts signal transduction and causes cell lysis. Another theory that has been put forth is that AgNPs interfere with DNA replication and ATP synthesis by affecting the bacteria's ribosomes and deoxyribonucleic acid. AgNPs have better antimicrobial activity against *A. hydrophila* than against *F. aquidurense. A. hydrophila* is more effectively inhibited by silver nanoparticles than *F. aquidurense*. The disruption of plasma membrane done by release of silver ions due to silver nanoparticles as shown in Fig. 16.3 is probably responsible for the possible antibacterial effect. The sticking of ions to the bacterial cell wall as it is negatively charged, cause it to burst and denaturing of the proteins that are responsible for killing of the bacterial cell (Nugroho et al. 2024). By employing nanoparticles either by themselves or in conjunction with widely used medications, a new phase of antimalarial treatment is developing, potentially involving an inventive therapeutic approach to the treatment of malaria (Rai et al. 2017).

All the aforementioned studies have demonstrated that silver nanoparticles have great potential to inhibit the growth of bacteria.

16.4 Factors Affecting Silver Nanoparticles

(a) Role of PVP (Polyvinylpyrrolidone)

Additionally, the different factors influencing silver nanoparticles size were investigated. Reducing agents, dispersants, their concentrations and different kinds of amines have been described to have an impact on particle size. As a surfactant, polyvinylpyrrolidone (PVP) is crucial to the production of silver nanoparticles because PVP covers the upper or external surface layer of nanoparticles to produce unchanging colloid, which is helpful in manufacturing in industries as it can preserve nanoparticles in AgNPs preparation. The amount of PVP used as a dispersant affect how big sized silver nanoparticles can be synthesized. Agglomeration occurs when insufficient PVP is applied because AgNPs are not completely covered with PVP. To create small-sized AgNPs, the ideal PVP/$AgNO_3$ weight ratio should be used. In this instance, PVP is applied on AgNPs' whole surface. As a result, PVP keeps the silver particles apart, preventing agglomeration. But as more PVP is added, the size of the metal nanoparticles becomes larger suggesting that the colloid's size is not further decreased by additional PVP. Overuse of PVP either doesn't seem to improve the silver colloid's surface coverage or might prevent the nanoparticles from forming (Natsuki et al. 2015). Large particle sizes, small-size AgNPs are far more toxic (Scherer 2019) as with small diameters they can act directly in cells by passing through the cell membrane, whereas the entry or exit of silver nanoparticles with greater diameters may be through ion channels. However, small particles of silver nanoparticles can better function because they have larger specific surface area (Nie and Xu 2023). Nanomaterial shapes are distinct including rings, spheres, flakes, tubes, lines, and irregularities (Roy 2017). Particles with different shapes interact differently with cells and plasma membranes as well as DNA due to variations in their precise surface area, surface charge, zeta potential, and chemical interfaces. Additionally, the chances of damage in the cells are much higher in irregular particles (Choi 2021). Various factors like temperature, surfactant and dispersing agents affect the shape of silver nanoparticles.

(b) Temperature

Higher temperature generated irregular shaped nanoparticles while lower temperature spherical shaped and small sized nanoparticles (Yakoob et al., 2020). Ait was observed when reaction mixture swiftly became brown at 50 °C, and the end product was collected as a black powder. The TEM picture of AgNPs revealed the formation of massive silver particles that appear to be connected rather than spherical. This is most likely the result of cooperation between the PVP-covered silver particle surface at high temperatures. This suggests that when there is a rise in temperature the rate of reaction also increases considerably, leading to synthesis of bigger particles and happening of undesired reaction. When the temperature was around 10 °C then the reaction proceeded in an extremely sluggish manner and there was no alteration in the colour of solution for about three hours (Natsuki, 2011). Silver ion reduction in aqueous solution is significantly aided by the amine. Na_3Ct is a reducing agent that may be employed as a dispersion agent. It is a poor reductant, though, and the pH of the solution affects how much it can reduce. Higher temperatures (100 °C) are sometimes needed to enhance reduction when Na_3Ct is utilized alone as a reducing agent. However, in our investigation, the existence of DMAE causes the reactions to happen quickly at room temperature (Natsuki et al. 2015). The reaction can only take place with a very tiny amount of the reactant. Ag^{++} will not be successfully reduced if there is an increase in the concentration of reactant then accumulation may become apparent. The amount of time that the reaction mixture is stirred is another crucial element for synthesis. The amount of time needed for the synthesis of silver nanoparticles, beginning with the addition of the reactant to the beaker, is known as the reaction stirring time.

(c) Reducing Agents

Out of various factors affecting silver nanoparticles (Fig. 16.4), the interaction between the silver salt and the reducing complex plays a major role. The right interaction between the silver salt and the reducing complex components in the test leaf extract depends on the stirring time. Silver salt will be reduced more quickly by the plant with higher quantity of phytochemicals; in other words, it will take longer for the plant with fewer reduced compounds to reduce silver salt (Lade and Shanware 2020).

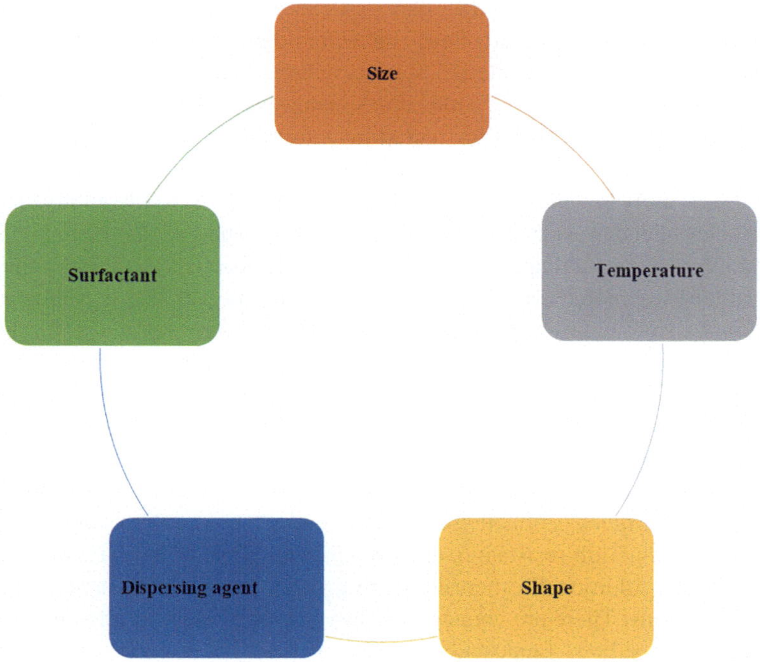

Fig. 16.4 Factors affecting silver nanoparticles

16.5 Applications of Silver Nanoparticles

Nanoparticles made from silver have varied applications in different fields, as shown in Fig. 16.5. Numerous commercial products, such as apparel and textiles, furniture, cosmetics, appliances in the home, contain these nanoparticles. AgNPs have strong antimicrobial and antifungal qualities, which have led to their high degree of commercialization; however, many manufacturers assert that possible AgNP toxicity is negligible. It is well known that silver ions and nanoparticles have such significant qualities that inhibit the microbial growth. Silver effectively eliminates existing bacteria while also having a suppressive effect on bacterial growth. According to recent reports, AgNPs have toxic effects on the majority of pathogenic bacteria that are threat to human health during wound healing as well as recovery period following surgery. The bacteria which are responsible for causing the majority of infections which are food and water borne, similar outcomes were noted. Bactericide effects were noted in *Salmonella, Choleraesuis, E. coli, P. aeruginosa, S. aureus, Bacillus cereus and Listeria innocua.* (Pantic 2014).

Antimicrobial qualities are also present in silver nanoparticles. Ag in nano forms is reportedly thought to cause less toxicity than Ag^{+} ions). Due to their antibiotic qualities, silver nanoparticles and other nano metals are used in a wide range of industrial, agricultural, packaging and health-related fields and cosmetics (Pattabi and Pattabi 2013).

The imaging applications of silver nanoparticles over the past 20 years has been a growing advancement in the field of nano biotechnology (Choi and Wang 2011). With their many useful optical characteristics, silver nanoparticles have made it possible to use them in sensing and imaging applications in new ways. They provide a variety of modes that are helpful in detection, including colorimetric, scattering, MEF (Metal Enhanced Fluorescence), SERS (Surface Enhanced Raman Spectroscopy), techniques at incredibly low limits of detection (Caro et al. 2010). Applications for silver nanoparticles include vast areas like healing of wound. The first commercially available wound dressing, Acticoat, having two layers composed of membranes made up of polyamide ester coated with silver nanoparticles (Reidy et al. 2013). Nanoparticles of silver are also found to help keratinocytes proliferate and migrate, inhibit fibroblasts from forming collagen, and alter the quantity of cytokines generated (Wong and Liu 2010). Additionally, it was noted that the dressing containing silver nanoparticles has the advantageous ability to shield the damaged tissue from bacterial infection (Sibbald et al. 2007).

It has been observed that the methyl methacrylate bone cement equipped with 1% silver nanoparticles is effective enough to hinder the growth of *Staphylococcus* epidermidis, methicillin-resistant *S. epidermidis*, and methicillin-resistant *S. aureus* (Alt et al. 2004). Additionally, there are some significant uses for silver nanoparticles in dental materials. There are greater chances of having bacterial infections during bone grafting procedures. Large and the extremely ruined bone defects are frequently replaced or restored with bone grafts (Burdușel et al. 2018). AgNPs directly affect the bacteria by interacting with the cell wall's peptidoglycan and plasma membrane,

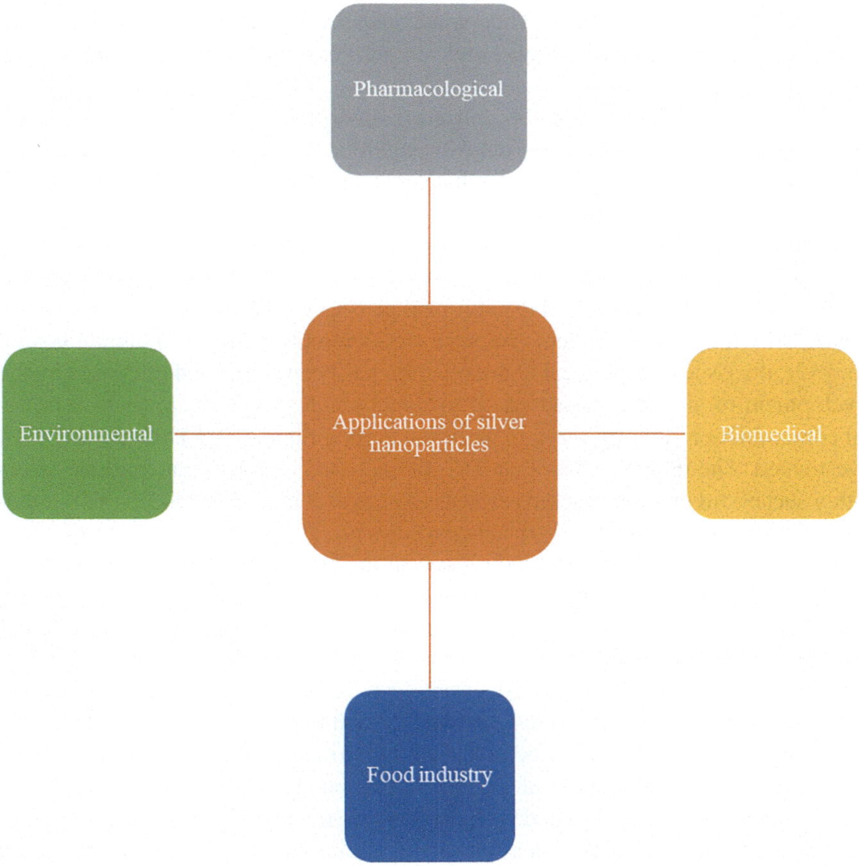

Fig. 16.5 Applications of silver nanoparticles in various fields

which prevents or reduces the growth of new or mature biofilm (Guo et al. 2019). Furthermore, AgNPs can prevent cell respiration and damage DNA via two methods which include Reactive Oxygen Species also denoted as ROS and non-ROS pathways (Khorsandi et al. 2021). According to a study, if silver nanoparticles are added to a composite resin at concentrations which vary considerably has positive mechanical effects together with significant antimicrobial potential (Cheng et al. 2012). When used in endodontic filling materials, NSPs significantly enhanced the anti-bactericidal effect against various bacteria *Enterococcus faecalis*, *S. aureus* and *Streptococcus milleri* (Magalhaes et al. 2012).

Additionally, further use of silver nanoparticles can be seen as drug carriers in drug delivery, as implants in cardiovascular, anti-inflammatory medications, contact lenses, cancer testing and treatment (Liu et al. 2012). Because local drug targeting raises local drug concentrations, it has certain drawbacks; therefore, require some more potent drug delivery techniques. These tactics are made possible by particular particles found in nanoparticles. Their small size enables them to bind and stabilize proteins, penetrate cell membranes, and release lysosomes following endocytosis (De et al. 2008).

Silver nanoparticles can also be utilized in environmental treatment. Numerous microorganisms in the air are extremely dangerous to humans. Airborne particles with biological origins are known as bio aerosols. These include bacteria, fungi, and viruses, all of which are capable of producing allergic and contagious illnesses. According to a WHO report, air handling systems are responsible for 50% of the microbiological hazards in indoor air, and air filters have been found to harbour harmless microorganisms like bacterial and fungal pathogens. The mycotoxins which are produced by different pathogenic bacteria are extremely harmful to human health. However, the issue of biological contamination has been somewhat mitigated by the development of air filters based on silver nanoparticles. AgNPs' impact on bacterial contamination removal through activated carbon filters (ACF) has already been investigated. The findings show that ACF filters based on silver nanoparticles were highly successful in removing bio aerosols (Agrawal et al. 2018). Due to their ability to inhibit microbial growth, silver nanoparticles are also being utilized to preserve food or extend its shelf life. In food preservation, silver nano particle coated paper may be helpful in halting microbial growth for extended time duration. The sono chemical coating, a straightforward technique for creating a coating of colloidal silver on paper utilising ultrasonic radiation was presented (Gottesman et al. 2011).

Nanomaterials can be employed as catalysts in power generation devices and automotive catalytic converters to react with certain harmful gases like CO and NO preventing pollution of environment caused by burning coal and gasoline (Sharma and Bhargava 2013; Sharma et al. 2025b). Additionally, scrub suits for healthcare professionals that are impregnated with silver can be made using silver nanoparticles. Freeman and his colleagues looked into the antimicrobial properties of scrub suits impregnated with silver that were utilized in a veterinary hospital (Freeman et al. 2012).

Paints containing Ag-NPs are particularly intriguing because of their possible bactericidal properties. Investigations were conducted into the eco–friendly chemical method of creating metal nanoparticles that could be embedded into paint instead of using regular household paint in a single step (Kumar et al. 2008). Consequently, the surfaces painted with nanoparticles made from silver exhibited antimicrobial qualities against human pathogens which include *E. Coli* and *S. aureus* (Agrawal et al. 2018). Water is basic necessity even then a huge number of people do not have enough resources to utilize safe and clean drinking water. Using 3-aminopropyltriethoxysilane (APTES) that acts as a linking molecule, the AgNPs created by the chemical process can be accumulated onto the porous materials made up of ceramics to create the Ag-NPs-porous ceramic composite. The sterilizing properties of the AgNPs-porous ceramic composite water filters were evaluated. It was discovered that when the input water had a bacterial load of approximately 105 CFU/ml, the output count of *E. coli* was zero at a flow rate of 0.01 l/min (Yakub et al. 2012).

16.6 Properties of Silver Nanoparticles

A thorough understanding of AgNPs' characteristics is essential for optimizing their potential uses. To understand how AgNPs affect the environment, their benefits and drawbacks must be precisely measured and qualified (Syafiuddin et al. 2017). When assessing AgNPs' biological interactions and effects, certain physicochemical characteristics—such as size, shape, charge as well as coating of the surface, agglomeration, and dissolution rate—are crucial. The tiny silver nanoparticles having larger surface area are much toxic (Johnston et al. 2010). It is commonly known that the physico- chemical characteristics of silver nanostructures can be significantly impacted by their shape. In the biomedical industry, silver nanostructures such as spherical nanoparticles, wires made from nano metals, nanorods, and nanocubes are commonly utilised (Rycenga et al. 2011). Accordingly, the various surface charges of AgNPs' coatings influence how they interact with living systems and determine their biological effects. Most engineered nanoparticles are known to agglomerate. It was demonstrated that AgNPs aggregate in culture media as well as in the cell's cytoplasm and nuclei of $HepG_2$ cells (Wei et al. 2015). By a process called the excitation of localized surface plasmon resonance, the electromagnetic radiations are absorbed by AgNPs in the visible spectrum between 380 and 450 nm. The surface plasmon resonance of 7 nm spherical AgNPs is at 410 nm, whereas that of 29 nm spherical AgNPs is at 425 nm. Furthermore, silver nanoparticles with a diameter of 89 nm showed a broader band that peaked at 490 nm. The width of surface plasmon resonance was found to be correlated with the size distributions of the nanoparticles (Syafiuddin et al. 2017). Their special qualities have recently been used to their advantage as sensors because of their sensitive colorimetric recognition of chromium in vegetable samples and surface waters together with industrial wastewater (Shrivas et al. 2016). Additionally, the crucial micelle concentration of cationic surfactants was determined using their superior properties (Salem et al. 2016). Additionally, silver nano-based sensor offers way to conduct ultrasensitive biodetection tests using incredibly lightweight, compact, reliable, and inexpensive equipment. AgNPs have been used as efficient catalytic agents to reduce distinct dyes, including methyl orange, eosin, rose bengal, methylene blue, yellow-12 and 4-nitrophenol (Syafiuddin et al. 2017). AgNPs can be used in electronic devices because of their special electrical characteristics. AgNPs grown in glass-ceramic with different sizes alternating from 4 to 12 nm were tested for electrical conductance (Roy and Chakravorty 1990). As an alternative, AgNPs were used in electronically conductive adhesives (ECAs) as conductive fillers. In contrast to the medium without AgNPs, it was discovered that the existence of AgNPs in ECAs reduced the resistivity, suggesting that it highly conductive. A forementioned results demonstrate that AgNPs may be used in electrical devices (Syafiuddin et al. 2017).

16.7 Fish Pathogens

There are a variety of bacterial species can be found in fish living in their natural habitats. Constant exposure to tainted water can cause bacterial colonization on fish skin, gills and adulterated feed or water can have an impact on the digestive tract. Muscle contamination in fish is also feasible in cases of compromised immunological resistance (Guzman et al. 2004).

The bacteria which are present on fish skin are commonly believed to be the same as those found in the contaminated water. These bacteria include members of the genera and species *Aeromonas (including A. hydrophila, A. bestiarum, A. schubertii, A. caviae,* and *A. veronii, Flexibacter, Proteus, Providencia, Psychrobacter, Moraxella, P. fluorescens, A. johnsonii, Alcaligenes piechaudii, Enterobacter aerogenes, E. coli, Micrococcus luteus, and Vibrio fluvialis* (Novoslavskij et al. 2016).

Numerous seafood items like fish, crabs, clams, oysters, lobsters, squid, cuttlefish mussels and octopuses, were found to contain *Salmonella.* Salmonella's capacity to form biofilms may also help it survive in freshwater reservoirs and other environments. The enteric red mouth disease is caused by pathogenic bacteria known as *Yersinia ruckeri* and is one of the species in the genus. Foodborne botulism is frequently linked to *Clostridium botulinum,* a member of the genus *Clostridium. C. botulinum* is present in soil and water environments and is widely distributed in nature. Because *C. botulinum* produces botulinum neurotoxin, it is the causative agent of botulism. *Salmonella, Listeria monocytogenes, Vibrio* spp., *Clostridium botulinum* and *Yersinia* spp. are among foodborne pathogens that frequently contaminate fish, demonstrating the microbiota of the surrounding water. Fish together with products made from fish can be a great source of bacteria having devastating effect to humans, so contaminating their natural habitat may have an impact on fish stocks as well as public health. A number of factors may impact the occurrence of the pathogenic bacterial strains in the fish, including human, contaminated water sources together with inadequate hygiene during fish capture, handling, and transportation. The unique capacity of these bacteria to endure in the environment raises the risk of these microorganisms (Novoslavskij et al. 2016).

16.7.1 Efficacy of Silver Nanoparticles Against Fish Pathogens

The usage of antibiotics was effective to treat bacterial infections. When there is an excessive usage of antibiotics, it causes bacteria to develop resistance against the antibiotics. Tetracycline was discovered to be maximum used antibiotic in fish farms. Hence, development of bacteria resistant to antibiotics and the excessive usage of antibiotics in aquaculture are strongly correlated. Numerous resistant bacteria have been observed to emerge, such as methicillin-resistant *Staphyloccus aureus* and multi- and super-drug resistant strains (Kumar et al. 2022a, 2022b, 2022c; Shaalan et al. 2016).

Aeromonas hydrophila isolates from tilapia in culture exhibited resistance to broad spectrum antibiotics, including erythromycin, streptomycin, and tetracycline (Son et al. 1997). *Yersinia ruckeri, Edwardsiella species, Aeromonas salmonicida, Vibrio, Listeria, Pseudomonas,* and *Photobacterium damselae subsp. piscicida* were also found to be resistant to antibiotics. It has been studied how to use nanoparticles as substitute antimicrobials to stop the development of resistant microorganisms in aquaculture against the antibodies (Seil and Webster 2012). Strong antimicrobial activity against viruses, fungi, and bacteria has been demonstrated by metal nanoparticles.

16.8 Conclusion

Silver nanoparticles are synthesised using various methods, out of which the most eco-friendly method i.e. biological method has proved to be effective against various fish pathogens. Now a day, it has become more difficult to treat fishes using drugs as they have become resistant to various medicines. Therefore, silver nanoparticles possessing antibacterial activities are effective enough to cure multiple diseases. AgNPs containing bactericidal properties find their utility in various fields making nanotechnology as a powerful tool to the present as well as upcoming world.

16.9 Future Prospectus

Silver nano particles have wide applications in distinct fields so it requires further explorations about their environmental impacts. Synthesis methods need further optimization to get consistent shape, size and stability. Their dose must be optimized so that they can be utilized efficiently in treating fishes by minimising dose of antibiotics. Toxicity of dose must be investigated to ensure safe usage of silver nano particles.

Acknowledgement I would like to express my sincere gratitude to Dr. R.K. Gupta, Dr. Pankaj Sharma, Sunidhi, Vijay Khushwaha and Nikki who encouraged me during the preparation of this book chapter. I also extend my appreciation to the editors and publishers for providing me the opportunity to contribute to this work. Lastly, I am grateful to my family for their constant support and motivation throughout this journey.

References

Agrawal S, Bhatt M, Rai SK, Bhatt A, Dangwal P, Agrawal PK (2018) Silver nanoparticles and its potential applications: a review. J Pharmacogn Phytochem 7(2):930–937

Alt V, Bechert T, Steinrücke P, Wagener M, Seidel P, Dingeldein E, Domann E, Schnettler R (2004) An in vitro assessment of the antibacterial properties and cytotoxicity of nanoparticulate silver bone cement. Biomaterials 25(18):4383–4391

Beer C, Foldbjerg R, Hayashi Y, Sutherland DS, Autrup H (2012) Toxicity of silver nanoparticles—nanoparticle or silver ion? Toxicol Lett 208(3):286–292

Burduşel AC, Gherasim O, Grumezescu AM, Mogoantă L, Ficai A, Andronescu E (2018) Biomedical applications of silver nanoparticles: an up-to-date overview. Nano 8(9):681

Caro C, Castillo PM, Klippstein R, Pozo D, Zaderenko AP (2010) Silver nanoparticles: sensing and imaging applications. Silver Nanoparticles 201

Cheng L, Weir MD, Xu HH, Antonucci JM, Lin NJ, Lin-Gibson S, Xu SM, Zhou X (2012) Effect of amorphous calcium phosphate and silver nanocomposites on dental plaque microcosm biofilms. J Biomed Mater Res B Appl Biomater 100(5):1378–1386

Choi D, Hwang J, Bang J, Han S, Kim T, Oh Y, Hwang Y, Choi J, Hong J (2021) In vitro toxicity from a physical perspective of polyethylene microplastics based on statistical curvature change analysis. Sci Total Environ 752:142242

Choi J, Wang NS (2011) Nanoparticles in biomedical applications and their safety concerns. Biomed Eng Theory Appl 29:486

Das CA, Kumar VG, Dhas TS, Karthick V, Govindaraju K, Joselin JM, Baalamurugan J (2020) Antibacterial activity of silver nanoparticles (biosynthesis): a short review on recent advances. Biocatal Agric Biotechnol 27:101593

De Jong WH, Borm PJ (2008) Drug delivery and nanoparticles: applications and hazards. Int J Nanomedicine 3(2):133–149

Dube E (2024) Antibacterial activity of engineered nanoparticles against fish pathogens. Aquac Rep 37:102240

Gottesman R, Shukla S, Perkas N, Solovyov LA, Nitzan Y, Gedanken A (2011) Sonochemical coating of paper by microbiocidal silver nanoparticles. Langmuir 27(2):720–726

Guo J, Qin S, Wei Y, Liu S, Peng H, Li Q, Luo L, Lv M (2019) Silver nanoparticles exert concentration-dependent influences on biofilm development and architecture. Cell Prolif 52(4):e12616

Guzmán MC, de los Angeles Bistoni M, Tamagnini LM, González RD (2004) Recovery of Escherichia coli in fresh water fish, Jenynsia multidentata and Bryconamericus iheringi. Water Res 38(9):2368–2374

Huang S, Wang L, Liu L, Hou Y, Li L (2015) Nanotechnology in agriculture, livestock, and aquaculture in China. A review. Agron Sustain Dev 35:369–400

Johnston HJ, Hutchison G, Christensen FM, Peters S, Hankin S, Stone V (2010) A review of the in vivo and in vitro toxicity of silver and gold particulates: particle attributes and biological mechanisms responsible for the observed toxicity. Crit Rev Toxicol 40(4):328–346

Keat CL, Aziz A, Eid AM, Elmarzugi NA (2015) Biosynthesis of nanoparticles and silver nanoparticles. Bioresour Bioprocess 2:1–11

Khandelwal N, Kaur G, Kumar N, Tiwari A (2014) Application of SIlver nanoparticles in viral inhibition: a new hope for antivirals. Dig J Nanomater & Biostructures (DJNB) 9(1)

Khorsandi K, Keyvani-Ghamsari S, Khatibi Shahidi F, Hosseinzadeh R, Kanwal S (2021) A mechanistic perspective on targeting bacterial drug resistance with nanoparticles. J Drug Target 29(9):941–959

Kim JS, Kuk E, Yu KN, Kim JH, Park SJ, Lee HJ, Cho MH (2007) Antimicrobial effects of silver nanoparticles. Nanomedicine 3(1):95–101

Krutyakov YA, Kudrinskiy AA, Olenin AY, Lisichkin GV (2008) Synthesis and properties of silver nanoparticles: advances and prospects. Russ Chem Rev 77(3):233

Kumar Rajesh S, Tharani M, Rajeswari VD, Alharbi NS, Kadaikunnan S, Khaled JM et al (2021) Synthesis of greener silver nanoparticle-based chitosan nanocomposites and their potential antimicrobial activity against oral pathogens. Green Process Synth 10(1):658–665

Kumar A, Vemula PK, Ajayan PM, John G (2008) Silver-nanoparticle-embedded antimicrobial paints based on vegetable oil. Nat Mater 7(3):236–241

Kumar R, Gupta R, Dahiya T, Yadav R, Saifi R, Yodha K, Kumari B, Singh P, Shefali. (2022a) Surveillance of *Heteropneustes fossilis* (stinging catfish) diseases and health management practices in Haryana. Appl Ecol Environ Sci 10(4):261–267. https://doi.org/10.12691/aees-10-4-12

Kumar R, Gupta RK, Yadav R, Saifi R, Yodha K, Pooja. (2022b) *Eisenia fetida* as protein source for growth enhancement of *Heteropneustes fossilis*. Egypt J Aquat Biol Fish 26(2):577–588. https://doi.org/10.21608/ejabf.2022.234553

Kumar R, Yadav R, Kumar Gupta R, Pooja (2022c) A sustainable way for fish health management by replacement of chemical and drugs by earthworm. In: Environmental degradation in Asia: land degradation, environmental contamination, and human activities. Springer International Publishing, Cham, pp 329–352. https://doi.org/10.1007/978-3-031-12112-8_16

Lade BD, Shanware AS (2020) Phytonanofabrication: methodology and factors affecting biosynthesis of nanoparticles. In: Smart nanosystems for biomedicine, optoelectronics and catalysis. IntechOpen

Liu J, Zhao Y, Guo Q, Wang Z, Wang H, Yang Y, Huang Y (2012) TAT-modified nanosilver for combating multidrug-resistant cancer. Biomaterials 33(26):6155–6161

Magalhaes APR, Santos LB, Lopes LG, Estrela CRDA, Estrela C, Torres ÉM, Bakuzis AF, Cardoso PC, Carrião MS (2012) Nanosilver application in dental cements. Int Sch Res Not 2012(1):365438

Natsuki J, Abe T (2011) Synthesis of pure colloidal silver nanoparticles with high electroconductivity for printed electronic circuits: the effect of amines on their formation in aqueous media. J Colloid Interface Sci 359(1):19–23

Natsuki J, Natsuki T, Hashimoto Y (2015) A review of silver nanoparticles: synthesis methods, properties and applications. Int J Mater Sci Appl 4(5):325–332

Nie P, Zhao Y, Xu H (2023) Synthesis, applications, toxicity and toxicity mechanisms of silver nanoparticles: a review. Ecotoxicol Environ Saf 253:114636

Novoslavskij A, Terentjeva M, Eizenberga I, Valciņa O, Bartkevičs V, Bērziņš A (2016) Major foodborne pathogens in fish and fish products: a review. Ann Microbiol 66:1–15

Nugroho RA, Hindryawati N, Aryani R, Manurung H, Sari YP, Nurhadi M, Nurti DD, Vieraldi M, Rudianto R, Prahastika W (2024) In vivo and in vitro assays using biosynthesized silver nanoparticles on Aeromonas hydrophila-infected Clarias gariepinus. J Appl Aquac 36(1):170–192

Pantic I (2014) Application of silver nanoparticles in experimental physiology and clinical medicine: current status and future prospects. Rev Adv Mater Sci 37:15–19

Pattabi RM, Pattabi M (2013) Antibacterial applications of silver nanoparticles. In: Materials science forum, vol 754. Trans Tech Publications Ltd., pp 131–142

Rai M, Ingle AP, Paralikar P, Gupta I, Medici S, Santos CA (2017) Recent advances in use of silver nanoparticles as antimalarial agents. Int J Pharm 526(1–2):254–270

Rai M, Yadav A, Gade A (2009) Silver nanoparticles as a new generation of antimicrobials. Biotechnol Adv 27(1):76–83

Ramya JR, Ali S, Vijayalakshmi R, Gajendiran J, Gnanam S, Ramachandran K (2024) Antimicrobial efficiency against fish pathogens on the green synthesized silver nanoparticles. Microb Pathog 193:106725

Reidy B, Haase A, Luch A, Dawson KA, Lynch I (2013) Mechanisms of silver nanoparticle release, transformation and toxicity: a critical review of current knowledge and recommendations for future studies and applications. Materials 6(6):2295–2350

Roy B, Chakravorty D (1990) Electrical conductance of silver nanoparticles grown in glass-ceramic. J Phys Condens Matter 2(47):9323

Roy E, Patra S, Saha S, Kumar D, Madhuri R, Sharma PK (2017) Shape effect on the fabrication of imprinted nanoparticles: comparison between spherical-, rod-, hexagonal-, and flower-shaped nanoparticles. Chem Eng J 321:195–206

Rycenga M, Cobley CM, Zeng J, Li W, Moran CH, Zhang Q, Qin D, Xia Y (2011) Controlling the synthesis and assembly of silver nanostructures for plasmonic applications. Chem Rev 111(6):3669–3712

Salem JK, El-Nahhal IM, Najri BA, Hammad TM (2016) Utilization of surface Plasmon resonance band of silver nanoparticles for determination of critical micelle concentration of cationic surfactants. Chem Phys Lett 664:154–158

Scherer MD, Sposito JC, Falco WF, Grisolia AB, Andrade LH, Lima SM, Caires AR (2019) Cytotoxic and genotoxic effects of silver nanoparticles on meristematic cells of Allium cepa roots: a close analysis of particle size dependence. Sci Total Environ 660:459–467

Seil JT, Webster TJ (2012) Antimicrobial applications of nanotechnology: methods and literature. Int J Nanomedicine:2767–2781

Shaalan M, El-Mahdy M, Theiner S, Dinhopl N, El-Matbouli M, Saleh M (2018) Silver nanoparticles: their role as antibacterial agent against Aeromonas salmonicida subsp. salmonicida in rainbow trout (Oncorhynchus mykiss). Res Vet Sci 119:196–204

Shaalan M, Saleh M, El-Mahdy M, El-Matbouli M (2016) Recent progress in applications of nanoparticles in fish medicine: a review. Nanomedicine 12(3):701–710

Shahverdi AR, Fakhimi A, Shahverdi HR, Minaian S (2007) Synthesis and effect of silver nanoparticles on the antibacterial activity of different antibiotics against Staphylococcus aureus and Escherichia coli. Nanomedicine 3(2):168–171

Sharma MMM, Kapoor D, Loyal A, Kumar R, Sharma P, Husen A (2025a) Positive and negative impact of carbon-based nanomaterials on the plant growth performance. In: Husen A (ed) Emerging carbon nanomaterials for sustainable agricultural practices: synthesis, plant growth, performance, production and protection. Springer Nature Singapore, pp 63–83

Sharma MMM, Kapoor D, Loyal A, Kumar R, Sharma P, Husen A (2024a) Effect of gold nanoparticles on seed germination, plant growth, and plant protection. In: Husen A (ed) Plant response to gold nanoparticles. Springer, pp 155–175. https://doi.org/10.1007/978-981-97-7174-5_8

Sharma MMM, Kapoor D, Loyal A, Kumar R, Sharma P, Husen A (2025b) Plant response to silver nanoparticles in terms of growth, development, production, and protection: an overview. In: Husen A (ed) Plant response to silver nanoparticles. Springer, Singapore, pp 1–22. https://doi.org/10.1007/978-981-97-7352-7_1

Sharma MMM, Kapoor D, Loyal A, Kumar R, Sharma P, Husen A (2024b) Environmental toxicity of engineered carbon nanoparticles. In: Bachheti AJ, Bachheti RK, Husen A (eds) Carbon-based nanomaterials: synthesis, agricultural, biomedical, and environmental interventions. Springer, pp 337–353. isbn:978–981–97-0239-8. https://doi.org/10.1007/978-981-97-0240-4_16

Sharma PAWAN, Bhargava MJRD (2013) Applications and characteristics of nanomaterials in industrial environment. Res Dev (IJCSEIERD) 3(4):63–72

Shrivas K, Sahu S, Patra GK, Jaiswal NK, Shankar R (2016) Localized surface plasmon resonance of silver nanoparticles for sensitive colorimetric detection of chromium in surface water, industrial waste water and vegetable samples. Anal Methods 8(9):2088–2096

Sibbald RG, Contreras-Ruiz J, Coutts P, Fierheller M, Rothman A, Woo K (2007) Bacteriology, inflammation, and healing: a study of nanocrystalline silver dressings in chronic venous leg ulcers. Adv Skin Wound Care 20(10):549–558

Son R, Rusul G, Sahilah AM, Zainuri A, Raha AR, Salmah I (1997) Antibiotic resistance and plasmid profile of Aeromonas hydrophila isolates from cultured fish, Telapia (Telapia mossambica). Lett Appl Microbiol 24(6):479–482

Syafiuddin A, Salmiati, Salim MR, Beng Hong Kueh A, Hadibarata T, Nur H (2017) A review of silver nanoparticles: research trends, global consumption, synthesis, properties, and future challenges. J Chin Chem Soc 64(7):732–756

Tang S, Zheng J (2018) Antibacterial activity of silver nanoparticles: structural effects. Adv Healthc Mater 7(13):1701503

Wei L, Lu J, Xu H, Patel A, Chen ZS, Chen G (2015) Silver nanoparticles: synthesis, properties, and therapeutic applications. Drug Discov Today 20(5):595–601

Wong KK, Liu X (2010) Silver nanoparticles—the real "silver bullet" in clinical medicine? MedChemComm 1(2):125–131

Wright GD (2005) Bacterial resistance to antibiotics: enzymatic degradation and modification. Adv Drug Deliv Rev 57(10):1451–1470

Yakub I, Soboyejo WO (2012) Adhesion of *E. Coli* to silver-or copper-coated porous clay ceramic surfaces. J Appl Phys 111(12)

Zhang XF, Liu ZG, Shen W, Gurunathan S (2016) Silver nanoparticles: synthesis, characterization, properties, applications, and therapeutic approaches. Int J Mol Sci 17(9):1534

Chapter 17
Role of Nanotechnology in Sustainable Aquaculture

Preeti Kumari, Rohit Sain, Tulsi Kumar, and Minakshi Sharma

Abstract Aquaculture is an integral system of food production that provides beneficial animal fats and proteins that contribute a lot in terms of economy and nutrition. The growth potential of aquatic life is highly affected due to various abiotic stressors involving pollution, climate change, water scarcity and enhancement in disease susceptibility due to various new pathogen's emergence. All these factors act as a big hurdle in the way of aquaculture sustainability. Nowadays, nanotechnology is an innovative approach in aquaculture that improves nutrition, pathology, genetics, biotechnology, environmental quality and reproduction. Due to the outstanding physiochemical properties of nanoparticles, they offer significant advantages for fish culture. Nanomaterials help in combating a wide variety of pathogens due to their antimicrobial properties. Nanoparticles act as efficient carriers, that's why used in the delivery of therapeutic agents, immune stimulants, and vaccines to fishes that improve fish health, immune system and reproduction. Nowadays, DNA nano vaccines, nano sensors, nano-tagging, nano-inspired genes, drug delivery, delivery of genes and preservation of seafood are also popular. The feed enriched with nanoelements is helpful in fish growth, prevent diseases, improve appearance and help in becoming healthier and stronger. The nanomaterials also help in effectively overcoming the pollutants and also reduce the treatment cost. This book chapter discusses nanotechnology applications in aquaculture including fish medicine, water treatment, feed ingredients. The present book chapter is providing in depth screening of various versatile nanotechnology applications in aquaculture sustainability including its prospects and challenges.

Keywords Aquaculture · Fish medicine · Nanoelements · Sustainability

P. Kumari · R. Sain · M. Sharma (✉)
Department of Zoology, Maharshi Dayanand University, Rohtak, Haryana, India
e-mail: sminakshi.2007@rediffmail.com

T. Kumar
Department of Microbiology, Maharshi Dayanand University, Rohtak, Haryana, India

R. Kumar et al. (eds.), *Sustainability in Aquaculture and Fisheries*,
Springer Water, https://doi.org/10.1007/978-3-032-16605-0_17

17.1 Introduction

Now a days, there is a need for food production enhancement, which is a major challenge to meet the demand of burgeoning human population. Culture of fishes is a supporting means to overcome this problem by providing valuable, inexpensive and rapidly digestible protein (Abbas 2021; Kumar et al. 2022a, b). Aquaculture is one of the rapid growing agricultural domains having ample growth potential. Aquaculture is a good source for healthy fat, easily digestible protein and a repository for wide variety of essential micronutrients across the world for billions of people (Hua et al. 2019). It also acts as a crucial platform for livelihood, employment for rural people and also acts as an excellent export earning potential that assists our national GDP via a good source of income (Phillips et al. 2016).

Now, shellfish and farmed fin fish production has been increased via intensive and improve aquaculture practices (Callier et al. 2018). Aquaculture farming systems have grown in number, which has reduced dependency on wild seedstock and fish supplies (Naylor et al. 2000). The food and agriculture organization research states that between 1961 and 2016, the yearly average consumption of fishes increased by 3.2% worldwide. Fish food considered as almost 17% animal protein that consumed in the world in 2015 and provided nearly 20% average animal protein ingested per person for nearly 3.2 billion population (Sarkar et al. 2022). An overview of aquaculture is representing in Fig. 17.1.

In comparison to the other animal producing industries, the aquaculture is undeniably a very promising activity as part of sustainable culture. Aquaculture faces significant practical obstacles that lead to a multitude of negative outcomes at different

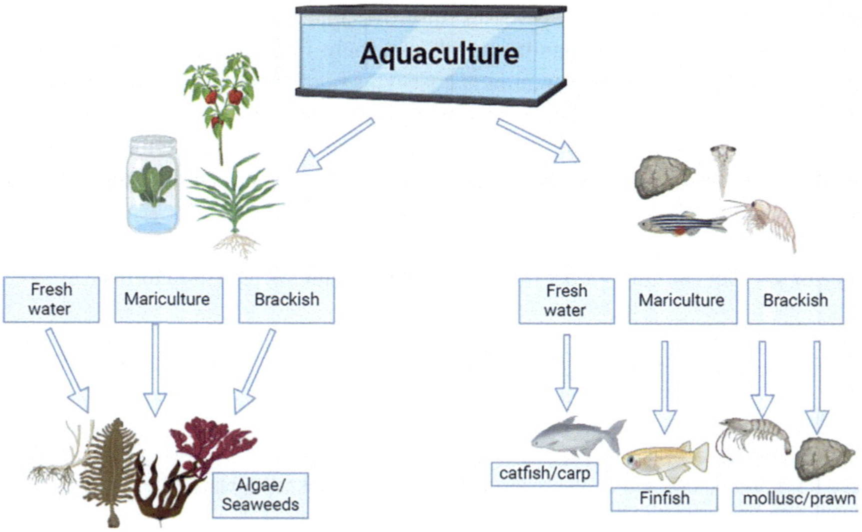

Fig. 17.1 Figure representing an overview of aquaculture

levels, including biofouling, pond contamination, water infection, acute and chronic illnesses, and postharvest preservation (Tayel et al. 2019).

17.2 Nanotechnology

The "nano," a Greek word meaning dwarf, is utilized to describe materials with one or more dimensions particles (Fig. 17.2) between one and one hundred nm (Abbas 2021). Nonetheless, nanotechnology is: "Understanding and control of matter at dimensions of roughly 1–100 nm where unique phenomena enable novel applications" via U.S. NNI that is National Nanotechnology Initiative (Rather et al. 2022; Sharma et al. 2025). For further preciseness, it is described as "the study, creation, design, manipulation, synthesis and application of the functional devices, materials and systems via matter control at the nanometer scale, which is at the atomic and molecular levels and is measured at 1×10^{-9} of a meter: 1–100 nm) and with exploitation of the properties and phenomena of the matter at that scale (Shah and Mraz 2020). Nanotechnology keeps great promise for fields like: materials science, aquaculture, electronics, animal husbandry, human health and biological/biomedical sciences. It can be used for biomolecule analysis, creation of the (non-viral) vectors that used in therapy of cancer, gene therapy, clinical diagnostics and therapeutics (Prasad et al. 2024). The maximum fund weightage allocated to nanomaterials research are focused on nanomedicine, nanoelectronics and nano-pharmaceutics. As a result, there are many nanotechnologies as well as nanoproducts available in various fields like: aquaculture, animal husbandry and agriculture compared to other sectors where nanotechnology investment is higher (Ahmad et al. 2022; Sharma et al. 2024). Nonetheless, nanotechnology has enormous promise in this crucial area, which is necessary for a revolution in the socioeconomic standing of a sizable population. The fields like: materials research, electronics, animal feed, human health and agriculture can all benefit greatly from nanotechnology. There are several indications of potential nanotechnological uses in aquaculture for water treatment, management of fish health, animal breeding/harvest/postharvest technologies, but still more work is needed to improve the research development in this area (Tayel et al. 2019).

17.3 Nanotechnology Applications in Aquaculture Sustainability

Nanotechnology has shown great promise in transforming many industries, including aquaculture and fisheries management (Sreadah 2023) (Fig. 17.3).

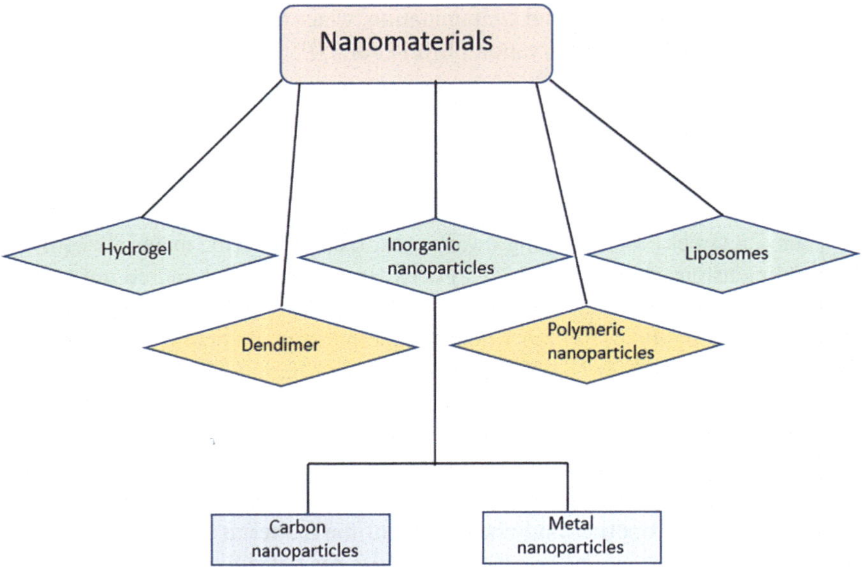

Fig. 17.2 Figure representing different kinds of nanoparticles

17.3.1 Nanoparticles as Feed Additives in Aquaculture

Now a days, fish farming has a significant and vital aspect in the global fishing sector. In their natural habitat, fish can obtain food from smaller animals and phytoplankton. However, in fish cultures, supplemental food ingredients are required to support rapid and healthy growth. These ingredients may include protein, carbohydrates, fats, fibers and dietary energy (Abbas 2021). It has been discovered that the addition of nanoparticles to fish food will improve aqua feeds via boosting nutrients amount that enter in fish through the gut tissue and are not wasted (Chris et al. 2018). The impact of various nanomaterials used for animal growth and development is shown in Table 17.1.

17.3.2 Role of Nanotechnology in Fish Breeding

Fish breeding and reproduction is an important area of aquaculture. Nanoparticles are used in fisheries to administer different hormones and steroids to increase gonadal growth and reproductive production (Haldar and Nath 2020). Using these strategies, fish farming can become more sustainable and an improved way to meet the seafood demand.

An attempt was made to implant a hormone pellet beneath the skin. The use of nano-encapsulated hormone administration was discovered to be a more effective

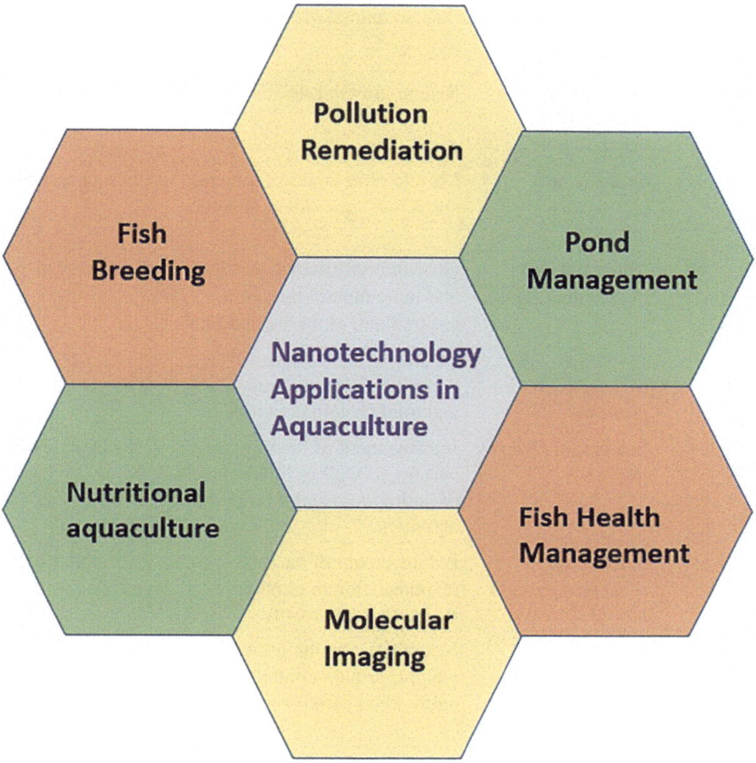

Fig. 17.3 Nanotechnology in aquaculture sustainability

method. Because of their high retention time and slower release, nanocarriers loaded with hormones can be injected for induction of maturation under the fish's skin during pre-spawning. The basic problem is LHRH (Luteinizing hormone releasing hormone) half-life in the blood circulation was shown to be resolved by an enhanced and regulated nano-delivery system, eliminating the need for repeated injections in fish. When tested with female *Cyprinus carpio*, chitosan as well as nanoconjugates of chitosan/gold/salmon LHRH showed that even nanoconjugate half dose might show greater reproductive effectiveness in comparison to the LHRH full dose. Minerals and vitamins are also important for managing broodstock (Sarkar et al. 2022).

17.3.3 Nanotechnology in Fish Health Management

The emergence of variety of pathogens that result in serious health hazards acts as a big menace to aquaculture system. To maintain the fish sustainability, it is crucial to prevent, cure fishes from pathogens and diagnose and treat diseases as soon as

Table 17.1 Impact of various nanomaterials on animal growth and development

Nanoparticles	Animal	Effects	References
Nano-selenium	Nile tilapia (*Oreochromis niloticus*)	Rise in growth rate	Chris et al. (2018)
	Crucian carp (*Carassius auratus*)	Muscle protein, development, and a few antioxidant enzymes	Chris et al. (2018)
	Common carp (*Cyprinus carpio*)	Growth performance metrices and increment in the antioxidant, globulin, and total protein activities	Ashouri et al. (2015)
	Mahseer fish *(Tor putitora)*	Amplify globulin, antioxidant, and total protein activities	Khan et al. (2017)
	Sea bream *(Pagrus major)*	Improvement of the immune system, as well as higher levels of antioxidant and total protein activities	Faiz et al. (2015)
Nano-zinc	Grass carp *(Ctepharyngodon idella)*	Enhancement of haematological response, immunological response, and growth	Faiz et al. (2015)
	Prawn *Macrobrachium rosenbergii (M. rosenbergii)*	Notable rise in the prawn's final weight, protein content, and antioxidant activity	Muralisankar et al. (2014)
	Grey Mullet *(Liza ramada)*	Weight and particular growth rate were significantly improved	Shukry et al. (2022)
	O. niloticus	Maximum survival rate, elevation of IL 8, IL1 genes, and increases in lysozyme, antioxidant potential and total protein. Increased responses to inflammation and immunological and antioxidant system impairment were caused by a dose of >60 mg/kg diet	Awad et al. (2019)
	Labeo rohita	Significant increases were also observed in the activity of digestive, haematological, and biochemical enzymes	Thangapandiyan and Monika (2020)

(continued)

Table 17.1 (continued)

Nanoparticles	Animal	Effects	References
Nano-iron	Freshwater prawn (*M. rosenbergii*)	Enhancement in fish biochemical and haematological parameters, growth, survival, and digestive enzymes activities in comparison to base diet	Srinivasan et al. (2016)
	Juvenile goldfish (*C. auratus*)	Elevation of the growth factor genes, alkaline phosphatase activity, and specific growth rate	Akbary and Jahanbakhshi (2019)
	O. mossambicus	Adverse alterations in haematology and physiology, as well as oxidative stress	Karthikeyeni et al. (2013)
	Major Carp (*L. rohita*) *Labeo rohita*	Adverse alterations in oxidative stress and physiological states	Remya et al. (2015)
	Bagridae catfish (*Clarias batrachus*)	Fish muscle was shown to have a larger protein and the lipid content. Cholesterol, Serum total protein and triglycerides also increased	Akter et al. (2018)
Nano-chitosan	Zebrafish (used to deliver ascorbic acid)	Raises fish ascorbic acid levels (because nano-chitosan has a strong penetration potential)	Jiménez-Fernández et al. (2014)
	Catfish	Adjust the impact of some mycotoxins	Abdel-Wahhab et al. (2016)
	O. niloticus	Improvement in growth as well as in immune system	Abd El-Naby et al. (2020), Abdel-Tawwab et al. (2019)
Copper nanoparticles	Common carp	Boosted the ash, fat, and protein contents of the carcass in a dose-dependent way; decreased the FCR and improved growth performance using linear and quadratic models	Dawood et al. (2020)

possible. Nanotechnology significantly contributes to disease management by novel means (Nasr-Eldahan et al. 2021).

Aquaculture has become increasingly important in recent decades in meeting enormous animal protein demand requirements and result in contribution to food security. Nonetheless, illness incidence and environmental degradation are regarded as crucial issues facing the industry (Nasr-Eldahan et al. 2021).

17.3.3.1 Recent Estimates of Financial Losses

- **Global Impact**: Disease outbreaks have been estimated to cost the global aquaculture industry around **US$6 billion per year**. This represents a major risk, especially considering over 90% of aquaculture occurs in developing countries.
- **India**: A recent 2025 study estimated the economic burden of diseases on India's aquaculture sector at **US$2.48 billion**, which is nearly 15% of the annual production value. Shrimp farming faced the highest cost per tonne, followed by marine fish. Major contributors included production loss (24%), prophylactic expenses (50%), and therapeutic costs (17%). Specific diseases causing major losses include *Enterocytozoon hepatopenaei* (EHP) in shrimp and bacterial *hemorrhagic septicemia* in finfish (Patil et al. 2025).
- **USA (Catfish)**: *Edwardsiellosis* remains a significant problem. A 2024 study estimated direct farm-level economic losses from this disease alone ranged from **US$5.2 to US$17.6 million per year**, with the total negative economic impact (including lost revenue) estimated at **US$15.5 to US$45.9 million per year**. Another study focusing on Alabama (2015–2021) found total annual disease-related financial losses were **US$11.1 million** (about 9.5% of food-size catfish sales), mainly due to bacterial diseases like *Aeromonas hydrophila* and columnaris.
- **Other Regions/Species**:
  - **Vietnam**: Estimated losses of US$800 million to the shrimp industry in 2011 due to AHPND alone.
  - **Brazil**: A 2017 study estimated annual losses in freshwater fish farms at US$84 million.
  - **Salmon**: Sea lice have caused significant losses, estimated at US$436 million annually for the Norwegian industry (Abolofia et al. 2017). Infectious Salmon Anaemia Virus (ISAv) cost the Chilean industry US$2 billion starting in 2007."

In aquaculture, disease (Fig. 17.4) issues are the main source of financial losses. Producers of channel catfish lost more than 100 million fish, valued at around $11 million, in 1988. Even greater losses are predicted by estimates for 1989. In 1988, the trout business reported losing more than 20 million fish, valued at more than $2.5 million. There are no statistics on the losses incurred by shellfish producers (Kumar et al. 2024a, b; Raja and Jithendran 2015; Tavares-Dias and Martins 2017). Aquaculture is negatively impacted by pathogenic germs and heavy metal deposition. These effects can be lessened by selenium (Se) nanoparticles' active surface. In this work, biological Se nanoparticles (Bio-SeNPs) were created using Se-resistant Bacillus subtilis AS12. The double-edged Bio-SeNPs were evaluated for their capacity to lessen the detrimental impacts of bacterial load and heavy metals on *Nile tilapia* (*Oreochromis niloticus*) as well as their respective effects on fish health, growth, and behavior. Two major problems affecting the aquaculture sector were addressed by the Bio-SeNPs (3–5 μg ml^{-1}) in the current study. First, the number of harmful bacteria, particularly *Aeromonas hydrophilia*, in fish organs decreased by 28–45%.

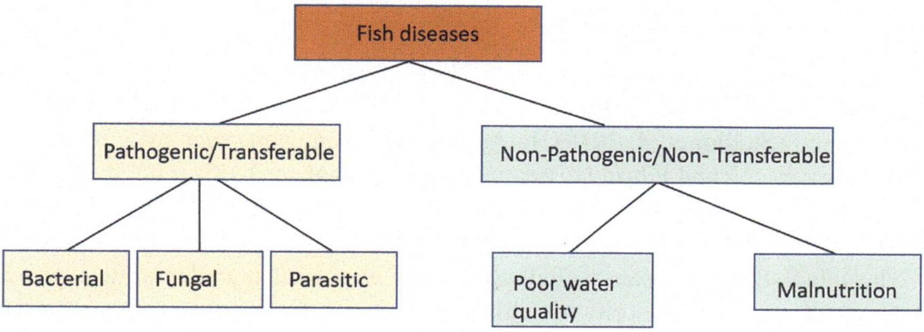

Fig. 17.4 Figure representing pathogenic/non-pathogenic cause of fish infections

Second, Bio-SeNPs (3–5 μg ml^{-1}) decreased the levels of heavy metals (Cd and Hg) at two levels (1 and 2 μg ml^{-1}) by 50–87% and 57–73%, respectively. As a result, inner immunity was increased and liver function parameters decreased. Applying Bio-SeNPs (3–5 μg ml^{-1}) to fish improved their gut health, growth, and behavior. Compared to controls, fish gained 36–52% more weight and had a 40% specific growth rate. In fish medium or organs, bio-SeNPs can decrease the bacterial load and lessen the buildup of heavy metals in a concentration-dependent way (Saad et al. 2022).

In order to properly address these difficulties, new technological avenues have been paved. Among these, nanotechnology has wider application range and enormous aquaculture potential as a revolutionary and cutting-edge instrument (Nasr-Eldahan et al. 2021). High antibacterial activity was demonstrated by nanoparticles (NPs) against the majority of microorganisms, including viruses, fungi, and bacteria. Nano-based techniques, with their unique chemical, physical, and biological characteristics, have been considered a promising option for aquaculture, especially as a fish medicinal agent (Das et al. 2020).

In fish health management, nanotechnology can be used in two ways: either directly to destroy pathogens or to detect them. By adopting these technique, a safe, sustainable and nutritious sea food is produced.

(A) ***Indirectly (As Diagnostic tool)***

Scientists studying aquaculture nowadays are searching for a simple yet effective technique for routinely diagnosing and detecting serious fish infections early on. The WSSV that is white spot syndrome virus strain in prawns targeted via antibody (Ab) based extremely sensitive immunodiagnosis procedure that attaches nano gold to alkaline phosphatase that is conjugated via secondary Ab (Thiruppathiraja et al. 2011). Fish pathogen identification using antibody-coated Au-NPs (*A. salmonicida*) from fish tissues is thought to provide a more accurate, efficient, and economical immunoassay (Saleh et al. 2011).

(B) ***Directly (To kill pathogens)***

To lessen pathogenic burden in aquaculture system, nanomaterials with anti-microbial and preventive qualities, such as nanosilver and ZnO nanoparticles, are already used (Siddiqi et al. 2018). Integral aquatic pathogens such as *S. aureus, P. aeruginosa, E. coli* and *Vibrio harveyi* inhibited by graphene oxide (GO), both alone and in presence of metal/polymeric/metal oxide nanoparticles (Shulga and Shulga 2020).

The most significant cause of illness issues across all manufacturing types is bacterial infections. In majority of cultivated species, gram-negative bacteria are the source of epizootics (Haenen et al. 2023). The second most significant cause of losses is fungi, particularly in the culture of salmon and crustaceans. Fungal infections have the potential to severely restrict both wild and captive aquatic animal populations (Pang et al. 2021). Unfortunately, our understanding of fungal illnesses is rather basic and mostly limited to the pathophysiology and identification of the etiological agents. Only gross pathologic observation, histological analyses, and conventional mycological isolation and identification techniques are necessary for the diagnosis of fungal infections (Leaño 2001). Aquatic fungus are usually thought of as secondary tissue invaders that develop as a result of pathogenic agents, traumatic injuries, or environmental factors like low water quality or cold temperatures (Qadir et al. 2024).

Silver nanoparticles (Ag-NPs) are considered the top choice for fighting bacteria. These tiny silver particles can stop harmful microbes in several ways. They can attach to the genetic material inside bacteria, damage the outer layers of bacterial cells, stop bacteria from multiplying, and interfere with how bacteria make important protein (Durán et al. 2016). Generally, AgNPs exhibit interaction with the sulfur accommodating proteins that resulting in blocking of cellular functions which cause bacterial cell death (Chen et al. 2011).

Gold nanoparticles (AuNPs) exhibiting antimicrobial properties through three primary mechanisms: (1) AuNPs have ability to disrupt the potential of bacterial membrane, that ultimately interfere the process of oxidative phosphorylation and consequently reducing ATP synthesis and metabolic activity; (2) they can inhibit protein synthesis by blocking ribosomal subunits and impeding tRNA binding; and (3) AuNPs can affect bacterial chemotaxis, altering the microbes' ability to detect and respond to chemical gradients in their environment (Cui et al. 2012). In this way nanotechnology reduces the chances of disease occurrence and enhances the food sustainability.

Detecting fish pathogens using antibody-coated gold nanoparticles (for *A. salmonicida*) in fish tissues is considered a more accurate, efficient, and cost-effective immunoassay method. This technique offers improved sensitivity and speed compared to traditional tests, potentially revolutionizing diagnosis of diseases in aquaculture (Saleh et al. 2011). Gold nanoparticles have shown promise in fighting various fish diseases, including those caused by *Salmonella typhi, A. salmonicida, and E. coli* (Lima et al. 2013). Zinc oxide nanoparticles work differently, breaking open bacterial cell walls and causing internal contents to leak out. Unlike silver nanoparticles, zinc oxide particles mainly affect microbes by creating harmful oxygen compounds, damaging cell membranes, producing hydrogen peroxide, and causing overall cell damage (Sirelkhatim et al. 2015). Zinc oxide nanoparticles can effectively fight various harmful microbes, including *Pseudomonas aeruginosa, E. coli, Aspergillus flavus, Candida albicans and Aeromonas hydrophila* (Jayaseelan et al. 2012). In fish farming the nanoparticles of TiO_2 shows strong antibacterial properties. Researchers suggest using these particles could help reduce fish deaths caused by bacterial infections (Jovanović et al. 2015). To combat Columnaris, a highly infectious disease occurred in farmed tilapia that happened due to *Flavobacterium columnare*, scientists developed a special nanovaccine that sticks to fish tissues. This vaccine uses chitosan, a substance that helps it adhere better (Kitiyodom et al. 2021). A novel mucoadhesive nanoencapsulated vaccine was developed by incorporating Flavobacterium oreochromis antigen into cationic lipid-based nanoparticles. When administered to Lates calcarifer, this enhanced vaccine formulation elicited a stronger immune response, characterized by elevated levels of Flavobacterium oreochromis-specific antibodies (Bunnoy et al. 2022). In a parallel study on *Labeo rohita*, researchers engineered a nanoDNA vaccine that encode gene of glyceraldehyde-3-phosphate dehydrogenase of *Edwardsiella tarda*. This vaccine synthesized through conjugation via a polymer complex of: poly (lactic-co-glycolic acid)-chitosan. Upon administration, the vaccine induced significant immunological improvements, including increased total serum protein and globulin concentrations, ultimately resulting in reduced mortality rates when challenged via *E. tarda* (Leya et al. 2021). This mode of mortality reduction contributes aquaculture sustainance.

17.3.4 Nanotechnology in Treatment of Water

Aquaculture operations are affected by water quality parameters that can be categorized into four primary classifications (Fig. 17.5): (1) physicochemical factors (including temperature, pH, salanity, dissolved oxygen concentration); (2) anthropogenic organic pollutants; (3) biochemical toxins (exemplified by cyanobacterial metabolites); and (4) biological contaminants (encompassing various pathogenic microorganisms) (Su et al. 2020). Table. 17.2. showing properties and applications of various nanomaterials that helpful in water quality improvement.

The development of sophisticated water treatment systems for aquaculture has made use of nanomaterials. Nanomaterials possess ability to improve soil and water quality by eliminating contaminants such as undesired microbes and heavy metals. Nanomaterials are investigated as a way that enhance the quality of water for aquaculture methods that are sustainable. Nanomaterials are highly reactive and also having several special qualities due to small size and wide surface-area (e.g., long-lasting nutrient release, improved cation exchange capacity and nutrient delivery) (Raja and Husen 2020; Nisha et al. 2025). Sustainable fish farming facilitates good water and soil quality with enhanced production.

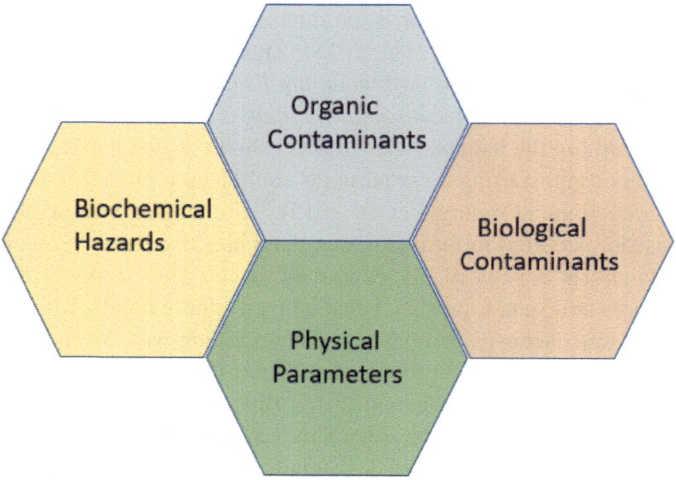

Fig. 17.5 Figure representing parameters that affect water quality

The use of nanobubble technology has shown great promise in improving the efficiency of ozone and oxygen in water treatment and pathogen management. The potential use of oxygen and ozone nanobubbles increase the hatching efficiency of common carp eggs is identified in the study. After 72 h, the three treatments employing O_2NB and O_3NB had greater larval survival rates than the control, with the maximum effectiveness being $94.07 \pm 0.90\%$. According to our research, common carp hatch more efficiently when O_2NB and O_3NB are applied at the right developmental phases (Trang et al. 2025).

Nano-sensors and Monitoring Systems

The creation of sophisticated sensors/monitoring systems that helpful in real-time water quality monitoring of indicators including temperature, pH and dissolved levels of oxygen (Fig. 17.6) has been made possible by nanotechnology (Manjakkal et al. 2023).

This technology has been primarily utilized for identification as well as detection of various water organic pollutants, especially PAHs that is polycyclic aromatic hydrocarbons, pyrene and anthracene, naphthalene and pyrene, and phenanthrene and fluorene (Gu et al. 2013; Péron et al. 2009; Xie et al. 2011). For the purpose of detecting PAHs, an example dimer: β-CD on the AgNPs that embedded via silica nanoparticles is created (Hahm et al. 2016). Biosensors are small analytical instruments that were primarily created for the measurement of microorganisms and toxins. They might be more desirable for on-site and instantaneous measurement. Optical and electrochemical sensors are the most often used biosensor forms for aquatic water analysis. Optoelectronic sensor that uses quantum dots (QDs) and DNA aptamer to detect toxins present in freshwater/marine aquatic bodies which produce via microorganisms like: cyanobacteria, diatoms and dinoflagellates) in real time on

Table 17.2 Table showing properties and applications of various nanomaterials that helpful in water quality improvement

Nanomaterial	Properties	Applications	References
AgNPs	Antifungal, antiviral, and antibacterial properties	Water disinfectants reduce bacterial activity and are extremely hazardous to *E. coli* and *Pseudomonas aeruginosa*	Gogoi et al. (2006)
Titanium oxide	Inexpensive, semiconducting, non-toxic, energy converting, gas detecting, electronic, photocatalytic and having ability to generate extremely reactive oxidants	Management of waste, surface self-cleaning, ustilization as a photocatalyst for treating water and microorganism disinfection for viruses, bacteria, algae and fungus	Cho et al. (2005), Li et al. (2008a, b)
MnO doped titanate nanotubes	Large specific surface area, dispersion of active species and active redox activity	Catalytic reduction	Guerra et al. (2018)
Iron and iron oxides NPs	Elemental iron (FeO, sometimes known as "zerovalent") makes up the core of a core–shell structure, while mixed valent oxides: Fe(II)/ Fe(III) form shell	Various heavy metals removal, including Cu^{2+}, Co^{2+}, Ni^{2+}, and $Cd^{2+}$and cleaning up of organic solvent s that have been chlorinated	Hooshyar et al. (2013), Guo et al. (2017)
Carbon based nanomaterials	Characteristics of adsorption and photocatalysis	Useful in remediation of environment from heavy metals along with organic compounds that are chlorinated via generating superoxide/ hydroxyl radicals, as well as the removal of inorganic/organic contaminants from the aqueous contaminations	Ren et al. (2011), Di Paola et al. (2012), Guerra et al. (2018)
Silica-based nanomaterials	Simple surface modification, a large and adjustable pore capacity, and an elevated level surface volume ratio	Eliminate contaminants, organic dyes, and metal ions like: Al^{2+}, Co^{2+}, Cd^{2+}, Cr^{3+} Zn^{2+}, Cu^{2+}, Ni^{2+}, from wastewater	Guerra et al. (2018), Tsai et al. (2016)
Polymer-based nanomaterials	Emulsifiers, stabilising agents, and surfactants	Withdrawl of ions (Ag^{+}, Cu^{2+}, Au^{2+}, etc.), dyes, and microbes. gases, heavy metals, organic and pathogenic contaminants remediation from the polluted locations	Guerra et al. (2018), Diallo et al. (2005), Mittal et al. (2015)

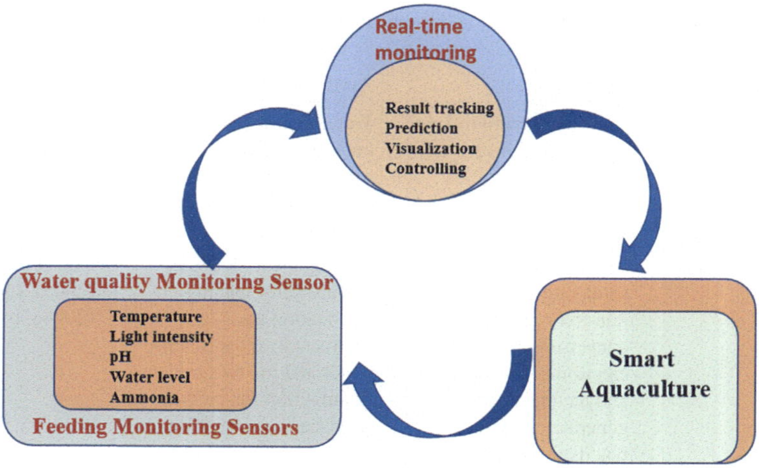

Fig. 17.6 Figure representing schematic representation of water quality monitoring of nanosensors

the spot (Cunha et al. 2018). Electrochemical immunosensor, specifically capacitive, for wide-spectrum identification for the class of poisonous cyclic peptides known as microcystins (Lebogang et al. 2014). To monitor freshwater supplies, a nucleic acid sensor assay was developed that identify the cyanopeptolin coding region of the single cynobacteria genome (Xie et al. 2016).

17.4 Green Nanotechnology as an Environmentally Friendly Approach for Sustainable Aquaculture

The majority of metal nanomaterials' problems can be solved in an economical, environmentally beneficial, and efficient manner with green nanotechnology, or the biosynthesis of nanoparticles. This includes the use of microbes, natural plants, plant extracts, or other materials like cellulose, chitosan, or tree gums. Additionally, a good substitute for metal nanoparticles is chitosan, a versatile natural polymer (Abbas 2021). Some aquaculture benefits obtained via green nanotechnology are: The use of bio-selenium nanoparticles (Bio-SeNPs) are very helpful in reducing the harmful bacteria especially *Aeromonas hydrophilia,* in fish organs, which was reduced in fish organs. Bio-SeNPs, also minimizes the heavy metals: Cd and Hg concentration and improve inner immunity. The application of Bio-SeNPs improved fish gut health, growth, and behavior, resulting in fish higher weight gain. In conclusion, Bio-SeNPs can mitigate the accumulation of heavy metals and reduce the bacterial load in a concentration dependent manner, either in the fish media or fish organs. The accumulation of heavy metals in fish organs causes oxidative stress by producing ROS, which damages cells and causes a low survival rate. Additionally, pathogenic bacteria accumulate in the gut and interfere with beneficial microorganisms, causing malnutrition, which affects growth performance. In this study, Bio-SeNPs synthesized by *Bacillus subtilis* AS12 isolate were proven with maximum efficiency as a remedy for heavy metal treatment and pathogen accumulation in fish organs and media. The SeNPs exhibited antioxidant, antimicrobial, antiviral, and anti-tumor activities. They also have insecticidal activity and have antifouling against *Pseudomonas aeruginosa*, *Staphylococcus aureus*, and *Candida albicans* (Saad et al. 2022).

Another eco-friendly approach includes anti-biofouling effectiveness in shrimp culture ponds. The potential of a chitosan nanocomposite containing bioactive metabolites isolated from *V. negundo* to prevent biofouling in prawn production ponds. The growth of biofouling was significantly inhibited by the nanocomposite. Treatment with chitosan nanocomposite showed an anti-biofouling effect of above 90%. Furthermore, the chitosan nanocomposite demonstrated non-toxic behaviour on zebrafish embryos (Karthick Raja Namasivayam et al. 2025).

Silver nanoparticles derived from Origanum vulgare leaf extract had cytotoxic and antimicrobial qualities. Using cashew nut shell liquid, silver and gold nanoparticles were produced in an environmentally friendly manner and demonstrated bactericidal effectiveness against a variety of fish illnesses. In young *Feneropenaeus indicus*, it was discovered that silver nanoparticles derived from tea leaf extract (Camellia sinensis) had a bactericidal impact on *Vibrio harveyi*. Zinc oxide nanoparticles (ZnO-NPs) were produced environmentally by using a broth derived from aloe leaf extract, which showed greater bactericidal action than nanoparticles produced using conventional chemistry. A novel method for biologically synthesising zinc oxide nanoparticles uses the reducing agent *Aeromonas hydrophila* bacteria. This process is economical and environmentally beneficial.

Significant antibacterial and antifungal properties against bacterial and fungal diseases of *Nile tilapia* and *Sea bass* were reported for a spherical green synthesised nanoparticle derived from the solvent extraction of *Origanum vulgare* leaves. The antibacterial properties of 10 μL nanoparticles/disk were found to exhibit a zone of inhibition between 23.7 and 31.3 mm against *Streptococcus agalactiae, Aeromonas hydrophila*, and *Vibrio alginolyticus*. Similarly, when treated with the same concentration (10 μL/disk), the fungal pathogens *Aspergillus flavus*, *Fusarium moniliforme*, and *Candida albicans* displayed an inhibitory zone of 11–18 mm (Okeke et al. 2022).

Heavy metals in fishpond sediments can be effectively immobilised by zinc oxide (Kumar et al. 2022c; Yadav et al. 2022a) and silicon nanoparticles made from *Azolla pinnata* and *Equisetum arvense*, which lowers the metals' bioavailability and possible environmental hazards. In addition to reducing heavy metal contamination, the use of these green-synthesised nanoparticles improves the sediments' nutritional content, which makes them more appropriate for use as fertilisers and soil conditioners. This double advantage demonstrates how these nanoparticles may be used to manage contaminated fishpond sediments in a sustainable manner while also boosting agricultural output.

The green nanotechnology make aquaculture more sustainable via eco-friendly means and making aqua food more safe and nutritious for consumption.

17.5 Challenges and Future Directions

17.5.1 *Toxicity and Environmental Impact*

Before being utilised in aquaculture, nanoparticles must first be properly tested for their potential toxicity and the effects they may have on aquatic environments and creatures. Due to fast nanotechnology development, there is a growing danger that materials/products based on it will be exposed to humans and the environment (Corsi et al. 2023). Since aquatic resources are especially susceptible to both direct and indirect nanomaterial pollution, its important to evaluate all possible toxicological and environmental nanomaterials impact on aquatic life (Wang et al. 2008). TiO_2 nanoparticles are found in many items, such as paint, building supplies, sunscreen, and cosmetics (Yadav et al. 2022b, 2023).

TiO_2 nanoparticles having concentration: 100 and 200 mg/L resulting in significantly increment in LPO levels in the tissues as well as significantly decreased SOD, CAT, and POD activities, indicating the fishes exposure to these doses of TiO_2 NPs experienced oxidative stress (Kumar et al. 2024a, b; Smith et al. 2007). In Medaka fish that are deficient in selenium, the comparison of toxicities of nano-Se with selenium was done. Comparing nano-Se to selenite, it was discovered that there were up to six times as many hyper-accumulated variations in the liver (Li et al. 2008a, b). Commercial products use silver nanoparticles because of their antifungal and antibacterial qualities. It's possible that some of these products will cause silver nanoparticles to enter the aquatic environment.

Examining cytotoxicity as well as genotoxicity of Ag nanospheres with a diameter of 30 nm on fish cells (Wise et al. 2010). SWCNTs that is single walled carbon nanotubes act as respiratory toxin in case of trout. Although fish can tolerate oxidative stress and osmoregulatory disruptions, additional cellular diseases give rise to worries regarding neurotoxicity and cell cycle problems (Can et al. 2011; Smith et al. 2007).

17.5.2 *Scalability and Cost-Effectiveness*

To be economically feasible for commercial use, aquaculture uses of nanotechnology must be made more scalable and cost-effective.

17.5.3 *Public Acceptance*

It is necessary to increase public acceptability of aquaculture's use of nanotechnology through education and awareness initiatives.

17.6 Conclusion

The productivity, sustainability, and efficiency of aquaculture could all be greatly increased by nanotechnology. Nonetheless, it is imperative to tackle the obstacles linked to its implementation in aquaculture. Toxicological and environmental effects of nanoparticles, their cost-effectiveness and scalability, the establishment of regulatory frameworks, and public acceptance all require more investigation.

17.7 Future Prospects

Aquaculture is an excellent source of nutritious food and also a way for generation of funds that helps in nation benefit. Lot of nanotechnology aspects have been used to improve aquaculture. There is a need to find more antifungal and antiviral properties of nanoparticles against the infections of fishes. There is also a need to investigate fish health related nanomaterials that will be helpful in diagnosis of fish diseases and therapeutics. In future it is necessary to focus on generating non-toxic, environmental-friendly nanomaterials production via simple means. Artificial intelligence and nanotechnology combination can allow for real-time aquaculture system control and monitoring. Creation of novel nanomaterials with particular characteristics for use in aquaculture. Commercialization will depend on the creation of scalable production technologies for nanomaterials.

Acknowledgements Author **Preeti Kumari [File no. 82-44/2020(SA-III)] and** author **Rohit Sain [NTA Ref. No.: 221610184504]** are thankful to the University Grants Commission (**UGC**), New Delhi-110012 for the award of UGC-JRF and UGC NFOBC SRF respectively.

References

Abbas WT (2021) Advantages and prospective challenges of nanotechnology applications in fish cultures: a comparative review. Environ Sci Pollut Res 28(7):7669–7690. https://doi.org/10.1007/s11356-020-12166-0

Abd El-Naby AS, Al-Sagheer AA, Negm SS, Naiel MAE (2020) Dietary combination of chitosan nanoparticle and thymol affects feed utilization, digestive enzymes, antioxidant status, and intestinal morphology of Oreochromis niloticus. Aquaculture 515(August 2019):734577. https://doi.org/10.1016/j.aquaculture.2019.734577

Abdel-Tawwab M, Razek NA, Abdel-Rahman AM (2019) Immunostimulatory effect of dietary chitosan nanoparticles on the performance of Nile tilapia, Oreochromis niloticus (L.). Fish Shellfish Immunol 88(January):254–258. https://doi.org/10.1016/j.fsi.2019.02.063

Abdel-Wahhab MA, Aljawish A, Kenawy AM, El-Nekeety AA, Hamed HS, Abdel-Aziem SH (2016) Grafting of gallic acid onto chitosan nano particles enhances antioxidant activities in vitro and protects against ochratoxin A toxicity in catfish (Clarias gariepinus). Environ Toxicol Pharmacol 41:279–288. https://doi.org/10.1016/j.etap.2015.12.005

Abolofia J, Asche F, Economics JW-MR (2017) The cost of lice: quantifying the impacts of parasitic sea lice on farmed salmon. J Uchicago Educ 32(3):329–349. https://doi.org/10.1086/691981

Ahmad F, Ogra AK, Ahmad S (2022) Nanomanufacturing of biomedicines: current status and future challenges. Nanomanufacturing Nanomater Des 95–136. https://doi.org/10.1201/9781003220602-7/NANOMANUFACTURING-BIOMEDICINES-AHMAD-OGRA-AHMAD

Akbary P, Jahanbakhshi A (2019) Nano and macro iron oxide (Fe_2O_3) as feed additives: effects on growth, biochemical, activity of hepatic enzymes, liver histopathology and appetite-related gene transcript in goldfish (Carassius auratus). Aquaculture 510(May):191–197. https://doi.org/10.1016/j.aquaculture.2019.05.052

Akter N, Alam MJ, Jewel MAS, Haque MA, Khatun S, Akter S (2018) Evaluation of dietary metallic iron nanoparticles as feed additive for growth and physiology of Bagridae catfish Clarias batrachus (Linnaeus, 1758). Int J Fisheries Aquatic Stud 6(3):371–377. www.fisheriesjournal.com

Ashouri S, Keyvanshokooh S, Salati AP, Johari SA, Pasha-Zanoosi H (2015) Effects of different levels of dietary selenium nanoparticles on growth performance, muscle composition, blood biochemical profiles and antioxidant status of common carp (Cyprinus carpio). Aquaculture 446:25–29. https://doi.org/10.1016/j.aquaculture.2015.04.021

Awad A, Zaglool AW, Ahmed SAA, Khalil SR (2019) Transcriptomic profile change, immunological response and disease resistance of Oreochromis niloticus fed with conventional and Nano-Zinc oxide dietary supplements. Fish Shellfish Immunol 93(June):336–343. https://doi.org/10.1016/j.fsi.2019.07.067

Bunnoy A, Thangsunan P, Chokmangmeepisarn P, Yata T, Klongklaew N, Pirarat N, Kitiyodom S, Srisapoome P, Rodkhum C (2022) Mucoadhesive cationic lipid-based Flavobacterium oreochromis nanoencapsulation enhanced the efficacy of mucoadhesive immersion vaccination against columnaris disease and strengthened immunity in Asian sea bass (Lates calcarifer). Fish Shellfish Immunol 127(June):633–646. https://doi.org/10.1016/j.fsi.2022.06.059

Callier MD, Byron CJ, Bengtson DA, Cranford PJ, Cross SF, Focken U, Jansen HM, Kamermans P, Kiessling A, Landry T, O'Beirn F, Petersson E, Rheault RB, Strand Ø, Sundell K, Svåsand T, Wikfors GH, McKindsey CW (2018) Attraction and repulsion of mobile wild organisms to finfish and shellfish aquaculture: a review. Rev Aquac 10(4):924–949 (Wiley-Blackwell). https://doi.org/10.1111/raq.12208

Can E, Kizak V, Kayim M, Can SS, Kutlu B, Ates M, Kocabas M, Demirtas N (2011) Nanotechnological applications in aquaculture-seafood industries and adverse effects of nanoparticles on environment. J Mater Sci Eng 5:605–609

Chen M, Yang Z, Wu H, Pan X, Xie X, Wu C (2011) Antimicrobial activity and the mechanism of silver nanoparticle thermosensitive gel. Int J Nanomed 6:2873–2877. https://doi.org/10.2147/ijn.s23945

Cho M, Chung H, Choi W, Yoon J (2005) Different inactivation behaviors of MS-2 phage and Escherichia coli in TiO_2 photocatalytic disinfection. Appl Environ Microbiol 71(1):270–275. https://doi.org/10.1128/AEM.71.1.270-275.2005

Chris UO, Singh NB, Agarwal A (2018) Nanoparticles as feed supplement on Growth behaviour of Cultured Catfish (Clarias gariepinus) fingerlings. Mater Today: Proc 5(3):9076–9081. https://doi.org/10.1016/j.matpr.2017.10.023

Corsi I, Venditti I, Trotta F, Punta C (2023) Environmental safety of nanotechnologies: the eco-design of manufactured nanomaterials for environmental remediation. Sci Total Environ 864:161181. https://doi.org/10.1016/j.scitotenv.2022.161181

Cui Y, Zhao Y, Tian Y, Zhang W, Lü X, Jiang X (2012) The molecular mechanism of action of bactericidal gold nanoparticles on Escherichia coli. Biomaterials 33(7):2327–2333. https://doi.org/10.1016/j.biomaterials.2011.11.057

Cunha I, Biltes R, Sales MGF, Vasconcelos V (2018) Aptamer-based biosensors to detect aquatic phycotoxins and cyanotoxins. Sensors (Switzerland) 18(7):1–34. https://doi.org/10.3390/s18072367

Das S, Aswani R, Midhun SJ, Radhakrishnan EK, Mathew J (2020) Advantage of zinc oxide nanoparticles over silver nanoparticles for the management of Aeromonas veronii infection in Xiphophorus hellerii. Microb Pathog 147:104348. https://doi.org/10.1016/j.micpath.2020.104348

Dawood MAO, Eweedah NM, Moustafa EM, El-Sharawy ME, Soliman AA, Amer AA, Atia MH (2020) Copper nanoparticles mitigate the growth, immunity, and oxidation resistance in common carp (Cyprinus carpio). Biol Trace Elem Res 198(1):283–292. https://doi.org/10.1007/s12011-020-02068-0

Di Paola A, García-López E, Marcì G, Palmisano L (2012) A survey of photocatalytic materials for environmental remediation. J Hazard Mater 211–212:3–29. https://doi.org/10.1016/j.jhazmat.2011.11.050

Diallo MS, Christie S, Swaminathan P, Johnson JH, Goddard WA (2005) Dendrimer enhanced ultrafiltration. 1. Recovery of Cu(II) from aqueous solutions using PAMAM dendrimers with ethylene diamine core and terminal NH 2 groups. Environ Sci Technol 39(5):1366–1377. https://doi.org/10.1021/es048961r

Durán N, Durán M, de Jesus MB, Seabra AB, Fávaro WJ, Nakazato G (2016) Silver nanoparticles: a new view on mechanistic aspects on antimicrobial activity. Nanomed: Nanotechnol Biol Med 12(3):789–799. https://doi.org/10.1016/j.nano.2015.11.016

Faiz H, Zuberi A, Nazir S, Rauf M, Younus N (2015) Zinc oxide, zinc sulfate and zinc oxide nanoparticles as source of dietary zinc: Comparative effects on growth and hematological indices of juvenile grass carp (Ctenopharyngodon idella). Int J Agric Biol 17(3):568–574. https://doi.org/10.17957/IJAB/17.3.14.446

Gogoi SK, Gopinath P, Paul A, Ramesh A, Ghosh SS, Chattopadhyay A (2006) Green fluorescent protein-expressing Escherichia coli as a model system for investigating the antimicrobial activities of silver nanoparticles. Langmuir 22(22):9322–9328. https://doi.org/10.1021/la060661v

Gu X, Tian S, Zhou Q, Adkins J, Gu Z, Li X, Zheng J (2013) SERS detection of polycyclic aromatic hydrocarbons on a bowl-shaped silver cavity substrate. RSC Adv 3(48):25989–25996. https://doi.org/10.1039/c3ra43442g

Guerra FD, Attia MF, Whitehead DC, Alexis F (2018) Nanotechnology for environmental remediation: materials and applications. Molecules 23(7):1–23. https://doi.org/10.3390/molecules23071760

Guo M, Weng X, Wang T, Chen Z (2017) Biosynthesized iron-based nanoparticles used as a heterogeneous catalyst for the removal of 2,4-dichlorophenol. Sep Purif Technol 175:222–228. https://doi.org/10.1016/j.seppur.2016.11.042

Haenen OLM, Dong HT, Hoai TD, Crumlish M, Karunasagar I, Barkham T, Chen SL, Zadoks R, Kiermeier A, Wang B, Gamarro EG, Takeuchi M, Azmai MNA, Fouz B, Pakingking R, Wei ZW, Bondad-Reantaso MG (2023) Bacterial diseases of tilapia, their zoonotic potential and risk of antimicrobial resistance. Rev Aquac 15(S1):154–185. https://doi.org/10.1111/raq.12743

Hahm E, Jeong D, Cha MG, Choi JM, Pham XH, Kim HM, Kim H, Lee YS, Jeong DH, Jung S, Jun BH (2016) β-CD Dimer-immobilized Ag assembly embedded Silica nanoparticles for sensitive detection of polycyclic aromatic hydrocarbons. Sci Rep 6(April):1–7. https://doi.org/10.1038/srep26082

Haldar C, Nath S (2020) Nanotechnology: a novel technique for aquaculture and fisheries development. Int J Fauna Biol Stud 7(6):23–27. https://www.researchgate.net/publication/351356703_Nanotechnology_A_novel_technique_for_aquaculture_and_fisheries_development

Hooshyar Z, Rezanejade Bardajee G, Ghayeb Y (2013). Sonication enhanced removal of nickel and cobalt ions from polluted water using an iron based sorbent. J Chem 2013(Ii). https://doi.org/10.1155/2013/786954

Hua K, Cobcroft J, Cole A, Condon K, Earth DJ-O (2019) The future of aquatic protein: implications for protein sources in aquaculture diets. One Earth. https://www.cell.com/one-earth/fulltext/S2590-3322(19)30132-0

Jayaseelan C, Rahuman AA, Kirthi AV, Marimuthu S, Santhoshkumar T, Bagavan A, Gaurav K, Karthik L, Rao KVB (2012) Novel microbial route to synthesize ZnO nanoparticles using Aeromonas hydrophila and their activity against pathogenic bacteria and fungi. Spectrochimica Acta Part A: Mol Biomol Spectrosc 90:78–84. https://doi.org/10.1016/j.saa.2012.01.006

Jiménez-Fernández E, Ruyra A, Roher N, Zuasti E, Infante C, Fernández-Díaz C (2014) Nanoparticles as a novel delivery system for vitamin C administration in aquaculture. Aquaculture 432:426–433. https://doi.org/10.1016/j.aquaculture.2014.03.006

Jovanović B, Whitley EM, Kimura K, Crumpton A, Palić D (2015) Titanium dioxide nanoparticles enhance mortality of fish exposed to bacterial pathogens. Environ Pollut 203:153–164. https://doi.org/10.1016/j.envpol.2015.04.003

Karthick Raja Namasivayam S, Arvind Bharani RS, Samrat K, Kavisri M, Moovendhan M (2025) Anti-biofouling efficacy of chitosan nanocomposites in shrimp culture ponds: a green nanotechnology solution. J Environ Chem Eng 13(2). https://doi.org/10.1016/j.jece.2025.115577

Karthikeyeni S, Vijayakumar S, Sakthivel V, Ganesh A, Manimegalai M, Subramanian P (2013) Biosynthesis of Iron oxide nanoparticles and its haematological effects on fresh water fish Oreochromis mossambicus. J Acad Ind Res 1(May 2014):645–649

Khan KU, Zuberi A, Nazir S, Ullah I, Jamil Z, Sarwar H (2017) Synergistic effects of dietary nano selenium and vitamin C on growth, feeding, and physiological parameters of mahseer fish (Tor putitora). Aquac Rep 5:70–75. https://doi.org/10.1016/j.aqrep.2017.01.002

Kitiyodom S, Trullàs C, Rodkhum C, Thompson KD, Katagiri T, Temisak S, Namdee K, Yata T, Pirarat N (2021) Modulation of the mucosal immune response of red tilapia (Oreochromis sp.) against columnaris disease using a biomimetic-mucoadhesive nanovaccine. Fish Shellfish Immunol 112(December 2020):81–91. https://doi.org/10.1016/j.fsi.2021.02.017

Kumar R, Gupta R, Dahiya T, Yadav R, Saifi R, Yodha K, Kumari B, Singh P, Shefali (2022a) Surveillance of *Heteropneustes fossilis* (Stinging Catfish) diseases and health management practices in Haryana. Appl Ecol Environ Sci 10(4):261–267. https://doi.org/10.12691/aees-10-4-12

Kumar R, Gupta RK, Yadav R, Saifi R, Yodha K, Pooja (2022b) *Eisenia fetida* as protein source for growth enhancement of *Heteropneustes fossilis*. Egypt J Aquatic Biol Fisheries 26(2):577–588. https://doi.org/10.21608/ejabf.2022.234553

Kumar R, Yadav R, Kumar Gupta R, Pooja (2022c) A sustainable way for fish health management by replacement of chemical and drugs by earthworm. In: Environmental degradation in Asia: land degradation, environmental contamination, and human activities. Springer International Publishing, Cham, pp 329–352. https://doi.org/10.1007/978-3-031-12112-8_16

Kumar G, Engle C, Aarattuthodi S, van Senten J, Hegde S, Khoo L, Hanson L, Peterman M, Dorman L (2024a) Economic impact of Edwardsiellosis on the U.S. catfish industry. Aquac Econ Manag 28(3):415–438. https://doi.org/10.1080/13657305.2024.2319083

Kumar R, Yadav R, Gupta RK, Pal A, Yodha K, Kumar A (2024b) Antioxidant enzyme activities and markers of oxidative stress in the life cycle of different Earthworm species. Indian J Exp Biol 62(07):578–583. https://doi.org/10.56042/ijeb.v62i07.12084

Leaño EM (2001) Fungal diseases. In: Health management in aquaculture, vol 2, pp 43–53. https://doi.org/10.1016/B978-0-7020-8225-2.00077-9

Lebogang L, Hedström M, Mattiasson B (2014) Development of a real-time capacitive biosensor for cyclic cyanotoxic peptides based on Adda-specific antibodies. Anal Chim Acta 826(1):69–76. https://doi.org/10.1016/j.aca.2014.03.028

Leya T, Ahmad I, Valappil RK, Kurcheti PP, Tripathi G, Sharma R, Bedekar MK (2021) Development of species-specific IgM antibodies and elevation of mucosal immune response in Labeo rohita using recombinant bicistronic nano DNA vaccine priming. Fish Shellfish Immunol 113(April):185–195. https://doi.org/10.1016/j.fsi.2021.04.008

Li H, Zhang J, Wang T, Luo W, Zhou Q, Jiang G (2008a) Elemental selenium particles at nano-size (Nano-Se) are more toxic to Medaka (Oryzias latipes) as a consequence of hyper-accumulation of selenium: a comparison with sodium selenite. Aquat Toxicol 89(4):251–256. https://doi.org/10.1016/j.aquatox.2008.07.008

Li Q, Mahendra S, Lyon DY, Brunet L, Liga MV, Li D, Alvarez PJJ (2008b) Antimicrobial nanomaterials for water disinfection and microbial control: potential applications and implications. Water Res 42(18):4591–4602. https://doi.org/10.1016/j.watres.2008.08.015

Lima E, Guerra R, Lara V, Guzmán A (2013) Gold nanoparticles as efficient antimicrobial agents for Escherichia coli and Salmonella typhi. Chem Cent J 7(1):1–7. https://doi.org/10.1186/1752-153X-7-11

Manjakkal L, Lorenzelli L, Willander M (2023) Sensing technologies for real time monitoring of water quality. In: Sensing technologies for real time monitoring of water quality. https://doi.org/10.1002/9781119775843

Mittal H, Maity A, Ray SS (2015) Synthesis of co-polymer-grafted gum karaya and silica hybrid organic-inorganic hydrogel nanocomposite for the highly effective removal of methylene blue. Chem Eng J 279:166–179. https://doi.org/10.1016/j.cej.2015.05.002

Muralisankar T, Bhavan PS, Radhakrishnan S, Seenivasan C, Manickam N, Srinivasan V (2014) Dietary supplementation of zinc nanoparticles and its influence on biology, physiology and immune responses of the freshwater prawn, Macrobrachium Rosenbergii. Biol Trace Element Res 160(1):56–66. https://doi.org/10.1007/s12011-014-0026-4

Nasr-Eldahan S, Nabil-Adam A, Shreadah MA, Maher AM, El-Sayed Ali T (2021) A review article on nanotechnology in aquaculture sustainability as a novel tool in fish disease control. Aquac Int 29(4):1459–1480. https://doi.org/10.1007/s10499-021-00677-7

Naylor RL, Goldburg RJ, Primavera JH, Kautsky N, Beveridge MC, Clay J, Folke C, Lubchenco J, Mooney H, Troell M (2000) Effect of aquaculture on world fish supplies. Nature 405:1017–1024

Nisha, Anshu, Rani S, Kumar S, Kumar R, Kumar A (2025) Recent technologies and strategies for bioremediation of toxic and hazardous waste. In: Dubey KK, Shah MP, Yadav A (eds) Challenges and sustainable solutions in bioremediation. CRC Press, pp 1–16. https://doi.org/10.1201/9781003407317-1

Okeke ES, Chukwudozie KI, Nyaruaba R, Ita RE, Oladipo A, Ejeromedoghene O, Atakpa EO, Agu CV, Okoye CO (2022) Antibiotic resistance in aquaculture and aquatic organisms: a review of current nanotechnology applications for sustainable management. Environ Sci Pollut Res 29(46):69241–69274 (Springer Science and Business Media Deutschland GmbH). https://doi.org/10.1007/s11356-022-22319-y

Pang KL, Hassett BT, Shaumi A, Guo SY, Sakayaroj J, Chiang MWL, Yang CH, Jones EBG (2021) Pathogenic fungi of marine animals: a taxonomic perspective. Fungal Biol Rev 38:92–106. https://doi.org/10.1016/j.fbr.2021.03.008

Patil PK, Geetha R, Mishra SS, Abraham TJ, Solanki HG, Krupesha Sharma SR, Pradhan PK, Manna SK, Avunje S, Abhinaya D, Felix KT, Vinay TN, Paniprasad K, Paria A, Raja SA, Saraswathy R, Sahoo SN, Rathod R, Rameshkumar P et al (2025) Unveiling the economic burden of diseases in aquatic animal food production in India. Front Sustain Food Syst 9. https://doi.org/10.3389/FSUFS.2025.1480094/PDF

Péron O, Rinnert E, Lehaitre M, Crassous P, Compère C (2009) Detection of polycyclic aromatic hydrocarbon (PAH) compounds in artificial sea-water using surface-enhanced Raman scattering (SERS). Talanta 79(2):199–204. https://doi.org/10.1016/j.talanta.2009.03.043

Phillips M, Subasinghe R et al (2016) Aquaculture big numbers. FAO Fisheries Aquac 1–25. https://search.proquest.com/openview/787d460622b747a4bdca4fd6d92646de/1?pq-origsite=gscholar&cbl=237320

Prasad RD, Prasad NR, Prasad RB, Prasad RY, Prasad SR, Gour M, Vaidya AK, Patil M, Patil S, Karmuse S, Athane A, Shaikh A, Shaikh YI, Kanthe A, Patil PD, Sarvalkar P (2024) A critical review on bio-mimetic synthesis of transition metal nanoparticles: their biomedical applications in veterinary medicine. ES Food Agrofor 15. https://doi.org/10.30919/esfaf1027

Qadir MS, Salem HS, Jalal TK, Mohammed SM (2024) Aquatic fungi and fungal diseases of fish: a review. Proc Zool Soc 1–13 (Springer). https://doi.org/10.1007/s12595-024-00553-1

Raja RA, Jithendran KP (2015) Aquaculture disease diagnosis and health management. In: Advances in marine and brackishwater aquaculture, pp 247–254. https://doi.org/10.1007/978-81-322-2271-2_23

Raja MA, Husen A (2020) Role of nanomaterials in soil and water quality management. In: Nanomaterials for agriculture and forestry applications. INC. https://doi.org/10.1016/B978-0-12-817852-2.00020-2

Rather MA, Agarwal D, Kumar S, Sundaray JK (2022) Nanobiotechnology: prospects and applications in aquaculture. In: Advances in fisheries biotechnology, pp 197–210. https://doi.org/10.1007/978-981-16-3215-0_14

Remya AS, Ramesh M, Saravanan M, Poopal RK, Bharathi S, Nataraj D (2015) Iron oxide nanoparticles to an Indian major carp, Labeo rohita: impacts on hematology, iono regulation and gill Na^+/K^+ ATPase activity. J King Saud Univ Sci 27(2):151–160. https://doi.org/10.1016/j.jksus.2014.11.002

Ren X, Chen C, Nagatsu M, Wang X (2011) Carbon nanotubes as adsorbents in environmental pollution management: a review. Chem Eng J 170(2–3):395–410. https://doi.org/10.1016/j.cej.2010.08.045

Saad AM, Sitohy MZ, Sultan-Alolama MI, El-Tarabily KA, El-Saadony MT (2022) Green nanotechnology for controlling bacterial load and heavy metal accumulation in Nile tilapia fish using biological selenium nanoparticles biosynthesized by Bacillus subtilis AS12. Front Microbiol 13. https://doi.org/10.3389/fmicb.2022.1015613

Saleh M, Soliman H, Haenen O, El-Matbouli M (2011) Antibody-coated gold nanoparticles immunoassay for direct detection of Aeromonas salmonicida in fish tissues. J Fish Dis 34(11):845–852. https://doi.org/10.1111/j.1365-2761.2011.01302.x

Sarkar B, Mahanty A, Gupta SK, Choudhury AR, Daware A, Bhattacharjee S (2022) Nanotechnology: a next-generation tool for sustainable aquaculture. Aquaculture 546(August 2021):737330. https://doi.org/10.1016/j.aquaculture.2021.737330

Shah BR, Mraz J (2020) Advances in nanotechnology for sustainable aquaculture and fisheries. Rev Aquac 12(2):925–942. https://doi.org/10.1111/raq.12356

Sharma MMM, Kapoor D, Loyal A, Kumar R, Sharma P, Husen A (2024) Mitigation of drought stress in plants using gold nanoparticles. In: Husen A (ed) Plant response to gold nanoparticles. Springer, pp 193–209. https://doi.org/10.1007/978-981-97-7174-5_10

Sharma P, Kapoor D, Kumar R, Husen A (2025) Rhizosphere and osmolytes coordination in plant systems under adverse situations. In: Husen A (ed) Roles of osmolytes in changing environment. Academic Press, pp 101–117

Shukry M, Albogami S, Gewaily M, Amer AA, Soliman AA, Alsaiad SM, El-Shehawi AM, Dawood MAO (2022) Growth performance, antioxidative capacity, and intestinal histomorphology of Grey Mullet (Liza ramada)–Fed dietary Zinc nanoparticles. Biol Trace Elem Res 200(5):2406–2415. https://doi.org/10.1007/s12011-021-02844-6

Shulga NN, Shulga IS (2020) Antibacterial properties of tea. Eurasianunionscientists 1(70):31–33

Siddiqi KS, Husen A, Rao RAK (2018) A review on biosynthesis of silver nanoparticles and their biocidal properties. J Nanobiotechnol 16(1). https://doi.org/10.1186/s12951-018-0334-5

Sirelkhatim A, Mahmud S, Seeni A, Kaus NHM, Ann LC, Bakhori SKM, Hasan H, Mohamad D (2015) Review on zinc oxide nanoparticles: antibacterial activity and toxicity mechanism. Nano-Micro Lett 7(3):219–242. https://doi.org/10.1007/s40820-015-0040-x

Smith CJ, Shaw BJ, Handy RD (2007) Toxicity of single walled carbon nanotubes to rainbow trout, (Oncorhynchus mykiss): respiratory toxicity, organ pathologies, and other physiological effects. Aquat Toxicol 82(2):94–109. https://doi.org/10.1016/j.aquatox.2007.02.003

Sreadah MA (2023) Emerging trends in nanotechnology for sustainable aquaculture and fishery management. 05(02):10–17

Srinivasan V, Saravana Bhavan P, Rajkumar G, Satgurunathan T, Muralisankar T, Bhavan S (2016) Effects of dietary iron oxide nanoparticles on the growth performance, biochemical constituents and physiological stress responses of the giant freshwater prawn Macrobrachium rosenbergii post-larvae. Int J Fisheries Aquatic Stud 4(2):170–182. www.fisheriesjournal.com

Su X, Sutarlie L, Loh XJ (2020) Sensors, biosensors, and analytical technologies for aquaculture water quality

Tavares-Dias M, Martins ML (2017) An overall estimation of losses caused by diseases in the Brazilian fish farms. J Parasitic Dis 41(4):913–918. https://doi.org/10.1007/S12639-017-0938-Y
Tayel AA, Elsaied BE, Diab AM (2019) Nanotechnology for aquaculture. In: Nanoscience for sustainable agriculture. Springer International Publishing, pp 479–544. https://doi.org/10.1007/978-3-319-97852-9_20
Thangapandiyan S, Monika S (2020) Green synthesized zinc oxide nanoparticles as feed additives to improve growth, biochemical, and hematological parameters in freshwater Fish Labeo rohita. Biol Trace Elem Res 195(2):636–647. https://doi.org/10.1007/s12011-019-01873-6
Thiruppathiraja C, Kumar S, Murugan V, Adaikkappan P, Sankaran K, Alagar M (2011) An enhanced immuno-dot blot assay for the detection of white spot syndrome virus in shrimp using antibody conjugated gold nanoparticles probe. Aquaculture 318(3–4):262–267. https://doi.org/10.1016/j.aquaculture.2011.06.008
Trang VT, Giang PT, Stejskal V, Sivaramasamy E, Thanh PT, Nguyen NT, Binh PT, Lua DT (2025) Nanobubbles of oxygen and ozone are effective tools for pathogen treatment and for improving the survival rates of common carp larvae (Cyprinus carpio)—A case study. Aquaculture 601. https://doi.org/10.1016/j.aquaculture.2025.742292
Tsai CH, Chang WC, Saikia D, Wu CE, Kao HM (2016) Functionalization of cubic mesoporous silica SBA-16 with carboxylic acid via one-pot synthesis route for effective removal of cationic dyes. J Hazard Mater 309:236–248. https://doi.org/10.1016/j.jhazmat.2015.08.051
Wang J, Zhang X, Chen Y, Sommerfeld M, Hu Q (2008) Toxicity assessment of manufactured nanomaterials using the unicellular green alga Chlamydomonas reinhardtii. Chemosphere 73(7):1121–1128. https://doi.org/10.1016/j.chemosphere.2008.07.040
Wise JP, Goodale BC, Wise SS, Craig GA, Pongan AF, Walter RB, Thompson WD, Ng AK, Aboueissa AEM, Mitani H, Spalding MJ, Mason MD (2010) Silver nanospheres are cytotoxic and genotoxic to fish cells. Aquat Toxicol 97(1):34–41. https://doi.org/10.1016/j.aquatox.2009.11.016
Xie Y, Wang X, Han X, Song W, Ruan W, Liu J, Zhao B, Ozaki Y (2011) Selective SERS detection of each polycyclic aromatic hydrocarbon (PAH) in a mixture of five kinds of PAHs. J Raman Spectrosc 42(5):945–950. https://doi.org/10.1002/jrs.2818
Xie P, Zhu L, Shao X, Huang K, Tian J, Xu W (2016) Highly sensitive detection of lipopolysaccharides using an aptasensor based on hybridization chain reaction. Sci Rep 6(July):1–8. https://doi.org/10.1038/srep29524
Yadav R, Gupta RK, Kumar R, Kaur T, Moun R (2022a) "Biochemical response of earthworm, *Eisenia fetida* to heavy metals toxicity. Journal of Applied and Natural Science 14(3):990–998
Yadav R, Gupta RK, Kumar R, Kaur T (2022b) Assessment of toxicity of lead and nickel on the biochemical and immunological parameters of Earthworm, Eudrilus Eugeniae. Agric Sci Dig 44(5):922–929
Yadav R, Kumar R, Gupta RK, Kaur T, Yodha K, Kour A, Kaur S, Rajput A (2023) Heavy metal toxicity in earthworms and its environmental implications: a review. Environ Adv 100374

Chapter 18
Recirculating Aquaculture System (RAS): A Sustainable Approach Towards Land-Based Aquaculture System

Sudha Shahi, Qurratul Ain Qureshi, Isabel Kachinda, Medha Shahi, Sonal Pathak, and Vibhu Vaid

Abstract Recirculatory Aquaculture System (RAS) is an advanced aquaculture system where fishes are raised in controlled environment in indoor tanks. The RAS uses minimum water replacement and sequential mechanical and biological filtration is done for maintenance of water quality. Fish culture in RAS promotes sustainability in aquaculture practices through utilisation of adequately available resources of land and water. As the water is treated regularly and exposure to external environment is minimised, it also promotes biosecurity measures in aquaculture practices. The main challenge of using this system is a lack of adequate knowledge about use of system, high initial investment, poor management and occurrence of disease and its control. The recent technological advancement in RAS includes use of denitrification

S. Shahi (✉)
Department of Aquaculture, Chaudhary Charan Singh Haryana Agricultural University, Hisar, Haryana, India
e-mail: sudhashahi252@gmail.com

Q. A. Qureshi
Department of Aquatic Animal Health Management, Chaudhary Charan Singh Haryana Agricultural University, Hisar, Haryana, India
e-mail: qurratulain2207@gmail.com

I. Kachinda
Ministry of Youth, Sports and Arts, Lusaka, Zambia
e-mail: isabelkachinda@gmail.com

M. Shahi
State Rural Development Mission, Dehradun, Uttarakhand, India
e-mail: medhashahi99@gmail.com

S. Pathak
Department of Agriculture Extension and Communication, Govind Ballabh Pant University of Agriculture and Technology, Pantnagar, Uttarakhand, India
e-mail: sonalpathak231997@gmail.com

V. Vaid
Department of Zoology, Punjab Agricultural University, Ludhiana, Punjab, India
e-mail: vibhuvaid30@gmail.com

R. Kumar et al. (eds.), *Sustainability in Aquaculture and Fisheries*,
Springer Water, https://doi.org/10.1007/978-3-032-16605-0_18

reactors, microalgae-based water treatment. Although the system generates a higher production in comparison to conventional aquafarming practices, its acceptability is lesser in small and marginal farmers of India due to higher initial investment and management regimes and water refreshing rates due to lack of adequate successful farms and training opportunities. Monitoring of this system prior to stocking of fish and maintaining record of challenges during the culture period of different fishes will help in facing these challenges. To improve equipment performance by conducting large-scale commercial research, there is a need to further investigate the best device combinations as per requirement.

Keywords Advancement · Biosecurity · Microalgae · Water replacement · Fish production

18.1 Introduction

The aquaculture industry is expanding rapidly, driven by the increasing demand for nutrient rich food to support the growing global population. Since traditional farming systems and catch fisheries have either reached or surpassed the maximum sustainable yields, there is little scope for sustainable production growth (Manjeet et al. 2022; Brown et al. 2024). RAS technology has developed steadily in last decade for rearing, raising of broodstock development targeting a number of species and successfully contributed in diversification of species in aquaculture rearing system. The treated water is recirculated within the system, with a maximum of 10% being replaced daily to account for evaporation and waste removal.

The main objective of this system is to conserve water and generate less effluent and maintain the environment sustainability (Zimmermann et al. 2023; Kumar et al. 2022a, b). A key trend in Recirculating aquaculture systems operations (RAS) is the reduction of water usage per kilogram of feed, resulting in a higher percentage of water being recycled. The quantities of fish feed play crucial roles in determining the ammonia removal capacity in recirculating aquaculture systems (RAS). The total ammonia nitrogen (TAN) production in fish is influenced by both the feeding intervals and the digestibility of the feed, and it can rise if the amino acid composition of the fish feed is unbalanced. Distributing the feed, setting up of feeding schedules, throughout the day in RAS is advantageous as it helps maintain a consistent ammonia load in the bioreactors (Pedersen et al. 2015; Kumar et al. 2024). Over the last decade RAS technology has advanced significantly and is now commonly used in broodstock management, hatcheries, and increasingly for the rearing of high-value fish species. Biosecurity, low water consumption, less area requirement and high production are the main features of this rearing system (Shefali et al. 2021; Murray et al. 2014). The system cleans the water by passing it through a series of biological and mechanical filters, allowing it to be recycled back into the fish culture tanks. A more than 90% of the water is recirculated through the culture units and rest is lost in consumption and evaporation losses. Moreover, the limited water usage simplifies and lowers cost of fish waste management, as the volume of discharged water is considerably smaller compared to traditional fish farms (Sugita et al. 2005).

Recirculatory Aquaculture Systems (RAS) are expanding rapidly across various areas of fish farming, from large-scale facilities producing significant quantities of fish for consumption to smaller, advanced systems designed for culture of coldwater species used in restocking or conserving endangered species. It occupies 1% of land for the high density rearing of aquaculture species in comparison to the traditional pond culture methods (Timmons 2002). Recirculation in aquaculture systems can be managed at varying intensities, with different degree of water recirculation, depending on the farm's capacity and the amount of water reused. Some farms operate as highly intensive systems within enclosed, insulated buildings, utilizing a minimum of 300 L of water and less for production of a kilogram of fish annually near the urban areas. Sustainable use of water in recirculation systems offers significant advantageous, especially since water has become a scarce resource in many areas. The tail water of RAS includes particulate matter, phosphorus, nitrite, nitrate and hazardous substances produced by residual feed, faeces deteriorate the water quality that induces changes in fish behaviour such as locomotion and movement. The use of water treatment unit is essential for creating ideal growth conditions for farmed organisms by filtering or breaking down pollutants in aquaculture wastewater and recycling the clean water, which can be reused for fish culture (Kumar et al. 2022c; Li et al. 2023). Moreover, there are several advantages in terms of area utilisation, productivity and environmental sustainability of this system, there are challenges of reuse and maintenance of water quality, specifically on the function and capability of equipment, pipeline and other important infrastructure and production facilities (Zimmermann et al. 2023).

The climate, water temperature, dissolved oxygen content, and the possibility of undesired items floating downstream and obstructing input screens are all major factors affecting the environment in traditional fish farming. Depending on the system type, recirculation level, and facility design, recirculating aquaculture systems (RAS) can completely remove the influence of these external factors. This approach enables fish farmers to have complete control over all aspects of production, making management of the culture system just as important as fish care. In India main species which are found suitable for RAS are as follows:

- Baramundi/Asian Seabass/Bhetki (*Lates calcarifer*)
- Tilapia (*Oreochromis niloticus*)
- Pangasius (*Pangasianodon hypophthalmus*)
- Rainbow Trout (*Oncorhynchus mykiss*)
- Indian Pompano (*Trichinotus Blochii/T. mookalee*)
- Pearl spot (*Etroplus suratensis*).

18.2 Components of Recirculatory Aquaculture System

There is need of basic design of Recirculatory Aquaculture System as shown in Fig. 18.1.

18.2.1 Design of RAS

A standard recirculating aquaculture system consists of several subsystems, with the selection of components for each one influenced by factors such as cost and operational efficiency. The main goal is to create a facility with a standardized process that can be easily replicated, while allowing flexibility for adjustments to' suit specific site conditions. The basic components include culture tanks, mechanical filter, biological filter, disinfection by chlorination or ultraviolet rays and degassing and oxygenation. Site-specific factors, including climate, reliable water and power supply, and water chemistry, are key considerations that impact project planning. Water flows from the fish culture tank to a mechanical filter, then to a biological filter, before being aerated, stripped of carbon dioxide, and returned to the tanks. Additional features, such as oxygenation to meet higher oxygen demands, ultraviolet light or ozone for disinfection, automatic pH regulation, can be added based on specific needs. For the design of CO_2 degassers, calculations for packing height, flow distribution plate area, and required counter current airflow are measured as per the requirement (Summerfelt

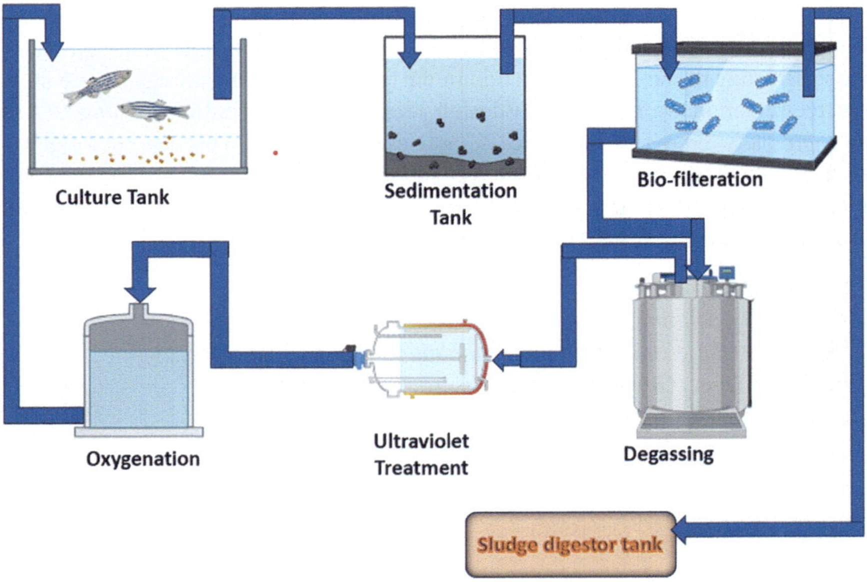

Fig. 18.1 Basic design of recirculatory aquaculture system

Table 18.1 Components of recirculatory aquaculture system

Sl. No.	Component	Material and type of component	Design	References
1	Fish culture tank	Cement, fiber, metal	Circular, rectangle	Ranjan et al. (2022)
2	Mechanical filters	Sand filters, drum filter	Fiber and metal	Timmons (2002)
3	Biological filters	Cement, fiber, metal	Cone, cylindrical	Tilley et al. (2002)
4	Aeration system	5 LPM to 15 LPM	Air blowers, oxygen concentrators and nanobubblers	Ranjan et al. (2022)
5	Degassing unit	Metal, fiber	As per availability	Summerfelt et al. (2000)
6	Storage sump	Cement, fiber	Circular	Tilley et al. (2002)
7	Disinfection	Metal	UV and ozone	Malone (2013)

et al. 2000). There are various component in RAS which are made of different shape, design and material as shown in Table 18.1.

18.2.2 Wastewater Management in Recirculatory Aquaculture System

18.2.2.1 Mechanical Filtration

In a recirculating system, continuous water treatment is important to remove waste from the fish tank and simultaneous oxygenation of treated water ensures both fish survival and efficient system operation. Solid waste particles in aquaculture recirculating systems (RAS) is most effectively removed through mechanical filtration methods. In mechanical filtration, the output water of tank is filtered using microscreens equipped with filter cloths that range in size from 40 to 100 microns. Microscreen filtration stabilizes the conditions for nitrification in subsequent biofiltration processes and lowers the organic load of biofilter. Drum filter is the most widely used microscreen type that gently removes coarse particles during the mechanical filtration (Ranjan et al. 2022). However, while selecting a filter design for solid waste removal in aquaculture, critical considerations such as energy consumption, backwash requirements, overall footprint, and performance must be assessed. Sand filters are known for their excellent performance in removing solid particles larger than 20 μm and where total suspended solids (TSS) concentrations are expected to be very low (Timmons 2002). Recently, the use of floating granular media as a filtration medium has gained considerable interest as an alternative to sand filter media,

Table 18.2 Based on type of system and intensity in recirculatory aquaculture system

System	Water consumption (/kg fish production) (m^3)	Degree of recirculation (%)	New water consumption per hour (m^3/h)	Volume of water consumption per day (%)	References
Flow through RAS	30	0	1712	102	Davidson et al. (2019)
Low level recirculation	3	95.9	171	103	Murray et al. (2014)
Intensive RAS	0.3	99.6	17	10	Dutta et al. (2018)
Phosphorus and nitrogen removal in RAS	0.03	99.9	1.7	1	Martins et al. (2010)

and addresses the drawbacks of sand while providing comparable filtration performance. The twin capabilities of floating media filters which can physically remove particles and convert total ammonia nitrogen into nitrates have drawn more attention in recent years (Malone and Beecher 2000). According to a study, the filter can raise the percentage of dissolved oxygen by 54.23%, decrease the concentration of total ammonia by 92.82%, and lower the concentration of unionized ammonia by 89.12% (Zahidah et al. 2018). Based on type of system and intensity in Recirculatory aquaculture system are categorised as illustrated in Table 18.2.

18.2.2.2 Biological Filtration

Among the most challenging tasks in a recirculating aquaculture system (RAS) are solids management and biofilter operation, which are often the primary causes of system failures (Badiola et al. 2012). Not all organic waste is eliminated by mechanical filters, dissolved substances like nitrogen and phosphate can also get through, as can finer particles. Ionised ammonia is harmful to the circulation loop and must be converted into nitrate within the biofilter through a sequential nitrification process (Pedersen et al. 2015). Bacteria oxidize organic waste, heterotrophic bacteria produce sludge, carbon dioxide, and ammonia while using oxygen. Nitrifying bacteria like *Nitrosomonas* then convert ammonia into nitrite, and ultimately into nitrate. It takes a few weeks to months to establish or colonize a biofilter (Masser et al. 2000). The primary factors affecting biofiltration efficiency are pH, oxygen and organic matter, alkalinity, and water temperature. For optimal nitrification, water temperatures should be maintained between 10 and 35 °C, and pH levels should be kept within the range of 7–8. The effectiveness of nitrification is decreased when alkalinity levels below 100 mg calcium carbonate levels, and low oxygen levels (Suhr and Pedersen 2010). In a water recirculation system, the pH is influenced by two main factors: the acid produced during the nitrification process and the carbon dioxide produced by microorganisms in biofilter respiration and biological activity of fish. Carbon dioxide is removed through aeration of water which promotes degassing. Acid (H^+) is produced during the nitrification process, which lowers pH (Suhr and Pedersen 2010). A base, usually lime, sodium hydroxide, or another appropriate base, must be added in order to stabilize the pH. Ammonia makes up the majority of the total ammonia nitrogen (TAN—total ammonium nitrogen), which is a mixture of ammonia and ammonium that fish excrete. However, the pH level, which controls the balance between ammonia and ammonium, affects the amount of ammonia in the water (Bregnballe 2022a, b).

18.2.2.3 Degassing and Aeration

To ensure the safety of the fish, all stored harmful gases must be removed before the water is reintroduced into the fish tanks. Biofiltration eliminates ammonia and dissolved organics, but it necessitates constant circulation between the filtration components and the tank. Lastly, aeration and degasification procedures balance dissolved gases, including carbon dioxide and oxygen (Malone 2013). The system's carbon dioxide (CO_2) levels are raised by fish respiration and bacterial activity, and they can rise even when pure oxygen is employed. This excess carbon dioxide is removed through a process called stripping, achieved by providing aeration in water. Another consequence of the ammonification and nitrification processes is free nitrogen (N_2) (Ranjan et al. 2022). The welfare and growth of fish will suffer if these carbon dioxide and nitrogen gas levels build up. Hydrogen sulfide (H_2S) can be created anaerobically, particularly in saltwater environments. Even at low concentrations, this gas is exceedingly hazardous to fish, and if hydrogen sulfide is produced in the system, fish will perish. Oxygenation and other cutting-edge technologies are being used more and more in recirculating systems to increase cost effectiveness. By encouraging higher stocking densities and feed rates and lowering the requirement for water flow, the use of pure oxygen lowers the expenses associated with producing fish. As a result, pumps, culture tanks, water reuse systems, and facility infrastructure become smaller and less expensive (Summerfelt et al. 2000).

18.2.2.4 Disinfection

Disinfection is the process of deactivating, or killing dangerous microbes which successfully targets harmful bacteria and single-celled organisms in aquaculture by employing light wavelengths that harm genetic material of biological organisms. UV light is widely used due to its simple installation and operation. Effective UV treatment relies on appropriate installation, lights must be submerged in the water to function effectively in contrary, light reflection renders them ineffective when placed above the water's surface. Controlling the bacterial population in conventional fish farms is often ineffective, as bacteria rapidly proliferate in organic matter. In order to effectively remove organic matter from the process water and make the UV radiation function successfully, the optimum control is obtained when mechanical filtration and comprehensive bio-filtration are coupled. Microwatt seconds per cm^2 ($\mu W\ s/cm^2$) is a unit of measurement for UV dose. The size, kind, and water turbidity of the target organisms affect efficacy of disinfection. To effectively eliminate pathogens, water must be treated with approximately 2,000–10,000 $\mu W\ s/cm^2$ to eradicate 90% of bacteria and viruses, 10,000–100,000 $\mu W\ s/cm^2$ for fungi, and 50,000–200,000 $\mu W\ s/cm^2$ to target microscopic parasites (Ranjan et al. 2022). Ozone, on the other hand, is used in larger facilities where the higher installation and operational costs can be managed by trained technical staff. Ozone is a disinfectant that poses significant risks to fish farming operations as it can kill undesirable organisms by oxidation of organic material. Risks include the possibility of overdosing and causing health problems for indoor facility workers through respiratory exposure to it (Malone 2013).

18.3 Feeding Management in RAS

The culture species are fed a high-protein diet that completes the nutritional needs of reared fishes and contains all necessary minerals and vitamins. The resulting biomass output is largely influenced by feeding strategies adopted in the system. When the given feed is consumed and digested, it is utilized by the metabolism of fish to support growth, physiological functions, and energy production (Bregnballe 2022a, b). When food is limited, such as during feed restriction, fish can rely on their body reserves to sustain themselves (Dar et al. 2018). These mobilizations, including the breakdown of lipids in the liver and adipose tissue, the utilization of muscle protein as a primary energy source, and the conversion of muscle and hepatic glycogen into glucose for energy, are crucial survival mechanisms during periods of feed restriction or environmental stress. However, prolonged reliance on these reserves can have negative consequences on fish health and performance (Li et al. 2018). The reared fishes exposed to various feed restriction and re-feeding methods may also have alterations in their haematological, immunological, and osmoregulatory systems (Silva et al. 2019). The undigested feed is expelled into the water as faeces commonly called as suspended solids and organic materials. Moreover, the gills discharge ammonia and carbon dioxide into the water. Fish therefore consume food and oxygen, which pollutes the water supply with ammonia, carbon dioxide, and excrement. As a result, in a recirculating system, only dry feed can be suggested. Its use is safe, and it has the added benefit of being made to precisely match the fishs biological requirements. The materials in dry fish feed can be blended to create specialized feeds for different stages of the fish life cycle, such as fry, broodstock, and grow-out, with dry feed provided in various pellet sizes tailored to each stage of fish. The management of feeding improves the feed conversion rate, and ensures maximum production with less requirement of feed. As a result, farmers achieve higher production yields with less pressure on the filtration system. Based on the type of culture system, different fish species shows different feed conversion efficiency

Table 18.3 Evaluation of Feed conversion efficiency in different culture systems

Fish species	Earthen pond	RAS system	Cage culture	References
Tilapia (*Oreochromis* sp.)	0.8–3.5	1.0–2.2	>1.5	El-Sayed (2006)
Rainbow trout (*Oncorhynchus mykiss*)	1.0	0.8–1.1	1.1–1.3	Bureau et al. (2003)
Asian Seabass (*Lates calcarifer*)	1.5–2.2	0.8–1.1	1.6–2	Vandeputte and Labbé (2012)
Cobia (*Rachycentron canadum*)	1.4	1.1	1.5–1.8	Benetti et al. (2008)

as illustrated in Table 18.3. Uneaten feed not only wastes money but also places an unnecessary burden on the filter. Henceforth, specialized feeds are formulated to enhance protein absorption in fish, thereby reducing ammonia excretion into the water. Some of the following characteristics should be included in the feed:

- Feed specific to requirement of reared species should be given.
- Feeding should be done at a rate of 3–5% of the body weight, depending on the quality and protein content of the meal.
- Faster development rates and a higher feed conversion ratio are the outcomes of more frequent feedings.

18.4 Status of Disease and Biosecurity in RAS

In RAS system, where healthy microbial community aids in water purification and purity, controlling disease outbreaks presents unique difficulties. The system may get accumulated with minerals, drug residues, toxic feed components, and metabolites (Martins et al. 2010). To limit the occurrence of recognized obligatory fish infections, the quality of intake water plays a crucial role in fish health management within RAS. Opportunistic pathogen groups, such as *Saprolegnia* and *Flavobacteria* species, might present serious problems, though, because traditional chemotherapy may harm the biofilter and deteriorate the quality of the water (Holan et al. 2020). It is possible for diseases to get into the system through transporting water, the fish themselves, or nets, baskets, gloves, and other items that are moved from one tank to another. Most pathogens are opportunistic, causing disease primarily in fish with weakened immune systems. However, when pathogen levels become sufficiently high, they can also infect healthy fish and continuous water flow within a system can facilitate the rapid spread of these pathogens. Fish should be quarantined, checked for diseases, and provided any care they require (Yanong 2003). Before transferring equipment between tanks, it should be sanitized (for example, by dipping it in chlorine). Each tank should ideally have its own specialized equipment, such as nets and baskets, to reduce the danger of cross-contamination. However, pathogens can quickly spread throughout the entire system in systems where tanks share equipment or when water passes through common parts like sumps, particle filters, or biofilters. Pathogens in a recirculating aquaculture system (RAS) can be transmitted through various routes, including:

- via the water,
- direct fish-to-fish contact,
- through vectors and fomites, and
- through contaminated feed.

Biosecurity plays a vital role in RAS by preventing disease outbreaks, protecting fish health, and supporting the system's overall performance. RAS operates in a closed or semi-closed loop, where water is recirculated and reused, making it easier to control environmental factors but also heightening the risk of disease outbreaks if biosecurity measures are not strictly enforced (Yanong 2003). Biosecurity strategies encompass various practices such as surveillance, quarantine, fish husbandry, preventive disease treatment, environmental monitoring, pest control, and the chemical management of existing diseases. Hence, the type and frequency of biosecurity measures implemented are influenced by factors such as farm size, fish species cultured, the source of incoming water, and the sociodemographic characteristics of the site operators (Delabbio et al. 2004). Since the environment in which fish are raised might serve as an incubator for the illness, it is frequently exceedingly difficult to eradicate a pathogen once it has established itself in a recirculating system. Furthermore, a lot of chemical treatments that are commonly utilized in flow-through systems to treat illnesses are not appropriate for use in recirculation systems. This is because these treatments can harm both the beneficial bacteria essential to the system and the targeted pathogens (Schwartz et al. 2000). Some of the common biosecurity measures beneficial for any intensive farm is:

1. Collection and discarding dead fish
2. Vaccination
3. Quarantine
4. Record keeping
5. Consultation with aquatic animal health specialist
6. Water quality monitoring
7. Cleaning of tanks
8. Use of chemical treatment.

18.5 Recent Advancement in RAS Technology

In today's time there is a lot of modification which is done and found suitable for more production and less handling of different stages manually. Major technical interventions in recirculating aquaculture systems (RAS) include the use of denitrification reactors, sludge thickening technologies, advanced filtration methods such as UV irradiation, ozone treatment, chlorination, and integrated systems like constructed wetlands and microalgal system (Martins et al. 2010). The technologies for thickening sludge are used to lower the amount of solids in the RAS system (Schneider et al. 2006; Rani et al. 2025; Nisha et al. 2025; Kiran et al. 2025). The installation of diffusers to supply oxygen directly to each tank during emergencies should also be considered. Microalgae-based water treatment effectively removes chemical oxygen demand (COD), biological oxygen demand (BOD), nutrients, heavy metals, pathogens, and facilitates the anaerobic digestion of algal–bacterial biomass (Munoz and Guieysse 2006a, b). Combining these technologies may result in a better sustainable integrated production.

18.5.1 Denitrification Reactors

These reactors are used to remove toxic products like ammonia and nitrites in which nitrification in aerobic biological filters primarily turns ammonia into nitrate and denitrification process converts nitrated into nitrogen gas. Denitrification can counteract high nitrate concentrations, and the reactors used for RAS are designed differently (van Rijn et al. 2006). The solids removal unit traps inorganic compounds, bacterial flocs, and dissolved and particulate fecal organic waste into the cylindric anoxic sludge blanket denitrification reactor. The reactor's bottom center is where the waste flow enters. The upflow velocity of the reactor is designed to be lower than the settling velocity of most particulate waste, allowing a sludge bed to form at the bottom. The denitrifying bacteria in the sludge bed break down the fecal particle matter (Eding et al. 2009). The decrease in water consumption is one of the main benefits of incorporating denitrification into RAS for environmental sustainability (Klas et al. 2006).

18.5.2 Sludge Thickening Technologies

This technology is employed to reduce the volume of solid waste generated as a byproduct of fish farming, thereby minimizing the labor, storage, transportation, and disposal costs associated with sludge discharge from RAS (Schneider et al. 2006). Belt filtration systems and geotextile bags or tubes are examples of thickening technology that can help reduce this problem as well as belt filtration systems. These systems decrease the volume of solids produced by enabling the sludge to be dewatered through the use of technology (Ebeling et al. 2006).

In RAS, geotextile bag filters are a useful pretreatment technique, especially when total suspended solids (TSS) need to be dewatered before disposal (Ebeling et al. 2006). Sludge that is released from the denitrification reactor once a week can be processed using geotextile tubes in RAS systems that are coupled with denitrification reactors. Adding polymer as a coagulation and flocculation aid to this sludge helps concentrate the waste solids after dewatering (Eding et al. 2009). The coagulation/flocculation process enhances the removal of nitrogen (64%), total ammonia nitrogen (50%), nitrite (68%), nitrate and total nitrogen (87%) in wastewater treatment. Activated sludge membrane bioreactors (MBRs) represent a promising waste treatment technology for managing solids-rich aquaculture effluents. They use fine-pore membranes (typically < 0.2 μm) to produce a semi-purified filtrate, while integrated aerobic and anoxic processes within the activated sludge promote both nitrification and denitrification (Davidson et al. 2019).

18.5.3 Ozone and Ultraviolet Irradiation

Ozone treatment in RAS is used to prevent infections and facilitate the oxidation of nitrite to nitrate, as well as to break down organic matter, total ammonia nitrogen, and small suspended particles (Tango and Gagnon 2003). Ozonation improves microscreen filter performance and minimizes the accumulation of dissolved matter affecting the water colour (Summerfelt et al. 2009). In order to maintain acceptable water quality and fish health, a wide range of 3–24 g of ozone per kilogram of feed to a RAS is typically mentioned in the literature (Bullock et al. 1997). It has been demonstrated that using ozone in conjunction with UV radiation to disinfect recirculated water is especially effective at inactivating bacteria. A combination of a moderate ozone dose and a high UV irradiation dose (ranging from 42.5 to 112.7 mJ/cm^2) effectively inactivated nearly all total heterotrophic and coliform bacterial populations in a freshwater RAS system (Sharrer and Summerfelt 2009). On the other hand, byproducts of ozonation might be dangerous, one of these potentially harmful byproducts is bromate concentrations in the ozonated RAS are harmful for fish health. Thus, more research should be done to determine how using ozone in RAS affects fish (Tango and Gagnon 2003).

18.5.4 Integrated System in RAS

In RAS system, naturally occurring processes such as algae or macrophyte formation, sedimentation, nitrification, denitrification, acidification, phosphate precipitation, aerobic and anaerobic decomposition, fish production, and temperature regulation are all managed within various reactors. However, the overall production capacity of the system is enhanced when some of these processes are compartmentalized and placed in reactors outside the fish tank. Despite these benefits, efficiency of phototrophic reactors treatment is still poor, which leads to a large disparity often by an order of magnitude between the surface areas needed for phototrophic reactors and those needed for fish production. Reuse of biomass as feed, further lowers the effectiveness of the treatment process by as much as 90%. Wetlands and algal ponds have recently gained significant attention as water treatment units in RAS due to their ability to enhance water reuse within these systems (Schneider et al. 2001).

18.5.4.1 Wetlands

Wetlands have been utilized to treat various forms of wastewater, including acid mine drainage, stormwater runoff, municipal wastewater, industrial effluent, and agricultural runoff from livestock operations (Lin et al. 2002). After aquaculture effluents have been concentrated, wetlands are typically utilized to treat them. At this stage, they are regarded as an affordable and practical biological treatment technique (Tavares and Braga 2008). Removal rates for organic suspended particles, phosphates, and total ammonia nitrogen in this pond-wetland system are over 90%, while removal rates for inorganic nitrogen compounds, total nitrogen, and total phosphorus are between 65 and 80% (Kerepeczki et al. 2003). Most artificial wetlands used in aquaculture are soil-based, horizontal subsurface flow systems. While these wetlands are highly effective in removing organic matter and BOD, nitrogen (N) and phosphorus (P) removal typically ranges from 30 to 50%. Additionally, nitrification is almost absent in these horizontal subsurface flow systems (Brix et al. 2007). Vertical flow constructed wetlands with partial recirculation have been introduced as innovative technologies to enhance the removal of total ammonium nitrogen and nitrate. Partial recycling of the effluent stabilizes system performance and enhances nitrogen removal via denitrification (Arias et al. 2005). Non-concentrated aquaculture effluents often result in less effective nutrient removal in artificial wetlands compared to concentrated effluents. In a constructed wetland, trout raceway effluents showed average removal rates of 68% for chemical oxygen demand, 58% for total phosphorus, and 30% for total nitrogen when hydraulic retention was applied (Schulz et al. 2003). Therefore, wetlands are found valuable in increasing fish production per cubic meter of water entering the farm, as well as in removing and concentrating solids from the recirculating flow (Sindilariu et al. 2009).

18.5.4.2 Algal Controlled Systems

Microalgae Availability

Aquaculture ponds are typically eutrophic, with primary production levels ranging from 1 to 3 g C/m^2/day in temperate regions and 4–8 g C/m^2/day in tropical and subtropical areas. Most of the algae in these ponds are decomposed, and the use of aqua feed also acts as a form of fertilizer. The amount of fertilizer needed to sustain productivity would be excessively high if all primary production were constantly harvested from the ponds (Cadoret and Bernard 2008), However, direct algae harvesting remains challenging, and innovations such as flocculation techniques may offer a breakthrough in making algae harvesting more feasible (Lee et al. 2009).

a. **Microalgae based water treatment**

Microalgae are crucial in wastewater treatment, helping to reduce chemical and biochemical oxygen demand, nutrients, heavy metals, and pathogens, while the anaerobic digestion of algal–bacterial biomass can generate biogas (Munoz and Guieysse 2006a, b). Algal ponds effectively process dissolved aquaculture waste, with the resulting algal biomass serving as a food source for certain aquatic species, while a commercial integrated culture system has demonstrated that combining shrimp, algae, and oyster production optimizes water use and treatment, turning it into a profitable activity (Wang 2003). Algal ponds are effective in processing dissolved aquaculture waste, with the resulting algal biomass providing a food source for specific aquatic species. Additionally, a commercial integrated culture system has shown that combining shrimp, algae, and oyster production enhances water use and treatment, making it a profitable venture. The high density of *Chaetoceros* also helps in reduction pathogens such as *Vibrio vulnificus*, benefitting shrimp health provided an overview and comparison of other phototrophic conversion systems (Sharma et al. 2024, 2025; Schneider et al. 2005). High-rate algal ponds (HRAPs) are specifically designed to align algal growth and oxygen production with the biochemical oxygen demand of incoming wastewater. These systems can achieve biochemical oxygen demand (BOD) removal rates of up to 175 g per cubic meter per day, significantly higher than the 5–10 g BOD removal rate of traditional waste stabilization ponds.

b. **Solids removal efficiency of RAS**

In general, modern RAS improve at dealing with nitrogenous waste and gas exchange, but they challenge with solid waste management. Optimizing nutrient sources, including feed types and feeding techniques, can help with some of these issues. Improving the hydraulic flow, tank design, and efficiency of solids removal systems is another potential optimization area. The solid waste should be controlled during production through reduction of feed spillage and improve feed efficiency, optimization of the consistency, water stability, and composition of the faeces. The primary goal is to develop fish feed that produces feces easily removed from the water, generates fewer fine solids, and is readily fermented by microbial communities within the system. Additionally, the degree of gelatinization in the diet influences feces removal rates, while insoluble non-starch polysaccharides, such as cellulose, enhance fecal recovery by improving particle removal efficiency in RAS (Amirkolaie et al. 2006). Drum filters or similar devices are frequently used in freshwater recirculating aquaculture systems (RAS) to remove bigger solid particles from tank effluents. Then, fixed-bed bio-filters are used to catch finer solids. To improve the removal of fine suspended materials, foam fractionation devices, including protein skimmers, are frequently used in combination with these filtering systems in marine RAS. Hydraulic dynamics, removal technologies, and tank design should be the focus of future developments in this field. Ozone is commonly utilized to enhance water quality in maritime systems which alters the properties of fine solids, enhancing the performance of foam fractionation systems (Tango and Gagnon 2003). The efficiency of nitrification in these systems is often constrained by organic matter entering the biofilters, which fosters the growth of both autotrophic and heterotrophic bacteria A key challenge lies in reducing the organic carbon content in the influent to promote chemoautotrophic conditions within the nitrification reactors (Eding et al. 2009).

c. **Phosphate removal efficiency of RAS**

Phosphate continues to accumulate in the system water and raise the phosphorus content of effluents because most Recirculating Aquaculture Systems (RAS) lack specialized phosphate removal systems. This makes it challenging to improve the efficiency and affordability of phosphate removal (Martins et al. 2009). The following techniques can be used to regulate phosphate levels:

- Improving the phosphorus retention of reared fish species
- Rapid solids removal from the water to prevent phosphorus from leaking from the surface of organic matter
- Aquaculture system with recirculation, integrated multitrophic aquaculture through phosphorus recycling can be used along with chemical flocculation for dephosphatation.

18.6 Ecological Sustainability of Recirculating Aquaculture Systems (RAS)

The sustainable approach in improving the efficiency of waste removal, particularly for solids, nitrogen, and phosphate ensures ecological balance in environment. The RAS system fulfils the food production requirement, maintain the optimum environment conditions, contain minimum ecological impact to environment and reduces the chances of pathogen susceptibility due to controlled environment (Martins et al. 2010). The aquaculture industry plays important role in provision of food security, nutritional fulfilment and providing employment opportunities to many people worldwide. In efforts to boost productivity, tanks are often stocked beyond optimal levels, leading to overcrowding, higher disease risks, and other management challenges. The adoption of sustainable culture techniques aligns with several United Nations Sustainable Development Goals (SDGs), making significant contributions to SDG 1 (No Poverty), SDG 2 (Zero Hunger), SDG 8 (Decent Work and Economic Growth), and SDG 14 (Life Below Water) (Zhang et al. 2024). Sustainable aquaculture practices are crucial to maintaining a balance between minimizing environmental impact and maximizing production. However, the high initial investment and the requirement for high stocking densities can be barriers to widespread RAS adoption. Moreover, impact of disease outbreak is lowered with the limited use of water and its effective management and provision of biosecurity during the culture period. Variability in species, diversification and conservation are practiced with freshwater, brackishwater and marine water fish species are successfully cultured by utilisation of all trophic levels in this system. By recirculating and treating water within a closed-loop system, RAS reduces water consumption, minimizes nutrient

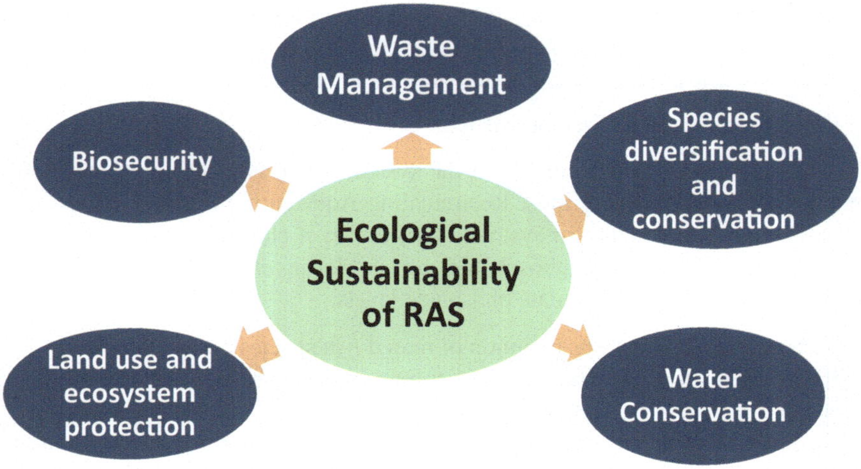

Fig. 18.2 Ecological sustainability of RAS system in fish production

discharge significantly and enables better management of culture conditions leading to improving fish health, productivity, and food quality (Ende et al. 2024). There are various advantages of maintaining ecological sustainability in fish culture system as shown in Fig. 18.2.

18.7 Conclusion

Due to limited supply of production from commercial fishing, aquaculture is essential to meet the growing demand for fish and fill the supply gap in the fish and seafood market. Water treatment technology through sequential filtration along with advanced treatment techniques is the best way of waste management, which is emerging rapidly in recirculation aquaculture system. The adoption of suitable treatment method depends on factors like location of recirculating system, water availability, waste discharge regulation limit and availability of land. Along with these factors market value of cultured organisms, proper cost benefit analysis and production intensity of the system make the system sustainable and useful for marginal farmers. However, economical benefit is accomplished only by the use of simple water treatment techniques with proper monitoring. The solid waste management is one of the challenges in modern RAS, which can be addressed by optimisation of feeding strategies, nutrient source and feed types, upgrading tank design, hydraulic flow, and by increasing efficiency of solids removal systems.

The integration of microalgae and wetland systems enhances sustainability by assimilating dissolved nutrients such as nitrogen and phosphorus, improving water quality through algal metabolism, while oxygenation via photosynthesis further supports the health and growth of the reared species. The proposed integration of microalgae cultivation offers potential to reduce environmental impacts through processes such as chemical remediation, carbon sequestration, and resource recovery via bioconversion. In outdoor RAS, waste is typically managed within the system by leveraging the nutrient uptake capabilities of phototrophic and heterotrophic organisms. In contrast indoor RAS uses alternative strategies like solid waste removal and the biological conversion of ammonia to nitrate, and onsite treatment of concentrated effluents is done before discharge to reduce the impact on environment. Advancement in cost-effective and eco-friendly waste management techniques is crucial for expanding the practical application of these systems. Since fish production is closely tied to water quality, species that are sensitive to environmental fluctuations especially in subtropical regions stand to benefit significantly from RAS, which offer consistent, optimized rearing conditions throughout the year. RAS facilities near urban areas provides additional advantages, including the supply of fresh, high-quality fish, reduced carbon emissions due to shorter transportation routes, and improved national food self-sufficiency. Despite its benefits, the adoption of RAS is currently constrained by high initial investment costs, significant energy requirements, and the need for skilled labor to manage its technical operations. Recirculatory aquaculture system is not just a technological alternative but a strategic method for sustainable aquaculture. As the pressure on natural resources is increasing and the demand for safe and quality aquatic food continues to rise, recirculatory aquaculture system offers a scalable, ecologically efficient, and resilient approach to sustainable aquaculture production in the twenty-first century.

18.8 Future Prospects

The adoption of RAS system for fish cultivation is increasing steadily, driven by the growing demand for healthy, fresh fish and the need for year-round availability for consumption. Although the core principles of Recirculating Aquaculture Systems (RAS) may seem conventional, advancing the technology to support a wider variety of species, environmental conditions, and life stages is crucial. The innovations in culture system are focused on making RAS more cost-effective, energy-efficient, and environmentally sustainable by designing systems that operate in a more closed-loop and cradle-to-cradle manner where waste is not discarded but repurposed into valuable resources or products. In today's era of rapid technological progress, the integration of digitalization, automation, and machine learning is transforming aquaculture. Artificial intelligence is increasingly becoming a vital tool for optimizing system performance. AI-driven solutions such as those capable of monitoring fish swimming behaviour for early detection of stress or forecasting feeding requirements are emerging as valuable assets for improving operational efficiency and productivity in fish farming. With growing global demand for seafood and increasing public concern over issues such as overfishing, water conservation, pollution, animal welfare, and ethical farming practices, both research and commercial interest in RAS are expected to expand significantly. The future of RAS lies in the continued development of cost-effective, low-energy waste treatment technologies, the refinement of integrated multi-trophic aquaculture models, and the incorporation of renewable energy to minimize the environmental footprint. Additionally, strong policy support, robust institutional frameworks, and active collaboration between public and private sectors will be critical in accelerating the adoption of RAS, particularly for small and medium scale producers. Together, these efforts will help position RAS as a cornerstone of sustainable, high-efficiency aquaculture for the future.

References

Amirkolaie AK, Verreth JA, Schrama JW (2006) Effect of gelatinization degree and inclusion level of dietary starch on the characteristics of digesta and faeces in Nile tilapia (Oreochromis niloticus (L.)). Aquaculture 260(1–4):194–205

Arias CA, Brix H, Marti E (2005) Recycling of treated effluents enhances removal of total nitrogen in vertical flow constructed wetlands. J Environ Sci Health 40(6–7):1431–1443

Badiola M, Mendiola D, Bostock J (2012) Recirculating aquaculture systems (RAS) analysis: main issues on management and future challenges. Aquac Eng 51:26–35

Benetti DD, Orhun MR, Sardenberg B, O'Hanlon B, Welch A, Hoenig R, Cavalin F (2008) Advances in hatchery and grow-out technology of cobia Rachycentron canadum (Linnaeus). Aquac Res 39(7):701–711

Bregnballe J (2022a) A guide to recirculation aquaculture: an introduction to the new environmentally friendly and highly productive closed fish farming systems. Food and Agriculture Organization

Bregnballe J (2022b) A guide to recirculation aquaculture—an introduction to the new environmentallyfriendly and highly productive closed fish farming systems. FAO and Eurofish International Organisation, Rome. https://doi.org/10.4060/cc2390en

Brix H, Schierup HH, Arias CA (2007) Twenty years experience with constructed wetland systems in Denmark—what did we learn? Water Sci Technol 56(3):63–68

Brown AR, Wilson RW, Tyler CR (2024) Assessing the benefits and challenges of recirculating aquaculture systems (RAS) for Atlantic Salmon production. Rev Fish Sci Aquac 1–22

Bullock GL, Summerfelt ST, Noble AC, Weber AL, Durant MD, Hankins JA (1997) Ozonation of a recirculating rainbow trout culture system I. Effects on bacterial gill disease and heterotrophic bacteria. Aquaculture 158(1–2):43–55

Bureau DP, Gunther SJ, Cho CY (2003) Chemical composition and preliminary theoretical estimates of waste outputs of rainbow trout reared in commercial cage culture operations in Ontario. N Am J Aquac 65(1):33–38

Cadoret JP, Bernard O (2008) Lipid biofuel production with microalgae: potential and challenges. J Soc Biol 202(3):201–211

Dar SA, Srivastava PP, Varghese T, Rasool SI, Anand G, Gupta S et al (2018) Regulation of compensatory growth by molecular mechanism in Labeo rohita juveniles under different feeding regimes. Gen Comp Endocrinol 261:89–96

Davidson J, Summerfelt S, Schrader KK, Good C (2019) Integrating activated sludge membrane biological reactors with freshwater RAS: preliminary evaluation of water use, water quality, and rainbow trout Oncorhynchus mykiss performance. Aquac Eng 87:102022

Delabbio J, Murphy BR, Johnson GR, McMullin SL (2004) An assessment of biosecurity utilization in the recirculation sector of finfish aquaculture in the United States and Canada. Aquaculture 242(1–4):165–179

Dutta S, Manojkumar B, Pillai D (2018) Design of a simple trickling biofilter for recirculating aquaculture system (RAS). Fish Technol 55(3)

e Silva WDS, Hisano H, Mattioli CC, Torres IFA, de Oliveira Paes-Leme F, Luz RK (2019) Effects of cyclical short-term fasting and refeeding on juvenile Lophiosilurus alexandri, a carnivorous Neotropical catfish. Aquaculture 505:12–17

Ebeling JM, Welsh CF, Rishel KL (2006) Performance evaluation of an inclined belt filter using coagulation/flocculation aids for the removal of suspended solids and phosphorus from microscreen backwash effluent. Aquac Eng 35(1):61–77

Eding E, Verdegem M, Martins C, Schlaman G, Heinsbroek L, Laarhoven B et al (2009) Tilapia farming using recirculating aquaculture systems (RAS)—case study in the Netherlands. In: Handbook for sustainable aquaculture. Project no. COLL-CT-2006-030384. http://www.sustainaqua.org. Accessed 22 June 2009

El-Sayed AFM (ed) (2006) Tilapia culture. CABI Publishing

Ende S, Henjes J, Spiller M, Elshobary M, Hanelt D, Abomohra A (2024) Recent advances in recirculating aquaculture systems and role of microalgae to close system loop. Bioresour Technol 131107

Holan AB, Good C, Powell MD (2020) Health management in recirculating aquaculture systems (RAS). In: Aquaculture health management. Academic Press, pp 281–318

Kerepeczki É, Gál D, Szabó P, Pekár F (2003) Preliminary investigations on the nutrient removal efficiency of a wetland-type ecosystem. Hydrobiologia 506(509):665–670

Kiran, Kumar N, Sindhu S, Gawaria B, Ansari A, Dhariwal A, Sharma P, aKumar R (2025) GWAS approaches in medicinal plants for enhanced secondary metabolite production. In: Chen JT (ed) Biotechnology, multiple omics, and precision breeding in medicinal plants. CRC Press, pp 119–134. https://doi.org/10.1201/9781003475491-10

Klas S, Mozes N, Lahav O (2006) A conceptual, stoichiometry-based model for single-sludge denitrification in recirculating aquaculture systems. Aquaculture 259(1–4):328–341

Kumar R, Gupta R, Dahiya T, Yadav R, Saifi R, Yodha K, Kumari B, Singh P, Shefali (2022a) Surveillance of *Heteropneustes fossilis* (stinging catfish) diseases and health management practices in Haryana. Appl Ecol Environ Sci 10(4):261–267. https://doi.org/10.12691/aees-10-4-12

Kumar R, Gupta RK, Yadav R, Saifi R, Yodha K, Pooja (2022b) *Eisenia fetida* as protein source for growth enhancement of *Heteropneustes fossilis*. Egypt J Aquat Biol Fish 26(2):577–588. https://doi.org/10.21608/ejabf.2022.234553

Kumar R, Yadav R, Gupta RK, Pal A, Yodha K, Kumar A (2024) Antioxidant enzyme activities and markers of oxidative stress in the life cycle of different earthworm species. Indian J Exp Biol 62(07):578–583

Kumar R, Yadav R, Kumar Gupta R, Pooja (2022c) A sustainable way for fish health management by replacement of chemical and drugs by earthworm. In: Environmental degradation in Asia: land degradation, environmental contamination, and human activities. Springer International Publishing, Cham, pp 329–352. https://doi.org/10.1007/978-3-031-12112-8_16

Lee AK, Lewis DM, Ashman PJ (2009) Microbial flocculation, a potentially low-cost harvesting technique for marine microalgae for the production of biodiesel. J Appl Phycol 21:559–567

Li H, Cui Z, Cui H, Bai Y, Yin Z, Qu K (2023) Hazardous substances and their removal in recirculating aquaculture systems: a review. Aquaculture 569:739399

Li H, Xu W, Jin J, Yang Y, Zhu X, Han D et al (2018) Effects of starvation on glucose and lipid metabolism in gibel carp (Carassius auratus gibelio var. CAS III). Aquaculture 496:166–175

Lin YF, Jing SR, Lee DY, Wang TW (2002) Nutrient removal from aquaculture wastewater using a constructed wetlands system. Aquaculture 209(1–4):169–184

Malone R (2013) Recirculating aquaculture tank production systems: a review of current design practice

Malone RF, Beecher LE (2000) Use of floating bead filters to recondition recirculating waters in warmwater aquaculture production systems. Aquac Eng 22(1–2):57–73

Manjeet, Airon A, Kumar R, Saifi R (2022) Temporal and spatial impact of lockdown during COVID-19 on air quality index in Haryana, India. Sci Rep 12:20046. https://doi.org/10.1038/s41598-022-20885-2

Martins CIM, Eding EH, Verdegem MC, Heinsbroek LT, Schneider O, Blancheton JP et al (2010) New developments in recirculating aquaculture systems in Europe: a perspective on environmental sustainability. Aquac Eng 43(3):83–93

Martins CI, Pistrin MG, Ende SS, Eding EH, Verreth JA (2009) The accumulation of substances in recirculating aquaculture systems (RAS) affects embryonic and larval development in common carp Cyprinus carpio. Aquaculture 291(1–2):65–73

Masser MP, Rakocy J, Losordo TM (2000) Recirculating aquaculture tank production systems. Southern Regional Aquaculture Center

Munoz R, Guieysse B (2006a) Algal–bacterial processes for the treatment of hazardous contaminants: a review. Water Res 40(15):2799–2815

Munoz R, Guieysse B (2006b) Algal-bacterial processes for the treatment of hazardous contaminants: a review. Water Res 40:2799–2815

Murray F, Bostock J, Fletcher D (2014) Review of recirculation aquaculture system technologies and their commercial application

Nisha A, Rani S, Kumar S, Kumar R, Kumar A (2025) Recent technologies and strategies for bioremediation of toxic and hazardous waste. In: Dubey KK, Shah MP, Yadav A (eds) Challenges and sustainable solutions in bioremediation. CRC Press, pp 1–16. https://doi.org/10.1201/9781003407317-1

Pedersen LF, Oosterveld R, Pedersen PB (2015) Nitrification performance and robustness of fixed and moving bed biofilters having identical carrier elements. Aquac Eng 65:37–45

Rani S, Sharma S, Singh D, Sheoran R, Kumar R (2025) Improved production of anti-cancer compounds from medicinal plants through molecular and biotechnological mutagenesis strategies. In: Chen JT (ed) Biotechnology, multiple omics, and precision breeding in medicinal plants. CRC Press, pp 287–297. https://doi.org/10.1201/9781003475491-21

Ranjan R, Sampath K, Raju SN, Bathina C, Avadhanula RK (2022) Recirculating aquaculture system engineering: design, components and construction

Schneider O, Blancheton JP, Varadi L, Eding EH, Verreth JAJ (2006) Cost price and production strategies in European recirculation systems. In: Conference: aqua 2006: linking tradition and technology highest quality for the consumer, Firenze (Florence), Italy, 9–13 May 2006, p 855

Schneider O, Eding EH, Verreth JAJ (2001) Introduction of a framework of zero nutrient discharge aquaculture by farming in integrated recirculating systems in Asia

Schneider O, Sereti V, Eding EH, Verreth JAJ (2005) Analysis of nutrient flows in integrated intensive aquaculture systems. Aquac Eng 32(3–4):379–401

Schulz C, Gelbrecht J, Rennert B (2003) Treatment of rainbow trout farm effluents in constructed wetland with emergent plants and subsurface horizontal water flow. Aquaculture 217(1–4):207–221

Schwartz MF, Bullock GL, Hankins JA, Summerfelt ST, Mathias JA (2000) Effects of selected chemotherapeutants on nitrification in fluidized-sand biofilters for coldwater fish production

Sharma MMM, Kapoor D, Rohilla R, Loyal A, Sharma P, Husen A (2024) Role of essential minerals and rhizosphere in plant under changing environmental conditions. In: Husen A (ed) Essential minerals in plant-soil systems. Elsevier, pp 235–252. https://doi.org/10.1016/B978-0-443-16082-0.00002-3

Sharma P, Kapoor D, Kumar R, Husen A (2025) Rhizosphere and osmolytes coordination in plant systems under adverse situations. In: Husen A (ed) Roles of osmolytes in changing environment. Academic Press, pp 101–117. https://doi.org/10.1016/B978-0-443-26549-5.00020-5

Sharrer MJ, Summerfelt S (2009) Evaluation of geotextile filtration applying coagulant and flocculant amendments for aquaculture biosolids dewatering and phosphorus removal. Aquac Eng 40(1):1–10

Shefali, Kumar R, Sankhla MS, Kumar R, Sonone SS (2021) Impact of pesticide toxicity in aquatic environment. Biointerface Res Appl Chem 11(3):10131–10140. https://doi.org/10.33263/BRIAC113.1013110140

Sindilariu PD, Brinker A, Reiter R (2009) Factors influencing the efficiency of constructed wetlands used for the treatment of intensive trout farm effluent. Ecol Eng 35(5):711–722

Sugita H, Nakamura H, Shimada T (2005) Microbial communities associated with filter materialsin recirculating aquaculture systems of freshwaterfish. Aquaculture 243(1–4):403–409

Suhr KI, Pedersen PB (2010) Nitrification in moving bed and fixed bed biofilters treating effluent water from a large commercial outdoor rainbow trout RAS. Aquac Eng 42(1):31–37

Summerfelt ST, Sharrer MJ, Tsukuda SM, Gearheart M (2009) Process requirements for achieving full-flow disinfection of recirculating water using ozonation and UV irradiation. Aquac Eng 40(1):17–27

Summerfelt ST, Vinci BJ, Piedrahita RH (2000) Oxygenation and carbon dioxide control in water reuse systems. Aquac Eng 22(1–2):87–108

Tango MS, Gagnon GA (2003) Impact of ozonation on water quality in marine recirculation systems. Aquac Eng 29(3–4):125–137

Tavares LHS, Braga FMDS (2008) Constructed wetland in wastewater treatment. Acta Sci: Biol Sci 261–265

Tilley DR, Badrinarayanan H, Rosati R, Son J (2002) Constructed wetlands as recirculation filters in large-scale shrimp aquaculture. Aquac Eng 26:81–109

Van Rijn J, Tal Y, Schreier HJ (2006) Denitrification in recirculating systems: theory and applications. Aquac Eng 34(3):364–376

Vandeputte M, Labbé L (2012) Cultured aquatic species information programme. Salmo trutta. FAO Cultured Species Information Programme

Wang JK (2003) Conceptual design of a microalgae-based recirculating oyster and shrimp system. Aquac Eng 28(1–2):37–46

Yanong RP (2003) Fish health management considerations in recirculating aquaculture systems—part 2: pathogens. Circular 121:1–8

Zahidah H, Rjibbusalam IM, Yuli A (2018) Novel mechanical filter for reducing ammonia concentration of Silver Barb culture in a recirculating aquaculture system (RAS). Res J Chem Environ 22:319–324

Zhang R, Chen T, Wang Y, Short M (2024) An optimisation approach for the design and operation of recirculating aquaculture systems integrated with sustainable hybrid energy systems. J Clean Prod 477:143860

Zimmermann S, Kiessling A, Zhang J (2023) The future of intensive tilapia production and the circular bioeconomy without effluents: biofloc technology, recirculation aquaculture systems, bio-ras, partitioned aquaculture systems and integrated multitrophic aquaculture. Rev Aquac 15:22–31

Chapter 19
Automation's Role in Sustainable Aquaculture Practices

Shivam Pandey, Gajender Singh, R. K. Gupta, Tejpal Dahiya, Reetu, Nisha Kumari, and Mayur Mukut Murlidhar Sharma

Abstract The demand for seafood is driving the fast expansion of the aquaculture industry as a food production sector. Sustainability in this industry is crucial, and automation plays a key role in achieving it. Automation technologies improve the efficiency of water, feed, and energy use, reduce waste and enhance overall operations. Technologies like water recirculation, automated feeding systems, and real-time monitoring promote responsible practices by optimizing resources, conserving water, improving energy efficiency, and managing waste more effectively. As the industry advances, automation powered by AI, machine learning, and IoT will enable precise, data-driven aquaculture management, including real-time adjustments and advanced fish health monitoring. Closed-loop systems and sustainable feed production will help reduce waste and dependence on overfished resources. Automation will also assist in climate adaptation through real-time monitoring. These technologies, becoming more affordable, will allow smaller farms to adopt sustainable practices, fostering long-term economic and environmental benefits for the sector.

Keywords Aquaculture automation · Sustainable aquaculture · AI in aquaculture · Real-time monitoring · Resource optimization

S. Pandey (✉)
Department of Aquatic Animal Health Management, College of Fisheries Science, Chaudhary Charan Singh Haryana Agricultural University, Hisar, Haryana, India
e-mail: karateshivam97@gmail.com

G. Singh · R. K. Gupta · T. Dahiya · Reetu
Department of Zoology, Chaudhary Charan Singh Haryana Agricultural University, Hisar, Haryana, India

N. Kumari
Department of Commerce, Gurugram University, Gurugram, India

M. M. M. Sharma
Department of Surgery, University of Illinois at Chicago, Chicago, USA
e-mail: mayurmms@uic.edu

R. Kumar et al. (eds.), *Sustainability in Aquaculture and Fisheries*,
Springer Water, https://doi.org/10.1007/978-3-032-16605-0_19

19.1 Introduction

The process of employing technology to do tasks with minimal assistance from people is called automation. The manufacturing, robotics, automotive, and IT systems sectors are the most common uses of automation, while it can be used to any company that handles repetitive tasks (Benanav 2020). The term "automation" was first introduced in 1946 to describe the growing use of automated controls and devices in production lines, with D.S. Harder, an engineering manager at Ford Motor Company, credited for coining it. While "automation" is primarily linked to manufacturing, its principles also apply to various systems where mechanical, electrical, or computerized processes replace human labor and decision-making significantly (Groover 2024). In essence, automation encompasses technology that enables processes to be executed through scripted commands and automatic feedback control, ensuring adherence to these instructions without human intervention. The development of automation technology has increasingly depended on computers and related technologies, resulting in more sophisticated systems that often surpass human capabilities in specific tasks (Au-Yong-Oliveira et al. 2019). Automation technologies are developed to the extent that a variety of technologies have occurred, with all detection and meaning. A notable technology is robotics, a special field within automation where machines have human characteristics. The most common anthropomorphic feature of modern industrial robots is your mechanical arm. This robotic arm can be programmed to load, unload, or programmed to apply spotch wide seams to the body components of the car during assembly. These examples illustrate how industrial robots are used primarily to replace human workers in manufacturing processes (Khatwani et al. 2023).

19.2 The Evolution of Automation

Since the first mechanized machines centuries ago, automation, robots, and artificial intelligence have been a part of human history. Ancient Greek water clocks were among the first automated devices ever discovered. The development of machinery to automate manual labor in factories during the Industrial Revolution increased productivity and efficiency. George Devol created unimate, the first industrial robot, in 1954, marking the beginning of the contemporary concept of robots. Productivity in the automotive industry was greatly increased by this robot, which was mostly utilized in manufacturing. The science fiction author Isaac Asimov coined the term "robot" and developed the Three Laws of Robotics in addition to researching robotic ethics. According to these laws, human safety comes first, robots must follow human instructions unless doing so jeopardizes human safety, and robots must defend their own lives unless doing so goes against the first two laws (Sauter 2007). Christopher Strachey created the first artificial intelligence program in 1951. It was a checkers game for a Ferranti Mark I computer. However, with the advent of new algorithms and machine learning techniques in the 1960s and 1970s, AI research made tremendous strides. A common belief is that the official study of AI began in 1956 with the Dartmouth Conference, where scientists first talked about the potential of the technology. These days, automation and artificial intelligence permeate every aspect of daily life, impacting everything from social media feeds and search engine algorithms to robotics in warehouses and factories. Further integration of AI and automation is anticipated to result in advanced robotics, driverless cars, and adaptive intelligent systems in the future. Even while automation has many benefits, there are worries about how it will affect jobs and the economy, therefore it must be carefully considered to optimize its advantages (Groover 2024).

19.3 Evolution of Robotics

The two main technologies that form the basis of robotics are teleoperators and numerical control. Using numerical codes that were first stored on punched paper tape or other media, numerical control (NC) is the process of controlling the motions of machine tools. The first NC machine tool was shown at the Massachusetts Institute of Technology (MIT) in 1952, marking the beginning of this technology's development in the late 1940s and early 1950s. Subsequent research at MIT produced the APT (Automatically Programmed Tools) language, which made machine tool programming easier (Garcia et al. 2007).

Teleoperators are mechanical manipulators controlled by humans from a distance, a concept that originated in the early 1940s during efforts to manage radioactive materials. In a standard teleoperator setup, a person operates a mechanical hand and arm from one location, while a manipulator at a different site mimics these movements. Both numerical control and teleoperator technology contribute to the development of industrial robotics. Teleoperator technology introduces the concept of a mechanical arm for performing tasks, while numerical control focuses on programmable machines. The first industrial robot, designed to unload parts from a die-casting machine, was installed in 1961 thanks to the collaboration of businessman Joseph F. Engelberger and inventor George C. Devol. Devol created the programmed manipulator and received a U.S. patent for it in the same year. Engelberger worked alongside Devol to advance robotics technology and founded Unimation, Inc., which became the first robotics company (McGrew 2021).

19.4 History of Artificial Intelligence in Aquaculture

Artificial intelligence (AI) is becoming increasingly important in aquaculture research and production, which allows both aspiring and established companies to develop innovative AI control solutions for the sector. The application of AI in aquaculture can be attributed to the introduction of fuzzy logic. In his influential work

"Fuzzy Set" in 1965, UC Berkeley's Lotfi A. Zadeh introduced the concept of a "continuum of membership classes." This basic idea paved the way for fuzzy logic and highlighted that many entities do not suit different categories properly (Lutz 2024). A surprising study by Lee et al. An intact logic control system for denitrification within closed recirculation systems investigated from the 2000 medical branch of the University of Texas Galveston University. Her research focused primarily on computer-controlled denitrifying bioreactors developed for Inkfish housing systems used in biomedical research. In this study, fuzzy logic was used to process real-time data from sensor pH monitoring, dissolved oxygen and potential for oxidation reduction. This allows for accurate adjustment of the pump speed of the bioreactor and carbon supply (Lutz 2024).

"The theory behind fuzzy logic acknowledges the nuances of human perception, where things often exist on a spectrum between 0 and 1, rather than being strictly true or false." True or false and binary values of 0 or 1 were the foundation of traditional computing. Instead of being completely binary, fuzzy logic acknowledged that human perception frequently incorporates degrees of truth, with values falling between 0 and 1. Fuzzy logic has been included into control systems for appliances like photocopiers, dishwashers, and automobile transmissions over time. It has also been used in a variety of computer applications, such as image processing and illness detection (Lutz 2024). Figure 19.1 illustrates the progression of aquaculture through four distinct technological generations, from Aquaculture 1.0 to Aquaculture 4.0. This evolution reflects a shift from traditional, manual practices to highly automated, data-driven systems.

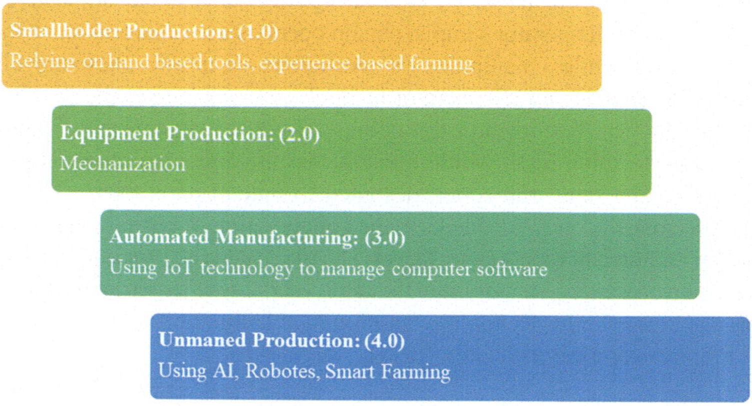

Fig. 19.1 Progression of aquaculture from generation 1.0 to 4.0

19.5 Basic Components of Automation

19.5.1 Power Source

Automated systems require a power supply to perform their designated tasks. Although various power sources are available, electricity is the most commonly used in modern automated systems due to its flexibility. Electrical power can be converted into mechanical, hydraulic, and pneumatic energy from multiple sources, including fossil fuels, hydroelectric power, solar energy, and nuclear energy. Additionally, electrical energy can be stored in durable, high-performance batteries (Groover 2024).

Automated systems typically carry out two main types of actions: (1) processing and (2) transfer and positioning. Processing involves applying energy to perform operations on an entity. This could involve data processing in a computer system, electrical signal switching in a communication system, molding plastic, or shaping metal. Energy is used in each of these processes to change the entity's condition to one that is more useful.

The second type of action, **transfer and positioning**, is particularly evident in automated manufacturing systems. In this context, products must be moved between different locations for various processing steps, often requiring precise placement. This concept also applies to automated communications and information systems, where it refers to the transfer and positioning of data or electrical signals between processing units. Additionally, it encompasses the delivery of information to output devices, such as printers or video display units, for human interpretation and use.

19.5.2 Feedback Controls

A feedback control system typically consists of five essential components: (1) input, (2) the controlled process, (3) output, (4) sensing elements, and (5) the controller along with actuating devices. This configuration is commonly referred to as a closed-loop feedback control system.

The input to the system, known as the reference value or set point, indicates the desired output level. For instance, in a heating system, the target room temperature serves as the input. The output is generated by the controlled process, which in this case is the furnace. In other contexts, this process could involve a manufacturing operation, rocket engines, or an automobile's cruise control system. The output variable is what is being monitored and compared against the input; in the heating example, it is the actual room temperature.

The equipment that keeps an eye on the output variable is called a sensing element. In the example of heating, a bimetallic strip is employed. It can monitor temperature because this strip, which is composed of two metals with varying rates of thermal expansion, bends in response to temperature variations. Feedback control systems are capable of using a wide variety of sensors.

To minimize the difference between the measured output and the reference input, the controller and actuating devices must adjust the process. Components such as motors, valves, solenoid switches, piston cylinders, and gears are designed to modify this process. In a heating system, for instance, the switch connected to the bimetallic strip acts as both the actuator and the controller. It turns on the heater when the ambient temperature falls below a specified level and turns it off when the temperature exceeds that level.

19.5.3 Machine Programming

The sequence of actions performed by an automated system is dictated by machine programming. This programming outlines the specific tasks the system must execute and how its various components should collaborate to achieve the desired outcome. The complexity of the program is significantly influenced by the system's design and requirements. A program may consist of a few clearly defined activities that are constantly performed in a predetermined order in simpler systems. On the other hand, more complicated systems could include a lot of commands with specific instructions, and the program might let you change things if the raw materials or other conditions change (Gottschlich et al. 2018).

In an automated system, feedback control is directly related to programming commands. For the system's feedback control loops, the program establishes the set points, or sequence of input values. A feedback loop's set point, which controls a particular activity, may be defined by each instruction. In essence, feedback control confirms that every programmed action has been carried out according to plan. The feedback control system, for example, makes sure that the arm moves as intended when a robot controller commands it to travel to a particular location.

Certain programmed orders can be carried out in an open-loop fashion, which means that feedback is not necessary to verify that they are being carried out. For instance, feedback verification might not be required for a command to turn an electrical switch. Feedback control, however, becomes essential when changes in raw materials or other circumstances need system modifications to preserve process quality. Without feedback, the system would find it difficult to efficiently regulate the output's quality (Soori et al. 2024).

Various media, including as computer memory, punched paper tape, magnetic tape, magnetic disks, and mechanical devices (such cams and linkages), can be used to store and run programs. Today, computer storage technology is commonly used for storing and managing programmed commands, allowing for easier updates and improvements. Changing a program stored on mechanical cams, by contrast, involves substantial effort (Ginzburg et al. 2013).

19.6 Role of Automation in Aquaculture Sustainability

Compared to other sectors such as agriculture, the performance of commercial aquaculture technology is relatively underdeveloped. Aquaculture farmers encounter numerous challenges, including manual water quality testing, resource management, and sudden climate variations. Conducting regular manual assessments of thousands of cultured organisms for their behavior and health is particularly challenging. The term automation refers to the shift from manual to automatic control or operation of labor processes, methods, or equipment components. As automation and intelligent technologies have progressed, aquaculture has gradually adopted a more intelligent and intensive approach. This evolution is a vital aspect of the third green revolution, aimed at tackling issues in fisheries development and boosting production. Machine learning technology, combined with advanced computing power, can analyze complex and high-dimensional data, offering solutions for intelligent aquaculture and ushering in a new era for the fishing industry. Historically, aquaculture has been viewed as both an art and a science, often relying more on the intuitive judgment of farm managers than on a comprehensive understanding of the physiology, ecology, and behavior of the species being cultivated. This reliance has made managers hesitant to adopt automated management solutions (Antonucci and Costa 2020).

However, recent advancements in research and business practices are transforming aquaculture into a more scientific discipline by encouraging the adoption of new technologies. Managers can now obtain valuable insights into the biological and physical conditions within their facilities through real-time monitoring of system parameters. This improvement is particularly significant given the labor-intensive nature of data collection, entry, graphing, and reporting in manually monitored systems. While such technology greatly benefits recirculating aquaculture systems, it also offers substantial advantages for pond and offshore systems (Abdullahi et al. 2015).

It is crucial to use cutting-edge technical strategies to address these issues. Comparatively speaking, the aquaculture industry has not evolved as far as the agricultural sector in integrating cutting-edge technologies like sensors, robots, data analysis, drones, IT devices, and autonomous tractors to improve farm management's sustainability and efficiency (Abdullahi et al. 2015).

The global aquaculture industry has been steadily advancing towards a more intensive and intelligent approach, driven by developments in automation and intelligent technologies. As a result of this change, sustainable aquaculture systems have been developed, greatly increasing industry efficiency. Aquaculture is still a labor-intensive industry, though, and managing agricultural organisms, habitats, and other aspects is difficult. As a result, problems like fish nutrition, disease control, and water pollution have arisen due to aquaculture's explosive growth. Intelligent aquaculture aims to address these issues and raise overall productivity in fisheries development as part of the third green revolution (Li 2018).

19.7 Automation in Different Ways of Aquaculture

1. Smart Feeding Systems in Aquaculture
2. Water Quality Management through Automation
3. AI-Driven Stock Monitoring and Health Management
4. Robotic Solutions for Harvesting and Sorting
5. Aquatic Drone Technology
6. Automation in Aquaculture Waste Management
7. Sustainability and Automation in Modern Aquaculture

19.7.1 Smart Feeding Systems in Aquaculture

Approximately 40% of all expenses in fish farming are attributed to feed, making it the largest operational cost in aquaculture. Estimates suggest that over 60% of the feed used in these systems breaks down into particulate matter. This degradation adversely affects aquatic life by depleting oxygen levels and generating harmful substances such as hydrogen sulfide and ammonia. Feeding rhythms significantly influence feed conversion rates (FCR) and the nutritional composition of fish flesh, highlighting the importance of proper feeding practices for effective feed application.

Aquaculture fish production heavily relies on water quality and is often associated with elevated levels of organic matter from feed and fertilizers. Excess feed can accumulate at the bottom of tanks, and if not removed, its decomposition can lead to the release of toxic compounds like hydrogen sulfide and ammonia while also reducing oxygen levels. To tackle challenges related to fish feeding, various types of feeders and methods for monitoring feed consumption have been explored in scientific research (Table 19.1).

In order to maximize feeding, this system took temperature and dissolved oxygen levels into account. These factors effect fish physiology, growth, and economic production in intensive aquaculture, which in turn affects feed conversion rate. For many control applications, fuzzy-logic control has been acknowledged as a practical approach that provides a means of creating nonlinear controllers with heuristic data. This study involved the creation and evaluation of a novel fuzzy-logic controlled feeder for tilapia (*O. niloticus*) production. The system adjusts feeding rates based on inputs such as temperature, dissolved oxygen, fish age, and body weight. The main goals of this feeder are to: (1) establish an innovative feeding system that optimizes food management, reduces waste, and enhances feed conversion rates; and (2) minimize water contamination from excess feed when feeding conditions are suboptimal.

Table 19.1 Different experiments by researchers for development of smart feeding system for aquaculture

Development of smart feeding system	References
In sea cages, where the presence of food dictated feed availability, used a hydro-acoustic sensor to detect food pellets	Juell et al. (1993)
Counted pellets and evaluated the quantity of food supplied using an underwater camera equipped with image analysis tools. Mal suggested a non-electric, automatic food dispenser that floats in the tank and distributes food horizontally by means of a pulley system that rotates a wheel	Foster et al. (1995)
Developed an innovative data collection system that enabled the monitoring of feeding schedules and quantities while integrating this information with various environmental factors	Fast et al. (1997)
Utilized underwater cameras to observe fish behaviors and pellet waste before, during, and after feeding, aiming to analyze fish feeding patterns based on general behaviors noted in various field conditions	Ang and Petrell (1997)
Employed the pseudo green water method to design and theoretically assess a fuzzy-logic control system aimed at estimating the daily feeding requirements of sea bream (*Sparus aurata*) larvae in intensive environments. This approach was intended to show that using floating feed could prevent feeding from continuing before it contaminates the water	Papandroulakis et al. (2000)
Utilized a reflecting photoelectric sensor to detect eel aggregation behavior and incorporated it into a feedback system	Fang et al. (2002)
Developed an intelligent control system for indoor eel farming	Chang et al. (2005)

19.7.1.1 Automatic or Demand Feeders

Fish species like rainbow trout (*Salmo gairdneri*), flatfish (*Limanda limanda*), blenny (*Blenniuspholis*), and goldfish (*Carassius auratus*) have all been the subject of numerous research that have investigated teaching them to self-feed by pressing a lever. By keeping track of the number of lever presses per unit of time, researchers can examine how environmental elements such as light temperature, food quality (energy content, palatability) and feeding schedules affect voluntary eating behavior. Free-running feeding cycles that correspond to the amount of time needed for absorption and digestion develop under constant light and temperature conditions. The periodicity of feeding patterns is revealed by statistical analysis of demand feeder activations. A known quantity of food is delivered with each lever press, making it possible to track food consumption over time. Food that hasn't been consumed is taken out and the food dispensing record is updated. Because random fish movements can lead to spurious lever activations-accounting for up to 10% of all activations-these must be taken into consideration (Tanveer et al. 2018; Kumar et al. 2022a, b).

A growth and environmental impact model for the California sea cucumber (*Parastichopus californicus*) was developed to analyze the performance of deposit feeders in integrated multi-trophic aquaculture (IMTA), using mathematical modeling approaches. The ORGANIX model was employed to simulate the organic particle loading resulting from both finfish cages and suspended shellfish cultures (Cubillo et al. 2016). Meanwhile, Mattos et al. (2016) investigated the ability of Pirarucu (*Arapaima gigas*) to feed using self-demand feeders, examining their locomotor activity and daily feeding patterns. In their study, self-feeders and occupancy sensors connected to a computer were installed in the tanks to monitor movement and feeding behavior.

A straightforward, inexpensive, non-electric feed dispenser designed for skipjack tuna live bait was developed using the clepsydra (water clock) principle, requiring no electrical components while maintaining simplicity in construction and operation. In contrast, a more advanced fish feeding system incorporated mechanical and electrical components-including a stepper motor for precise food delivery-controlled by Arduino-based programming to automate dispensing schedules and quantities, significantly reducing manual labor (Baldwin 1983; Swarnakar et al. 2017).

19.7.1.2 Benefits of Automatic Fish Feeders

Currently, there are no affordable, power-free automatic fish feeders available in India. However, Baldwin developed a simple, electrical-free automated feed dispenser at the Hawaii Institute of Marine Biology (Baldwin 1983). This type of mechanical fish feeder could gain popularity in India and other developing countries due to several potential advantages:

- It helps with labor shortages by guaranteeing uniform feed distribution.
- It is probably less expensive than alternative feeding techniques.
- By modifying the feeder, feed may be added to pond systems at the best periods.
- Because it doesn't require expensive materials or complex parts, it is comparatively simple to construct and run.

19.7.2 Water Quality Management Through Automation

Advanced computer-controlled intensive culture systems and automated remote monitoring are anticipated to be features of aquaculture in the future. Aquaculture management in the present day depends on efficient water quality monitoring. By maintaining environmental parameters within acceptable ranges, proper water quality regulation helps fish grow more quickly, reduce the occurrence of major fish diseases, and enhance nutritional efficiency. Accurately controlling water quality at the appropriate time and location is practically impossible without Knowledge of the physical and chemical characteristics of water quality and their ecological effects (Stigebrandt et al. 2004).

19.7.2.1 Waterproof Temperature Sensors

Waterproof temperature sensors measure the temperature of the submerged water body. These sensors typically operate within a temperature range of −55 °C to + 125 °C and are equipped with data transmission, power, and ground wires, along with a temperature limit warning system (Barman et al. 2015). Temperature sensing methods can be categorized into electric and non-electric devices. Thermocouples, which utilize the Seebeck effect, require nearby electrical instrumentation to function. They are commonly made from metals such as copper, nickel, iron, platinum, rhodium, and their alloys, consisting of two connected metal conductors. Thermocouples are compact, cost-effective, and highly durable (Zhu et al. 2010).

19.7.2.2 pH Sensor

A pH sensor uses the pH scale, which ranges from 0 to 14, to assess the alkalinity of a specific body of water or solution. Typically employed for measuring water pH, these sensors are constructed with Ag/AgCl reference electrodes and pH glass electrodes (Aakash 2019).

19.7.2.3 Dissolved Oxygen Sensor

This sensor measures the oxygen levels in a body of water to determine whether there is sufficient oxygen for aquatic life to thrive. It achieves this through electrochemical sensors, including polarographic and galvanic types. These sensors utilize dissolved oxygen from the water, which diffuses across an oxygen-permeable membrane, to evaluate the electrical signals generated by the chemical reduction reaction with a NaOH solution.

Polarographic and optical dissolved oxygen (DO) sensors are currently the two primary varieties. The rate of molecular oxygen diffusion across a film is measured by polarographic sensors, which are based on the Clark DO electrode. Although this approach is straightforward and minimally intrusive, it needs regular maintenance and has a short lifespan. Conversely, fluorescence quenching of oxygen for luminous materials is used in optical DO sensors. This technique, which offers great accuracy, stability, and a longer service life, evaluates DO by examining fluorescence intensity or lifetime (Menon et al. 2021).

A conditioning circuit is used to first transform the data from the dissolved oxygen and pH sensors into current signals. After being converted into voltage signals using a current/voltage converter, these signals are transmitted to a microcontroller. Utilizing this information, the microcontroller controls water quality parameters, uploads data, and displays it on a host computer. The device can automatically change the temperature and, if any irregularities are found, send out an SMS alert if the pH or temperature of the water rises above predetermined levels. Furthermore, in case the levels of dissolved oxygen deviate from the norm, the system triggers an oxygenation pump automatically to guarantee proper oxygenation, achieving the Chunxia and Qiuchan study (2020) on real-time controlling and monitoring of aquaculture environmental parameters.

19.7.2.4 Nitrite Ion Sensor

Nitrite ion sensors measure the concentration of nitrite ions in a water sample or body of water using specialized nitrite ion-selective electrodes. Nitrates can be easily detected through a biosensing approach, which involves combining biological materials with a detection device and a signal-conditioning circuit to determine the concentration of specific ions in a sample. The sample solution is directly exposed to the biosensor and the sensing material, allowing the biological material to facilitate interaction with the target ions. The resulting electrical signal—such as voltage, current, or impedance—from this interaction is then used to calculate the ion concentration. The effectiveness of the biosensing technique relies on the ability to convert the sensing signal into a measurable electrical output (Alahi and Mukhopadhyay 2018).

19.7.2.5 Ammonium Ion Sensor

An ammonium ion sensor measures the concentration of ammonium ions in a solution or body of water. Typical sensors include Ag/AgCl electrodes with specially composed filler solutions, often oxidized bromides which guarantee ammonium ions will interact with the electrode. Ammonia electrodes assess ammonia levels in water and use a pH glass electrode as the indicator and a silver-silver chloride electrode as the reference. This apparatus case contains electrodes and gas permeable membrane sleeves housing 0.1 mol/L ammonium chloride solution. Ammonium salts can convert to ammonia when the sample solution pH is raised to 11 or higher by ionic strength addition. Ammonia's diffusion through the permeable membrane increases the pH and alters the balance. All of these changes are detected by the pH glass electrode. With constant controls of ionic strength, temperature, and electrode features, ammonia concentration and electromotive force in water samples obey the Nernst equation. Recorded voltage signals establish a standard curve for unknown substance concentration determination (Kusumah et al. 2020).

19.7.3 AI-Driven Stock Monitoring and Health Management

The health monitoring of fish in aquaculture facilities can be enhanced significantly with the use of artificial intelligence. These techniques can help in proactive actions by analyzing diverse data and identifying patterns, assessing fish health, and in taking timely action by early identifying potential health issues.

19.7.3.1 Computer Vision and Image Recognition

Fish diseases often show symptoms such as lesions, discolored patches, or unusual behavioral patterns that can be seen externally. Fish can be captured on images or videos which can then be processed by image processing methods such as object detection, computer vision, or even deep learning techniques such as CNN, to detect disease indicators automatically. These algorithms are capable of making decisions based on the provided datasets and thus assist in proactive actions (Zhang et al. 2020).

19.7.3.2 Anomaly Detection and Pattern Recognition

Changes such as fish health behavior monitoring feed rates, and water conditions like temperature, pH, ammonia, and dissolved oxygen level are critical in determining the health status of fish. AI models can monitor these indicators and identify anomalies and deviations from the set parameters. Clustering and autoencoders as other forms of unsupervised learning are capable of marking unusual activities in fish populations which are likely to be under some form of stress or disease (Kaur et al. 2023).

19.7.3.3 Support Vector Machines Classification Systems

Classification algorithms, including SVMs, utilize health-related information to classify fish into distinct disease categories. By utilizing training data from healthy and diseased fish, these algorithms can distinguish between different diseases, including those involving water quality measurement, behavioral aspects of fish, or even molecular indicators. This makes treatment and diagnosis more accurate (González et al. 2023).

19.7.3.4 Assessment and Prediction Models of Time-Series Data

Fish populations suffering from disease may have trends and patterns as time progresses. Certain time-series analytical techniques such as autoregressive integrated moving average (ARIMA) and recurrent neural networks (RNN) or long short-term memory (LSTM) models can forecast future outbreaks by analyzing historical health data. These models offer insights into effective disease management and prevention by considering numerous factors like previous incidences, water quality, and environmental variables.

Systems that rely on AI are best suited when utilized alongside seasoned aquaculture professionals or veterinarians. Maintaining real-time observance through monitoring systems, such as sensors or underwater cameras, enables uninterrupted surveillance. This, coupled with the assistance of AI, enhances the ability to identify diseases in their nascent stages, thereby reducing the likelihood of incurring losses in aquaculture (Kumar et al. 2022a, b).

19.7.4 Robotic Solutions for Harvesting and Sorting

The use of robotics in fishing and aquatic life farming (aquaculture) has brought in new technological advancements which assures accuracy, effectiveness, and ecological balance. Robotics innovations help in checking the fish fry, monitoring the environment, checking for diseases, and feeding (Fig. 19.2). Robotics in the aquaculture sector is aimed at optimizing the production processes, improving waste effectiveness, and improving the health of the living aquatic organisms (Antonucci and Costa 2020).

19.7.4.1 Observation and Surveillance

Robotic systems fitted with cameras, sensors, and underwater drones have revolutionalized monitoring of aquaculture. These systems can monitor fish behavioral patterns, environmental conditions, and water quality in real time. Operators are able to make timely interventions to address any abnormalities and mitigate potential challenges (Ubina and Cheng 2022).

19.7.4.2 Reduction of Labor and Enhancement of Workforce Safety

The automation of repetitive tasks like data capturing and feeding results to a manual labor reduction. This improves the costs and enhances the aquatic environment safety of the workers as they don't have to work in dangerous environments (Ubina and Cheng 2022).

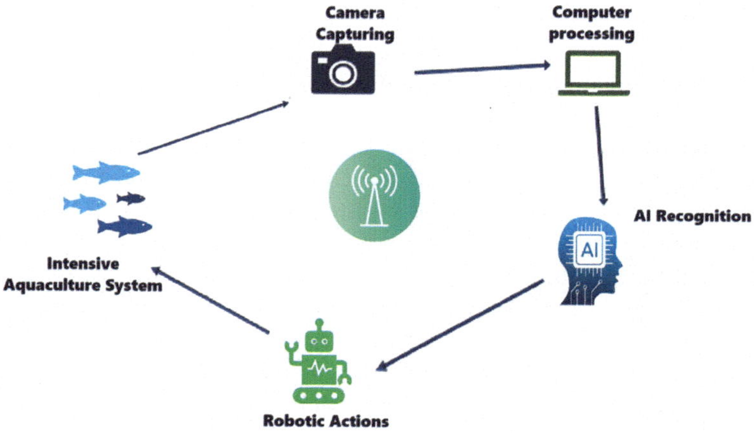

Fig. 19.2 AI and robotics managed smart aquaculture farms

19.7.4.3 Continuous Monitoring and Prompt Response

Robotic systems equipped with sensors and cameras can operate around the clock, providing continuous real-time monitoring of critical parameters. This constant surveillance enables immediate detection of anomalies, such as changes in water quality or fish behavior. Operators can then respond quickly to address issues and prevent problems from escalating, thereby avoiding potential stock losses.

Intelligent equipment in modern aquaculture integrates traditional fishery tools with advanced information and intelligent manufacturing technologies to enable precise operations in smart fish farms. This equipment is categorized into mobile and fixed types (Ubina and Cheng 2022).

19.7.4.4 Mobile Equipment Devices

These device operate autonomously include remotely operated underwater vehicles (ROVs), unmanned surface boats (USVs), unmanned ground vehicles (UGVs), and unmanned aerial vehicles (UAVs).

19.7.4.5 Fixed Equipment

Fixed Equipment encompasses systems that remain stationary, including water quality monitoring and control devices, feeding systems, and oxygenation equipment. It also includes harvesting tools like fish suction pumps and sorting equipment (Ubina and Cheng 2022).

19.7.4.6 Automatic or Robotic Patrol System

Unmanned aerial vehicles (UAVs), unmanned surface vessels (USVs), and robotic fish are examples of patrol equipment used in intelligent fish farms. Fish activity monitoring, video data collecting, daily underwater management, and ecological environment surveillance are all made easier by these techniques. The flight control module of the UAV controls pitch, rotation, hovering, and vertical movement using fuzzy and PID control algorithms. The automatic navigation system tracks the UAV's speed, course, height, location, and distance using GNSS, machine vision, and 5G technology. This allows for exact path planning and intelligent obstacle avoidance for precise positioning in challenging conditions. For thorough environmental and feeding data collecting, the monitoring system includes cameras, hyperspectral sensors, and machine vision-equipped remote sensing instruments (Ubina and Cheng 2022).

In open-pen sea cage aquaculture, 5G technology supports high-throughput HD video transmission, addressing challenges related to the high costs and difficulties of optical cable installation. Future UAVs, powered by AI algorithms, will include advanced airborne edge computing modules for autonomous takeoff, landing, cruising, and monitoring. Additionally, UAVs will be equipped with automatic charging controllers for truly autonomous operation.

USVs, guided by GPS or Beidou navigation systems, manage speed, course, position, and operational path planning. They can carry underwater sonar for fish stock tracking and biomass estimation and interact with ROVs. USVs can also serve as platforms for other aquaculture equipment, including aerators, bait feeders, harvesting tools, and as launching platforms for UAVs and biomimetic robotic fish (Kelasidi and Svendsen 2023).

19.7.4.7 Robotic Harvesting System

In the fish breeding cycle, the robotic harvesting system is the last step that makes it easier to get aquatic products—with or without water—to market. At the moment, the most effective fishing technique is trawling. To guarantee accurate fishing in intelligent fish farms, this method uses sonar, underwater cameras, and net positioning tools. The efficiency of production is greatly increased when aquatic items are transported by ships and unmanned vehicles. This strategy encourages high efficiency, conserves energy, and lessens the amount of work required by fisherman(Wang et al. 2021).

Robots that harvest sea cucumbers, for instance, use mechanical servo-drive technology, computer vision, and Doppler-GPS integrated navigation to collect them automatically. The navigation system of the robot organizes the route, collects position data, and manages the robot's motion. According to Qiao et al. (2019), the machine vision system takes pictures of sea cucumbers, analyzes them, identifies them, and determines their locations.

With the assistance of cutting edge technologies such as automation, robotics and smart processing, aquaculture is evolving world-wide. The automated fish farming techniques are meant to improve efficiency, production, and to eliminate the dependence on human labor. Nowadays robots, unmanned vessels, and even robotic submarines are employed to turn aquafarms into autonomous facilities capable of operating hands-free (Wang et al. 2021).

19.7.5 Aquatic Drone Technology

In the present day, drones and remote controlled flying devices have become a powerful tool enabling marine behavioral research. Their advanced applications have improved observational capabilities for tracking fish movements extending to social interactions among species like sharks, sea turtles and whales. With extensive data collection being renewable with drones, the potential for future developments in marine ecology is limitless. New UAV applications are highly anticipated as fresh ideas are constantly being developed by researchers (Farinha et al. 2022).

Drones equipped with cameras and sensors are capable of monitoring aquaculture from above while capturing measurements of turbidity, temperature, pH, and dissolved oxygen levels. Besides monitoring, drones equipped with specialized technology can increase feeding efficiency by delivering feed at calculated intervals (Farinha et al. 2022).

Drones with cameras and computer vision technology can also monitor the environment such as the weather, plant or invasive species distribution, locate potential pollution sources, and assess the impact of aquaculture associated activities on the surrounding ecosystems. Early detection of disease outbreaks is crucial in aquaculture. Drones with thermal imaging cameras can spot water temperature abnormalities which could indicate the presence of a disease. Drones can help mitigate the risk posed by birds and other pests on aquaculture operations. Besides providing alternatives to conventional aerial scanning, modern LIDAR outfitted drones offer the ability to shoot lasers to measure distances while producing detailed three-dimensional maps of the aquatic environment. This resourceful approach allows for non-invasive, cost-effective collection of current fish population data (Carrivick and Smith 2019).

From above, drones equipped with sensors and cameras can see shrimp ponds and fish farms. This lets farmers watch above the health and activity of aquatic animals. Early on, surveillance drones may also identify issues including disease outbreaks, predatory threats, or poor water quality, so enabling prompt action (Fish 2020).

i. Drones fitted with sensors may track temperature, pH, turbidity, and dissolved oxygen, hence providing water quality analysis. Using this understanding, farmers may make wise choices regarding water treatment and management (Carrivick and Smith 2019).
ii. Automated drone systems can sprinkle feed pellets on aquaculture ponds at prearranged periods. This improves feeding techniques and lowers manual labor in addition to guaranteeing more precise feeding (Carrivick and Smith 2019).
iii. According to Duan et al. (2019), drones fitted with cameras and computer vision technology may monitor and count aquatic species, therefore assisting in harvest time estimation and inventory management.
iv. Drones can track weather, potential sources of pollution, and consequences of aquaculture activities on nearby ecosystems (Carrivick and Smith 2019).
v. Drones can generate complex maps and three-dimensional (3D) representations of aquaculture sites, therefore supporting facility layout and planning and enabling infrastructure design and space usage optimization (Duan et al. 2019).
vi. Early disease detection is crucial in aquaculture. Drones equipped with thermal imaging cameras may detect temperature fluctuations in the water that could indicate sickness (Fish 2020).
vii. To evaluate the nutritional content and water quality of a pond or fish farm, drones can collect water samples from several sites (Fish 2020).

viii. Birds and other pests threaten aquaculture activities, which may be avoided with drones equipped with non-lethal deterrents like lights or loud sound (AVPL International 2023).

ix. **Emergency Response**: Drones can aid operators in making well-informed judgments and successfully responding to emergencies, such as chemical spills or breaches in containment structures, by providing a brief overview of the situation (AVPL International 2023).

Overall, by increasing productivity, cutting expenses, and promoting sustainability through improved environmental management and monitoring, drones have the potential to revolutionize aquaculture. Adherence to rules, meticulous planning, and training are necessary for the successful integration of this technology.

19.7.6 Automation in Aquaculture Waste Management

Automation in aquaculture waste management is crucial to improve the sustainability and efficiency of aquaculture systems. Ongoing monitoring and management of water quality, waste disposal, and nutrient cycling are made easier by automated systems, all of which are necessary for guaranteeing a suitable habitat for aquatic life. Mostly made of fish waste, uneaten feed, and metabolic byproducts, aquaculture waste can damage fish health and compromise water quality. Time-consuming and inadequate are aquaculture waste management methods including water exchange and manual cleaning. Automation in this sector aims to fix these problems by improving the efficiency and efficacy of waste collection.

19.7.6.1 Waste Management Automated Systems

i. **RAS, or recirculating aquaculture systems**, is a significant advancement in aquaculture trash management. It allows for the reuse and treatment of water, therefore lowering demand for large volumes of fresh water and so lowering trash dump into the environment. Among the water quality indicators automatically sensors monitor are ammonia, nitrate, and oxygen levels, triggering waste treatment operations including biofiltration and filtration (Martins et al. 2010).

ii. **Automated Sludge Removal Systems**: Waste items including feces and extra feed accumulate at the bottom of ponds and tanks to create "sludge." Automated sludge removal systems detect garbage accumulation and activate pumps to remove the sludge without need for human interaction (Gutierrez-Wing and Malone 2006). These tools help to maintain the optimal level of water quality and prevent the development of hazardous bacteria and algae.

iii. **Real-time monitoring systems** monitor variables like temperature, pH, and dissolved oxygen in real-time. Real-time feedback from the system helps to more quickly and precisely control waste and water quality by beginning reactions like turning on waste filtration equipment or boosting aeration when water quality declines (Davidson and Summerfelt 2005).

iv. **Robotic cleaning devices** are increasingly being used to eliminate waste from pond liners and tank surfaces. Equipped with sensors that can recognize biofilm and waste buildup, these robots enable surface cleaning without disturbing fish activity. Lowered personnel expenditures guarantee constant garbage management.

19.7.6.2 Benefit Automation Offers Waste Management

- Better Water Quality: Automation ensures quick garbage removal, therefore maintaining water quality and lowering disease danger.
- Automated systems improve waste management efficiency by operating constantly without human interaction.
- Automation helps to make aquaculture systems more environmentally sustainable by reducing water usage and waste generation.
- Enhancing the sustainability and efficiency of aquaculture businesses depends on automating waste management. Waste can be properly controlled using automated sludge removal, recirculating systems, and real-time monitoring to improve water quality and aquatic life health.

19.7.7 Modern Aquaculture's Sustainability and Automation

Among the food production categories with the fastest growth rate in the world is aquaculture; as demand for seafood increases, assuring the sustainability of this industry is becoming more and more vital. Automation helps to solve the environmental problems brought up by traditional aquaculture techniques by eliminating trash, using less resources, and raising general efficiency, therefore guaranteeing a more sustainable future for the industry.

By maximizing the usage of resources including water, feed, and energy and minimizing environmental effect and garbage, automation in aquaculture considerably improves sustainability. More environmentally benign and effective methods made possible by modern technologies—including water recirculation systems, automated feeding systems, and real-time monitoring—can help to satisfy the increasing worldwide need for seafood.

- One of automation's main advantages for aquaculture sustainability is resource optimization. Automated systems, especially those that monitor environmental factors like fish behavior and water quality, make precise control over inputs like feed and water achievable. Automated feeding systems utilizing sensors to monitor fish activity can, for instance, significantly reduce overfeeding, therefore lowering waste output and feed costs (Torrissen et al. 2011). These technologies help to reduce nutrient pollution in close environments and relieve the pressure on water treatment facilities by minimizing garbage (Martins et al. 2010).

- Water use by aquaculture carries a significant environmental hazard. Because they require regular water exchanges to keep water quality, traditional open systems utilize a great deal of freshwater and run the risk of polluting natural water bodies. Automation, especially with Recirculating Aquaculture Systems (RAS), which constantly treat and reuse water inside the system, solves this problem. Real-time filtration and aeration system adjustments are facilitated by automated sensors that monitor water characteristics like ammonia, pH, and oxygen levels. Utilizing up to 99% less water than traditional methods, this closed-loop approach is very environmentally friendly.
- Automation increases the energy efficiency of aquaculture as well. By changing tasks such water pumps, aerators, and illumination in accordance with present demands rather than running continually at fixed rates, automated systems can maximize energy use. Integrating smart technologies helps to reduce total consumption by making sure energy is only used as needed. By including renewable energy sources like solar or wind into these automated systems, one may also reduce the environmental footprint of aquaculture activities (Crampton et al. 2021).
- Another important field automation improves sustainability is waste management by efficiently eliminating and treating both liquid and solid waste, thereby preventing hazardous substances from building up in water (Bostock et al. 2010). Robotic sludge removal systems, for instance, keep sediment from accumulating in ponds and tanks, so ensuring better habitats for aquatic life. To promote circularity and minimize the ecological effect of waste disposal, certain advanced systems might also recycle nutrients from waste goods for subsequent application as feed or fertilizer in further aquaculture activities.
- Automation helps aquaculture companies to minimize their environmental footprint by improving energy efficiency, lowering waste, and maximizing resource use; therefore, lowering the negative effects related to intensive aquaculture, which have formerly included habitat destruction, water contamination, and depletion of natural resources; hence, automated aquaculture systems match production methods with environmental conservation goals, hence supporting the long-term sustainability of the sector (Valenti et al. 2018).
- Automation helps sustainability in aquaculture to a large degree; additionally, it increases resource efficiency, saves water, lowers energy usage, and reduces environmental effects. As the industry grows, automated systems will become even more important since they will enable aquaculture to be a more moral and ecologically friendly approach of food production.

19.8 The Economic and Environmental Advantages of Automation

- Automated aquaculture operations have significant advantages for both the economy and the ecology. Automation first of all enables exact control of the use of resources such water and feed, therefore minimizing trash and increasing resource efficiency. This lowers farmers' operating costs (Crampton et al. 2021) and lessens the environmental impact of activities, hence making aquaculture more financially sustainable.
- Automation also reduces the chance of human mistake in complex aquaculture system management, improving the overall health and productivity of the farmed species. Automation also causes a reduced demand for labor, which is another important element.
- Many aquaculture companies would find automation to be a worthwhile investment because of the long-term labor cost savings and higher production efficiency even if the first investment in automated equipment might be significant. Moreover, by helping smaller farms to embrace sustainable methods, these cost savings democratize technology in the industry.

19.9 Conclusion

In essence, for sustainability in the sector the incorporation of automation systems in aquaculture methods is imperative. Aquaculture companies can greatly lower their environmental impact by utilizing cutting-edge technologies including automated feeding systems, real-time monitoring, and water recirculation. Automation increases overall operations by reducing waste, improving water, feed, and energy consumption effectiveness. Moreover, it fosters responsible habits through water preservation, enhanced energy efficiency, and perfect water quality maintenance. Embracing automation in aquaculture is crucial to guarantee a sustainable and environmentally conscious food production industry as seafood demand keeps growing. Adopting automation will enable the aquaculture sector to lessen its environmental footprint, raise profitability, and help to create a more food-secure future.

19.10 Future Prospectives

Opportunities Ahead As population and demand for seafood increase, aquaculture's contribution to the global food supply grows and automation powered by artificial intelligence (AI), machine learning, and Internet of Things (IoT) devices is essential for guaranteeing sustainable development without exacerbating environmental issues (Valenti et al. 2018). Real-time data analysis and predictive management enabled by artificial intelligence and machine learning help to optimize feeding, water quality, and health monitoring while lowering resource consumption and waste (Martins et al. 2010). Robotic and autonomous systems now handle cleaning, health monitoring, even disease detection, therefore minimizing manual labor and improving fish welfare. IoT devices connect sensors and equipment for continuous, remote monitoring of environmental and operational parameters, supporting timely interventions and efficient resource use (Crampton et al. 2021). Automation also supports closed-loop, zero-waste systems by recycling waste into valuable byproducts and improving water conservation. AI-driven feed optimization uses real-time behavioral and environmental data to reduce reliance on wild fishmeal and promote sustainable alternatives. By giving early alerts and automatic changes to keep ideal conditions, these technologies assist farms to adjust to climate change. Small and medium-sized farms profit from higher efficiency and sustainability as automation becomes more available, so democratizing best practices throughout the sector.

References

Aakash P (2019) Water quality monitoring system using RC boat with wireless sensor network. Int J Res Appl Sci Eng Technol 7(5):1–6

Abdullahi HS, Mahieddine F, Sheriff RE (2015) Technology impact on agricultural productivity: a review of precision agriculture using unmanned aerial vehicles. In: Wireless and satellite systems: 7th international conference, WiSATS2015, Bradford, UK, July 6–7, 2015. Revised selected papers 7. Springer International Publishing, pp 388–400

Alahi MEE, Mukhopadhyay SC (2018) Detection methods of nitrate in water: a review. Sens Actuators, A 280:210–221

Ang KP, Petrell RJ (1997) Control of feed dispensation in sea cages using underwater video monitoring: effects on growth and food conversion. Aquac Eng 16(1–2):45–62

Antonucci F, Costa C (2020) Precision aquaculture: a short review on engineering innovations. Aquac Int 28(1):41–57

Au-Yong-Oliveira M, Canastro D, Oliveira J, Tomás J, Amorim S, Moreira F (2019) The role of AI and automation on the future of jobs and the opportunity to change society. In: New knowledge in information systems and technologies, vol 3. Springer International Publishing, pp 348–357

AVPL International (2023, October 7) Aquaculture advances with drone technology. AVPL International—Medium. Medium. https://medium.com/@info_24967/aquaculture-advances-with-drone-technology-fa7ce25aa022

Baldwin WJ (1983) The design and operation of an automatic feed dispenser. Aquaculture 34(1–2):151–155

Barman P, Partha B, Mondal KC, Mohapatra PKD (2015) Water quality improvement of Penaeus monodon culture pond for higher productivity through biomediation. Acta Biologica Szegediensis 59(2):169–177

Benanav A (2020) Automation and the future of work. Verso Books

Carrivick JL, Smith MW (2019) Fluvial and aquatic applications of structure from motion photogrammetry and unmanned aerial vehicle/drone technology. Wiley Interdiscip Rev Water 6(1):e1328

Chang CM, Fang W, Jao RC, Shyu CZ, Liao IC (2005) Development of an intelligent feeding controller for indoor intensive culturing of eel. Aquac Eng 32(2):343–353

Chunxia JIN, Qiuchan BAI (2020) The monitoring system of aquaculture environment. In: 2020 13th international symposium on computational intelligence and design (ISCID). IEEE, pp 184–187

Crampton V, Howell B, Hastein T, Krogdahl Å (2021) Advances in aquaculture: sustainable feed ingredients, health, and energy management. J Aquac Res Dev 12(7):1–15

Cubillo AM, Ferreira JG, Robinson SM, Pearce CM, Corner RA, Johansen J (2016) Role of deposit feeders in integrated multi-trophic aquaculture—a model analysis. Aquaculture 453:54–66

Davidson J, Summerfelt ST (2005) Solids removal from a coldwater recirculating system—comparison of a swirl separator and a radial-flow settler. Aquac Eng 33(1):47–61

de Mattos BO, Nascimento Filho ECT, Barreto KA, Braga LGT, Fortes-Silva R (2016) Self-feeder systems and infrared sensors to evaluate the daily feeding and locomotor rhythms of Pirarucu (*Arapaima gigas*) cultivated in outdoor tanks. Aquaculture 457:118–123

Duan Z, Li Y, Wang J, Zhao G, Svanberg S (2019) Aquatic environment monitoring using a drone-based fluorosensor. Appl Phys B 125:1–8

Fang W, Chang CM, Shyu CZ, Liao IC (2002) Development of an intelligent feeding system for indoor intensive culturing of eel. World Aquaculture Beijing

Farinha AT, Di Tria J, Reyes M, Rosas C, Pang O, Zufferey R et al (2022) Off-shore and underwater sampling of aquatic environments with the aerial-aquatic drone MEDUSA. Front Environ Sci 10:1023269

Fast AW, Qin T, Szyper JP (1997) A new method for assessing fish feeding rhythms using demand feeders and automated data acquisition. Aquac Eng 16(4):213–220

Fish FE (2020) Advantages of aquatic animals as models for bio-inspired drones over present AUV technology. Bioinspir Biomim 15(2):025001

Foster M, Petrell R, Ito MR, Ward R (1995) Detection and counting of uneaten food pellets in a sea cage using image analysis. Aquac Eng 14(3):251–269

Garcia E, Jimenez MA, De Santos PG, Armada M (2007) The evolution of robotics research. IEEE Robot Autom Mag 14(1):90–103

Ginzburg SA, Lekhtman IY, Malov VS (2013) Fundamentals of automation and remote control: international series of monographs in automation and automatic control, vol 7. Elsevier

González C, Astudillo CA, López-Cortés XA, Maldonado S (2023) Semi-supervised learning for MALDI-TOF mass spectrometry data classification: an application in the salmon industry. Neural Comput Appl 35(13):9381–9391

Gottschlich J, Solar-Lezama A, Tatbul N, Carbin M, Rinard M, Barzilay R et al (2018, June) The three pillars of machine programming. In: Proceedings of the 2nd ACM SIGPLAN international workshop on machine learning and programming languages, pp 69–80

Groover MP (2024) Automation. In: Technology, types, rise, history, and examples. Encyclopedia Britannica. https://www.britannica.com/technology/automation

Gutierrez-Wing MT, Malone RF (2006) Biological filters in aquaculture: trends and research directions for freshwater and marine applications. Aquac Eng 34(3):163–171

Juell JE, Furevik DM, Bjordal Å (1993) Demand feeding in salmon farming by hydroacoustic food detection. Aquac Eng 12(3):155–167

Kaur R, Kumar R, Gupta M (2023) Deep neural network for food image classification and nutrient identification: a systematic review. Rev Endocr Metab Disord 24(4):633–653

Kelasidi E, Svendsen E (2023) Robotics for sea-based fish farming. Encyclopedia of smart agriculture technologies. Springer International Publishing, Cham, pp 1–20

Khatwani R, Raghuram G, Mishra M (2023) Automation, technological employment, and society: a review. In: AIP conference proceedings, vol 2914, no 1. AIP Publishing

Kumar R, Gupta R, Dahiya T, Yadav R, Saifi R, Yodha K, Kumari B, Singh P, Shefali (2022a). Surveillance of *Heteropneustes fossilis* (stinging catfish) diseases and health management practices in Haryana. Appl Ecol Environ Sci 10(4):261–267

Kumar R, Gupta RK, Yadav R, Saifi R, Yodha K, Pooja (2022b) Eisenia fetida as protein source for growth enhancement of Heteropneustes fossilis. Egypt J Aquat Biol Fish 26(2):577–588

Kusumah BR, Kostajaya A, Supriadi D, Nugraha EH, Siskandar R (2020) Engineering of automatically controlled energy aeration systems for fisheries cultivation pools. Aquacultura Indonesiana 21(2):74–81

Li D (2018) Unmanned fish farming—the 4th generation of aquaculture [PowerPoint presentation]. https://www.itu.int/en/ITU-D/RegionalPresence/AsiaPacific/SiteAssets/Pages/E-agriculture-Solutions-Forum-2018/unmanned%20fish%20farming.pdf

Lutz CG (2024, April 23) The rise of AI in aquaculture. The Fish Site. https://thefishsite.com/articles/the-rise-of-ai-in-aquaculture-artificial-intelligence

Martins CIM, Eding EH, Verdegem MC, Heinsbroek LT, Schneider O, Blancheton JP et al (2010) New developments in recirculating aquaculture systems in Europe: a perspective on environmental sustainability. Aquac Eng 43(3):83–93

McGrew G (2021, April 15) The evolution of robotics. https://www.amtonline.org/article/the-evolution-of-robotics

Menon AG, Prabhakar M (2021, March) IoT-based automated pond water quality monitoring system for aquaculture farms. In: 2021 8th international conference on computing for sustainable global development (INDIACom). IEEE, pp 287–293

Papandroulakis N, Markakis G, Divanach P, Kentouri M (2000) Feeding requirements of sea bream (Sparus aurata) larvae under intensive rearing conditions: development of a fuzzy logic controller for feeding. Aquac Eng 21(4):285–299

Qiao X, Bao J, Zhang H, Wan F, Li D (2019) Underwater sea cucumber identification based on principal component analysis and support vector machine. Measurement 133:444–455

Sauter T (2007) The continuing evolution of integration in manufacturing automation. IEEE Ind Electron Mag 1(1):10–19

Smart Agriculture Market Size, Share and Industry Outlooks. (n.d.). DataHorizzon Research. https://datahorizzonresearch.com/smart-agriculture-market-2294

Soori M, Jough FKG, Arezoo B, Dastres R (2024) Robotical automation in CNC machine tools: a review. Acta Mechanica et Automatica 18:434–450

Stigebrandt A, Aure J, Ervik A, Hansen PK (2004) Regulating the local environmental impact of intensive marine fish farming: III. A model for estimation of the holding capacity in the modelling—ongrowing fish farm—monitoring system. Aquaculture 234(1–4):239–261

Swarnakar R, Jayarajan A, Nirwan S, Shah P (2017) The development of automatic fish feeder system using Arduino UNO. J Aquac Res Dev 8(5):1–5

Tanveer M, Balasubramanian S, Sivakumar M, Manimehalai N, Jagan P (2018) A technical review on feeders in aquaculture. Int J Fish Aquat Stud 6(4):305–309

Torrissen O, Olsen RE, Toresen R, Hemre GI, Tacon AG, Asche F et al (2011) Atlantic salmon (Salmo salar): the "super-chicken" of the sea? Rev Fish Sci 19(3):257–278

Ubina NA, Cheng SC (2022) A review of unmanned system technologies with its application to aquaculture farm monitoring and management. Drones 6(1):12

Valenti WC, Kimpara JM, Preto BDL, Moraes-Valenti P (2018) Indicators of sustainability to assess aquaculture systems. Ecol Indic 88:402–413

Wang C, Li Z, Wang T, Xu X, Zhang X, Li D (2021) Intelligent fish farm—the future of aquaculture. Aquac Int 1–31

Yue K, Shen Y (2022) An overview of disruptive technologies for aquaculture. Aquac Fish 7(2):111–120

Zhang S, Yang X, Wang Y, Zhao Z, Liu J, Liu Y et al (2020) Automatic fish population counting by machine vision and a hybrid deep neural network model. Animals 10(2):364

Zhu X, Li D, He D, Wang J, Ma D, Li F (2010) A remote wireless system for water quality online monitoring in intensive fish culture. Comput Electron Agric 71:S3–S9

Chapter 20
Achieving Balance: Strategies for Sustainable Fisheries Supply and Conservation

Avisek Banerjee, Avishek Das, Pradip Kumar Kar, and Triparna Sen

Abstract Global fisheries and aquaculture are undergoing a pivotal transformation, driven by the need to balance escalating demands for aquatic protein with the imperative of environmental stewardship. With wild capture fisheries operating at or beyond their maximum sustainable yields, the sector is shifting toward innovative practices in aquaculture, recreational fishing, and governance. This chapter examines the complex interplay between ecological vulnerability and economic sustainability, with case studies ranging from the recovery of overexploited precious corals to the environmental and market dynamics in the Atlantic sea scallop fishery. The role of integrated assessment models that connect ecological trends with economic outcomes is also explored. Particular emphasis is placed on sustainable aquaculture, especially in the context of feed innovation. The ecological drawbacks of traditional fishmeal and fish oil are contrasted with alternative protein sources such as plant-based feeds and insect meals, which offer environmentally responsible solutions within a circular bioeconomy framework. The chapter also reviews emerging technological tools such as satellite and drone monitoring, selective fishing gear, and

A. Banerjee (✉) · A. Das · P. K. Kar
Department of Zoology, Ramakrishna Mission Vidyamandira, Belur Math, Howrah, West Bengal, India
e-mail: avisek_banerjee@yahoo.com

A. Das
e-mail: dasavishekdas2012@gmail.com

P. K. Kar
e-mail: karpradip@gmail.com

T. Sen
Department of Zoology, Cooch Behar Panchanan Barma University, Cooch Behar, West Bengal, India

Internal Medicine at The Ohio State University and Lung Cancer Preclinical Therapeutics Platform at the OSUCCC-James, The Ohio State University, Columbus, OH, USA

T. Sen
e-mail: triparnasen@gmail.com; sen.149@osu.edu

R. Kumar et al. (eds.), *Sustainability in Aquaculture and Fisheries*,
Springer Water, https://doi.org/10.1007/978-3-032-16605-0_20

blockchain-enabled traceability systems that are reshaping modern fisheries management. Finally, it emphasizes the importance of ecosystem-based approaches, participatory governance, and the incorporation of traditional and Indigenous knowledge systems in promoting long-term sustainability.

Keyword Sustainable aquaculture · Fisheries management · Alternative fish feed · Ecosystem-based approach

20.1 Introduction

The global fisheries and aquaculture sectors are experiencing significant need to balance increasing demand for aquatic protein and the imperative to conserve environmental integrity. Wild capture fisheries have approached maximum sustainable yields and therefore innovations are necessary in aquaculture and recreational fishing practices to meet productivity with sustainability. In aquaculture, feed remains a major cost factor, comprising up to 60% of operational expenses (Singh et al. 2024a). This economic burden is combined by sustainability concerns surrounding traditional feed components like fishmeal (Salick et al. 2007) and fish oil (Tran et al. 2022; Kumar et al. 2022a). While plant protein alternatives offer a more affordable and sustainable option, they are hindered by imbalances in essential amino acids. In recreational fishing, extensive use of ground baits (GBs) is a growing concern. GBs have evolved from being traditionally cereal-based to complex ingredients derived both from plant and animal sources. A study performed on *Cyprinus carpio* for six-week revealed that while animal-derived GBs promoted higher growth rates, plant-based GBs yielded superior nitrogen utilization efficiency, producing greater biomass per unit nitrogen input. Interestingly this study found that there were no significant differences between the two groups in hepatosomatic indices or inflammatory gene expression though, the presence of antinutritional factors such as phytic acid and protease inhibitors in plant-based GBs calls for targeted processing techniques to enhance digestibility and performance (Fazekas et al. 2024). Insect meal has emerged as a promising fish feed (Singh et al. 2024b; Kumar et al. 2022b). For example, insect protein derived from *Tenebrio molitor* (mealworm) and *Acheta domesticus* (house cricket), presents a high nutritional profile and shows efficient conversion of organic waste into biomass (Rumpold and Schlüter 2013). A study from 2013 to 2022 highlights an exponential increase in research on insect-based feeds in aquaculture (Raghuvaran et al. 2024).

Genetic innovations have also contributed to sustainable aquaculture, particularly in species such as *Micropterus salmoides* (largemouth bass), where all-female populations are favored for their enhanced growth performance (Gu et al. 2022; Kumar et al. 2022c; Zhang et al. 2022). Hormonal treatments involving 17α-methyltestosterone (MT) and aromatase inhibitors like letrozole have proven effective in producing functional pseudo-males. Notably, an M20 dosage (20 mg/kg MT) achieved a 95% masculinization rate and yielded 92% female offspring. This method is supported by shifts in gene expression (upregulation of *dmrt1*, downregulation of *cyp19a1a*) and altered endocrine profiles, offering a controlled and efficient approach to monosex breeding (Zhang et al. 2024).

As aquaculture intensifies, stress-related challenges such as oxidative damage pose threats to fish health and performance. Selenium (Se) is an essential trace element that plays role in antioxidative defense through glutathione peroxidase activity (Raymond and Ralston 2009). Through nanotechnology selenium nanoparticles (Se-NPs) offers enhanced stability, absorption, and immunostimulatory properties making Se-NPs as promising agents for mitigating stress and improving resilience in cultured species (Nisa et al. 2024).

20.2 Challenges in Achieving Balance Between Supply and Conservation

Ecosystem-based fisheries management (EBFM) has transformed our approach to managing marine resources (Pikitch et al. 2004). Unlike traditional strategies that focus on individual fisheries, EBFM emphasizes a broader, more integrated perspective in policy evaluation and development, a shift that has already led to advancements in ecological models and methods (Kroetz et al. 2019). While ecological linkages within an ecosystem such as intricate food-web dynamics are acknowledged (Stouffer et al. 2012), economic connections also play a significant role, especially when fishers are involved in multiple fisheries simultaneously. A study finds that catch-share programs in Alaska can lead to cross-fishery participation spillovers and alter the economic connectivity among fisheries, highlighting the need to consider these economic linkages when developing and assessing EBFM policies (Kroetz et al. 2019). One of the key challenges to achieving sustainable fisheries is the inability of management systems to effectively address the inherent risks and uncertainties in marine fisheries. Several major sources of uncertainty have been outlined, including: (i) fluctuations in resource abundance across space and time; (ii) limitations in model structures and parameter estimates used to assess stock dynamics; (iii) difficulties in accurately modeling fishing effort dynamics; and (iv) unpredictability in future environmental and economic conditions. These uncertainties are rarely quantified, complicating the management process (Caddy and Seijo 2005).

20.3 Impacts of Overfishing and Unsustainable Practices

The intensification of global fishing pressure, compounded by climate-induced ocean change, has led to widespread declines in marine biodiversity such as precious corals and commercially valuable bivalves. We discuss the ecological and economic impacts of overfishing, emphasizing *Corallium rubrum* and *Placopecten magellanicus* as focal taxa.

20.3.1 Vulnerability and Recovery of Precious Corals

Corallium rubrum has long been harvested for its rigid, calcareous skeletons prized in jewelry. This exploitation, compounded by environmental stressors such as ocean warming and acidification, has resulted in marked reductions in biomass, colony size, and three-dimensional habitat complexity, key indicators of ecological degradation in benthic environments. As ecosystem engineers, *C. rubrum* colonies support a variety of associated marine life, and their decline disrupts not only coral populations but also the broader community structure. The species' vulnerability is heightened by its slow growth, long lifespan, and limited larval dispersal, which constrain natural recovery processes. Longitudinal photographic surveys conducted over 5 to 7 years on two populations near Marseilles have provided rare before-and-after data following an unplanned harvesting event (Montero-Serra et al. 2015; Hart and Chute 2009). These time-series analyses tracked critical demographic responses, including colony survival, regrowth of damaged individuals, and new recruitment, yielding valuable insights into resilience trajectories. Model projections of biomass recovery under different exploitation scenarios revealed that heavily disturbed populations exhibited limited short-term resilience, particularly where prior harvesting had already reduced structural and reproductive capacity. These cumulative impacts underscore the necessity of considering exploitation history when assessing coral recovery potential.

20.3.2 Environmental and Economic Vulnerabilities in the Atlantic Sea Scallop Fishery

The Atlantic Sea scallop (*Placopecten magellanicus*) represents one of the most economically significant single-species fisheries in the United States. Found along the northeastern continental shelf from the mid-Atlantic Bight to Georges Bank, this species inhabits a broad thermal range (0 to 17 °C) and exhibits region-specific differences in growth rates and maximum size. Its life history traits, including a 4 to 7 weeks planktonic larval stage, sexual maturity at around two years, and typical harvest at age four, render the population particularly responsive to both environmental conditions and management interventions (MacDonald and Thompson 1986). The adoption of rotational closures and effort-limited harvest strategies has contributed to short-term stability and biomass recovery in the fishery. However, climate change particularly ocean acidification and warming are among growing challenges. Anthropogenic CO_2 emissions have already reduced global surface ocean pH by approximately 0.1 units since pre-industrial times, a trend expected to accelerate under high-emission scenarios. Consequently, declining pH levels will impose higher energetic demands for biomineralization for calcifying organism like *P. magellanicus* which relies on sufficient carbonate ion availability for shell formation. Experimental evidence from a related species *P. maximus* under acidified conditions has shown reduced growth rates and compromised metabolic efficiency (Cooley et al. 2015).

20.3.3 Integrated Assessment Models-Bridging Ecology and Economics

Conventional fisheries assessments have historically prioritized biological stock status and short-term yield optimization, often at the expense of incorporating slower-developing but pervasive environmental changes such as climate-driven ocean acidification and warming. While socioeconomic factors are frequently acknowledged, they remain underrepresented in most operational management frameworks. To address this gap, Integrated Assessment Models have emerged as powerful tools that synthesize ecological, physical, and economic dimensions to provide long-term, climate-informed projections for fisheries systems.

A simplified IAM developed for the U.S. Atlantic Sea scallop (*P. magellanicus*) fishery exemplifies this integrative approach. It merges three respective interconnected modules: an ocean box model simulating depth-stratified pH and temperature variations under changing climate scenarios, a stochastic demographic model that captures recruitment fluctuations and fishing mortality, and an economic module projecting revenue and cost trajectories through mid-century. When applied under a high-emissions scenario coupled with assumptions of steady economic growth, the model forecasts notable declines in biomass and fishery revenues by 2050, even in the presence of consistent, effective harvest regulations (Cooley et al. 2015). These outcomes suggest that traditional benchmarks of management success may obscure underlying ecological and economic vulnerabilities driven by long-term climate pressures. By explicitly quantifying trade-offs among harvest intensity, economic returns, and environmental sustainability, IAMs provide a robust decision-support framework for anticipating and navigating future risks. Their capacity to simulate multiple interacting stressors makes them especially relevant for adaptive fisheries governance in an era of climate uncertainty. Ultimately, IAMs bridge the often-separated domains of ecological forecasting and economic planning, offering a holistic foundation for more resilient and forward-looking fisheries management.

20.4 Environmental Impacts of Fisheries

At a broader scale, fisheries contribute to accelerated biodiversity loss, particularly through unsustainable harvesting of keystone species and habitat engineers. Recent estimates indicate that vertebrate extinction rates are occurring at 100 to 1000 times background levels, driven in part by direct exploitation and indirect ecological disruptions. While iconic megafauna such as African elephants garner attention, declines in lesser-known marine taxa ranging from sessile invertebrates to apex predators can alter community composition, trophic interactions, and even long-term evolutionary trajectories (Ceballos and Ehrlich 2023).

20.4.1 Habitat Destruction and biodiversity Loss

Modern fishing practices impose substantial physical and chemical pressures on marine ecosystems, leading to widespread habitat degradation and declines in biodiversity. Among the most persistent yet underrecognized threats is ghost fishing the continued capture and killing of marine life by lost or abandoned fishing gear such as gillnets and traps. These derelict gears can entangle a wide range of non-target species, including fish, crustaceans, sea turtles, and marine mammals (Richardson et al. 2019). Beyond their direct lethality, they contribute to chronic habitat damage, particularly in structurally complex ecosystems like coral reefs, sponge gardens, and seagrass meadows. As these habitats degrade, their capacity to support diverse and productive marine communities diminishes, contributing to broader ecological instability and fragmentation. Chemical pollution compounds these threats. The widespread use of aquatic herbicides in freshwater and nearshore environments introduces another layer of stress. Frequently employed in agricultural and urban weed control, aquatic herbicides often enter aquatic systems through runoff, particularly in areas adjacent to aquaculture zones and capture fisheries. Experimental studies have demonstrated that sublethal concentrations of 2,4-dichlorophenoxyacetic acid, an aquatic herbicide can impair fish metabolism, disrupt endocrine function, and reduce reproductive success, ultimately lowering survival rates over time (Anton et al. 2025). These physiological stress responses can ripple through fish populations, leading to population declines and reduced resilience.

20.4.2 Bycatch and Mitigation Strategies

Bycatch is unintentional capture of non-target species during fishing operations. Its ecological footprint spans taxa and regions, frequently impacting threatened or poorly studied aquatic species and undermining broader conservation efforts. For example, in Mediterranean Sea, the loggerhead sea turtle (*Caretta caretta*) suffers particularly high bycatch mortality across various fishing gear types, including bottom trawls, pelagic longlines, and set nets (Lucchetti et al. 2017). Despite being listed as "Least Concern" by the IUCN, an estimated 44,000 individuals are incidentally captured each year, raising serious concerns about long-term population viability (Casale 2011). Promising mitigation tools such as circle hooks, turtle excluder devices (TEDs), and illuminated nets are currently being trialed, though their success varies depending on gear design and regional fishery practices. In the Bay of Bengal, particularly along the Bangladeshi coast, elasmobranchs (sharks and rays) represent a large proportion of the bycatch, especially in Lakkha nets known for their poor species selectivity. Conservation efforts have largely focused on the economically vital *Tenualosa ilisha* (Hilsa), leaving regulatory gaps for elasmobranch protection (Das et al. 2024). Here, Productivity-Susceptibility Analysis (PSA) has emerged as a valuable tool to assess risk levels across species and guide the development of sustainable gear modifications. Odontocetes (toothed whales, dolphins, and porpoises) are especially vulnerable to gillnet entanglement (Read et al. 2006). Annual bycatch mortality likely numbers in the thousands, with some populations, like the vaquita (*Phocoena sinus*) in the northern Gulf of California, teetering on the brink of extinction, current estimates suggest only 5–15 individuals remain (Jaramillo-Legorreta et al. 2017). This alarming scenario is compounded by synergistic pressures such as prey depletion, chemical contamination from persistent organic pollutants (e.g., PCBs), and noise pollution. Mitigation strategies under exploration include acoustic deterrents (pingers), modified gear designs, and community-based conservation initiatives, though results are often site-specific and species-dependent. In the Northeast Atlantic, pelagic longline fisheries targeting swordfish (*Xiphias gladius*) and blue shark (*Prionace glauca*) inadvertently capture other vulnerable species, including shortfin mako (*Isurus oxyrinchus*) and loggerhead turtles (Lima et al. 2023).

20.5 Principles of Sustainable Fisheries Management

20.5.1 Ecosystem-Based Fisheries Management (EBFM)

EBFM is a holistic, climate-adaptive framework that seeks to balance ecological resilience with human livelihoods (Fogarty 2014). By integrating species interactions, habitat dependencies, and socioeconomic drivers, EBFM addresses a wide array of stressors, from overexploitation and habitat loss to climate-induced environmental shifts. EBFM recognizes that fish populations do not exist in isolation but are embedded within dynamic food webs and human systems. This systems-based perspective is demonstrated by the Eastern Bering Sea case, where a two-million-ton cap on groundfish harvests has played a critical role in sustaining commercially important species such as walleye pollock and Pacific cod (Stram and Evans 2009; Holsman et al. 2020). Here, multi-species models and ecosystem indicators guide management decisions, allowing for more adaptive responses to ecological variability. A key element of EBFM is the implementation of catch-share programs, which allocate fixed portions to individual fishers or communities to curb overcapacity. While EBFM have demonstrated success in stabilizing stocks and improving economic outcomes, it has challenges. Effort leakage, including unintended or unauthorized shift of fishing effort is a documented risk. A network analysis of six major Alaskan catch-share fisheries (1991 to 2015) revealed that although not all programs cause leakage, some have significantly altered participation patterns and inter-fishery

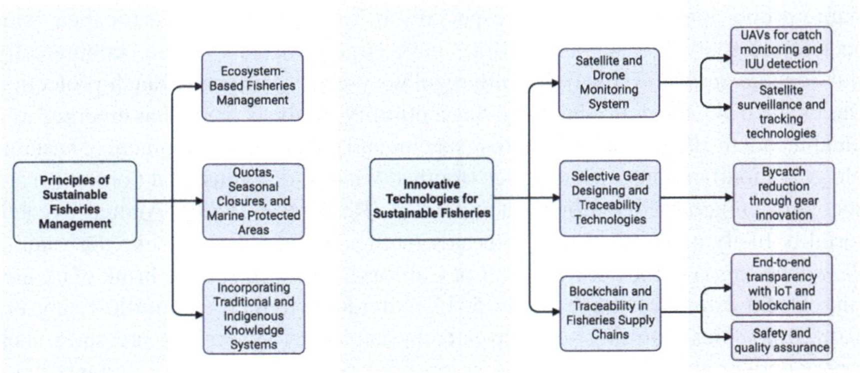

Fig. 20.1 Integrating principles and innovative technologies for sustainable fisheries management

dynamics (Kroetz et al. 2019). This highlights the importance of understanding cross-fishery spillover effects when evaluating the true ecosystem impact of management interventions. Ultimately, EBFM represents a progressive step toward resilient and adaptive fisheries governance. However, its success depends on coupling ecological modeling with robust policy enforcement, stakeholder engagement, and climate mitigation strategies. As marine environments continue to shift under anthropogenic influence, the need for integrated, forward-looking management has never been more urgent (Fig. 20.1).

20.5.2 Quotas, Seasonal Closures and Marine Protected Areas

As part of the broader framework of ecosystem-based fisheries management, quotas, seasonal closures, and marine protected areas (MPAs) remain foundational tools in mitigating overexploitation and conserving marine biodiversity. Yet their success depends on context-specific implementation and integration with ecological modeling frameworks. To address the complexity of multi-species dynamics and food web interactions, a trophic-level (TL) approach rooted in ecological theory has been introduced. This framework uses Trophic Transfer Efficiency typically around 10% to model biomass flow from lower trophic levels, such as zooplankton (TL2), to top predators (TL5) (Lindeman 1942). By incorporating Ecopath modeling and production-to-biomass (P/B) ratios, managers can estimate sustainable fishing mortalities, particularly in pelagic ecosystems where species interactions are critical to ecosystem stability (Hodgson 2022). Among spatial and temporal regulatory tools, quotas remain a widely used mechanism to limit total allowable catches. They are particularly effective when stock assessments are robust and regularly updated.

However, for highly mobile or data-poor species, quota effectiveness can be undermined by uncertainties in abundance estimates and shifting distributions driven by climate change.

Seasonal closures, designed to protect vulnerable life stages or aggregation events (e.g., spawning), offer more targeted protection. Their success hinges on accurate spatiotemporal data to align closures with biologically critical periods. For example, closures during known spawning windows can significantly enhance recruitment rates, provided enforcement mechanisms are in place (Grüss et al. 2011).

MPAs serve as spatial refuges that limit or prohibit extractive activities to preserve ecosystem function and biodiversity. While MPAs can effectively safeguard benthic habitats and sedentary species, their performance declines for migratory or wide-ranging taxa unless these areas are part of larger, interconnected networks (La Manna et al. 2024). To optimize placement, Species Distribution Models (SDMs) such as MaxEnt are increasingly used to map high-suitability habitats based on environmental parameters (Melo-Merino et al. 2020). A case in point is Sardinia's Capo Caccia–Isola Piana MPA, where SDM-guided zoning has aligned protection with core habitat areas, improving ecological outcomes (Di Franco et al. 2016).

20.5.3 Incorporating Traditional and Indigenous Knowledge Systems (TEK)

TEK like indigenous fishing practices like selective gear use, timing harvests, and breeding ground protection enhance sustainability derived from Indigenous and local community practices, aligns with EBFM's ecosystem-centered ethos. For example, in Brazil's Amarante community, spiritual beliefs tied to mythological water guardians shape respectful fishing behaviors (da Silva et al. 2019), suggesting that these culturally embedded practices may function as informal regulatory mechanisms.

20.6 Innovative Technologies for Sustainable Fisheries

20.6.1 Satellite and Drone Monitoring Systems

Emerging digital technologies like unmanned aerial vehicles (UAVs) and satellite-based monitoring systems are transforming the landscape of marine resource governance. These tools are increasingly integral to sustainable fisheries management, offering better surveillance, enhanced data accuracy, and fresh approaches to enforce compliance. Yet, their adoption remains uneven, especially in small-scale and tropical coastal fisheries, where technological and economic barriers persist. Drones, or UAVs, provide high-resolution imagery at relatively low operational cost. They have shown significant potential in improving catch documentation, stock assessments, and the monitoring of illegal, unreported, and unregulated (IUU) fishing. Case studies from Australia, the USA, Brazil, and Spain illustrate how UAVs are used for vessel tracking, species behavior analysis, and bycatch reduction. In countries like Norway and regions of Australia, artisanal fishers are beginning to integrate drones into daily operations, citing gains in efficiency, safety, and environmental monitoring (McCauley et al. 2018). However, in tropical regions such as Kerala, India, UAV adoption remains limited. Fieldwork conducted in Alappad, a coastal village in Kerala, explored the perceptions of local fishers toward drone integration (Lukambagire et al. 2024). While there was general interest in drone-enabled monitoring, successful implementation requires culturally sensitive, economically viable, and locally co-designed solutions. Engaging stakeholders through participatory approaches ensures that modern technologies align with traditional knowledge systems and support the broader "blue transformation" agenda one that seeks to couple ecological resilience with livelihood security (Caldeira et al. 2025). The challenge of IUU fishing, estimated to cause US$10–24 billion in annual global losses, remains a formidable threat to ocean sustainability (Agnew et al. 2009). Its mobile and often offshore nature complicates enforcement. In response, some countries have developed Monitoring, Control, and Surveillance (MCS) systems supported by advanced digital infrastructure. Satellite-based technologies, including Vessel Monitoring Systems (VMS) and Satellite Automatic Identification Systems (S-AIS), are at the forefront of this shift. While VMS offers government-regulated tracking, platforms like Global Fishing Watch leverage S-AIS to provide publicly accessible, near-real-time vessel data, enabling detection of unauthorized activity in marine protected areas (MPAs) and revealing inconsistencies in catch reporting (Watson et al. 2018; Coro et al. 2023).

20.6.2 Selective Gear Design and Traceability Technologies

In parallel, advancements in gear selectivity and traceability technologies contribute significantly to sustainable fisheries. Selective fishing gear, designed to reduce bycatch and protect non-target species, plays a critical role in ecosystem-based fisheries management (Dempster and Taquet 2004). Moreover, occupational safety is increasingly recognized as a priority. A study on commercial fishing fatalities highlighted vessel incidents, gear entanglements, and mechanical failures as primary causes of injury (Victoroff et al. 2023). Future gear innovations must therefore balance ecological functionality with ergonomic safety and automation to mitigate operational risks (Fig. 20.1).

20.6.3 Blockchain and Traceability in fisheries Supply Chains

Blockchain, a cornerstone of Industry, has emerged as a powerful tool for enhancing traceability and sustainability across fisheries supply chains. When integrated with Internet of Things sensors and smart tags, blockchain enables end-to-end transparency from catch to consumer effectively combating IUU fishing by providing immutable transaction records. This digital traceability enhances supply chain integrity, ensures regulatory compliance, and improves consumer trust. The COVID-19 pandemic further underscored the importance of traceable, safe, and transparent food systems (Alam et al. 2022; Manjeet et al. 2022). In response, technologies such as Radio Frequency Identification (RFID), blockchain, and wireless sensor networks (WSN) have been employed to ensure cold-chain integrity, reduce contamination risks, and bolster supply chain resilience. RFID, particularly when paired with IoT, facilitates real-time monitoring of seafood products, enabling detailed tracking of environmental conditions during storage and transport (Rahman et al. 2021). Studies have demonstrated the applicability of RFID-based systems in both capture fisheries and aquaculture. For instance, a study developed a sturgeon monitoring platform combining RFID and QR codes (Zhang et al. 2019). Interestingly, RFID data loggers demonstrated that improve seafood quality and traceability during logistics (Trebar et al. 2013), while another study reported up to 95% increases in efficiency through WSN use in aquaculture (Parreño-Marchante et al. 2014).

20.7 Conclusion

Effective fisheries governance today requires integrated, adaptive, and inclusive policy frameworks that align scientific knowledge with regulatory enforcement and participatory stakeholder engagement. Global case studies highlight the critical role of multi-actor collaboration. The South African hake trawl fishery stands out as a successful example, where coordination among government agencies, scientific bodies, industry representatives, and organizations such as the Marine Stewardship Council and World Wide Fund for Nature has led to notable improvements in sustainability standards and certification. In contrast, the mackerel scad (*Decapterus macarellus*) fishery in Cabo Verde illustrates the consequences of weak monitoring and insufficient regulatory frameworks, underscoring the importance of localized capacity-building and effective enforcement (Delgado et al. 2024). Broader systemic frameworks, such as the Driver-Pressure-State-Impact-Response (DPSIR) model used in coastal Ghana, provide insights into the socio-ecological impacts of overfishing, pollution, and aquaculture mismanagement (Faseyi et al. 2023). Emerging ecological indicators such as parasite community structures in species like the black-spotted croaker (*Protonibea diacanthus*) in northern Australia offer novel tools for environmental monitoring (Porter et al. 2024). The COVID-19 pandemic further exposed the socioeconomic fragility of fishing communities, as seen in the Philippines, where poverty and child labor increased among fisheries-dependent households (Macusi et al. 2022).

20.8 Future Prospects

Community-Based Fisheries Management (CBFM) continues to gain traction as a governance model that integrates conservation goals with local knowledge and participation. However, challenges persist in scaling CBFM while accounting for complex cultural, institutional, and governance contexts. The PROMIS framework (PRactice-Oriented Multi-level perspective on Innovation and Scaling), originally developed in agricultural research, offers a useful strategy for promoting stewardship, equity, and transformative system change in fisheries governance (Steenbergen et al. 2022). Future policy directions must prioritize inclusive and participatory governance structures, particularly emphasizing gender equity. Despite increasing community involvement in monitoring and decision-making, deeply rooted gender disparities continue to limit meaningful participation. A reflexive governance approach one that critically examines the quality and impact of participation is essential to shift existing power dynamics and ensure genuinely inclusive outcomes. Overall, effective fisheries governance in the coming decades will depend on the ability to build adaptive, ecologically grounded, and socially embedded management systems. Strengthening local institutions, fostering equitable participation, and embracing context-sensitive strategies will be central to reconciling biodiversity conservation with the livelihood needs of coastal and fishing communities.

Acknowledgements This work was partially supported by Gennova Biopharmaceuticals Ltd., Pune, Maharashtra, India, through corporate social responsibility (CSR) funding to AB.

References

Agnew DJ, Pearce J, Pramod G, Peatman T, Watson R, Beddington JR, Pitcher TJ (2009) Estimating the worldwide extent of illegal fishing. PLoS One 4:e4570

Alam GM, Nazirul MM, Sarker I, Gatto M, Bhandari H, Naziri d (2022) Impacts of COVID-19 on the fisheries and aquaculture sector in developing countries and ways forward. Sustainability 14:1071

Anton BJ, Oguchi Y, White AM, Karasov WH, Dehnert GK (2025) The impacts of 2,4-D herbicide DMA® 4 IVM on reproductive health and gene expression along the hypothalamic-pituitary-gonad-liver [HPGL] axis in the fathead minnow (*Pimephales promelas*). Chemosphere 371:143994

Caddy JF, Seijo JC (2005) This is more difficult than we thought! The responsibility of scientists, managers and stakeholders to mitigate the unsustainability of marine fisheries. Philos Trans R Soc Lond Ser B Biol Sci 360:59–75

Caldeira M, Sekinairai AT, Vierros M (2025) Weaving science and traditional knowledge: toward sustainable solutions for ocean management. Mar Policy 174:106591

Casale P (2011) Sea turtle by-catch in the Mediterranean. Fish Fish 12:299–316

Ceballos G, Ehrlich PR (2023) Mutilation of the tree of life via mass extinction of animal genera. Proc Natl Acad Sci USA 120:e2306987120

Cooley SR, Rheuban JE, Hart DR, Luu V, Glover DM, Hare JA, Doney SC (2015) An integrated assessment model for helping the United States Sea scallop (*Placopecten magellanicus*) fishery plan ahead for ocean acidification and warming. PLoS One 10:e0124145

Coro G, Sana L, Ferrà C, Bove P, Scarcella G (2023) Estimating hidden fishing activity hotspots from vessel transmitted data. Front Sustain Food Syst 7:2023

da Silva A, Bastos JB, Lopes LS, Figueiredo RF, de Barros M, Souto WMS, Alencar NL, Lopes CGR (2019) Water spirits within the fishers' worldview: implications for fishing management in Northeast Brazil. J Ethnobiol Ethnomedicine 15:70

Das R, Faruque MH, Sakib S, Ahmad MT, Seba RN, Zahid MA, Yeasmin MN, Islam MM (2024) Assessing the vulnerability of elasmobranch species in the bay of Bengal: insights from Lakkha gill net fishery of Bangladesh. Heliyon 10:e37331

Delgado K, Osemwegie I, Medina A, Luz A, Kubik Z, Kouamelan E (2024) Ex-post evaluation of fishery management policies on wild fisheries production in northern Cabo Verde: an example of mackerel scad (Decapterus macarellus, Carangidae). J Fish Biol 105:1212–1226

Dempster T, Taquet M (2004) Fish aggregation device (FAD) research: gaps in current knowledge and future directions for ecological studies. Rev Fish Biol Fish 14:21–42

Franco D, Antonio PT, Di Carlo G, Dimitriadis C, Francour P, Gutiérrez NL, Jeudy A, de Grissac D, Koutsoubas MM, del Mar M, Otero CP, Plass-Johnson J, Sainz-Trapaga S, Santarossa L, Tudela S, Guidetti P (2016) Five key attributes can increase marine protected areas performance for small-scale fisheries management. Sci Rep 6:38135

Faseyi CA, Miyittah MK, Yafetto L (2023) Assessment of environmental degradation in two coastal communities of Ghana using driver pressure state impact response (DPSIR) framework. J Environ Manag 342:118224

Fazekas DL, Antal L, Ardó L, Halasi-Kovács B, Sándor ZJ, Pirger Z, Tóth F, Vitál Z, Specziár A, Mozsár A (2024) Effects of angler's groundbaits on fish physiology and growth. Sci Rep 14:28541

Fogarty MJ (2014) The art of ecosystem-based fishery management. Can J Fish Aquat Sci 71:479–490

Grüss A, Kaplan DM, Guénette S, Roberts CM, Botsford LW (2011) Consequences of adult and juvenile movement for marine protected areas. Biol Conserv 144:692–702

Gu J, Liang H, Ge X, Xia D, Pan L, Mi H, Ren M (2022) A study of the potential effect of yellow mealworm (Tenebrio molitor) substitution for fish meal on growth, immune and antioxidant capacity in juvenile largemouth bass (Micropterus salmoides). Fish Shellfish Immunol 120:214–221

Hart DR, Chute AS (2009) Estimating von Bertalanffy growth parameters from growth increment data using a linear mixed-effects model, with an application to the sea scallop Placopecten magellanicus. ICES J Mar Sci 66:2165–2175

Hodgson BR (2022) An analytical solution to ecosystem-based FMSY using trophic transfer efficiency of prey consumption to predator biological production. PLoS One 17:e0276370

Holsman KK, Haynie AC, Hollowed AB, Reum JCP, Aydin K, Hermann AJ, Cheng W, Faig A, Ianelli JN, Kearney KA, Punt AE (2020) Ecosystem-based fisheries management forestalls climate-driven collapse. Nat Commun 11:4579

Jaramillo-Legorreta A, Cardenas-Hinojosa G, Nieto-Garcia E, Rojas-Bracho L, Ver Hoef J, Moore J, Tregenza N, Barlow J, Gerrodette T, Thomas L, Taylor B (2017) Passive acoustic monitoring of the decline of Mexico's critically endangered vaquita. Conserv Biol 31:183–191

Kroetz K, Reimer MN, Sanchirico JN, Lew DK, Huetteman J (2019) Defining the economic scope for ecosystem-based fishery management. Proc Natl Acad Sci USA 116:4188–4193

Kumar R, Gupta R, Dahiya T, Yadav R, Saifi R, Yodha K, Kumari B, Singh P, Shefali (2022a) Surveillance of *Heteropneustes fossilis* (stinging catfish) diseases and health management practices in Haryana. Appl Ecol Environ Sci 10(4):261–267. https://doi.org/10.12691/aees-10-4-12

Kumar R, Gupta RK, Yadav R, Saifi R, Yodha K, Pooja. (2022b) *Eisenia fetida* as protein source for growth enhancement of *Heteropneustes fossilis*. Egypt J Aquat Biol Fish 26(2):577–588. https://doi.org/10.21608/ejabf.2022.234553

Kumar R, Yadav R, Kumar Gupta R, Pooja (2022c) A sustainable way for fish health management by replacement of chemical and drugs by earthworm. In: Environmental degradation in asia: land degradation, environmental contamination, and human activities. Springer International Publishing, Cham, pp 329–352. https://doi.org/10.1007/978-3-031-12112-8_16

La Manna G, Ronchetti F, Moro Merella M, Vargiu R, Perretti F, Ceccherelli G (2024) Using spatial distribution modeling of commercial species to inform management of small-scale fisheries in a mediterranean marine protected area. Ecol Evol 14:e70169

Lima FD, Parra H, Alves RB, Santos MAR, Bjorndal KA, Bolten AB, Vandeperre F (2023) Effects of gear modifications in a North Atlantic pelagic longline fishery: a multiyear study. PLoS One 18:e0292727

Lindeman RL (1942) The trophic-dynamic aspect of ecology. Ecology 23:399–417

Lucchetti A, Vasapollo C, Virgili M (2017) An interview-based approach to assess sea turtle bycatch in Italian waters. PeerJ 5:e3151

Lukambagire I, Agrah T, Sophie von Lieres J, Matovu B, Bhavani RR (2024) Fishermen's attitudes towards drone use for sustainable fishing in a coastal community in Kerala, India: an exploratory qualitative study. Sustain Futur 7:100225

MacDonald BA, Thompson RJ (1986) Influence of temperature and food availability on the ecological energetics of the giant scallop Placopecten magellanicus. Mar Biol 93:37–48

Macusi ED, Rafon JKA, Macusi ES (2022) Impact of COVID-19 and closed fishing season on commercial fishers of Davao gulf, Mindanao, Philippines. Ocean Coast Manag 217:105997

Manjeet, Airon A, Kumar R, Saifi R (2022) Temporal and spatial impact of lockdown during COVID-19 on air quality index in Haryana, India. Sci Rep 12:20046. https://doi.org/10.1038/s41598-022-20885-2

McCauley DJ, Jablonicky C, Allison EH, Golden CD, Joyce FH, Mayorga J, Kroodsma D (2018) Wealthy countries dominate industrial fishing. Sci Adv 4:eaau2161

Melo-Merino SM, Reyes-Bonilla H, Lira-Noriega A (2020) Ecological niche models and species distribution models in marine environments: a literature review and spatial analysis of evidence. Ecol Model 415:108837

Montero-Serra I, Linares C, García M, Pancaldi F, Frleta-Valić M, Ledoux JB, Zuberer F, Merad D, Drap P, Garrabou J (2015) Harvesting effects, recovery mechanisms, and management strategies for a long-lived and structural precious coral. PLoS One 10:e0117250

Raghuvaran N, Varghese T, Jana P, Brighty RJA, Sethupathy AM, Sudarshan S, Alrashdi YBA, Ibrahim AE, El Deeb S (2024) Current status and global research trend patterns of insect meal in aquaculture from scientometric perspective: (2013-2022). Aquac Nutr 2024:5466604

Nisa S, Fatima M, Shah SZH, Khan N, Aftab B, Ali W, Maryam SS, Fatima A (2024) Evaluation of growth, antioxidant status, hepatic enzymes and immunity of Nanoselenium-fed Cirrhinus mrigala. PLoS One 19:e0308761

Parreño-Marchante A, Alvarez-Melcon A, Trebar M, Filippin P (2014) Advanced traceability system in aquaculture supply chain. J Food Eng 122:99–109

Pikitch EK, Santora C, Babcock EA, Bakun A, Bonfil R, Conover DO, Dayton P, Doukakis P, Fluharty D, Heneman B, Houde ED, Link J, Livingston PA, Mangel M, McAllister MK, Pope J, Sainsbury KJ (2004) Ecosystem-based fishery management. Science 305:346–347

Porter M, Barton DP, Williams J, Randall J, Ovaskainen O, Crook DA, Shamsi S (2024) Influence of body size and environmental conditions on parasite assemblages of the black-spotted croaker (Protonibea diacanthus) (Teleostei: Sciaenidae) in northern Australia. Parasitology 151:864–874

Rahman LF, Alam L, Marufuzzaman M, Sumaila UR (2021) Traceability of sustainability and safety in fishery supply chain management systems using radio frequency identification technology. Foods 10

Raymond LJ, Ralston NVC (2009) Selenium's importance in regulatory issues regarding mercury. Fuel Process Technol 90:1333–1338

Read AJ, Drinker P, Northridge S (2006) Bycatch of marine mammals in U.S. and global fisheries. Conserv Biol 20:163–169

Richardson K, Asmutis-Silvia R, Drinkwin J, Gilardi KVK, Giskes I, Jones G, O'Brien K, Pragnell-Raasch H, Ludwig L, Antonelis K, Barco S, Henry A, Knowlton A, Landry S, Mattila D, MacDonald K, Moore M, Morgan J, Robbins J, van der Hoop J, Hogan E (2019) Building evidence around ghost gear: global trends and analysis for sustainable solutions at scale. Mar Pollut Bull 138:222–229

Rumpold BA, Schlüter OK (2013) Nutritional composition and safety aspects of edible insects. Mol Nutr Food Res 57:802–823

Salick J, Amend A, Anderson D, offmeister K, Gunn B, Zhendong F (2007) Tibetan sacred sites conserve old growth trees and cover in the eastern Himalayas. Biodivers Conserv 16:693–706

Singh SK, Pawar L, Thomas AJ, Debbarma R, Biswas P, Ningombam A, Devi AG, Waikhom G, Patel AB, Meena DK, Chakraborty G (2024a) The current state of research and potential applications of insects for resource recovery and aquaculture feed. Environ Sci Pollut Res Int 31:62264–62282

Singh SK, Pawar L, Thomas AJ, Debbarma R, Biswas P, Ningombam A, Devi AG, Waikhom G, Patel AB, Meena DK, Chakraborty G (2024b) The current state of research and potential applications of insects for resource recovery and aquaculture feed. Environ Sci Pollut Res 31:62264–62282

Steenbergen DJ, Song AM, Andrew N (2022) A theory of scaling for community-based fisheries management. Ambio 51:666–677

Stouffer DB, Sales-Pardo M, Sirer MI, Bascompte J (2012) Evolutionary conservation of species' roles in food webs. Science 335:1489–1492

Stram DL, Evans DCK (2009) Fishery management responses to climate change in the North Pacific. ICES J Mar Sci 66:1633–1639

Tran HQ, Nguyen TT, Prokešová M, Gebauer T, Van Doan H, Stejskal V (2022) Systematic review and meta-analysis of production performance of aquaculture species fed dietary insect meals. Rev Aquac 14:1637–1655

Trebar M, Lotrič M, Fonda I, Pleteršek A, Kovačič K (2013) RFID data loggers in fish supply chain traceability. Int J Antennas Propag 2013:875973

Victoroff TM, Case SL, Robertson LD, Syron LN (2023) Workplace injuries caused by commercial fishing winches-Alaska, 2000-2020. J Agromedicine 28:433–443

Watson JT, Haynie AC, Sullivan PJ, Perruso L, O'Farrell S, Sanchirico JN, Mueter FJ (2018) Vessel monitoring systems (VMS) reveal an increase in fishing efficiency following regulatory changes in a demersal longline fishery. Fish Res 207:85–94

Zhang D, Li S, Tian T, Du J, Lei C, Zhu T, Han L, Song H (2024) Effects of 17α-methyltestosterone and letrozole on growth and gonadal development in largemouth bass (*Micropterus salmodies*). Front Physiol 15:1444918

Zhang Q, Bian Y, Zhao Y, Yang X, Jiankai W, Wang D, Wang J, Wang A, Qi Z (2022) Replacement of fishmeal by fermented silkworm pupae meal in diets of largemouth bass (*Micropterus salmoides*): effects on growth performance and feed utilization. J Appl Ichthyology

Zhang Y, Wang W, Yan L, Glamuzina B, Zhang X (2019) Development and evaluation of an intelligent traceability system for waterless live fish transportation. Food Control

Chapter 21
Organic Aquaculture Practice Towards Establishing Sustainability in Aquaculture

Shubham Janbandhu, Panchakarla Sedyaaw, Smruthi Hareendran, Priya Singh, and Vyankatesh Tekade

Abstract One area of farming that has a lot of potential to embrace modernity is aquaculture. By developing appropriate procedures for fish farming, it tackles a number of problems, including animal abuse, consumer health, and environmental degradation. In this way, it does not employ artificial chemicals, antibiotics or genetically modified organisms (GMOs), but provides natural alternatives such as probiotics, organic feeds and appropriate management of water. It minimizes the risk of oxygen depletion and nutrients pollution since it incorporates waste recycling and the introduction of aquaponics systems. In economic terms, there are interesting developments in organic aquaculture as new markets are developing as customers are demanding seafood that is of high quality, obtained in sustainable ways and more importantly traceable. The sector has provided a doable model on how to ensure sustainability in producing seafood in the world in spite of challenges such as higher cost of operation and more comprehensive guidelines. Organic aquaculture can, therefore, serve as a form of diversified land use with beneficial effects on all three pillars—environmental, social, and economic—provided it integrates some elements of traditional skilled aquaculture.

S. Janbandhu (✉)
Department of Fisheries Resource Management, College of Fisheries, Mangalore, Karnataka Veterinary Animal and Fisheries Sciences University, Bidar 575002, India
e-mail: shubhamjanbandhu18@gmail.com

P. Sedyaaw
Department of Fish Processing Technology, College of Fisheries, Ratnagiri, Maharashtra, Dr. Balasaheb Sawant Konkan Krishi Vidyapeeth, Dapoli, Maharashtra 415612, India

S. Hareendran
Department of Aquaculture, Kerala University of Fisheries and Ocean Studies, Kochi, India

P. Singh
Department of Aquaculture, College of Fisheries Science, Chaudhary Charan Singh Haryana Agricultural University, Hisar, Haryana, India

V. Tekade
Department of Fish Processing Technology, College of Fisheries, Mangalore, Karnataka Veterinary Animal and Fisheries Sciences University, Bidar 575002, India

R. Kumar et al. (eds.), *Sustainability in Aquaculture and Fisheries*,
Springer Water, https://doi.org/10.1007/978-3-032-16605-0_21

Keywords Probiotics · Eco-labeling · Productivity · Aquaponics

21.1 Introduction

Over the past few years, organic aquaculture has seen a sharp increase in food output. Although it currently represents small amounts globally, it has nearly increased since 2014. In 2016, around 415 thousand metric tonnes of food were produced globally using organic aquaculture (Willer et al. 2023; 2024). With almost 74% of all organic aquaculture products coming from China, the region with the largest production was Asia, continuing its long-standing strong position. Europe came next, with around 20% of the total. Ireland was the biggest producer in Europe, ranking second globally, and it made up about 10% of the total. It also had a notable 31% increase between 2015 and 2016. A form of supply-demand contradiction characterises organic aquaculture in emerging nations, which are mostly represented by Asian nations (Gambelli et al. 2019).

In addition to organic agricultural land, there are additional organic land regions used for various purposes. The majority of them are beekeeping and wild gathering sites. Aquaculture, woods, and grazing regions on non-agricultural land are additional non-agricultural areas. The combined area of all organic areas was 108.3 million hectares, and these areas accounted for 31.8 million hectares (Kumar et al. 2023; IFOAM 2023). Organic aquaculture accounts for a meager 4.7% of the total aquaculture production in the EU, which includes six major species: carp, salmon, trout, sea bream, sea bass, and mussels (Lembo et al. 2019).

Capture and culture fisheries production was at 178 million tonnes in 2020. Despite this, the capture industry is growing steadily, generating around 90.3 million tonnes, of which 11.5 million come from inland waterways and 78.8 million from marine waters. About 88 million tonnes of seafood, including fish, shellfish, and bivalves, have been shared among the major producers in the cultural sector. Thus, the capture sector contributed about 51% in global production while the culture sector is prospering to satisfy the increasing demand for seafood and shares remaining 49%. Considering the awareness about the importance of seafood in human nutrition, global per capita consumption has reached to 20.2 kg. (FAO 2022) In the year 2021, the fish production in India has reached to 16.248 million tons from marine and inland fisheries sector. Although the marine sector has contributed to 41.27 lakh tons, but the inland sector showed a rise up to 121.21 lakh tons (MPEDA 2022).

In 2021–2022, India exported 13,69,264 tonnes of seafood, of which 48.63% came from the aquaculture sector and 51.37% from the catch fisheries. Thus, the Indian economy benefited from the 57,586.48 crores that the fishing industry provided. 5.89 lakh tonnes were generated by inland and marine fisheries in Maharashtra state, which has a 720 km coastline and 173 landing places. 1.57 lakh tonnes came from the inland sector, whereas 4.33 lakh tonnes came from the maritime sector (MPEDA 2022). In the food industry, one of the sectors in the world that is expanding fastest is aquaculture (Yılmaz et al. 2022). Because of the increased need for high-quality food items, aquaculture output has increased significantly in recent years. Data shows that during the past 30 years, aquaculture output has increased fourfold, from 21.8 million tonnes in the 1990s to 87.5 million tonnes in 2020 (FAO 2022).

According to Lembo and Mente (2019), a systematic strategy for farm management and food production, organic aquaculture demands optimum fish welfare to meet certain consumer preferences while including the best environmental practices, biodiversity preservation, and natural resource conservation. Organic fish farming is an ecosystem-based managed farm that aims to avoid any type of chemical inputs other than nature or some diseases control, which produced less environmental impacts as well makes an effort for making it sustainable (IFOAM 2014; Naturland 2020). Given that output from capture fisheries has stagnated, aquaculture will have to increase in order for food production to keep up with the expanding global demand for seafood and fish as humans continue growing their numbers. This means that any upcoming increase in fish production will likely be challenged by a range of environmental problems, as aquaculture is constrained to those regulations and therefore inherently unsustainable. Organic aquaculture methods will reduce the impact of this method on environment. The purpose is to increase fish production, which will be achieved using strategies such as sustainable intensification across a range of productive systems (ecosystem-based polyculture, integrated aquaculture and organic Aquaculture).

In recent years it has been pointed out the need of feed raw materials and their sources in relation to the extent of sustainable aquaculture practised. Due to the shrinking resources of natural fisheries, aquaculture requires probably alternative sources of new raw material (Nizza 2012; Shefali et al. 2021). Open-access human activities such as aquaculture are growing rapidly each day, and inescapably they have implications on the environment that contribute to threatening food security. Consequently, there is now an urgent need to use human food quality and aquaculture-green compatible sources for the feeds used.

The International Federation of Organic Agriculture Movements (IFOAM) lists health, ecology, justice and compassion as the four main tenets of organic agriculture (Gould et al. 2019). Organic aquaculture's growth may be attributed to the acceptance of principles such as animal welfare, value addition, traceability (a way to track a product and certify if there are any issues), environmentally friendly practices in particular for feeds), eco-labeling (deciding which products have no negative impacts on nature) or consumer awareness. Leading to the consumer expectations are one of some important factors behind organic aquaculture production (Ahmed et al. 2020).

Economic, ecological, as well as socially responsible production methods are provided by organic aquaculture. Organic aquaculture incorporates a number of factors and takes into account the health and welfare of the organisms that are raised in order to guarantee that farming techniques are in balance with the environment. Due to market instability, customer preference differences, and difficulties procuring organic feed and feedstuffs, organic fish production has advanced more slowly than organic agriculture (Prein et al. 2010). After the aforementioned issues were resolved, the company has, however, made significant progress. The expansion of organic aquaculture is largely dependent on the use of organic feed and feed ingredients. Organic feed is getting easier to get, and there are more and more sources available. The availability of organic feed is becoming simpler and the variety of sources is growing. Comprehensive information regarding organic aquaculture is provided in this chapter. The biggest cause of biodiversity loss is unsustainable agriculture, which also pollutes soil, threatens rural livelihoods, and jeopardises food and nutrition security (IFOAM 1981).

Increased efficiency, lower resource consumption, avoiding chemicals and medications, generating less waste, and recycling materials and nutrients are the goals of organic aquaculture (Folke et al. 1992). Due to concerns about aquaculture‘s sustainability, regional and international organisations suggested Best Management Practices (BMP) to improve sustainable output (FAO 1997; Kumar et al. 2022a; Kumar et al. 2022b). BMPs aim to make aquaculture socially and economically sustainable while also making it ecologically responsible. A number of indicators were created to assess aquaculture‘s sustainability and resource use efficiency (Boyd et al. 2007).

“Organic aquaculture“has gained attention due to a number of factors, including diminishing fisheries harvests, food safety problems with wild fish, environmental concerns, rising fish consumption, and the growing market share of organic goods (Dube et al. 2012). The importance of organic aquaculture and beekeeping cannot be quantified in hectares, and other metrics (such as productivity and the number of beehives) are more significant than the area.

21.2 Aquaculture Production Systems

21.2.1 Flow-Through Systems

Currently, the most popular methods for intensively raising a wide range of farmed species are flow-through systems. Conventional land-based flow-through farms rely on pumps or gravity to draw water from a nearby lake or river. Before being released into the receiving waters downstream, the farm water is filtered mechanically (using micro-sieves and settling ponds) and biologically using a variety of water treatment units. The water may stay in the system for a variety of lengths of time.

Type 1 fish farm is a relatively big farm where mechanical water treatment is employed, and the fish stocking density is rather low. The farm does not water recycle much, hence the need for efficient internal (natural) nutrient digestion. While some methods of purifying water include sludge basins, plant lagoons, micro-sieves/contact filters, and sludge cones, natural processes also play a role (Jokumsen and Svendsen 2010), and these methods obviously adhere to the ideals of organic agriculture.

Compared to farm type I, farm type 2 is more intense and the volume of water used per AUM is less, while also a much higher percentage (almost all) will be reused. It also carries both mechanical and biological water treatment. In addition to mineralising nutrients, water is cleansed through sludge cones, micro-sieves, biofilters and two stage lagoons of greater than 2 hectares with a minimum hydraulic loading rate; this does not require plant lagoons. This sort of farm could still be considered organic farming, if it met certain requirements.

Although categorized as FT due to its much-reduced water use (less than 3.6 m3 of water per kilogramme, i.e., kg or L) compared to traditional flow-through systems, the hybrid farm type is not really an apiary-based system). This kind of agricultural undertaking utilizes a great deal of recirculating systems in which there is conservation of water such as in the recycling system and in treating used water as this reduces the dependency of such system on the environment (Jokumsen and Svendsen 2010).

21.2.2 Recirculating Aquaculture Systems (RAS)

Recently, organic aquaculture has emerged as a solution to a number of problems. It is a healthy substitute that lowers pollution and creates safer food. Recirculating aquaculture systems (RAS) were developed in recent decades with the goal of preserving pond water quality by reducing water exchange and emphasising water reuse following treatment. The technical feasibility of using RAS in organic aquaculture systems was examined in this study. Widespread and semi-intensive cultures will cost less, while intensive systems with higher stocking densities will provide higher yields. The waste water of organic aquaculture can be used to improve sedimentation efficiency and remove more nutrients. Because of this, recirculating aquaculture systems may be easily modified for organic aquaculture (Beg et al. 2024).

RAS are mostly utilised for indoor or outdoor finfish production in order to produce fish for human consumption. Since 2004, Denmark uses semi-intensive commercial RAS called "Model-Trout-Farms" to boost land-based rainbow trout (Oncorhynchus mykiss) production while conforming to the EU Water Framework Directive and minimizing damage on neighbouring aquatic habitats (Dalsgaard et al. 2013; EU-Water Frame-Directive 2000; Jokumsen and Svendsen 2010). Smolt grown in RAS grows better and cause fewer difficulties with infestation of sea-lice in cages later on (Clarke and Bostock 2017). Investments in RAS have increased during the past 10 years. In Chile and northern Europe, these RAS are increasingly being used to produce Atlantic salmon smolt before moving them to sea cages (EUMOFA 2020).

21.2.3 Extensive Pond Systems

In contrast, extensive aquaculture, where it is an ecosystem part and relies on the environment for sustaining the quality of water, providing most of the animals' feed and other needs, is completely different (Soltan 2016). In many places, due to their presence and often great sizes, fish ponds have become a significant factor in rural development as well as looking naturally integrated into the surroundings (Lembo et al. 2019). Probably the most old-fashioned large system of aquaculture development is composed of fish ponds, man-made water bodies which constitute a natural feature of the area (Kořínek et al. (1987); Adámek et al. 2012).

According to Kořínek et al. (1987), Adámek et al. (2012), a fish pond is the most conventional big aquaculture system and the primarily artificial aquatic habitat that is seen as part of the landscape. Fish ponds may be useful for controlling water resources. Fishponds offer lands to not only produce fish but also, catch and store water when rainfall is scarce for watering purposes. Ponds also help to preserve the surrounding biotopes and wildlife. The organic matter and nutrients are accumulated in fish pond sediments, which provides an agricultural fertilizer that can be collected between growing when ponds were dried over as part of extensive farming. This means that fish farming allows horticulture (culture of other terrestrial plants as well) to be added on top as well instead, providing a semi-terrestrial and aquatic production system (Kumar et al. 2022c; Lembo et al. 2019).

21.2.4 Cage Systems

Floating cages vary in size and the intended use in open water from small and weak ones to hundreds of cubic meters in capacity (Soltan 2016). Cage farming has several advantages over any other method of protein source production, including cheap investment and energy costs, besides having a very minimal carbon foot print cages offers low consumption of natural resources of space and natural flushing (Angel et al. 2019). Marine protected areas and no-take zones are other names for sea cage farms. By serving as a safeguard against overfishing and other types of fishery mismanagement, these would provide safe havens for wild fish while halting the detrimental effects of fishing operations on fish populations. Although this is often linked to adverse effects (International Union for Conservation of Nature 2007; Grigorakis and Rigos 2011), sea cages may help the local and regional ecosystem by providing a place for wild fish to live and feed as a hatchery and nursery (Ozgul and Angel 2013).

21.2.5 Integrated Multi-trophic Aquaculture (IMTA)

Recent concept of Integrated multi-trophic aquaculture emerged in Europe that strives to take advantage of extractive species like shellfish and macro algae ability to absorb and uptake particulate and dissolved material from aquaculture effluents. Ecologically efficient, environment friendly, product-diversified, rewarding, and societally helpful goals can be achieved in integrated multitrophic aquaculture by raising fed species-such as finfish that are fed sustainable commercial diets-side-by-side with extractive species that use the inorganic (like seaweeds) and organic (like suspension-

and deposit-feeders) excess nutrients of aquaculture for growth (Chopin and Thierry 2010; Kumar and Singh 2022). One of the high-tech solutions proposed for public acceptance is to promote greater support of the industry and its special safe products, as well as for economic stability product diversification and environmental sustainability through biomitigation. By converting the waste of one crop (fed animals) into fertiliser, food, and energy for the other crops (extractive plants and animals), which can then be marketed, the goal is to increase long-term sustainability and profitability per cultivation unit rather than per species in isolation as is done in monoculture.

21.3 Rationale of Organic Aquaculture

Sustainable practice like organic fish farming emerged. In other words, organic fish farming, which is a management method that simply aimed at minimizing the negative impacts of fish farming practices whereby every effort possible, is made. However, economic transition from aquaculture production practices to organic aquaculture is more or less problematic, costly and requires patience. The existence of the organic aquaculture certification bodies will foster growth of the industry. The switch to organic aquaculture also has some positive effects on the environment. Largely owing to much lower output level of organic aquaculture as compared to traditional systems, incorporation of this system in food production will add minimally to enhancing food security conditions worldwide (Ahmed et al. 2020). Given that the harvest of wild fisheries has reached a limit and cannot fulfill the increasing requirement for fish and seafood of the ever-expanding population, aquaculture must also produce more food. Indisputably, environmental dilemmas will arise from the increase of fish production levels in aquaculture because this will always be subject to limitations of the surrounding environment.

Promoting organic aquaculture principles could mitigate the harm that aquaculture inflicts on natural systems. Production methods such as polyculture, integrated aquaculture and organic aquaculture in our view can progressively be enhanced to improve fish production sustainably. According to Lembo and Mente; and Mente et al. (2019), organic aquaculture is a holistic, encompassing methodology that incorporates agriculture, preservation and management of the farms. As noted by Gould et al. (2019), the organic aquaculture has four guiding principles: health, ecology, and social justice and compassion. The five components aquaculture practice consists of is site and farm design; breeding; feeding; animal welfare and health; harvesting and product quality. These components work together to influence all the principles simultaneously, as each principle will be addressed below.

21.4 Principles of Organic Aquaculture

Organic aquaculture adheres to principles rooted in organic farming, focusing on sustainability, animal welfare, and ecological balance. These principles are aimed at minimizing environmental impact, enhancing biodiversity, and promoting the health and well-being of aquatic species.

21.4.1 Ecological Sustainability

Natural ecosystems are intended to coexist peacefully with organic aquaculture methods. Instead of using artificial chemicals, antibiotics, or genetically modified organisms (GMOs), farmers employ natural approaches to preserve water quality and stop disease outbreaks. Probiotics, herbal extracts, and biological controls are examples of natural compounds that are frequently utilised to prevent disease and support health. Nutrient recycling and waste reduction are made possible by integrated multi-trophic aquaculture (IMTA) systems, which co-culture species from many trophic levels.

21.4.2 Biodiversity Conservation

To preserve genetic variety and the integrity of regional ecosystems, organic aquaculture promotes the use of native species and shuns the use of genetically modified organisms. In organic systems, IMTA and polyculture systems, multiple species cultivated together-are prevalent because they support ecological balance and lessen the environmental impact of farming.

21.4.3 Animal Welfare

By guaranteeing proper stocking densities, preserving species-specific environmental conditions, and refraining from using artificial growth stimulants, organic aquaculture places a high priority on animal welfare (European Union 2021). Rather, the emphasis of organic techniques is on reducing stress and providing natural conditions, which improve animal health and lower the likelihood of disease outbreaks.

21.5 Key Practices in Organic Aquaculture

Organic aquaculture practices are characterized by sustainable resource use, careful management of farmed species, and an emphasis on natural biological processes. Several key practices stand out.

21.5.1 Sustainable Feed Management

Purchasing and handling feed is a vital part of aquaculture that is organic. Feed components for organic systems must originate from sustainable sources. Plant-based feed must be organic, and fishmeal and fish oil must come from wild fisheries that are sustainably managed or from organic production systems (Willer and Lernoud 2019; Yadav et al. 2022). Genetically engineered substances and synthetic additives are not allowed. In preserving the nutritional requirements of farmed animals, this helps lessen the environmental effect of feed production. Table 21.1 presents a comparison between conventional and organic feed.

According to the FAO (2022a, 2022b), the production volume of organic aquaculture varies by species. The highest production volume falls under the category "Other/ no details," with 241,112 metric tons. This is followed by salmon, with a production volume of 25,546 metric tons, and mussels, at 25,419 metric tons. Shrimps account for 7382 metric tons, while sturgeon contributes 2520 metric tons. Rainbow trout have a production volume of 1821 metric tons, and carps have the lowest reported volume at 761 metric tons.

Table 21.1 Comparison of conventional and organic feed in aquaculture

Aspect	Conventional feed	Organic feed	Reference
Feed ingredients	Fishmeal, fish oil, soy, grains, poultry by-products	Organic-certified plant sources, fish trimmings from sustainable fisheries	Mente et al. (2011)
Additives	Synthetic amino acids, growth promoters, antibiotics	Natural additives, probiotics, herbal supplements	NRC (2011)
Consumer preference	Widely accepted; not always traceable	Preferred for health-conscious markets; traceable supply chain	FAO (2018)
Regulation	General feed safety standards	Must comply with organic aquaculture regulations (e.g., EU Reg. 2018/848)	EU Commission (2018)

Table 21.2 Organisms experimented in IMTA

Types of fish	Species	Reference
Finfish	*Salmo, Oncorhynchus, Scophthalmus, Dicentrarchus, Gadus, Anoplopoma, Hippoglossus, Melanogrammus, Paralichthys, Pseudopleuronectes,* and *Mugil* sp.	Yokoyama (2013), Cubillo et al. (2016), Zhou et al. (2006), Zamora et al. (2018), Fang et al., (2016) and Shpigel et al.(2016)
	Most preferred: Red seabream, salmon, carp and catfish	Kibria and Haque (2018)
	India: Cobia, Pearspot, mullet (Mugil cephalus, Liza parsia), milkfish, Rohu	Viji et al. (2014), Johnson et al. (2019), Biswas et al. (2019) and Nath et al. (2021)
Organic extractive species/ Suspension feeders (shellfish)	Mollusks are the most tested organisms in the IMTA context	Granada et al. (2016)
	Mussels (Mytilidae) such as *Mytilus edulis* or *Mytilus trossulus* are particularly effective in coastal temperate areas and present high potential to be used as secondary species in IMTAs	Ren et al. (2012)
	Haliotis, Crassostrea, Pecten, Argopecten, Placopecten, Mytilus, Choromytilus and *tapes*	Soto (2009)
	Crustaceans-*Penaeus* and *Homarus*	Soto (2009)
	India- Green mussel (*Perna viridis*), *Crassostrea madrasensis, C. Cuttackensis, Lamellidens marginalis*	Viji et al. (2014), Nath et al. (2021) and Biswas et al. (2019)
DepositFeeders	Echinoderms: Strongylocentrotus, Paracentrotus, Psammechinus, Loxechinus, Cucumaria, Holothuria, Stichopus, Parastichopus, Apostichopus, and Athyonidium	Soto (2009)

21.5.2 Water Quality Management

The focus of organic aquaculture is on using natural processes to maintain water quality. In order to control waste and lower nutrient loads, biofilters, artificial wetlands, and vegetative buffers are frequently employed. Organic farms must periodically check water characteristics, such as pH, ammonia, and dissolved oxygen levels, to make sure that environmental regulations are being followed.

21.5.3 Polyculture and IMTA

In organic systems, polyculture and integrated multi-trophic aquaculture (IMTA) are crucial techniques. One species' waste becomes a resource for another, lowering the demand for outside inputs and enhancing the quality of the water (Troell et al. 2021). An illustration of an IMTA system utilised in organic aquaculture may be seen in Table 21.2. There are few examples of IMTA System in Organic Aquaculture as given in Table 21.3.

21.6 Organic Aquaculture Production in India and the World: Latest Data and Insights

In order to produce fish, shellfish, and other aquatic species without the use of synthetic pesticides, antibiotics, or genetically modified organisms (GMOs), organic aquaculture is a sustainable farming method. The principles of ensuring animal welfare, maintaining the natural balance of ecosystems, and advancing long-term environmental sustainability form the foundation of organic aquaculture. Due to rising consumer awareness of environmental and health concerns, the market for organic aquaculture products has expanded significantly over the past ten years.

21.6.1 Global Organic Aquaculture Production

The aquaculture industry has grown significantly on a global scale. According to estimates from the Food and Agriculture Organisation (FAO 2023), around 3% of global aquaculture production is organic. According to FAO (2023), the worldwide market for organic aquaculture was valued at USD 8.6 billion in 2022 and is projected to rise at a compound annual growth rate (CAGR) of 10.2% between 2023 and 2028.

Table 21.3 IMTA system in organic aquaculture

System type	Species combination	Location	Key benefits	Reference
Marine cage culture	Atlantic salmon, blue mussels, kelp (*Saccharina latissima*)	Bay of Fundy, Canada	Nutrient recycling, mussel and kelp growth enhanced	Chopin et al. (2008)
Brackishwater system	Tiger shrimp (*Penaeus monodon*), mullets, estuarine oyster, seaweed	Sundarbans, India	Biodiversity conservation, ecological sustainability	Patra et al. (2019)
Tropical marine cage	Donkey's ear abalone (*Haliotis asinina*), Gracilaria, Eucheuma seaweeds	Cebu, Philippines	Seaweed used as feed and biofilter, improved abalone survival	Hurtado et al. (2015)
Freshwater polyculture	Nile tilapia (*Oreochromis niloticus*), freshwater prawn, azolla, duckweed	Thailand	Nutrient recycling, water purification	Edwards (2009)
Coastal Sea cage	Cobia (*Rachycentron canadum*), *Kappaphycus alvarezii* seaweed	Tamil Nadu, India	Economic upliftment, organic certification potential	Gopakumar et al. (2019)

21.6.1.1 Key Countries Contributing to Organic aquaculture Production Include

Europe: Particularly in nations like Norway, Ireland, and Germany, the European Union is a pioneer in organic aquaculture. About 50,000 metric tonnes of organic salmon were produced in Norway in 2022, making up over 15% of the nation's total salmon output (European Commission 2023).

Asia: China is a major contributor in organic aquaculture, specialising in the production of seaweed, tilapia, and prawns. China produced over 200,000 metric tonnes of organic aquaculture in 2022 (China Aquaculture Association 2023).

North America: The organic aquaculture sector is growing in the U.S. and Canada, with salmon and mussels being the primary organic products. In 2022, the U.S. organic aquaculture production was around 10,000 metric tons (NOAA 2023).

Table 21.4 Key challenges for transforming from conventional to organic aquaculture

Feature	Challenge in organic aquaculture	References
Culture species	Native species are preferred as exotic species may have adverse ecological effects	Censkowsky and Altena (2013)
Farming system	Monoculture is allowed, but polyculture is recommended	Naturland (2020)
Farming intensity	Extensive farming is suggested, not intensive farming	
Hatchery operation	Application of antibiotics and hormones are prohibited	
Fish stocking	Not exceed carrying capacities, maximum stocking densities: Salmon and tropical freshwater fish 10 kg/m^3, shrimp 15 $postlarvae/m^2$	Biao (2008)
Fertilization	Using synthetic fertilizers is incompatible	Ahmed et al. (2020)
Feeding	Application of certain feed additives are not allowed	Green (2022)
Disease treatment	Antibiotics and chemo-therapeutic treatments are not permitted	Taşbozan (2023)
Productivity	Annual carp and shrimp productivity should not exceed 1500 kg/ha and 1600 kg/ha, respectively	
Certification	Cost of certification is high for small-scale producers	Lieke et al. (2020)

21.6.2 Organic Aquaculture Production in India

Organic aquaculture is one of the growing sectors in India. The huge coastline and inland water of the country have given much scope for organic prawn and fish farming. Encouragement towards organic aquaculture is being taken by the Marine Products Export Development Authority (MPEDA 2023) by offering financial assistance to the farmers and technical assistance to them. A huge portion of the organic shrimps of India is produced in the coastal region, where organic aquaculture is followed.

An estimated 35,000 metric tonnes of organic aquaculture were produced in India in 2022; 80% of the production was accounted for by organic prawns (MPEDA 2023). In addition, organic fish species such as catfish and tilapia are increasingly being farmed, especially in Tamil Nadu and Andhra Pradesh. The main export markets for organic aquaculture products from India are Japan, the United States, and the European Union.

21.7 Organic Aquaculture and the Standards

The history of organic aquaculture started with the principles of organic farming and up to these days has maintained influences from them (IFOAM EU Group 2010; Ahmed et al. 2020). Organic aquaculture is a broad approach which focuses on management techniques and practices that sustain and improve ecosystem quality. Generally speaking, organic production requires strict production standards intended to actualise ecologically, socially, and commercially sustainable production methods. As a result, items made organically are different from those made commercially. Natural or conventional production techniques are the foundation of organic farm management. Similar to organic agricultural systems, the foundation of organic aquaculture is ecology and the health of humans and animals (Rani et al. 2025; Nisha et al. 2025; Kiran et al. 2025; Ahmed et al. 2020). As a result, an ecologically responsible, sustainable aquaculture strategy that doesn't rely on artificial raw materials is known as organic aquaculture.

The first known organic aquaculture standard for the carp species, *Cyprinus carpio*, was created in Austria in the beginning of the 1990s. Through the Organic Salmon Project, this made its way to Ireland. The first national general criteria pertaining to goods from organic fisheries were developed in France and the UK in 2000. Moreover, IFOAM drafted the first-ever worldwide organic aquaculture guidelines in the year 2000. Led by a team consisting of Lembo et al. 2019 the publication explains the organic aquaculture practices adopted by various countries (2007). Five years later, Council of Europe provided legal framework to the sector of organic aquaculture by creating regulation on quite sustainable organic farming practices for the region under EC No 884 of year 2007.

Regulation (EC) No 834/2007, (FIBL and IFOAM 2022), deals with the goals and underlying principles exceeded this 2017 year. Case of Organic Farming is a regulation of the Organic Production, labelling and Marketing of Organic Produce being Sisal products acts but also foreign trade of organic commodities.

At the beginning, the provision for organic aquaculture had been covered in a separate Commission rule (Commission Regulation EC No 889/2008) of which aquaculture was not part. Therefore, to replace the previous regulation, regulation EEC 710/2009, provision of organic aquaculture was also executed. In that way, the integrated schemes for the biological production of fish and the harvesting of marine algae were approved. That is, regulation EEC 889/2008 was not only updated but also overridden after 2009 by some Commission regulations which were particularly concerned with aquaculture, such as: Regulation No 505/2012, Regulation No 1030/2013, Regulation No 1364/2013, and Regulation No 1358/2014, concerning 673 of 2016 and 838 of 2017 and finally 1584 of 2018 (Busacca and Lembo 2019).

21.8 Opportunities in Organic Aquaculture

Demand for fish is growing. There are concerns, which are reasons for increased consumption of fish: the popularity of organic foods, the decrease in catch rates of fisheries, and concerns over the safety of wild fish for human consumption and for the environment. "Organic aquaculture"is now in the eyes of the customers. China dominates the market of organic aquaculture with 77%, Europe has 22%, and Indonesia, Vietnam, Brazil, and Thailand come afterwards. Common fish types include carp, prawns, salmon, and mussels. Although this is a specialized industry, the growth witnessed in this industry was at 8% in 2015. This could perhaps address some of the problems in aquaculture. The selling price of some species is lower than that of the production. The world prices are decreasing due to increased productivity and the shift in land use.

Moreover, all the above-mentioned instruments—ponds, cages, raceways, net pens, and RAS—can be effectively applied in organic aquaculture. Even though it has an extremely high degree of "non-nature proximity" which renders it unsuitable for organic aquaculture standards, it is still possible to refer to RAS as organic system which provides almost all fish with a healthy and secure environment (Kerr and Potthast 2018; Meisch and Stark 2019). As more of Naturland (2020) asserts, the inclusion of other elements such as livestock, grains, and vegetables into organic aqua farming, it implies earthy and extremely laborious inter-cultivation practices.

Since polyculture makes far better use of the natural food webs and space, organic fish farming may support monoculture (Xie et al. 2013). Through aquaculture, polyculture diversifies species (Naturland 2020). Organic farming aims to prevent organic aquaculture activities from affecting the surrounding ecosystems. As a result, a number of agricultural standards have been established, such as utilising native species with organic fry raised in hatcheries and moving from wild to organic broodstock.

In organic aquaculture, up to a certain extent participants can gather wild post larval shrimp or broodstocks into fish and shrimp hatcheries, only that ponds are allowed to refill non drenched. The natural reproduction as opposed to induced with antibiotics or pituitary hormones, is organic aquaculture, as the hatchery system is unable to support them. Organic aquaculture is also, confined from incorporating organisms that have been genetically modified such as prepared feeds (Naturland 2020). Overstocking leads to misuse of eco-system resources (Censkowsky and Altena 2013). Such as the quality of water and respect for ecological cycles in order for fishes to feed, spawn, and grow must be upheld (Naturland 2020). The transition from conventional to organic aquaculture carries with it complexities especially in matters related to food safety, issues related to animal the care of animals, environmental and economic factors. Employing suitable feeds and organic manures enhances plankton production in fish for extensive and semi-intensive systems.

Raising *Macrobrachium rosenbergii*, *Penaeus monodon*, and *Litopenaeus vannamei* at organic aquaculture shall not exceed densities of 1600 kg/ha/year (Naturland 2020). According to Stanciu et al. (2015), organic carp can reach a production rate of 1500 kg/ha on an annual basis. For the low intensity production scale, as less scale production will like accepted in, organic aquaculture will not be widely practiced. Over 90 countries have adopted organic guidelines (Gould et al. 2019). Other than that, because organic systems have strict limitations for the controlled markets, it is essential for the advancement of organic aquaculture to involve active programs of third party certifications with trustworthy organic brands. Organic aquaculture involves a farm to market third party verification much the same as organic horticulture.

Organic aquaculture is capital intensive and thus small holder farmers are likely to face these challenges. The difficulty is in generating financial difficulties (Ha et al. 2012). The Global South finds it challenging to adopt third party certification of origin as it is premised on Western techno-scientific norms and ideals (Hatanaka 2010). Marschke and Wilkings (2014) suggest that smaller producers might fall short of aquaculture norms. It is now encouraged that small scale industries in developing countries may be able to overcome such challenges through group certification. The group certification scheme should be encouraged among the small scale organic aquaculture producers primarily to reduce certification and administrative costs, enhance domestic local registration networking, maintain fair competition in the local markets, and make retail markets less congested. There has to be a thorough understanding of this and proper integration into how group certification is carried out thanks to the concerted efforts of organization, business and other parties undertaking group certification.

21.9 Organic Aquaculture Practices and Challenges

Accessories such as raceways, cages, ponds, net-pens, tanks, and RAS are all applicable in organic aquaculture (Angel et al. 2019). However, in terms of animal welfare and biosafety, much as RAS doesn't promote 'closeness to nature' (Kerr and Potthast 2018)—a higher fish stocking RAS could still be looked at in organic production. (Stark and Meisch 2019). This is in spite of the fact that organic fish production monoculture can be practiced i.e. only one species cultivated, in reality, two or more species are advisable. Polyculture is one of the methods used to diversify aquaculture species, and is therefore promoted (Naturland 2020).

The core objective of organic farming is to ensure that the activities conducted in organic aquaculture farms do not harm the surrounding ecosystems. Therefore, several guidelines have been formulated to outline many aspects of farming activity. The best stock has to be natural native species and organic fry that are brought up in hatcheries (Naturland 2020). Organic broodstock is therefore recommended as an alternative source to wild broodstocks (IFOAM EU Group 2010). It is, however also acceptable under organic aquaculture to use wild caught prawn post larvae and capture wild broodstocks for fish and prawn hatchery operations if the input is passive when ponds are refilled.

Furthermore, because hatcheries rely on natural reproduction, organic aquaculture does not support hatchery operations that employ pituitary hormone and antibiotics. Organic aquaculture is prohibited from using GMOs, particularly those found in compounded feed. Since there is a chance of some ecological repercussions, stocking density shouldn't exceed the water body's carrying capacity, according to Centskowsky and Altena, (2013). For fish to feed, reproduce, nidify, and grow-out, biological processes and water quality must be preserved (Naturland 2020). The main obstacles to switching from conventional to organic aquaculture are outlined in Table 21.4.

21.10 Environmental Sustainability of Organic Aquaculture

Positive effects on the Organic aquaculture have various benefits (see Fig. 21.1). The IFOAM EU Group (2010) sees organic fish farming as an environmentally friendly management approach that enhances and protects aquatic ecosystems. In addition, it alleviates the negative impacts on the environment that are caused by the practice of intensive production, to a large extent. Organic management is aimed at enhancing biodiversity, unlike the practices of the conventional farming system (Tuck et al. 2014). When compared to conventional farming, organic management has been shown to have a favourable effect on biotic abundance and richness; on average, species richness and organism abundance increase by 30% and 50%, respectively (Smith et al. 2019). In organic aquaculture, producers must do all in their power to preserve biodiversity (Naturland 2020).

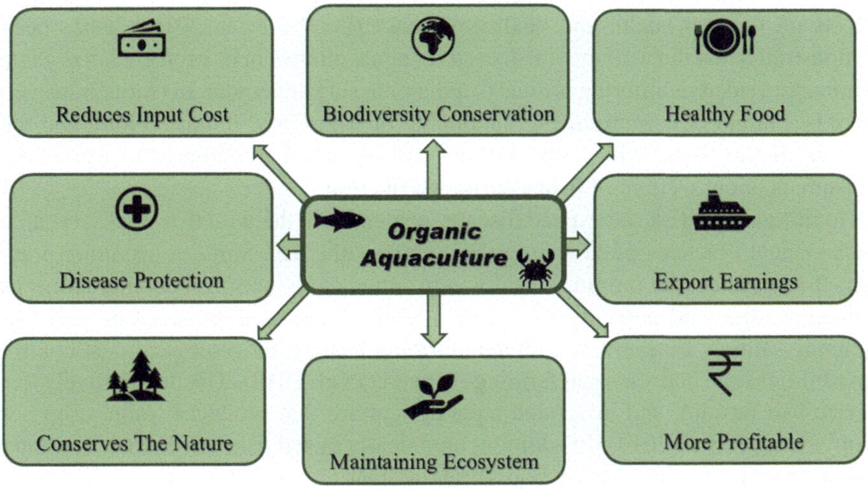

Fig. 21.1 Benefits of organic aquaculture. *Source* Author

Centroskowsky and Altena (2013) have noted that green-pond dikes which are afforested are crucial for maintaining organic fish farming and contribute rapidly to the local biodiversity as well. Organic agriculture, in contrast to traditional cropping practices, has improved soil fertility and water quality, nutrient management, and reduced the dependency on fossil energy sources (Reganold and Wachter 2016). As a result, it offers a variety of ecological functions. It is therefore frequently more sustainable than traditional farming as it lowers environmental externalities (Reganold and Wachter 2016). For instance, it has been demonstrated that organic farming enables a 97% reduction in the use of pesticides and a 34%–53% reduction in the usage of fertiliser and energy.

Organic farming obviously has a lower overall environmental effect and improves the environment when less fertiliser and pesticide are used. Compared to conventional farming, organic farming appears to have reduced greenhouse gas (GHG) emissions per unit land area (Skinner et al. 2014). Research has proven that adding more organic matter inputs in organic farming systems stimulates carbon sequestration. In Spain, organic agriculture led to a decrease of 56% of the greenhouse gases emitted per land unit, and 39% for every product unit, compared to the traditional agriculture practices. Moreover, it has been proved that the same applies to organic aquaculture.

Traditional intensive aquaculture uses antibiotics and fertilisers, which increases the danger of antibiotic resistance and contaminates water (Brunton et al. 2019). Given that it reduces water pollution and preserves water quality, organic farming is preferable to conventional production in this regard. However, chemical-free organic fish farming has less of an adverse effect on water quality because it depends on clean water (IFOAM EU Group 2010). Organic fish culture attempts to shield fish from injury, environmental threats, parasite infestations, and disease outbreaks by using minimum fish stocking. This is advantageous for the preservation and improvement of water quality as well as animal welfare (Bergleiter and Meisch 2015).

It is believed that low stocking density is appropriate because of its beneficial effects on the fish health and quality of water (Lembo et al. 2018). It has been demonstrated that the welfare of fish used for aquaculture can be protected in organic farming provided monitoring is done to adherer to set parameters. In reality however, there is a threat of compromising the quality of water used in aquaculture and the welfare of the fishes with high stocking density which mimmics the worries of consumers about welfare of fishes during production.

To mitigate the risk of pest and disease incidences in the wild fish stocks, organic management practices often restrict the escape of the fish. Some aquaculture practices, for instance, the farming of salmon in submersible nets, has distinct issues as it threatens the wild population of salmon with diseases as a result of the farming escaping salmon. Furthermore, organic farming has greater biological pest control possibilities than conventional farming (Muneret et al. 2018). This imperatively ties to rice-fish farming and integrated aqua agriculture that produces 'more crop per drop' (Ahmed et al. 2014). In addition, organically raised fish, due to their breeding techniques are healthier and more disease resistant.

21.11 Confronting Environmental Challenges

As a result of the insufficient understanding of additional ecological factors, farmers from developing countries tend to practice aquaculture in an organic way but without pesticides (FAO 2019a, 2019b). This leads to the engagement of these farmers in practices which are not friendly to the environment. For example, in Vietnam integrated mangrove shrimp farms are mostly farmed in an organic way by farmers who do not have any certification. While farmers may start importing post larvae of prawn from certified organic farms thus beginning the process of migrating from non-certified organic aquaculture to certified organic aquaculture, this does not assure an improvement of the aquaculture activities and their associated environmental degradation (Jonell and Henriksson 2015). Moreover, certain organic aquaculture regulations (as in the case of salmon farming) were also incapable of addressing the harmful impacts of ocean acidification, water eutrophication, and marine toxic pollution.

It is quite clear that the already recognized organic aquaculture standards can be further harmonized and more efforts within the producers, the consumers, and the market participants would assist the growth of the business exponentially. However, this means a qualitatively new level for organic aquaculture—for consumers' protection, for better environmental conditions and for benevolence towards farmers. Of course, organic certification also requires particular feed. However, life cycle assessment studies have shown that often organic salmon farming does not actually decrease the environmental burden associated with aqua feed production, including oil and meal and other ingredients derived from crops. The issue of deploying different kinds of feed for organic aquaculture is however of paramount importance, particularly for ecological reasons (Mente et al. 2019).

Usage of plant protein and fat in feeding of salmon as there is a shortage of fishmeal and fish oil have became common in aquaculture. Carnivorous fish, such as salmon, require diets that are highly lipid, fatty acid, and amino acid rich. This technique can be applied to other amino acids such as histidine or compounds obtained from fermenting food micronutrients from naturally occurring bacteria not derived from genetic modification. This technique has been recognized by the European legislation.

Organic aquaculture proposes additional feed, the single cell proteins (SCP), which are proteins produced from microbial and algal cells and organic soluble fish protein concentrate (Sharma et al. 2024; Sharma et al. 2025; Ritala et al. 2017; Mente et al. 2019). There are some species-specific alternative protein sources that could be adopted by organic aquaculture, such as processed animal protein (PAP) with organic inclusions in the diets containing up to 45–90% of protein sourced from organic agriculture-based insect meals.

It is further evident that organic feeds are more expensive to produce, have limited availability, and therefore carry the entire burden of certifying all their ingredients. Culture of organic farming encourages utilizing more land in a unit area for cropping, but not for a unit product thus tempering agriculture with aquaculture. It offers clean, healthy nutrition without any synthetic chemicals in fertilizers, pesticides or otherwise, thus heightening interior design; (Pacanoski 2009).

Due to the extra space and labor associated with generating human-edible food, organic agriculture is less popular than conventional farming. Assimilating organic farming practices will pose a threat to the non-agricultural land, in attempt to sustain conventional agricultural crop yields. This will elevate the worries of biodiversity and forest cover loss and reduce the merits of organic agriculture. This begs the question of if it is really feasible to convert a significant portion of land to organic farming as this could create a negative impact on the environment. A more sensible approach would be to integrate organic farming concepts into conventional agriculture instead.

In view of the fact that organic techniques have higher carbon stocks and sequester more carbon than conventional terrestrial agriculture, this approach is said to decrease the prospects of global crop production due to increased deficit in output per acre of land cultivated (Gattinger et al. 2012; Skinner et al. 2014). Organic farming cannot become a serious weapon to combat global warming through soil carbon storage, as it requires converting additional carbon-rich land into cultivated land to produce food for the population of the world (Leifeld et al. 2013). The changing environment and rising human numbers necessitate the adoption of a more environmentally friendly agricultural system that is less.

21.12 Limitations of Organic Aquaculture

Farmers in developing countries engage in nonorganic aquaculture practices without pesticides, which help protect ecosystems, however, they are involved in environmental degrading practices because they do not understand other limits of the ecosystem (FAO 2019a, 2019b). In Vietnam, developers of mangrove swamps often raise the shrimp Tarabaevi without any proper documentation. In the case of organic conversion of shrimp farming practice, substituting organic prawn post larvae, for instance, is the first step, but does not guarantee better aquaculture practices and environmental impacts (Jonell and Henriksson 2015). Salmon culture was subjected to one of the standards for organic aquaculture but it did not stop acidifying or eutrophying or to the ecotoxicity of the surrounding marine environment (Pelletier and Tyedmers 2007). In the case of Germany, the organic farmed carp systems may be more acidic and more eutrophic than the conventional makes (Biermann and Geist 2019). If the organic aquaculture standards are harmonised and the relationships of the producers with the consumers and the other actors of the market strengthened, the development of organic aquaculture will be possible (Mente et al. 2011).

A life cycle analysis showed that organic farming of salmon cannot mitigate the negative effects of harvesting fish oil, fish meal or any other organic aqua feed applicable to more than aquaculture of fish (Pelletier and Tyedmers 2007). Due to ecological issues, organic aquaculture has to have different feed (Mente et al. 2019). Even though oil is scarce, salmon farms can use both plant protein and oil in diet for the fish (Olesen et al. 2011). Carnivorous fish require high levels of lipids, amino acids and fatty acids in their diet. Most fish protein goes into soluble concentrates and single cell protein (SCP) produced from microorganisms and algae which are hostile to organic farming (Ritala et al. 2017; Mente et al. 2019).

Diets including insects and "processed animal protein" (PAP), with protein levels reaching up to 90% (Mente et al. 2019). Organic aquaculture may in turn make use of species-specific protein ingredients that have been recently developed. Within this context, there is a possibility of certifiable feed production, although the uhc is still in the experimental stage and is expensive and inaccessible compared to conventional non-organic certified feed. Certifying organic aquaculture may call to question some of the environmental benefits of organic agriculture to aquaculture because organic agriculture believes in the environmental advantages per unit area but not per unit productivity. Also, since organic farming does not apply any chemical pesticides or fertilizers, it cannot guarantee consumers availability of safe and sufficiently nutritious foods (Pacanoski 2009). More land and effort is required in organic production of food for human consumption as compared to the alternative approach of conventional farming (Floros et al. 2010). If more opted for organic farming, there will be little land that will remain unfarmed for other purposes such as ecosystems and wildlife habitats. The problems associated with deforestation and loss of biodiversity will outweigh the benefits that organic farming brings (Trewavas 2001; Pacanoski 2009).

As it is not necessary to excessivly organicise cultivation of all zones of agriculture, which as a result of large scale buffering of the land may in fact enhance the environment, it is recommended to assimilate the principles of organic farming into the processes of farming. As farming resolves larger carbon stocks and carbon sink, organic farming is anticipated to have a less output of agricultural produce in the world relative to its land area yield (Gattinger et al. 2012; Skinner et al. 2014).

The situation would improve in carbon emissions below level with organic farming diversity practices as the latter introduces some level of oils corbon sequestration in the soils. This is although exacerbating problems of global warming and not alleviating them.

21.13 Conclusion

In light of the fact that production from catch fisheries has remained constant, aquaculture needs to suffice in boosting the supply of global fish and seafood to cater for the growing food needs of the population. However, aquaculture is not without its challenges and the current production of fish and seafood is bound to face many challenges over time. As a response to the existing environmental limitations of contemporary aquaculture, organic aquaculture has been developed. Organic aquaculture is a management system that contrasts with industrial fish farming in that it is focused on the health of ecosystems, animals, people, and their relationships with one another. Transitioning to organic aquaculture offers numerous environmental benefits, including improved input and output, healthy soil, clean water, infection resistance, and preservation of bio-diversity and ecosystem. However, making the transition from inorganic to organic aquaculture is a challenging process with several challenges. It posits that what the future holds for aquaculture is that the level of production should be improved sustainably. In order to mitigate environmental damage, there has to be an increase in fish production through the sustainable rehabilitation of integrated aquaculture and organic systems. There is need for research in practice in order to find out how much the impact of using other types of feeds, reducing the environmental burden of feeds, lessening the costs of certification and other administrative processes. Organic aquaculture of high productive capacity should aim at providing healthy above organic food and at contributing to sustainability and food security.

21.14 Future Prospects

The future of organic aquaculture holds great potential for ensuring the long-term viability of global aquaculture systems. As worries about environmental degradation, chemical usage, and food safety rise, organic aquaculture emerges as a feasible alternative that meets the needs of eco-friendly and health-conscious consumers. Organic techniques emphasise the use of natural inputs, the avoidance of synthetic chemicals and antibiotics, and the upholding of high animal welfare and environmental standards. These principles not only assist to safeguard aquatic ecosystems from pollution and overexploitation, but they also encourage biodiversity by encouraging sustainable feed sources and minimising pressure on wild fish stocks. Furthermore, organic aquaculture systems often incorporate holistic techniques including polyculture, low-density stocking, and the utilisation of renewable resources, all of which improve ecosystem resilience. As consumer knowledge grows, market demand for organic seafood is likely to climb, pushing more producers to follow organic standards and practices. Government regulations, certification programs, and international collaboration will all play an important role in developing and promoting the growth of this industry. Investment in R&D is also critical, particularly in optimising organic feed formulations, disease prevention without chemicals, and increasing the cost-effectiveness of organic operations. Furthermore, educating stakeholders—farmers, consumers, and policymakers—on the benefits of organic aquaculture will help to increase its adoption and implementation. Although challenges remain, such as higher production costs, limited availability of certified organic inputs, and the need for consistent global standards, the long-term environmental and socioeconomic benefits make organic aquaculture an important step towards a more sustainable and responsible aquaculture industry. With joint efforts and new solutions, organic aquaculture has the potential to become a mainstream practice that not only serves rising seafood demand but also protects the environment for future generations.

References

Adámek Z, Linhart O, Kratochvíl M, Flajšhans M, Randák T, Policar T, Masojídek J, Kozák P (2012) Aquaculture The Czech Republic in 2012: modern European prosperous sector based on thousand-year history of pond culture. Aquac Eur 37:5–14

Ahmed N, Thompson S, Turchini GM (2020) Organic aquaculture productivity, environmental sustainability, and food security: insights from organic agriculture. Food Secur:1253–1267. https://doi.org/10.1007/s12571-020-01090-3

Ahmed N, Ward JD, Saint CP (2014) Can integrated aquaculture-agriculture (IAA) produce "more crop per drop"? Food Security:767–779

Angel D, Jokumsen A, Lembo G (2019) Aquaculture production systems and environmental interactions. In: Organic aquaculture: impacts and future developments, pp 103–118

Beg, MM, Roy SM, Kar A, Mukherjee CK, Bhagat SK, Tanveer M (2024). Study on recirculating aquaculture system (RAS) in organic fish production. In IOP conference series: earth and environmental science Vol. 1391, No. 1, p. 012013). IOP Publishing. https://doi.org/https://doi.org/10.1088/1755-1315/1391/1/012013

Bergleiter S, Meisch S (2015) Certification standards for aquaculture products: bringing together the values of producers and consumers in globalised organic food markets. J Agric Environ Ethics:553–569

Biao X (2008) The development of and prospects for organic aquaculture worldwide. Outlook Agric 37:255–260. https://doi.org/10.5367/000000008787167718

Biermann G, Geist J (2019) Life cycle assessment of common carp (Cyprinus carpio L.)–a comparison of the environmental impacts of conventional and organic carp aquaculture in Germany. Aquaculture 501:404–415. https://doi.org/10.1016/j.aquaculture.2018.10.019

Biswas G, Kumar P, Kailasam M, Ghoshal TK, Bera A, Vijayan KK (2019) Application of integrated multi trophic aquaculture (IMTA) concept in Brackishwater ecosystem: the first exploratory trial in the Sundarban, India. J Coast Res 86(SI):49–55

Boyd DE, Tucker C, McNevin A, Bostick K, Clay J (2007) Indicators of resource use efficiency and environmental performance in fish and crustacean aquaculture. Rev Fish Sci 15:327–360. https://doi.org/10.1080/10641260701624177

Brunton LA, Desbois AP, Garza M, Wieland B, Mohan CV, Häsler B, Tam CC, Le PNT, Phuong NT, Van PT, Nguyen-Viet H (2019) Identifying hotspots for antibiotic resistance emergence and selection, and elucidating pathways to human exposure: application of a systems-thinking approach to aquaculture systems. Sci Total Environ 687:1344–1356. https://doi.org/10.1016/j.scitotenv.2019.06.134

Busacca E, Lembo G (2019) EU regulation on organic aquaculture. Org Aquac: Impacts Futur Dev:23–39. https://doi.org/10.1007/978-3-030-05603-2_2

Censkowsky U, Altena A (2013) Scoping study on organic aquaculture in 5 east African countries. International Federation of Organic Agriculture Movements, Bonn

China Aquaculture Association. (2023). Annual report on organic aquaculture production in China. https://www.china-aquaculture.cn

Chopin Thierry (2010) Integrated multi-trophic aquaculture, pp 195–217

Chopin T, Robinson SMC, Troell M, Neori A, Buschmann AH, Fang J (2008) Integrated multi-trophic aquaculture (IMTA)—a responsible practice for sustainable aquaculture. World Aquac 39(3):27–30

Clarke R, Bostock J (2017) Regional review on status and trends in aquaculture development in Europe-2015. Food Agric Organ United Nations

Cubillo AM, Ferreira JG, Robinson SM, Pearce CM, Corner RA, Johansen J (2016) Role of deposit feeders in integrated multitrophic aquaculture—a model analysis. Aquaculture 453:54–66

Dalsgaard J, Lund I, Thorarinsdottir R, Drengstig A, Arvonen K, Pedersen PB (2013) Farming different species in RAS in Nordic countries: current status and future perspectives. Aquac Eng 53:2–13. https://doi.org/10.1016/j.aquaeng.2012.11.008

Danilo G, Vairo D, Solfanelli F, Zanoli R (2019) Economic performance of organic aquaculture: a systematic review. Mar Policy 108:103542. https://doi.org/10.1016/j.marpol.2019.103542

Dube K, Thongam C I (2012) Organic aquaculture: way to sustainable production. Adv Fish Res:219–229

Edwards P (2009) Traditional Asian aquaculture. In: Baird DJ et al (eds) Aquaculture in the ecosystem. Springer, pp 29–52

EU Commission (2018) Regulation (EU) 2018/848 on organic production and labelling of organic products. Off J Eur Union

EUMOFA (2020) Recirculating aquaculture Sstems. Eur Mark Obs Fish Aquac Prod:45

European Commission (2023) Organic Aquaculture in the EU: A Growing Industry. https://www.ec.europa.eu

Union E (2021) Organic production and labelling of organic products in aquaculture. EU Org Aquac Regul

EU-Water-Frame-Directive (2000) Ditective 2000/60/EC of the European Parlement and of the council of 23 Octover 2000 establishing a framwork for community action in the field of water policy, OJ L, vol 327, pp 1–40

Fang J, Zhang J, Xiao T, Huang D, Liu S (2016) Integrated multi-trophic aquaculture (IMTA) in Sanggou Bay, China. Aquac Environ Interact 8:201–205

FAO (1997) Aquaculture development. FAO technical guidelines for responsible fisheries. 1020–5292, no 5. FAO., Fisheries Department, Rome, p. 40

FAO (2018) The state of world fisheries and aquaculture 2018—meeting the sustainable development goals, Rome

FAO (2019a) Organic Agriculture. Food and Agriculture Organization of the United Nations, Rome. http://www.fao.org/organicag/oafaq/oa-faq/en/. Accessed 10 Jan 2019

FAO (2022a) The state of world fisheries and aquaculture 2022. Towards blue transformation. Rome, pp. 19–20 & 80

FAO 2022b. FAO fishery and aquaculture statistics, 2022 yearbook. https://www.fao.org/3/cc0461en/cc0461en.pdf. Accessed 15 Dec 2022

FAO (2019b) Organic agriculture. Food and Agriculture Organization of the United Nations, Rome. www.fao.org/organicag/oafaq/oa-faq/en/. Accessed 10 Jan 2019

FiBL and IFOAM (2022) The world of organic agriculture, statistics, emerging trends 2022. http://www.organic world.net/yearbook/yearbook-2022.html. Accessed 15 Dec 2025

Floros JD, Newsome R, Fisher W, Barbosa-Cánovas GV, Chen H, Dunne CP, German JB, Hall RL, Heldman DR, Karwe MV, Knabel SJ (2010) Feeding the world today and tomorrow: the importance of food science and technology: an IFT scientific review. Compr Rev Food Sci Food Saf 9:572–599

Folke C, Kautsky N (1992) Aquaculture with its environment: prospects for sustainability. Ocean Coast Manag 17:5–24

Food and Agriculture Organization (2023) World aquaculture production report 2022. https://www.fao.org

Gattinger A, Muller A, Haeni M, Skinner C, Fliessbach A, Buchmann N, Mäder P, Stolze M, Smith P, Scialabba NEH, Niggli U (2012) Enhanced top soil carbon stocks under organic farming. Proc Natl Acad Sci 109:18226–18231. https://doi.org/10.1073/pnas.1209429109

Gopakumar G et al (2019) CMFRI technology—integrated multi-trophic aquaculture (IMTA). In: CMFRI Booklet Series No. 11/2019. https://eprints.cmfri.org.in/13366/

Gould D, Compagnoni A, Lembo G (2019) Organic aquaculture: principles, standards and certification. Impacts and future developments, Organic aquaculture, pp 1–22

Granada L, Sousa N, Lopes S, Lemos MF (2016) Is integrated multitrophic aquaculture the solution to the sectors' significant challenges? –a review. Rev Aquac 8(3):283–300

Green BW (2022) Fertilizer use in aquaculture. In feed and feeding practices in aquaculture. Woodhead Publishing, pp 29–63. https://doi.org/10.1016/B978-0-12-821598-2.00012-6

Grigorakis K, Rigos R (2011) Aquaculture effects on environmental and public welfare—the case of Mediterranean mariculture. Chemosphere 855:899–919. https://doi.org/10.1016/j.chemosphere.2011.07.015

Ha TTT, Bush SR, Mol APJ, van Dijk H (2012) Organic coasts? Regulatory challenges of certifying integrated shrimpmangrove production systems in Vietnam. J Rural Stud 28:631–639. https://doi.org/10.1016/j.jrurstud.2012.07.001

Hatanaka M (2010) Governing sustainability: examining audits and compliance in a third-party-certified organic shrimp farming project in rural Indonesia. Local Environment15:233–244. https://doi.org/https://doi.org/10.1080/13549830903575588

Hurtado AQ, Critchley AT, Neish IC (2015) Tropical seaweed farming trends. Elsevier, Problems and Opportunities. In Seaweed Sustainability

IFOAM (2014) The IFOAM norms for organic production and processing. International Federation of Organic Agriculture Movements, Bonn

IFOAM EU Group (2010) Organic aquaculture: EU regulations (EC) 834/2007, (EC) 889/2008, (EC) 710/2009: background, assess ment, interpretation. IFOAM EU Group, Brussels

International Union for Conservation of Nature (2007) Guide for the sustainable development of Mediterranean aquaculture. Interaction between aquaculture and the environment. IUCN, Gland. https://cmsdata.iucn.org/downloads/acua_en_final.pdf. Accessed 27 Sep 2012

Johnson B, Nazar AKA, Jayakumar R, Tamilmani G, Sakthivel M, Ramesh Kumar P, Anikuttan KK, Samal AK, Sankar M (2019) Integrated multi-trophic aquaculture (IMTA): a technology for uplifting rural economy and bio-mitigation for environmental sustainability. Aquac Spectr 2(5):9–18

Jokumsen A, Svendsen LM (2010) Farming of freshwater rainbow trout in Denmark., DTU Aqua Report no. 219-2010. DTUAqua- Technical University Denmark, pp 4–43

Jonell M, Henriksson PJG (2015) Mangrove–shrimp farms in Vietnam—comparing organic and conventional systems using life cycle assessment. Aquaculture 447:66–75.https://doi.org/10.1016/j.aquaculture.2014.11.001

Kerr M, Potthast T (2018) As close as possible to nature: possibilities and constraints for organic aquaculture systems. In Professionals in food chains. Wageningen academic publishers, pp 279-289. https://doi.org/10.3920/978-90-8686-869-8

Kibria ASM, Haque MM (2018) Potentials of integrated multitrophic aquaculture (IMTA) in freshwater ponds in Bangladesh. Aquac Rep 11:8–16

Kiran, Kumar N, Sindhu S, Gawaria B, Ansari A, Dhariwal A, Sharma P, Kumar (2025). GWAS approaches in medicinal plants for enhanced secondary metabolite production. In J. T. Chen (Ed.), Biotechnology, multiple omics, and precision breeding in medicinal plants (pp. 119–134). CRC Press. https://doi.org/https://doi.org/10.1201/9781003475491-10

Korinek V, Fott J, Fuksa J, Lellák J, Prazakova M (1987) Carp ponds of Central Europe. In: IN: managed aquatic ecosystems. Ecosystems of the world, 29. Elsevier Science Publishing Co., New York, pp 29–62

Kumar R, Yadav R, Gupta RK, Kataria SK, Kadyan P, Kaur S (2023) The earthworms: Charles Darwin's ecosystem engineer. In: Hakeem KR (ed) Organic Fertilizers. Intech Open, pp 1–7. https://doi.org/10.5772/intechopen.1001339

Kumar R, Singh D (2022) Insecticide- tolerant bacterial population in *Eisenia fetida*'s gut and Vermicast exposed to Chlorantraniliprole and Fipronil. Appl Biol Res 24(3):273–279

Kumar R, Gupta R, Dahiya T, Yadav R, Saifi R, Yodha K, Kumari B, Singh P, Shefali. (2022a) Surveillance of *Heteropneustes fossilis* (stinging catfish) diseases and health management practices in Haryana. Appl Ecol Environ Sci 10(4):261–267. https://doi.org/10.12691/aees-10-4-12

Kumar R, Gupta RK, Yadav R, Saifi R, Yodha K, Pooja. (2022b) *Eisenia fetida* as protein source for growth enhancement of *Heteropneustes fossilis*. Egypt J Aquat Biol Fish 26(2):577–588. https://doi.org/10.21608/ejabf.2022.234553

Kumar R, Yadav R, Kumar Gupta R, Pooja (2022c) A sustainable way for fish health management by replacement of chemical and drugs by earthworm. In: Environmental degradation in Asia:

land degradation, environmental contamination, and human activities. Springer International Publishing, Cham, pp 329–352. https://doi.org/10.1007/978-3-031-12112-8_16

Leifeld J, Angers DA, Chenu C, Fuhrer J, Kätterer T, Powlson DS (2013) Organic farming gives no climate change benefit through soil carbon sequestration. Proc Natl Acad Sci 110:E984–E984. https://doi.org/10.1073/pnas.1220724110

Lembo G, Jokumsen A, Spedicato MT, Facchini MT, Bitetto I (2018) Assessing stakeholder's experience and sensitivity on key issues for the economic growth of organic aquaculture production. Mar Policy 87:84–93. https://doi.org/10.1016/j.marpol.2017.10.005

Lembo G, Mente E (2019) Organic aquaculture. Springer International Publishing, Cham, Switzerland

Lieke T, Meinelt T, Hoseinifar SH, Pan B, Straus DL, Steinberg CE (2020) Sustainable aquaculture requires environmental-friendly treatment strategies for fish diseases. Rev Aquac 12:943–965. https://doi.org/10.1111/raq.12365

Marine Products Export Development Authority (MPEDA) (2023). India's organic aquaculture progress report. https://www.mpeda.gov.in

Marschke M, Wilkings A (2014) Is certification a viable option for small producer fish farmers in the global south? Insights from Vietnam. Mar Policy 50:197–206. https://doi.org/10.1016/j.marpol.2014.06.010

Meisch S, Stark M (2019) Recirculation aquaculture systems: sustainable innovations in organic food production? Food Ethics 4:67–84. https://doi.org/10.1007/s41055-019-00054-4

Mente E, Jokumsen A, Carter CG, Antonopoulou E, Tacon AG (2019) Nutrition in relation to organic aquaculture: sources and strategies. In: Organic aquaculture: Impacts and future developments, pp 141–188. https://doi.org/10.1007/978-3-030-05603-2_8

Mente E, Karalazos V, Karapanagiotidis IT, Pita C (2011) Nutrition in organic aquaculture: an inquiry and a discourse. Aquac Nutr 17:e798–e817. https://doi.org/10.1111/j.1365-2095.2010.00846.x

MPEDA (2022) The marine products export development authority. (Ministry of Commerce & industry). Government of India, pp 37–41

Muneret L, Mitchell M, Seufert V, Aviron S, Djoudi EA, Pétillon J, Plantegenest M, Thiéry D, Rusch A (2018) Evidence that organic farming promotes pest control. Nat Sustainability 1:361–368. https://doi.org/10.1038/s41893-018-0102-4

Nath K, Munilkumar S, Patel AB, Kamilya D, Pandey PK, Sawant PB (2021) *Lamellidens* and *Wolffia canopy* improve growth, feed utilization, and welfare of *Labeo rohita* (Hamilton, 1822) in an integrated multi-trophic freshwater aquaculture system. Aquaculture 534:7360207

National Oceanic and Atmospheric Administration (NOAA) (2023) U.S. organic aquaculture: market insights. https://www.noaa.gov

Naturland (2020) Naturland standards for organic aquaculture. Naturland, Gräfelfing

Nisha A, Rani S, Kumar S, Kumar R, Kumar A (2025) Recent technologies and strategies for bioremediation of toxic and hazardous waste. In: Dubey KK, Shah MP, Yadav A (eds) Challenges and sustainable solutions in bioremediation. CRC Press, pp 1–16. https://doi.org/10.1201/9781003407317-1

Nizza A (2012) What future for organic aquaculture? J Aquac Res Dev 3:1. https://hdl.handle.net/11588/570859

NRC (National Research Council) (2011) Nutrient requirements of fish and shrimp. The National Academies Press

Olesen I, Myhr AI, Rosendal GK (2011) Sustainable aquaculture: are we getting there? Ethical perspectives on salmon farming. J Agric Environ Ethics 24:381–408. https://doi.org/10.1007/s10806-010-9269-z

Ozgul A, Angel DL (2013) Aggregations of wild fish at net-cage fish farms in the Gulf of AqabaRed Sea; implications for fisheries management and conservation. Aquac Environ Interact 4:135–145

Pacanoski Z (2009) The myth of organic agriculture. Plant Prot Sci 45:39–48

Patra BC, Mitra A, Bhattacharya A (2019) Application of integrated multi-trophic aquaculture (IMTA) concept in Brackishwater aquaculture in Indian Sundarbans. J Coast Res SI 86:43–51

Pelletier N, Tyedmers P (2007) Feeding farmed salmon: is organic better? Aquaculture 272:399–416. https://doi.org/10.1016/j.aquaculture.2007.06.024

Prein M, Bergleiter S, Ballauf M, Brister D, Halwart M, Hongrat K, Kahle J, Lasner T, Lem A, Lev O, Morrison C (2010) Organic aquaculture: the future of expanding niche markets. In farming the waters for people and food. Proceedings of the global conference on aquaculture, pp 549-567

Rani S, Sharma S, Singh D, Sheoran R, Kumar R (2025) Improved production of anti-cancer compounds from medicinal plants through molecular and biotechnological mutagenesis strategies. In: Chen JT (ed) Biotechnology, multiple omics, and precision breeding in medicinal plants. CRC Press, pp 287–297. https://doi.org/10.1201/9781003475491-21

Reganold JP, Wachter JM (2016) Organic agriculture in the twenty-first century. Nat Plants 2:1–8. https://doi.org/10.1038/nplants.2015.221

Ren JS, Stenton-Dozey J, Plew DR, Fang J, Gall M (2012) An ecosystem model for optimizing production in integrated multitrophic aquaculture systems. Ecol Model 246:34–46

Ritala A, Häkkinen ST, Toivari M, Wiebe MG (2017) Single cell protein—state-of-the-art, industrial landscape and patents 2001–2016. Front Microbiol 8:2009. https://doi.org/10.3389/fmicb.2017.02009

Sharma MMM, Kapoor D, Rohilla R, Loyal A, Sharma P, Husen A (2024) Role of essential minerals and rhizosphere in plant under changing environmental conditions. In: Husen A (ed) Essential minerals in plant-soil systems. Elsevier, pp 235–252. https://doi.org/10.1016/B978-0-443-16082-0.00002-3

Sharma P, Kapoor D, Kumar R, Husen A (2025) Rhizosphere and osmolytes coordination in plant systems under adverse situations. In: Husen A (ed) Roles of Osmolytes in changing environment. Academic, pp 101–117. https://doi.org/10.1016/B978-0-443-26549-5.00020-5

Shefali K, Sankhla MS, Kumar R, Sonone SS (2021) Impact of pesticide toxicity in aquatic environment. Biointerface Res Appl Chem 11(3):10131–10140. https://doi.org/10.33263/BRIAC113.1013110140

Shpigel M, Ari TB, Shauli L, Odintsov V, Ben-Ezra D (2016) Nutrient recovery and sludge management in seabream and grey mullet co-culture in integrated multi-trophic aquaculture (IMTA). Aquaculture 464:316–322

Skinner C, Gattinger A, Muller A, Mäder P, Flieβbach A, Stolze M, Ruser R, Niggli U (2014) Greenhouse gas fluxes from agricultural soils under organic and non-organic management—a global meta-analysis. Sci Total Environ 468:553–563. https://doi.org/10.1016/j.scitotenv.2013.08.098

Smith OM, Cohen AL, Rieser CJ, Davis AG, Taylor JM, Adesanya AW, Jones MS, Meier AR, Reganold JP, Orpet RJ, Northfield TD (2019) Organic farming provides reliable environmental benefits but increases variability in crop yields: a global meta-analysis. Front Sustain Food Syst 3:82. https://doi.org/10.3389/fsufs.2019.00082

Soltan MA-H (2016) Aquaculture systems:27. https://www.researchgate.net/publication/308928410

Soto D. (2009) Integrated mariculture - A global review. FAO Fisheries and Aquaculture Technical Paper No. 529

Stanciu S, Radu RI, Vîrlănuță FO (2015) The development of the organic aquaculture. Case study: Romania. SEA–practical application of. Science 3:99–107

Taşbozan O (2023) Organic aquaculture and organic feeds. Turk J Agric-Food Sci Technol 11:774–779. https://doi.org/10.24925/turjaf.v11i4.774-779.5927

Trewavas A (2001) Urban myths of organic farming. Nature 410:409–410. https://doi.org/10.1038/35068639

Troell M, Chopin T, Reid G (2021) Integrated multi-trophic aquaculture: ecological benefits and challenges. Aquac Environ Interact 13:95–110

Tuck SL, Winqvist C, Mota F, Ahnström J, Turnbull LA, Bengtsson J (2014) Land-use intensity and the effects of organic farming on biodiversity: a hierarchical meta-analysis. J Appl Ecol 51:746–755. https://doi.org/10.1111/1365-2664.12219

Viji CS, Chadha NK, Kripa V, Prema D, Prakash C, Sharma R, Jenni B, Mohamed KS (2014) Can oysters control eutrophication in an integrated fish-oyster aquaculture system? J Mar Biol Assoc India 56(2):67–73

Willer H, Lernoud J (2019) The world of organic agriculture. Stat Emerg Trends 2019

Willer H, Schlatter B, Trávníček J (2023) The world of organic agriculture. Stat Emerg Trends 2023:1–358

Xie B, Qin J, Yang H, Wang X, Wang YH, Li TY (2013) Organic aquaculture in China: a review from a global perspective. Aquaculture 414:243–253. https://doi.org/10.1016/j.aquaculture.2013.08.019

Yadav R, Gupta RK, Kumar R, Kaur T, Moun R (2022) Biochemical response of earthworm, *Eisenia fetida* to heavy metals toxicity. J Appl Nat Sci 14(3):990–998

Yılmaz S, Ergün S, Yiğit M, Yılmaz E (2022) An extensive review on the use of feed additives against fish diseases and improvement of health status of fish in Turkish aquaculture sector. Aquac Stud 22. https://doi.org/10.4194/AQUAST710

Yokoyama H (2013) Growth and food source of the sea cucumber *Apostichopus japonicas* cultured below fish cages—potential for integrated multi-trophic aquaculture. Aquaculture 372:28–38

Zamora LN, Yuan X, Carton AG, Slater MJ (2018) Role of deposit-feeding sea cucumbers in integrated multitrophic aquaculture: progress, problems, potential and future challenges. Rev Aquac 10(1):57–74

Zhou Y, Yang H, Hu H, Liu Y, Mao Y, Zhou H, Xu X, Zhang F (2006) Bioremediation potential of the macroalga *Gracilaria lemaneiformis* (Rhodophyta) integrated into fed fish culture in coastal waters of North China. Aquaculture 252(2–4):264–276

Chapter 22
Sustainable Live Feed Production for Enhancing Aquaculture Output

Farhat Jahan and Haroon Ali

Abstract Sustainable production of high-quality live feed is a significant challenge and a key factor for success in aquaculture. This chapter presents a comprehensive, research-driven analysis of good strategies for sustainable live feed production, with the aim of continually improving aquaculture yields while minimizing environmental impacts. Aquaculture is a critical sector for global food security. Its ever-increasing consumption and global growth are dependent on a few key and reliable factors. Successful aquaculture business depends on the availability of nutritionally adequate feed sources for the development of the delicate larval and juvenile stages of aquatic organisms. Live feed has historically played an important role in aquaculture due to its superior nutritional profile, digestibility, and often conventional live feed production methods face limitations related to high cost, nutritional variability, disease risk, and climate change. This book chapter critically discusses the importance of live feed in aquaculture and the challenges associated with conventional production practices. It then presents an analysis of new and emerging technologies for nutritionally sustainable live feed production based on principles such as biofloc technology, use of insect larvae, and single-cell protein integration. Furthermore, this chapter discusses the methods for enhancing the nutritional value of live feed through enrichment and sustainable live feed production in diverse aquaculture systems. Through a number of case studies, the chapter addresses the challenges and future directions for research and development in this important area of aquaculture and highlights the successful implementation of this area. Finally, this chapter highlights the critical role of sustainable live feed production in ensuring long-term viability, environmental responsibility and economic efficiency required for the aquaculture industry. For commercial aquaculture, larvae of some insect species are produced in the early to late juvenile stages. We also introduce important live feeds such as microalgae, rotifers (*Brachionus* spp.), brine shrimp (Artemia spp.), and copepods.

F. Jahan (✉)
Department of Zoology, Hansraj College, University of Delhi, Delhi, India
e-mail: farhatjahan927@gmail.com

H. Ali
Department of Public Health, College of Applied Medical Sciences, Khamis Mushait Campus, King Khalid University (KKU), Abha, Kingdom of Saudi Arabia
e-mail: haalahmad@kku.edu.sa

R. Kumar et al. (eds.), *Sustainability in Aquaculture and Fisheries*, Springer Water, https://doi.org/10.1007/978-3-032-16605-0_22

And we critically review existing studies on the cultivation of these organisms, with an emphasis on innovative and efficient techniques. Additionally, we examine recent advances in nutrition enrichment using sustainable ingredients and methods and discuss the integration of live feed production within more comprehensive aquaculture systems such as integrated multi-trophic aquaculture (IMTA). This chapter draws on specific research findings, including experimental designs and results, to develop robust, biosafe and environmentally friendly live feed production systems that enhance efficiency and sustainability in aquaculture and to demonstrate best practices

Keywords Sustainable production · Live feed · Aquaculture · Nutritional value · Environmental impact · Biofloc technology · Insect larvae · Single cell protein · Cultivation · Aquatic organisms

22.1 Introduction

In global food security, there is an urgent need to adopt environmentally sustainable practices that can meet growing demands without depleting natural resources or causing significant ecological damage (FAO 2022). A critical aspect of this transformation is the reliable and sustainable supply of high-quality feeds, especially live feeds, which are vital for the sensitive larval and early juvenile stages of many economically important farmed aquatic species (Tacon and Metian 2013). Ongoing world-wide research on different taxa shows that live feeds introduced at early stages of development of various finfish, crustaceans and molluscs (Kullman et al. 2021; Yufera and Darias 2007; Shefali et al. 2021) result in better nutrition, higher digestibility and stronger feeding response compared to inert, formulated diets. However, conventional live feed production methods are resource-intensive and pose environmental risks.

This chapter presents a comprehensive synthesis of ongoing research that has shown to increase aquaculture yields by maintaining ecosystem integrity. It also presents an overview of evidence-based in-depth evaluation of sustainable strategies designed for live feed production. We will analyse some of the scientific literature that examines the cultivation, enrichment and integration of live feed organisms required for aquaculture systems. This step towards increasing sustainable production of live feeds, which are vital for enhancing aquaculture productivity, is crucial for the growth and survival of various aquatic species. To increase the production of live feed, we can increase the aquaculture yield by adopting efficient and environmentally friendly methods and reducing environmental impacts. Especially if we talk about the early larval and juvenile stages of aquaculture, it becomes important to pay attention to the important facts given below.

22.1.1 Importance of Live Feed in Larval and Juvenile Stages

Live feed while supporting the healthy growth of aquatic organisms, then it includes microalgae, copepods, rotifers, and Artemia and all these provide important nutrients like protein, lipids and vitamins to the larval and juvenile stages. Providing live feed develops high digestive capacity. Larvae and young of aquatic organisms can easily consume it due to its appropriate size, nutrient composition and motility. Using live feed helps aquatic organisms meet specific dietary requirements which increases survival rates during critical developmental stages. The larvae of most aquaculture species have critical nutritional requirements in their early stages of juvenile development which can only be met through live feed. Live organisms provide a safe natural delivery system of essential nutrients such as protein which is a great source of highly unsaturated fatty acids (HUFAs) such as docosahexaenoic acid (DHA) and eicosapentaenoic acid (EPA) in live feeds for larval fish (Lavens and Sorgeloos 1996; Kumar et al. 2022a, b). Early stages of aquatic organisms' ontogeny in their digestive systems often involve a period of underdeveloped enzymatic activity. While in live feeds, these endogenous enzymes are already available which aid in food breakdown, facilitating easy absorption of nutrients by the developing larvae (Kolkovski 2001). Live prey exhibit chemo-attractiveness which attracts larvae to capture and consume it quickly. Proper size and density of live feed ensures adequate larval feeding rate. Live feeds are a good source of important bioactive compounds, including abundant phospholipids, free amino acids, and immunostimulants. These bioactive compounds contribute significantly to improving larval health and increasing immunity to diseases (Sargent et al. 1997). For successful first feeding of larvae during their juvenile stage, it is very important to match the size of live feed to the larval mouth gape (Henry et al. 2015a, b).

22.1.2 Nutritional Advantages of Live Feed Over Artificial Diets

Live feed gives better nutritional profile containing a balanced amount of all the essential nutrients important for larval development (Kolkovski 2001). The digestive enzymes available in live feed often provide better digestibility and nutrient profile than processed feeds. The presence of natural growth-promoting substances in larval live feed positively affects larval growth (Dhont and Lavens 1996). Some important and sustainable live feed production systems, such as biofloc technology, can contribute to improving aquaculture water quality through nutrient cycling (Conceição et al. 2010a).

22.1.3 *Challenges and Limitations of Conventional Live Feed Production*

Conventional live feed production requires significant resources such as labour, energy and water (Dhont and Lavens 1996). Timely availability of feed sources to maintain consistent nutritional quality in live feed production can be very difficult. Contaminated live feed can cause various diseases in aquaculture systems. Live feed supply may become unstable due to lack of consistent resource availability, which may require dependence on external sources. Wastewater produced from aquaculture facilities can raise environmental concerns that need to be remedied.

22.1.4 *Need for Sustainable and Innovative Approaches*

Ensuring and expanding long-term aquaculture growth and sustainable live feed production is critical for the environmental and economic viability of the aquaculture industry. Environmentally friendly production systems are needed to optimally utilize resources and reduce waste. Technologies that increase the efficiency and predictability of live feed production also need to be developed.

22.2 Traditional Live Feeds in Aquaculture: Current Practices and Limitations

Live feed plays an important role in modern aquatic agriculture, especially during the early development stages of Finfish and Crustacean. Natural method makes the ideal for consumption of larvae. These larvae of aquatic agriculture do not like to eat passive diet. In this section, we discuss the applications and related boundaries of the widely used traditional live feeds, including microalgae, rotifers, artemia and zooplankton and phytoplankton culture (Table 22.1).

22.2.1 *Microalgae: Culture Technology, Nutrition Profile and Application*

Microalgae's common species such as isocysis, nanocoloropsis, tetraselmis and chlorella filter-feeding larvae act as super feeds. Microalgae is used directly to cultivate bivalve and crustaceans. They are also used for the promotion of nurture for the species of Microalgae, rotifers and zooplankton like Artemia. This increases their nutritional quality (Muller-Feuga 2000; Sharma et al. 2024). Microalgae are grown in nutrient-rich media such as F/2 or Walne medium, and for better development of

Table 22.1 Showing comparison of traditional live feeds in aquaculture

Live feed type	Culture techniques	Nutritional value	Applications	Limitations	References
Microalgae	Grow in batch/ semi-continuous/ continuous in nutrient-rich media (e.g. F/2) control of light, CO_2, temp and pH	Rich in PUFAs, vitamins, pigments e.g. EPA from Nannochloropsis, DHA from Isochrysis	Direct feed for larvae, enrichment of zooplankton (e.g. rotifers, Artemia)	Requires strict control, potential contamination	Lavens and Sorgeloos (1996), Muller-Feuga (2000)
Rotifers	Static or continuous systems, fed with microalgae or yeast	Require enrichment, lack sufficient EPA/ DHA naturally	First exogenous feed for marine fish larvae	Nutritional variability, prone to contamination	Lubzens et al. (2001), Conceicao et al. (2010b)
Artemia	Incubated at 30–35 ppt salinity, aeration, 28 °C for 24–36 h hatching	Nutrient content enhanced post-hatching with microalgaeor fatty acid emulsions	Versatile feed for various larvae, enrichment increases quality	Harvesting raises ecological concerns, and availability fluctuates	Lavens and Sorgeloos (1996), Dhont and Sorgeloos (2002)
Copepods	Cultured in controlled tank systems, sensitive to environment	High DHA content, good digestibility and prey stimulation	Used in marine larviculture, preferred for sensitive larvae	Sensitive to environmental changes	Støttrup (2000), Lee (2005)
Moina & Daphnia	Cultured in freshwater with organic/inorganic fertilization	Rich in fatty acids, suitable for freshwater species	Feed for carp, catfish, ornamental freshwater fish	Eutrophication risk, inconsistent nutritional quality	Sharma and Sharma (2014), Saha and Datta (2005)

this process, controlled supply of light, temperature, pH and co_2 is required (Kumar et al. 2022c; Sharma et al. 2024). Nannochloropsis is primarily the main source of EPA, while DHA is found in high amounts in isochrysis (Brown et al. 1997).

22.2.2 Rotifers: Follow-Up, Nutrition Promotion and Utility

Two Important Species of Rotifers *Brachionus plicallis* and *B. rotundiformis* are considered suitable feeds in the initial stages for the larvae of sea fishes because they are small and rapid. The rotifers are followed in stable or processed systems, where they are fed with special food Microalgae and yeast. For their optimal increase, the temperature should be between 25 and 28 °C and salinity between 15 and 20 PPT (Dhont and Lavens 1996). Microalgae-focused foods and fish oil-based emulsions (e.g., DHA selco) or liposome are used to meet the lack of EPA and DHA required in the promotion of rotifers (such as DHA selco) or liposome (Kolkovski et al. 1993a, b). In the time of making a rotifer, as a live feed in the form of sea feeders, as a live feed, as a live feed, as a live feed. The larvae of clownfish etc. are given in the early stages of life (Conceição et al. 2010b).

22.2.3 *Artemia: Hatching Process, Enrichment Methods and Challenges of Durability*

Artemia or Brine Shrimp is a very popular and large-scale live feed used for larvae nutrition of various aquatic species. In the hatching process, hatching occurs in 24–36 h at 28°–30 °C, usually at 28–30 °C (Sorgeloos et al. 2001a), after keeping the cyst of artemia in plenty and constant aeration with about 30–35 ppt salinity (Sorgeloos et al. 2001b). The amount of DHA and EPA can be increased (Dhont and Lavens 1996). Environmental anxiety: Excessive use of Artemia cysts negatively affects natural ecosystems. Such as the instability and prices fluctuations of their production in USA's Great Salt Lake and Iran's Urmia Lake are worrying (FAO 2011a, b).

22.2.4 *Other Live Feed: Copepods, Moina and Daphnia*

Many organisms such as Copepods, Moina and Daphnia are rich in nutrition and are considered highly important for larvae of aquatic species. The essential fatty acids present in these are quite high in especially DHA. Creation of copy requires a controlled tank or Mesocosum system, while Moina and Dafania can be produced in large quantities on a large amount in tanks with low resources (Støttrup 2000). Copepods are used in the early stages of the larvae of marine fish because they are naturally dynamic but the tendency to take food in the larvae is activated. At the same time, freshwater fish such as carp, catfish and decorative fish are given as Moina and Daphnia live feeds (Evjemo et al. 2003; Sharma and Sharma 2014; Saha and Datta 2005).

22.2.5 *Traditional Live Feed Limits*

Live feeds also show many negative aspects even after widely used, such as experts require high costs and labor requirements to maintain regular living feed production (Lavens and sorgeloos 1996). The nutritional quality of these living feeds may be reduced with ecological effects that can affect larvae growth (Conceição et al. 2010a). These traditional live feed systems are high in microbial infection and pollution risk, which can be harmful to larvae. Therefore, now scientific biofloc, insect-based feeds, and single-cell protein are moving towards alternate, and more permanent solutions.

22.3 Principles of Sustainable Live Feed Production

In aquaculture production, sustainable live feed is a responsible cornerstone.

22.3.1 Methods of Sustainable Live Feed Production

Production of rotifers and copepods can be boosted by cultivating microalgae. Microalgae are a key food source for many live feed organisms (such as rotifers and copepods) using the following sustainable methods, as shown in Fig. 22.1.

Fig. 22.1 Aspects for sustainable live food production methods

Live Food and its Culture

Live food plays a very important role in aquaculture, especially for the early developmental stages of many aquaculture species. Live food viz., zooplankton and phytoplankton are suitable starter feed for many cultivable crustaceans and fish larvae. These are suitable starter feed because of their easy digestibility and nutritional value. They are called as "living capsule of nutrition" for their high nutritional value, both macro and micronutrients are present. The jerking movement of zooplankton in water column influences the feeding behavior of larvae. Larvae can easily detect and digest the live zooplankton compared to the compound commercial diets. Therefore, it is essential to culture the zooplankton to meet the requirements of the Aquaculture industry. Among various zooplankton, Rotifers, Cladocera and Copepods are preferable food for many aquaculture species. Zooplankton are cultured using organic manures viz., cattle manure, poultry droppings and mustard oil cake. The cattle manure, poultry droppings and mustard oil cake are mixed at the ratio of 1:1:1 and the dose is 0.526 kg/m. These manures were decomposed for initial 10 days and then the zooplankton are inoculated in the culture unit. After 7 days of inoculation, the bloom of zooplankton appears. The water containing zooplankton is passed through the plankton nets of various mesh sizes. Collected in a bucket and should be cleaned with water. In the culture unit, fresh cattle manure may be supplied at a weekly interval for the maintenance of zooplankton. The culture may be continued for 4–6 months. Then the culture unit should be cleaned and re-start with culture following the same method. Schematic presentation of the method of zooplankton culture are shown in Fig. 22.2.

The key principles governing the sustainable production of live feeds such as rotifers, Artemia, copepods and microalgae without causing economic or social harm to the environment.

22.3.2 Scientific Basis of the Importance of Live Feed in Early Aquaculture

A number of researchers have documented the critical role of live feed in the complex processes of larval development. Studies on larval fish show that the use of live prey in the larval and juvenile stages of aquatic organisms activates the synthesis and secretion of specific digestive enzymes that improve the efficiency of nutrient absorption which enhances the growth rate and survival of aquatic organisms (Gisbert et al. 2004). In addition, by examining the fatty acid profiles of aquatic organisms fed live feeds, scientists also found that nutrient-rich copepods and Artemia contained high concentrations of beneficial compounds HUFAs such as EPA and DHA, which are highly important for the development of the nervous and visual systems of fish larvae (Villeneuve et al. 2005). Studies on crustacean larvae have shed light on how live feed supplies hormones necessary for successful metamorphosis and chitinolytic

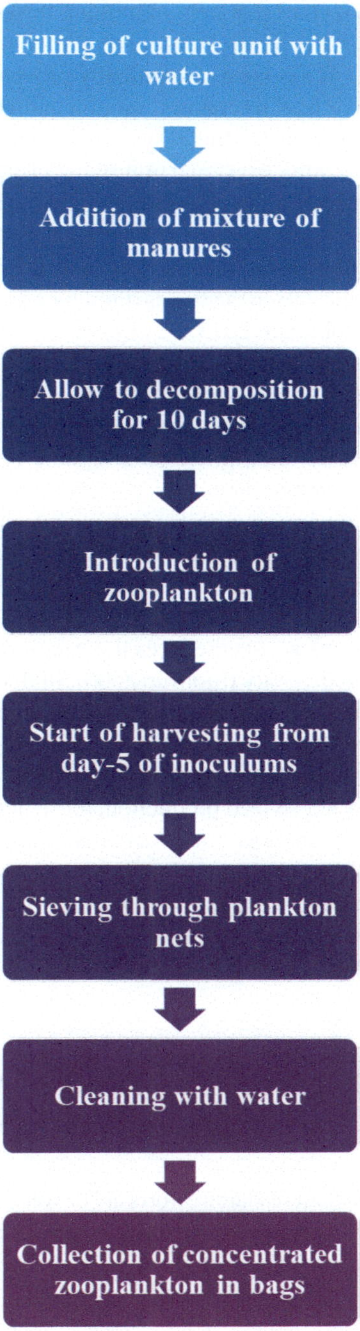

Fig. 22.2 Schematic presentation of the method of zooplankton culture

enzymes that help in molting through various developmental stages (Kurmaly et al. 1989).

It has been scientifically proven that live feed is essential in aquatic life because live feed has natural motion and chemo-attractant properties which stimulate more vigorous feeding responses in larvae and juveniles than inert feeds, thereby reducing weaning stress and improving overall feed intake (Kolakowski et al. 1993). Adopting recirculating aquaculture systems (RAS) can reduce water and energy use to conserve nutrients present in live feed systems (Timmons and Ebeling 2013). To prevent eutrophication, it is essential to develop algae-based systems as they recycle dissolved nutrients in the system thereby promoting circular nutrient economy (Borowitzka and Mohimani 2013).

Energy-efficient aeration and lighting systems such as using LED lights for algae growth also contribute to reducing the carbon footprint of live feed production and improve nutrient management. Waste minimisation by recycling the remaining wastewater, processing organic residues from live feed cultures into fertilizers or microbial inoculants can also reduce overall waste generation and convert biomass into protein-rich feed (Avnimelech 2009). Continuous harvesting of rotifer cultures reduces biomass loss and helps maintain system efficiency and reduce the need for excessive inputs (Dhont and Lavens 1996). Reducing the ecological footprint is a major goal of sustainable live feed production. Cultivation of indigenous copepods or microalgae reduces the risk of introducing non-native species and preserves local biodiversity (Støttrup 2000). Reducing dependence on fishmeal and fish oil by incorporating insect larvae, single-cell proteins, or plant-based diets in rotifer and Artemia cultures can improve sustainability metrics (Henry et al. 2015a, b). For economic viability, especially for small and medium scale hatcheries, cost reductions through in-house feed production, reuse of culture media, and automation can make live feed systems cost-effective (Conceição et al. 2010a, b). Scalability is essential for aquaculture businesses. Modular culture systems allow gradual expansion without compromising sustainability. As such, periphyton-based units and decentralized microalgae bioreactors can be adapted to various scales with minimal input costs (Azim et al. 2005).

22.3.3 Detailed Research Advances in Sustainable Live Feed Production for Aquatic Organisms

Recent research efforts are devoted to developing innovative and sustainable methods for large-scale production of important live feed organisms, with special focus on minimizing resource use and reducing environmental impact

22.3.3.1 Microalgae

Studies on microalgae indicate that there is a strong need to introduce closed photobioreactor systems to promote microalgae cultivation for aquatic life. These systems are essential to better control key environmental factors such as temperature, light intensity and CO_2 levels, reduce the risk of contamination from unwanted organisms and significantly improve volumetric productivity compared to traditional open pond systems (Enzing et al. 2014). Research has been done on activities such as harnessing CO_2 emitted from industrial sources to promote algae growth for aquatic life, recovering essential nutrients from wastewater discharged from aquaculture, supporting the increase of algae biomass production and using energy-efficient LED lighting (Chemodanov et al. 2020; Marko and Georgakakis 2011). Furthermore, substantial research efforts are focused on selecting and genetically modifying microalgal strains. Algae with improved nutritional properties are particularly used to increase the levels of specific long-chain of polyunsaturated fatty acids (HUFAs) and essential vitamins required for the larval aquaculture, (Ryckebosch et al. 2012).

22.3.3.2 Rotifers

Research based on the cultivation of rotifers has focussed on sustainable production, optimising high-density culture systems through advanced aeration and the techniques used in water management. These techniques, particularly those employed in recirculating aquaculture systems (RAS), aim to minimise water exchange and enhance utilisation of resources (Hoff and Snell 2020). Extensive investigations are underway to identify alternative sources of feeding that make the rotifer cultivation cost-effective and environmentally friendly. This is particularly important as traditional microalgae monocultures can be expensive and may not be sustainable. Researchers are exploring specific strains of heterotrophic yeasts, bacteria grown on readily available organic wastes, and carefully formulated microparticulate diets enriched with specific nutrients. They are essential for rotifer reproduction and quality (Park et al. 2016; Suantika et al. 2019). Furthermore, research is investigating the use of probiotics and bacteriophages as biocontrol agents to manage harmful microbial contaminants in rotifer cultures. This approach aims to reduce the need for conventional chemical treatments and antibiotics, thereby improving biosafety and sustainability (Gomez-Gil et al. 2000).

22.3.3.3 Artemia (Brine Shrimp Farming)

Brine shrimp, small crustaceans, are commonly used as a live food source in aquaculture. Fish larvae are particularly vulnerable to this issue. Research efforts to improve the sustainability of Artemia use focus on creating natural harvesting sites and implementing enhanced management strategies. These strategies include scientifically based harvesting quotas and active habitat conservation methods to prevent overexploitation, ensuring the long-term sustainability of these natural resources. (Van Stappen 2002) Also, studies optimizing controlled Artemia culture systems are intensively exploring precisely tailored feeding strategies that use specific microalgae and specialized artificial feeds, as well as advanced methods to improve the growth rate and nutritional quality of the biomass produced (Sánchez-Saavedra et al. 2016). Additionally, ongoing research in cryopreservation is focused on improving protocols for the long-term storage of Artemia cysts and nauplii, with the aim of establishing reliable seed banks and guaranteeing a constant supply of quality Artemia for aquaculture hatcheries, which would help reduce dependence on variable natural harvests (Clegg 2005).

22.3.3.4 Copepods

Copepods are used as a food source for larvae of a variety of marine fishes for improved nutrition. Recognizing the benefits of this, especially during the critical early developmental stages, current research has been devoted to creating scalable and economically viable sustainable culture methods for these important zooplankton (Shields 2001). Ongoing investigations are based on optimizing semi-continuous and continuous culture systems. Investigating how different microalgal diets affect the nutritional composition of copepods (particularly HUFA content) and their reproduction rates such as essential factors of fertility and hatching success (Ohs 2015). In addition, novel studies are exploring the potential of integrating the culture of copepods with microalgal production and larval rearing systems to develop more efficient and self-sustaining aquaculture setups (Norsker and Olsson 2007). Simultaneously, significant attention is being paid to enhancing cryopreservation techniques applicable to different copepod life stages, to establish reliable seed banks and ensure a continuous supply of high-quality copepods for aquaculture (Marcus 2005).

22.3.4 Extensive Research on Sustainable Nutritional Enrichment

Research focused on improving the nutritional value of live feeds for aquatic organisms through sustainable practices is a major area of study. Several investigations are evaluating the effectiveness of bioencapsulation techniques that use specially formulated microparticulate diets containing microalgal oils rich in essential HUFAs, plant-based sources of omega-3 fatty acids, and vitamins and carotenoids important for larval growth and pigmentation (Bell et al. 2003; Olson et al. 1999). Additionally, research is closely analyzing the optimal timing and duration of enrichment strategies to deliver these essential nutrients to aquatic organism larvae while minimizing stress or negative impacts on the viability and nutritional quality of live feed organisms (Tocher et al. 2005). There is also a focus on strategically incorporating immunostimulants and probiotics into live feeds to enhance the disease resistance of aquatic organisms and the overall health of aquaculture larvae (Gatesoupe 1999).

22.3.5 Social Responsibility and Ethical Considerations

Fair labour practices, training opportunities, and benefiting local communities from aquaculture ventures are essential in maintaining social sustainability in live feed cultures. Women's involvement in hatchery operations and microalgae farming, especially in South Asia and Africa, enhances rural livelihoods and gender equality (FAO 2020). Ethical considerations need to include humane treatment of live feed organisms and responsible sourcing. Avoiding wild Artemia cyst collection from endangered ecosystems such as Lake Urmia or the Great Salt Lake supports broader ecological ethics (Sorgeloos et al. 2001a).

22.4 Innovative and Emerging Technologies for Sustainable Live Feed Production

Different innovative technologies for live feed production are used as summary in Table 22.2.

Table 22.2 Summary of innovative technologies for live feed production

Technology	Key features	Benefits	References
Biofloc technology (BFT)	Microbial aggregates in high C:N environment	Enhances water quality, microbial feed, rich in proteins	Avnimelech (2009),Crab et al. (2012)
Periphyton systems	Natural bio-film on substrates	Cost-effective, protein-rich, eco-friendly	Azim et al. (2001), Kumar et al. (2007)
Insect larvae (BSF, Mealworms)	Organic waste-based larvae culture	High protein and fat content, sustainable	Makkar et al. (2014), Henry et al. (2015a, b)
Single-cell proteins (SCPs)	Yeast, algae, bacteria biomass	Rich in essential amino acids, low environmental impact	Ravindra (2000), Anupama and Ravindra (2000)
Automation & AI	Real-time monitoring, smart sensors	Reduces labor, increases consistency	Molina Grima et al. (2003), Ramos et al. (2021)

22.4.1 *Biofloc Technology (BFT) for Live Feed Production*

The flocs of bacteria (biofloc) are a nutrient-rich (protein-lipid source) available as fish food throughout the day made by a complex interaction between living microbes, planktons, organic matter, substrates and grazers, such as rotifers, ciliates, protozoa and copepods which serves as secondary source of food. Biofloc is a Profitable Method of Fish Farming. It has become very popular all around the world as an alternative to open pond fish farming. Biofloc fish farming is a low-cost way of fish farming in which toxic materials for the fish such as Ammonia, Nitrate, and Nitrite can be converted into feed. The principle of this technique is to recycle nutrients. Biofloc also helps in cleaning the culture water of the fish while giving an additional source of feed. It is a sustainable and eco-friendly process. Management of Artemia (Brine Shrimp): Artemia is preferred due to its nutritional benefits and resilience. Sustainable practices emphasize reducing dependence on wild stocks and encouraging production of Artemia in captivity. Optimizing the yield of nauplii from cysts through improved hatching technology. Promoting integrated aquaculture production by culturing Artemia with other aquaculture species in polyculture systems.

22.4.1.1 Principles and Mechanisms of BFT

The basis of biofloc technology is to encourage the growth of heterotrophic microorganisms while maintaining a high carbon and nitrogen ratio (C:N) in aquaculture. In this process, a group of microorganisms known as biofloc, consisting of bacteria, protozoa, algae, and organic particles combine together to form a nutrient-rich mixture. This mixture is highly useful as feed. The basic objective of this technology is to convert nitrogenous wastes into microbial biomass, which can be consumed directly by filter feeders or used to enhance the nutrition of other live feeds (Avnimelech 2009).

22.4.1.2 Enhancing Live Feed Production Within BFT Systems

The inclusion of BFT in rotifer and Artemia culture increases nutritional richness and also improves water quality. Biofloc particles act as a dietary supplement and probiotic that enhances the trophic properties of live feed organisms (Emerenciano et al. 2013).

22.4.1.3 Nutritional Quality of BFT-Derived Live Feed

Bioflocs are rich in protein, essential amino acids, fatty acids (PUFAs) and vitamins. The microbial biomass present in BFT enhances larval growth and survival in several fish species. (Kuhn et al. 2009).

22.4.1.4 Case Studies and Applications in Aquaculture

Successful applications in tilapia, shrimp and catfish culture showed higher survival and growth rates when biofloc-enriched live feed was given (Crab et al. 2012).

22.4.2 Periphyton-Based Systems for Live Feed Generation

22.4.2.1 Cultivation of Periphyton as a Natural Food Source

Periphyton—biofilm communities associated with submerged surfaces—are grown on artificial substrates such as bamboo or PVC. The ecosystem consists of algae, bacteria, detritus, and small invertebrates, which serve as direct food sources (Azim et al. 2001).

22.4.2.2 Enhancing the Nutritional Value of Periphyton

The nutritional value of periphyton larvae can be enhanced by promoting algae growth through fertilization strategies and availability of sunlight which provides proteins, carbohydrates and pigments required for its growth (Kumar et al. 2007).

22.4.2.3 Integration with the Aquaculture Systems

Studies have shown that integrating grass carp and tilapia culture into freshwater fish farming systems leads to improved feed conversion ratios and cost savings (Azim and Little 2008).

22.4.3 Insect Larvae Are a Sustainable Live Feed Alternatives

22.4.3.1 Nutritional Composition of Common Insect Larvae

Black soldier fly (BSF) larvae and mealworms are rich in protein (40–60%), which can be used for aquaculture. BSF larvae are highly nutritious and contain fat and minerals and essential amino acids, lauric acid and antimicrobial peptides (Makkar et al. 2014).

22.4.3.2 Mass Production Technology for Insect Larvae

For scale-up using controlled environment chambers and mechanized harvesting systems, rearing insects on organic waste substrates makes them environmentally friendly (van Huis et al. 2013).

22.4.3.3 Processing and Applications in Aquaculture Diets

Using insect larvae as paste or powder to enrich live feeds such as Artemia or rotifers showed improved larval growth and stress resistance (Henry et al. 2015a, b).

22.4.3.4 Regulatory and Consumer Acceptance

Globally, despite the acceptance of insect meal, regulatory hurdles remain in some countries. Awareness about its sustainability is needed to improve consumer acceptance (OECD 2021).

22.4.4 Single-Cell Protein (SCP) as a Live Feed Supplement

22.4.4.1 Production of SCP from Bacteria, Yeast and Algae

Some yeast and algae such as Chlorella, Spirulina and Saccharomyces cerevisiae are SCPs derived from microbial biomass. Using industrial wastes to culture them can provide a sustainable protein source (Anupama and Ravindra 2000).

22.4.4.2 Nutritional Value and Potential as Live Feed Enrichment

SCPs contain more than 50–80% protein and essential amino acids, B-vitamins and nucleotides. They can be used to enrich Artemia or rotifers by feeding them to larvae (Beccatorou et al. 2006).

22.4.4.3 Incorporation Strategies in Live feed Cultures

SCPs can be incorporated as enrichment emulsions in the media of live feed systems. They enhance the nutritional quality while reducing the dependence on fish meal (Ravindra 2000).

22.4.5 Automation and Precision Aquaculture in Live Feed Production

22.4.5.1 Automated Culture Systems for Microalgae and Zooplankton

Automated photobioreactors and climate-controlled systems improve the sustainability and efficiency of microalgae and zooplankton production (Molina Grima et al. 2003).

22.4.5.2 Sensor Technologies for Monitoring and Control

Sensors for pH, temperature, turbidity and nutrient levels enable real-time optimization of live feed cultures and reduce risks and labor costs (Timmons and Ebeling 2010).

22.4.5.3 Data Analytics for Optimizing Production Efficiency

Algorithms from AI and machine learning are being introduced to predict culture crashes, optimize harvesting times, and streamline production cycles (Ramos et al. 2021).

22.5 Enhancing the Nutritional Value of Live Feeds Through Enrichment

Live diets, while inherently providing nutritional advantages over inert diets, their nutritional profiles can be further enhanced to meet the specific requirements of the target aquaculture species at different life stages. Enrichment techniques aim to replenish essential nutrients in live feed organisms that are not available from inert diets.

22.5.1 Bio-encapsulation Techniques for Delivering Essential Nutrients

Bio-encapsulation techniques involve the incorporation of essential nutrients within the live feed organism through the culture medium making them readily available. Emulsions or liposomes containing nutrients such as fatty acids, vitamins are used to formulate live feeds with a particle size that can be easily ingested by the live feed. Rotifers and Artemia nauplii readily ingest these particles while filter-feeding and the nutrients are effectively absorbed in their body tissues (Dhont et al. 1993). Some microalgal species are also cultured with specific nutrient enrichments, which are then consumed by zooplankton such as rotifers or copepods (Sargent et al. 1999). This "trophic transfer" found in natural form of bio-encapsulation. Cultured yeast and bacteria as genetically modified vectors can be used to deliver targeted nutrients in live feed as an immuno-stimulant. These microorganisms are fed first to live feed and then to aquaculture larvae (Olsson et al. 1999).

22.5.2 Enrichment Strategies for Certain Fatty Acids (e.g., DHA, EPA)

Highly unsaturated fatty acids (HUFA), especially docosahexaenoic acid (DHA) and eicosapentaenoic acid (EPA), are important for the growth, development and survival of many marine fish larvae (Sargent et al. 1997).

Enrichment with oil emulsions rich in DHA and EPA is a widely used technique for live feed. This emulsion technique is first stabilized with emulsifiers and then can be administered for different periods depending on the desired level of enrichment (McEvoy et al. 1998).

Microalgae rich in HUFAs are used for high DHA and EPA content e.g., Isochrysis, Nannochloropsis whereby cultivation of live feed on specific microalgae species can enrich zooplankton naturally (Benemann 1992).

Commercial enrichment products: Several commercial enrichment products are available that have been formulated to provide a balanced profile of essential fatty acids and other nutrients. These often come as self-emulsifying concentrates that are easily dispersed in the culture water (Coutteau and Sorgeloos 1992).

22.5.3 Vitamin and Mineral Fortification of Live Feed

Vitamins and minerals play important roles in various metabolic processes, immune function and skeletal development in aquaculture larvae. But enrichment can ensure that live feed will provide these micronutrients in adequate quantities. Adding water-soluble vitamins and trace minerals to the culture medium of live feed can lead to their bioaccumulation within the organism (Merchie et al. 1995).

Incorporation into enrichment emulsions: Fat-soluble vitamin fatty acids can be effectively delivered through oil emulsions used for enrichment (Watanabe et al. 1983). Similarly, minerals can be incorporated in a fine particulate form within these emulsions for easy ingestion.

Use of vitamin-enriched microalgae or yeast: Similar to fatty acid enrichment, enrichment of live feed on microalgae or yeast grown in vitamin- or mineral-fortified media can be an effective strategy (Brown et al. 1997).

22.5.4 *Crucial Role of Probiotics and Prebiotics in Live Feed Nutrition*

In addition to essential nutrients, consuming the microbial environment within the live feed can further enhance the nutritional value for the health of the larvae. Consuming the microbial environment within the live feed, in addition to essential nutrients, can further enhance the nutritional value for the health of the larvae. Probiotics, containing live microorganisms, provide health benefits to gut of the host when administered in adequate amounts. (FAO/WHO 2001) The nutritional profile of the live feed can be improved by improving its own health through the addition of probiotics and certain vitamins in the live feed culture.

22.6 Integration of Sustainable Live Feed Production Into Aquaculture Systems

Integrating sustainable live feed production into aquaculture systems has proven to be an important step towards improving overall efficiency, reducing costs and minimising environmental impacts. Traditional aquaculture systems often rely on external sources for live feed which makes them subject to disruptions in consistent availability, fluctuations in cost and variable nutritional quality. By integrating live feed production within or around aquaculture systems, producers can ensure a consistent and high-quality feed supply while promoting circular resource use.

22.6.1 *Integrated Multi-trophic Aquaculture (IMTA)*

IMTA is an approach that interlinks species from different trophic levels in the same system. In integrated multi-trophic aquaculture, the waste of one species is used as nutrients for another species. For example, waste from finfish serves to provide nutrients for filter feeders such as microalgae and mussels. The microalgae feed the rotifers which serve to enrich the live feed. IMTA enhances resource efficiency by reducing nutrient pollution (Muhhammad Hanif Azhar and Mamis 2023).

22.6.2 *On-Site Microalgae and Zooplankton Production*

Setting up the hatchery near microalgae and zooplankton culture units ensures constant live feed availability. Species such as Chlorella and Nannochloropsis should be cultivated in photobioreactors to ensure a constant supply of nutritionally balanced feed. Integration of LED lighting and CO_2 injection has significantly improved yields and DHA content in these systems (Spolaore et al. 2006).

22.6.3 *Biofloc Integration with Live Feed Systems*

Biofloc technology (BFT) has grown rapidly in larviculture systems. Biofloc-enriched water serves as a culture medium for zooplankton such as rotifers and Artemia, thereby improving survival rates and immune responses in larvae while reducing live feed input costs (Avnimelech 2009; Sinha et al. 2020).

22.6.4 *Recirculating Aquaculture System (RAS) with Live Feed Unit*

RAS is a closed loop system that recycles water while maintaining its quality. System resilience can be increased by producing live feed such as copepods and microalgae in RAS. Waste from fish tanks, which is rich in nutrients, is fed to microalgae or periphyton, which feed on the microalgae and feed on them for zooplankton growth (National Fisheries Development Board).

22.6.5 *Aquaponics and Live Feed Synergy*

Aquaponics, a system that combines aquaculture and hydroponics. By adding live feed production to the aquaponic loop, waste from the fish provides nutrition for plants and microalgae. This reduces the need for external feed inputs, increasing overall productivity (Zhang et al. 2017).

22.7 Case Studies and Best Practices in Sustainable Live Feed Production

The growing global demands for aquaculture products have forced live feed production towards sustainable solutions. Various innovative approaches being used across the world to boost aquaculture production, which have demonstrated success in optimising cost, sustainability and nutritional value, are highlighted. This chapter discusses key case studies and best practices from key aquaculture research and commercial settings.

22.7.1 Successful Implementation of Biofloc in Larviculture

The most transformative breakthrough has been the development of biofloc technology (BFT) to improve water quality in larviculture systems and reduce feed costs. The best example is the shrimp hatcheries of India and Indonesia where successful implementation of biofloc in larviculture has been done by maintaining high C:N ratio through carbon supplementation (e.g., molasses or rice husk) to allow heterotrophic bacteria to thrive, which assimilate toxic nitrogen compounds to form microbial flocs rich in protein and other micronutrients (Avnimelech 2009).

In a trial conducted by ICAR-CIBA in Chennai, a 25% increase in survival rate of Litopenaeus vannamei post-larvae was observed when reared in biofloc-integrated nurseries compared to conventional systems and improved feeding behaviour of larvae due to the continuous availability of natural particulate nutrition (Sinha et al. 2020).

22.7.2 Using Insect Larvae as a Cost-Effective Feed Source

Insect-based live feeds such as black soldier fly larvae (BSFL) and mealworms have shown a way forward by providing high protein content and sustainable options through efficient bioconversion of organic wastes. In a Kenyan tilapia farm, BSFL production units were integrated using poultry manure and vegetable scraps to produce low-cost, high-nutrient live feed, providing a highly valuable alternative to larval production as live feed. These larvae were fed directly to juvenile tilapia, reducing dependence on commercial fishmeal by 40% (Makkar et al. 2014).

Studies conducted in Europe revealed that insect larvae have a desirable amino acid profile and lipid content (lauric acid), which supports larval immunity and gut health (Henry et al. 2015a, b). As regulatory frameworks evolved, the European Union has approved the cultivation of live feeds for larvae of several insect species for use in aquafeed under strict hygienic conditions.

22.7.3 Innovative Approaches to Microalgae Production for Aquaculture

Microalgae remain a cornerstone of hatcheries in aquaculture as filter feeders and feed for zooplankton. Innovations in photobioreactor design have enabled high-density cultures of microalgae such as Nannochloropsis, Isochrysis and Tetraselmis with controlled CO_2 injection, LED lighting and nutrient recycling (Spolaore et al. 2006).

Another practical example from the Netherlands used flat-panel photobioreactors to continuously culture Isochrysis galbana, reducing contamination and yielding high DHA-rich biomass. Additionally, closed systems improved the quality of microalgal feed by reducing the risk of protozoan contamination, resulting in improved production reliability of oyster larvae and rotifers.

22.7.4 Examples of Integrated Live Feed Production Systems

Integrated systems linking live feed production with aquaculture units have gained success. A small-scale hatchery in Vietnam incorporated periphyton-based substrates within larval rearing tanks and this substrate then facilitated the growth of biofilms composed of algae, bacteria and detritus, which were used as supplemental feed for fish larvae, reducing feed inputs by 20% (Azim and Little 2006).

In an integrated model, waste nutrients from shrimp tanks were redirected into algae ponds, enabling Chlorella cultivation without additional fertilisers. This closed-loop design reduced nutrients discharge into surrounding water bodies by supporting rotifer enrichment, which demonstrated improved environmental compliance (Zhou et al. 2017).

22.7.5 Comprehensive Research on the Integration of Live Feed Production Within Aquaculture Systems

Research is increasingly focusing on strategically integrating live feed production units within aquaculture systems to promote sustainability and enhance efficiency. Studies on integrated multi-trophic aquaculture (IMTA) systems have revealed that nutrient-rich wastewater from aquaculture can be an economical and accessible method for growing microalgae. These microalgae can then be utilised as a healthy feed for rotifers or other filter-feeding organisms, establishing a closed-loop system that recycles nutrients and minimises waste (Renaudeau et al. 2011). Ongoing research is meticulously exploring the design and optimisation of on-site feed production systems that integrate live feed production units into hatchery systems. This approach helps to reduce transportation costs, prevents loss of nutritional quality during shipping, and guarantees a steady supply of fresh, nutritious feed, exactly when and where it is needed (Holm et al. 2016). Another important area of research involves using advanced automation techniques and sophisticated sensors to monitor and control key environmental and operational factors in live feed culture systems for aquatic organisms. The goal of this study is to significantly increase production efficiency, maintain consistent feed quality and reduce the labor required for intensive live feed production (Benetti et al. 2010). Different case studies and best practices of sustainable live feed production are shown in Table 22.3.

Table 22.3 Case studies and best practices of sustainablelive feed production

Case study focus	Location/example	Key innovation	Impact/benefit	References
Biofloc in larviculture	ICAR-CIBA, India	C:N ratio adjustment to promote biofloc growth and nutrient recycling	25% increase in larval survival, improved water quality	Avnimelech (2009), Sinha et al. (2020)
Insect larvaeas feed	Tilapia farms, Kenya	Use of BSFL and mealworms for direct feeding and waste bioconversion	Reduced fishmeal dependency by 40%, cost-effective	Makkar et al., (2014), Henry et al. (2015a, b)
Microalgae production	Hatchery, Netherlands	Photobioreactors with CO_2 injection and LED lighting	High-quality DHA-rich algae, reduced contamination	Spolaore et al. (2006)
Integrated live feed systems	Small-scale hatcheries, Vietnam	Periphyton substrates and algae-bacteria integration	Reduced feed input by 20%, improved sustainability	Azim and Little (2006), Zhou et al. (2017)

22.8 Challenges and Future Directions in Sustainable Live Feed Production

Sustainable live feed production remains a cornerstone of modern aquaculture, especially in larviculture systems where the nutritional needs of young fish are critical. While traditional methods of producing live feeds such as microalgae, rotifers, and Artemia have supported hatchery operations for decades, a range of challenges still hampers their large-scale, consistent, and environmentally responsible use.

22.8.1 Key Challenges in Current Practices

- High Production Costs: Live feed culture requires specialized infrastructure, high-quality inputs (nutrients, fertilizers, or enrichment materials), and skilled labor. The daily maintenance of culture conditions adds significantly to operational expenses (Lavens and Sorgeloos 1996).
- Nutritional Inconsistencies: The nutrient composition of live feeds often varies with the culture conditions, feed inputs, and species used. This variability significantly impacts larval growth and survival (Conceição et al. 2010b).
- Live feed cultures are highly susceptible to microbial contamination, which can introduce pathogens into larval systems and pose a risk of contamination and disease. This is particularly critical in high-density hatchery environments, as highlighted by Dhont and Lavens (1996).
- Conventional live feed systems also have an environmental impact. They often consume large amounts of water and energy, and the harvesting of natural sources like Artemia cysts contributes to habitat degradation (FAO 2011a, b).
- Furthermore, the use of insect larvae or genetically modified microalgae faces regulatory restrictions in many countries. Scaling up production without compromising feed quality also poses logistical challenges.

22.8.2 Emerging Solutions and Innovations

22.8.2.1 Closed-Loop Systems

Closed-loop Systems such as Biofloc (BFT) and Integrated Multi-Trophic, are technologies that utilise closed-loop systems. Aquaculture (IMTA) enable nutrient recycling and reduce water usage, offering eco-friendly alternatives (Rani et al. 2025; Avnimelech 2009).

22.8.2.2 Alternative Protein Sources

Insect larvae, such as the black soldier fly e.g.,Hermetia illucens, are gaining recognition as sustainable feed sources due to their rapid growth on organic waste and their high protein contents (Kiran et al. 2025; Henry et al. 2015a, b).

22.8.2.3 Automation and Artificial Intelligence

Precision aquaculture tools, such as sensor-based systems and machine learning algorithms, are being developed to monitor live feed cultures and predict optimal harvest times (Nisha et al. 2025; Goddek et al. 2019).

22.8.2.4 Genetically Engineered Microorganisms

Genetically Engineered Microorganisms such as strains of microalgae and yeast, are being modified to enhance fatty acid profiles and reduce the reliance on enrichment protocols (Spolaore et al. 2006).

22.8.3 Future Research Directions and Perspectives

- **Standardisation of Culture Protocols** The development of universal guidelines for the culture, enrichment, and storage of live feeds can ensure consistent quality and mitigate risks.
- **Eco-certification and Sustainability Labels** As consumer demand for environmentally conscious seafood grows, certification systems for sustainable live feed production could play a major role in eco-certification and sustainability labels.
- **Research and Education** Investment in the research and development (R&D) and capacity building is crucial for expanding knowledge about novel species, cultural methods, and integration strategies.
- **Decentralised production models** which encourage small-scale, local production units, can improve access to live feeds and reduce the carbon footprint associated with transportation and storage.

Future research should focus on these aspects

- Optimizing closed photobioreactor systems for cost-effective microalgae production.
- Developing robust culture systems for copepods to meet the demands of aquaculture.
- Identifying sustainable nutritional enrichment materials from non-traditional sources.

- Understanding the nutritional requirements of larval aquaculture species for tailored feed enrichment.
- Developing standardised protocols for integrating live feed production units into aquaculture systems.
- Conducting comprehensive life cycle assessments to enhance environmental sustainability.

Through research on sustainable live feed production and investment in innovation, the aquaculture industry can transition towards a more responsible and resilient future. This will ensure a consistent supply of high-quality seafood while safeguarding the environment.

Prospects in Aquaculture

- Enhancing live feed production with suitable strategies.
- Integrating aquaculture waste into live feed production to develop circular systems for integrated system improvement requirements.
- Genetic enhancement of live feed organisms through biotechnology can improve yield and nutritional value

Sustainable live feed production is crucial for enhancing aquaculture efficiency while simultaneously minimising environmental impacts. By adopting these strategies, we can effectively strike a balance between the increasing demand for seafood and the imperative to safeguard natural ecosystems.

22.9 The Future of Sustainable Live Feeds in a Growing Aquaculture Industry

As global demand for aquatic food continues to increase, sustainable live feed production emerges as a pivotal pillar for ensuring the health, growth, and survivability of fish and shellfish during their early life stages. The advancement of aquaculture hinges not only on the quantity but also on the quality and environmental responsibility of feed inputs. This chapter encapsulates key insights from previous discussions and presents a forward-looking perspective on the sustainable transformation of live feed systems through interdisciplinary innovations.

22.9.1 Recapitulation of Key Findings

Traditional live feeds like microalgae, rotifers, Artemia, and cladocerans have supported hatchery operations for decades. However, they face challenges such as fluctuating nutritional profiles, contamination risks, and high operational costs (Conceição et al. 2010a, b; Lavens and Sorgeloos 1996). New technologies like biofloc systems, periphyton-based setups, and the use of insect larvae and single-cell proteins offer scalable, eco-friendly alternatives with enhanced nutritional value (Avnimelech 2009; Henry et al. 2015a, b; Spolaore et al. 2006).

Moreover, automation and precision aquaculture technologies have facilitated enhanced control over live feed parameters, thereby improving efficiency and reducing labour dependency (Goddek et al. 2019). The integration of these systems within closed-loop or circular aquaculture frameworks enables nutrient recycling, waste minimization, and resource optimisation.

22.9.2 The Importance of Interdisciplinary Approaches

Sustainable live feed for aquaculture production cannot evolve independently. It necessitates collaboration among microbiologists, aquaculture engineers, nutritionists, ecologists, and policymakers. Advancements in microbial biotechnology like genetically enhanced microalgae, environmental modelling, bioinformatics, and artificial intelligence are transforming the cultivation, enrichment, and delivery of live feeds to aquatic organisms (Tacon and Metian 2015; FAO 2022).

For instance, combining entomology with aquaculture has led to the rapid emergence of black soldier fly larvae as a promising live and processed feed. Likewise, the intersection of waste management and aquafeed research has enabled the conversion of agro-industrial residues into valuable feed biomass through microbial fermentation (Makkar et al. 2014).

Education and capacity building are equally crucial. Farmers and hatchery operators must be trained not only in the technical aspects of live feed culture but also in environmental stewardship and ethical aquaculture practices.

22.9.3 Outlook for Sustainable and Resilient Aquaculture Through Innovative Live Feed Solutions

Looking ahead, the future of sustainable aquaculture will hinge on the ability to make live feed systems more resilient, affordable, and ecologically sound. Key strategies include:

- **Diversifying Feed Sources** Reducing dependency on single species like Artemia by promoting alternatives such as copepods, Moina, and periphyton.
- **Decentralised Production Units** Promoting small-scale, modular systems to enhance accessibility and minimise transportation-related carbon emissions.
- **Policy Support and Incentives** Governments and agencies must allocate substantial resources to research and development (R&D), implement standardised quality control measures, and provide financial incentives for sustainable feed innovations.

The alignment of technological, ecological, and socio-economic priorities will determine the extent to which the aquaculture industry can transition towards a future where live feed is not only nutritionally optimal but also sustainably produced.

22.10 Conclusion

The integration of sustainable live feed production for aquaculture systems presents numerous advantages, such as diminished reliance on external feed, enhanced nutrient recycling, and heightened system efficiency. These innovations contribute to long-term sustainability and resilience in aquaculture practices. Sustainable live feed production in aquaculture has transcended the experimental realm and become a practical necessity. These types of case studies highlight the potential of biofloc, insect larvae, advanced algae culture, and integrated systems to transform larviculture. By drawing inspiration from successful implementations, hatchery operators and researchers can adopt optimal practices tailored to their ecological and economic circumstances. To meet the escalating demands of sustainable aquaculture, the live feed sector must undergo a transformative evolution through a synergistic interplay of innovation, policy support, and industry collaboration. A paradigm shift towards precision, circularity, and diversity in live feed production can effectively contribute to the assurance of food security, economic viability, and environmental stewardship.

Sustainable live feed production transcends mere operational necessity; it constitutes a fundamental pillar for responsible aquaculture development. Through unwavering innovation, cross-disciplinary collaboration, and global cooperation, the industry can fortify its future while simultaneously mitigating environmental impact and bolstering food security. In-depth research conducted into sustainable live feed production for aquaculture has yielded substantial advancements in enhancing practices. This transformation towards meticulously controlled farming systems, the pursuit of alternative feed sources, and the refinement of nutritional enrichment techniques all play pivotal roles in adopting more environmentally friendly and economically viable methods within aquaculture. Integrating live feed production units within aquaculture systems, adhering to circular economy principles, and ensuring on-demand feed availability presents substantial potential to enhance efficiency and reduce environmental footprint.

22.11 Future Prospects

There is a need for Interdisciplinary approaches for sustainable live food production. New innovative technologies that combine the aquatic plants, insects and earthworms to use these organisms as protein sources as well as for medicinal properties in aquaculture. New polices that not only sustain the ecosystem but also increase the farmer economy. Education and capacity building are equally crucial. Farmers and hatchery operators must be trained not only in the technical aspects of live feed culture but also in environmental stewardship and ethical aquaculture practices.

References

Anupama, Ravindra P (2000) Value-added food: Single-cell protein. Biotechnol Adv 18(6):459–479

Avnimelech Y (2009) Biofloc technology: a practical guide book. The World Aquaculture Society, Baton Rouge, Louisiana, USA

Azim ME, Wahab MA, Verdegem MCJ, Beveridge MCM (2001) Periphyton-based aquaculture: ecological perspective. NAGA WorldFish Center Q 24(4):15–19

Azim ME, Verdegem MCJ, van Dam AA, Beveridge MCM (2005) Periphyton: ecology, exploitation and management. CABI Publishing

Azim ME, Little DC (2006) Intensifying aquaculture production through new approaches to periphyton-based systems. CAB Rev Perspect Agric Vet Sci Nutr Nat Resour 1(62)

Azim ME, Little DC (2008) The biofloc technology (BFT) in indoor tanks: water quality, biofloc composition, and growth and welfare of Nile tilapia. Aquaculture 283(1–4):29–35

Bekatorou A, Psarianos C, Koutinas AA (2006) Production of food grade yeasts. Food Technol Biotechnol 44(3):407–415

Bell JG, Farndale BM, Pradhan AD, Sargent JR, Henderson RJ (2003) Effects of dietary lipid source on fatty acid composition of larval turbot (Scophthalmus maximus). Aquaculture 217(1–4):515–531

Benemann JR (1992) Microalgae aquaculture. In: Richmond A (ed) CRC handbook of applied phycology: algal biotechnology. CRC Press, pp 251–336

Benetti DD, Olson CD, Ojha T, Ojha S, Ojha R, Ojha B, Ojha M (2010) Intensive larval rearing of Pacific threadfin (Polydactylus sexfilis) using automated feeding and water quality management systems. Aquaculture 306(1–4):220–227

Borowitzka MA, Moheimani NR (2013) Sustainable biofuels from algae. Mitig Adapt Strat Glob Change 18(1):13–25

Brown MR, Jeffrey SW, Volkman JK, Dunstan GA, Garland CD (1997) Nutritional properties of microalgae for mariculture. Aquaculture 151(1–4):315–331

Chemodanov AA, Zamyatin NA, Baranov VV, Solovchenko AE, Chekanov KV (2020) Microalgae cultivation using flue gas CO_2: a review of current technologies and future perspectives. Renew Sustain Energy Rev 125:109834

Clegg JS (2005) Cryptobiosis in embryos of Artemia salina: a review. J Exp Mar Biol Ecol 321(1):3–19

Conceição LEC et al (2010a) Nutritional physiology of fish larvae: state of the art and future perspectives. Aquaculture 302(1–2):2–10

Conceição LEC, Yufera M, Makridis P, Morais S, Dinis MT (2010b) Live feeds for early stages of fish rearing. Aquac Res 41(5):613–640. https://doi.org/10.1111/j.1365-2109.2009.02355.x

Coutteau P, Sorgeloos P (1992) Quality improvement of live food for fish larvae. Aquaculture 107(1–2):179–191

Crab R, Defoirdt T, Bossier P, Verstraete W (2012) Biofloc technology in aquaculture: Beneficial effects and future challenges. Aquaculture 356–357:351–356

Dhont J, Støttrup JG, Sorgeloos P (1993) Rotifers in aquaculture: biology, culture and use. Aquaculture 126(1–4):47–63

Dhont J, Lavens P (1996) Artemia: in manual on the production and use of live food for aquaculture. FAO Fisheries Technical Paper No. 361, pp 79–106

Dhont J, Sorgeloos P (2002) Applications of Artemia. In: Live feeds in marine aquaculture

Emerenciano M, Gaxiola G, Cuzon G (2013) Biofloc technology (BFT): a review for aquaculture application and animal food industry. In: Biomass now—Cultivation and utilization. InTech.

Enzing C, Barbier M, Wijffels RH, Trzcinski T, Slegers PM, Jørgensen U (2014) Microalgae-based products for the food and feed sector: an outlook for Europe. Eur J Lipid Sci Technol 116(12):1378–1391
Evjemo JO, Reitan KI, Olsen Y (2003) Copepods as live food organisms in marine larviculture. Aquac Res 34(11):987–993
FAO (2011a) Artemia: the production of Artemia cysts. Food and Agriculture Organization of the United Nations
FAO (2011b) The state of world fisheries and aquaculture. Food and Agriculture Organization of the United Nations
FAO (2020) The State of World Fisheries and Aquaculture 2020. Sustainability in action. Rome. https://doi.org/10.4060/ca9229en
FAO (2022) The state of world fisheries and aquaculture 2022: towards blue transformation. Food and Agriculture Organization of the United Nations
FAO/WHO (2001) Health and nutritional properties of probiotics in food including powder milk with live lactic acid bacteria. Report of a Joint FAO/WHO Expert Consultation. Córdoba, Argentina
Gatesoupe FJ (1999) The use of probiotics in aquaculture. Aquaculture 180(1–2):147–165
Gisbert E, Gómez-Gil B, Castelló-Orvay F, Lin S (2004) Ontogeny of the digestive system in larval gilthead seabream (Sparus aurata) and the effect of weaning on digestive enzyme activity. Aquaculture 237(1–4):255–276
Goddek S, Joyce A, Kotzen B, Burnell GM (eds) (2019) Aquaponics food production systems: combined aquaculture and hydroponic production technologies for the future. Springer
Gomez-Gil B, Roque A, Turnbull JF (2000) The use of probiotics in aquaculture: a review. Aquaculture 197(1–4):1–23
Henry M, Gasco L, Piccolo G, Fountoulaki E (2015a) Insect meals for aquaculture: a review of current evidence. Anim Feed Sci Technol 205:1–22
Henry M, Gasco L, Piccolo G, Fountoulaki E (2015b) Review on the use of insects in the diet of farmed fish: past and future. Anim Feed Sci Technol 203:1–22
Hoff FH, Snell TW (2020) Plankton culture manual, 6th edn. Florida Aqua Farms Inc.
Holm J, Øie G, Vadstein O, Olsen Y (2016) On-demand production of live feed for marine fish larvae: challenges and perspectives. Rev Aquac 8(3):207–221
Kiran, Kumar N, Sindhu S, Gawaria B, Ansari A, Dhariwal A., Sharma P, Kumar R (2025) GWAS approaches in medicinal plants for enhanced secondary metabolite production. In: Chen JT (ed) Biotechnology, multiple omics, and precision breeding in medicinal plants. CRC Press, pp 119–134. https://doi.org/10.1201/9781003475491-10
Kołakowski E, Szybowicz Z, Raczek K (1993) Comparison of minced flesh yield from several fish species using a Baader-694 separator. J Food Qual 16(3):209–221. https://doi.org/10.1111/j.1745-4549.1993.tb00227.x
Kolkovski S, Arieli A, Tandler A (1993a) Visual and chemical cues stimulate microdiet ingestion in seabream larvae. Aquacult Int 1(4):275–282
Kolkovski S, Yufera M, Olsen Y, Seraj H (1993b) The effect of live feed and artificial diets on the growth and survival of larval gilthead seabream (Sparus aurata). Aquaculture 110(2):169–179
Kolkovski S (2001) Digestive enzymes in fish larvae and juveniles—Implications and applications to formulated diets. Aquaculture 200(1–2):181–201. https://doi.org/10.1016/S0044-8486(01)00603-1
Kuhn DD, Boardman GD, Lawrence AL, Marsh L, Flick GJ Jr (2009) Microbial floc meals as a replacement ingredient for fish meal and soybean protein in shrimp feed. Aquaculture 296(1–2):51–57
Kullman SW, Schkoda SM, Rice JA, Schreiber DLS, Buchwalter DB (2021) Developmental toxicity and respiratory physiology in fish larvae exposed to environmental stressors. Sci Rep. https://doi.org/10.1038/s41598-021-00000-0
Kumar K, Mukherjee SC, Jana RK (2007) Enhancing periphyton productivity through nutrient management for fish culture in inland water bodies. Fish Manage Ecol 14(5):365–374

Kumar R, Gupta R, Dahiya T, Yadav R, Saifi R, Yodha K, Kumari B, Singh P, Shefali (2022a) Surveillance of *Heteropneustes fossilis* (Stinging Catfish) Diseases and Health Management Practices in Haryana. Appl Ecol Environ Sci 10(4):261–267. https://doi.org/10.12691/aees-10-4-12

Kumar R, Gupta RK, Yadav R, Saifi R, Yodha K, Pooja (2022b). Eisenia fetida as protein source for growth enhancement of Heteropneustes fossilis. Egypt J Aquat Biol Fisheries 26(2):577–588. https://doi.org/10.21608/ejabf.2022.234553

Kumar R, Yadav R, Kumar Gupta R, Pooja (2022c) A sustainable way for fish health management by replacement of chemical and drugs by earthworm. In: Environmental degradation in Asia: land degradation, environmental contamination, and human activities. Springer International Publishing, Cham, pp 329–352. https://doi.org/10.1007/978-3-031-12112-8_16

Kurmaly K, Jones DA, Iyah Y (1989) Biochemical changes during larval development of Penaeus monodon (Fabricius). Aquaculture 79(1–4):307–315

Lavens P, Sorgeloos P (1996) Manual on the production and use of live food for aquaculture. FAO Fisheries Technical Paper 361

Lee CS (2005) Copepods in aquaculture. Blackwell Publishing

Lubzens E et al (2001) Rotifers as live food in aquaculture. Hydrobiologia 446–447:145–153

Makkar HPS, Tran G, Heuzé V, Ankers P (2014) State-of-the-art on use of insects as animal feed. Anim Feed Sci Technol 197:1–33

Marko J, Georgakakis D (2011) Economic and technical aspects of aquaculture development. J Appl Ichthyol (FAO Regional Report). https://doi.org/10.1111/j.1439-0426.2011.01700.x

Marcus NH (2005) Diapause in copepods. Adv Mar Biol 48:141–247

McEvoy LA, Naess T, Bell JG, Lie Ø (1998) Lipid and fatty acid metabolism during early development of marine fish larvae. Prog Lipid Res 37(2–3):131–165

Merchie G, Lavens P, Dhont J, Sorgeloos P (1995) State of the art of Brachionus culture in aquaculture. In: Schram E, Knutsen J, Olsen Ø (eds) Rotifera VII. Kluwer Academic Publishers, pp 1–16

Molina Grima E, Belarbi EH, Acién Fernández FG, Robles Medina A, Chisti Y (2003) Recovery of microalgal biomass and metabolites: process options and economics. Biotechnol Adv 20(7–8):491–515

Muhhammad Hanif Azhar, Mamis D (2023) Application of the IMTA (integrated multi-trophic aquaculture) system in freshwater, brackish and marine aquaculture. Aquatic Sci Eng 38(2):106–121 https://doi.org/10.26650/ASE20231252136

Muller-Feuga A (2000) The role of microalgae in aquaculture: situation and trends. J Appl Phycol 12(3–5):527–534

Nisha, Anshu, Rani S, Kumar S, Kumar R, Kumar A (2025) Recent technologies and strategies for bioremediation of toxic and hazardous waste. In Dubey KK, Shah MP, Yadav A (eds) Challenges and sustainable solutions in bioremediation. CRC Press, pp 1–16. https://doi.org/10.1201/978 1003407317-1

Norsker NH, Olsen Y (2007) Copepod cultivation: status and perspectives. Aquaculture 266(1–3):1–13

OECD (2021) Insects as sustainable feed ingredients: production, processing and policy. OECD Publishing, Paris

Ohs CL, D'Abramo LR, Buddington RK (1999) Evaluation of the partial or complete substitution of live Artemia nauplii with an artificial diet for the pond culture of freshwater prawn Macrobrachium rosenbergii larvae. J World Aquaculture Soc 29(4):488–495. https://doi.org/10.1111/j.1749-7345.1998.tb00673.x

Ohs CL (2015) Effects of different microalgae diets on the growth and fatty acid composition of the calanoid copepod Paracartia grani. Aquaculture 438:1–8. https://doi.org/10.1016/j.aquaculture.2014.12.034

Olsson JC, Kjørsvik E, Westerdahl G, Vadstein O (1999) Influence of cultivation parameters on the bacterial flora associated with Brachionus plicatilis. Aquaculture 170(3–4):223–234

Park HG, Do HG, Devaraja TN, Song YC, Kim JH, Yoo JS (2016) Evaluation of different diets for mass culture of the rotifer Brachionus plicatilis. J World Aquac Soc 47(6):815–825. https://doi.org/10.1111/jwas.12344

Ramos FM, Valerio E, Soares M (2021) Artificial intelligence in aquaculture: a review. Aquac Rep 20:100711

Rani S, Sharma S, Singh D, Sheoran R, Kumar R (2025) Improved production of anti-cancer compounds from medicinal plants through molecular and biotechnological mutagenesis strategies. In: Chen JT (ed) Biotechnology, multiple omics, and precision breeding in medicinal plants. CRC Press, pp 287–297. https://doi.org/10.1201/9781003475491-21

Ravindra P (2000) Single cell protein: promising protein for future. Curr Sci 81(5):515–516

Renaudeau S, Mikkelsen S, Angel D, Ferreira JG, Pérez JP (2011) Integrated multi-trophic aquaculture (IMTA) in Europe: a review. Aquacult Int 19(4):611–630. https://doi.org/10.1007/s10499-010-9393-1

Ryckebosch E, Muylaert K, Foubert I (2012) Optimization of lipid production in microalgae for biodiesel: a review. Appl Energy 89(1):271–282

Saha A, Datta NC (2005) Culture of Daphnia magna and Moina micrura in organic and inorganic media. J Inland Fisheries Soc India 37:12–18

Sánchez-Saavedra MP, Milán-Aldaco M, Hernández-Llamas H, Durazo-Amador D, Hernández-Montaño DA, Durazo-Amador DA (2016) Effect of different diets on growth and biochemical composition of Artemia franciscana nauplii. Lat Am J Aquat Res 44(1):117–126. https://doi.org/10.3856/vol44-issue1-fulltext-11

Sargent JR, Bell MV, McEvoy L, Tocher DR, Estevez A (1999) Recent developments in the essential fatty acid nutrition of fish. Aquaculture 177(1–4):191–210

Sargent JR, McEvoy LA, Bell MV, Tocher DR (1997) Lipids and essential fatty acids in cultured marine fish larvae. Aquaculture 155(1–4):117–129

Sharma BK, Sharma S (2014) Freshwater zooplankton diversity and distribution: Moina and Daphnia. Int J Aquatic Biol 2:1–10

Sharma MMM, Kapoor D, Rohilla R, Loyal A, Sharma P, Husen A (2024) Role of essential minerals and rhizosphere in plant under changing environmental conditions. In: Husen A (eds) Essential minerals in plant-soil systems, pp 235–252

Shefali, Kumar R, Sankhla MS, Kumar R, Sonone SS (2021) Impact of pesticide toxicity in aquatic environment. Biointerface Res Appl Chem 11(3):10131–10140. https://doi.org/10.33263/BRIAC113.1013110140

Shields RJ (2001) Mass production of the calanoid copepod Acartia tonsa for use as a live food for marine fish larvae. Aquaculture 201(3–4):315–327

Sinha S, Thirunavukkarasu AR, Ghoshal TK (2020) Application of biofloc technology in indian aquaculture. CIBA Technical Bulletin

Sorgeloos P, Dhert P, Candreva P (2001a) The use of the brine shrimp, Artemia spp., in aquaculture. In: Newkirk G, лица JS (eds) Finfish hatchery management, 2nd edn. Blackwell Publishing, pp 205–232

Sorgeloos P, Lavens P, Léger P, Tackaert W, Versichele D (2001b) Manual for the culture and use of brine shrimp Artemia in aquaculture. Artemia Reference Center, Ghent University

Spolaore P, Joannis-Cassan C, Duran E, Isambert A (2006) Commercial applications of microalgae. J Biosci Bioeng 101(2):87–96

Støttrup JG (2000) The elusive copepods: their production and suitability in marine aquaculture. Aquac Res 31(8–9):703–711

Suantika G, Astuti NN, Nursid NA, Nurilmala A, Andriyono S (2019) Growth and fatty acid profile of rotifer (Brachionus plicatilis) fed with different types of yeast. Aquac Res 50(11):3181–3188. https://doi.org/10.1111/are.14272

Tacon AGJ, Metian M (2011) Global overview on the use of fish meal and fish oil in industrially compounded aquafeeds: Trends and future prospects. Aquaculture 311(1-4):1–10. DOI: https://doi.org/10.1016/j.aquaculture.2010.11.046

Tacon AGJ, Metian M (2013) Feed matters: towards responsible aquaculture. Rev Aquac 5(1):1–17

Tacon AGJ, Metian M (2015) Feed matters: satisfying the feed demand of aquaculture. Rev Fisheries Sci Aquac 23(1):1–10

Timmons MB, Ebeling JM (2010) Recirculating aquaculture. Cayuga Aqua Ventures, Ithaca, NY

Timmons MB, Ebeling JM (2013) Recirculating aquaculture. Ithaca Publishing Company

Tocher DR, Agaba MK, Skiba-Cassy S, Hamilton RJ, Dick JR, Houlihan DF (2005) Effects of dietary docosahexaenoic acid on the fatty acid composition and expression of genes involved in lipid metabolism in larval turbot (Scophthalmus maximus). Comp Biochem Physiol B: Biochem Mol Biol 141(3):356–363. https://doi.org/10.1016/j.cbpc.2005.04.007

Van Huis A, Van Itterbeeck J, Klunder H et al (2013) Edible insects: future prospects for food and feed security. FAO Forestry Paper 171, Food and Agriculture Organization of the United Nations

Van Stappen G (2002) состояние of Artemia resources in natural salt lakes. Hydrobiologia 467(1–3):117–131

Villeneuve LA, Ruzicka D, Penney T, Murray C, Hori R, Guderley H (2005) Effects of dietary docosahexaenoic acid on brain fatty acid composition and visual performance in larval Atlantic salmon (Salmo salar). Lipids 40(5):501–509

Watanabe T, Kitajima C, Fujita S (1983) Nutritional values of live organisms used in Japan for mass propagation of fish: a review. Aquaculture 34(1–2):115–143

Yúfera M, Darias MJ (2007) The onset of exogenous feeding in marine fish larvae. Aquaculture 268(1-4):53–63. https://doi.org/10.1016/j.aquaculture.2007.04.050

Zhang Y, Hu H, Zhang X (2017) Co-cultivation of microalgae in aquaponic systems. Biores Technol 245:117–123. https://doi.org/10.1016/j.biortech.2017.08.123

Zhou Q, Zhang X, Xie X (2017) Integrated nutrient management through algal-bacterial systems in aquaculture. Aquacult Eng 78:1–10

Chapter 23
Sustainable Management of Freshwater Ornamental Fish Culture: Practices and Challenges

Annu Malik, Dharambir Singh, and Amit Kour

Abstract Ornament fishes are now considered a luxury with goldfish and koi carp as the fish of choice by many enthusiasts. Advanced artificial propagation practices trigger ornamental fish reproduction in a way to ensure enough fry for aquaculture necessary to help and strategize the economy and maintain the balance among practicality, social values, and protection of the environment by breeding in eco-friendly ways. Encouraging artificial spawning using bio-filters, RAS, and artificial feeds prevents over-nutrition and eutrophication along with policies implementation on wastewater, good sanitation, quarantine, and antibiotic use leading to less reliability on wild-caught brood stock which in turn help wild fish conservation. Meanwhile, challenges like poor efforts to prevent overfishing and ruining the natural habitats of wild fishes as well as lack of industrial knowledge and high cost of RAS and bio-filters hinders the use of environment friendly measures. A collaborative multidisciplinary approach involving policymakers, researchers, industry practitioners, and other stakeholders is essential to tackle challenges by offering guidelines for technical support, financial incentives, and effective conservation strategies.

Keywords Ornamental fish · Aquaculture · Sustainability · Practices · Challenges

23.1 Introduction

Aquaculture encompasses the cultivation, breeding, and harvesting of aquatic species, both animal and plant, within controlled environments for commercial and recreational purposes. It emphasizes food production, habitat restoration, and the conservation of endangered and threatened species (Bostock et al. 2010; Froehlich et al. 2017). A notable segment of aquaculture is ornamental fish culture, which concentrates on the breeding, rearing, and marketing of visually appealing fish species intended for decorative use (Pouil et al. 2020). These extraordinary aquatic

A. Malik · D. Singh · A. Kour (✉)
Department of Zoology, COBS&H, CCS Haryana Agricultural University, Hisar, Haryana, India
e-mail: akour625@gmail.com

R. Kumar et al. (eds.), *Sustainability in Aquaculture and Fisheries*,
Springer Water, https://doi.org/10.1007/978-3-032-16605-0_23

creatures, often referred to as "living gems", are kept primarily in aquariums and water gardens and play an important role in a multi-billion dollar global market (Biondo and Burki 2020). Freshwater species are leaders in this field due to their strong adaptability, simple breeding process and low maintenance requirements. The ornamental fish market is present in more than 100 countries, with around 1 billion fish sold annually, a large portion of which comes from developing countries (Shefali et al. 2021; Kumar et al. 2022a; Jain and Karunasagar 2025). Major part of the population in developing countries is becoming interested in ornamental fish culture for their appearance and convenience over regular pets providing therapeutic advantages and fosters environmental consciousness, however, in developing countries like India, this hobby is practiced by only 10% of the population, apart from that India is likely to become an important source of new ornament fish culture resources (Watson 2000; Evers et al. 2019; Pailan et al. 2022). The ornamental fish industry is expected to become a major player in the fish market by providing financial provisions and substantiating ancillary sectors such as aquarium manufacturing and fish feed production if resources are managed well (Raja et al. 2019; Pountney 2023). Nonetheless encountering obstacles such as the overexploitation of wild fish populations, habitat degradation, and the risk of invasive species are major challenges. To palliate these challenges, robust regulations with growing precedence on sustainable practices circumscribing eco-friendly breeding gumptions, waste recycling methods, and the corroboration of native species in aquaculture and community help is required (Arthington et al. 2016).

23.2 Ornamental Fish Resources in India

Freshwater ornamental fish farming is an expanding and vital part of the world's aquaculture, helping to sustain jobs, businesses, and the environment (Wuyep and Rampedi 2018; Durrah and Mall 2020; Bhattacharya et al. 2025). The business of ornamental fish is worth billions of dollars, and India, Thailand, Singapore, Brazil, and the Philippines are leading countries helping with foreign exchange and offering job opportunities to millions, especially in rural parts of the world (Monticini 2010). The accessibility and hobbyist culture of freshwater ornamental fish have spurred the growth of a global hobbyist culture, driving demand for aquarium manufacturing, ornamental plant trade, and fish feed production (Pouil et al. 2020; Froehlich et al. 2017). India possesses rich resources of marine ornamental fishes such as the lagoons and coral reefs of Lakshadweep and Minicoy islands, Andaman and Nicobar Islands, Gulf of Kutch, Coast of Kerala, Gulf of Mannar and Palk Bay are abounded with highly attractive and varied species (Pailan et al. 2022; Pandey and Mandal 2017). The potential with regard to the freshwater ornamental fishes inhabiting the hill streams, major river systems, reservoirs and lakes are immense. Rivers of north-eastern states and their Himalayan streams have an abundant variety of ornamental fish species. In India, the north-eastern states play leading role in ornamental fish trade. India has recorded at least 150 commercially important ornamental fish species and trade mainly indigenous freshwater species, collected from rivers, streams, wet lands, etc. Prominent among the freshwater Indian ornamental fish are loaches, eels, barbs, catfish and goby. About 90% of ornamental fish is traded from Kolkata port followed by 8% from Mumbai and 2% from Chennai. India's share to global ornamental fish trade is less than one percent but still she is projected as a "sleeping giant" because of yet untapped potential resources (Pandey and Mandal 2017; Sandilyan 2016; Shyam et al. 2013).

23.3 Practices for Sustainable Management

Freshwater ornamental fish farming is an important part of the aquaculture industry, securing livelihoods and facilitating world trade. However, unsustainable practices can lead to environmental damage and biodiversity loss. Sustainable management practices focus on balancing ecological health, economic feasibility, and social responsibility. Some of the key practices in sustainable ornamental fish culture includes.

23.3.1 Eco-Friendly Breeding and Stock Management

Eco-friendly breeding programs in ornamental fish culture prioritize sustainability, biodiversity conservation, and minimal environmental impact. Efforts are underway to breed ornament fish by using effective and environment friendly methods and approaches enabling people to reduce the use of wild fish in order to protect both the environment and the aquatic fish inhabiting (Abisha et al. 2022). The important ideas in eco-friendly ornament fish breeding are about making sure this practice favours people, fish and environment.

(a) Focusing on sustainability to give wild fish populations a chance to recover by growing fish in captivity and using eco-friendly sources like feed, water, and energy without causing harm to the environment (Baruah et al. 2017; Saini et al. 2024).
(b) Focusing on biodiversity conservation by protecting native or endangered species by their captive breeding, help inflow of invasive species harmful to endemic species (Ebner et al. 2020).
(c) Implementing efficient water and waste handling methods, like using systems that recirculate water, thus, help save water and prevent environment pollution. Bio-filters and integrated systems help clean fish waste and turn it into useful nutrients for the plants growing in an aquaponics system (Balami 2021; Nair et al. 2025) (Table 23.1).

Table 23.1 Ornamental fish breeding practices and their benefits

Approaches	
Captive breeding programs	Helps to conserve wild species and safekeeping wild population by preventing over harvesting of wild species as that might disrupt the environment and reduce species
Selective breeding	Changing genetic make-up for production of disease resistance species helping farmers to use less chemicals and better results
Avoiding hybridization	Hybridization discouragement as it may lead to erosion or partial loss to pure breeds/genetic lines

23.3.2 Water Quality Management

The use of bio-filters and mechanical systems to recycle water and cut down significantly on wastewater discharge to the environment, thus, guarantees environment stability in terms of pH, temperature, oxygen, and less use of harmful nitrogenous compounds, which is good for sensitive species.

a. **Recirculating Aquaculture Systems (RAS)**: RAS stands for an inventive way to cultivate ornamental fish in a sustainable way, mainly through better water quality, more effective resource usage, and reducing the effects on the environment (Roy et al. 2025; Halachmi 2006; Elalouf et al. 2018). The growth of solar technology, along with the partnership of aquaponics and RAS aquaculture, makes it easier for nutrients to be recycled and helps sensitive aquatic species to thrive (Yusoff et al. 2024; Gupta et al. 2024; Turlybek et al. 2025). These systems also helps to use resources more efficiently by using less fresh water and having a smaller impact on the environment compared to basic regular ways of ornament fish culture, furthermore, prevents diseases by employing biosecurity protocols thus reducing pathogen load (Subasinghe et al. 2023). It is clear from the environment's standpoint that RAS knock-down discharge is important, since it allows for the use of captive-reared species instead of harming wild ones, lower cost and disease reduction, offering more benefits, make water reuse useful and possible in urban and peri-urban settings with cheaper transportation at the same time helping small-scale farmers to earn sustainable incomes (Aly and Fathi 2024). In defiance of the existing certain challenges, exaltations in technology amalgamated with reductions in costs have the potential to meliorate the accessibility of RAS, thereby furthering its broader adoption.
b. **Bio-filtration**: Bio-filtration is a continuously developing artistic and practical way of growing ornamental fish with a minimum impact on the nature. By removing dangerous ammonia and nitrates from the water, bio-filters help maintain a safe environment for fish and prevent eutrophication in surrounding ecosystems (Gutierrez-Wing and Malone 2006; Palao et al. 2020). Advances in bio-filtration technology are essential for preserving water quality in ornamental fish culture, fostering healthy fish habitats, and reducing environmental impacts. Cutting edge bio-filtration systems, like fluid bed's filters and MBBR systems

are engineered to maximize surface area for bacteria to grow on successfully, metabolize harmful ammonia and nitrites into less harmful nitrates by using nitrifying bacteria and to promote water movement to encourage efficient biochemical reactions even in high density aquaculture rearing operations (Tripathi and Hussain 2022; Fadhil et al. 2011). Denitrification filters reduce nitrate build-up again by encouraging anaerobic bacteria which in turn reduces the risk of nitrate poisoning. Bio-filtration improves the health and survival rates of fish and makes ornamental fish farming more sustainable by ensuring stable and safe water conditions (Irhayyim and Fotedar 2019; Preena et al. 2021). An efficient integration of bio-filtration with the aquaculture system prevents nutrient-packed effluent from entering surrounding ecosystems and the eutrophication of water bodies is minimized, thereby mitigating the risk to aquatic biodiversity.

c. **Regular Monitoring**: Advanced sensors and manual testing of parameters like pH, dissolved oxygen, and temperature help maintain optimal water quality and reduce stress on fish and use of IoT and AI for real-time monitoring of water quality (Mandal and Ghosh 2024).

23.3.3 Feed Management

a. **Low-Impact Feed Formulations**: Sustainable feeds incorporate natural ingredients, such as algae, and insect protein into feed, which is less environmentally damaging than fishmeal and fish oil (Hameed et al. 2022; Hua et al. 2019). Sustainable low-impact feeds are changing the way ornamental fish culture practices sustainability by decreasing the reliance on fishmeal or fish oil (which might be linked to overfishing and environmental damage) as sustainable feeds now include certified algae, insect meal and plant-based proteins, all providing high nutritional value and a much smaller environmental impact (Ramasamy 2025; Vala et al. 2024). Some species of algae, for example, contain such valuable ingredients as omega-3 fatty acids and fish-colouring pigments, an important selling-point in the ornamental fish sector (Rout et al. 2013). Insects' proteins, such as that of black soldier fly larvae, are also high in amino acids; they can also be made from organic waste, according to circular economy principles. These creative feed types are not only environmentally sound for the aquaculture they are used for, but also to improve feed efficiency and fish health. Diversifying away from finite marine resources, low-impact feeds may thus contribute to the sustainable development of ornamental fish culture, as well as being in accordance with the worldwide campaign to preserve aquatic environment (Little et al. 2016).
b. **Optimized Feeding Practices**: Preventing overfeeding can minimize waste build-up that degrades water quality and promotes harmful algal growth. Feeding strategy is very important in optimizing the advantages conferred by low-impact feeds in order to promote an environmentally friendly ornamental fish industry (Moorhead and Zeng 2010). Good feeding practices (e.g., feeding rations, correct

feeding time and automatic feeding) improve feed utilization and reduce feed waste and nutrient loss. Feeding practices should be adjusted based on the dietary requirements and feeding habits of ornamental species of fish to minimize overfeeding, as well as uneaten feed accumulation, which can deteriorate water quality and result in a higher filtration requirement (Sales and Janssens 2003). Pellet design and processing procedures which provide optimum nutrient absorption enable a higher level of consumability of low impact feed formulations such as algae or insect protein based feed (Munguti et al. 2014). Routine surveillance of fish health and their growth is known to fine-tune feeding practices so as to fine tune feeding conversion ratios without compromising the health and vigour of ornamental fish. By integrating efficient feed formulations complemented with efficient feeding procedures, the ornamental fish culture can have an increased efficiency and lesser environmental sustainability (Huntington and Hasan 2009).

c. **Utilization of Alternative Protein Sources**: Reuse of by-products as alternative sources of proteins from other industries (poultry or soybean) provides lower ecological footprint in ornamental fish culture as revolutionary method of attaining sustainability (Kari et al. 2023). Proteins that result from algae, insect meal and by-products of agriculture provide eco-friendly and nutritive sources of proteins capable of promoting the growth and health of fish for example, algae-based proteins contain sufficient amounts of essential amino acids and bioactive compounds which improve immunity and coloration in ornamental species (Hua et al. 2019; Kumar et al. 2022b, 2022c). In addition, plant-based proteins including soybean meal if subjected to a proper processing to eliminate anti-nutritional factor can be a good supplement in aqua feeds. These other sources of proteins limit the environmental footprint of the aquaculture by lessening the effect on the wild fish reserves as well as reducing the greenhouse gases emission from traditional feed production (Aragão et al. 2022). Including such innovations is not only in line with the global sustainability goals but also improves economic sustainability of the ornamental fish culture with cost-effective and renewable feed solutions (Hua et al. 2019).

23.3.4 Waste Management

a. **Sludge Recycling**: Fish tanks' solid waste can be treated and used as agriculture fertilizer, thus closing the nutrient cycles. Sludge recycling from aquaculture waste to ornamental fish tanks is a sustainable approach that converts the aquaculture waste into useful agricultural products thus closing nutrient cycle and minimizing the stress on the environment (Zhanga et al. 2021). Solid waste produced from fish excreta, uneaten feed and bio-filter residue as they contain high amount of organic matter and nutrients (like nitrogen and phosphorus) can be used as a fertilizer crop. Biological activities of treating sludge using such measures as compositing, anaerobic digestion, or dewatering increase usability and safety of the sludge by stabilizing nutrients and destroying pathogens. When pouring this recycled sludge on agricultural fields, its application increases soil fertility, plants growth and lowers the dependence on chemical fertilizers, which are usually costly and harmful to the environment (Chiquito-Contreras et al. 2022; Sharma et al. 2024, 2025; Nisha et al. 2025). This practice does not only eliminate the problems of waste disposal that accompany aquaculture waste but also helps to foster a symbiotic relationship between fish farming and agriculture by promoting the use of efficient resources and circular economy principles. Implementing sludge recycling practice, ornamental fish culture operations can reduce wastes and minimize the ecological footprint and develop sustainable food and ornamental plant production systems (Campanati et al. 2022).

b. **Effluent Treatment Systems**: The pre-dispose treatment of waste reduces the risk of contamination of the surrounding water bodies. Ornamental fish culture effluent treatment systems must be introduced in order to minimize the effects of the wastewater disposal on the environment and ensure the sustainability of fish cultures (Turcios and Papenbrock 2014). These systems are meant to treat wastewater, so organic matter, nutrients, and potential pollutants—such as ammonia, nitrates, and phosphates—that can harm the aquatic system near the facility are excluded (Rani et al. 2025a, b; Tom et al. 2021). Sedimentation tanks, mechanical filtration, and advanced bio-filtration practices guarantee the removal of the solid and dissolved waste. Disinfection techniques such as UV sterilization or ozonation destroy pathogens hence minimizing incidences of diseases transmission to wild populations (Lieke et al. 2020). For nutrient removal, constructed wetlands or denitrification filters have been used more and more to reduce nitrogen and phosphorus levels that help to avoid eutrophication of water bodies in the vicinity. Effluent treatment does not only protect the local biodiversity but also reflects the standards of regulatory requirements thus making the environmental compliance of aquaculture operations better (Kiran et al. 2025; Reid et al. 2013). Treating wastewater successfully, the ornamental fish farmers help the long-term sustainability of their enterprise and ensure the health and balance of aquatic ecosystems (Campanati et al. 2022).

23.3.5 Disease Management

a. **Preventative Health Measures**: Quarantine policies and procedures for new stock minimize the entry of diseases by optimized preventative health measures as crucial components in the management of diseases and sustainability of ornamental fish culture (Sampaio et al. 2015). It is important to keep newly bought fish in different tanks or systems for two to four weeks to watch for and treat any diseases before adding them to the main tanks while monitoring fish for stress, diseases and parasites, and if any are present, get preventative treatments, such as treatment with saline solution or medicated feed (Mocho et al. 2022). The risk of water systems picking up diseases from abroad is limited by this approach, helping to maintain water quality and avoiding increasing cases of antimicrobial resistance. Aquaculture operations in quarantine tends to increase biosecurity and strengthen the overall safety of these operations helping ornamental fish farmers to achieve good health for their culture, greater yields, and a more environment friendly business ventures (Mishra et al. 2017).

b. **Biological Control**: Relying on addition of beneficial microbes in culture tanks for its health means reducing the dependence on chemical antibiotics by managing the health of tanks (Cardoso et al. 2019; Sihag and Sharma 2012). Probiotics made up of live beneficial bacteria that include *Lactobacillus*, *Bacillus*, and *Pseudomonas* species are being introduced into aquaculture systems in order to improve water quality and improve fish health. These microorganisms also compete with the injurious pathogens by colonizing the surfaces and invading nutrients ultimately inhibiting the growth of disease causing organisms (Saha et al. 2022). Probiotics also fortify the digestion and immunity of fish in generating enzymes and bioactive ingredients that would strengthen gut health and stress resistance. Likewise, the bio-floc technology utilizes microbial communities to convert waste products such as ammonia and nitrites to microbial protein, which can be used as additional feed (Ogello et al. 2021; Rani et al. 2025a, b). These biological control methods limit the use of chemical that would pollute the environment and cause risk of antimicrobial resistance. Thus, the ornamental fish farming acts as productive systems with compatible sustainable aquaculture practices by encouraging balanced microbial ecosystems.

c. **Limited Use of Chemicals**: The sustainable systems promote vaccines and natural treatments rather than excessive prescription of antibiotics or pesticides and avoidance of drug resistance and environmental pollution (Assefa and Abunna 2018). Limited application of chemicals in the cultivation of ornamental fish is an integral part of sustainable management, as practices do not involve the use of chemicals instead, they are based on preventative measures, such as vaccination, natural remedies, etc. to promote fish health. With the oral vaccinations, the fish is protected from certain diseases, thus minimizing the use of therapeutic antibiotic treatments that may give rise to drug resistance and residue contamination towards the aquatic systems (Radhakrishnan et al. 2023). Natural cures such as the extracts of herbs, probiotics, and immune-booster supplements can be safe and environmentally friendly substitutes of disease prevention and maintenance (Hoseinifar et al. 2020; Hodar et al. 2021). Such practices reduce the number of harmful chemicals, including antibiotics and pesticides, released into the environment, where they are not able to concentrate in the water bodies and lead to the development of resistant microbial strains. Sustainable systems give the priority to combined strategies of health management, which include biosecurity practices, stress avoidance tips and natural health promoters to promote the balanced ecosystem (Lieke et al. 2020). By eliminating chemical dependency, the ornamental fish farmers are able to maintain healthy stock, abide by environment regulation and are able to achieve long-term sustainability in aquiculture enterprises.

23.3.6 Habitat and Biodiversity Conservation

a. **Avoiding Wild-Caught Specimens**: Sustainable practices also prevent massive capture of wild fish, especially the endangered, and vulnerable as an important component of sustainable practices in ornamental fish culture as it will assist in protecting the endangered and threatened species, as well as lowering the burden on the natural ecosystems (Evers et al. 2019). Large scale capture of wild fish to serve the ornamental trade destroys and upsets biodiversity and also can encourage over-exploitation of sensitive species. Sustainable aquarium fish culture encourages breeding and multiplication of fish under captivity so that only fishes from farms are traded thus mitigating the need for wild-caught fish (King 2019). This apparent holds natural habitats to be maintained and mitigates the risk of transferring diseases from wild populations to the farmed stocks and aid in the conservation of the endangered species. Additionally, the farmed fish are usually bred in highly controlled condition thus making them more disease resistant and meeting their specific needs of the aquarium trade. Captive breeding given top priority can make ornamental fish culture a part of the conservation measures, reduce the impact on environment, sustain the process long-term (King 2019).
b. **Restocking Programs**: Restocking programs are one of the critical components under sustainable management in ornamental fish culture, especially where the populations of such species have been depleted as a result of either overfishing or loss of habitat (Domínguez and Botella 2014). Many fish farms dedicated to the ornamental fish also take part in such programs and breed and release farmed fish back into ecosystems so that they could restore the equilibrium and send supplemental stocks of the fish under the recovery of the wild stocks. These restocking initiatives are usually well planned and supervised so that the species can survive and continue to be genetically pure (Buckley et al. 2024). Reintroducing captive-bred fish into wild also contributes to the replenishing of natural stocks, biodiversity promotion, and lessening the effects of the ornament fish trade on imperiled species through these programs. The process of restocking also helps to achieve the long-term conservation goals because of the increased resilience of the ecosystem and decreased necessity to explore the wild populations further (Shraborni et al. 2024). Ornamental fish culture can then play a decisive role in ensuring the sustainability of the industry and protecting the aquatic biodiversity by responsible breeding and restocking (Trushenski et al. 2010).
c. **Preserving Natural Habitats**: It is important to preserve breeding grounds and reduce the level of habitat destruction for long-term sustainability. Care for breeding areas and reduction of habitats destruction is fundamental for the long-term sustainability of ornamental fish culture. Destruction of these important ecosystems, leading to pollution, coastal development, and irresponsible fishing practices destroy these species and the ecological equilibrium threatening the survival of such species (Reid et al. 2013). The sustainable ornamental fish

culture highlights the need for preservation of habitat and ushers in practices that do not impact the environment badly and realize the need to conserve existing natural breeding grounds. Such measures as restoration of habitat and creation of marine protected areas, as well as environmentally friendly aquaculture methods are applied for the protection of these crucial systems (Froehlich et al. 2017).

23.3.7 Energy and Resource Efficiency

a. **Renewable Energy Use**: Reduction in the level of greenhouse gas emission occurs as a result of solar powered pumps and energy efficient aerators. The incorporation of renewable energy in ornamental fish culture especially through solar energy pumps or energy efficient aerators tremendously increases sustainability by mitigating greenhouse gas emissions and de-emphasizing dependence on non-renewable forms of energy (Scroggins et al. 2022). Solar-powered pumps utilize the sun to move water and ensure the required flow rates, therefore, reducing the dependency on electricity drawn from fossil fuels (Pringle et al. 2017). On the same lines, energy efficient aerators maximize the oxygenation without wasting power, which in turn lowers the energy footprint of the overall aquaculture setup (Badiola et al. 2018; Troell et al. 2004).
b. **Water Conservation**: Reducing water use is essential for sustainability in ornamental fish culture, RAS and similar closed systems designed are modern approach to reduce water usage by far compared to traditional types of systems. While typical open systems keep losing water by always exchanging it, RAS collects and treats the water internally reducing amount of water used. The filtration and treatment technologies of RAS allow maintaining an optimal water quality by disposing of the waste, keeping the ammonia and nitrate on a balanced level and providing a stable environment for the fish (Sicuro et al. 2024). This efficiency is not only saving on the freshwater resources but also minimizing the risk of human-induced pollution of the environment through wastewater abstraction. Such water-saving systems can help ornamental fish farms increase their ecological footprint while lowering the costs of operation and complying with regulations on water management (Elalouf et al. 2018; Turlybek et al. 2025). The transformation towards water-efficient technologies means that ornamental fishes' industry can grow sustainably even in water-deficient parts of the world, while advancing efforts from around the globe to conserve the world's precious water resources (Ahmed and Turchini 2021).
c. **Reducing Carbon Footprint**: The reduction in distance to transport feed and materials reduces the industry's net impact on the environment. Decreasing the level of carbon footprint is an important aspect of strengthening the sustainability of ornamental fish culture and minimizing travel distances for feeds and supplies is an important factor in the process (Li et al. 2025). Locally sourced feed ingredients, equipment, and other necessary materials not only minimize the greenhouse gas emission incurred from transport across long distances, but

also promote regional economies as well as lessen reliance of global supply chains (Darras 2020). With local procurement in mind, ornamental fish farms can reduce consumption of fuel and cost in terms of logistics, as well as we can guarantee the freshness and high quality of inputs. In addition, on-site or nearby production of feed and supplies like bio-floc and live food production minimizes transportation requirements and also increases resource efficiency. It keeps with environmentally friendly processes, contributing to ensuring that the industry of ornamental fish can have a reduced impact on the environment and contribute to global climate change fighting. Local sourcing can become part of the operations of aquaculture practitioners, and they can build a more sustainable and robust production system (Patel et al. 2023).

23.4 Challenges to Implementation Practices

The adoption of sustainable freshwater ornamental fish culture practices is characterized by challenges that should be taken care for wide spread adoption. The barriers to the adoption of sustainable practices for this type of fish culture include a number of ecological, economical, and logistical reasons (Table 23.2).

23.5 Future Prospects

- **Exploration of alternative feed sources**

The search for alternative sources of feed is critical to sustainable management of ornamental fish culture since it seeks to solve the ecological and economic problems related to the culture of this type of fish. Sustainable substitutions like plant-based ingredients (soybean meal, rice bran, and spirulina) are laden with proteins, vitamins and antioxidants, insect meals (black soldier fly larvae (BSFL), mealworms and crickets) show some traction as high-protein feed alternatives, and microbial proteins (single-cell proteins (SCP) from yeast, bacteria or microalgae) are getting more attention in terms of cost effectiveness and sustainability so as to lessen dependency on traditional sources and prey on nutritional requirements of ornamental fish (Fantatto et al. 2024). Microalgae such as *Chlorella* and *Haematococcus* are especially of interest to ornamental fish diet owing to high carotenoid content thereby contributing to brilliant coloration (Hoseinifar et al. 2023; Hemaiswarya et al. 2011; Shah et al. 2018). Securing feed processing advances in the forms of extrusion and fermentation helps ensure better digestibility and palatability of plant-based feeds and makes it suitable for use with ornamental species (Reverter et al. 2021; Goyal et al. 2021). In addition, the potential of circular economy associated with the insect farming—based on the use of organic waste as feedstocks conforms to sustainable aquaculture practices (Fantatto et al. 2024).

Table 23.2 Challenges to implementation of sustainable freshwater ornamental fish culture practices

Challenges	Remarks
Knowledge gaps	Limited access to education and training
Market constraints	Demand for cheap ornamental fish leading to unsustainable practices to maximize profit
Weak enforcement	Poor regulatory oversight allowing overexploitation and habitat degradation
Environmental degradation	Habitat destruction, water pollution, wild stock overharvesting leads to natural resource depletion making it difficult to source brood stock sustainably
Lack of awareness	Lack of knowledge of stakeholders like farmers leading to reliance on unsustainable collection methods or improper farming techniques
High initial costs	High upfront costs of RAS, water quality management and energy-efficient equipment can be prohibitive for small-scale operators
Water quality management	Optimal water management posing a challenge for farmers with limited resources or technical knowledge
Market constraints	Limited demand for sustainably bred ornamental fishes by consumers due to unawareness of ecological impacts
Regulatory and policy gaps	Inadequate regulations and enforcement, and illegal wildlife trade
Disease management	Sustainable practices with focus on minimizing chemical treatment can make disease control challenging
Climate change	Disruption in aquaculture operations with changes in water temperature and availability due to climate change affecting viability or certain species in specific regions

In addition, aqua feed companies are also carrying out experiments on waste by-products of agriculture and food industries like the poultry by-product meal and fish processing waste to be able to come up with cost effective and nutrient dense feeds sparing resource efficiency. Microencapsulation and encapsulation technologies are also being implemented to enhance the delivery of nutrients to ensure that the feed is catering for specific dietary needs of different ornamental species. Adoption of these alternative sources of feed does not only alleviate the environmental impacts but also makes ornamental fish culture economically sustainable as the dependency on costly traditional feed is lowered. Further research and further cooperation of feed manufacturers and researchers with aquaculture scientists will be vital for optimizing these feed formulations and for their generalization in the ornamental aquaculture industry.

- **Development of more sustainable breeding techniques**

Sustainable breeding helps protect the environment and the business of ornamental fish farming. Breeding one type of species can create problems as in terms of inbreeding depression, reduced diversity among the species, a greater chance of inbreeding, and lack of healthy ecosystems. Hybridization has been used on purpose to create ornamental fish that have brighter colors, cooler patterns, and can survive in different habitats, but it's important to make sure these hybrids don't escape out into the wild (Rahman et al. 2018). To solve such problems, advanced genetic management instruments, including marker-assisted selection (MAS), pedigree analysis, etc. should be implemented in a program of development to increase desirable characteristics while ensuring the diversity of genetics (Maqsood and Ahmad 2017; Alam et al. 2024). Hormonal techniques that induct species with complex species needs have allowed year round breeding thus limiting dependence on wild populations (Zamri et al. 2022).

Cryopreservation of gametes and embryos has become a sustainable way to store genetic resources and conduct selective breeding programs without putting a lot of pressure on live broodstock (Martínez-Páramo et al. 2017). Genomic advances such as CRISPR-Cas9 gene editing provide precision breeding possibilities for enhancing such things as disease resistance, stress tolerance, but this will be dependent on adhering to ethical and ecological reasons (Gutási et al. 2023; Coogan et al. 2022). Moreover, the enriched breeding conditions can copy natural ecosystems to increase reproductive success and quality of offspring in captivity (Alam et al. 2024). Combination of such techniques with sustainable aquaculture systems like RAS increases water efficiency, reduces waste discharge and continues contributing to environmental sustainability. Certification programs like those offered by the Marine Aquarium Council (MAC) guarantee that breeding follows worldwide sustainability standards, with ethical breeding and trading.

23.6 Conclusion

To combat the challenges faced by sustainable ornamental fish culture, several practices such as eco-friendly breeding and stock management, water quality management, feed management, waste management, disease management, habitat and biodiversity conservation, sustained energy and resource efficiency are being amended to ensure sustainability at global level. However, there is a need to come together and work as a community comprising of education, reforms in policy formulation, financial aid and technological interventions, to make the freshwater ornamental fish culture industry more sustainable capable of withstanding shocks. The presented, scientifically viable approaches in combination of will help ornamental fish industry to minimize its ecological footprint while satisfying the increased demand for diverse and attractive species.

References

Abisha R, Krishnani KK, Sukhdhane K, Verma AK, Brahmane M, Chadha NK (2022) Sustainable development of climate-resilient aquaculture and culture-based fisheries through adaptation of abiotic stresses: a review. J Water Clim Chang 13(7):2671–2689

Ahmed N, Turchini GM (2021) Recirculating aquaculture systems (RAS): environmental solution and climate change adaptation. J Clean Prod 297:126604

Alam M, Abbas K, Zehra Z, Kamil F (2024) Genetic advancement, global trade dynamics, persistent challenges and future prospects in ornamental fish culture. Asian J Res Zool 7(1):32–46

Aly SM, Fathi M (2024) Advancing aquaculture biosecurity: a scientometric analysis and future outlook for disease prevention and environmental sustainability. Aquac Int 32(7):8763–8789

Aragão C, Gonçalves AT, Costas B, Azeredo R, Xavier MJ, Engrola S (2022) Alternative proteins for fish diets: implications beyond growth. Animals 12(9):1211

Arthington AH, Dulvy NK, Gladstone W, Winfield IJ (2016) Fish conservation in freshwater and marine realms: status, threats and management. Aquat Conserv Mar Freshwat Ecosyst 26(5):838–857

Assefa A, Abunna F (2018) Maintenance of fish health in aquaculture: review of epidemiological approaches for prevention and control of infectious disease of fish. Vet Med Int 2018(1):5432497

Badiola M, Basurko OC, Piedrahita R, Hundley P, Mendiola D (2018) Energy use in recirculating aquaculture systems (RAS): a review. Aquac Eng 81:57–70

Balami S (2021) Recirculation aquaculture systems: components, advantages, and drawbacks. Trop Agroecosystems 2(2):104–109

Baruah K, Norouzitallab P, Pal AK (2017) Development of low cost and eco-friendly feed for various candidate species for the sustainability of commercial aquaculture and reduction of aquatic pollution. In: Developing new functional food and nutraceutical products. Academic Press, pp 441–453

Bhattacharya S, Das BK, Ekka A, Parida PK, Roy A, Lianthuamluaia L, Saha A (2025) Empowering women through alternative sustainable livelihood: a case study of ornamental fish farming in India. Front Sustain Food Syst 9:1507917

Biondo MV, Burki RP (2020) A systematic review of the ornamental fish trade with emphasis on coral reef fishes—an impossible task. Animals 10(11):2014

Bostock J, McAndrew B, Richards R, Jauncey K, Telfer T, Lorenzen K, Corner R (2010) Aquaculture: global status and trends. Philos Trans R Soc B: Biol Sci 365(1554):2897–2912

Buckley SJ, Brauer C, Lamin C, Rose P, Vornicu DE, Beheregaray LB (2024) A community-driven captive-breeding and reintroduction program maintains genetic diversity in a threatened freshwater fish. Conserv Sci Pract 6(1):e13054

Campanati C, Willer D, Schubert J, Aldridge DC (2022) Sustainable intensification of aquaculture through nutrient recycling and circular economies: more fish, less waste, blue growth. Rev Fish Sci Aquac 30(2):143–169

Cardoso PHM, Moreno AM, Moreno LZ, Oliveira CHD, Baroni FDA, Maganha SRDL, Balian SDC (2019) Infectious diseases in aquarium ornamental pet fish: prevention and control measures. Braz J Vet Res Anim Sci 56(2):1–16

Chiquito-Contreras RG, Hernandez-Adame L, Alvarado-Castillo G, Martínez-Hernández MDJ, Sánchez-Viveros G, Chiquito-Contreras CJ, Hernandez-Montiel LG (2022) Aquaculture—production system and waste management for agriculture fertilization—a review. Sustainability 14(12):7257

Coogan M, Alston V, Su B, Khalil K, Elaswad A, Khan M, Dunham RA (2022) CRISPR/Cas-9 induced knockout of myostatin gene improves growth and disease resistance in channel catfish (Ictalurus punctatus). Aquaculture 557:738290

Darras AI (2020) Implementation of sustainable practices to ornamental plant cultivation worldwide: a critical review. Agronomy 10(10):1570

Domínguez LM, Botella AS (2014) An overview of marine ornamental fish breeding as a potential support to the aquarium trade and to the conservation of natural fish populations. Int J Sustain Dev Plan 9(4):608–632

Duarah JP, Mall M (2020) Diversified fish farming for sustainable livelihood: a case-based study on small and marginal fish farmers in Cachar district of Assam, India. Aquaculture 529:735569

Ebner B, Millington M, Holmes B, Wilson D, Sydes T, Bickel TO, Morgan DL (2020) Scoping the biosecurity risks and appropriate management relating to the freshwater ornamental aquarium trade across northern Australia

Elalouf H, Kaspi M, Elalouf A, Halachmi I (2018) Optimal operation policy for a sustainable recirculation aquaculture system for ornamental fish: simulation and response surface methodology. Comput Oper Res 89:230–240

Evers HG, Pinnegar JK, Taylor MI (2019) Where are they all from?—sources and sustainability in the ornamental freshwater fish trade. J Fish Biol 94(6):909–916

Fadhil R, Endan J, Arziman SNBS, Taip FS, Ja'afar MSBH (2011) Effectiveness of floating microbead bio-filter for ornamental fish in a recirculating aquaculture system. Fish Aquac J 1:1–8

Fantatto RR, Mota J, Ligeiro C, Vieira I, Guilgur LG, Santos M, Murta D (2024) Exploring sustainable alternatives in aquaculture feeding: the role of insects. Aquac Rep 37:102228

Froehlich HE, Gentry RR, Halpern BS (2017) Conservation aquaculture: shifting the narrative and paradigm of aquaculture's role in resource management. Biol Cons 215:162–168

Goyal S, Ott D, Liebscher J, Höfling D, Müller A, Dautz J, Reuss R (2021) Sustainability analysis of fish feed derived from aquatic plant and insect. Sustainability 13(13):7371

Gupta S, Makridis P, Henry I, Velle-George M, Ribicic D, Bhatnagar A, Netzer R (2024) Recent developments in recirculating aquaculture systems: a review. Aquac Res 2024(1):6096671

Gutási A, Hammer SE, El-Matbouli M, Saleh M (2023) Recent applications of gene editing in fish species and aquatic medicine. Animals 13(7):1250

Gutierrez-Wing MT, Malone RF (2006) Biological filters in aquaculture: trends and research directions for freshwater and marine applications. Aquac Eng 34(3):163–171

Halachmi I (2006) Systems engineering for ornamental fish production in a recirculating aquaculture system. Aquaculture 259(1–4):300–314

Hameed A, Majeed W, Naveed M, Ramzan U, Bordiga M, Hameed M, Rana N (2022) Success of aquaculture industry with new insights of using insects as feed: a review. Fishes 7(6):395

Hemaiswarya S, Raja R, Ravi Kumar R, Ganesan V, Anbazhagan C (2011) Microalgae: a sustainable feed source for aquaculture. World J Microbiol Biotechnol 27:1737–1746

Hodar AR, Vasava R, Mahavadiya D, Joshi N, Nandaniya V, Solanki H (2021) Herbs and herbal medicines: a prominent source for sustainable aquaculture. J Exp Zool India 24(1)

Hoseinifar SH, Sun YZ, Zhou Z, Van Doan H, Davies SJ, Harikrishnan R (2020) Boosting immune function and disease bio-control through environment-friendly and sustainable approaches in finfish aquaculture: herbal therapy scenarios. Rev Fish Sci Aquac 28(3):303–321

Hoseinifar SH, Maradonna F, Faheem M, Harikrishnan R, Devi G, Ringø E, Carnevali O (2023) Sustainable ornamental fish aquaculture: the implication of microbial feed additives. Animals 13(10):1583

Hua K, Cobcroft JM, Cole A, Condon K, Jerry DR, Mangott A, Strugnell JM (2019) The future of aquatic protein: implications for protein sources in aquaculture diets. One Earth 1(3):316–329

Huntington TC, Hasan MR (2009) Fish as feed inputs for aquaculture—practices, sustainability and implications: a global synthesis. FAO Fish Aquac Tech Pap 518:1–61

Irhayyim T, Fotedar R (2019) Effects of fish size and biofiltration techniques on water quality and nitrogen removal efficiency in recirculating aquaculture systems. Aquac, Aquar, Conserv Legis 12(5):1606–1616

Jain AK, Karunasagar I (2025) Ornamental fish and aquarium trade in the globalized world. In: Ornamental fisheries and aquarium keeping. CRC Press, pp 1–70

Kari ZA, Mat KHAIRIYAH, Kabir MA, Zal WA, Munir MB, Wei LS, & Téllez-Isaías GUILLERMO (2023) Soybean by-product: as an alternative to fish meal as protein source for aquaculture industry. J Sustain Sci Manag 18(5):179–202

King TA (2019) Wild caught ornamental fish: a perspective from the UK ornamental aquatic industry on the sustainability of aquatic organisms and livelihoods. J Fish Biol 94(6):925–936

Kiran, Kumar N, Sindhu S, Gawaria B, Ansari A, Dhariwal A, Sharma P, Kumar R (2025) GWAS approaches in medicinal plants for enhanced secondary metabolite production. In: Chen JT (ed) Biotechnology, multiple omics, and precision breeding in medicinal plants. CRC Press, pp 119–134. https://doi.org/10.1201/9781003475491-10

Kumar R, Gupta R, Dahiya T, Yadav R, Saifi R, Yodha K, Kumari B, Singh P, Shefali (2022a) Surveillance of *Heteropneustes fossilis* (stinging catfish) diseases and health management practices in Haryana. Appl Ecol Environ Sci 10(4):261–267. https://doi.org/10.12691/aees-10-4-12

Kumar R, Gupta RK, Yadav R, Saifi R, Yodha K, Pooja (2022b) *Eisenia fetida* as protein source for growth enhancement of *Heteropneustes fossilis*. Egypt J Aquat Biol Fish 26(2):577–588. https://doi.org/10.21608/ejabf.2022.234553

Kumar R, Yadav R, Kumar Gupta R, Pooja (2022c) A sustainable way for fish health management by replacement of chemical and drugs by earthworm. In: Environmental degradation in Asia: land degradation, environmental contamination, and human activities. Springer International Publishing, Cham, pp 329–352. https://doi.org/10.1007/978-3-031-12112-8_16

Li H, Zhou X, Gao L, Liang J, Liu H, Li Y, Liang S (2025) Carbon footprint assessment and reduction strategies for aquaculture: a review. J World Aquac Soc 56(1):e13117

Lieke T, Meinelt T, Hoseinifar SH, Pan B, Straus DL, Steinberg CE (2020) Sustainable aquaculture requires environmental-friendly treatment strategies for fish diseases. Rev Aquac 12(2):943–965

Little DC, Newton RW, Beveridge MCM (2016) Aquaculture: a rapidly growing and significant source of sustainable food? Status, transitions and potential. Proc Nutr Soc 75(3):274–286

Mandal A, Ghosh AR (2024) Role of artificial intelligence (AI) in fish growth and health status monitoring: a review on sustainable aquaculture. Aquac Int 32(3):2791–2820

Maqsood HM, Ahmad SM (2017) Advances in molecular markers and their applications in aquaculture and fisheries. Genet Aquat Org 1(1):27–41

Martínez-Páramo S, Horváth Á, Labbé C, Zhang T, Robles V, Herráez P, Cabrita E (2017) Cryobanking of aquatic species. Aquaculture 472:156–177

Mishra SS, Rakesh D, Dhiman M, Choudhary P, Debbarma J, Sahoo SN, Mishra CK (2017) Present status of fish disease management in freshwater aquaculture in India: state-of-the-art-review. J Aquac Fish 1(003):14

Mocho JP, Collymore C, Farmer SC, Leguay E, Murray KN, Pereira N (2022) FELASA-AALAS recommendations for biosecurity in an aquatic facility, including prevention of zoonosis, introduction of new fish colonies, and quarantine. Comp Med 72(3):149–168

Monticini P (2010) The ornamental fish trade: production and commerce of ornamental fish: technical-managerial and legislative aspects, vol 102. Food and Agriculture Organization of the United Nations, Rome, Italy

Moorhead JA, Zeng C (2010) Development of captive breeding techniques for marine ornamental fish: a review. Rev Fish Sci 18(4):315–343

Munguti JM, Musa S, Orina PS, Kyule DN, Opiyo MA, Charo-Karisa H, Ogello EO (2014) An overview of current status of Kenyan fish feed industry and feed management practices, challenges and opportunities

Nair CS, Manoharan R, Nishanth D, Subramanian R, Neumann E, Jaleel A (2025) Recent advancements in aquaponics with special emphasis on its sustainability. J World Aquac Soc 56(1):e13116

Nisha A, Rani S, Kumar S, Kumar R, Kumar A (2025) Recent technologies and strategies for bioremediation of toxic and hazardous waste. In: Dubey KK, Shah MP, Yadav A (eds) Challenges and sustainable solutions in bioremediation. CRC Press, pp 1–16. https://doi.org/10.1201/978 1003407317-1

Ogello EO, Outa NO, Obiero KO, Kyule DN, Munguti JM (2021) The prospects of biofloc technology (BFT) for sustainable aquaculture development. Sci Afr 14:e01053

Pailan GH, Banu H, Manna S, Singh DK (2022, August) Ornamental fish culture for enhancing livelihood of coastal farming communities. In: Transforming coastal zone for sustainable food and income security: proceedings of the international symposium of ISCAR on coastal agriculture, March 16–19, 2021. Springer International Publishing, Cham, pp 403–417

Palao C, Marquez GA, Ibasco K, Ferrer LC, Sagge P, Mendoza MTB (2020) Renewable waste water filtration system with phytoremediation used in aquaculture of freshwater ornamental fish. In: Proceedings of the international conference on industrial engineering and operations management, pp 1333–1334

Pandey PK, Mandal SC (2017, May) Present status, challenges and scope of ornamental fish trade in India. In: Conference: aqua aquaria India, Mangalore, pp 1–10

Patel AB, Kumar G, Debbarma S, Mutum D, Debnath S, Yadav NK (2023) Live duckweed-based circular aquaculture for climate resilience and carbon footprint reduction of fed aquaculture. In: Outlook of climate change and fish nutrition. Springer Nature Singapore, Singapore, pp 337–351

Pouil S, Tlusty MF, Rhyne AL, Metian M (2020) Aquaculture of marine ornamental fish: overview of the production trends and the role of academia in research progress. Rev Aquac 12(2):1217–1230

Pountney SM (2023) Survey indicates large proportion of fishkeeping hobbyists engaged in producing ornamental fish. Aquac Rep 29:101503

Preena PG, Rejish Kumar VJ, Singh ISB (2021) Nitrification and denitrification in recirculating aquaculture systems: the processes and players. Rev Aquac 13(4):2053–2075

Pringle AM, Handler RM, Pearce JM (2017) Aquavoltaics: synergies for dual use of water area for solar photovoltaic electricity generation and aquaculture. Renew Sustain Energy Rev 80:572–584

Radhakrishnan A, Vaseeharan B, Ramasamy P, Jeyachandran S (2023) Oral vaccination for sustainable disease prevention in aquaculture—an encapsulation approach. Aquac Int 31(2):867–891

Rahman MA, Lee SG, Yusoff FM, Rafiquzzaman SM (2018) Hybridization and its application in aquaculture. Sex Control Aquac 163–178

Raja K, Aanand P, Padmavathy S, Sampathkumar JS (2019) Present and future market trends of Indian ornamental fish sector. Int J Fish Aquac Stud 7(2):6–15

Ramasamy SNR (2025) Sustainable feed: insects, algae, and plant-based alternatives in aquaculture. In: Developments in aquaculture: trends, challenges, and applications, vol 1, p 175

Rani S, Sharma S, Singh D, Sheoran R, Kumar R (2025a) Improved production of anti-cancer compounds from medicinal plants through molecular and biotechnological mutagenesis strategies. In: Chen JT (ed) Biotechnology, multiple omics, and precision breeding in medicinal plants. CRC Press, pp 287–297

Rani S, Sharma S, Singh D, Sheoran R, Kumar R (2025b) Improved production of anti-cancer compounds from medicinal plants through molecular and biotechnological mutagenesis strategies. In: Chen JT (ed) Biotechnology, multiple omics, and precision breeding in medicinal plants. CRC Press, pp 287–297. https://doi.org/10.1201/9781003475491-21

Reid GM, Contreras MacBeath T, Csatádi K (2013) Global challenges in freshwater-fish conservation related to public aquariums and the aquarium industry. In: International zoo yearbook, vol 47, no 1, pp 6–45

Reverter M, Tapissier-Bontemps N, Sarter S, Sasal P, Caruso D (2021) Moving towards more sustainable aquaculture practices: a meta-analysis on the potential of plant-enriched diets to improve fish growth, immunity and disease resistance. Rev Aquac 13(1):537–555

Rout J, Sharma B, Swain S, Mishra S (2013) Algae in nutrition and colouration of ornamental fish: a review. Renew Sustain Energy 14(2):578–597

Roy D, Saha PK, Sarker S, Parves MA, Zavyalov AP, Latifa GA (2025) Culture practice of Oreochromis niloticus through recirculating aquaculture system (RAS): a modern and growth-optimized approach of fish culture by maintaining water quality with proper fish stocking density in Bangladesh. Environ Sci Pollut Res 32(18):11749–11766

Saha MK, Bhattacharya M, Patra BC (2022) The progressive development of probiotics, prebiotics, and synbiotics research and its multipurpose use in the ornamental fishery. In: Prebiotics, probiotics and nutraceuticals. Springer Nature Singapore, Singapore, pp 95–111

Saini VP, Singh AK, Paul T, Thakur A, Singh M (2024) Advances in environment management for sustainable fisheries and livestock production

Sales J, Janssens GP (2003) Nutrient requirements of ornamental fish. Aquat Living Resour 16(6):533–540

Sampaio FD, Carolina AF, Vinícius MST, Jean RV, Luís FF (2015) The precautionary principle and its approach to risk analysis and quarantine related to the trade of marine ornamental fishes in Brazil. Mar Policy 51:163–168

Sandilyan S (2016) Occurrence of ornamental fishes: a looming danger for Inland fish diversity of India. Curr Sci 2099–2104

Scroggins RE, Fry JP, Brown MT, Neff RA, Asche F, Anderson JL, Love DC (2022) Renewable energy in fisheries and aquaculture: case studies from the United States. J Clean Prod 376:134153

Shah MR, Lutzu GA, Alam A, Sarker P, Kabir Chowdhury MA, Parsaeimehr A, Daroch M (2018) Microalgae in aquafeeds for a sustainable aquaculture industry. J Appl Phycol 30:197–213

Sharma MMM, Kapoor D, Rohilla R, Loyal A, Sharma P, Husen A (2024) Role of essential minerals and rhizosphere in plant under changing environmental conditions. In: Husen A (ed) Essential minerals in plant-soil systems. Elsevier, pp 235–252. https://doi.org/10.1016/B978-0-443-16082-0.00002-3

Sharma P, Kapoor D, Kumar R, Husen A (2025) Rhizosphere and osmolytes coordination in plant systems under adverse situations. In: Husen A (ed) Roles of osmolytes in changing environment. Academic Press, pp 101–117. https://doi.org/10.1016/B978-0-443-26549-5.00020-5

Shefali, Kumar R, Sankhla MS, Kumar R, Sonone SS (2021) Impact of pesticide toxicity in aquatic environment. Biointerface Res Appl Chem 11(3):10131–10140. https://doi.org/10.33263/BRIAC113.1013110140

Shraborni A, Mandal SC, Parhi J (2024) Freshwater ornamental fishes of India: sustainable management and conservation. In: Aquaculture and conservation of inland coldwater fishes. Springer Nature Singapore, Singapore, pp 155–173

Shyam SS, Sirajudeen T, Biju Kumar A, Pramod Kiran RB, Premdev KV (2013) Harnessing ornamental fisheries resources for sustainable growth and development: a trade perspective from Kerala, India. J Aquat Biol Fish 1(1 & 2):155–168

Sicuro B, Akther KR, Tafuro V, Roy NC (2024) Exploring the intersection of aquaponics and ornamental fish culture in Italy's aquaculture evolution: a review of innovative integration. Am J Multidiscip Res Innov 3(2):7–13

Sihag RC, Sharma P (2012) Probiotics: the new ecofriendly alternative measures of disease control for sustainable aquaculture. J Fish Aquat Sci 7(2):72–103

Subasinghe R, Alday-Sanz V, Bondad-Reantaso MG, Jie H, Shinn AP, Sorgeloos P (2023) Biosecurity: reducing the burden of disease. J World Aquac Soc 54(2):397–426

Tom AP, Jayakumar JS, Biju M, Somarajan J, Ibrahim MA (2021) Aquaculture wastewater treatment technologies and their sustainability: a review. Energy Nexus 4:100022

Tripathi S, Hussain T (2022) Biofiltration treatment of wastewater through microbial ecology. In: An innovative role of biofiltration in wastewater treatment plants (WWTPs). Elsevier, pp 19–44

Troell M, Tyedmers P, Kautsky N, Rönnbäck P (2004) Aquaculture and energy use. In: Encyclopedia of energy, vol 1, pp 97–108

Trushenski J, Flagg T, Kohler C (2010) Use of hatchery fish for conservation, restoration, and enhancement of fisheries. In: Inland fisheries management in North America, 3rd edn. American Fisheries Society, Bethesda, Maryland, pp 261–293

Turcios AE, Papenbrock J (2014) Sustainable treatment of aquaculture effluents—what can we learn from the past for the future? Sustainability 6(2):836–856

Turlybek N, Nurbekova Z, Mukhamejanova A, Baimurzina B, Kulatayeva M, Aubakirova KM, Alikulov Z (2025) Sustainable aquaculture systems and their impact on fish nutritional quality. Fishes 10(5):206

Vala RB, Borichangar RV, Solanki HG, Patel MR, Parmar AM, Varma A (2024) Insect meal: an emerging eco-friendly approach for sustainable aquaculture feed industry. J Exp Zool India 27(2)

Watson I (2000) The role of the ornamental fish industry in poverty alleviation. Natural Resources Institute, Kent, UK, p 120

Wuyep SZ, Rampedi IT (2018) Urban fish farming in Jos, Nigeria: contributions towards employment opportunities, income generation, and poverty alleviation for improved livelihoods. Agriculture 8(7):110

Yusoff FM, Umi WA, Ramli NM, Harun R (2024) Water quality management in aquaculture. Camb Prism: Water 2:e8

Zamri AS, Zulperi Z, Esa Y, Syukri F (2022) Hormone application for artificial breeding towards sustainable aquaculture—a review. Pertanika J Trop Agric Sci 45(4)

Zhanga H, Gaoa Y, Liua J, Lina Z, Leeb CT, Hashimb H, Lia C (2021) Recovery of nutrients from fish sludge as liquid fertilizer to enhance sustainability of aquaponics: a review. Chem Eng 83

Chapter 24
Indicators for Sustainability in Aquaculture

Sudhir Kumar Kataria, Mohit Deswal, Geetanjli Sahni, Jaya Saini, Renu, Jyoti Chahal, Nisha Kumari, Pooja Kadyan, and Mayur Mukut Murlidhar Sharma

Abstract Aquaculture is one of the fastest-growing sectors in global food production, now accounting for more than half of the fish consumed worldwide. While this growth presents significant opportunities for enhancing global food security and reducing overfishing pressures on wild populations, the sustainability of aquaculture practices has raised concerns in both research and industry practice. Sustainability relates to taking care of the environment by also supporting economic and social well-being, and looking out for the needs of future generations. In sustainable aquaculture, one aims to harvest aquatic life in an environmentally friendly manner that protects the environment and works hand-in-hand with the local community. We want to generate and share benefits in an effective way and help profit the company. It should use resources in a balanced way without damaging the lands and surroundings. It should create more jobs for people living nearby, give them a better way of life, and help them keep their own traditions alive. It is important to check that aquaculture doesn't harm the environment or the people who do it, so we need to make sure it will work well all the time. Here, we talk about how to check and figure

S. K. Kataria · M. Deswal · J. Saini · Renu · P. Kadyan (✉)
Department of Zoology, Maharshi Dayanand University, Rohtak, Haryana, India
e-mail: pooja.kadyan546@gmail.com

S. K. Kataria
e-mail: sudhir.zoology24@mdurohtak.ac.in

Renu
e-mail: renugill1212@gmail.com

G. Sahni
Department of Zoology, Arya P.G College, Panipat, Haryana, India

J. Chahal
Department of Zoology, Hindu Girls College, Sonipat, Haryana, India

N. Kumari
Department of Commerce, Gurugram University, Gurugram, Haryana, India

M. M. M. Sharma
J1 Research Collaborator, Department of Surgery, University of Illinois at Chicago, Chicago, USA
e-mail: mayurmms@uic.edu

R. Kumar et al. (eds.), *Sustainability in Aquaculture and Fisheries*,
Springer Water, https://doi.org/10.1007/978-3-032-16605-0_24

out how well aquaculture systems are doing in terms of sustainability. This chapter looks at how aquaculture affects the economy, the environment, and society, and then talks about how we can check and get better at keeping it sustainable. The problems with keeping things sustainable as people need more and new technology come up are also talked about. Lastly, we will mention some basic things that need to be done to make sure fish farming is done in a better, more careful way everywhere.

Keywords Aquaculture · Biodiversity · Sustainable development · Environment protection · Aquaculture sustainability

24.1 Introduction

Sustainable aquaculture can be defined as the aquaculture practice which focuses on environmental, economic, and social sustainability to improve capacity building and utilize land effectively for the aquaculture sector (FAO 2020). Sustainable aquaculture involves the efficient production of aquatic organisms, ensuring minimal environmental degradation, economic viability, and social equity. Sustainable aquaculture as a system that optimizes the use of natural resources without degrading the surrounding environment. Sustainable practices include responsible water usage, efficient feed conversion, and minimizing pollutants. Additionally, sustainable aquaculture supports social and economic benefits by promoting fair labor practices, enhancing food security, and contributing to the livelihoods of coastal communities. Sustainable practices include integrated multi-trophic systems, recirculating aquaculture systems, and the use of plant-based or responsibly sourced feed. Through support of small economies and adoption of eco-friendly practices, aquaculture is vital for feeding today's world without harming nature for those who come after us (Kumar et al. 2020; Valenti et al. 2011).

It is important to realize that aquaculture can affect the environment, even if these effects are not included in the market price. Examples of these negative effects are pollution in water supplies, destroying habitats, and sourcing feed that is unsustainable (Garlock et al. 2024; Kumar and Singh 2021; Kumar et al. 2022a). For this reason, checking sustainability means looking at all the aspects, including environmental, social, and economic in aquaculture (Kumar et al. 2022c; Valenti et al. 2011). Through the use of tough policies, green technologies, and the cooperation of all stakeholders, aquaculture can help meet the rising need for seafood while protecting oceans and other bodies of water.

24.2 Key Dimensions of Aquaculture Sustainability

Using fish farms also means focusing on advancing sustainability by using only the amount needed to keep farming, limiting damage to nature, and upholding social benefits. Aquaculture sustainability should be reviewed by examining the economic, environmental, and social effects (Valenti et al. 2018). Sustainability and efficiency in aquaculture are largely affected by various key aspects represented in Figs. 24.1, 24.2 and Table 24.1. The plan confirms the importance of aquaculture for specific communities and values their way of life. These include:

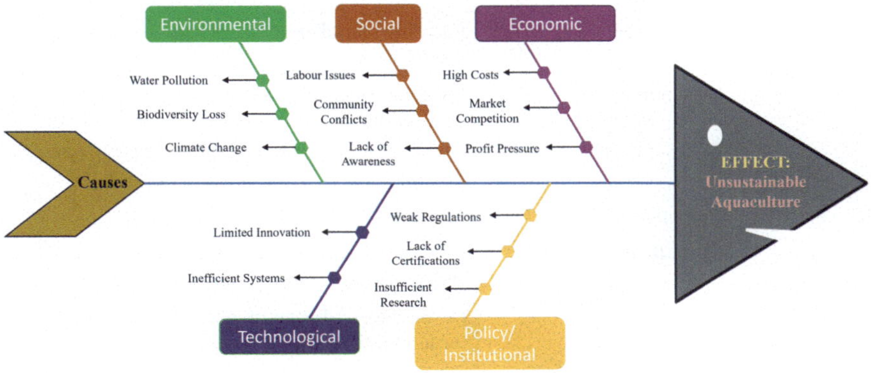

Fig. 24.1 Key drivers of unsustainability in aquaculture (*Source* Author)

Fig. 24.2 Dimensions of aquaculture sustainability (*Source* Author)

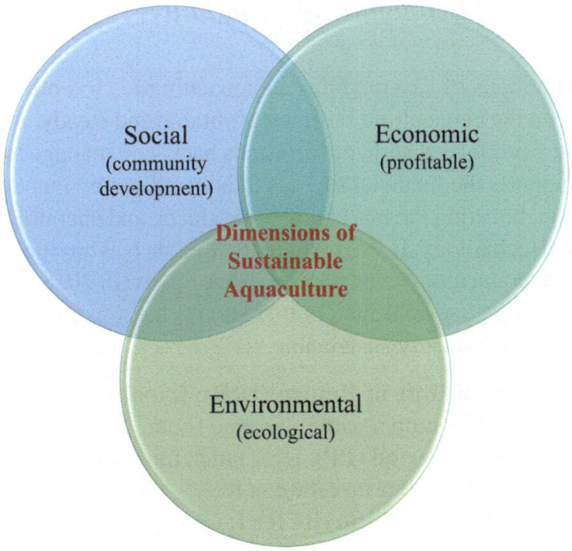

Table 24.1 Key dimensions of sustainable aquaculture

Dimension	Aspect	Focus areas
Environmental	Resource efficiency	Efficient use of water, feed, and energy
	Pollution management	Reducing waste, controlling nutrient run-off, and mitigating water pollution
	Climate impact	Lowering greenhouse gas emissions and using renewable energy
	Biodiversity conservation	Protecting wild species, avoiding overfishing, and preventing escapes of farmed species
Social	Employment and livelihoods	Creating jobs and supporting local economies
	Worker safety and welfare	Ensuring fair labor practices and safe working conditions
	Community engagement	Involving stakeholders in planning and decision-making
	Cultural values	Respecting traditional practices and local cultural dynamics
Economic	Profitability	Achieving sustainable financial returns
	Cost efficiency	Reducing production costs through technological and operational improvements
	Market access	Developing trade opportunities and infrastructure for fair pricing
	Resilience	Adapting to economic and environmental challenges

24.2.1 Economic Sustainability

The focus of sustainability in aquaculture is to optimize using capital, which helps keep the aquaculture business profitable and steady. Aquaculture systems succeed if they produce wealth, put proceeds back, and manage to survive economic changes. It supports the financial success of aquaculture by increasing the efficiency of production, boosting the marketing of products, and encouraging the use of new techniques and technology. It also includes factors such as buying and selling, earning profits, and how aquaculture can handle various issues. In 2018, researchers Valenti et al. (2011) mentioned several factors that should be considered to check whether ecosystems are economically sustainable.

- **Internal Rate of Return (IRR)**: It looks at how profitable an investment could become within a particular time frame, keeping expenses and pollution in mind.
- **Payback Period (PP)**: How much time it will take for an aquaculture business to break even after investing in it.
- **Benefit–Cost Ratio (BCR)**: It looks at whether the aquaculture system generates enough benefits to cover its costs, in addition to any negative impacts on the environment.

For an economy to be sustainable, businesses must find a way to remain viable for multiple generations.

24.2.2 *Environmental Sustainability*

This means taking good care of the environment and the water when running aquaculture activities to ensure things remain healthy and productive over the long term. Aquaculture is known to harm the environment, mainly due to excessive usage of resources and the release of pollutants into the water systems. The main idea of environmental sustainability is to reduce the impact of humans on nature by filing the waste, limiting pollution, protecting nature, and restoring habitats. It covers details on using alternatives to wild fish in feed and implementing solar power in farming (Fig. 24.2).

According to Valenti et al. (2018), some of these factors are:

- **Water and Energy Use**: Using water and energy more efficiently during aquaculture helps protect the environment. How much water and energy is used for each product is an important sign of sustainability.
- **Nutrient Management**: Although nitrogen and phosphorus are vital in fish farming, too much of them can cause aquaculture ponds to become polluted with algae and other harmful components. It is necessary to observe how much and often these nutrients are used so that the balance of the environment is not disturbed.
- **Pollutant Release**: One way to reduce the environmental effects of aquaculture is to keep the amount of pollutants to a minimum.

Practices like rice-fish farming have succeeded in making better use of resources and lessening the effects on the environment (Valenti et al. 2011).

24.2.3 *Social Sustainability*

Aquaculture should be socially sustainable, which means generating good living conditions and meeting ethical principles for all who are involved. The primary focus includes the needs, concerns, and lives of humans and their communities when it comes to aquaculture. Sustainable aquaculture systems also need to offer employment, ensure everyone earns fairly, and make food more accessible, all the while respecting local customs and practices.

- **Job Creation**: One way to assess social sustainability is to look at the number of jobs created by aquaculture systems. Greater job opportunities in a field are associated with more social sustainability (Valenti et al. 2011).

- **Income Distribution**: Paying workers a fair wage and offering social benefits along with other income benefits in aquaculture can help promote social welfare (Valenti et al. 2018).
- **Community Development**: Purchasing goods and services locally is good for aquaculture systems that want to support the local economic growth (Valenti et al. 2011).

Sustainable farms for fish and shellfish should watch over the security of the staff and give staff members opportunities to progress (Valenti et al. 2018).

24.3 Indicators of Aquaculture Sustainability

Making sure that aquaculture is sustainable focuses on taking care of the environment, people, and the economy. The sustainability indicators (seen in Fig. 24.3 and Table 24.2) are used for measuring and assessing the different kinds of impacts from aquaculture. Such indicators are used by stakeholders to assess and lead the way in sustainable aquaculture development. Here are some of the main factors used to judge the sustainability of aquaculture:

24.3.1 Environmental Indicators

Looking at how effectively resources are managed and how much environmental pollution is reduced is a common approach to sustainability. It is mainly nitrogen and phosphorus in wastewater that help determine the level of nutrient pollution.

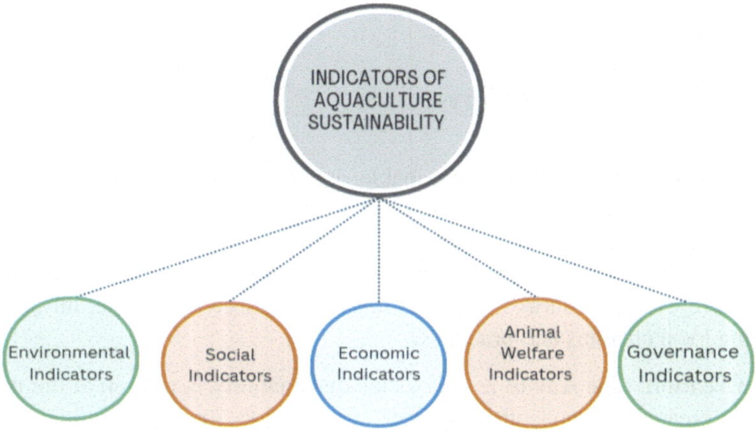

Fig. 24.3 Indicators of aquaculture sustainability (*Source* Author)

Table 24.2 Indicators of sustainability in aquaculture

Category	Indicators	Metric/unit	References
Environmental	Water quality	Nutrient levels, dissolved oxygen, pH levels	Boyd and Tucker (1998)
	Effluent management	Pollutant discharge, suspended solids	Hall et al. (2011)
	Resource use efficiency	Feed conversion ratio (FCR), water use	Tacon and Metian (2008)
	Biodiversity impact	Interaction with wild populations, habitat alteration	Naylor et al. (2000)
	Chemical use	Antibiotic and pesticide levels, Trace metals	Boyd and McNevin (2015)
	Carbon and energy footprint	GHG emissions, energy use	Parker et al. (2018)
	Disease and pathogen management	Disease outbreak records, Pathogen load in effluents	Rico et al. (2013)
	Land use and habitat conversation	Area of land used, rehabilitation efforts	Boyd and McNevin (2015)
	Regulatory compliance	Adherence to local and international environmental standards	FAO (1999)
	Socio-economic co-benefits		Naylor et al. (2000)
Social	Employment and income generation	Job creation, income levels, economic diversity	Brugère and Ridler (2004)
	Working conditions and labor rights	Fair labor practices, health and safety standards, Job security	Brugère and Ridler (2004)
	Community engagement and participation	Local stakeholder involvement, community consultation, support for local culture and traditions	Bush et al. (2013); Bush et al. (2013)
	Access to resources and land rights	Equitable resource distribution, land use and ownership, water access	Bush and Marschke (2014)
	Food security and nutrition	Contribution to local food supply, Affordability	Beveridge et al. (2013)
	Education and skill development	Training programs, knowledge transfer	Bush and Marschke (2014)

(continued)

Table 24.2 (continued)

Category	Indicators	Metric/unit	References
	Equity and inclusion	Gender equality, social inclusivity	Belton et al. (2017)
	Health and well-being of communities	Impact on public health, livelihood resilience	Kabir et al. (2020)
	Social license to operate	Public perception, trust and transparency	Bush and Marschke (2014)
	Contribution to sustainable development Goals (SDGs)	Alignment with SDGs	Belton et al. (2017); Bush and Marschke (2014)
Economic	Profitability	Net profit margin, return on investment, economic Efficiency	Troell et al. (2009)
	Production and yield	Annual production volume, growth rates, yield per unit area or volume	Badiola et al. (2012); Troell et al. (2009)
	Market competitiveness	Market share, price stability, product diversification	Asche et al. (2018)
	Export and trade contribution	Export revenues, trade balance, global competitiveness	Asche et al. (2018); Kumar et al. (2018)
	Economic impact on local economies	Local income generation, Economic spillover effects, multiplier effect	Badiola et al. (2012)
	Employment and job creation	Number of jobs created, quality of employment, Skill development and training	Boyd et al. (2020)
	Cost of production	Operational costs, feed conversion ratio (FCR), energy use and resource costs	Asche et al. (2018)
	Risk management and economic resilience	Insurance and risk mitigation strategies, diversification of revenue streams, adaptability to market changes	Kumar et al. (2018)
	Investment in infrastructure	Capital expenditure (CAPEX), maintenance and upkeep costs	Badiola et al. (2012)
	Economic contribution to sustainable development	Alignment with sustainable economic growth, Economic value added (EVA)	Belton et al. (2021)

(continued)

Table 24.2 (continued)

Category	Indicators	Metric/unit	References
Animal welfare	Health status and disease management	Disease incidence, mortality and survival rates, use of vaccines and prophylactic treatments	Ashley (2007)
	Stocking density	Space allocation per animal, behavioral indicators of stress	Turnbull et al. (2005)
	Water quality and environmental conditions	Oxygen levels, temperature control, pH and ammonia levels	Boyd et al. (2016)
	Feeding practices	Diet composition, Feeding frequency and method, Natural feeding behaviors	Glencross et al. (2007)
	Behavioral health	Opportunities for enrichment, reduced aggressive behaviors	Barton (2002); Huntingford et al. (2006)
	Handling and transport	Minimized handling, Stress-free transport, post-transport recovery	Lines et al. (2004)
	Breeding practices	Avoidance of genetic manipulation, selective breeding for robustness	Lines et al. (2004)
	Slaughter methods	Humane slaughter techniques, Training for slaughter practices	Lines et al. (2004)
	Pain and suffering management	Recognition of pain indicators, appropriate medical treatment	Barton (2002)
	Certification and compliance	Adherence to welfare standards, third-party audits, public reporting	Ellis et al. (2012)
	Training and worker awareness	Staff training, empathy and ethical responsibility	Lines et al. (2004), Näslund and Johnsson (2016)
Governance	Policy and regulatory frameworks	Existence of comprehensive regulations, Alignment with international standards, adaptability of policies	Tacon and Metian (2013)
	Enforcement and compliance	Monitoring and inspection, sanctions for Non-compliance, effectiveness of enforcement agencies	Buschmann et al. (2009)

(continued)

Table 24.2 (continued)

Category	Indicators	Metric/unit	References
	Licensing and permits	Transparent licensing processes, Environmental impact assessments, Periodic renewal of license	Bush, et al. (2013), Bush et al. (2013)
	Stakeholder engagement and participation	Involvement of local communities, consultative committees, feedback mechanisms	Pomeroy et al. (2016)
	Transparency and accountability	Public reporting, open access to information, independent audits	Molnar et al. (2008)
	Conflict resolution mechanisms	Dispute resolution systems, mediation and arbitration, community rights protections	Ratner et al. (2013)
	Interagency coordination	Collaboration between government agencies, regional and international cooperation	FAO (1995)
	Sustainability certification programs	Adoption of certification standards, support for small-scale producers	Bush et al. (2013), Bush et al. (2013)
	Capacity building and education	Training programs for regulators and industry, research and development funding	Grafton et al. (2007)
	Funding and investment incentives	Subsidies for sustainable practices, Access to credit for sustainable initiatives, public–private partnerships	Grafton et al. 2007)
	Risk management and preparedness	Disaster preparedness plans, CONTINGENCY funds, climate change adaptation policies	Ashley (2007)
	Evaluation and improvement mechanisms	Regular policy reviews, feedback loops, indicators of success	Molnar et al. (2008)

The environmental impact of aquaculture is generally assessed by using LCA to look at the water, energy, and waste used during aquaculture (Kiran et al. 2025; Nisha et al. 2025; Rani et al. 2025; Valenti et al. 2011). People often argue that using external feeds in aquaculture is harmful to the environment and not very efficient. It's important to add fishmeal and fish oil to the feeds used in aquaculture. As a result, wild fish are removed from their habitat, which hurts the environment. They make sure that carrying out the activity won't harm the environment to the extent that it would endanger its use for years.

There are some key environmental indicators of sustainable aquaculture:

i. **Water Quality**

- **Nutrient Levels (e.g., nitrogen, phosphorus):** This makes it possible to watch for the early appearance of eutrophication.
- **Dissolved Oxygen (DO):** Needs to be present for healthy aquatic organisms and warns about stress in an ecosystem.
- **pH Levels:** Helps detect the pH level in water that can be important for the fish and environmental balance (Boyd and Tucker 1998).

ii. **Effluent Management**

- **Pollutant Discharge:** Checking the release of waste, particularly its chemicals, antibiotics, and feed residues, helps make sure that standards are met.
- **Suspended Solids:** By monitoring, scientists are able to notice the changes happening in near-by waters, and any sedimentation.

iii. **Resource Use Efficiency**

- **Feed Conversion Ratio (FCR):** It measures the amount of feed used compared to animal weight, with a lower FCR signifying better sustainability.
- **Water Use:** Number of aquaculture products produced, while paying attention to not wasting water and reusing it (Tacon and Metian 2008).

iv. **Biodiversity Impact**

- **Interaction with Wild Populations:** Looking at how the escape of farm fish affects wild animals and their environment.
- **Habitat Alteration:** Accessing the extent of the destruction of mangrove forests or the seabed from aquaculture operations (Naylor et al. 2000).

v. V. Chemical Use

- **Antibiotic and Pesticide Levels:** By keeping track of fields, farmers can use fewer pesticides, which reduces both antibiotic resistance and toxic impacts on other animals.
- **Trace Metals:** It is necessary to control copper and zinc from aquaculture because if not, they can end up bioaccumulating.

vi. **Carbon and Energy Footprint**

- **Greenhouse Gas (GHG) Emissions:** Taking into account emissions from the ways things are manufactured and delivered.
- **Energy Use:** Often shoppers are rewarded for being regulars and this is where Coles's loyalty program comes in.

vii. **Disease and Pathogen Management**

- **Disease Outbreak Records:** Seeing how outbreaks are handled is important for making sure all safety and sustainability measures are working properly.
- **Pathogen Load in Effluents:** Can cause risks to both wild and farmed populations of fish (Rico et al. 2013).

viii. **Land Use and Habitat Conservation**

- **Area of Land Used:** Trying to use as little space as possible and bypassing vulnerable areas.
- **Rehabilitation Efforts:** Reforestation of mangroves and healing natural areas should be performed after any operation (Boyd and McNevin 2015).

ix. **Regulatory Compliance**

Full compliance with the specifications put forth by the local and international Environmental Standards. Following FAO standards and regulations (FAO 1999) in the operations.

x. **Socio-Economic Co-Benefits**

Although the main focus is on the environment, it is also important to look at how sustainability relates to people and resources (Naylor et al. 2000).

They are important for those involved in aquaculture, as they help them manage farming practices to protect the environment and ensure future availability of resources.

24.3.2 Social Indicators

Sustainability measures used in social context focus on how many jobs and fair pay are provided in the aquaculture industry to local communities. Socially, projects are considered sustainable when they give jobs and make income distribution fair (Valenti et al. 2011). The number of jobs made available, the share of the local labor involved, and the earnings from aquaculture for the community are some vital social markers. Often, export-minded aquaculture businesses focus more on earning profits than ensuring nearby people have enough to eat, which can result in disputes over resources and inequality in the community. Social indicators for aquaculture sustainability include factors that reveal the effects on the community, people's lives, and society at large. Thanks to these indicators, it is possible to tell if farming fish is helping people grow economically, get fair working conditions, and enjoy community benefits (Valenti et al. 2018). Here are some of the primary indicators for judging how sustainable aquaculture practices are:

i. **Employment and Income Generation**

- **Job Creation**: The total of jobs created in the local area thanks to aquaculture and the impact it has on the economy.

- **Income Levels**: Actions designed to increase income at the local and area level and ensure workers are fairly paid.
- **Economic Diversity**: All in all, developing aquaculture strengthens local communities by earning more money and making them less dependent on just one form of income (Boyd et al. 2016).

ii. **Working Conditions and Labor Rights**

- **Fair Labor Practices:** Keeping up with all rules about work, avoiding any kind of unfair treatment, and making sure no children or adults do work against their will.
- **Health and Safety Standards**: Steps that are taken so workers don't get injured while doing their job.
- **Job Security**: Prospects and promises of better lives for people who work in fishing and farming companies (Boyd et al. 2016).

iii. **Community Engagement and Participation**

- **Local Stakeholder Involvement**: Local populations playing a role when decisions about aquaculture projects are made.
- **Community Consultation**: Inviting the community to share their views on the design and running of projects.
- **Promotion and Upkeep of Key Culturally and Traditionally Tied Activities**: Taking into account the customs and practices of the community while performing aquaculture (Bush et al. 2013a; b).

iv. **Claim to Resources and Land**

- **Equitable Resource Distribution**: Noticing that what is used for aquaculture is not taken from the fishers and farmers in local communities.
- **Land Use and Ownership**: See to it that buying land is just and remain sensitive to claims made by local and indigenous communities.
- **Water Access**: Guaranteeing that the amount of fresh and clean water available to town residents isn't diminished by the aquaculture (Bush and Marschke 2014).

v. **Food Security and Nutrition**

- **Helping to Sustain the Local Community with Food**: Bringing aquaculture products to markets helps boost the total range of food and improves nutrition among members of the community.
- **Affordability**: Local people can get and use products from fish farming, which helps them have enough good food to eat, as Beveridge and his colleagues explain in their study (Beveridge et al. 2013).

vi. **Education and Skill Development**

- **Training Programs**: Arrange training sessions where workers and people living near the farm can pick up new skills and knowledge on aquaculture and environmental protection.

- **Knowledge Transfer**: To keep good results, programs support learning of new techniques and technologies that people can use going forward (Bush and Marschke 2014.

vii. **Equity and Inclusion**

- **Gender Equality**: Taking steps to ensure women and marginalized people have the same chances of getting work in aquaculture.
- **Social Inclusivity**: Constructing projects in aquaculture in a way that all groups of people gain advantages and can share in the benefits (McRea et al., 2020).

viii. **Health and Well-being of Communities**

- **Impact on Public Health**: Looking at potential health concerns with aquaculture, for example, waterborne diseases, and chemicals ending up in the water.
- **Livelihood Resilience**: If local people can use aquaculture for various types of income, they are better prepared to survive hard times, just as noted in the study by (Kabir et al. 2020).

ix. **Social License to Operate**

- **Public Perception**: If local people are happy with aquaculture, it tends to become successful in achieving its goals.
- **Trust and Transparency**: The extent to which people in the aquaculture industry show what they are doing and how it affects the environment (Bush and Marschke (2014).

x. **Implementing the United Nations' Sustainable Development Goals** helps countries focus on key areas like ending hunger, ensuring clean water, and making sure people have better health.

- **Alignment with SDGs**: Studying how Red Cross uses aquaculture to support their goals like helping people out of poverty, making sure there is equal treatment for men and women, giving people fair and good jobs, and supporting communities in a way that lasts.

24.3.3 Economic Indicators

To check whether an investment plan is economically viable, many people use traditional measures such as internal rate of return, payback period, and benefit–cost ratio (Valenti et al. 2011). Even so, they should take into account the after-effects on nature and communities to get the true picture of sustainable aquaculture. Take, for example, the way aquaculture can grow near seashores, conflict with tourism or fishing, and result in challenges for both society and the economy (Garlock et al. 2024). They allow experts to check if aquaculture is financially profitable, saves money, and leads to a stable economy in the future. The following economic indicators can show whether aquaculture is sustainable:

i. **Profitability**

- **Net Profit Margin:** The amount of profit left after all expenses have been deducted in aquaculture. If a company has a positive net profit margin, it is likely economically sustainable.
- **Return on Investment (ROI):** The difference between how much money you make from aquaculture and how much you spend setting it up.
- **Economic Efficiency:** Producing more fish from less energy, feed, and labor, which indicates efficient costs for the operation (Troell et al. 2009).

ii. **Production and Yield**

- **Annual Production Volume:** How many products do we produce through aquaculture every year.
- **Growth Rates**: Checking if production is on the rise over the years to determine how scalable and sustainable the business will be.

iii. The amount of wheat produced for each square meter or cubic meter. It considers how efficiently and effectively a space is used (Badiola et al. 2012; Troell et al. 2009).

iv. **Market Competitiveness**

- **Market Share:** How much of the overall aquaculture sales is due to a specific operation or sector.
- **Price Stability:** Monitoring changes in prices to find out if aquaculture products generally stay the same or vary in the market.
- **Product Diversification:** To make the economy more secure and the market stronger, offer items such as several types of fish and shellfish (Asche et al. 2018).

v. **Export and Trade Contribution**

- **Export Revenues:** Income earned from international sales helps add to a country's foreign exchange earning.
- **Trade Balance:** When the country has more aquaculture exports than imports, it is good for the domestic economy.
- **Global Competitiveness:** Meeting international trade standards allows aquaculture to succeed more easily on the global market (Asche et al. 2018; Kumar et al. 2018).

vi. **Economic Impact on Local Economies**

- **Economic Spillover Effects:** When events use local services, they boost the local economy.

- **Multiplier Effect:** How much developing the aquaculture industry helps support the same industries related to economic growth (Badiola et al. 2012).

vii. **Employment and Job Creation**

- **Number of Jobs Created:** Both direct and indirect jobs and employment prospects connected to working in aquaculture.
- **Quality of Employment:** As a result of such benefits, it gets easier for employees to look after themselves and their families in the future.
- **Skill Development and Training**: Offering training and education to employees can boost both the quality and productivity of workers within the company (Boyd et al. 2020).

viii. **Cost of Production**

- **Operational Costs:** Checking the expenses for feed, energy, and care for your animals helps you spot ways to cut down on costs for the farm.
- **Feed Conversion Ratio (FCR):** How much feed is needed to help the chicken grow and produce eggs is an economic indicator of feed efficiency. A low FCR is ideal for a farm as it means they save on feed and make extra profit.
- **Energy costs associated with making products:** When you try to use less energy, you can keep your monthly bills down (Asche et al. 2018).

ix. **Risk Management and Economic Resilience**

- **Insurance and Risk Mitigation Strategies:** Use of financial tools and practices to protect your business from losing money because of things like a disease, bad weather, or sudden changes in the market.
- **Diversification of Revenue Streams:** Generating income from more than one stream of business (like making products as well as offering processing services) can help lower the financial risks that come with running a business.
- **Adaptability to Market Changes:** Ability to make changes in how goods are made or what is being made in order to meet what customers want or need (Kumar et al. 2018).

x. **Investment in Infrastructure**

- **Capital Expenditure (CAPEX):** Investment in things like green buildings and new technology that can help people work more effectively in the future.
- **Maintenance and Upkeep Costs:** Regular reinvestment in facilities and equipment so that the factory can keep making products well.

xi. **Economic Contribution to Sustainable Development**

- **Alignment with Sustainable Economic Growth:** Support for national and regional plans that help grow the economy in ways that work for the long term and take care of people and the environment.
- **Economic Value Added (EVA):** Measurement of things other than just regular costs and investments, helping to show how much the business is actually helping the economy grow (Belton et al. 2021).

These economic indicators help us see if aquaculture is making money now as well as if it's possible for the business to keep growing in a way that helps all involved, including the people who invest, the people who work there, and the people who live nearby.

24.3.4 Animal Welfare Indicators

Health, well-being, and overall handling of farmed aquatic animals are monitored in assessing the sustainability of aquaculture. Using these indicators ensures the industry works ethically and contributes to its productivity and sustainability. Those are some main animal welfare benchmarks to show if aquaculture is sustainable:

i. **Health Status and Disease Management**

   - **Disease Incidence**: The rate and importance of diseases that break out among farmed fish or shellfish.
   - **Mortality and Survival Rates**: Keeping an eye on these rates reflects on how well the stock is performing, both physically and mentally.

ii. **Immunization and use of medications** aimed at disease prevention are important parts of public health care. Helping animals escape illnesses by using vaccines is a sign of caring for their welfare and future (Ashley 2007).

iii. **Stocking Density**

   - **Space Allocation per Animal**: Guiding the number of animals in a habitat to ensure natural actions, little stress, space for all, and a low level of aggression.
   - **Behavioral Indicators of Stress**: Seeing fish that swim strangely, gulp air often, or behave aggressively toward each other is often a sign of a full tank (Turnbull et al. 2005).

iv. **Water Quality and Environmental Conditions**

   - **Oxygen Levels**: If the oxygen is too low, fish suffer from stress and do not reach their full growth potential.
   - **Temperature Control**: Optimal temperatures for each type of animal lessens stress and makes them feel happier.
   - **pH and Ammonia Levels**: Managing chemical levels to protect the fish from any harm (Boyd et al. 2016).

v. **Feeding Practices**

- **Diet Composition**: Ensuring that your pet's diet contains all the necessary nutrients and is not full of additives supports their growth and health.
- **Feeding Frequency and Method**: The right meals are important for animals since giving them too much food or not enough can be bad for their health.
- **Natural Feeding Behaviors**: Animals need to be able to forage and feed normally to keep themselves healthy (Glencross et al. 2007).

vi. **Behavioral Health**

- **Opportunities for Enrichment**: Putting in places that allow fish to hide or flee, as well as to interact with each other.
- **Reduced Aggressive Behaviors**: Tank set-up should be planned properly, the amount of fish should be right for the size of the tank, and their species need to get along.
- **Indicators of Chronic Stress**: Noticing changes in behavior and measuring cortisol in the animal are both ways to identify persistent stress (Barton 2002; Huntingford et al. 2006).

vii. **Handling and Transport**

- **Minimized Handling**: Using modern and automated technology helps reduce the handling and stress involved.
- **Stress-Free Transport**: It's important to protect live animals from extreme heat or cold, as well as from a lack of oxygen during transport.
- **Post-Transport Recovery**: Observing any changes in the animals' health and whether they continue to live after the trial (Lines et al. 2004).

viii. **Breeding Practices**

- **Avoidance of Genetic Manipulation**: Ensuring that no activities happen that can cause bodily deformities or a loss of fitness.
- **Selective Breeding for Robustness**: Work should be done to breed cows more resistant to disease and to maintain good treatment of cows during their reproductive phase (Lines et al. 2004).

ix. **Slaughter Methods**

- **Humane Slaughter Techniques**: Abattoirs are allowed to use electrical stunning or sound pulses to ease the animal's stress and pain, which satisfies animal welfare regulations.
- **Training for Slaughter Practices**: Having all those who handle animals be trained in how to take care of them properly.

x. **Pain and Suffering Management**

- **Recognition of Pain Indicators**: Informing and preparing all my staff to handle feelings of pain, pain, or distress.

- **Appropriate Medical Treatment**: Ensuring that animals receive regular health care from a veterinarian, and only using medication when necessary (Barton 2002).

xi. **Certification and Compliance**

- **Adherence to Welfare Standards**: Adhering to industry standards and certification programs like ASC and Global GAP, both of which outline how to care for farm animals.
- **Third-Party Audits**: Having independent bodies audit welfare practices on a regular basis to ensure they follow best practices.
- **Public Reporting**: Part of sustainability reporting ought to include details about welfare and the indicators used for this purpose.

xii. **Training and Worker Awareness**

- **Staff Training on Animal Welfare**: Ongoing training is provided to ensure all workers understand how to best look after the fish.
- **Empathy and Ethical Responsibility**: Ensuring that the culture in the workplace places importance on empathy and good care for animals (Lines et al. 2004; Näslund and Johnsson 2016).

xiii. **Through these animal welfare indicators**: aquaculture is ensured to be sustainable financially, environmentally, and also in taking care of the animals involved. Using better welfare methods makes animals stronger, the production process more reliable, and consumers more likely to have a positive image of the meat products.

24.3.5 Governance Indicators

Governance indicators of aquaculture sustainability look at how well rules, policies, and organizations help make sure that aquaculture is done in a way that protects the environment and is good for the future. These indicators help make sure that aquaculture is run in a way that takes care of the environment, looks after the community, and supports a healthy economy while also keeping things open, fair, and involved everyone who is affected. Here are some main signs to look for to check if aquaculture is being done in a sustainable way:

i. **Policy and Regulatory Frameworks**

- **Existence of Comprehensive Regulations:** Clear, straight-forward rules and guidelines that help us take care of the environment, treat animals in a good way, keep our food safe, and make sure everyone working in aquaculture is looked after.

- **Alignment with International Standards:** Consistency with global rules, like the ones made by organizations such as the Food and Agriculture Organization (FAO) and the Aquaculture Stewardship Council (ASC).
- **Adaptability of Policies:** The way countries' rules and standards can change depending on if new science or technology shows up, or if the environment goes through changes.

ii. **Enforcement and Compliance**

- **Monitoring and Inspection:** Regular check-ups of aquaculture farms are important so that the people running them follow the laws and protect the environment.
- **Sanctions for Non-Compliance:** Presence of clear penalties for those who don't follow the rules helps stop people from doing actions that can be bad for the environment.
- **Effectiveness of Enforcement Agencies:** Adequately supported and trained agencies should be available to help make sure the rules are followed and obeyed (Buschmann et al. 2009).

iii. **Licensing and Permits**

- **Transparent Licensing Processes:** Clear and fair steps for getting licenses that allow the public to give their opinions and make sure everything is done honestly and without any bribes.
- **Environmental Impact Assessments (EIA):** Requirement for doing thorough EIAs to see if giving a permit would cause only a bit of harm to the environment or people.
- **Periodic Renewal of Licenses:** Regular checks and updates so that the company keeps doing the right things for the environment (Bush et al. 2013a, b).

iv. **Stakeholder Engagement and Participation**

- **Involvement of Local Communities:** Provisions to help local communities, indigenous people, and small-scale farmers have a chance to share their views on making policies.
- **Consultative Committees:** Setting up advisory boards made up of people from different areas like business, government, charity groups, and schools.
- **Feedback Mechanisms:** Means for stakeholders to speak up about what they think and what issues concern them when it comes to how aquaculture is run and controlled.

v. **Transparency and Accountability**

- **Public Reporting:** There must be clear and easy-to-find information shared about how the aquaculture works, how it affects the environment, and whether the farm is following all the rules.
- **Open Access to Information:** Anyone can get access to the information on permits, regulations, and inspections carried out by agencies.

- **Independent Audits:** Third-party reports are important to check how effectively the company complies with good governance standards (Molnar et al. 2008).

vi. **Conflict Resolution Mechanisms**

- **Dispute Resolution Systems:** Clearly outlined steps for dealing with issues that come up from sharing or using land and water by various parties.
- **Mediation and Arbitration:** There is the option to use neutral third-party mediators who can help resolve conflicts properly and quickly.
- **Community Rights Protections:** Making certain that local people and their users are able to protect what belongs to them and the resources they rely on (Ratner et al. 2013).

vii. **Interagency Coordination**

- **Collaboration Between Government Agencies:** Working together with all the relevant government departments (e.g., environment, fisheries, and trade) to prevent overlap or conflicts in their policies.
- **Regional and International Cooperation**: Cooperating with other nations on shared resource management and aquaculture effects (FAO 1995).

viii. **Sustainability Certification Programs**

- **Adoption of Certification Standards:** Requiring or suggesting to farmers to obtain approved sustainability certificates (like ASC or Global GAP).
- **Support for Small-Scale Producers**: Offering help to small-scale fisheries to fulfill quality certification standards with proper instruction and materials (Bush et al. 2013a, b).

ix. **Capacity Building and Education**

Training programs are available for financial regulators and for individuals working in the industry. Professional development for government, business, and community leaders in the field of sustainable practices and compliance.

- **Research and Development Funding:** Investments by the government or groups in R&D to achieve sustainability in aquaculture (Grafton et al. 2007).

x. **Funding and Investment Incentives**

- **Subsidies for Sustainable Practices**: Incentives for farms that work toward eco-friendly practices and new technology. Sources of funding for aquaculture businesses that wish to invest in sustainability.
- **Public–Private Partnerships:** Many areas now depend on the teamwork between governments and private companies in driving sustainable projects (Grafton et al. 2007).

xi. **Risk Management and Preparedness**

- **Disaster Preparedness Plans**: Steps that can be taken to lessen the problems caused by natural calamities, diseases, and other hazards for fish farmers.
- **Contingency Funds:** How to use money to help with emergencies and the process of recovery.
- **Climate Change Adaptation Policies:** Policies that look at how climate change may affect aquaculture are necessary (Ashley 2007).

xii. **Evaluation and Improvement Mechanisms**

- **Regular Policy Reviews:** Review the governance of aquaculture on a regular basis to look for areas that can be made better.
- **Feedback Loops**: Strategy for including stakeholder comments and feedback during policy updates.
- **Indicators of Success**: Assessing how much the company is lowering impacts on nature and how it benefits the communities (Molnar et al. 2008).

As a result of using these signs, aquaculture becomes safer because all important interests are taken care of and protected. In aquaculture, trust, a decrease in conflicts, and sustained industry success are all possible thanks to good governance. By watching these indicators, stakeholders can see where there is room for growth, meet global sustainability targets, and help make the industry greener in the long term. By using these indicators together, the big picture of aquaculture sustainability can be checked and supported, making sure it follows the required standards. Sustainable ways of doing things make communities and our environment stronger, kinder, and help them benefit in the long run.

24.4 Challenges in Achieving Aquaculture Sustainability

Evaluating whether aquaculture is sustainable requires us to look at the systems from every angle. Ecoreint and LCA offer much information about the environment, but they mostly neglect significant economic and social issues (Valenti et al. 2011).

According to (Valenti et al. 2018), a combination of indicators evaluates the sustainability of a strategy in all three aspects. These indicators help researchers, policy makers, and investors see what is good and what needs to improve in aquaculture systems, and also to see how different systems measure up over time and in different places. Nevertheless, finding important data can be tricky, and new efforts should be made to streamline how sustainability is measured. Aquaculture encounters many obstacles to sustainability, which influence environmental, economic, and social factors (Fig. 24.4). Below are key challenges:

Fig. 24.4 Challenges in aquaculture sustainability (*Source* Author)

24.4.1 Environmental Challenges

Sustainable aquaculture can be easily challenged by the possibility of harming the ecosystem.

Because high-intensity systems can produce big amounts of waste, it can lead to eutrophication in nearby water bodies and a decrease in biodiversity.

- Harvesting wild fish for aquafeeds can negatively affect marine ecosystems and disturb the balance between different creatures in the water (Garlock et al. 2024; Ratner et al. 2013).
- So far, alternatives to animal feed using plants and insects are still being developed (Ratner et al. 2013).
- Water temperature, salinity, and dramatic changes in the weather brought about by climate change pose significant problems for fish farming. Such changes often lower the productivity of farming sea animals and make their activities more harmful to the environment (Boyd et al. 2016).
- By doing this, aquaculture can lead to the destruction of different habitats (for example, the destruction of mangroves for shrimp farming), water pollution from

nutrients on land, and species loss both among farmed and wild fish because escaped fish from aquaculture could compete or interbreed with them (Boyd and McNevin 2015).

- Having too many fish in a small space in aquaculture makes it much easier for diseases and parasites to spread, negatively impact productivity and increase the use of antibiotics and chemicals (Ponte 2012).

Economic Challenges

- Aquaculture's sustainability may be weakened by changes in the market and the economy. As a result of fishmeal and fish oil becoming more expensive, it is now harder for small producers to stay economically afloat (Valenti et al. 2011).
- In developing countries, aquaculture ventures often face difficulties with access to markets and financing, hindering their investment in making the industry more sustainableAsche et al. 2018).
- Getting a sustainability certification is often too expensive for smaller farmers, which means they can't sell their products in valuable markets (Bondad-Reantaso et al. 2005).
- Places for aquaculture growth are found only after evaluating their suitability, following the rules, and avoiding conflicts with users of the coastal area (Bush et al. 2010).

24.4.2 *Social Challenges*

i. Among the social issues in aquaculture are exploitation of workers, gaining advantages over others, and conflict with people in the area (Bush et al. 2013a, b).
ii. At times, people lose their traditional jobs in fishing or farming because of large-scale fish farming, which then causes both social and financial complications (Brugère and Ridler 2004).
iii. In some cases, if aquaculture remains in a single area, it favors the growth of wealth for rich parties over the locals living nearby (Garlock et al. 2024).
iv. The growth of aquaculture often leaves small-scale and indigenous farmers at a disadvantage because they lack the finances or resources the bigger companies have (Aguilar-Manjarrez et al. 2017).

24.5 Future Approaches to Enhancing Aquaculture Sustainability

Various future strategies for sustainable aquaculture development are shown in the Fig. 24.5.

24.5.1 Policy and Regulatory Frameworks

Running good governance is essential to encourage sustainable fish farming. Introducing certifications and market incentives makes economies more interested in following green and social policies. Such labels for seafood encourage fisheries to use sustainable methods. It is possible for governments to help by imposing regulations that reduce aquaculture's effect on the environment. For example, they might restrict the use of wild fish in feed or limit the growth of aquaculture in particularly sensitive places (Garlock et al. 2024). Below are key policy measures:

i. **National Aquaculture Development Plans (NADPs)**

   In many cases, national plans about the fisheries include rules for shaping sustainable growth in aquaculture, and they cover environmental, social, and economic issues (European (European Commission (2013).

ii. **Disease Control and Biosecurity Protocols**

   To prevent and manage diseases in aquaculture, regulations require biosecurity controls, specific procedures for vaccination, and watch for diseases (Subasinghe et al. 2001).

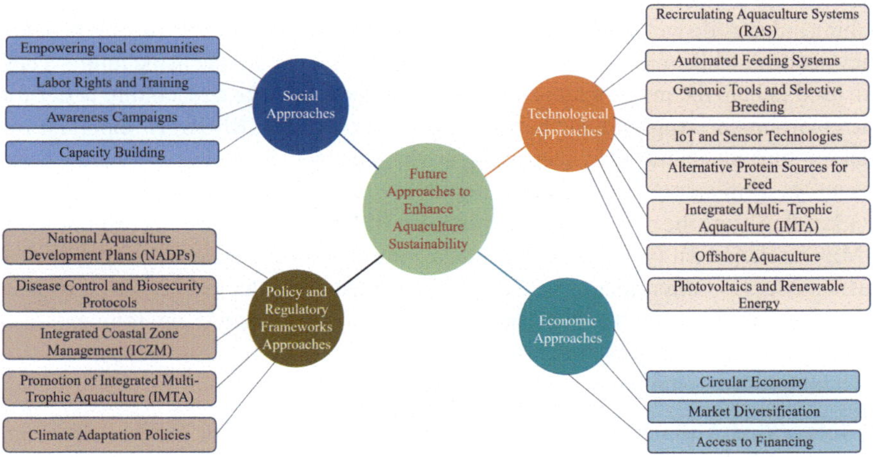

Fig. 24.5 Future Strategies for sustainable aquaculture Development (*Source* Author)

iii. **Integrated Coastal Zone Management (ICZM)**

Through ICZM, aquaculture and other activities by the sea or rivers are balanced so that marine and freshwater environments are safeguarded (R.W. et al., 1998).

iv. **Improving Integrated Multi-Trophic Aquaculture (IMTA)**

Curricula support IMTA practices by giving out technical advice, research grants, and financial support to decrease pollution and encourage other types of cultivation (Troell et al. 2009).

v. **Climate Adaptation Policies**

Several strategies are designed by governments to help aquaculture cope with rising seas, warmer temperatures, and extra-heavy weather (Barange et al. 2018).

24.5.2 Community Engagement and Capacity Building

Getting local people involved in aquaculture management helps make the industry more sustainable socially. Involving people from the local community in making decisions can ensure that all members gain from aquaculture. Providing training and support to local small-scale producers can help aquaculture become steadier financially because it lets farmers use their resources better and find a wider range of buyers. People are given a chance to voice their views and needs so that the development strategy can be shaped by their culture and area (FAO 2022).

24.6 Best Practices for Sustainable Aquaculture

Strategies for sustainable aquaculture help keep the environment healthy, help the fishing industry do well financially, and make sure people can keep getting the benefits they need, all while making sure things work out well for everyone in the future. They are used to help fix any problems that might happen and make aquaculture better for the environment. Applying the best management practices (BMPs) is important for making sure aquaculture can keep going in a healthy way. Reducing how hard we use resources from nature, making sure animals are properly taken care of, and treating workers fairly are the most important things for BMPs (Valenti et al. 2011). Following BMPs on their own is not enough to make sure that a landscape stays healthy and sustainable. It is important to watch what's happening in the market, think of new ways to improve when needed, and make sure to care for the environment so a business can succeed in the long run.

Examples of sustainable practices include:

- **Integrated Aquaculture:** Integrating aquaculture with other kinds of farming, like rice-fish farming, can help you use resources more efficiently and lower the effects on the environment (Valenti et al. 2011).

- **Efficient Feed Use:** Changes that make it easier for animals to get the most from the food they eat and help them waste less, can reduce some of the negative effects that farming fish and seafood can have on the environment (Valenti et al. 2018).
- **Water Quality Management**: Continuous monitoring of water quality so we can stop it from getting too dirty, and to check that the plants and animals in the stream are doing okay. Closed-loop systems have been put to use on farms so that less water gets wasted, there is less pollution, and the place where they are used stays safe.
- **Community Engagement**: Involving local communities in planning and making decisions so the needs and wishes of local people are listened to. It also means giving workers a safe place to work, paying them a fair amount, and taking care of their rights, to help make things more equal for everyone.
- **Supply Chain Efficiency**: Streamlining processes from producing food to getting it out to stores helps cut down on waste, makes it easier to track where food comes from, and keeps the farm affordable by finding ways to reduce costs, try new ideas, and try different ways to make money.
- **Advanced Monitoring Systems**: Using remote sensing, internet-of-things, and artificial-intelligence technologies to monitor water quality, fish health, and how fish farms are working in real time (Moghadam et al. 2019).
- **Genetic Improvement:** Breeding programs for wild rice that is hardier, grows fast, and has less impact on the environment (Houston et al. 2020).
- **Adherence to Regulations**: Complying with national and international rules on the environment, social issues, and safety, while also keeping people informed about what is being done, how things are affecting the environment, and how things could be made better by putting out regular reports (Buschmann et al. 2009).

Globally, people are trying to use ways of raising fish in the water that are better for the environment. However, it is still hard to keep these farming methods working as more people need food, while also making sure they cost less and can be done in a way that is better for the environment.

24.7 Case Study: The Aquaculture Stewardship Council Certification in Andhra Pradesh

All over the world, moves are being made to make aquaculture more sustainable. However, it is still challenging to quickly increase these methods to meet more demand without affecting their economy or harming the environment (Valenti et al. 2018). Even so, expanded aquaculture activities have brought up concerns about harming the environment, adding pollutants to the water, and excess use of antibiotics. Various shrimp farms in Andhra Pradesh gained the ASC Certification to address these matters, as the certification supports eco-friendly aquaculture ways (Kumar and Singh 2021, Paul 2024).

24.8 Sustainability Practices Implemented

- Farms started using water recycling and recirculation systems to help save freshwater. Regular water quality testing helped make sure that the wastewater was clean enough to meet the rules for discharging it back into the environment.
- Farmers started to use feed that comes from sources that are better for the environment, and this helped ease the demand for wild fish to be caught. Improved ways to feed animals were put in place to make sure less food went uneaten.
- Farms started using different practices like keeping certain ponds under control, to stop diseases from spreading.
- Reduced reliance on antibiotics to help slow down the growth of resistance to these drugs.
- Buffer zones were made around farms so that mangroves and other natural areas would be kept safe from any mistakes or accidents.
- Efforts were made to help fix and restore damaged parts of the mangrove forests.
- Worker safety and rights were made a top priority, and everyone got paid fairly and had the proper training they needed.
- Community engagement programs helped the company get along better with the local people.

Outcomes

- A decrease in the amount of wastewater discharged resulted in cleaner local water.
- Security of mangroves reduced erosion in the coast and enhanced the region's variety of life.
- Farms that earned the ASC certification were able to sell in top international markets.
- Because the value of their products increased, the income of farmers went up.
- The reforms improved the comfort and safety of workers in the aquaculture industry.
- As trust in the community increased, so did people's acceptance of how resources are used.

Challenges

- The high prices for being certified prevented small farmers from getting involved.
- It took a lot of resources to give proper training and capacity-building to farmers and workers.

24.9 Conclusion

Even though aquaculture can meet the world's needs for food, it must be managed well in order to continue being sustainable. The framework provided by (Valenti et al. 2018) helps determine and address the sustainability issues in aquaculture. Aquaculture can develop even more while having a smaller effect on the environment and local people by choosing best practices, being innovative, and following sustainability. Balancing the goals of sustainability, profit, and well-being is hard work, but it can be done in aquaculture. Embracing digital solutions and engaging with the people living nearby can allow the aquaculture industry to grow sustainably and fairly distribute its benefits. Concern for the environment, fairness at work, and limited use of resources play a key role in supporting food security and building up communities. When aquaculture adopts new technologies, sources alternative feed solutions, and uses inclusive policies, it can reduce its negative effects on nature and increase its benefits for people and the economy. Aquaculture will only remain sustainable in the long run if there is a dedication to continuous improving, full transparency, and teamwork among all involved parties. For aquaculture to continue providing food sustainably, it needs to focus equally on the three pillars of sustainability.

24.10 Future Prospects

Thanks to new technology, aquaculture can be made more sustainable. Using integrated multi-trophic aquaculture (IMTA), which brings together several types of species, has helped improve both the environment and the bottom line. Using plant proteins or insect meal as feed can give the industry a way to use less wild-caught fish. Aquaculture can be made more sustainable and viable thanks to development in technology that leads to better use of resources, fewer negative effects on the environment, and economical operations. These are the main improvements and their importance for sustainable development:

i. **Recirculating Aquaculture Systems (RAS)**

   By using a closed-loop system, RAS removes wastewater and lowers the amount of clean water needed, as well as reduces waste released into the environment. Because of these systems, plants can be grown for the entire year without changes in weather (Martins et al. 2010).

ii. **Automated Feeding Systems**

   Automated systems ensure food is only given as needed, saving on wasted feed and helping fish to gain weight faster. They also bring down the costs of labor and make activities less harmful to the environment (Buentello et al. 2009).

iii. **Genomic Tools and Selective Breeding**

   Using genomic techniques, breeders can focus on traits such as disease resistance and faster and better-converting feed. Using genetic markers can make breeding more efficient and keep the genetic material of the population diverse (Houston et al. 2020).

iv. **IoT and Sensor Technologies**

IoT devices and sensors keep an eye on water quality (like pH, oxygen, temperature) all the time, giving farmers the ability to manage things ahead of time and save losses from bad conditions (Moghadam et al. 2019).

v. **Integrated Multi-Trophic Aquaculture (IMTA)**

Fish, shellfish, and seaweed are frequently co-cultivated in IMTA systems, which allows them to support and maintain an ecosystem where one species' waste is used by another. It cuts down on nutrients put into the system and offers farmers an additional source of money (Troell et al. 2009).

vi. **Alternative Protein Sources for Feed**

Switching to algae, insects, and single-cell proteins, rather than using fishmeal and fish oil, is helping to cut back on products made from wild fish (Turchini et al. 2019).

vii. **Offshore Aquaculture**

By using submersible cages, fishing vessels can farm farther offshore, which frees space by the coast and lessens the impact on vulnerable habitats.

viii. **Photovoltaics and Renewable Energy**

Adding solar or wind energy to farms in aquaculture helps cut down the use of fossil fuels and reduces the amount of carbon the farms produce (Ahmed et al. 2017).

Acknowledgements The author Pooja Kadyan is thankful to CSIR, India, for the award of Senior Research Fellowship.

References

Aguilar-Manjarrez J et al (2017) Mapping the potential for sustainable development of aquaculture. FAO Fish Aqua Proc 44

Ahmed N et al (2017) The integration of renewable energy in aquaculture systems: a technological innovation for sustainability. J Clean Prod 162:1394–1403

Asche F et al (2018) Market shocks in aquaculture: the case of tilapia in the US. Aquac Econ Manag 22(2):141–156

Ashley PJ (2007) Fish welfare: current issues in aquaculture. Appl Anim Behav Sci 104(3–4):199–235

Badiola M et al (2012) Recirculating aquaculture systems (RAS) analysis: main issues on management and future perspectives. Aquac Eng 51:26–35

Barange M et al. (2018) Impacts of climate change on fisheries and aquaculture. FAO Fish Aquac Tech Paper No. 627

Barton BA (2002) Stress in fishes: a diversity of responses with particular reference to changes in circulating corticosteroids. Integr Comp Biol 42(3):517–525

Belton B et al (2017) Equitable development for aquaculture-dependent communities: a framework for social responsibility in the Asian aquaculture industry. Aquac Econ Manag 21(2):201–220

Belton B et al (2021) Small-scale aquaculture and its role in rural livelihoods. World Dev 147:105650

Beveridge MCM et al. (2013) Aquaculture and food security, poverty alleviation, and nutrition in Bangladesh. *WorldFish Report*, explores how aquaculture impacts local food security

Bondad-Reantaso MG et al (2005) Disease and health management in Asian aquaculture. Vet Parasitol 132(3–4):249–272

Boyd CE et al (2016) Water quality in aquaculture. J World Aquaculture Soc 47(1):20–32

Boyd CE et al (2020) Regulatory frameworks for sustainable aquaculture. J World Aquaculture Soc 51(5):898–920

Boyd, C. E., & McNevin, A. A. (2015). Aquaculture, resource use, and the environment. *Wiley-Blackwell.*

Boyd CE, Tucker CS (1998) Pond aquaculture water quality management. Springer Sci Business Media

Brugère C, Ridler N (2004) Global aquaculture outlook in the next decades: an analysis of national aquaculture production forecasts to 2030. FAO Fisheries Circular No. 1001, FAO.

Buentello JA et al (2009) Advances in automation and control of aquaculture feeding. Aquac Res 40(9):1193–1202

Buschmann AH et al (2009) Salmon aquaculture and coastal ecosystem health in Chile: analysis of regulations, monitoring, and enforcement. Mar Pollut Bull 58(10):1767–1775

Bush SR et al (2010) The "devils triangle" of aquaculture development: Sustainability, equity, and resilience. Aquac Econ Manag 14(4):231–247

Bush SR et al (2013) The "devils triangle" of MSC certification: Balancing credibility, accessibility, and continuous improvement. Mar Policy 37:288–293

Bush SR, Marschke M (2014) Making social sense of aquaculture transitions. Mar Policy 45:153–160

Bush SR et al. (2013). Governing sustainable seafood. WorldFish Report, discusses the benefits of aquaculture for local infrastructure and community resilience

Ellis T et al (2012) A review of the welfare issues in commercial aquaculture. Aquaculture 358–359:2–12

European Commission (2013) Strategic guidelines for the sustainable development of EU aquaculture

FAO (2020) The State of World fisheries and aquaculture 2020. Sustainability in action. Rome.https://doi.org/10.4060/ca9229en

FAO (1995) Code of conduct for responsible fisheries. food and agriculture organization of the United Nations

FAO (1999) Sustainable aquaculture development and the code of conduct for responsible fisheries

Food and Agriculture Organization (FAO) (2022) Sustainable aquaculture development

Garlock TM, Asche F, Anderson JL et al (2024) Environmental, economic, and social sustainability in aquaculture: the aquaculture performance indicators. Nat Commun 15:5274

Glencross BD et al (2007) A review of recent developments in understanding the nutritional requirements of farmed aquatic species. Rev Aquac 10(1):79–89

Grafton RQ et al (2007) Adaptive management: principles and practices. Environ Manage 39(6):726–739

Hall SJ et al. (2011) Blue frontiers: Managing the environmental costs of aquaculture. WorldFish Center Report, focuses on waste management and strategies to mitigate impacts

Houston RD et al (2020) Harnessing genomics to fast-track genetic improvement in aquaculture. Nat Rev Genet 21(7):389–409

Huntingford FA et al (2006) Current issues in fish welfare. J Fish Biol 68(2):332–372

Kabir MH et al (2020) Health, safety, and well-being of aquaculture workers in Bangladesh: challenges and solutions. Aquaculture Reports 18:100437

Kiran, Kumar N, SindhuS, Gawaria B, Ansari A, Dhariwal A, Sharma P, Kumar R (2025) GWAS approaches in medicinal plants for enhanced secondary metabolite production. In J. T. Chen (Ed.), Biotechnology, multiple omics, and precision breeding in medicinal plants, pp 119–134. CRC Press. https://doi.org/10.1201/9781003475491-10

Kumar G et al (2018) The role of aquaculture in global food security: analysis of opportunities and challenges. Food Security 10(6):1309–1321

Kumar P, Singh S (2021) Sustainable aquaculture in India: case studies and insights. J Environ Manage 283:111–120

Kumar S, Sankhla MS, Kumar R, Swaroop S Sonone (2020) Impact of pesticide toxicity in aquatic environment. Biointerface Res Appl Chem 11(3):10131–10140. https://doi.org/10.33263/BRIAC113.1013110140

Kumar R, Gupta R, Dahiya T, Yadav R, Saifi R, Yodha K, Kumari B, Singh P, Shefali (2022a) Surveillance of *Heteropneustes fossilis* (Stinging Catfish) Diseases and Health Management Practices in Haryana Appl Ecol Environ Sci 10(4):261–267. https://doi.org/10.12691/aees-10-4-12

Kumar R, Gupta RK, Yadav R, Saifi R, Yodha K. Pooja (2022b) Eisenia fetida as protein source for growth enhancement of Heteropneustes fossilis. Egypt J Aquatic Biol Fish 26(2):577–588. https://doi.org/10.21608/ejabf.2022.234553

Kumar R, Yadav R Kumar Gupta R, Pooja (2022c) A sustainable way for fish health management by replacement of chemical and drugs by earthworm. In Environmental degradation in Asia: land degradation, environmental contamination, and human activities, pp 329–352. Springer International Publishing, Cham. https://doi.org/10.1007/978-3-031-12112-8_16

Lines JA et al (2004) Electric stunning of fish: The relationship between the electric field strength and water conductivity. Aquaculture 241(1–4):219–234

Madin EM, Macreadie PI (2015) Incorporating carbon footprints into seafood sustainability certification and eco-labels. Mar Policy 57:178–181

Martins CI et al (2010) New developments in recirculating aquaculture systems in Europe: a perspective on environmental sustainability. Aquacult Eng 43(3):83–93

Moghadam AA et al (2019) IoT-based smart aquaculture systems: a review of recent advancements. Journal of Aquatic Systems Research 11(1):45–59

Molnar JL et al. (2008) The global distribution and effectiveness of marine protected areas.PLoS ONE 3(7):e2939

Naylor RL et al. (2000) Effect of aquaculture on world fish supplies. Nature 405(6790):1017–1024, discusses aquaculture's impact on wild fish populations and biodiversity

Nisha A, Rani S, Kumar S, Kumar R, Kumar A (2025) Recent technologies and strategies for bioremediation of toxic and hazardous waste. In Dubey KK, Shah MP, Yadav A (Eds.), Challenges and sustainable solutions in bioremediation, pp 1–16. CRC Press. https://doi.org/10.1201/9781003407317-1

Näslund J, Johnsson JI (2016) Environmental enrichment for fish in captive environments: effects on cognition, behavior, and fitness. Front Psychol 6:1893

Parker RW et al (2018) Measuring the environmental performance of blue foods: An international perspective. Environ Res Lett 13(11):114002

Paul R (2024) AMR in aquaculture: enhancing Indian shrimp exports through sustainable practices and reduced antimicrobial usage (No. 209). Research Paper

Pomeroy R et al (2016) Stakeholder participation in coastal zone management. Ocean Coast Manag 134:100–110

Ponte S (2012) The marine stewardship council (MSC) and the making of a market for "sustainable fish". J Agrarian Change 12(2–3)

Pullin RS, Froese R, Pauly D (2007) Indicators for the sustainability of aquaculture. Ecological and genetic implications of aquaculture activities. Springer, Netherlands, Dordrecht, pp 53–72

Rani S, Sharma S, Singh D, Sheoran R, Kumar R (2025) Improved production of anti-cancer compounds from medicinal plants through molecular and biotechnological mutagenesis strategies. In Chen JT (Ed.), Biotechnology, multiple omics, and precision breeding in medicinal plants, pp 287–297. CRC Press. https://doi.org/10.1201/9781003475491-21

Ratner BD et al (2013) Addressing conflict through collective action in natural resource management. Int J Commons 7(1):139–161

Rico A et al (2013) Use of veterinary medicines, feed additives, and probiotics in four major internationally traded aquaculture species farmed in Asia. Aquaculture 412:231–243

Subasinghe R et al (2001) Biosecurity in aquaculture: principles and best practices. FAO Aquaculture Newsletter 28:9–15
Tacon AGJ, Metian M (2008) Global overview on the use of fish meal and fish oil in industrially compounded aquafeeds: trends and future prospects. Aquaculture 285(1–4):146–158
Tacon AGJ, Metian M (2013) Aquaculture feed and food safety: the role of policy and regulation. Aquaculture 467:320–329
Troell M et al (2009) Future of aquaculture and the environment: a sustainable balance? Rev Environ Sci Biotechnol 8(3):297–307
Troell M et al (2009) Ecological engineering in aquaculture: Use of integrated multi-trophic aquaculture (IMTA) to improve environmental sustainability. Rev Aquac 1(1):27–41
Turchini GM et al (2019) Fish oil replacement in finfish nutrition. Rev Aquac 11(3):645–664
Turnbull JF et al (2005) Stocking density and welfare of farmed fish. Fish Physiol Biochem 31(2–3):123–145
Valenti WC, Kimpara JM, Preto BL (2011) Measuring aquaculture sustainability. World Aquaculture 42(3):26–30
Valenti WC, Kimpara JM, Preto BDL, Moraes-Valenti P (2018) Indicators of sustainability to assess aquaculture systems. Ecol Ind 88:402–413

Chapter 25
Mitigating Microplastic Pollution in Aquaculture: Impacts on Food Systems

Pooja Kadyan, Jaya Saini, Mohit Saharan, Divya Vashisth, Poonam Yadav, Sunita Dalal, Ajay Kumar, Sudhir Kumar Kataria, and Sindhu Sheoran

Abstract Concerns are being raised globally about the substantial amount of microplastic contamination in the environment today as a result of the pervasive usage of plastic and its long-lasting ecological repercussions. Aquatic animals in aquatic habitats consume microplastics, which has an impact on their flourishment and maturation and cause change of trophic level to higher living organisms in the food cycle. Therefore, eating aquatic products is one of the main ways that people are exposed to microplastics. The production of fish raised in aquaculture has increased dramatically recently and surpasses that of captured fish. Since aquaculture products are a major source of protein in the human diet, they have a direct and qualitative impact on food safety. This chapter compiles and summarizes the documented research on the origins and routes of microplastic pollution in aquatic environments, the effects it has on aquatic life and ecosystems, and the implications it has for food sustainability and safety. External importation, which includes plastic litter and waste from the terrestrial ecosystem, atmospheric deposition, transit, and shipping, is the main cause of microplastics in aquaculture environments. Higher levels of microplastic buildup are also caused by the utilization of plastic tools and gear, health and nourishing items for aquaculture, and specialized settings for aquaculture. We also discuss the harm that microplastics cause to farmed species and the

P. Kadyan · J. Saini · M. Saharan · D. Vashisth · P. Yadav · S. K. Kataria (✉)
Department of Zoology, Maharshi Dayanand University, Rohtak, Haryana, India
e-mail: sudhir.zoology24@mdurohtak.ac.in

P. Kadyan
Department of Zoology, Sant Shiromani Shri Sain Bhagat Government College, Jind, Haryana, India

S. Dalal
Department of Biotechnology, Kurukshetra University, Kurukshetra, India

A. Kumar
Department of Zoology, Maharaja Neempal Singh Government College, Bhiwani, India

S. Sheoran
Senior Research Program Coordinator, Hawkesbury Institute for the Environment, Western Sydney University, Penrith, Australia
e-mail: S.Sheoran@westernsydney.edu.au

R. Kumar et al. (eds.), *Sustainability in Aquaculture and Fisheries*,
Springer Water, https://doi.org/10.1007/978-3-032-16605-0_25

potential health risks they pose to humans by moving up the food chain. This chapter emphasizes negative consequences of microplastic pollution in aquatic life, especially environmental harm that microplastic pollution causes to aquaculture species and the resulting in disease susceptibility to humans. It also advocates for controlling microplastics in aquaculture ecosystems to a limited extent.

Keywords Microplastic pollution · Aquaculture · Sustainability · Aquatic organisms · Contaminated fish

25.1 Introduction

Plastic goods' advantageous characteristics and affordable price have greatly enhanced modern life's convenience. The longevity of plastic products is one of its best features. Although, plastic garbage is becoming one of the biggest environmental problems because of limited biodegradability and poor management (Krüger et al. 2020). Based on estimations, in 2019 just 9% of plastic garbage gets recycled, 19% is burned, and 50% is dumped into the environment and 22% is mismanaged, leading to environmental leakage (Vuppaladadiyam et al. 2024). The world's growing usage of plastics has led to large areas of floating plastic garbage, sometimes referred to as "plastic soup," in the oceans. The buoyant plastic waste is breaking down into small fragments over time, which will eventually turn into microplastics and nanoplastics. These might be in the shape of beads, spheroids, pellets, or flakes (Cowger et al. 2020; Geremia et al. 2023; Hale et al. 2020).

The term "plastic" describes a broad category of polymer elements that may be crafted under particular temperature and pressure conditions and have varied characteristics based on the needs of the final commodities. Plastic has a broad range of consumer and industrial applications and is either synthetic or semi-synthetic. The majority of these synthetic materials are made of polymers composed mostly of chains of carbon atoms with gaps between them filled with hydrogen, oxygen, sulphur, and nitrogen (Shrivastava 2018). Plastics are still produced worldwide today, having grown over the last few years. These are the numbers provided by the plastics industry itself, however, since synthetic fibres are not included, the actual output is far greater. By 2050, output is predicted to have quadrupled (in comparison to 2019) to reach around 1480 million tons. This is nearly three times the current global population's weight (Plastic Atlas, Asia edition 2021).

Plastics may be divided into three main categories: thermosets, thermoplastics, and elastomers. Thermosets, such as polyester resins, epoxy resins, and Bakelite, polyurethane (PU) never soften once they are moulded, in contrast to thermoplastics, which Polymers like polyethylene (PE), polyethylene terephthalate (PET), polytetrafluoroethylene (Teflon), polyvinyl chloride (PVC), polystyrene (PS), polypropylene (PP), polyamide (PA), and which solidify when cold and soften when heated. Elastomers, such as rubber and neoprene, are polymers with an elastic quality that allows the material to revert to its initial shape after stretching. Around twenty different groups make up the highly varied class of plastic polymers (Mills et al. 2020; Mohamadi 2023).

Microplastics (MPs), generally refer to plastics with a size range of 0.1–5000 μm, or 5 mm (Frias and Nash 2019). These days, plastic products come in a wide variety of forms and dimensions, using the ranges of microns and submicrons referred to as microplastics, or primary microplastics (directly produced MPs) (Tirkey and Upadhyay 2021). Primary microplastics are produced on purpose for a variety of uses, such as for the creation of major plastic products, microplastic powders, and microbeads used in abrasive scrubbers and cosmetics (Auta et al. 2017; Duis and Coors 2016). Pharmaceutical drug delivery systems, and biomedical goods are among the products made from plastics that are generated at the nanoscale (Koelmans et al. 2015). However, secondary microplastics are produced when large plastic objects break down into tiny, micron-sized pieces and end up in the atmosphere. This disintegration can happen through different type of processes, including chemical, biological, or mechanical breakdown (Fig. 25.1) (Bacha et al. 2021; Hale et al. 2020). Road wear, washing synthetic materials, physical deterioration, and tire abrasion of huge objects are examples of mechanical breakdown. Exposure to acids or alkalis causes chemical breakdown, whereas non-ionizing UV light causes UV fragmentation. In addition, microorganisms such as mealworms, waxworms, and other living things with the ability to consume and break down plastics can cause biological breakdown (Bacha et al. 2021).

Microplastics have the potential to be mistakenly consumed, breathed, and bioaccumulated by organisms because of their tiny size, posing a health concern (Cole et al. 2015; Wright et al. 2013). Consequently, microplastic become more and more frequent, garnering interest on a global scale in recent years. Every kind of environment on Earth has been discovered to include microplastics, including the atmosphere and aquatic ecosystems (reservoirs, rivers, lakes, ponds, and sea) and terrestrial ecosystems (beach, land and island) (Chen et al. 2021). Oceans and the

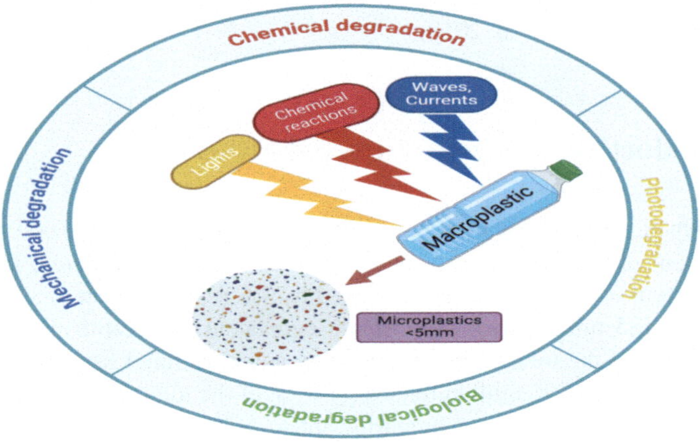

Fig. 25.1 Formation of microplastics (secondary microplastics) by breakdown mechanism

ecosystem are seeing a sharp rise in plastic pollution because of inadequate management (Campanale et al. 2020). Between 4.8 and 12 million tons of plastic debris are dumped into the ocean annually (Jambeck et al. 2015). There are some statistics available, especially on the marine environment, about microplastics in food. Biota may ingest microplastics, which may then build up in digestive tracts of them. This have been identified in a different type of organisms, mainly in whales, fish, bivalves and zooplanktons (Harmon et al. 2024; Kumar et al. 2022a; Rezania et al. 2018). Fish exhibit large quantities of microplastics, but as these particles are mostly found in the stomach and intestines, they are often eliminated before reaching consumers. However, there is a chance of exposure because one might consume the digestive tract while eating crustaceans and bivalve molluscs like mussels and oysters. There have also been reports of microplastics in table salt, beer, and honey. Once organisms consume microplastics, several harmful physiological consequences may manifest, which may impact behaviour and feeding, impede flourishing and development, and result in immune system toxicity, genetic damage, and reproductive toxicity (Harmon et al. 2024).

One major concern is that both macroplastics and a variety of chemicals are added during the process of production to create microplastics and they may also absorb or adsorb environmental pollutants. These additional chemicals include flame retardants, plasticizers, oligomers, and monomers (Botterell et al. 2019; Issac and Kandasubramanian 2021). On the other hand, environmental contaminants that have been absorbed include bacteria, trace metals, PAHs, PCBs and chlorinated pesticides, that belong to the category of persistent organic pollutants (POPs). POPs are also categorized as harmful compounds that persist and accumulate over time (PBTs). Additionally, substances used in packaging, such as bisphenol A (BPA), may have left behind traces (Issac and Kandasubramanian 2021; Rochman 2015; Yadav et al. 2022a; Yadav et al. 2022b; Yadav et al. 2023). According to certain research, microplastics have the potential to penetrate and accumulate in tissues, intensifying their harmful effects on living things. Microplastics and other pollutants can spread and bioaccumulate in high-trophic organisms through the food chain, creating major health hazards (Carbery et al. 2018).

25.2 Production of Aquaculture and Fisheries

Fish and aquatic organism rearing has increased eightfold since the Food and Agricultural Organization (FAO) began collecting data from member nations in 1950. In 2022, the output of aquatic animals increased by 4% year over year to a new global maximum of 185 million tons (live weight equivalent). An estimated 94 million tonnes, or 51% of the total, were produced by aquatic animal farming, more than the first-ever output of 91 million tonnes (or 49% of the total) from catch fisheries. 62% of the total, were produced in marine areas; 31% came from aquaculture and 69% from catch fisheries. Inland waterways contributed 70 million tons (38 percent of overall production), with aquaculture accounting for 84% and catch fisheries for

16% of the production (Kumar et al. 2022b; Kumar et al. 2022c). Approximately, 164.6 million tonnes, or 89%, of the 185 million tonnes of aquatic animals grown in 2022 were intended for human consumption, or an average of 20.7 kg per person. Fishmeal and fish oil (17 million tonnes, or 83 percent) were the primary non-food applications for the remaining 20.8 million tonnes.

Seafood consumption has been rising in line with the growth of the global inhabitants. In response to the increased consumer need for fisheries, aquaculture was established, and production techniques have continued to advance. Fisheries and aquaculture have developed significantly under the influence of plastics industry. The industrial synthesis of plastics led to advancements in fishing tactics, and most equipment used today is comprised of synthetic or semisynthetic materials. When synthetic fibre ropes were first presented, they not only reduced the net's total weight but also gave superior strength and longevity over organic ropes. Plastic is used to make fish cages and boxes, fish crates, fish tanks, pond liners, and the packaging and shipping of seafood (Lusher et al. 2017). Aquaculture environments will unavoidably get contaminated with microplastics because of the use of these products and therefore, pose a health risk to aquatic organisms as well as other living organisms throughout the food chain (Shefali Kumar et al. 2021; Wu et al. 2023).

25.3 Sources of Microplastic pollution

Microplastics can enter as primary or secondary plastics to the environment, as was previously discussed (Fig. 25.2). These kinds of polymers are found in marine, freshwater, and terrestrial environments. Wind, currents, and other environmental factors carry microplastics across great distances, dispersing them globally in the sky, on land, in lakes, rivers, estuaries, and in deep seas (Souza Machado et al. 2018).

Plastic pellets pollution is primary source of microplastic pollution, is closely associated with spills that occur during manufacture and/or transportation (Karlsson et al. 2018). In addition to the likelihood of plastic particulates being carried by industry emissions, locations near to the plastic industry are more exposed to microplastic contamination (Kurniawan et al. 2021). It can also utilized as components in PCPs,examples of these include fibres for clothes and exfoliants for soap, skin cleansers, liquid baths. Approximately 1000–19,000 microplastic particles per millilitre are present in face washes, and they vary in size, color, and shapes (Napper et al. 2015). Microbeads are discharged from PCPs to the aquatic environment and leachate in landfills. Since microbeads are smaller than filter pores, many microbead particles can get through filtration and added in the aquatic environment. Because of their small enough sizes, aquatic life can readily take them and amass them via the food chain (Duis and Coors 2016).

In techniques known as air blasting and industrial grating technology, condensed air creates air-driven velocity that pushes rough material via nozzles onto the surface.

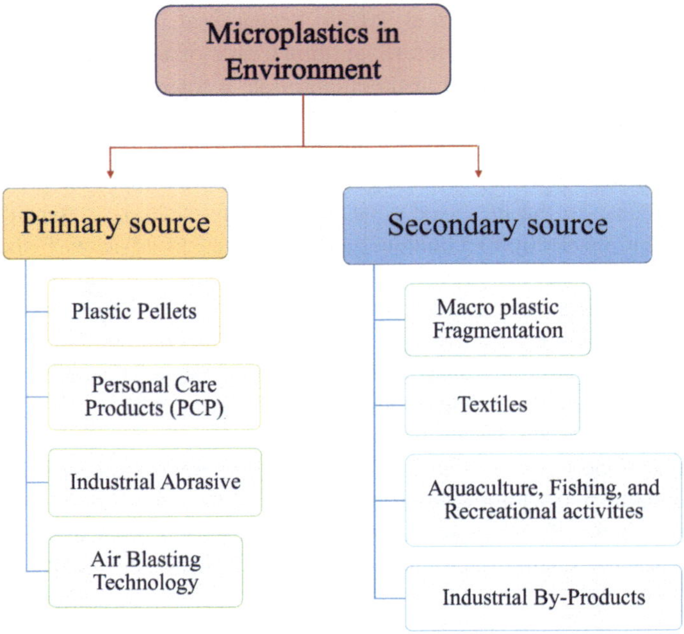

Fig. 25.2 Common sources of microplastic pollution

Before applying fresh coatings via the media blasting process, industrial grating technology and pneumatic cleaning are frequently used to remove rust, colour, and other impurities from surface of steel like, ship hulls, machines, engines, and walls (Dhir et al. 2017). The automotive, aerospace, oceanic, communications, and industrial sectors all make extensive use of the air blasting technology. Approximately, 0.2–2 mm microplastic particles are utilized as an abrasive medium, which is the key source of microplastic. This issue becomes a major worry for harbors due to the large tankers being painted white and the untreated effluent being poured into the ocean (Waldschläger et al. 2020). Rubber is also classified as a plastic, in addition to thermoplastics like PE, PP, PVC, and PET. One of the primary sources of rubber release is tires of automobile. Tires generate a lot of power and debris that ranges in size from nano to micrometers. Particulate particles may be released from the road as a result of the abrasion action of tires (Galafassi et al. 2019).

Human activity, industrial effluents, sewage treatment facilities, and agricultural practices are among the sources of river microplastics, rivers are mostly contaminated with fibrous microplastics due to industrial effluent, especially textile wastewater (Kumar et al. 2021). In the clothing industry, washing a single fabric product produces 900particles in the form of microfiber (Napper and Thompson 2016). Microplastics are also outcome of the regular washing of synthetic garments made of fibres. Statistics show that in a single wash, polyester fleece textiles shed an average of 7360 fibres/m^2/L (Carney Almroth et al. 2018). Clothes made of nylon, acrylic, and when cleaned, different synthetic substances also release various concentration of microplastics. Home sewerage will carry these microplastics into the river. Due to the massive sewage discharge, there is still a significant amount of microplastics entering rivers even with the present wastewater treatment procedure having some treatment impact (Sun et al. 2019). Numerous plastic items, such as plastic sheets and pipes, are utilized in agricultural production. As these polymers deteriorate, tiny plastic particles will be produced, which wind and rain will carry into waterways (Carney Almroth et al. 2018).

Microplastics will also inevitably find their way into the aquatic ecology through fishing gear, fish nets, boats, fences, fish packaging, transport materials, and aquatic vessels. Microplastics can be produced by UV catalytic degradation will expedite the deterioration of plastic products because of prolonged immersion, erosion, wear, and collision. (Song et al. 2017). The two materials associated with fishing that are most frequently discovered garbage are nylon nets and fibrous ropes. (Karbalaei et al. 2020). In aquaculture, fishing equipment would also be lost for a variety of causes, such as being carried away by the waves or falling off the ropes (Chaves and Silveira 2016), and the breakdown of these items produces microplastic particles.

25.4 Pathways to Get into the Environment

25.4.1 Microplastics in the Atmosphere

Plastic is light weight, a lot of microparticles are suspended and move as "urban dust" through the atmosphere. These can lead to the deposition of microplastics into aquatic or terrestrial ecosystems. They are frequently derived from paint and tire fragments found on roadways, as well as fibers from synthetic, especially those present in soft furnishings. While cities and other densely populated areas are the primary locations where urban dust is created, plastic particles can be carried far from the source by wind and air currents (Abbasi et al. 2019; Kitahara and Nakata 2020).

25.4.2 Microplastics on Land

Products intended for consumer usage employ the vast majority of plastics produced. Improper waste management, which includes spills from industry, landfill releases, and losses within the waste disposal chain, is mostly responsible for the rise of plastics on land. The environment may eventually break down these materials to create secondary microplastics. Along with sewage sludge that is spread as fertilizer on arable land, microplastics can also be discharged directly onto land (He et al. 2019; Yadav et al. 2022a).

25.4.3 Microplastics in the Aquatic System

Microplastics are becoming more common in aquatic systems, which is concerning because of the potential harm they might do to a variety of organisms (Gall and Thompson 2015). The primary ways it get to marine system are from terrestrial sources, such as unintentional plastic loss or mishandling, plastic waste dumping, and poorly managed landfills—for example, lacking a fence around the landfill or using the wrong kind of synthetic material to cover trash (Duis and Coors 2016). After rain washes the land, large plastic garbage penetrates into aquatic ecosystem. An estimated millions of tons of plastic are generated in the Ganges River basin each year. WWTPs are seen as a significant MPs gateway into the aquatic habitats and add to these entry pathways (Lebreton et al. 2017).

25.4.3.1 Microplastics in the Freshwater Environment

When it comes to the movement and retention of microplastics, freshwater sources including ditches, streams, lakes, rivers and ponds constitute the bulk of complicated systems. They break down bigger things to make microplastics, and they also receive plastics from the land. Bigger plastic objects can enter freshwater habitats as a result of inappropriate garbage disposal, which can happen by surface runoff or wind carrying them off the land, or through littering or loss from landfills (Fahrenfeld et al. 2019; Li et al. 2020).

25.4.3.2 Microplastics in the Marine System

Since seas are believed as the world's largest sink for plastic, there are many sources of microplastics in marine ecosystems. Plastics will also directly reach seas through improper waste management (Coyle et al. 2020; Priya et al. 2022). For example, misplaced fishing equipment or inadvertent cargo loss, and unlawful dumping, besides the inflows from rivers (Coyle et al. 2020; Li et al. 2020). This will most likely take the shape of trash made of plastic, this will decompose to produce microplastics in maritime system (Tang et al. 2021). Worldwide, Microplastics are found in many different contexts including marine life. Certain regions of accumulation, like the well-known "Great Pacific Garbage Patch," are caused by currents in the ocean (Vivekanand et al. 2021).

25.4.4 Microplastics in Aquaculture

The worldwide aquaculture ecosystem and aquaculture goods are heavily impacted by the widespread presence of microplastic contamination. Microplastics that brought in from rivers, oceans, atmosphere, land and microplastics introduced throughout the process are two keys of microplastics in aquaculture. The effects of microplastics in rivers on aquaculture environments are numerous because marine and freshwater aquaculture ecosystems are strongly related to rivers (Fig. 25.3). Different findings have shown a link between elevated concentrations of microplastics to nearby rivers and aquaculture sites (Lam et al. 2022; Ma et al. 2020). Consequently, it is vital to treat rivers carefully as they are a main source of microplastic contamination in aquaculture.

25.5 Effects of Microplastics on Aquatic Organisms

Research has emphasised primarily on marine life as it is well known that a major portion of microplastic contamination is deposited at the water's surface. Since plastic is constituted mostly of large-molecule carbon monomers, it is a biochemically unreactive substance that takes much time to break down in the environment and has negative impacts (Yu et al. 2020). Plastic particle size has an inverse relationship with plastic toxicity (Rillig et al. 2019). Because they can be ingested by organisms and increase their toxicity, microplastics are more harmful than macroplastics (Guerranti et al. 2019).

There have been reports of MPs entering aquatic animals via a variety of routes. Trophic transfer, suspension feeding, direct ingestion, and filter feeding are a few of these, but they're not the only ones (Nelms et al. 2018). Microplastic density determines which kinds of microplastics are present in saltwater. In surface waters, plastics including PE, PP, and PS are frequently encountered by pelagic fish, Demersal fish, on the other hand, are more likely to encounter thick microplastics such as PVC and PET in deep waters (Wang et al. 2020; Yu et al. 2020). In ocean system, microplastics eventually find their way into sediments. Additionally, surface biofilm development and detritus hetero-aggression can alter the density of microplastics (Carbery et al. 2018; Wang et al. 2020). MPs being consumed by living organism with chemoreceptor-based prey selection is enhanced when a biofilm forms on a microplastic surface. Microplastics are engulf by organisms and pass via the digestive tract to produce mechanical disruptions to the system, including penetration into the intestinal wall and blockage of the intestines (De-la-Torre 2020). Because being a polymer it cannot be disrupted by an organism's enzymes, which lowers the amount of feeding activity. These issues then result in changes in the ratio and distribution of cholesterol, weight loss, depletion of energy reserves, disruption of reproduction, nutritional deficiencies, and slower growth (Eltemsah and Bøhn 2019; Weber et al. 2021).

In prior research, fish shown MPs required about twice as long as the non-exposed control group to eat the same quantity of zooplankton (Carbery et al. 2018). Reduced fertilityand fecundity could result from diminished growth (Yu et al. 2020). Allergies are brought on by the GI tract's microplastic accumulation. An allergy reaction, an augmented immunological response, and a change in the metabolic profile are

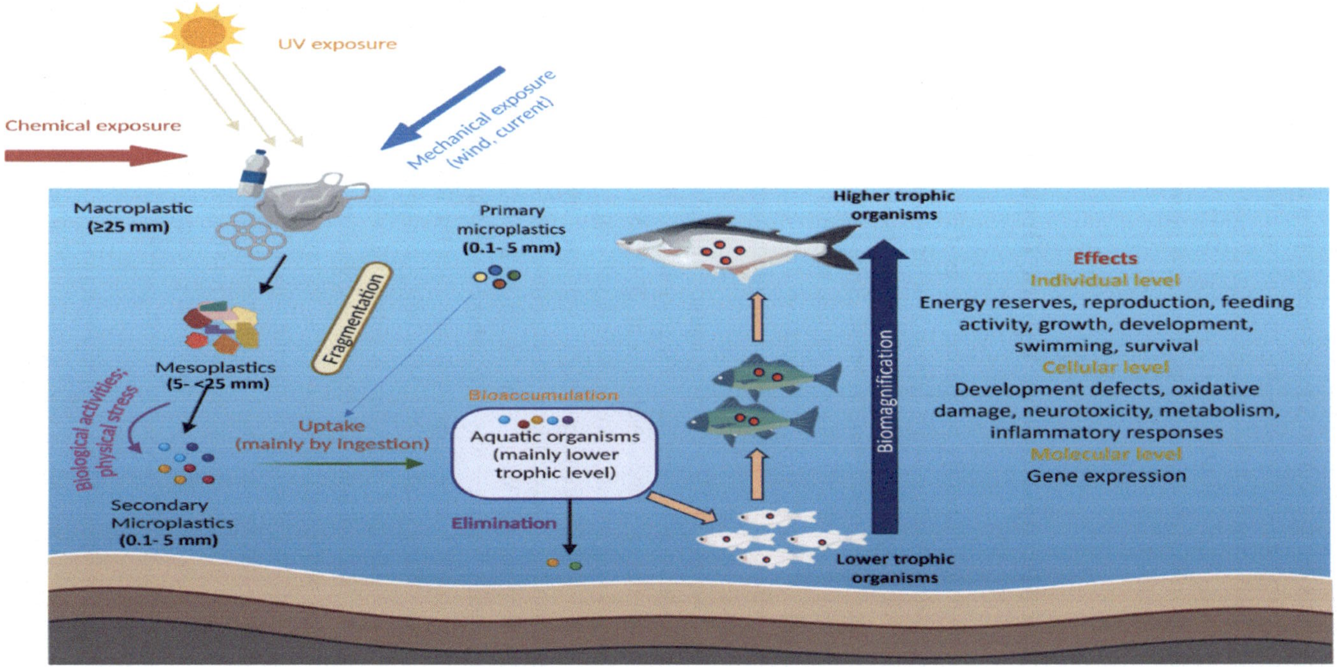

Fig. 25.3 Possible trophic transmission, MPs in the aquatic systems and consequences

brought on by the accumulation of microplastic in the gastrointestinal system (Wang et al. 2020). Plastics can release harmful compounds into the body and have both immediate and long-term effects in acidic stomach conditions (Bosker et al. 2019).

Three main mechanisms—ingestion, bioaccumulation, and biomagnification—are involved in the trophic transfer of microplastics. Ingestion mainly occurs for microplastic particles. Lower trophic-level organisms are typically the ones that directly consume microplastics. These organisms, particularly non-selective feeders, inadvertently ingest microplastics or them in combination with naturally occurring prey items that are comparable in size to plankton and sand grains (Wang et al. 2020). On the other hand, when predators eat contaminated prey, this is known as indirect ingestion. Bioaccumulation is a process through which pollutants can be deposited inside an organism's body. Microplastic particle bioaccumulation occurs predominantly in the gastrointestinal tract. The process known as "biomagnification" describes the buildup of additional particles inside the body of an organism because of predatory behavior within the food chain. Consuming lower-trophic-level species which is already collected microplastics within their bodies can expose higher-trophic-level organisms to a greater quantity of microplastics (Nelms et al. 2018). As microplastics may enter aquatic, benthic, and pelagic food webs, there is rising worry about potential trophic transfer. The consumption of microplastic-contaminated prey by predatory creatures may lead to the indirect buildup of microplastics and bioaccumulation at higher trophic levels. Microplastics can also be consumed by predators and detritivores when they scavenge microplastic-containing detrital debris. Lab tests show that when green crabs were given mussels comprising MPs, the particles ended up to digestive tracts (Besseling et al. 2014). Furthermore, it has been noted that various fish species, such as Atlantic salmon, bleak, crucian carp), Rudd, tench, and northern pike, experience trophic transfer to the planktonic water fleafrom green algae. Also, fluorescent polystyrene microspheres are exchanged from zooplankton to shrimp (Besseling et al. 2014; Watts et al. 2015).

Several studies involving aquatic animals have shown that microplastics have negative, if not fatal, effects on individuals, cells, and molecules (Table 25.1) (Xu et al. 2020). Impacts at the individual level shown in *Aranecola marina*, where it has been discovered that exposure to microplastic reduced lugworm life by impeding its ability to feed and store energy (Fig. 25.4) (Besseling et al. 2013). Reduced rates of ingestion, feeding ability, energy budget, assimilation efficiency, and swimming speed have also been documented for the oyster (Gardon et al. 2018), the diving beetle (Kim et al. 2018), copepod (Cole et al. 2015), and the crab (Watts et al. 2015).

When exposed to PS microplastics, the green alga *Scenedesmus obliquus* showed inhibited growth and a drop in chlorophyll-a levels (Besseling et al. 2014). PS microplastics have been demonstrated to decrease the number and body size of newborns in the the first three sets of *D. magna.* The research also observed a reduction in the neonates' body size and many abnormalities, such as a bent shell spine, short antenna setae, vacuoles surrounding the ovary (Rehse et al. 2016). Microplastics significantly hinder quality of gametes, gametogenesis, and efficacy of oyster reproduction. This results in reduced production of oocyte, formation of sperm velocity, and shell deformities (Sussarellu et al. 2016).

Table 25.1 Impacts of microplastics in lab-cultured aquatic organisms

Studied organisms	Microplastics		Exposure		Effects	References
	Size (μm)	Type	Duration	Concentration		
Pomatoschistus microps (Goby fish)	1–5 μm	PE	96 h	0.184 mg/L	Neurotoxicity	Oliveira et al. (2012)
Zooplankton taxa	0.4–7.3 μm	PS	24 h	3000 beads/ml	Ingestion rate decreased	Cole et al. (2013)
Mytilus edilus L. (blue mussel)	<80 μm	HDPE	96 h	2.5 g/L	Granulocytomas formation increased	Moos et al. ((2012)
Danio rerio (zebrafish)	5 and 20 μm	PS	7 days	20 mg/L	Disturbed the lipid metabolism, change of metabolic profiles in liver	Lu et al. (2016)
Dicentrarchuslabrax (European sea bass)	10–45 μm	PS	846 h	10–100 particles/mg of diet	Mortality and initiation of cytochrome P450	Mazurais et al. (2015)
Mytilus galloprovincialis (Marine mollusc)	<100 μm	PS and PE	168 h	2×10^3 mg/L	Alterations in enzymes	Avio et al. (2015)
Scrobicularia plana (clam)	20 μm	PS	14 days	4 particles/mL	oxidative damage, neurotoxicity and Antioxidant capacity changed	Ribeiro et al. (2017)
Tripneustesgratilla (sea urchin)	10–45 μm	PE	120 h	3×10^5 particles/L	development without affecting its survival and Influence larval growth	Kaposi et al. (2014)
Paracentrotuslividus (sea urchin)	0.04–0.05 μm	PS	6–48 h	1–5 mg/L	Up-regulation of the Abcb 1 gene responsible for protection and multi-drug resistance	Della Torre et al. (2014)
Arenicola marina (Blow lungworm)	< 100 μm	PE	336 h	1.1×10^5 particles/L	Energy consumption increased	Cauwenberghe et al. (2015)

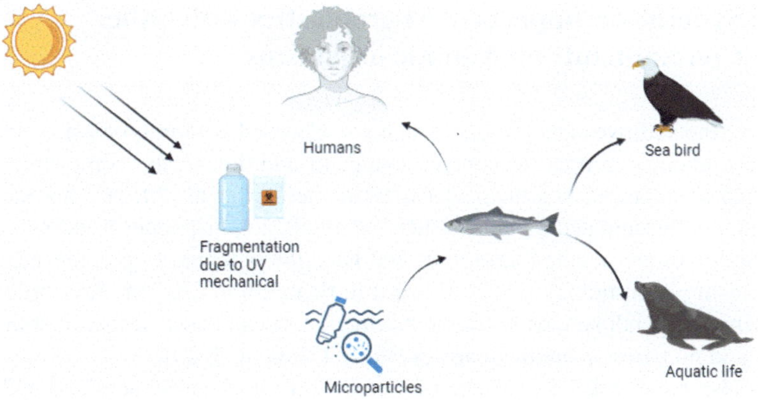

Fig. 25.4 Microplastic effects on environment, aquatic animals and on sea birds

Exposure to PS microplastics treated with amino acids, sea urchin *P. lividus* embryos displayed abnormal proliferation and thickened ectodermal membranes at 6 h post-fertilization (hpf). At 48 hpf, the larvae displayed severe malformation, incomplete skeletal rods, reduced arm length and fractured ectoderm (Della Torre et al. 2014). The zebrafish and brine shrimp treated with microplastics displayed aberrant intestinal epithelial cell ultrastructure, including an increase in mitochondria, a decrease in microvilli, and the splitting of enterocytes and villi (Wang et al. 2019). Microplastics cause neurotoxicity, inflammatory reactions, and oxidative damage in crabs, coral, rotifers, fish, and mussels in addition to developmental abnormalities. Moreover, microplastics suppress the immune system and detoxifying processes of the *Pocillopora damicornis*, a scleractinian coral (Tang et al. 2018).

Microplastics modify the activity of genes linked with an extensive range of behaviors at the molecular level. Like, in young Chinese mitten crabs, microplastics boost activity of genes associated with the stress, worm *C. elegans*, and gilthead seabream *Sparus aurata* L. When juvenile *E. sinensis*is exposed to microplastics, ROS production triggers the MAPK signaling pathway, suggesting that genes associated with oxidative stress are more highly expressed (Yu et al. 2018). Coral *P. damicornis* may show up- or down-regulation of genes linked to the EGF-ERKand JNK signaling pathway and zymogen granulesin response to exposure to microplastics. According to (Tang et al. 2018), this shows that the particles, in this species, may work as an environmental hormone regulating processes. Exposure of Mytilus to microplastics has been shown to modify gene expression which is related with lipid metabolism signaling pathways DNA repair, Biotransformation, immunology and stress response (Brandts et al. 2018).

25.6 Synergistic Impacts of Microplastics with Other Contaminants on Aquatic organisms

Effects of the additives and pollutants that are adsorbed with microplastics are also a concern about microplastic contamination, in addition to the composition and properties of microplastics themselves (Bouwmeester et al. 2015). This includes triclosan, polybrominated diphenyl ethers, nonylphenol, bisphenol Aandphthalates. These additives are deemed hazardous and have the potential to permanently alter an organism's morphology if it is exposed to them while it is still developing. By interfering with endogenous hormone production, it can cause sexual dysfunction, cancer, and mutation in adult aquatic creatures (Yu et al. 2020).

MPs also have ability as carriers of metals, hazardous substances and additives from the environment to living things. Adsorbed chemical pollutants on microplastics develop harmful endocrine, immunological, and reproductive issues when they desorb into tissue (Table 25.1) (Bouwmeester et al. 2015; Nelms et al. 2018). It can disperse throughout body by translocating their harmful compounds or monomers into living cells, tissue, internal organs and the lymphatic and circulatory systems (Bouwmeester et al. 2015). MPs that are ingested can harm aquatic environment creatures both physically and toxicologically, which can then seriously endanger human health when digested.

25.7 Impact on Humans from Consuming MP-Contaminated fish

The existence of MPs in marine organisms is a significant factor in the intake of microplastics by humans (De-la-Torre 2020). Because microplastics can pass from herbivores to carnivores across trophic levels, they have accumulated and become more abundant in food chain (Tang et al. 2021). It is present in roughly 80% of dietary fish (Senathirajah et al. 2021; Walkinshaw et al. 2020). Compared to seafood processed by bypassing the digestive tract, seafood from the commercial market exposes people to a higher risk of consuming microplastics. This is especially true for seafood taken in its whole (Nelms et al. 2018). People may have negative consequences if they consume aquaculture goods polluted with microplastics. In addition to eating seafood, individuals may also consume microplastics through their skin (through hair follicles, open wounds, and sweat glands), ingestion (by sugar, honey, drinking water and table salt), and inhalation (from city dust and synthetic fabrics) (De-la-Torre 2020; Ma et al. 2020).

Despite increasing evidence of MPs in seafood, our understanding of how they affect human food safety is perhaps in early phases (Garrido Gamarro et al. 2020). Currently, it is known that eating seafood exposes humans to thousands of MP particles, the exact amount varies according to the types and frequency of marine items ingested in different nations (Cox et al. 2019). For instance, as comparison with diets which depends on gadoids, a meal heavy in species from sardines or mussel's results in higher MP intake (Galgani et al. 2015). Since the highest concentrations have been found in the ingested digestive tracts or gills, this would be related to both the higher concentrations found and the tendency for smaller fish (such as bivalves and anchovies) to be eaten whole which raises exposure for humans.

Microplastics can cross the placenta, cell membrane and blood–brain barrier. Once inside, they cause cell damage, inflammation, poor energy allocation, oxidative stress, and marine creatures are also similarly affected (Bouwmeester et al. 2015). Human health risks associated with microplastics include skin irritation, respiratory issues, cancer risk, cardiovascular illnesses, and reproductive issues (Fig. 25.5) (De-la-Torre 2020; Yu et al. 2020). Tan et al. (2020) study on the digestive tract demonstrated that microplastics might interact with lipases as well as lipid droplets. It also identified two processes which impact absorption and digestion: (i) Because it is highly hydrophobic, they reduce the bioavailability of lipid droplets by forming large lipid-microplastic hetero-aggregates, and (ii) Because microplastics disrupted the crucial open conformation and changed the secondary structure, they absorbed lipase and reduced its activity (Powell et al. 2010; Tan et al. 2020). Persorption and endocytosis are two of the most popular routes via which MPs reach the human body. In vitro experimentation with human gastric cancer and lung cancer cells exposed to PS has shown that microplastics can induce the generation of cytokinesis, leading to an inflammatory response in both cell types (De-la-Torre 2020). Microplastics can enter the lymphatic and circulatory systems from the gut cavity, upsetting the intestinal flora and causing lipid issues in the liver. This can lead to an immune reaction and inflammation (Bouwmeester et al. 2015; Wright and Kelly 2017; Yu et al. 2020). When it gets into the bloodstream, they also induce haemolysis. Following their entry into the circulatory system, microplastics will travel via blood to different organs inside the human body, therefore impacting those organs. Braun et al. (Braun et al. 2021) reported the detection of MPs, including PU, PE, PS, and PP in human placenta. These findings raise concerns about the possible health risks that microplastics may pose to developing foetuses (Braun et al. 2021). It is believed that microplastics possess teratogen potential. It has been proven that certain chemicals associated with microplastics seriously damage human health, including mutagenicity, carcinogenicity (Powell et al. 2010). Studies have revealed that approximately 1700 viruses are transported in aquatic ecosystems by microplastics (Li et al. 2022).

25.8 The Economic Impacts of Microplastic pollution

Microplastic contamination has several notable economic impacts. Microplastic exposure can have a wide range of consequences and endanger members of many species, their habitats, and eventually human health. Microplastics commonly found in food and water supplies, which has led to problem about possible health hazards and higher healthcare expenses for treating associated ailments. Aquaculture items,

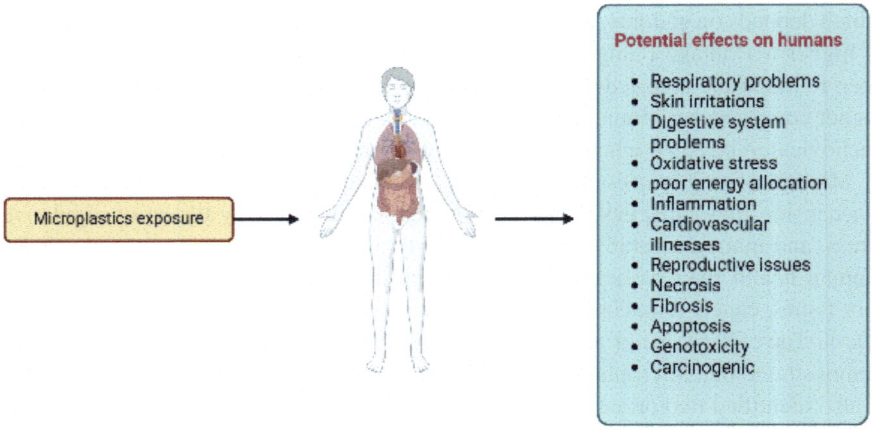

Fig. 25.5 Potential impacts of microplastics on humans

including fish, mussels, shrimp, and crabs, have been reported to contain microplastics (FAO 2024). Aquaculture products are susceptible to various toxicological consequences caused by microplastics, such as oxidative stress and negative impacts on behaviour, growth, and reproduction. These effects eventually lower the aquaculture products' economic advantages.

25.9 Conclusion

There are also some harmful effects of MPs on the microbial growth in aquatic environment. For instance, during the production process, some polymers are supplemented with antibacterial compounds. (Sui et al. 2020). Microplastics can release these antibacterial substances, and aquatic microorganisms in aquatic environments may develop drug resistance, hence increasing the danger of infectious disease transmission (Ta and Babel 2020). Antibiotic resistance genes may become more prevalent in aquaculture environments due to microplastics, which also raises the possibility that antibiotics may lose their potency. (Lu et al. 2019). Additionally, a lot of viruses get loaded onto microplastics. The environment will deteriorate as a consequence for aquatic life, decrease the number of items produced by aquaculture, and decrease its financial advantages.

Moreover, during the photocatalytic breakdown, certain chlorine-containing polymers (like PVC) may emit HCl, which would acidify the surrounding water body (Gewert et al. 2015). Harmful effects of MPs and additives on microalgae can disrupt the harmony of the aquaculture ecosystem. Microplastics are distinguished by their huge specific surface area when compared to larger plastic contaminants. Furthermore, microplastics can carry heavy organic compounds (HOCs) like PCBs, PAHs, and organochlorine insecticides due to their surface hydrophobicity (Cai et al. 2017;

Yadav et al. 2022a; Yadav et al. 2023). Simultaneously, microorganisms have the ability to develop biofilms on the surfaces of microplastics, which can facilitate the discharge of HOCs into the aquatic system (Rummel et al. 2017). This impacts fish populations and reduces the quality of catches which in turn affects the income of fishermen and the aquaculture industry, leading to increased costs for cleaning and monitoring. Pollution reduction initiatives need labour and costly technology.

25.10 Future Prospects

Several studies have evidence of microplastic contamination, particularly in the oceans. The advancement in technology needed to combat the problem of plastic pollution, which has previously come out and persist to take place in the ecosystem, has not been extensively discussed. The majority of research concentrated on preventing further pollution, whereas very few examined the technologies utilized to repair the environment that was already contaminated.

Acknowledgements The authors are highly thankful to the respective university for guiding them to write this book chapter. The authors are also thankful to UGC for providing a fellowship for their research work.

References

Abbasi S, Keshavarzi B, Moore F, Turner A, Kelly FJ, Dominguez AO, Jaafarzadeh N (2019) Distribution and potential health impacts of microplastics and microrubbers in air and street dusts from Asaluyeh County, Iran. Environ Pollut 244:153–164

Auta HS, Emenike CU, Fauziah SH (2017) Distribution and importance of microplastics in the marine environment: a review of the sources, fate, effects, and potential solutions. Environ Int 102:165–176

Avio CG, Gorbi S, Milan M, Benedetti M, Fattorini D, d'Errico G, Regoli F (2015) Pollutants bioavailability and toxicological risk from microplastics to marine mussels. Environ Pollut 198:211–222

Bacha AUR, Nabi I, Zhang L (2021) Mechanisms and the engineering approaches for the degradation of microplastics. Acs Es&T Eng 1(11):1481–1501

Besseling E, Wegner A, Foekema EM, Van Den Heuvel-Greve MJ, Koelmans AA (2013) Effects of microplastic on fitness and PCB bioaccumulation by the lugworm Arenicola marina (L.). Environ Sci Technol 47(1):593–600

Besseling E, Wang B, Lurling M, Koelmans AA (2014) Nanoplastic affects growth of S. obliquus and reproduction of D. magna. Environ Sci Technol 48(20):12336–12343

Bosker T, Bouwman LJ, Brun NR, Behrens P, Vijver MG (2019) Microplastics accumulate on pores in seed capsule and delay germination and root growth of the terrestrial vascular plant *Lepidium sativum*. Chemosphere 226:774–781

Botterell ZL, Beaumont N, Dorrington T, Steinke M, Thompson RC, Lindeque PK (2019) Bioavailability and effects of microplastics on marine zooplankton: a review. Environ Pollut 245:98–110

Bouwmeester H, Hollman PC, Peters RJ (2015) Potential health impact of environmentally released micro-and nanoplastics in the human food production chain: experiences from nanotoxicology. Environ Sci Technol 49(15):8932–8947

Brandts I, Teles M, Gonçalves AP, Barreto A, Franco-Martinez L, Tvarijonaviciute A, Oliveira M (2018) Effects of nanoplastics on *Mytilus galloprovincialis* after individual and combined exposure with carbamazepine. Sci Total Environ 643:775–784

Braun T, Ehrlich L, Henrich W, Koeppel S, Lomako I, Schwabl P, Liebmann B (2021) Detection of microplastic in human placenta and meconium in a clinical setting. Pharmaceutics 13(7):921

Cai L, Wang J, Peng J, Tan Z, Zhan Z, Tan X, Chen Q (2017) Characteristic of microplastics in the atmospheric fallout from Dongguan city, China: preliminary research and first evidence. Environ Sci Pollut Res 24:24928–24935

Campanale C, Massarelli C, Savino I, Locaputo V, Uricchio VF (2020) A detailed review study on potential effects of microplastics and additives of concern on human health. Int J Environ Res Public Health 17(4):1212

Carbery M, O'Connor W, Palanisami T (2018) Trophic transfer of microplastics and mixed contaminants in the marine food web and implications for human health. Environ Int 115:400–409

Carney Almroth BM, Åström L, Roslund S, Petersson H, Johansson M, Persson NK (2018) Quantifying shedding of synthetic fibers from textiles; a source of microplastics released into the environment. Environ Sci Pollut Res 25:1191–1199

Chaves PDT, Silveira BED (2016) Artisanal fishing net float loss and a proposal for a float design solution. Braz J Oceanogr 64:89–94

Chen G, Li Y, Wang J (2021) Occurrence and ecological impact of microplastics in aquaculture ecosystems. Chemosphere 274:129989

Cole M, Lindeque P, Fileman E, Halsband C, Goodhead R, Moger J, Galloway TS (2013) Microplastic ingestion by zooplankton. Environ Sci Technol 47(12):6646–6655

Cole M, Lindeque P, Fileman E, Halsband C, Galloway TS (2015) The impact of polystyrene microplastics on feeding, function and fecundity in the marine copepod *Calanus helgolandicus*. Environ Sci Technol 49(2):1130–1137

Cowger W, Gray A, Christiansen SH, DeFrond H, Deshpande AD, Hemabessiere L, Primpke S (2020) Critical review of processing and classification techniques for images and spectra in microplastic research. Appl Spectrosc 74(9):989–1010

Cox KD, Covernton GA, Davies HL, Dower JF, Juanes F, Dudas SE (2019) Human consumption of microplastics. Environ Sci Technol 53(12):7068–7074

Coyle R, Hardiman G, O'Driscoll K (2020) Microplastics in the marine environment: a review of their sources, distribution processes, uptake and exchange in ecosystems. Case Stud Chem Environ Eng 2:100010

De Souza Machado AA, Kloas W, Zarfl C, Hempel S, Rillig MC (2018) Microplastics as an emerging threat to terrestrial ecosystems. Glob Change Biol 24(4):1405–1416

De-la-Torre GE (2020) Microplastics: an emerging threat to food security and human health. J Food Sci Technol 57(5):1601–1608

Della Torre C, Bergami E, Salvati A, Faleri C, Cirino P, Dawson KA, Corsi I (2014) Accumulation and embryotoxicity of polystyrene nanoparticles at early stage of development of sea urchin embryos *Paracentrotus lividus*. Environ Sci Technol 48(20):12302–12311

Dhir RK, de Brito J, Mangabhai R, Lye CQ (2017) Production and properties of copper slag. Sustain Construct Mater: Copper Slag 55:27–86

Duis K, Coors A (2016) Microplastics in the aquatic and terrestrial environment: sources (with a specific focus on personal care products), fate and effects. Environ Sci Eur 28(1):2

Eltemsah YS, Bøhn T (2019) Acute and chronic effects of polystyrene microplastics on juvenile and adult Daphnia magna. Environ Pollut 254:112919

Fahrenfeld NL, Arbuckle-Keil G, Beni NN, Bartelt-Hunt SL (2019) Source tracking microplastics in the freshwater environment. TrAC, Trends Anal Chem 112:248–254

Frias JP, Nash R (2019) Microplastics: Finding a consensus on the definition. Mar Pollut Bull 138:145–147

Galafassi S, Nizzetto L, Volta P (2019) Plastic sources: a survey across scientific and grey literature for their inventory and relative contribution to microplastics pollution in natural environments, with an emphasis on surface water. Sci Total Environ 693:133499

Galgani F, Hanke G, Maes T (2015) Global distribution, composition and abundance of marine litter. Marine Anthropogen Litter, pp 29–56

Gall SC, Thompson RC (2015) The impact of debris on marine life. Mar Pollut Bull 92(1–2):170–179

Gardon T, Reisser C, Soyez C, Quillien V, Le Moullac G (2018) Microplastics affect energy balance and gametogenesis in the pearl oyster Pinctada margaritifera. Environ Sci Technol 52(9):5277–5286

Garrido Gamarro E, Ryder J, Elvevoll EO, Olsen RL (2020) Microplastics in fish and shellfish–a threat to seafood safety? J Aquat Food Prod Technol 29(4):417–425

Geremia E, Muscari Tomajoli MT, Murano C, Petito A, Fasciolo G (2023) The impact of micro-and nanoplastics on aquatic organisms: Mechanisms of oxidative stress and implications for human health—a review. Environments 10(9):161

Gewert B, Plassmann MM, MacLeod M (2015) Pathways for degradation of plastic polymers floating in the marine environment. Environ Sci Process Impacts 17(9):1513–1521

Guerranti C, Martellini T, Perra G, Scopetani C, Cincinelli A (2019) Microplastics in cosmetics: environmental issues and needs for global bans. Environ Toxicol Pharmacol 68:75–79

Hale RC, Seeley ME, La Guardia MJ, Mai L, Zeng EY (2020) A global perspective on microplastics. J Geophys Res Oceans 125(1):e2018JC014719

Harmon SM, Chen Q, Ma C, Ji M, Yan X, Ji R, Shi H (2024) The effects of microplastic pollution on aquatic organisms. In Microplastic contamination in aquatic environments, pp 355–379. Elsevier

He P, Chen L, Shao L, Zhang H, Lü F (2019) Municipal solid waste (MSW) landfill: a source of microplastics?-Evidence of microplastics in landfill leachate. Water Res 159:38–45

Issac MN, Kandasubramanian B (2021) Effect of microplastics in water and aquatic systems. Environ Sci Pollut Res 28:19544–19562

Jambeck JR, Geyer R, Wilcox C, Siegler TR, Perryman M, Andrady A, Law KL (2015) Plastic waste inputs from land into the ocean. Science 347(6223):768–771

Kaposi KL, Mos B, Kelaher BP, Dworjanyn SA (2014) Ingestion of microplastic has limited impact on a marine larva. Environ Sci Technol 48(3):1638–1645

Karbalaei S, Golieskardi A, Watt DU, Boiret M, Hanachi P, Walker TR, Karami A (2020) Analysis and inorganic composition of microplastics in commercial Malaysian fish meals. Mar Pollut Bull 150:110687

Karlsson TM, Arneborg L, Broström G, Almroth BC, Gipperth L, Hassellöv M (2018) The unaccountability case of plastic pellet pollution. Mar Pollut Bull 129(1):52–60

Kim SW, Kim D, Chae Y, An YJ (2018) Dietary uptake, biodistribution, and depuration of microplastics in the freshwater diving beetle *Cybister japonicus*: effects on predacious behavior. Environ Pollut 242:839–844

Kitahara KI, Nakata H (2020) Plastic additives as tracers of microplastic sources in Japanese road dusts. Sci Total Environ 736:139694

Koelmans AA, Besseling E, Shim WJ (2015) Nanoplastics in the aquatic environment. Critical review. Marine Anthropogen Litter 325–340

Krüger L, Casado-Coy N, Valle C, Ramos M, Sánchez-Jerez P, Gago J, Sanz-Lazaro C (2020) Plastic debris accumulation in the seabed derived from coastal fish farming. Environ Pollut 257:113336

Kumar R, Sharma P, Manna C, Jain M (2021) Abundance, interaction, ingestion, ecological concerns, and mitigation policies of microplastic pollution in riverine ecosystem: a review. Sci Total Environ 782:146695

Kumar R, Gupta R, Dahiya T, Yadav R, Saifi R, Yodha K, Kumari B, Singh P, Shefali (2022a) Surveillance of *Heteropneustes fossilis* (Stinging Catfish) Diseases and Health Management

Practices in Haryana. Appl Ecol Environ Sci 10(4):261–267. https://doi.org/10.12691/aees-10-4-12

Kumar R, Gupta RK, Yadav R, Saifi R, Yodha K, Pooja (2022b) *Eisenia fetida* as protein source for growth enhancement of *Heteropneustes fossilis*. Egypt J Aquat Biol Fish 26(2):577–588. https://doi.org/10.21608/ejabf.2022.234553

Kumar R, Yadav R, Kumar Gupta R, Pooja (2022c) A sustainable way for fish health management by replacement of chemical and drugs by earthworm. In Environmental degradation in Asia: land degradation, environmental contamination, and human activities, pp. 329–352). Springer International Publishing, Cham. https://doi.org/10.1007/978-3-031-12112-8_16

Kurniawan SB, Said NSM, Imron MF, Abdullah SRS (2021) Microplastic pollution in the environment: Insights into emerging sources and potential threats. Environ Technol Innov 23:101790

Lam TWL, Fok L, Ma ATH, Li HX, Xu XR, Cheung LTO, Wong MH (2022) Microplastic contamination in marine-cultured fish from the Pearl River Estuary, South China. Sci Total Environ 827:154281

Lebreton LC, Van Der Zwet J, Damsteeg JW, Slat B, Andrady A, Reisser J (2017) River plastic emissions to the world's oceans. Nat Commun 8(1):15611

Li C, Busquets R, Campos LC (2020) Assessment of microplastics in freshwater systems: a review. Sci Total Environ 707:135578

Li HX, Shi M, Tian F, Lin L, Liu S, Hou R, Xu XR (2022) Microplastics contamination in bivalves from the Daya Bay: species variability and spatio-temporal distribution and human health risks. Sci Total Environ 841:156749

Lu Y, Zhang Y, Deng Y, Jiang W, Zhao Y, Geng J, Ren H (2016) Uptake and accumulation of polystyrene microplastics in zebrafish (*Danio rerio*) and toxic effects in liver. Environ Sci Technol 50(7):4054–4060

Lu J, Zhang Y, Wu J, Luo Y (2019) Effects of microplastics on distribution of antibiotic resistance genes in recirculating aquaculture system. Ecotoxicol Environ Saf 184:109631

Lusher A, Hollman P, Mendoza-Hill J (2017) Microplastics in fisheries and aquaculture: status of knowledge on their occurrence and implications for aquatic organisms and food safety. FAO

Ma J, Niu X, Zhang D, Lu L, Ye X, Deng W, Lin Z (2020) High levels of microplastic pollution in aquaculture water of fish ponds in the Pearl River Estuary of Guangzhou, China. Sci Total Environ 744:140679

Mazurais D, Ernande B, Quazuguel P, Severe A, Huelvan C, Madec L, Zambonino-Infante J (2015) Evaluation of the impact of polyethylene microbeads ingestion in European sea bass (Dicentrarchuslabrax) larvae. Mar Environ Res 112:78–85

Mills N, Jenkins M, Kukureka S (2020) Plastics: microstructure and engineering applications. Butterworth-Heinemann

Mohamadi M (2023) Plastic types and applications. Plastic Waste Treatment and Management: Gasification Processes, pp 1–19

Von Moos N, Burkhardt-Holm P, Köhler A (2012) Uptake and effects of microplastics on cells and tissue of the blue mussel *Mytilus edulis* L. after an experimental exposure. Environ Sci Technol 46(20):11327–11335

Napper IE, Thompson RC (2016) Release of synthetic microplastic plastic fibres from domestic washing machines: effects of fabric type and washing conditions. Mar Pollut Bull 112(1–2):39–45

Napper IE, Bakir A, Rowland SJ, Thompson RC (2015) Characterisation, quantity and sorptive properties of microplastics extracted from cosmetics. Mar Pollut Bull 99(1–2):178–185

Nelms SE, Galloway TS, Godley BJ, Jarvis DS, Lindeque PK (2018) Investigating microplastic trophic transfer in marine top predators. Environ Pollut 238:999–1007

Oliveira M, Ribeiro A, Guilhermino L (2012) Effects of short-term exposure to microplastics and pyrene on Pomatoschistus microps (Teleostei, Gobiidae). Compar Biochem Physiol A 163:S20

Organisation for Economic Co-operation and Development (2022) Global plastics outlook: economic drivers. OECD Publishing, Environmental Impacts and Policy Options

Powell JJ, Faria N, Thomas-McKay E, Pele LC (2010) Origin and fate of dietary nanoparticles and microparticles in the gastrointestinal tract. J Autoimmun 34(3):J226–J233

Priya KL, Renjith KR, Joseph CJ, Indu MS, Srinivas R, Haddout S (2022) Fate, transport and degradation pathway of microplastics in aquatic environment—A critical review. Region Stud Marine Sci 56:102647

Rehse S, Kloas W, Zarfl C (2016) Short-term exposure with high concentrations of pristine microplastic particles leads to immobilisation of *Daphnia magna*. Chemosphere 153:91–99

Rezania S, Park J, Din MFM, Taib SM, Talaiekhozani A, Yadav KK, Kamyab H (2018) Microplastics pollution in different aquatic environments and biota: a review of recent studies. Mar Pollut Bull 133:191–208

Ribeiro F, Garcia AR, Pereira BP, Fonseca M, Mestre NC, Fonseca TG, Bebianno MJ (2017) Microplastics effects in *Scrobicularia plana*. Mar Pollut Bull 122(1–2):379–391

Rillig MC, Lehmann A, de Souza Machado AA, Yang G (2019) Microplastic effects on plants. New Phytologist 223(3):1066–1070

Rochman CM (2015) The complex mixture, fate and toxicity of chemicals associated with plastic debris in the marine environment. Marine Anthropogen Litter, pp 117–140

Rummel CD, Jahnke A, Gorokhova E, Kühnel D, Schmitt-Jansen M (2017) Impacts of biofilm formation on the fate and potential effects of microplastic in the aquatic environment. Environ Sci Technol Lett 4(7):258–267

Senathirajah K, Attwood S, Bhagwat G, Carbery M, Wilson S, Palanisami T (2021) Estimation of the mass of microplastics ingested–a pivotal first step towards human health risk assessment. J Hazard Mater 404:124004

Shefali Kumar R, Sankhla MS, Kumar R, Swaroop S Sonone (2021) Impact of pesticide toxicity in aquatic environment. Biointerface Res Appl Chem11(3):10131–10140. https://doi.org/10.33263/BRIAC113.1013110140

Shrivastava A (2018) Introduction to plastics engineering. William Andrew

Song YK, Hong SH, Jang M, Han GM, Jung SW, Shim WJ (2017) Combined effects of UV exposure duration and mechanical abrasion on microplastic fragmentation by polymer type. Environ Sci Technol 51(8):4368–4376

Sui M, Lu Y, Wang Q, Hu L, Huang X, Liu X (2020) Distribution patterns of microplastics in various tissues of the Zhikong scallop (*Chlamys farreri*) and in the surrounding culture seawater. Mar Pollut Bull 160:111595

Sun J, Dai X, Wang Q, Van Loosdrecht MC, Ni BJ (2019) Microplastics in wastewater treatment plants: Detection, occurrence and removal. Water Res 152:21–37

Sussarellu R, Suquet M, Thomas Y, Lambert C, Fabioux C, Pernet MEJ, Huvet A (2016) Oyster reproduction is affected by exposure to polystyrene microplastics. Proc Natl Acad Sci 113(9):2430–2435

Ta AT, Babel S (2020) Microplastics pollution with heavy metals in the aquaculture zone of the Chao Phraya River Estuary, Thailand. Marine Pollut Bull 161:111747

Tan H, Yue T, Xu Y, Zhao J, Xing B (2020) Microplastics reduce lipid digestion in simulated human gastrointestinal system. Environ Sci Technol 54(19):12285–12294

Tang J, Ni X, Zhou Z, Wang L, Lin S (2018) Acute microplastic exposure raises stress response and suppresses detoxification and immune capacities in the scleractinian coral Pocilloporadamicornis. Environ Pollut 243:66–74

Tang Y, Liu Y, Chen Y, Zhang W, Zhao J, He S, Yang Z (2021) A review: Research progress on microplastic pollutants in aquatic environments. Sci Total Environ 766:142572

Tirkey A, Upadhyay LSB (2021) Microplastics: an overview on separation, identification and characterization of microplastics. Mar Pollut Bull 170:112604

Van Cauwenberghe L, Claessens M, Vandegehuchte MB, Janssen CR (2015) Microplastics are taken up by mussels (*Mytilus edulis*) and lugworms (*Arenicola marina*) living in natural habitats. Environ Pollut 199:10–17

Vivekanand AC, Mohapatra S, Tyagi VK (2021) Microplastics in aquatic environment: challenges and perspectives. Chemosphere 282:131151

Vuppaladadiyam SSV, Vuppaladadiyam AK, Sahoo A, Urgunde A, Murugavelh S, Šrámek V, Pant KK (2024) Waste to energy: trending key challenges and current technologies in waste plastic management. Sci Total Environ 913:169436
Wagner S, Klöckner P, Reemtsma T (2022) Aging of tire and road wear particles in terrestrial and freshwater environments–a review on processes, testing, analysis and impact. Chemosphere 288:132467
Waldschläger K, Lechthaler S, Stauch G, Schüttrumpf H (2020) The way of microplastic through the environment–Application of the source-pathway-receptor model. Sci Total Environ 713:136584
Walkinshaw C, Lindeque PK, Thompson R, Tolhurst T, Cole M (2020) Microplastics and seafood: lower trophic organisms at highest risk of contamination. Ecotoxicol Environ Saf 190:110066
Wang Y, Zhang D, Zhang M, Mu J, Ding G, Mao Z, Wang J (2019) Effects of ingested polystyrene microplastics on brine shrimp, Artemia parthenogenetica. Environ Pollut 244:715–722
Wang W, Ge J, Yu X (2020) Bioavailability and toxicity of microplastics to fish species: a review. Ecotoxicol Environ Saf 189:109913
Watts AJ, Urbina MA, Corr S, Lewis C, Galloway TS (2015) Ingestion of plastic microfibers by the crab Carcinusmaenas and its effect on food consumption and energy balance. Environ Sci Technol 49(24):14597–14604
Weber A, von Randow M, Voigt AL, von der Au M, Fischer E, Meermann B, Wagner M (2021) Ingestion and toxicity of microplastics in the freshwater gastropod Lymnaeastagnalis: no microplastic-induced effects alone or in combination with copper. Chemosphere 263:128040
Wright SL, Kelly FJ (2017) Plastic and human health: a micro issue? Environ Sci Technol 51(12):6634–6647
Wright SL, Thompson RC, Galloway TS (2013) The physical impacts of microplastics on marine organisms: a review. Environ Pollut 178:483–492
Wu H, Hou J, Wang X (2023) A review of microplastic pollution in aquaculture: Sources, effects, removal strategies and prospects. Ecotoxicol Environ Saf 252:114567
Xu S, Ma J, Ji R, Pan K, Miao AJ (2020) Microplastics in aquatic environments: occurrence, accumulation, and biological effects. Sci Total Environ 703:134699
Yadav R, Gupta RK, Kumar R, Kaur T, Moun R (2022a) Biochemical response of earthworm, *Eisenia fetida* to heavy metals toxicity. Journal of Applied and Natural Science 14(3):990–998
Yadav R, Gupta RK, Kumar R, Kaur T (2022b) Assessment of toxicity of lead and nickel on the biochemical and immunological parameters of earthworm, *Eudrilus eugeniae*. Agric Sci Digest 44(5):922–929
Yadav R, Kumar R, Gupta RK, Kaur T, Yodha K, Kour A, Kaur S, Rajput A (2023) Heavy metal toxicity in earthworms and its environmental implications: a review. Environ Adv 100374
Yu P, Liu Z, Wu D, Chen M, Lv W, Zhao Y (2018) Accumulation of polystyrene microplastics in juvenile *Eriocheir sinensis* and oxidative stress effects in the liver. Aquat Toxicol 200:28–36
Yu Q, Hu X, Yang B, Zhang G, Wang J, Ling W (2020) Distribution, abundance and risks of microplastics in the environment. Chemosphere 249:126059

Chapter 26
Plankton Diversity in Aquatic Ecosystem for Sustainability

Prapti Sudan, Aditya Kumar Upadhyay, and Ambrish Singh

Abstract To keep the water ecosystems in balance, plankton, small living things, drift through the water. The study examines phytoplankton and zooplankton to understand the diversity of plankton and how they help support the ecosystem. Because they are responsible for most photosynthesis in the oceans and are the base of the aquatic food web, phytoplankton impact many processes. Zooplankton take part in removing carbon from the ocean and in delivering energy from phytoplankton to larger species. Temperature, salinity, light and nutrient amounts are environmental conditions that play a role in determining the diversity of plankton, as explained in the document. When humans pollute the oceans with fertilizers and cause acidification or climate change, it often results in unsafe algae growth and problems for planktonic biodiversity. Remote sensing, analysis of DNA and microscopic tools are all used by scientists to study the diversity, amount and spread of planktons. Ecologists use the Continuous Plankton Recorder Survey and similar programs to observe how plankton communities have changed through the years. Preserving plankton diversity is crucial to help ecosystems, the long-term fishing industry and to control world climate change. To make aquatic ecosystems more resilience against climate change, this study advises that research on planktons can protect aquatic life through coordinated efforts and limiting pollution.

Keywords Plankton diversity · Phytoplankton · Zooplankton · Freshwater ecosystems · Biogeochemical cycles · Environmental factors

---

P. Sudan · A. K. Upadhyay (✉)
Department of Aquatic Environment Management, College of Fisheries, Guru Angad Dev Veterinary and Animal Sciences University, Ludhiana 141004, Punjab, India
e-mail: adityaupadhyay.nd@gmail.com

A. Singh
Department of Fisheries Resource Management, College of Fisheries Science, DUVASU, Mathura 281001, UP, India

R. Kumar et al. (eds.), *Sustainability in Aquaculture and Fisheries*,
Springer Water, https://doi.org/10.1007/978-3-032-16605-0_26

26.1 Introduction

The term "plankton" comes from the Greek word "planktos," meaning "drifter" in English, and describes the tiny creatures that float on the surface of water. Plankton, despite their small size, play an essential role in aquatic ecosystems as nutrient cyclers, primary producers, and health indicators. The delicate equilibrium of ecosystems, whether in freshwater or marine environments, depends on their variety and abundance. Photosynthesis by phytoplankton, a kind of plankton, accounts for approximately half of all photosynthetic activity on Earth and has a major impact on both oxygen generation and carbon sequestration. These minuscule creatures nourish a wide variety of marine life, from small zooplankton to massive whales, and they are the foundation of the aquatic food chain. The process goes on as zooplankton feed on phytoplankton, giving higher levels of life a key food source and enabling energy to be passed along in the ecosystem. The presence of different types of plankton helps keep aquatic ecosystems healthy by helping bacteria eat garbage and recycle nutrients (Hamilton and Duggan 2010).

The types of plankton found are influenced by many environmental elements, including light, temperature, salinity and nutrients. Changes in the ocean's acidity and rising temperatures may result in different patterns and types of plankton, so climate change is a serious danger to plankton variety. When pollution and eutrophication are caused by humans, they add to these issues and may lead to bad algal blooms that affect the food chain. It is necessary to keep track of and understand these factors in order to preserve and care for aquatic ecosystems. Experts use several ways to study plankton diversity, including remote sensing, using flow cytometry, microscopy and DNA sequencing. Studying the location, number and types of plankton helps researchers identify problems in ecosystems and decide how to conserve them. To discover shifts in the environment and design efficient strategies for fisheries, programs like the Continuous Plankton Recorder Survey are vital for monitoring changes in plankton communities through the years. There would be no ecological balance in freshwater and marine ecosystems without plankton. In the oceans, phytoplankton build food chains, take part in biogeochemical cycles and foster primary productivity, but also affect the water quality of rivers and lakes (Lodang and Kurnia 2019; Nisha et al. 2025). The ongoing sustainability of aquatic ecosystems and their useful services (including fisheries, storing carbon and producing oxygen) rely on healthy and varied plankton populations. A healthy aquatic environment needs a balanced plankton diversity. Protecting plankton and the ecosystems they support requires us to reduce nutrients, limit pollution and lower climate change. To ensure that aquatic ecosystems stay strong and sustainable, it is important to keep studying and monitoring how plankton change in response to environmental factors.

26.2 Classification and Types of Plankton

Plankton are small animals that drift along in the water in different aquatic environments. Although they are relatively small, these animals are crucial to balanced aquatic ecosystems because they eat, decompose waste and produce food. Three major aspects used to classify plankton are size, how long they live and their function.

26.2.1 Functional Classification

(a) **Phytoplankton**: Microscopic photosynthetic organisms called phytoplankton inhabit the sunlit parts of the ocean, lakes and rivers. Though they are small, they play a big role in the environment of Earth. Phytoplankton are responsible for making about half of the oxygen on Earth by photosynthesis, turning light, carbon dioxide and nutrients into organic matter (Kiran et al. 2025; Neculina et al. 2016). This is as important as all the plants living on land combined. Producers in aquatic food webs are these small animals. Phytoplankton is eaten by zooplankton and fish; whales and seabirds get their food from these zooplankton. This cascade demonstrates that phytoplankton supports marine biodiversity greatly. Among phytoplankton, cyanobacteria, dinoflagellates and diatoms are more recognizable. Rich nutrients in water make diatoms common which have shells made from silica. When specific conditions exist, dinoflagellates can make up algal blooms that are toxic and emit light, thanks to their two flagella (Kumar et al. 2023). Blue-green algae (cyanobacteria) were some of the first photosynthetic organisms and help with fixing nitrogen. Phytoplanktons are important for the global carbon cycle as well as for their ecological roles. To help with climate change, they assist by absorbing carbon dioxide during photosynthesis. Some of the remains of phytoplankton that sink to the seabed help to lock up carbon in deep-sea sediments. Phytoplanktons respond to changes happening around them. Ocean warming, acidification and nutrient pollution may affect both the productivity and the location of fish stocks. If there are changes in phytoplankton communities, it can lead to major changes in the ocean and climate, so scientists must watch for such changes (Baird et al. 2022).

(b) **Zooplankton**: Some zooplankton which have sizes from tiny protozoans to full-sized jellyfish, are also heterotrophic and drift around on the water. Because they are food for fish and whales, zooplankton help keep aquatic food chains active by feeding on phytoplankton (Nanda et al. 2020). This is why they are important for the spread of nutrients and energy in water-based systems. There are two main groups of zooplankton defined by their life cycles: meroplanktons which are only planktonic as young and holoplankton which stay in the plankton stage their whole life. Aquatic insects are influenced by things such as temperature, salinity and the presence of nutrients (Goździejewska et al. 2024;

Rani et al. 2025). Besides their role in ecosystems, zooplanktons affect the flow of carbon and other important chemical cycles. They help the biological pump move carbon from the upper ocean to the lower ocean. Carbon dioxide sequestration and controlling the climate are reliant on this process. Watching zooplankton helps determine the health of the environment in lakes, rivers and oceans. The arrangement of a coral reef community can reveal changes in environmental factors such as pollution or warmer oceans. To understand and control how changes in the environment affect freshwater and marine life, we need to research zooplankton (Singh et al. 2021).

(c) **Bacterioplankton**: Bacterioplankton serve to maintain the well-being and sustainability of freshwater and marine ecosystems. In spite of being very tiny, these organisms help manage nutrient cycling, organic material breakdown and control several biogeochemical processes. A main purpose of bacterioplankton is to decompose organic material. When something living dies, bacterioplankton work to break it down and simplify the organic substances in its remains. In doing so, it allows important elements such as nitrogen and phosphorus to be available to phytoplankton (Wetzel 2001). Keeping the base of the aquatic food web and productivity of the ecosystem rely on recycling nutrients. Apart from nutrient recycling, bacterioplankton are crucial for regulating carbon in aquatic systems. Part of the microbial loop involves bacteria consuming DOC which is later passed on to other organisms such as protozoans and small zooplankton when they prey on bacteria. In addition to transferring energy, this loop causes organic debris to sink to the ocean's bottom, where it takes carbon and aids in managing global carbon levels. Bacterioplankton take part in the circulation of other substances such as sulfur and silica. Some species of bacterioplankton oxidize sulfur compounds to affect how sulfur moves in the ocean and others facilitate silica release from diatom cells, improving the availability of silica to phytoplankton. Temperature, the amount of nutrients in the water and the amount of light allows through affect bacterioplankton diversity and activity. Bacterioplankton communities can be changed by factors like pollution and climate change which may upset nutrient cycles and stability in marine ecosystems. The health of aquatic ecosystems and their protection can be understood by monitoring bacterioplankton populations (Cordone et al. 2022).

(d) **Mycoplankton**: Mycoplankton are saprotropic members of the plankton community that live in freshwater and marine habitats. These communities are formed by free-living yeast and fungi, not by plankton or phytoplankton. These aquatic fungi, which are related to bacterioplankton, are crucial in the processes of nutrient cycling and heterotrophic mineralization. The maximum size for mycoplankton is 50 mm in length and 20 mm in diameter. There are around 103 to 104 fungal cells per milliliter of seawater. This figure is higher in estuaries and coastal ecosystems because of the nutritional runoff from land-based populations. The vertical profile is determined by the abundance of phytoplankton, and the most diverse and numerous mycoplankton species are located in surface waters (<1000 m). In addition, the availability of nutrients may cause this distributional difference to change with the seasons. Because of their inability to produce their own oxygen, aquatic fungi rely on two sources: turbulence-induced oxygen diffusion and oxygen produced by photosynthetic organisms (Adam and Magdalena 2022).

Aquatic fungi can be classified using three groups (Sridhar 2009):

- **Lower fungi**—adapted to marine environments (zoosporic fungi, encompassing mastigomycetes: oomycetes and chytridiomycetes)
- **Higher fungi**—Filamentous organisms adapted to a planktonic lifestyle (hyphomycetes, ascomycetes, basidiomycetes).
- **Terrestrial fungi**—include appendages of marine fungi (trichomycetes)

Mycoplankton primarily consists of higher-level fungal species, specifically those belonging to the phylums Ascomycota and Basidiomycota. Fossil evidence suggests that fungi have been around since the late Proterozoic period, between 900 and 570 million years ago. It is believed that mycoplankton likely originated from terrestrial fungi around 390 million years ago, during the Paleozoic period. It is still a mystery how and where terrestrial fungi adapted to life in water.

(e) **Virioplankton**: Viruses constitute the most prevalent biological entities in aquatic ecosystems, with an estimated global abundance of 10^{31} particles. They are essential in regulating microbial population dynamics and nutrient cycling. Viruses infect planktonic organisms, including phytoplankton and bacteria, causing cell lysis and releasing organic matter and nutrients into the environment, a phenomenon termed the "viral shunt." This mechanism redirects carbon and nutrients from elevated trophic levels to the microbial loop, thereby augmenting nutrient recycling and impacting biogeochemical cycles. Moreover, viral infections can regulate phytoplankton blooms, preserving microbial diversity and ecosystem stability. Viruses contribute to genetic diversity within microbial communities through horizontal gene transfer (Wommack and Colwell 2000).

26.2.2 Size-Based Classification

(a) **Picoplankton:**"Picoplankton are minuscule planktonic organisms ranging from 0.2 to 2 μm in size. They are prevalent in both marine and freshwater ecosystems and contribute significantly to global primary production, particularly in nutrient-deficient (oligotrophic) waters. In these environments, algal picoplankton may constitute as much as 90% of daily carbon production. These minute organisms encompass both prokaryotic and eukaryotic phototrophs and heterotrophs. Picoplankton, including Prochlorococcus and Synechococcus, are among the most prevalent photosynthetic organisms in the ocean, playing a crucial role in the global carbon cycle. Picoplankton possesses a high surface-area-to-volume ratio due to their diminutive size, enabling efficient nutrient absorption in nutrient-poor environments. This adaptation allows them to prosper in oligotrophic environments where larger phytoplankton may be scarce.

(b) **Nanoplankton (2–20 μm)**: Nanoplankton are diminutive phytoplankton, encompassing minute diatoms and flagellates. They conduct photosynthesis, significantly contributing to primary production and constituting an essential component of aquatic food webs.

c) **Microplankton (20–200 μm)**: Microplankton includes larger phytoplankton and protozoans. They are essential in photosynthesis and nutrient cycling, acting as a primary food source for numerous zooplankton species.

d) **Macroplankton (>200 μm)**: Macroplankton encompasses larger zooplankton such as krill and jellyfish. Clearly observable, they play a crucial role in transferring energy from lower to higher trophic levels within marine ecosystems.

This size classification helps in understanding the ecological roles and dynamics of different plankton groups."

26.2.3 Life Cycle-Based Classification

(a) **Holoplankton**: Holoplankton are organisms that remain planktonic for their entire life cycle, in contrast to meroplankton, which are planktonic only during certain life stages. This group includes both phytoplankton and zooplankton, comprising diatoms and copepods. Diatoms are unicellular algae characterized by silica-based cell walls, essential for primary production and oxygen generation. Many aquatic food webs depend on copepods, as they eat phytoplankton and are commonly eaten by much larger creatures (Dias and O, Bonecker ACT, de Castro MS, de Carvalho PF, Paranhos R, Bonecker SLC, 2020).

(b) **Meroplankton**: Meroplankton include organisms that are members of plankton for part of their lives only, mainly during the larval or earliest stages of their development. Following their planktonic phase, some animals migrate to live on the bottom or start moving freely in the water column. This collection includes the larvae of many marine creatures, including crabs, sea urchins, starfish, mollusks and fish. The main job of meroplankton is to be dispersed throughout the ocean. The use of ocean currents by larvae allows them to travel far from their home, helping found colonies in new areas and conserving new variations within species (Dias et al. 2020).

26.3 Plankton Biodiversity and Ecosystem Health

Plankton diversity measures the condition of the marine ecosystem, supports oceanic food chains and is very sensitive to changes in the environment. Photosynthetic phytoplankton and zooplankton, both microscopic organisms, help make oxygen, keep carbon dioxide out of the air and recycle important nutrients in the sea. The great variety in plankton, with many species having their own shape and way of working, makes them strong against potential changes in the environment. Still, issues like climate change, ocean acidification, eutrophication and pollution are now putting this biodiversity in greater danger because of humans. As a result of these threats, plankton populations are affected which can impact food webs in the oceans and lower the quality of services provided by these ecosystems to people. About 71% of Earth is covered by marine ecosystems and they are home to many different species, the most common and varied being plankton. Both phytoplankton (light-powered microbes) and zooplankton (miniature animals and protozoa) make up the base of the seafood web and are important for Earth's main biochemical activities. Each component of plankton biodiversity—taxonomy, genetics, function and phylogeny—strengthens the ability of ecosystems to respond to change (Falkowski 2012). The type and arrangement of plankton communities have a strong impact on marine ecosystems, changing primary production, nutrient cycles, carbon levels and food webs.

26.3.1 Plankton Biodiversity Patterns

The diversity of plankton shows detailed patterns in space and time that depend on the ocean environment, the availability of nutrients, temperature and interactions among different species. Frequently, there are many different species found in tropical and subtropical waters, while the polar areas tend to support communities with unusual adaptations. Microbial diversity among ocean plankton was recently revealed by the Tara Oceans expedition and similar studies, with hundreds of millions of distinct species now believed to exist globally. The fact that these creatures change and adapt quickly, have many mutations and live in different spots in the water makes them so diverse. The different ways plankton change and develop each season reflect the shifting environment of their ecosystems. These patterns help increase the variety of living species and the way ecosystems operate (Vargas et al. 2015; Irigoien et al. 2004).

26.3.2 Ecosystem Services Provided by Plankton

Phytoplankton play a key role, making up about half the world's primary production and releasing oxygen. They also provide nutrition for the ocean's food chains. Microscopic organisms that can photosynthesize are crucial in biogeochemical cycles, especially the biological carbon pump that brings carbon dioxide down to the ocean depths and at the bottom. Energy from producers is passed along to fish, marine mammals and seabirds by zooplankton acting as intermediaries. Moreover, plankton communities facilitate nutrient recycling via the microbial loop, in which dissolved organic matter is reintroduced into the food web through bacterial assimilation and subsequent predation. The functional diversity in plankton communities improves ecosystem stability via complementarity effects and response diversity to environmental changes. This resilience, driven by biodiversity, sustains ecosystem functions despite variable conditions (Naselli-Flores and Padisák 2023).

26.3.3 Threats to Plankton Biodiversity

Climate change constitutes a significant threat to plankton biodiversity, as increasing ocean temperatures modify species distributions, phenology, and community composition. Elevated water temperatures typically promote smaller phytoplankton species, which may diminish energy transfer efficiency within food webs. Ocean acidification disproportionately impacts calcifying plankton, such as coccolithophores, foraminifera, and pteropods, potentially hindering calcium carbonate formation and undermining skeletal integrity. These effects jeopardize not only individual species but also the ecosystem functions reliant on calcifying organisms. The nitrates and phosphates released by people often lead to eutrophication and the growth of dangerous algal blooms that decrease oxygen in the water and produce toxins for marine life and humans. Plankton communities are facing new threats from marine pollution thanks to microplastics which they can mistake for food or become exposed to their toxins.

26.4 Environmental and Anthropogenic Influences on Plankton Diversity

Phytoplankton and zooplankton, both members of plankton, play a major role in aquatic ecosystems by producing food and being part of the food web. Diversity and abundance in animals are determined by a mix of nature and human activities.

26.4.1 Environmental Influences on Plankton Diversity

Temperature, salinity, light conditions and nutrient levels in the water have a strong effect on plankton. Plankton adjust their energy use and reproduce in response to temperature, but their salt balance and spread change due to salinity. Photosynthetic activity in phytoplankton is controlled by the amount of light that reaches them because of water clarity and depth. Correct amounts of nutrients, mostly nitrogen and phosphorus, are needed for healthy growth and reproduction and a lack can result in eutrophication and intense algae blooms. The number of plankton species varies with each season. Studies suggest that both plankton density and biomass are usually higher prior to and after the monsoon, since the increased availability of nutrients and suitable conditions lead to these increases. Contrary to that, monsoon seasons usually have fewer plankton since there is more cloudy water and fewer nutrients in the sea. They illustrate that plankton communities are flexible and able to adjust to environmental changes throughout the year (Geng et al. 2022; Lim et al. 2025).

26.4.2 Anthropogenic Influences on Plankton Diversity

Environmental stresses caused by humans are significantly altering the lives of plankton communities. Fertilizer and waste from farming and wastewater commonly cause eutrophication which means too many nutrients in water bodies. Extra nutrients let phytoplankton grow very quickly, usually causing algal blooms that harm aquatic life by reducing the amount of oxygen in the water. The Pearl River Estuary in China, a large urbanized coastal area, demonstrates these patterns, since nutrient variations can influence phytoplankton communities and the way they photosynthesize. Chemical contamination with heavy metals such as copper, may cause changes in the structure of plankton communities. Science indicates that having either too much or too little copper can lead to algae overtaking the water, disturb plankton relationships and affect the area's ecological processes. The consequences of climate change caused by human activity are made worse when temperature, salinity and the way layers are arranged in water is altered in aquatic ecosystems. Modifications in the plankton community can have effects on both the variety of species and important services found in the ecosystem. Changes in phytoplankton species in the Arctic are being caused by increasing temperatures and less sea ice and this may lead to changes in the marine food web (Roy and Sarkar 2022).

26.4.3 Interactive Effects of Multiple Stressors

Combined effects of climate change, increasing acidity in oceans, eutrophication and metal pollution are particularly damaging to plankton communities. The combined impact of these stressors often leads to stronger effects on how plankton function, develop and reproduce. Planktonic communities are naturally disturbed in rich environments by metals because they can alter how nutrients are absorbed, raise the cost of respiration and increase toxic compound formation.

26.4.4 Monitoring and Conservation Implications

Since plankton communities react strongly to environmental changes, they are important signs of how the ecosystem is doing. Noticing shifts in the types and numbers of plankton can be among the first signs of environmental harm and can help managers plan their efforts. Steps to control the flow of nutrients, reduce chemical pollution and battle climate change are necessary for saving plankton diversity and upholding the health of aquatic habitats.

26.5 Methodologies for Assessing Plankton Diversity

Plankton diversity must be examined to help us understand aquatic ecosystems, since they perform important duties in food webs, biogeochemical cycles and the control of climate. For many years, ways to study plankton diversity have advanced by adding both modern and advanced lab tools to old favourites.

26.5.1 Traditional Sampling Methods

Traditional net-based sampling has long been important in plankton research since it directly allows for studying these microscopic organisms. In most cases, researchers use conical or cylindrical nets with mesh openings ranging from 50 to 300 μm to gather specified sizes of plankton (Frey 2025). Scientists pull the nets up or down to collect plankton which they later examine in the laboratory. The simplicity and low-cost nature of net sampling have made it a preferred method for many ecologists over a long time. This way of researching has its own limits. The size of the mesh in the nets can make some organisms escape, so they are not well represented in the catch. Dragging the net in the water can injure some delicate plankton which may affect the accuracy of biodiversity surveys. However, there is still an important place for traditional net sampling, when paired with more recent approaches. The combination of net sampling and environmental DNA analysis helps catch more species and explains plankton communities in more depth. Besides, progress in imaging helps us observe plankton in their natural habitats without surgery (Garcia et al. 2021).

26.5.2 Continuous Plankton Recorder (CPR)

The Continuous Plankton Recorder (CPR) helps monitor plankton species across the world's oceans by collecting valuable and extended data. Sir Alister Hardy invented the CPR in 1931 and it has since been vital for monitoring the seas, mainly in the North Atlantic and North Sea, through the Continuous Plankton Recorder Survey (Reid et al. 2003). The CPR is a straight device that follows behind a ship at around 10 m below the surface, so researchers can continue sampling along large chunks of ocean. As water enters the CPR, it carries plankton to a mesh that is 270 μm wide, then another silk layer is placed over it. The "sandwich" goes into a tank containing formalin which preserves the specimens until laboratory analysis. In the laboratory, the silk is cut into portions that stand for 10 nautical miles of tow. They do two basic analyses on each sample, using the Phytoplankton Colour Index to assess phytoplankton numbers and visually checking each sample on a microscope to identify and count the individuals. Because the CPR can operate in several sea conditions and be placed from opportunity vessels, it offers an affordable solution for long-term, large-scale ocean monitoring. The large amount of data collected has helped find changes in plankton distribution and numbers which are usually linked to climate and other environmental changes. While the CPR has positive points, it still does not collect all plankton because of the mesh and could damage delicate species. Its steady way of working over the years has produced many useful discoveries about marine biodiversity and ocean health, proving why it matters for oceanography (Batten and Burkill 2010).

26.5.3 Video Plankton Recorder (VPR)

The VPR is a device designed to image plankton up close in their regular habitats. The VPR, fitted on undersea equipment, supports real-time observations of plankton and their activities without harming their natural habitat. While this method makes plankton ecology clearer, it also needs a lot of space and power to process the data (Davis et al. 1992).

26.5.4 Environmental DNA (eDNA) Metabarcoding

Environmental DNA (eDNA) metabarcoding involves using water samples to detect the genetic material of various organisms. Using this method, scientists spot many different species, both seen and rare, without harming or trapping them. Checking for environmental DNA is a proven way to keep an eye on different species and locate invasive ones. Moreover, the data may show differences in population density soon after, due to the rapid deterioration of the materials in time (Wee et al. 2023).

26.5.5 Flow Cytometry

The paths of particles are studied as they move with the fluid while crossing a laser beam in flow cytometry. Researchers can measure and describe plankton using the Sable Scat Based Video Profiler by looking at their size, texture and fluorescence. Researchers study many forms of cells using this method, except they cannot always tell the difference between species if their light-based features are the same (Barendregt 2024).

26.5.6 Remote Sensing and Satellite Imaging

Remote Sensing and satellite pictures help researchers monitor and recognize different Earth surface features. Because of satellites, we can examine a significant area and measure chlorophyll as a sign of the abundance of phytoplankton. Plankton records show us the location and season where plankton is found in the ocean. Still, most of these models rely on surface populations and may not consider subsurface or zooplankton species (Revanna et al. 2017).

26.5.7 Statistical and Modeling Approaches

Scientists rely on statistical models and computer programs to study plankton information, recognize patterns and anticipate trends in the ecosystem. Principal component analysis (PCA) and cluster analysis help explain both the organization of communities and how the environment is impacted. Through model-based simulations, it becomes possible to understand plankton movements in many ecosystems which supports management and conservation projects (Acevedo-Trejos et al. 2022).

26.5.8 Integration of Multidisciplinary Methods

Combining several approaches enhances how well plankton diversity is studied. By adding eDNA analysis to other sampling and imaging approaches, scientists can discover both the types and numbers of organisms present in a sample. Studying plankton involves different fields and this approach is important when facing changes in the environment.

26.6 Plankton in Freshwater and Marine Ecosystems

Freshwater and marine systems always depend on the presence of plankton such as phytoplankton and zooplankton. Water currents move these small organisms which support the whole structure of aquatic food chains. Phytoplankton make food for themselves by photosynthesis, turning sunlight, carbon dioxide and nutrients into organic matter. As a result, they produce about half of the Earth's oxygen and are key producers in the lives of aquatic organisms. While phytoplankton provides nutrients, zooplankton obtain nutrients by eating phytoplankton. They are food for fish, crustaceans and whales. As well as being part of food webs, plankton play a major role in global biogeochemical cycles. Photosynthesis by phytoplankton helps to capture carbon dioxide in the biological carbon pump. As these species die, their bodies settle on the ocean floor which locks up carbon and reduces climate change. Furthermore, plankton contribute to nutrient cycling by using and recycling key elements such as nitrogen and phosphorus, important to aquatic animals. Tracker populations help us recognize if the environment has been altered. Abundance and distribution of certain types of fish may be impacted by the levels of water temperature, salinity, pH and nutrient supply (Kumar and Veerwal 2024).

Plankton can be found in both marine and freshwater places, forming communities that are not the same.

26.6.1 Freshwater Ecosystems

The variation of plankton in freshwater ecosystems depends on the diet, light availability and water-related conditions present. Various environmental elements cause green algae (from the Chlorophyta group) and cyanobacteria (blue-green algae) to usually be the most abundant organisms in these communities. The amount of nitrogen and phosphorus in the water helps determine what species of plankton live in a body of water. Excess amounts of nutrients in the water are usually due to runoff and wastewaters and they frequently cause eutrophication (Potharaju and Aruna 2024). As a result, there is more growth of cyanobacteria, some of which can create HABs which release toxins that harm aquatic creatures and people. On the other hand, green algae thrive where nutrients are just right which helps the plankton community stay diverse and steady. How much light is available is a key factor in deciding plankton diversity. By using specific pigments, cyanobacteria are able to survive and succeed in water that is not well lit. Alternatively, green algae thrive with high light intensity and are found in a lot more transparency and shallower waters. Plankton distributed and varied throughout the sea are greatly influenced by the movement and mixing of water. In these layered aquatic areas, uneven distribution of nutrients can influence the variety of species found in plankton. A stable aquatic network can cut down on nutrients at the surface, benefiting cyanobacteria that can only thrive in areas with nutrient shortages.

26.6.2 Marine Ecosystems

Diatoms and dinoflagellates, together known as phytoplankton, provide a base for the ocean food chain and strongly affect global biogeochemical cycles. Many factors affect fish distribution and their rate of growth, including ocean currents, the salt (salinity) content of the water and upwelling nutrients. Phytoplankton and nutrients are transported by ocean currents across large marine areas and this decides their spatial distribution (Chen et al. 2021). Along Australia's southeastern coast, the East Australian Current strongly affects the number and location of phytoplankton. In the Western Pacific Ocean, each phytoplankton species is affected by different factors;

diatoms respond strongly to changes in nutrients, while dinoflagellates react mainly to temperature changes. Changes in salinity shape the types of phytoplankton found in the ocean. If the salinity varies, it can affect which nutrients are available and the distribution of water layers which in turn can change phytoplankton growth and spread. As an example, an increase in salinity has been found to boost phytoplankton levels and encourage more blooms, probably due to the upwelling of richer ocean waters (Armbrecht et al. 2014). Upwelling of nutrients is very important for how phytoplankton function. Nutrient-rich deep water is brought up by upwelling, giving phytoplankton the nitrates, phosphates and silicates it needs for growth. Diatoms find this process helpful since they rely on silicate for their silicone-made cell walls. Even though upwelling regions cover only a small area, they are highly productive and play a key role in global marine productivity.

Understanding both the similarities and differences among plankton communities in these ecosystems is key for good aquatic ecosystem management.

26.7 Role of Plankton in Aquatic Sustainability

The survival of many aquatic ecosystems depends on plankton which consists of phytoplankton and zooplankton. Among their many duties are producing oxygen, absorbing carbon and forming the base for the food in oceans which makes them vital for keeping ecosystems balanced. Phytoplankton get their energy from sunlight by photosynthesis and use that to create organic matter, releasing oxygen into the air which makes up about 50% of Earth's oxygen. With photosynthesis, aquatic ecosystems stay healthy and climate change is reduced by plants absorbing carbon dioxide from the air. A part of that carbon is also carried down to the ocean floor, where it stays for a long time. Though zooplankton do not have the ability to make their own food, they bring phytoplankton and large marine animals together. By eating phytoplankton, they become prey for fish larvae, small fish and whales, helping to complete an essential link in the ocean's food network (Kumar et al. 2022a; Kumar et al. 2022b; Sahoo et al. 2024). The movement of food energy from plankton to fish is crucial for global fisheries, as fish populations needed for both daily nutrition and livelihoods depend on plankton. When plankton behavior is off-track such as changes in species or decrease in populations, the effects spread to affect fish recruitment and the diversity of species in the area. Larval fish need zooplankton which is why their appearance is important for a successful fishery. In addition, plankton are used as signs to monitor water quality. Since they quickly adjust to changes in nutrient levels, air or water temperatures or pollution, they are vital for understanding the condition of an ecosystem. Eutrophic conditions cause some species of phytoplankton to multiply which can cause dangerous algal blooms (such as red tides) and make ocean environments toxic for marine organisms. When pollution-sensitive plankton are not found, it suggests water pollution is increasing, so immediate actions are required to improve the water. In polluted urban lakes, rotifers and cladocerans are abundant, whereas uncontaminated lakes are home to a wide variety of copepods.

Plankton take organic matter and break it down to form the inorganic nutrients nitrates and phosphates. The presence of recycled nitrogen helps phytoplankton flourish which is important for the constant production of organic matter. Some species are able to turn atmospheric nitrogen into forms that help water fertility and encourage aquatic plants to grow. However, pollution from human sources like farms and factories introduces too many nutrients to the water which leads to eutrophication and hypoxia. Climate change impacts these issues by altering how heat is distributed in the water column which then affects where and how much plankton grows. Higher water temperatures encourage smaller types of phytoplankton, lower the efficiency of removing carbon through the oceans and reduce the ocean's ability to capture CO_2. Although they are very small, plankton have a big impact on the Earth's biogeochemical cycles. The carbon sequestration by trees regulates the amount of carbon dioxide in the air and the oxygen they produce is vital for the health of land and marine animals (Table 26.1 shows plankton diversity and their role in sustainability). Such innovations as iron fertilization aimed at boosting phytoplankton for better carbon storage suggest they may play a role in climate change; but, these plans should always be carefully examined to avoid unexpected environmental problems (Sharma et al. 2024; Sharma et al. 2025; Shefali et al. 2021).

Table 26.1 Plankton diversity and their role in sustainability

Type of plankton	Examples	Ecosystem role	Sustainability contribution	References
Phytoplankton	Diatoms, dinoflagellates, cyanobacteria	Primary producers that undergo photosynthesis	Produce ~50% of Earth's oxygen; are the foundation of aquatic food webs; sequester carbon dioxide	Falkowski et al. (1998); Field et al. (1998)
Zooplankton	Copepods, krill, jellyfish larvae	Feed on phytoplankton and are a prey for larger species	Facilitate energy transfer in marine food webs; fisheries; nutrient and carbon cycling	Calbet and Saiz (2005)
Bacterioplankton	Marine bacteria, archaea	Break down dead organic matter and recycle nutrients	Control nutrient availability; facilitate biogeochemical cycles such as nitrogen and carbon	Azam et al. (1983)
Mycoplankton	Marine fungi, yeasts	Decomposition of organic matter; parasitic or symbiotic function	Contribution to organic matter breakdown; influence on microbial interactions and carbon fluxes	Gutiérrez et al. (2011)
Virioplankton	Marine viruses	Contribute to infection of other plankton; control microbial populations	Contribute to microbial loop; increase nutrient regeneration; affect carbon and nutrient cycles	Suttle (2005)

26.8 Conclusion

Diversity among plankton is important for aquatic ecosystems because it supports many trophic levels, manages biogeochemical processes and notifies us about any changes in the environment. This study points out that, due to their tiny size, plankton—phytoplankton and zooplankton—have a large influence on the well-being of various water bodies. Phytoplankton help produce oxygen that the world needs and they also lock away carbon from the atmosphere. Because they take in carbon dioxide from the atmosphere and begin the process of food webs in aquatic habitats, they play a key role in ecological balance and climate maintenance. At the same time, zooplankton pass along energy from primary producers to fish, seabirds and marine mammals which occupy the top parts of the food chain. Both kinds of plankton contribute to recycling nutrients and to a healthy environment for many aquatic species. Many abiotic elements such as light, temperature, salinity and the number of nutrients, play a big role in shaping plankton diversity. Such factors have an impact on both how plankton population's function and where and how they distribute in the oceans. In this case, pressures caused by humans such as nutrient pollution, discharge from industry, too much plastic and changes in climate, especially warming sea water and ocean acidification, are major risks to plankton populations. Such disturbances often cause algae blooms, changes in food webs and shifts in the kinds of life found there. Plankton can be trusted to reveal the general health of water bodies. Because they react rapidly to changes in their surroundings, they are crucial for ongoing monitoring and management efforts. Plankton scientists use flow cytometry, remote sensing, microscopy and DNA sequencing to study changes in plankton over time. The Continuous Plankton Recorder Survey has highlighted the importance of continuous monitoring to track ongoing changes in aquatic ecosystems. Ensuring plankton diversity is key to keeping ecosystems alive and healthy catches, pure water and a safe climate. Healthy plankton populations make aquatic ecosystems more able to withstand changes from climate change and pollution. That's why we should use integrated management systems. Examples are less nutrient runoff, sticking with pollution control guidelines, better waste management and working harder to combat climate change. Plankton's ecological roles should be promoted in the community and to local leaders and environmentalists. While being barely visible, plankton help support life in and around oceans and lakes. Plankton can be protected by holding classes, getting people in the area to participate in protection work and formally including monitoring in environmental regulations.

26.9 Future Prospects

Phytoplanktons are the primary producers who contribute almost half of the global oxygen. They are at the bottom of the aquatic food chain. They have the power to absorb carbon dioxide. Their use in photosynthesis renders them unavoidable in the effort to combat climate change. Zooplankton play a vital role in the chain of energy transferring the energy between phytoplankton and other organisms. Such diversity gives stability to the environment changes and enhances many ecosystem services. Therefore, there is a need for innovative technologies that can use planktons more efficiently. All planktons are not discovered, there is a need for more research and identify such kinds of planktons that can replace 100 percent dependence on other protein source and medicinal for aquatic ecosystem.

In the future, the development of genomics, remote sensing, and AI-based monitoring will greatly improve our knowledge of plankton biodiversity. The technologies will help to achieve more accurate forecasts of the population dynamics in the changing environment. Restoration plans, mitigation of nutrient overloads, preservation of wetlands and marine ecosystems should represent critical steps that will make it possible to preserve the diversity of plankton.

The maintenance of plankton diversity will not only serve to make aquatic ecosystems healthier but will also benefit the global food security, climate control, and biodiversity. Scientific research, policy-making, and the awareness of the population should be incorporated into future strategies that are aiming at long-term sustainability of the aquatic environment.

References

Acevedo-Trejos E, Cadier M, Chakraborty S, Chen B, Cheung SY (2022) Modelling approaches for capturing plankton diversity (MODIV), their societal applications and data needs. PERSPECTIVE Article, 1–8. https://doi.org/10.3389/fmars.2022.975414

Adam C, Magdalena Ś (2022) Species diversity of mycoplankton on the background of selected indicators of water quality in stratified mesotrophic lakes. Int J Environ Res Public Health 19(20):13298. https://doi.org/10.3390/ijerph192013298

Armbrecht LH, Roughan M, Rossi V, Schaeffer A, Davies PL, Waite AM, Armand LK (2014) Phytoplankton composition under contrasting oceanographic conditions: upwelling and downwelling (Eastern Australia). Cont Shelf Res 75:54–67. https://doi.org/10.1016/j.csr.2013.11.024

Azam F, Fenchel T, Field JG, Gray JS, Meyer-Reil LA, Thingstad F (1983) The ecological role of water-column microbes in the sea. Mar Ecol Prog Ser 10:257–263. https://doi.org/10.3354/meps010257

Baird M, Dutkiewicz S, Hickman A (2022) Chapter 8—Modeling phytoplankton processes in multiple functional types. Adv Phytoplank Ecol Appl Emerging Technol 245–264. https://doi.org/10.1016/B978-0-12-822861-6.00016-9

Barendregt A (2024) The optics of flow cytometry. Misumi

Batten SD, Burkill PH (2010) The continuous plankton recorder: towards a global perspective. J Plankton Res 32(12):1619–1621. https://doi.org/10.1093/plankt/fbq140

Calbet A, Saiz E (2005) The ciliate-copepod link in marine ecosystems. Aquat Microb Ecol 38(2):157–167. https://doi.org/10.3354/ame038157

Chen Z, Sun J, Chen D, Wang S, Yu H, Chen H, Wang M (2021) Effects of ocean currents in the western pacific ocean on net-phytoplankton community compositions. Diversity 13(9):428. https://doi.org/10.3390/d13090428

Cordone A, D'Errico G, Magliulo M (2022). Bacterioplankton diversity and distribution in relation to phytoplankton community structure in the Ross sea surface waters. ORIGINAL RESEARCH Article 13:1–20. https://doi.org/10.3389/fmicb.2022.722900

Davis CS, Gallager SM, Berman MS, Haury LR, Strickler JR (1992) The Video Plankton Recorder (VPR): design and initial results. Arch Hydrobiol Beih Ergebn Limnol 36:67–81

de Vargas C, Audic S, Henry N, Decelle J, Mahé F, Logares R, Lara E, Berney C, Le Bescot N, Probert I, Carmichael M, Poulain J, Romac S, Colin S, Aury J-M, Bittner L, Chaffron S, Dunthorn M, Engelen S, Velayoudon D (2015) Eukaryotic plankton diversity in the sunlit ocean. Science 348(6237). https://doi.org/10.1126/science.1261605

Dias C. de O, Bonecker ACT, de Castro MS, de Carvalho PF, Paranhos R, Bonecker SLC (2020) Holoplankton and meroplankton of three western Atlantic sedimentary basins. Marine Biol Res 16(10):695–713. https://doi.org/10.1080/17451000.2021.1894341

Falkowski PG, Barber RT, Smetacek V (1998) Biogeochemical controls and feedbacks on ocean primary production. Science 281(5374):200–206. https://doi.org/10.1126/science.281.5374.200

Falkowski P (2012) Ocean science: the power of plankton. Nature 483(7387):S17–S20. https://doi.org/10.1038/483S17a

Field CB, Behrenfeld MJ, Randerson JT, Falkowski P (1998) Primary production of the biosphere: integrating terrestrial and oceanic components. Science 281(5374):237–240. https://doi.org/10.1126/science.281.5374.237

Frey EF (2025) Zooplankton Sampling Techniques. Simplyecologist. https://simplyecologist.com/zooplankton-sampling-techniques/?utm_source=chatgpt.com

Garcia TM, Santos NMO, Campos CC, Costa GAS, Belmonte G, Rossi S, Soares MO (2021) Plankton net mesh size influences the resultant diversity and abundance estimates of copepods in tropical oligotrophic ecosystems. Estuar Coast Shelf Sci 249:107083. https://doi.org/10.1016/j.ecss.2020.107083

Geng Y, Li M, Yu R, Sun H, Zhang L, Sun L, Lv C, Xu J (2022) Response of planktonic diversity and stability to environmental drivers in a shallow eutrophic lake. Ecol Ind 144:109560. https://doi.org/10.1016/j.ecolind.2022.109560

Goździejewska AM, Cymes I, Glińska-Lewczuk K (2024) Zooplankton functional diversity as a bioindicator of freshwater ecosystem health across land use gradient. Sci Rep 14:1–13. https://doi.org/10.1038/s41598-024-69577-z

Gutiérrez MH, Galand PE, Pantoja S (2011) Marine fungi in the Humboldt Current system and their potential role in the decomposition of sinking organic matter. Fungal Ecol 4(3):218–229. https://doi.org/10.1016/j.funeco.2011.01.006

Hamilton DP, Duggan IC (2010) Plankton. In Waters of the Waikato: ecology of New Zealand's Longest River, pp 117–132

Irigoien X, Huisman J, Harris RP (2004) Global biodiversity patterns of marine phytoplankton and zooplankton. Nature 429(6994):863–867. https://doi.org/10.1038/nature02593

Kiran, Kumar N, Sindhu S, Gawaria B, Ansari A, Dhariwal A, Sharma P, Kumar R (2025) GWAS approaches in medicinal plants for enhanced secondary metabolite production. In Chen JT (Ed.), Biotechnology, multiple omics, and precision breeding in medicinal plants, pp 119–134. CRC Press

Kumar PS, Dharani G, Santhanakumar J, Jha DK, Pandey V (2023) Assessment of phytoplankton diversity, distribution, and environmental variables along the southeast coast of India. Front Marine Sci 1–8. https://doi.org/10.3389/fmars.2023.1215627

Kumar R, Gupta R, Dahiya T, Yadav R, Saifi R, Yodha K, Kumari B, Singh P, Shefali (2022a) Surveillance of *Heteropneustes fossilis* (Stinging Catfish) diseases and health management practices in Haryana Appl Ecol Environ Sci 10(4):261–267. https://doi.org/10.12691/aees-10-4-12

Kumar R, Gupta RK, Yadav R, Saifi R, Yodha K, Pooja (2022b) *Eisenia fetida* as protein source for growth enhancement of *Heteropneustes fossilis*. Egypt J Aquatic Biol Fish 26(2):577–588. https://doi.org/10.21608/ejabf.2022.234553

Kumar S, Veerwal B (2024) Importance of plankton in aquatic ecosystem. Futur Trends Agric Eng Food Sci 3:1–6

Lim AS, Park NY, Choi HS, Kang SU, Ok JH, Yoo YD (2025) Influence of environmental parameters on marine plankton diversity in the southern coastal waters of Korea: emphasis on thermal stratification. Mar Environ Res 203:106829. https://doi.org/10.1016/j.marenvres.2024.106829

Lodang H, Kurnia N (2019) Distribution and abundance of plankton in the downstream of jeneberang river. IOP Conf Ser: J Phys: Conf Ser 1–8. https://doi.org/10.1088/1742-6596/1244/1/012011
Nanda P, Muthu J, Sinha B (2020). Zooplankton. Bull Pure Appl Sci Zool 39a(2):516–523. https://doi.org/10.5958/2320-3188.2020.00061.3
Naselli-Flores L, Padisák J (2023) Ecosystem services provided by marine and freshwater phytoplankton. Hydrobiologia 850(12–13):2691–2706. https://doi.org/10.1007/s10750-022-04795-y
Neculina A, Mih AM, Tening AS, Che CA (2016) Phytoplankton diversity and abundance in water bodies as affected by anthropogenic activities within the Buea municipality, Cameroon. J Ecol Nat Environ 8(7):99–114. https://doi.org/10.5897/JENE2016.0566
Nisha A, Rani S, Kumar S, Kumar R, Kumar A (2025) Recent technologies and strategies for bioremediation of toxic and hazardous waste. In Dubey KK, Shah MP, Yadav A (Eds.), Challenges and sustainable solutions in bioremediation (pp. 1–16). CRC Press. https://doi.org/10.1201/9781003407317-1
Potharaju R, Aruna M (2024) Phytoplankton diversity in a fresh water lake. Int J All Res Educ Scient Method 12(09):1–6. https://doi.org/10.56025/IJARESM.2024.12092401
Rani S, Sharma S, Singh D, Sheoran R, Kumar R (2025) Improved production of anti-cancer compounds from medicinal plants through molecular and biotechnological mutagenesis strategies. In: Chen JT (ed) Biotechnology, multiple omics, and precision breeding in medicinal plants. CRC Press, pp 287–297
Reid PC, Colebrook JM, Matthews JBL, Aiken J (2003) The Continuous Plankton Recorder: concepts and history, from Plankton Indicator to undulating recorders. Prog Oceanogr 58(2–4):117–173. https://doi.org/10.1016/j.pocean.2003.08.002
Revanna S, Deepa P, RVK (2017) Remote sensing satellite image processing techniques for image classification: a comprehensive survey. Int J Comput Appl 161(11):24–37. https://doi.org/10.5120/ijca2017913306
Roy S, Sarkar SK (2022) Anthropogenic impacts on plankton biodiversity and water quality parameters. IJEP2 42(12):1456–1463
Sahoo D, Bera SK, Pal PP, Tamili DT, Sarkar B, Madhu NR, Ghorai SK (2024) Plankton act as a key indicator of the health and stability of aquatic ecosystems: a review. In A basic overview of environment and sustainable development (Vol. 3, pp 257–272). International Academic Publishing House (IAPH). https://doi.org/10.52756/boesd.2024.e03.018
Sharma MMM, Kapoor D, Rohilla R, Loyal A, Sharma P, Husen A (2024) Role of essential minerals and rhizosphere in plant under changing environmental conditions. In Husen A (Eds.) Essential minerals in plant-soil systems, pp 235–252. Elsevier. https://doi.org/10.1016/B978-0-443-16082-0.00002-3
Sharma P, Kapoor D, Kumar R, Husen A (2025) Rhizosphere and osmolytes coordination in plant systems under adverse situations. In Husen A (Ed.) Roles of osmolytes in changing environment, pp 101–117. Academic Press. https://doi.org/10.1016/B978-0-443-26549-5.00020-5
Shefali KR, Sankhla MS, Kumar R, Swaroop SS (2021) Impact of pesticide toxicity in aquatic environment. Biointerface Res Appl Chem 11(3):10131–10140. https://doi.org/10.33263/BRIAC113.1013110140
Singh S, Kumari V, Monalisa, Gupta BK, Arif M (2021) Original articlestudy on zooplankton diversity in a fresh water pond (RajaBandh) of Jamtara, Jharkhand, India. Int J Adv Life Sci Res 4(2):5–13
Sridhar KR (2009) Aquatic fungi—are they planktonic? Plankton Dynamics of Indian Waters. Pratiksha Publications, Jaipur, India, pp 133–148
Suttle CA (2005) Viruses in the sea. Nature 437(7057):356–361. https://doi.org/10.1038/nature04160
Wee AKS, SalmoIII SG, Sivakumar K, Then AY-H, Basyuni M, Fall J, Habib KA (2023) Prospects and challenges of environmental DNA (eDNA) metabarcoding in mangrove restoration in Southeast Asia. Front Marine Sci 1–9. https://doi.org/10.3389/fmars.2023.1033258
Wetzel RG (2001) Bacterioplankton. In Limnology (Third Edition). https://doi.org/10.1016/B978-0-08-057439-4.50021-6

Wommack KE, Colwell RR (2000) Virioplankton: viruses in aquatic ecosystems. Microbiol Mol Biol Rev 64(1):69–114. https://doi.org/10.1128/mmbr.64.1.69-114.2000

Chapter 27
Earthworm: A Potential Replacement for Fish Meal

Nikki, Shikha Ahalavat, Rishu, Khushbu Sharma, Renu Yadav, Rameshwar Singh, and Sindhu Sheoran

Abstract The increasing demand for fish and other aquatic goods around the world is driving rapid growth in the aquaculture sector. Fish feeds, which constitute up to 70% of aquaculture's financial input, play a pivotal role in the industry's success. Fish meal has long been the main source of protein in aquatic organism feeds because of its high nutritional content, but overfishing, environmental concerns, and growing costs have made it unsustainable and expensive, which has led to a search for more sustainable protein sources. Researchers are now looking at alternative ingredients such as plant proteins, animal byproducts, and agricultural byproducts because of the decline of wild fish supplies and the detrimental effects that fish meal production has on the ecosystem. Earthworms, notably *Eisenia fetida*, have emerged as a promising possibility among these options because they have a great nutritional profile, with high-quality protein, essential amino acids, and omega-3 fatty acids that are good for you, similar to traditional fish meal. Earthworms are a useful and cheap way to make aquafeed, especially on small farms, because they are easy to cultivate and full of nutrients. Additionally, the integration of earthworm farming with aquaculture and agriculture (vermiculture) can improve nutrient recycling, reduce feed costs, and contribute to sustainable farming practices. This chapter examines the nutritional advantages, viability of production, and contribution to sustainable aquaculture practices of earthworm-based feeds as a potential replacement for fish meal in aquaculture.

Nikki (✉) · S. Ahalavat · Rishu · K. Sharma · R. Yadav
Department of Zoology, College of Basic Sciences and Humanities, Chaudhary Charan Singh Haryana Agricultural University, Hisar, Haryana, India
e-mail: nikki.hariyana@gmail.com

R. Singh
Chaudhary Charan Singh Haryana Agricultural University, Regional Research Station, Bawal, Rewari, Haryana, India

S. Sheoran
Senior Research Program Coordinator, Hawkesbury Institute for the Environment, Western Sydney University, Sydney, Australia
e-mail: S.sheoran@westernsydney.edu.au

R. Kumar et al. (eds.), *Sustainability in Aquaculture and Fisheries*,
Springer Water, https://doi.org/10.1007/978-3-032-16605-0_27

Keywords Earthworm meal · Disease-resistant · Fish meal · Medicine · Protein source

27.1 Introduction

As global food needs rise, aquaculture is increasingly contributing to worldwide production, providing a significant amount of protein for human consumption. It refers to the cultivation of diverse aquatic organisms—finfish, mollusks, crustaceans, and aquatic plants-within controlled culture systems (Nikki et al. 2023). The importance of aquaculture in supplying the expanding demand for fish and other seafood products is increasing (Ali et al. 2021). Aquaculture is essential for producing food on a global scale (FAO 2021). Freshwater aquaculture represents a significant segment of the aquaculture industry. Over the last decade, the aquaculture sector has exhibited remarkable expansion and is expected to continue its upward trend, fuelled by escalating worldwide demand for accessible and affordable animal protein (Dauchak and Nikki, 2024).

Fish feeds, which make up as much as 70% of all aquaculture-related financial inputs, are essential to aquaculture output but can pose serious problems for fish farming (FAO 2009). Fish meal‘s widespread use in aquafeeds is attributed to its optimal nutrient composition, including elevated protein content, essential amino acid balance, micronutrients and favourable digestibility characteristics (Tacon 1993; FAO 2017). However, the decrease in catch fisheries, fierce competition between the human and animal feed industries, rising global energy prices, and an unpredictable supply throughout the year have all contributed to the steady decline in fish meal availability and its increasing cost (FAO 2013). The ongoing reliance on fish meal in aquaculture has posed significant risks to the sustainability of fisheries ecosystems, while simultaneously driving up the demand for fish, thereby impacting the profit margins of fish farmers (Musyoka et al. 2019).

Fish meal, as a limited fishery resource, cannot provide the aquaculture sector with a sustainable supply of affordable protein over the long term (Ng et al. 2001). Furthermore, the increasing price of premium fish meal used to produce aqua-feed has forced fish nutritionists around the world to investigate and find novel, renewable, and unconventional alternative protein sources that are crucial for the continued expansion and sustainability of the aquaculture sector (Sales and Janssens 2003; Sogbesan et al. 2007). Studies are being conducted to determine and assess sustainable options such as plant proteins, grain legumes, and oilseed plants for use in animal feeds, as well as locally accessible protein sources. Plant proteins have already been tested as effective alternatives in fish diets (Zhou et al. 2011). Even while plant proteins may be good for organic fish farming, anti-nutritional factors might make it harder for fish to absorb nutrients, which can hurt their growth and health (EU, Council Regulation (EC), 2007). Also, studies have looked into the potential of meat and bone meal (Forster et al. 2003), earthworm (Kumar et al. 2022a; Kumar et al. 2022b) and chicken by-products as new sources of animal-derived protein to deal with sustainability issues in aquaculture nutrition (Shapawi et al. 2007). Some of these results were promising.

The rising costs and deficits of fish meal (FM) stem from its primary role as a protein source in aquaculture feeds. Demand is increasing day by day, due to animal protein consumption, aquatic environmental challenges, depletion of wild fish stocks and the competition from other feeders. On the other hand, the supply is limited which increases the competition amongst other feeders. Out of the many sources, FM is one of the most impacted regarding sustainability which considerably decreases the profits of many farmers, which is why there is abundant research for alternate sources of protein (Musyoka et al. 2019). Economically viable aquaculture protein sources should be explored with the hope of easing reliance on expensive protein feed resources. These sources can include plant and animal by-products, as well as fish discards (Davies et al. 2019; Mansour et al. 2021).

27.2 Fish Meal

Fish meal, which is mostly used as feed for farmed fish, is a commercially produced product made from entire wild-caught fish, by-catch, and other fish by-products. Both the growth of industrial farming and the number of animals raised for food have benefited greatly from it. Fish or fish trimmings are dried and ground to create fish meal, which is sold as either a powder or a cake. When using fatty fish, a large amount of the fish oil is first extracted by pressing the fish. The manufacture and extensive use of fish meal have generated debate. The profitable market for fish meal as animal feed incentivizes corporate fisheries to maximize their by-catch yields, resulting in ecosystem depletion, environmental degradation, and the potential collapse of local fisheries (Miles and Chapman, 2005). Advocates of fish meal emphasize its pivotal role in enhancing nutrition and breeding outcomes for millions of farmed animals, thereby supporting global food security. Fish meal's balanced amino acid profile, vitamin content, and growth-promoting factors render it an optimal protein source in aquaculture feed. Nonetheless, its sustainability as a limited fishery resource is a concern for long-term supply (Goddard, 2008).

In order to ensure the long-term viability and sustainability of the aquaculture industry, research into alternative, renewable, and unconventional protein sources for aquafeed production has been sparked globally by the rising costs of premium fish meal.

Three main sources of fish meal are as follows:

1. Fish that are tiny, bony, and oily—such as anchovies, horse mackerel, menhaden, capelin, sandeel, blue whiting, herring, and pollack—are the main target species for fishmeal extraction.
2. By-catch received from different types of fishing.

3. Offal and leftovers from preparing fish meant for human consumption that are either inedible or quickly deteriorate (FIN, 2008).

Fish meal is a premium feed ingredient that supports optimal growth and health by providing critical micronutrients, omega-3 fatty acids (EPA and DHA), and high-quality protein (IFOMA, 2001). The kind of raw materials used and the processing techniques used have an impact on the quality of fish meal.

27.2.1 Processing of Fish Meal

- Fish or fish by-products are cooked, pressed, dried, and ground into a solid during the production of fishmeal, as shown in (Figs. 27.1and 27.2).
- During this process, most of the water content and a portion or all of the oil is extracted.
- To produce one tone of dry fishmeal, approximately four to five tons of fish are required.
- Although there are several ways to make fishmeal from raw fish, the simplest method is to let the fish dry in the sun before grinding and pressing.
- This traditional method continues to be utilized in certain regions where processing facilities are lacking; however, the quality of the final product is inferior when compared to that produced by contemporary techniques (Miles and Chapman, 2015; Abbey et al. 2016).

27.2.2 Environmental Impacts

The production of fish meal plays a substantial role in overfishing, potentially leading to the depletion of fish stocks beyond their natural replenishment capacity. In certain regions, particularly in Western Fish meal production has significantly increased in Africa, which has had a negative impact on local fisheries and contributed to their decline (Jacobs and Andrew, 2017; Yadav et al. 2022a; Yadav et al. 2023).

27.2.3 Use of Earthworm as a Replacement for Fish Meal

The main challenges in aquaculture are associated with overreliance on fishmeal due to its exceptionally high nutritional value, scarcity, and high cost. Earthworm meal can be used in place of fishmeal because it does offer good nutritional value and palatability.

Earthworms, sometimes known as the “farmer’s friend,” play a key role in turning organic waste and litter into organic manure, which is a great natural fertilizer that

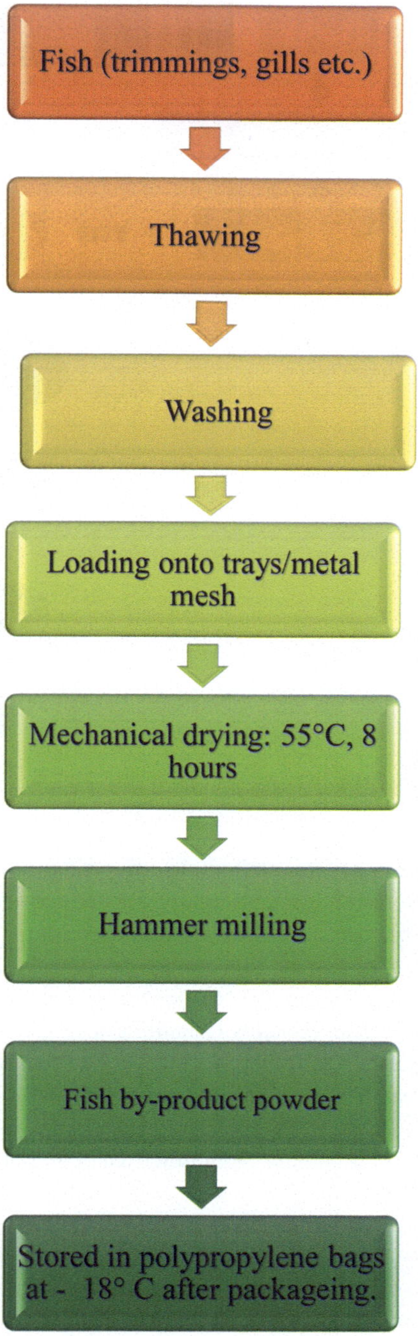

Fig. 27.1 Production flow chart for powdered fish byproducts

Fig. 27.2 Process of forming fishmeal

enriches soil with vital nutrients. Furthermore, earthworms can also be beneficial for aquaculture, particularly for fish farmers, when utilized effectively. They utilize in different form like earthworm custard, earthworm pellet, whole earthworm, earthworm extract, and earthworm powder as shown in Fig. 27.3. They provide a nutritious feed source for the ornamental fish industry, particularly for broodstock. Novel bacterial proteins (Khalil et al. 2021), insect meals (Abdel-Latif et al. 2021), and earthworm meals have all been investigated recently as prospective substitutes for fishmeal in aquafeed formulations (Parolini et al. 2020).

Earthworms is a source of dietary protein and can be used either independently or in combination with other feed components for formulation of fish diets (Sogbesan and Madu, 2008). Due to their nutrition profile, earthworms is a cost-effective protein source in aquaculture (MM Beg et al. 2016). Moreover, earthworms have variable lipid content, polyunsaturated fatty acids and also some concentration of ω-3 fatty acids which boosts immunity in many animals (Dynes RA, 2003). Lipid content in earthworms ranges between 5–10%, while the protein content varies between 50–70%. In addition, catfish and other species of tropical fish can utilize the sodium, calcium, and potassium provided by earthworms in appropriate physiological quantities (Sales and Janssens 2003; Sogbesan and Madu 2008). So, use of dry earthworm meal as substitute of fish meal is very effective.

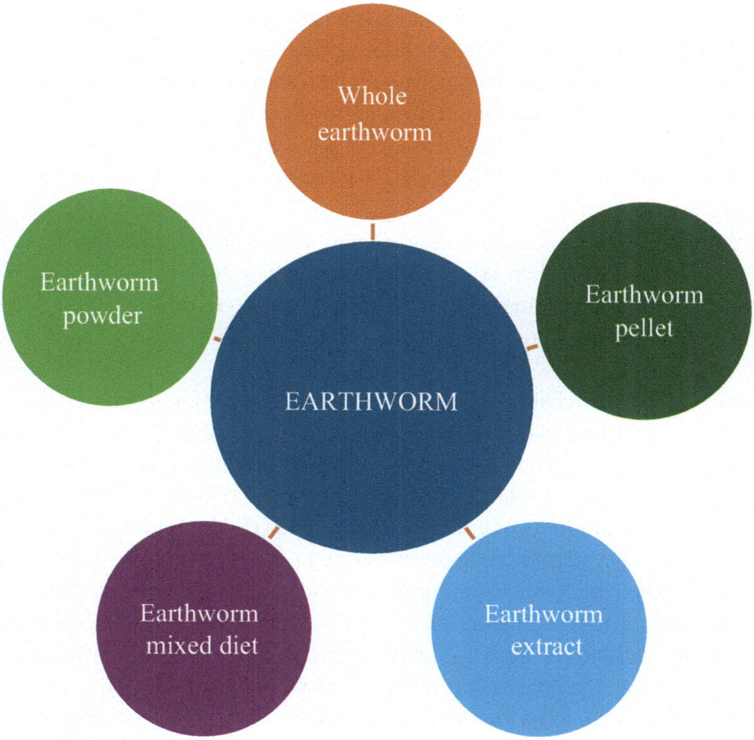

Fig. 27.3 Use of Earthworm in Different Form

African night crawler *Eudrilus eugeniae* and blueworm *Perionyx excavatus* contain nutrional profiles comparable to fish meal, thus fulfil the nutritional requirements for various species of fish (Pucher et al. 2014; Tram et al. 2005; Vodounnou et al. 2016).). However, both species exhibit significant resilience to stress along with various environmental conditions (Hasanuzzaman et al. 2010). *P. excavatus* has the capacity to support growth, protein efficiency, and energy retention in Common carp that is equal to or superior to rates observed in fishmeal containing diets (Ngoc et al. 2016). When earthworm meal is blended with 5 per cent sardine oil, it can replace fish meal in aquaculture and help in growth and development. For rohu advanced fry, two earthworm-based diets were formulated targeting nitrogen and energy content of 500 g/kg and 17.0 MJ/kg, respectively. Earthworm diets can be used in form of pelleted earthworm, earthworm custard, and whole earthworm. Advanced fry utilizing pelleted earthworm meal attained the highest growth rates (Beg et al. 2016). But use of unprocessed *P. excavatus* meal's contains anti-nutritional factors that risk fish growth and development. Therefore, gut-cleaned earthworms are used (Hasanuzzaman et al. 2010).

Among the earthworm species that may serve as a fishmeal substitute, one of the most tropical adapted species is the *E. fetida.* This earthworm is most well-known for ease of reproduction compared to other earthworm species, along with being short-lived, enabling quick maturation, and exhibiting high fecundity (Edwards et al. 2019). The species has a short maturation period and a relatively high rate of reproduction with about three hatchlings per egg. Reproduction flourishes at 20 °C–29 °C temperatures. It inhabits the surface of the soil as an epigeic earthworm and feeds primarily on organic debris and litter. Therefore, *E. fetida* excels in converting organic waste into beneficial organic fertilizers (Chattopadhyay, 2012; Vodounnou, 2016a, b). *Eisenia fetida's* capacity to regulate and enhance energy use efficiency is the basis for its relatively low mortality when compared to most other earthworm species. Due to this capacity, it can survive under a wide range of harsh conditions, including exposure to cold temperatures, toxic substances, and salty conditions (Sinha et al. 2009; Rorat et al. 2015).

They are beneficial to fish species due to their good nutritional composition which includes protein, fat, and essential amino acids comparable to fishmeal (Vodounnou et al. 2016a, b). *E. fetida* can replace conventional protein sources for growing and reproducing fish without hindering their growth (Staffod and Tacon 1985). *Eisenia fetida* contains high amounts of essential amino acids (EAA) and lipids, comparable to those in fishmeal. *E. fetida* crude protein content is within the range of 52–62% on a dry weight basis, which makes it eligible for use in aquafeeds (Gunya and Masika, 2021). Also, the protein composition of *E. fetida* is influenced by the production and processing techniques used to achieve specific dietary goals (Musyoka et al. 2020a, b). Moreover, the relatively low capital cost associated with culturing *E. fetida* reduces the risks associated with finding alternative worms as substitutes for fish feed. This species is particularly rich in Omega-3 fatty acids and other EAA (Dynes 2003). This species has remarkable versatility in employing a range of organic substrates, with biodegradation rates up to five times faster than similar processes (Tripathi and Bhardwaj 2004; Visvanathan and Trankler et al. 2005).

In the production of catfish, the protein obtained from *E. fetida* is more cost-effective than soybean and chicken waste meal (Zakaria et al. 2012). Consequently, *E. fetida* makes an appropriate feed source because it rarely competes directly with humans or most livestock for nutritional resources. Given the attributes of *E. fetida*, scientists are exploring its potential use in various applications, including replacing fish meal. This chapter assesses the primary factors determining *E. fetida* suitability for producing sustainable fish feed, such as production and processing.

27.3 Importance of Earthworm Meal

Rural farmers in developing nations have the opportunity to participate in semi-intensive aquaculture by incorporating earthworms into their supplement fish diets through the integration of earthworm farming with pisciculture. Earthworm protein, when partially substituted for fish meal protein in complete feeds, shown to improve growth outcomes in common carp and Rohu in controlled laboratory settings without natural food sources (Tuan, 2010; Deborah et al 2011). The fish fed with earthworm diets as a total replacement for fish meal exhibited comparable growth rates to those receiving fish meal-based diets (Tuan, 2010). Earthworm can be used in different form to feed the fishes like earthworm pellet, earthworm custard, extract, powder, and also whole earthworm as shown in (Fig. 27.3).

27.3.1 Whole Earthworm Diet

After being fasted for 48 hours, the entire earthworm was boiled in water with a tiny bit of regular salt. The worm was well cleaned, then cut with a sharp knife into pieces around 0.5 mm in size. It was then kept in a refrigerator at 4 °C for later usage (Bhuvaneshwaran et al. 2019).

27.3.2 Pelleted Earthworm Diet

(i) To prepare a pelleted diet for earthworms, the initial step involved boiling the earthworms in water (45 minutes), incorporating a small amount of common salt, followed by a thorough rinsing in clean water.
(ii) Following boiling, the earthworms were subjected to oven drying for 24 hours, after which they were processed into a fine powder using a kitchen grinder, yielding earthworm meal. The specified amounts of fish meal, groundnut oil cake, prawn meal, earthworm meal, mineral-vitamin supplement, and vegetable oil were thoroughly mixed using a kitchen mixer.
(iii) Using an electric heater, the necessary amount was progressively added to one liter of warm water (80 °C) in a beaker while being gently stirred to achieve gelatin disintegration. The gelatin was added to the feed mixture when it had fully dissolved.
(iv) To create a consistent dough, lukewarm water was added to the feed mixture and thoroughly blended. A hand pelletizer was then used to pelletize the dough, and the finished pellets were stored at 4 °C and dried at 60 °C (Bhuvaneshwaran et al. 2019).

27.3.3 Earthworm Custard

- To prepare earthworm custard, the initial step involves boiling the earthworm in water for 45 minutes, with the addition of a small amount of common salt. The earthworms were ground in a kitchen mixer grinder (Philips India Limited, Mumbai) after being thoroughly cleaned with sanitized water. To ensure even mixing, the resulting earthworm paste was next mixed with exact amounts of skim milk powder and entire hen's egg (albumin and yolk).

- A measured volume of water (1 l) was heated to 80 °C using an electric heater. Gelatin was then dissolved in the heated water with gentle agitation, followed by incorporation into the feed mixture. Subsequently, a vitamin and mineral blend is incorporated, and the entire mixture is uniformly blended using the kitchen mixer grinder.
- After that, the feed mixture is mixed with lukewarm water until a cohesive dough forms. To make the earthworm custard, the dough was moved to an aluminum container and steam-cooked in a home pressure cooker for 20 minutes.
- The earthworm custard was cooked and then allowed to cool to room temperature before being refrigerated at 4 °C. The experimental fish were fed custard portions cut into 0.5 mm pieces with a kitchen knife on a daily basis (Bhuvaneshwaran et al. 2019).

27.4 Bio-Magnification

Aquaculture has become more intense and commercialized as a result of its advancement, which also enhance risk for the establishment and spread of new fish diseases. In aquaculture, disease outbreaks are common and mostly caused by intensive farming methods meant to maximize profits (Kumar et al. 2022c). Eventually, this leads to environmental deterioration and stresses out populations of cultured fish. Overcrowding, poor nutrition, contaminated water, and high microbial loads are some of the factors that can turn opportunistic diseases into virulent agents and cause significant injury to fish (Mishra et al. 2015). These are the factors that contribute to the incidence of microbial diseases and poor conditions in aquaculture that hinder the growth of fish culture. Although drugs are used to control fish diseases, but it led to biomagnification of the aquatic ecosystem (Fig. 27.4A).

Earthworms can be used as an alternative of drugs to cure fish diseases (Fig. 27.4B). Various earthworm species have been utilized as therapeutic agents in the treatment of fish diseases in different Asian nations with the use of powder, extract,

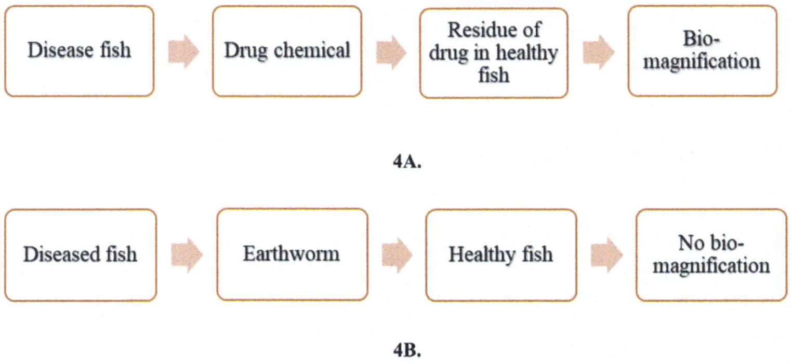

Fig. 27.4 Role of earthworms in preventing drug-biomagnification

paste, and live earthworms. Earthworms also have a two-pronged function by offering the necessary nutrients to enhance fish immunity and controlling disease outbreaks, thus ensuring long-term quality. Medicinal applications of earthworm products like powder, extract, and coelomocytes have been extensively studied (Urmila et al. 2025; Kumar et al. 2024).

27.5 Medicinal Properties of Earthworms

Earthworm species such as *Eisenia fetida*, *Eudrilus eugenie*, *Lumbricus rubellus* and *Pheretima* spp. are known for their numerous medicinal properties like analgesic, anti-inflammatory, antipyretic, skin wound healing, antioxidative, anticarcinogenic, hepatoprotective as shown in Fig. 27.5 and are widely used in natural medicine. Commonly referred to as "Earth Dragons" in countries like China, Korea and across much of Southeast Asia, these organisms have a long history of therapeutic use (Kumar et al 2025). In various tribal communities and remote villages, earthworms are traditionally employed to treat a range of ailments (El-Kamali, 2000). They are believed to boost overall health, and when consumed baked with bread, are said to help reduce bladder stones (Cooper et al. 2012).

27.5.1 Earthworms: A Rich Source of Bioactive Compounds

Bio-active molecule- Microbial activity produces a variety of substances, including organic acids, chemical compounds, and enzymes. Two significant sources of bioactive substances with possible therapeutic applications in both human and veterinary medicine are coelomic fluid and tissue homogenate derived from earthworms. Glycoproteins, polysaccharides, peptides, amino acids, glutathione, fibrinolytic enzymes, vitamins, and other bioactive compounds such as lumbrofebrin, lumbritin, purine, choline, microelements, fatty acids, terrestrolumbrolysin, and cholestrin make up the bioactive molecular profile of earthworms (Yadav et al. 2022b; Jamshidzadeh et al. 2018).

These bioactive compounds modulate interactions between host cellular proteins and pathogen-derived proteins, influencing disease progression, thereby playing an essential role in maintaining homeostasis, facilitating wound healing, and promoting cellular proliferation. Earthworms have emerged as a valuable resource for biologically active molecules, offering considerable biomedical potential for disease treatment as shown in flow chart below. Notably, these bioactive compounds have long been used by indigenous people around the world, particularly in Asian nations including China, India, Korea, Vietnam, and Myanmar (Kumar and Singh 2022; Kumar et al. 2023; Ranganathan 2006).

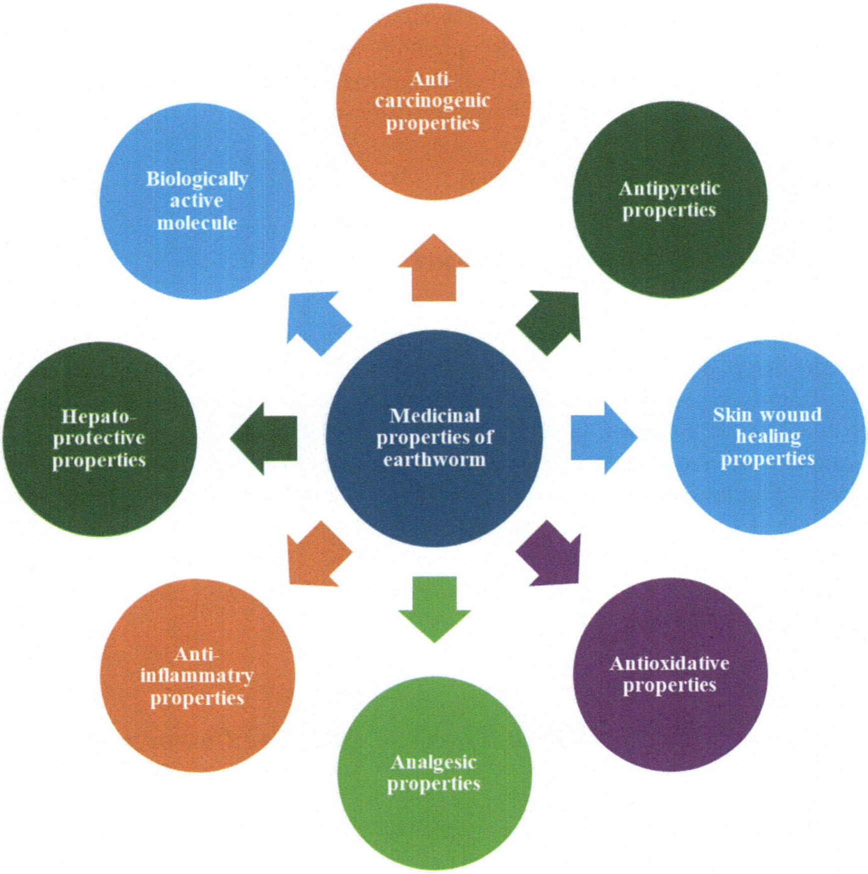

Fig. 27.5 Different medicinal properties of earthworm

27.5.2 Anticarcinogenic Properties

Coelomic fluid of earthworm contains many active molecules having anticarcinogenic properties. Enzymes like proteases and fibrinolytic enzyme present in earthworm extract have exceptional anticarcinogenic action in hepatoma cells. Metalloproteinase—2 facilitate in cancer cell invasion and metastasis and by inhibiting this enzyme, earthworm fibrinolytic enzyme protects cell (Hong C et al. 2007).

27.5.3 Antipyretic Properties

Antipyretic substances are those substances that lower body temperature and also limits bacterial growth. Arachidonic acid, antipurine, anti-toxins are among the antipyretic substances found in earthworm. Anti-inflammatory properties of earthworm extract may also reduce the high body temperature. Earthworm also contains antibacterial substances that prevents the growth of harmful bacteria (Waluyo et al. 2019).

27.5.4 Skin Wound Healing Properties

Wound healing is a complex biological process which includes many steps such as, inflammation, homeostasis, proliferation, granulation, matrix production and remodelling. Earthworm's coelomic fluid have certain factors that promote formation of growth factors which helps in wound healing (Metausic-Pisl et al. 2010). By improving granulation, epithelization, and collagen production, various extracts from *E. fetida* enhanced wound healing and minimize healing time (Grdisa, 2013).

27.5.5 Analgesic Properties

Earthworm extract also contains important analgesic properties that raise the pain threshold. Like other analgesic medicine such as aspirin, morphine, it additionally reduces the concentration of serum brain transmitter such as 5-hydroxytryptamine, nor-epinephrin and nitric oxide synthase (Luo et al. 2018).

27.5.6 Antioxidative Properties

Antioxidants are those substances that protect cell from harmful effects of free radicals and other oxygen species that are found in organism. Earthworm extract contains high quantities of antioxidants such as glutathione peroxidase, glutathione, catalase, and superoxide dismutase having exceptionally high antioxidative activities. Additionally, it also contains high concentration of phenolic compound that neutralize the free radicals due to presence of hydroxyl group (Kumar et al. 2024).

27.5.7 Anti-Inflammatory Property

TNF- α is a cytokine that regulate various functions related to systemic inflammation like inflammation, apoptotic cell death, inducing fever, stimulating acute-phase reactions etc. Earthworm extract exhibits anti-transudative, anti-proliferative, and anti-histaminic properties that inhibits TNF- α and other inflammatory mediators.

Different types of cells such as cardiac myocytes, endothelial cells, mast cells, fibroblasts, lymphoid cells, and adipose tissue release a high concentration of TNF in response of inflammatory agents (Kumar et al. 2022c).

27.5.8 *Hepato-Protective Property*

Hepato-toxic substances damage liver cells mostly through oxidative damage such as lipid peroxidation. Elevated blood enzyme activities such as bilirubin, aspartate amino transferase (AST), alanine amino transferase (ALT), alkaline phosphatase (ALP) and reduced liver antioxidants are sign of liver damage. Earthworm extract function as a hepatoprotective agent because it reduces the amount and activities of serum enzyme and enhance the amount of liver antioxidants (Singh et al. 2018).

27.6 Integration of Vermiculture into Small-Scale Farms: A Supplementary Farming Activity

Earthworms can be used by small-scale farmers in rural areas to produce additional feed for omnivorous fish species like common carp. One tactic to improve integrated farming systems in underdeveloped nations is the incorporation of vermiculture, or earthworm production technology, into agricultural techniques. By turning underutilized waste materials and low-quality by-products into high-value feed resources for livestock and aquaculture, as well as nutrient-rich vermicompost for horticulture and aquaculture, vermiculture promotes nutrient cycling on farms and provides financial advantages (Sogbesan et al. 2007; Chakrabarty et al. 2010).

A study on small-scale farming in northern Vietnam found that vermiculture, from underutilized on-farm waste like ruminant manure, can produce 6 to 36 kg of dry earthworm matter per year per farm (Müller et al. 2012). Adding earthworm meal to plant-based diets can increase common carp production and profitability of small-scale farmers by 18–112 kg/farm and 12–75% in yearly financial revenues (Pucher et al. 2013).

27.6.1 *Use of Vermicompost as Feed and Fertilizer*

A composite farm can have agriculture, aquaculture, and vermiculture function in harmony with each other. The water-based environment for aquaculture can be used as a sink for diverse organic materials, such as solid and liquid animal manure, silkworm pupa residues, earthworms, and human food grain residues. These materials can be effectively transformed into high-quality fish feed. Additionally, the water from the reservoir can be employed for irrigating crops grown along the reservoir's embankment, and this water can subsequently be recycled for aquaculture purposes.

The integration of vermiculture, sericulture, and carnivorous fish culture represents a significant methodological advancement. when *Clarias batrachu*, commonly known as Magur and *Heteropneustes fossilis*, commonly referred to as Singhi are fed earthworms and silkworm pupae, instances of cannibalism are notably reduced, even under conditions of high stocking density (Chakrabarty et al. 2010).

Organic manure is extensively utilized in aquaculture ponds to enhance productivity while simultaneously minimizing expenses associated with expensive feeds and fertilizers (Chakrabarty 1994; Brown and Gilloy 2003). In agricultural systems, it is estimated that nearly fifty percent of the total investment is allocated to fertilizers and feed. By substituting these artificial feeds with organic alternatives derived from earthworms, significant reductions in feed costs can be achieved.

Vermicompost serves as a viable food source for invertebrate organisms that are part of the fish diet, as well as for fish themselves. In polyculture systems where both artificial feed and natural food sources are used then it is very important that they supplement each other. Any nutrients that are lack in natural food should be provided by artificial feed and vice versa. The supplementary artificial feed must fulfill the overall nutritional requirements necessary for optimal fish growth and health.

Artificial feed must supplement and enhance other sources of fish nutrition derived from pond manuring and fertilization practices. These approaches promote the growth of detritus and natural food organisms for fish, such as phytoplankton, zooplankton, and another biota. Fish can easily eat vermicompost, which is high in bacterial mass and a significant source of protein. The remaining components of vermicompost dissolve in the water, supplying essential nutrients that foster phytoplankton growth. Consequently, vermicompost can function effectively as both an artificial feed and a fertilizer (Chakrabarty et al. 2007).

Establishing a feeding schedule for the young fish is essential when they are raised in the same aquatic environment as the adult fish. In such circumstances, vermiwash can be utilized as an outstanding direct application feed for the juvenile fish (Chakrabarty et al. 2009).

27.7 Conclusion

This chapter highlight the potential of earthworms as a fish meal alternative, calling for more research into its application as an environmentally friendly source of animal protein in aquaculture. Demand is anticipated to rise across a number of market categories due to consumers' increasing preference for fish that is produced organically. Given the growing customer demand, organic food marketing methods must be effective. In order to increase their production capacity, farmers must be made aware of the advantages of using earthworm meal as an affordable addition to fish meal. Additionally, the sustainability of fisheries and aquaculture activities could be greatly enhanced by the year-round production of this earthworm species using well-established mass culture techniques.

27.8 Future Prospects

Since fish meal is a major cost in aquaculture, earthworm meal shows promising potential as a replacement. Earthworm meal could improve profitability and sustainability. Earthworm meal has the strong potential as a sustainable and high-protein alternative to replace fish meal, especially in aquaculture. Advancements in vermiculture, bioprocessing, and waste-to-protein techniques can make earthworm meal an important feed component in the near future.

Acknowledgement The authors are grateful to Chaudhary Charan Singh Haryana Agricultural University and Western Sydney University, Australia, for their guidance and support.

References

Abbey L, Glover-Amengor M, Atikpo MO et al (2016) Nutrient content of fish powder from low value fish and fish byproducts. Food Sci Nutr 5(3):374–379

Abdel-Latif HMR, Abdel-Tawwab M, Khalil RH et al (2021) Black soldier fly (*Hermetia illucens*) larvae meal in diets of European seabass: effects on antioxidative capacity, non-specific immunity, transcriptomic responses, and resistance to the challengewith vibrio alginolyticus. Fish Shellfish Immunol 111:111–118

Ali A, El Sherif S, Abd Alla J et al (2021) Morphometric, histochemical, and ultrastructural analysis of the reproductive system and spermatogenic stages of male blue crab (*Callinectessapidus Rathbun*, 1896). J Mar Sci Eng 9:1105

Andrew J (2017) "China's appetite pushes fisheries to the brink". The New York Times

Beg MM, Mandal B, Moulick S (2016) Potential of earthworm meal as a replacement of fish meal for Indian major carps. Int J Fish Aquat Stud 4(3):357–361

Bhuvaneshwaran T, Sanjay G, Jayakumar N et al (2019) Potentiality of earthworm as a replacement for fish meal and its role in aquaculture. J Aqua Trop 34(1–2):115–127

Chakrabarty D, Das SK, Das MK et al (2010) Assessment of Vermicompost as direct application manure in fish farming ponds. Turk J Fish Aquat Sci 10:47–52

Cooper EL, Balamurugan M, Huang CY, Tsao CR, Heredia J, Tommaseo-Ponzetta M, Paoletti MG (2012) Earthworms dilong: ancient, inexpensive, noncontroversial models may help clarify approaches to integrated medicine emphasizing neuroimmune systems. Evid-based complement Altern med 1–12

Dauchak S, Nikki (2024) The impact of medicinal plants on fish immunity. Research and Reviews in Animal Science, vol IV, 1st edn. Bhumi Publishing, India, pp 1–12

Davies SJ, Laporte J, Gouveia A et al (2019) Validation of processed animal proteins (mono-PAPS) in experimental diets for juvenile gilthead sea bream (*Sparus aurata* L.) as primary fish meal replacers within a European perspective. Aquac Nutr 25:225–238

Deborah PT, Divya VV, Ulaganathan P et al (2011) Replacing fish meal with earthworm and mushroom meals in practical diets of *Labeo rohita* and *Hemigrammus caudovittatus* fingerlings. Indian J Anim Res 45:115–119

Dynes RA (2003) Earthworms–Technology information to enable the development of earthworm production. Kingston: RIRDC 1–33
El-Kamali HH (2000) Folk medicinal use of some animal products in Central Sudan. J Ethnopharmacol 72:279–289
EU, Council Regulation (EC) No. 834/2007 of 28 2007 On organic production and labelling of organic products and repealing regulation (EEC) no. 2092/91. Off J Eur Union Legis 2007 189(50):1–23
FAO (2009) Impact of rising feed ingredient prices on aquafeeds and aquaculture production. FAO Fisheries and aquaculture technical paper, Rome, Italy, p 541
FAO (2013) The state of food insecurity in the world. The multiple dimensions of food security. Food and Agriculture Organization, Rome
FAO (2017) Aquaculture newsletter. No. 56 (April). Rome: FAO
Forster IP, Dominy W, Obaldo L et al (2003) Rendered meat and bone meals as ingredients of diets for shrimp *Litopenaeus vannamei* (Boone, 1931). Aqua 219:655–670
Goddard S, Al-Shagaa G, Ali A (2008) Fisheries by-catch and processing waste meals as ingredients in diets for Nile tilapia, *Oreochromis niloticus*. Aquac Res 39:518–525
Grdisa M (2013) Therapeutic properties of earthworms. Bioremediat Biodivers Bioavailab 7(1):1–5
Gunya B, Masika PJ (2021) *Eisenia fetida* worm as an alternative source of protein for poultry: a review. Int J Trop Insect Sci:1–8
Hasanuzzaman KU, Ahamed NM, Rahmatullah N et al (2010) Plant growth characters and productivity of wetland rice (*Oryza sativa* L.) as affected by application of different manures. Emirate journal of food. Agriculture 22(1):46–58
Hong C, Takahashi S, Imamura M et al (2007) Earthworm fibrinolytic enzyme: anti-tumor activity on human hepatoma cellsin vitro and in vivo. Chin Med J 120(10):898–904
Jamshidzadeh A, Dabagh F, Farshad O et al (2018) Hepatoprotective properties of the glycolipoprotein extract from Eisenia foetida. Trends Pharm Sci Technol 4(3):149–160
Khalil HS, Momoh T, Al-Kenawy D et al (2021) Nitrogen retention, nutrient digestibility and growth efficiency of Nile tilapia (*Oreochromis niloticus*) fed dietary lysine and reared in fertilized ponds. Aquac Nutr 00:1–13
Kumar R, Yadav R, Gupta RK, Kiran, Kataria SK, Kadyan P, Pooja and Kaur S. 2023. The earthworms: Charles Darwin's ecosystem engineer. In: Organic Fertilizers Khalid Rehman Hakeem(eds) 1–7, Intech Open https://doi.org/10.5772/intechopen.1001339
Kumar R, Yadav R, Gupta RK, Pal A, Yodha K, Kumar A (2024) Antioxidant enzyme activities and markers of oxidative stress in the life cycle of different earthworm species. Indian J Exp Biol 62(07):578–583
Kumar R, Singh D, Yadav R (2025) Toxic effects of chlorantraniliprole and fipronil on *Eisenia fetida*: from growth and reproductive parameters to avoidance behavior. Environ Eng Manag J 24(7):1499–1508. https://doi.org/10.30638/eemj.2025.117
Kumar R, Singh D (2022) Insecticide- tolerant bacterial population in *Eisenia fetida*'s gut and Vermicast exposed to Chlorantraniliprole and Fipronil. Appl Biol Res 24(3):273–279
Kumar R, Gupta R, Dahiya T, Yadav R, Saifi R, Yodha K, Kumari B, Singh P, Shefali. (2022a) Surveillance of *Heteropneustes fossilis* (stinging catfish) diseases and health management practices in Haryana. Appl Ecol Environ Sci 10(4):261–267. https://doi.org/10.12691/aees-10-4-12
Kumar R, Gupta RK, Yadav R, Saifi R, Yodha K, Pooja (2022b) *Eisenia fetida* as protein source for growth enhancement of *Heteropneustes fossilis*. Egypt J Aquat Biol Fish 26(2):577–588. https://doi.org/10.21608/ejabf.2022.234553
Kumar R, Yadav R, Kumar Gupta R, Pooja (2022c) A sustainable way for fish health management by replacement of chemical and drugs by earthworm. In: Environmental degradation in Asia: land degradation, environmental contamination, and human activities. Springer International Publishing, Cham, pp 329–352. https://doi.org/10.1007/978-3-031-12112-8_16
Luo W, Deng ZH, Li R et al (2018) Study of analgesic effect of earthworm extract. Biosci Rep 38(1):BSR20171554

Mansour AT, Allam BW, Srour TM et al (2021) The feasibility of monoculture and polyculture of striped catfish and nile tilapia in different proportions and their effects on growth performance, productivity, and financial revenue. J Mar Sci Eng 9:586

Metausic-Pisl M, Cupic H, Kasuba V, Mikecin AM, Grdisa M (2010) Gliko-lipoprotein extract (G-90) from Eisenia foetida accelerated the rats wound healing. Eur Rev Med Pharm Sci 14:177–184

Miles RD, Chapman FA (2005) The benefits of fish meal in aquaculture diets. UF/IFAS Extension. Reviewed, Fisheries and Aquatic Sciences Department

Mishra SS, Dhiman M, Swain P et al (2015) Fish diseases and health management issues in aquaculture. ICAR-CIFA training manual no. 18, central Institute of Freshwater Aquaculture, Bhubaneswar, India

Müller JL, Pucher J, Tran TNT et al (2012) The potential of vermiculture to produce on-farm feed resources for aquaculture in mountainous areas of North Vietnam. In: Proceedings of the international scientific conference "sustainable land use and rural development in mountain areas". University of Hohenheim, Stuttgart, Germany, pp 113–114

Musyoka SN, Liti D, Ogello E et al (2020a) Using earthworm, *Eisenia fetida*, to bio-convert agro-industrial wastes for aquaculture nutrition. Bioresources 15:574–587

Musyoka SN, Liti D, Ogello EO et al (2020b) Earthworm, *Eisenia fetida*, bedding meal as potential cheap fishmeal replacement ingredient for semi-intensive farming of Nile tilapia, *Oreochromis niloticus*. Aquac Res 11

Musyoka SN, Liti DM, Ogello E et al (2019) Utilization of the earthworm, *Eisenia fetida* (Savigny, 1826) as an alternative protein source in fish feeds processing: a review. Aquac Res 50:2301–2315

Ng WK, Liew FL, Ang LP et al (2001) Potential of mealworm (*Tenebrio molitor*) as an alternative protein source in practical diets for African catfish, *Clarias gariepinus*. Aquac Res 32:273–280

Ngoc T, Pucher J, Becker K et al (2016) Earthworm powder as an alternative protein source in diets for common carp (*Cyprinus carpio* L.). Aquac Res 47:2917–2927

Nikki, Ahalavat S, Poonam (2023) Aquaculture industry—their present and future prospects. Research Trends in Animal Science 1st Edition, Bhumi Publishing, India, 43–52

Parolini M, Ganzaroli A, Bacenetti J (2020) Earthworm as an alternative protein source in poultry and fish farming: current applications and future perspectives. Sci Total Environ 734:139460

Pucher J, Ngoc T, ThiHanhYen T et al (2014) Earthworm meal as fishmeal replacement in plantbased feeds for common carp in semi-intensive aquaculture in rural northern Vietnam. Turk J Fish Aquat Sci 14:557–565

Pucher J, Steinbronn S, Mayrhofer R et al (2013) Improved sustainable aquaculture Systems for Small-Scale Farmers in northern Vietnam. In: Fröhlich HL, Schreinemachers P, Clemens G, Stahr K (eds) Sustainable land use and rural development in southeast. Innovations and Policies for Mountainous Areas. Springer Verlag, Asia, pp 281–317

Ranganathan LS (2006) Vermibiotechnology: from soil health to human health. Agrobios (India)

Rorat A, Suleiman H, Grobelak A et al (2015) Interactions between sewage sludge-amended soil and earthworms—comparison between *Eisenia fetida* and *Eisenia Andrei* composting species. Environ Sci Pollut Res 23(4):3026–3035

Sales J, Janssens PJG (2003) Nutrient requirements of ornamental fish. Aqua Liv Res 16:533–540

Shapawi R, Ng WK, Mustafa S (2007) Replacement of fish meal with poultry by- product meal in diets formulated for the humpback grouper, *Cromileptes altivelis*. Aquac 273:118–126

Singh K, Bhartiya DK (2018) Review on therapeutic use of earthworms. Int J Basic Appl Res

Sinha R, Herat S, Valani B et al (2009) Earthworms: the 'unheralded soldiers of mankind' and 'farmer's friend' working day and night under the soil: reviving the dreams of sir Charles Darwin for promoting sustainable agriculture. Am-Eurasian J Agric Environ Sci 5:01–55

Sogbesan AO, Ugwumba AAA, Madu CT (2007) Productivity potentials and nutritional values of semiarid zone earthworm (*Hyperiodrilus euryaulos*; Clausen, 1967) cultured in organic wastes as fish meal supplement. Pak J Biol Sci 10:2992–2997

Sogbesan OA, Madu CT (2008) Evaluation of earthworm (*Hyperiodrilus euryaulos*; Clausen, 1914; Oligocheata: Eudrilidae) meal as protein feedstuff in diets for *Heterobranchus longifilis*, Valenciennes 1840 (Teleostei Clariidae) fingerlings under laboratory condition. Res Jour of Envi Sci 2:23–31

Stafford EA, Tacon AJ (1985) The nutritional evaluation of dried earthworm meal (*Eisenia fetida*, Savigny 1826) included at low levels in production diets for rainbow trout. *Salmo gairdneri* Richardson Aquac Res 16(3):213–222

Tacon AGJ (1994) Feed ingredients for carnivorous fish species. Alternatives to fishmeal and other fishery resources (FAO of United Nations, fisheries circular no.881. FIRI/C881,), Rome 35

Tacon AJ (1993) Feed ingredients for warm water fish: fish meal and 3 other processed feedstuffs. FAO, fisheries circular 856. Rome: FAO

Tram ND, Ngoan LD, Ogle B (2005) Culturing earthworms on pig manure and the effect of replacing trash fish by earthworms on the growth performance of catfish (*Clarias macrocephalus* x *Clarias gariepinus*). Hue University of Agriculture and Forestry, Vietnam

Tripathi P, Bhardwaj B (2004) Earthworms in waste and environmental management. Bioresour Technol 92:275–283

Tuan NN (2010) Development of supplemental diets for carp in Vietnamese upland ponds based on locally available resources. (PhD thesis). PhD in Agriculture Sciences, Stuttgart,. University of Hohenheim

Urmila G, R. K., Sharma, S. and Kumar, R. (2025) Earthworms: the environmentally preferred bioengineers for managing heavy metals. In: Dey S, Bhattacharya S (eds) Biotechnological interventions in the removal of emerging pollutants. Springer Nature Singapore, Singapore, pp 365–384

Visvanathan C, Trankler J, Jospeh K et al (2005) Vermicomposting as an eco-tool in sustainable solid waste management. India, Asian Institute of Technology, Anna University

Vodounnou DS, Juste V, Kpogue DNS et al (2016a) Culture of earthworm (*Eisenia fetida*), production, nutritive value and utilization of its meal in diet for *Parachanna obscura* fingerlings reared in captivity. Int J Fish Aquat Stud 4(5):01–05

Vodounnou DS, Kpogue D, Tossavi NS et al (2016b) Effect of animal waste and vegetable compost on production and growth of earthworm (*Eisenia fetida*) during vermiculture. Int J Recycl Org Waste Agric 5:87–92

Waluyo J, Wahyuni D, Nuri N (2019) Antipyretic effects of dried earthworm (Pheretima javanica k.) in male white rat (Rattus norvegicus) with typhoid fever. Int J Sci Technol Res 8:243–246

Yadav R, Gupta RK, Kumar R, Kaur T, Moun R (2022a) Biochemical response of earthworm, *Eisenia fetida* to heavy metals toxicity. J Appl Nat Sci 14(3):990–998

Yadav R, Gupta RK, Kumar R, Kaur T (2022b) Assessment of toxicity of Lead and nickel on the biochemical and immunological parameters of earthworm. Eudrilus Eugeniae Agric Sci Dig 44(5):922–929

Yadav R, Kumar R, Gupta RK, Kaur T, Yodha K, Kour A, Kaur S, Rajput A (2023) Heavy metal toxicity in earthworms and its environmental implications: a review. *Environmental*. Advances p:100374

Zakaria Z, Salleh NM, Mohamed AR et al (2012) Optimization of protein content in earthworm-based fish feed formulation for catfish (*Clarius gariepinus*). Sains Malaysiana 41:1071–1077

Zhou F, Song W, Shao Q et al (2011) Partial replacement of fish meal by fermented soybean meal in diets for black seabream, *Acanthopagrus schlegelii*, juveniles. J World Aquacult Soc 42:184–197

Chapter 28
Prawn Culture and Beneficial Attributes of Medicinal Plants in Their Health

Vibhu Vaid, Nisha Vashishat, Rahul Kumar, Ajay, Monika Pilania, and Sudha Shahi

Abstract In recent years, aquacultural development has enhanced rapidly and has become the fastest- rising food production industry globally. Among different aquatic organisms, the culture of the crustaceans is one of fastest growing aquaculture worldwide. Out of these, *Macrobrachium rosenbergii* has gained a significant consideration due to its higher economical value, disease resistance, quality, larger abdomen, fecundity, environmental tolerance, flavour, rapid growth and development. However, external environmental stressors and diseases caused by pathogens are chief restrictions for the large scale development of aquacultural practices. For many years, antibiotics and other chemotherapeutic agents were utilised to regulate disease outbreaks in aquaculture. However, their adverse impacts on aquatic biota and their environment along with the development of anti-microbial resistance has led to the restriction of their utilisation. Therefore, medicinal plants are utilised as possible cheap and safe alternatives to different antibiotics to enhance growth, enhance immune

V. Vaid (✉) · N. Vashishat
Department of Zoology, Punjab Agricultural University, Ludhiana, Punjab, India
e-mail: vibhuvaid30@gmail.com

N. Vashishat
e-mail: nisha.vashishat@pau.edu

R. Kumar
Department of Zoology, Chaudhary Charan Singh Haryana Agricultural University, College of Agriculture, Bawal, Rewari, Haryana, India
e-mail: drrkr321@gmail.com

Ajay
Department of Agronomy, Chaudhary Charan Singh Haryana Agricultural University, Hisar, Haryana, India
e-mail: ajaykalaramna05@gmail.com

M. Pilania
Biotechnology Technician, Health and Life Sciences, Conestoga College, Kitchener, ON, Canada
e-mail: Monika5090@conestogac.on.ca

S. Shahi
Department of Aquaculture, Chaudhary Charan Singh Haryana Agricultural University, Hisar, Haryana, India
e-mail: sudhashahi252@gmail.com

R. Kumar et al. (eds.), *Sustainability in Aquaculture and Fisheries*,
Springer Water, https://doi.org/10.1007/978-3-032-16605-0_28

system and prevent diseases. Furthermore, they defend prawns against different environmental factors including hydrogen ion concentration (pH), water temperature, salinity, ammonia and carbon/nitrogen (C/N) ratio. The chapter aims to fill a knowledge gap regarding biological and environmental factors affecting prawn culture and beneficial roles of medicinal plants such as growth promotion, appetite stimulation, anti-stressor, immunostimulants, improve pigmentation and feed attractant in *M. rosenbergii*. In addition, the literature highlighted the general mode of action of medicinal plants, limitations in their application in prawn culture and recommendations for further progression in the field.

Keywords Aquaculture · Medicinal plants · Mode of action · Prawn culture

28.1 Introduction

Aquaculture refers to the production of aquatic animals and plants under semi-controlled or controlled environmental conditions. Currently, more than 202 countries are aquaculture producers contributing to food security, animal protein and economic development worldwide (Asaikkutti et al. 2018; Food and Agriculture Organization (FAO) 2018; Shefali et al. 2021; Chen-Fei et al. 2020). Aquaculture has progressed as the fastest developing field in the food production sector with approximately 5.3% annual increment in food production from 2001 to 2018 to meet the demands of inclining population worldwide (Food and Agriculture Organization (FAO) 2020). Furthermore, its diversification and production vary among countries (Cai et al. 2023) and is expected to grow globally by 53% in 2030 (Food and Agriculture Organization (FAO) 2020).

With aquaculture development, fish consumption has shown a constant incline out passing other sectors in providing a substantial amount of animal food (Beltrán et al. 2018). The main goal of aquaculture is to produce maximum weight per culture unit for molluscs, crustaceans or fish (Direkbusarakom 2004). It supplies about half of the worldwide demand for fishery products with more than 90% food coming from Asia. Additionally, world aquaculture is an economically vital source of foreign exchange for different developing countries (Food and Agriculture Organization (FAO) 2009). Among different countries, India holds second place after China, accounting about 3.12×10^6 tons aquaculture production (Shanthi et al. 2012). Furthermore, India and China made rapid strides with blue revolutions producing 7.76 and 61.35% aquaculture, respectively (Kanagarasu et al. 2017). Among cultured species, crustaceans including shrimps, lobsters, crabs, prawns and crayfish, is one of the rapidly growing animal production sectors having around 50,000 to 67,000 species worldwide (Radhakrishnan et al. 2012; Asaikkutti et al. 2018).

Rearing of crustaceans is 3rd largest sector in aquaculture having 11.2×10^6 tons production in 2022, which is 62% more in comparison to production in 2015 (Food and Agriculture Organization (FAO) 2022). Out of these, about 9.5×10^6 tons crustaceans were reared in Asian regions (Stentiford et al. 2017). However, with prawn culture intensification in Asian regions, the frequency of diseases also increased resulting in noteworthy economic losses (Chen-Feiet al. 2020). In addition, other major constraints including seed availability, nutrition, water quality and cost of feed account for fluctuation in prawn production (Food and Agriculture Organization (FAO) 2010). Disease challenges are key factors which hinders the expansion of aquaculture (Zhong et al. 2018; Kumar et al. 2022a, b), therefore, antibiotics are utilised for health management of aquaculture species (Van Boeckelet al. 2019; Liu et al. 2020). However, utilisation of antibiotics leads to accumulation of antibiotic residues in tissues of fauna. Furthermore, the action of antibiotics is not restricted to the target bacteria but impacts other microorganisms in aquatic fauna and environment (Martinez 2009). Moreover, with rise in the rate of utilisation of antimicrobial agents, antibiotic-resistant bacterial strains develop which further negatively affects human health and environment (Direkbusarakom 2004; Stratev and Odeyemi 2016).

Presently, utilisation of antibiotics has been banned for culture worldwide (Kumar et al. 2022c; Binh et al. 2018). Therefore, different ecofriendly and economically viable strategies are required to be implemented against disease challenge to ensure sustainable aquaculture production. Modulation of immune system of culture through prophylactic measures such as probiotics, medicinal plants, immunostimulants and vaccination play central role in aquaculture (Wang et al. 2018). Among these, herbal plants are rich source of oils and secondary metabolites, which are necessary for aquaculture's health. The substantial benefits of medicinal plants in different treatments include their easy availability, safety, low cost and efficacy leading to utilisation of herbs in animal feeds by both feed companies and researchers (Pandey et al. 2012). Therefore, the present chapter was designed to enlist beneficial roles of medicinal plants in prawn culture.

28.2 Prawns: Species, Trends and Their Culture

Aquatic term refers to variable water environments including brackish, marine and freshwater. Pioneering studies of Tukuji Fijumura and Shao Wen Ling, fathers of prawn culture, together made possible the development of freshwater prawn farming. Freshwater culture refers to faunal cultivation where the final product is raised in freshwater having less than 0.55% salinity including rivers, canals, lakes, groundwater, reservoirs and ponds (Farook et al. 2019). In India, freshwater resources include paddy fields (2.3×10^6 ha), ponds (2.25×10^6 ha), reservoirs and lakes (2.09×10^6 ha), and canals and rivers (0.16×10^6 ha) (Banerjee et al. 2024). Freshwater prawn fishery has the ability to transform the inland aquaculture, generate income and employment (Parameswaran 1994). Prawns refer to freshwater crustaceans who spend atleast juvenile and adult phases in freshwater bodies. Around the world, most farmed species belong to the genus *Macrobrachium*, which is native to Asia, America and Africa (Holthuis and Ng 2010).

From 1995 to 2003, prawn production of *M. rosenbergii* (Scampi) elevated from 5.00 to 28.00×10^4 metric tons throughout the world (Selvaraj and Kumar 2003). In wild and farm, freshwater prawns grow to 500 and 30–100 g, respectively (Tayamen 2005; Ragaza 2016). Different palemonid prawns including *M. lamarrei, M. rude, M. birmanicumchoprai, M. mirable, M. hendersonii, M. hendersonii, M. scabriculum, M. nobilii* and *M. villosimanus* are cultured worldwide to support local fisheries (Mariappan et al. 2003). In India, commercial species of freshwater prawns cultured are *M. gangeticum, M. malcolmsonii* (monsoon river prawn) and *M. rosenbergii.* Out of these, freshwater adapted macroura species *M. rosenbergii* has gained a significant attention due to its high economical value due to its larger abdominal portion, rapid development, acceptance of different formulated diets, fecundity, environmental tolerance, disease resistance, delicate flavour, meat quality and growth in China, America, Africa and Japan (Kumar and Singh 2022; Wei et al. 2021).

In addition, production of crustaceans has substantially reached 78.62×10^8 tons during the year 2010–2016 (Food and Agriculture Organization (FAO) 2018). Out of these, prawns and shrimps account for around 4,048 species worldwide (Radhakrishnan et al. 2012). In 2020, production of crustaceans was dominated by *M. rosenbergii, Procambarus clarkii* (red swamp crawfish), *Litopenaeus vannamei* (white leg shrimp), *Penaeus monodon* (giant tiger prawn), *M. nipponense* (Oriental river prawn) and *Eriocheir sinensis* (Chinese mitten crab) (Pourmozaffar et al. 2024). The technology of prawn culture was advanced to support high density prawn fishery in constructed ponds (Chen-Fei et al. 2020). Among different crustaceans, *M. rosenbergii,* belonging to family Palaemonidae, accounts for around 2.34×10^5 tons (~3%). In India, prawn fishery of different rivers including Mahanadi, Ganges, Godavari, Cauvery, Krishna, Hooghly and Parkas indicate that *M. rosenbergii* and *M. malcolmsonii* are major commercial species (Kanagarasu et al. 2017). In India, the production of *M. rosenbergii* elevated from ~ 7.15×10^3 tons during 1999–2000 to ~ 2.73×10^4 tons in 2007–2008 (Wang et al. 2022). In India, *M. rosenbergii* production was higher (4.28×10^4 tons) in 2005, which unceasingly declined during 2006–2009 (30.12–6.60 $\times 10^3$ tons) followed by dramatic increase in production (1.35×10^4 tons) (Food and Agriculture Organization (FAO) 2010). In 2010, annual production of *M. rosenbergii* was around 5.00×10^6 tons (Food and Agriculture Organization (FAO) 2012). Additionally, *M. malcolmsonii* is also extensively distributed throughout India and is 2nd largest freshwater prawn after *M. rosenbergii* (Hossain et al. 2012).

Out of these, *M. rosenbergii* is distributed in around 43 countries, with Asia producing about 98% of them (Mather and de Bruyn 2003). Among these, countries such as India, Vietnam, Thailand, Bangladesh, Taiwan, Myanmar and China are among the principal producers, with annual production of about 3.00×10^4tons (Food and Agriculture Organization (FAO) 2010; New and Nair 2012). Since 1999, the area under *M. rosenbergii* culture has elevated noticeably and touched over 4.00×10^6 ha and is still increasing in different Indian states. *M. rosenbergii* has expanded from 1.31 to 2.03×10^5 tons during 2000–2011 (Asaikkutti et al. 2016a). In 2004 and 2016, global *M. rosenbergii* culture reached over \$0.81and $1.90 \times 10^9$with production around 0.19 and 0.23×10^6 tons, respectively (Food and Agriculture Organization (FAO) 2004; Sui et al. 2019). In addition, *M. rosenbergii* is a good source of protein, carbohydrate, polyunsaturated fatty acids and essential amino acids, along with considerable amount of minerals, fibre, lipid and vitamins (Asaikkutti et al. 2016a). They are omnivorous scavengers and feed on decomposing animals, plants, worms, insects and their larva (Bhavan et al. 2010).

28.3 Environmental and Biological Factors Affecting Prawn culture

Prawn culture is challenged by host immunity; and multidisciplinary interaction between biological and chemical stressors originating from over-exploitation, aquatic pollution and other anthropogenic stresses which downregulates the expression of genes (Dawood et al. 2019). Therefore, for sustainable prawn culture, genes responsible for growth, immunity and other factors are required to be recognised and their expression should be quantitatively measured. Cellular genes including Toll, HSP70 and dorsal have crucial role in immune systems and physiological functions. Alterations in different environmental factors including water temperature, water and soil quality, hydrogen ion concentration (pH) and rainfall affect prawn culture (Kumar et al. 2022a, b; Banerjee et al. 2024).

28.3.1 Water Temperature

Among different physico-chemical parameters, optimum water temperature is a crucial environmental factor for the survivability of any aquatic fauna. Furthermore, stress induced water temperature stimulates innate immune (Nikoskelainen et al. 2004). Prawns such as *M. rosenbergii* can withstand a water temperature ranging 14.00–35.00 °C, whereas, Bir et al. (2024) and Chaitanawisuti et al. (2013) studied that 29.00–30.00 °C is ideal for the growth of prawns. In addition, virus multiply actively up to 30.00 °C, therefore, viral diseases prevail in prawn aquaculture between 25 and 30 °C. Moreover, at 35.50 °C viral infections drastically reduce, whereas, incidence of bacterial infections elevates at this temperature range. Furthermore, temperature between 35 and 40 °C, weakens the immune system of prawns along with growth of bacteria including *Enterococcus* sp. (Cheng and Chen 1998). A recent study by Tsai et al. (2025) proposed 15° and 35 °C as lower and upper critical temperatures for *M. rosenbergii*, respectively. The scientists also proposed optimal performance temperature (26 °C) and range (24–28 °C) for *M. rosenbergii*.

28.3.2 Salinity

Similarly, fluctuations in salinity directly relates with mortality of aquatic biota. Prawns including *M. rosenbergii* requires low salinity for optimum their growth, behaviour and health (Huang et al. 2019). Furthermore, they show the growth promotion and higher adaptability in the salinity range of 10–12% (Kumar et al. 2022c; Mulyasari et al. 2022). Chand et al. (2015) recorded that the median lethal of salinity was 24.6 ppt for prawn culture. In addition, rate of survival of prawns varied from 78 (at 20 ppt salinity) to 91% (at 0 ppt salinity). Though, prawns cultured at 0 ppt salinity were more prone to bacterial infections. According to Cheng and Chen (2000), the activity of phenoloxidase (PO) showed positive relation with salinity and a maximum activity of PO was recoded in range of 10–15 ppt. Additionally, salinity >8 ppt alters the rate of reproduction (Cheng and Chen 1998).

Bir et al. (2024) also examined the effect of different temperatures (22.00, 26.00, 30.00, 34.00 and 38.00 °C) and salinities (0, 5, 10, 15 and 20 ppt) on adolescent *M. rosenbergii* for 168 h. The results showed normal movement and behaviour at 10 ppt salinity in *M. rosenbergii* whereas unusual behaviour and mortality (50% within 72 h and 100% after 96 h) was observed at 20 ppt. In addition, the results indicated normal behaviour in prawns exposed to 22.00–30.00 °C whereas abnormal temperature and mortality were observed after 34.00 °C. Interestingly, survival rate of prawns declined to 33.33% at 38.00 °C within 72 h. Though, differential haemocyte count (DHC) and total haemocyte count (THC) showed non-significant alterations at temperature $\leq$30.00 °C and salinity $\leq$ 10 ppt whereas significant decrease was recorded at higher temperatures ($\geq$34.00 °C) and salinities ($\geq$15 ppt). Furthermore, lipid content showed a significant decline at salinity $\geq$ 15 ppt.

28.3.3 Light Spectrum and Photoperiod

Along with this, alterations in light spectrum and photoperiod also control the immunity and growth of different aquatic fauna (Choi et al. 2012). Wei et al. (2021) examined the impact of light spectrum (blue, yellow, green, white and red), light: dark cycle/photoperiod (0:24, 6:18, 12:12, 18:6 and 24:0 h) and different salinities (10, 13 and 16%) on the activities of enzymes and growth of larvae of Scampi. The results revealed that the studied factors significantly influenced metamorphosis, growth rate and activity of enzymes viz. lysozyme, catalase (CAT), superoxide dismutase (SOD) and alkaline phosphatase (ALP) in prawn larvae. In addition, the best performances were recorded at 24:0 h (light: dark), 13% salinity and under white or green light. Moreover, red light negatively affected the larval development of prawns.

28.3.4 pH Value

pH also directly impacts immunity and growth of a living biota. Extracellular pH impacts innate immune components such as decline in proliferation of polymorphs and lymphocytes, reduction in production of antibodies, alteration in chemotaxis and expression of immune genes in crustaceans (Lardner 2001; Sharma et al. 2009). Optimum pH for *M. rosenbergii* aquaculture is 7.00–8.00, where activity of PO and THC significantly rise between pH 7.00–7.70. Furthermore, Cathepsin L activity also significantly increased at pH 7.5 in *M. rosenbergii*, whereas, it significantly reduced at lower (pH 3.00–5.00) and higher (pH 9.00–10.00) pH values (Arockiaraj et al. 2012). The impact of different pH values (5.40 $\pm$ 0.20, 6.40 $\pm$ 0.50 and 7.70 $\pm$ 0.40) on the growth, development and feeding of larvae of *M. rosenbergii* was evaluated by Liew et al. (2022). The results exhibited negative effect of decline in pH on the larval species, with no larval survival at pH value 5.40 $\pm$ 0.20 after 48 h of exposure period. Along with this, pH value of 6.40 $\pm$ 0.50 also negatively affected survival, growth and feeding of freshwater prawn larvae compared to 7.70 $\pm$ 0.40. Furthermore, metamorphism to the post-larval stage at different pH values (pH 6.40 $\pm$ 0.50 and 7.70 $\pm$ 0.40) were observed at 30 and 23 days, respectively. The results emphasized the necessity of pH monitoring during the prawn culture.

28.3.5 Carbon/Nitrogen (C/N) Ratios

The effect of different C/N ratios (20/1, 15/1 and 10/1) was studied in ponds for *M. rosenbergii* culture by Asaduzzaman et al. (2010). The results exposed significant rise in the biovolume of crustaceans, rotifers and phytoplankton. In addition, positive relation was observed in the biovolume of crustaceans, rotifers and phytoplankton and increase in C/N ratio as follows: 15, 6 and 11%, respectively. In addition, significant decline was recorded in the biovolume of benthic macroinvertebrates indicated their grazing by *M. rosenbergii*. However, periphyton, microbial and plankton were under-utilized by *M. rosenbergii* at 20 C/N ratio.

28.3.6 Ammonia Concentration

Ammonia is a major nitrogenous compound released by decomposition of organic matter, fertilizers, faeces, pesticides and left over feed. Prawn culture is also affected by intensification of toxic materials including ammonia which affect production, growth and finally cause death (Boyd and Tucker 2012; Kumar et al. 2024, 2025). About 30 g ammonia is released by each kg of feed (20–40% crude protein) incorporated. According to New (2000), recommended levels of ammonia for *M. rosenbergii* is below 500 mgL^{-1} at 9.5 pH value. Mugnier et al. (2008) recorded the effect of rise in ammonia concentration on all the stages of shrimps especially their early post-moult stage. Similarly dissolved oxygen also influences growth performance of prawns. The scientists also observed that combination of hypoxia and ammonia toxicity result in more mortalities than single effects of ammonia toxicity and hypoxic condition. Moreover, positive relation between pH and ammonia toxicity was also studied by Mallasen and Valenti (2005) and Coyle et al. (2005) in larval and juvenile stages of *M. rosenbergii* and other crustaceans. Figueroa-Lucero et al. (2012) also studied that rise in the concentration of ammonia (12 to 20 mg/L) led to 100-fold rise in ammonium toxicity in *M. rosenbergii.*

28.3.7 *Hydrogen Sulphide (H_2S)*

H_2S is a toxic colourless gas produced by the decomposition of organic matter in H_2O or damp environment under limited O_2 by sulphate consuming bacteria in sewerage and petroleum industries. Additionally, lower water temperature and pH contributes to rise in H_2S levels, which further impacts aquatic fauna including fishes, shrimps and prawns by interfering reoxidation of cytochrome a_3 and irritating different tissues including hepatopancreas, intestine, gills and gut (Maji, 2024). Lalitha and Surendran (2004) stated that H_2S producing *Shewanella putrefaciens* impacts the spoilage of muscles of *M. rosenbergii.* Additionally, Jayamanne (1986) studied that0.26 and 0.43 mgL^{-1} are safer concentrations for H_2S in the juveniles of *M. rosenbergii.* Kang et al. (1995) also studied significant effects of the combination ofH_2S and hypoxic conditions on the survival of prawns in Lake Kojima (Japan).

28.3.8 *Pesticides*

Pesticides are one of the major pollutants affecting water, soil and different animals. Different pesticides utilised in agriculture reach prawn cultivation ponds and have various effects on the immunity and growth of prawns (Kumar and Singh, 2022). Rose et al. (2023) evaluated the effects of Imidacloprid, a broad-spectrum neonicotinoid insecticide on the quality of larvae, feeding behaviour, survival and metamorphosis in *M. rosenbergii* larvae for 48 h. The results disclosed significant ($p < 0.05$) decline in larval health, feeding rate and survival rate with rise in insecticide concentration, which led to delay in metamorphosis. Similarly, the effects of glyphosate (SUNPHOSATE-G®) and emamectin benzoate (EMACOT®) were investigated on *M. macrobrachion* juveniles in Benin by Houssou et al. (2024) for 28 days. The results indicated that different concentrations of EMACOT® prevented the moulting of prawns further hindering weight gain in comparison to control groups. On the contrast, treatment with glycophosate showed non-significant impact and hence lesser sensitivity of the juveniles of prawns.

28.3.9 *Heavy Metals*

Heavy metal pollution acts as a major environmental threat worldwide due to their release from municipal, industrial and agricultural sources, affecting histological, behavioural and physiological activities in aquatic organisms (Vaid and Hundal 2019; Yadav et al. 2022a, b, 2023). Out of these, behavioural changes are the earliest responses to heavy metal contaminant. Ahmed and Lodhi (2024) investigated behavioural changes induced by cadmium chloride ($CdCl_2$) and copper sulphate on *M. lamarrei*. Behavioural alterations such as aggregation, darkening of gills, avoidance response and excessive formation of mucus were recoded. This study indicated that initially $CdCl_2$ was more toxic for prawn culture but with time both chemically become equally toxic. The results suggested that crustaceans act as bioindicators of heavy metal contamination and emphasized to mitigate heavy metal pollution to protect ecosystem.

Similarly, the toxicological effects of heavy metals through sewages were studied by Onunkwor et al. (2022) on *M. macrobrachion* in Ogun River and Epe Lagoon (Nigeria). The results showed that physicochemical parameters (chemical and biochemical oxygen demand were above WHO permissible limits. The results indicated that heavy metals declined nutritional value and induced oxidative stress in *M. macrobrachion* from Epe lagoon and Ogun River. Likewise, Yamuna et al. (2009) stated that mercury chloride impacts hepatopancreas and gills in *M. rosenbergii*. Moreover, mercury toxification affects protein content in muscles of *M. rosenbergii* in duration and dose dependent manner (Desai and Tank 2008). Furthermore, even a trace concentration of mercury like 0.002 mg/L could alter protein content of muscles and hepatopancreas in prawns (Bhardwaj et al. 2014).

28.3.9.1 Diet

Formulation of better-quality feed is also crucial to improve the growth and quality of larvae of *M. rosenbergii*. These are omnivorous which allow them to feed on industrial by-products as ingredients of formulated feed (Mukhopadhyay et al. 2003). Dietary proteins are crucial to produce enzymes, energy, antibodies, hormones (Gimenez et al. 2009) and are essential component of crustaceans from eggs to adult phase especially younger phase (Dinakaran and Soundarapandian 2009). Furthermore, protein could decline nitrogen loading, increase production cost and growth (Tibbetts et al. 2000; Glencross 2006). In Juveniles of *P. monodon* and *P. japonicus*, around 40 to 52% proteins are required for survivability and growth, however, protein levels above 60% cause depressing effect on growth (Kumar et al. 2006). In addition, carbohydrates of plant origin are utilised as short-term energy source and facilitates prawn metabolism (Briggs 1991; Fegan 2004).

Along with this, requirement of dietary lipid in *M. rosenbergii* ranges from 6 to 8%, is an important component of biota and acts as energy reservoir. In addition, it is crucial for growth, precursor of steroids, part of cell membranes, moulting, weight gain and maintenance (Rangacharyulu 1999). They are deposited in hepatopancreas and play crucial role in immunity and lipid metabolism (Rőszer 2014). Moreover, vitamins and minerals in feeds lead to absorption of inorganic minerals (Mukhopadhyay et al. 2003). Disproportion of mineral-vitamin mixture in feed could negatively affect growth of prawns (Conklin 1997). Furthermore, dietary carotenoids contribute around 20% flesh pigmentation (White et al. 2002; Moretti et al. 2006). In *M. rosenbergii*, astaxanthin utilisation from dietary carotenoids contribute to muscle coloration (Lim et al. 2018).

28.3.9.2 Biological Factors

Continuous growth of breeding density and deterioration of water quality, has contributed to incidence of diseases caused by bacteria, fungi and virus; and related health issues, alterations in survivability of post-larvae along with other impacts on different life stages of prawn (Shailender et al. 2012). Among different pathogens, viruses and bacteria the main group that impacts prawn culture because of their lethality, infectiousness and pathogenicity (Bachère 2000). Furthermore, *Aeromonas* and vibriosis can cause death in early larvae of prawns as they are highly susceptible to diseases (Tonguthai 1997). White tail disease (WTD) of *M. rosenbergii* is the among the most studied diseases of freshwater prawns (Pillai and Bonami 2012). In Taiwan, during last few decades, prawn culture was negatively impacted with bacteria and yeasts in hot and cool seasons (Hsu and Liu 1994; Halim and Prajitno 2017). Different infectious diseases in prawns are listed below.

Although *M. rosenbergii* is considered to be hardy and is a highly resistant species, however, disease outbreaks caused by protists, viruses, fungi, bacteria and yeast had been observed particularly during larval and post-larval stages (Pillai and Bonami 2012). Few pathogens like infectious hypodermal, branchiostegal blister disease (BBD), haematopoietic necrosis virus (IHHNV), appendage deformity syndrome (ADS), *M.* hepatopancreatic parvo like virus (MHPV), extra small virus like particle (XSV), *M.* muscle virus (MMV), white spot syndrome virus (WSSV) and WTD caused by *M.* nodavirus (MrNV). Among the listed viruses, XSV and MrNV are responsible for greater economic losses causing mortalities in various life stages (larvae, post-larvae and juvenile) of *M. rosenbergii*, with clinical signs including whitish tail, telson degeneration and opaqueness in post-larval stage (Farook et al. 2016). Similarly, rearing of *M. malcolmsonii* was mostly affected by MMV, idiopathic muscle necrosis (IMB), IHHNV, slow growth syndrome (SGS), BBD and ADS resulting in alterations in muscular atrophy, colouration of cuticles, swimming behaviour and delayed growth (Sahoo 2008). Moreover, bacterial diseases including *Pseudomonas* spp., *Lactococcus garviae*, *Aeromonas* spp. and *Vibrio* spp. cause deaths in cultured prawns (Musthafa et al. 2016). Similarly, few protozoans have also been recorded from infected prawn species (Shailender et al. 2012).

28.4 Impacts of Antibiotics on Prawn culture

Antibiotics are generally utilised to treat bacterial infections; and are usually administered as therapeutics and prophylactics through injection, oral administration or baths in the range 200-700g/ton of culture production worldwide (Ng et al. 2009; Reverter et al. 2014).Likewise, they are effective against infectious diseases; however, their use has been banned globally due to their bioaccumulation and ability to lead bacterial-resistance towards antibiotic drugs in cultured organisms (Binh et al. 2018; Chen et al. 2018). Bermúdez-Almada and Espinosa-Plascencia (2012) reported that oxytetracycline, one of most administered antibiotics worldwide, accumulated in hepatopancreas and other tissues of shrimp when applied against bacterial infections. Therefore, development of non-antibiotic methods is required to improve shrimp culture (Guo et al. 2011). For instance, chloramphenicol and nitrofurans, used mostly in farms, caused serious issues such as aplastic anaemia and acted as carcinogens (Uddin and Kader 2006), therefore were banned worldwide for food production.

Paul et al. (2011) recorded chemicals and antibiotics utilised hatcheries and farms for prawn culture in north and south 24 Parganas (West Bengal). The researchers recorded the presence of different antibodies including erythromycin, nitrofurans, oxytetracycline and chloramphenicol in studied areas. Out of these, mostly chloramphenicol (40%) was used followed by 23 and 20% oxytetracycline and erythromycin, respectively. Along with this, disinfectants such as malachite green, methylene blue, potassium permanganate and formalin were also recorded. Additionally, synthetic antioxidants are also utilised for preservation and rotting of food. Furthermore, synthetic antioxidants reduce damage induced by free radicals and converse anti-aging effects. However, long term use of butylated hydroxytoluene (BHT), butylated hydroxyanisole (BHA) and propylgallate are harmful to consumers (Elshafie et al. 2012). Researchers evaluated that BHT adversely effects body weight and triggers lung tumour development (Meyer and Hansen 1980; Malkinson 1999), whereas, BHA induces stomach tumour development (Kahl and Kappus 1993).

28.5 Beneficial Roles of Medicinal Plants as Additives

The production cost and growth performance of aquaculture species depend upon the type and quality of feed provided. In addition, artificial feeds could account for as much as 40% operating cost in the culture of freshwater prawns (Mitra et al. 2005). However, the addition of herbs increases palatability and digestibility of feed (Poongodi et al. 2012). Around 75,000 medicinal plant species are recorded worldwide, still only few plants have been studied in different studies (Mustafa et al. 2017).In diet, the active bio-compounds of herbs viz. pigments, phenolics, flavonoids, alkaloids, terpenoids, essential oils, starch and steroids act as appetite stimulators, growth promoters, antioxidants; and increase the release of digestive enzymes, improve

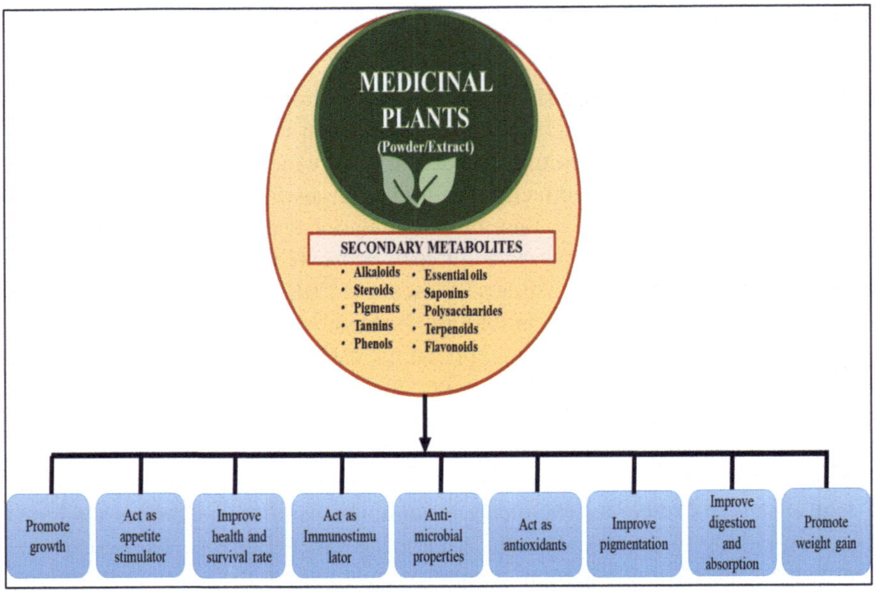

Fig. 28.1 Secondary metabolites of medicinal plants and their beneficial roles in prawn culture

food consumption and utilisation which further promotes growth by enhancing metabolism, protein synthesis and rate of transcription (Awad and Awaad 2017; Kawamura et al. 2019). Additionally, herbs comprise tannins, alkaloids, volatile oils, flavonoids, polysaccharides, minerals, organic acids, carbohydrates, amino acids and vitamins, which further improve enzyme activity, digestibility, protein synthesis and palatability of the feed (Shanthi et al. 2012; Pu et al. 2017). Furthermore, the market for medicinal plants is projected to meet $\$5 \times 10^{12}$ by 2050 worldwide (Harikrishnan et al. 2012). The secondary metabolites of medicinal plants and their beneficial roles are depicted in the Fig. 28.1.

Medicinal plants are usually directed as powder and extract; and the optimal concentration of these components depend on the type of extraction process, prawn species, plants and method of administration (Van Hai 2015). Survival and production of healthy larvae and post-larva stages are primary deterrent in the intensification of prawn culture (Prakash and Karmagam 2013). In addition, reduction in duration at larval stage enhance survivability and improve disease resistance in prawns. In common, larva and post larva of *M. rosenbergii* depend on live feed and pellets of different size groups, respectively (Amiruddin et al. 2021). Furthermore, feed accounts for maximum (about 40 to 60%) operational cost in production of prawns (Silva et al. 2007). The formulation of cheap, balanced diets from animal and plant by-products including ground nut oilcake, wheat brawn, poultry waste, rice brawn, blood meal, soy bean meal and fishmeal are essential for running aquaculture successfully (Langer et al. 2011). Now, scientists have focused on algae, herbal plants, waste of fruits and vegetables as cheaper food supplements for survival and growth of prawns (Rajkumar et al. 2017).

28.5.1 Medicinal Plants as Growth Promoter in *M. rosenbergii* Aquaculture

Feed additives significantly act as growth promoters, and hence improve health and animal performance leading to sustainable development of prawns (Awad and Awaad 2017). Compounds such as polysaccharides, fibres, carbohydrates and other components contribute to the increase digestibility and nutrient absorption in shrimps (Rani et al. 2025; Nisha et al. 2025; Kiran et al. 2025; Niu et al. 2015). Studies showed that inclusion of herbal power as feed additive in post larva of *M. rosenbergii* significantly improve levels of amino acids, digestive enzymes' activity and ultimately meat quality (Muralisankar et al. 2017). The protein, lipid, astanxanthin and carbohydate contents in different perennial grasses viz. *Cynodon dactylon* (Bermuda grass), *Eragrostis amabilis* (featherlove grass), *Eleusine indica* (goose grass), *Setaria pumila* (pigeon grass) and *Dactylocteum aegyptium* (crowfoot grass) were studied by (Banerjee et al. 2024). The results indicated that *C. dactylon* showed better results and hence selected as feed ingredient for *M. rosenbergii* for 180 days. The scientists observed that the feed condition index and growth (growth rate, survival rate and specific growth rate) were significantly more in prawns fed with *C. dactylon* formulated diet. A recent study conducted by Pelinson da Fonseca et al. (2025) also observed weight gain and improvement in physiological and metabolic activity in *M. rosenbergii* juveniles supplemented with *Curcuma longa* extract (0.05, 0.2 and 1%) for 60 days.

Interestingly, Muralisankar et al. (2018) also observed significant ($P < 0.05$) effect of leaf powder of *Phyllanthus amarus, Ocimum sanctum* and *Solanum trilobatum* on growth performance, digestive enzymes, vitamins, energy utilisation and survival on *M. rosenbergii.* The growth promotion in prawns supplemented with different leaf powders was in the order: *P. amarus* > *S. trilobatum* > *O. sanctum.* Similarly, El-Desouky et al. (2012) studied that inclusion of extracts of *Zingiber officinalis*(ginger) and *C. dactylon* increase food conversion ratio and growth (weight gain, specific growth rate and length gain) in *M. rosenbergii* juveniles; and juveniles fed with these herbal extracts showed better growth rate compared to control group. Rebecca (2011) also evaluated the impact of *Withania somnifera* and *O. sanctum* on survival, energy utilisation, biochemical constituents and growth of post larvae of *M. rosenbergii* for 45 days. The results indicated significant ($P < 0.05$) changes in studied parameters in feeds enriched with herbs. In addition, the researchers observed that post larvae fed with *W. somnifera* showed better growth promotion than *O. sanctum* enriched feed.

The role of seeds of different medicinal plants including *Ricinus communis* (caster), *P. emblica* (gooseberry), *Azadirachta indica* (neem) and *Syzygiumcumini* (jambolana) individually as feed additives at different concentrations (1, 3 and 5%)

were examined in early juveniles of *M. malcolmsonii* for 45 days (Bhavan et al. 2014). The researchers observed that seeds of studied plants acted as appetizer and significantly increased the activities of different digestive enzymes viz. amylase, lipase and protease) in the order: *S. cumini* (5%), *P. emblica* (5%), *A. indica* (1%) and *R. communis* (5%).In addition, the seeds increased consumption of food, which further significantly ($P < 0.05$) increased non-enzymatic antioxidants (vitamins E and C), amino acid, protein, lipid and carbohydrate levels and ultimately health and growth of *M. malcolmsonii* early larvae. Moreover, THC also displayed significant ($P < 0.05$) increase which showed presence of good non-specific immune mechanisms, whereas the levels of lipid peroxidation (LPO) and activities of CAT and SOD indicated non-significant change. The results highlighted that incorporation of seeds in feed improved health of *M. malcolmsonii* and suggested that they can be incorporated as additives in freshwater prawn culture.

The individual effects of various concentrations (1, 3 and 5%) of different parts viz. fruit powder of *Quercus infectoria* (gallnut), seed powder of *Myristica fragrans* (nutmeg) and stem powder of *Glycyrrhiza glabra* (liquorice) of herbs on the growth of post larvae of *M. rosenbergii* were investigated by Bhavan et al. (2013a) for 60 days. The results showed significant ($P < 0.05$) increase survival rate, weight gain, protein levels, non-enzymatic antioxidants (vitamins E and C), digestive enzymes (amylase, lipase and protease) and mineral (K^+ and Na^+) content were found in the order: *M. fragrans*, *G. glabra* and *Q. infectoria.* In addition, intensely stained protein bands (molecular weight: 116–14 kDa) were observed in muscles of exposed post larvae compared to control. The results suggested that medicinal herbs induce secretion of lipase, amylase and protease in *M. rosenbergii*, which further enhanced food consumption, digestion, absorption and growth of post-larval prawn.

Poongodi et al. (2012) also studied the single effects of *Z. officinale*, *Allium sativum* (garlic), *Trigonella foenum-graecum* (fenugreek) and *C. longa* (turmeric) on growth of post larvae of *M. rosenbergii.* The feeding rate and survival rate were recorded in the order: garlic > ginger > turmeric > fenugreek. However, feed conversion ratio, lipid content and condition factor were in the sequence: garlic < ginger < turmeric < fenugreek). Additionally, the levels of amino acid, protein, DNA, RNA, transcription rate, carbohydrate and RNA/DNA ratio were comparatively higher in treated groups. Moreover, Shanthi et al. (2012) also examined the effects of different concentrations (1, 3 and 5%) of medicinal herbs (*Andrographis paniculata, Cissus quadrangularis* and *Eclipta alba*) on energy utilisation, growth, biochemical constituents, digestive enzymes and antioxidants of *M. rosenbergii* post larvae for 90 days. The results showed significant improvement in growth, survival, energy utilisation, protein, lipid, carbohydrate, vitamin C, activity of digestive enzymes (amylase, lipase and protease) and vitamin E. In addition, the intensity of staining of some bands (46, 48 and 50kDa) were more in *C. quadrangularis* fed post larvae compared to *E. alba* and *A. paniculata* treated *M. rosenbergii.*

28.5.2 *Medicinal Plants as Immunostimulators and Antimicrobial Agents in M. rosenbergii Aquaculture*

Medicinal plants contain active components that stimulate non-specific immune system in animals, and hence act as immune stimulants in culture of different species (Sharma et al. 2024; Sharma et al. 2025; Simtoe et al. 2024). Immunostimulants refer to a naturally occurring compounds that modulate the immune response by enhancing the disease resistance against pathogens in host organisms (Bricknell and Dalmo 2005). Dietary immunostimulants act as effective and cheaper method for increasing cost-effectiveness and productivity of aquaculture by increasing stress tolerance, disease resistance and declining the requirement of expensive disease control measures (Ragaza 2016). The immune system of crustaceans lacks adaptive immune system and therefore depend on non-specific immune system. The non-specific immune response of crustaceans depends on haemocytes that perform humoral and cellular mechanisms including encapsulation, phagocytosis, formation of antimicrobial substances to fight different foreign and infectious agents. THC and DHC are important functional parameters for assessing the immune potential (Johansson and Söderhäll 1989).

The haemocytes (specialized blood cells) are essential for killing infectious agents, direct sequestration and synthesis of bioactive molecules (Smith and Chisholm 1992). In addition, haemocytes in crustaceans increase the expression of immune related genes against viruses and other pathogens (Lio-Pio and Inui 2010). Furthermore, haemocytes including granular, semi-granular and hyaline cells are essential in phagocytosis and melanin production through prophenoloxidase (proPO) system (Rodrıguez and Moullac 2000). Reactive oxygen species (ROS) are produced in healthy fauna in response to different environmental stimuli (hypoxia, elevated temperature and xenobiotics) leading to oxidative stress, cellular damage and death. ROS include hydroxyl radicals including superoxide anion (SOA), hydrogen peroxide (H_2O_2), hydroxide ion and oxygen (Beaulieu and Costantini 2014). Different antioxidants including CAT, SOD, glutathione and different peroxidases prevent the oxidation of non-self-antigens (Munoz et al. 2000). SOD play major defensive role by catalysing the breakdown of SOA to H_2O_2 and O_2. On the other hand, CAT results in the conversion of H_2O_2 to H_2O and O_2. Fortuitously, internal mechanisms of cells check ROS levels to maintain normal physiological activities (Karplus 2015). Glutathione peroxidase (GPx) is involved in the reduction ofH_2O_2 and lipid peroxide through its catalytic actions, and its ability to protect the cell against LPO can be assured (Stara et al. 2012). Furthermore, the decline in malondialdehyde (MDA) shows rise in antioxidants, therefore acts as direct evidence of toxicity posed by free radicals (Hegazi et al. 2010).

In addition, many medicinal have antifungal and antibacterial properties. For example, *O. basilicum*'s extract has an ability to control infectious diseases caused by *Fusarium oxyspoum* and *Aspergillus flavus* (Adigüzel et al. 2005). Poongodi et al. (2012) observed improvement in survivability (82 and 70%) of *M. rosenbergii* fed with 1% of garlic and 1% turmeric powder, respectively. Moreover, herbs have a potential to act as antiviral against shrimp and fish viruses. For instance, Diab et al. (2023) studied the increase in the tolerance of *Oreochromis niloticus* (Nile tilapia) against *Aeromonas hydrophila* after incorporation of methanolic extract of fenugreek. A study conducted by Balasubramanian et al. (2008) also showed antiviral activity of plant extract of *C. dactylon* against WSSV in *P. monodon*, where no mortality occurred in *P. monodon* incorporated with artificial feed having 2% *C. dactylon* extract. Similarly, Citarasu et al. (2003) examined the effects of mentholic extracts of different herbs (*Psoralea corylifolia, S. trilobatum* and *A. paniculata*) encapsulated in Artemia on post larvae of *P. monodon* aquaculture. The results showed that *P. corylifolia* resulted in comparatively major decline in bacterial load; and higher specific growth rate and survival rate than other experimental herbs. In crustaceans, immunostimulants stimulate the innate immune system by elevating the number of macrophages and phagocytes, and increasing the activity of granulocytes phagocytes (Yin et al. 2009).

The immunomodulatory action of herbal extract of *Hygrophila spinosa* supplemented diet (0.01, 0.1 and 1.0 g) on *M. malcolmsonii* against *Vibrio alginolyticus* was investigated by Devi et al. (2019). The activity of proPO and THC in prawns supplemented with 1.0 g *H. spinosa* supplemented diet showed significant increase compared to 0.01 and 0.1 g between 1 and 4 weeks. SOD and respiratory burst showed significant increase when fed 0.1 and 1.0 g herbal extract compared to 0.01 g. In addition, phagocytic activity showed significant increase at 1.0 gin comparison to 0.01 g extract enriched diet. Furthermore, the mortality of prawns was lower when fed 0.1 and 1.0 g extract compared to 0.01 g herbal extract. The effects of different concentrations of emodin (0, 25 and 50 mg/Kg) on antioxidation property, intestinal histology, biochemical constituents and gene expression of heat shock proteins of *M. rosenbergii* were examined by Tadese et al. (2022a) for 8 weeks. The results showed significant decrease in total cholesterol, SOD, triglycerides and alanine aminotransferase (ALT) whereas significant increase was observed in glucose, SOA and MDA followed emodin supplementation. In addition, the gene expression of HSP90 and HSP60 declined compared to control group. Furthermore, intestinal epithelial cells were well-organised, tightly affixed and complete in emodin remedy in comparison to slightly disorganised basement membrane in control.

28.5.3 Medicinal Plants as Feeding Attractants in *M. rosenbergii* Aquaculture

Medicinal plants as feed additives also act as feeding attractants which further promotes growth performance. Kawamura et al. (2019) assessed the feeding attractiveness of about 21 different herbs at different stages of *M. rosenbergii* and *L. vannamei* by studying behavioural changes. *Mentha piperita* (peppermint) and *Houttuynia cordata* (dokudami) were observed as feed attractants for all stages of *M.*

rosenbergii especially post larvae and juveniles whereas *A. sativum* (garlic) acted as attractant for juveniles of prawn. On the other hand, *Capsicum annuum* (red chilli) acted as best feeding attractant for *L. vannamei* while *M. piperita* and *Z. officinale* were observed as moderate feed attractants for juveniles and adults. Moreover, the results indicated that the addition of higher concentration of herbal extracts have negative effects on feeding because of the occurrence of feeding deterrents including saponins in herbs leading to decline in food palatability in living organisms. Saponins in plants are basically triterpene glycosides or steroids which cause lower food palatability in fishes, crustaceans and insects; and have toxic impacts on aquatic organisms due to alterations in respiratory epithelium (Makkar et al. 2007). Moreover, different herbs (*C. quadrangularis*, *Alteranthera sessilis* and *E. alba*) acted as appetizers and promoted growth in post larvae of *M. rosenbergii* after 90 days (Radhakrishnan et al. 2014). The researchers observed significant rise in specific growth rate, survival rate, protein efficiency ratio and weight gain in herb fed prawns. Additionally, the studied herbs significantly increased the activities of lipase, amylase and protease, which further significantly increased energy utilisation and biochemical constituents. Among three tested herbs, *C. quadrangularis* resulted in better growth, survival and production followed by other herbs including *E. alba* and *A. sessilis.*

Likewise, the individual effects of feeding attractants viz. *Zingiber officinale* (dried zinger), *Piper nigram* (black pepper) and *Piper longum* (long pepper) on biochemical constituents, growth, survival and digestive enzymes of post larvae of *M. rosenbergii* for 60 days were studied by Bhavan et al. (2013b). The results showed that survival, protein levels, mineral salts (potassium and sodium), activities of digestive enzymes (lipase, protease and amylase), growth (food conversion ratio, specific growth rate and weight gain), vitamins E and C significantly ($P < 0.05$) in post larvae fed with herbal feed in the order: *P. longum*, *P. nigram* and *Z. officinale.* In addition, comparatively a greater number of Coomassie blue stained protein bands in herbal-treated post larvae compared to control group suggested their effect on protein synthesis. The results showed that tested herbs acted as appetizers, increased the release of digestive enzymes, absorption of nutrients and hence enabled efficient digestion, health, survival and growth of post larvae of *M. rosenbergii.* Likewise, Rebecca and Bhavan (2014) also observed significant increase in growth of *M. rosenbergii* post larvae supplemented with different concentration (1, 3 and 5%) of *C. longa* (curcumin), *A. sativum* (garlic) and *Z. officinale* (ginger) for 90 days. The scientists suggested that medicinal plants could be utilised as safer and cheaper alternatives against antibiotics and synthetic hormones for survival and growth performance.

28.5.4 Medicinal Plants as Anti-Stressor Against Environmental Factors in M. rosenbergii Aquaculture

Success of prawn culture depend on various physicochemical-parameters including sunlight, water temperature, rate of evaporation, soil quality, pH, land elevation, water quality, meteorological variables such as rainfall, wind speed, transportation and marketing facilities (Banerjee et al. 2024). Alteration in these parameters could lead to negative impacts on growth and survival of prawn species. However, medicinal plants incorporated in feed antagonise stressful environmental conditions by inhibiting oxygen anions generation and scavenging free radicals. Moreover, secondary metabolites significantly reduce stress and improve immune system by modifying, catalysing and binding DNA and proteins (Tadese et al. 2022b). El-Desouky et al. (2012) also investigated the effects of 2% *C. dactylon* and 1.5% *Zingiber officinalis* on 9.5 pH levels in juveniles of *M. rosenbergii* for 5 weeks. The results exhibited significantly more length gain rate, feed conversion ratio, average body length, average body weight, condition factor, specific growth rate and weight gain rate in herbs incorporated groups in comparison to control group. Additionally, significant ($p < 0.05$) increase was recorded in haemocyte viability, phagocytic index (PI), phagocytic rate, DHC and THC in groups exposed to high pH and medicinal plants compared to control group after 4 and 5 weeks. No alterations were observed in histology of stomach and hepatopancreas of herb-exposed juveniles. The results highlighted that medicinal plants could control stress induced by alterations in pH levels in *M. rosenbergii*. Likewise, a study conducted by Liu et al. (2010) also studied the improvement in protein levels and the activity of ALP (haemolymph), lysozyme (haemolymph), CAT (liver) and SOD (liver) in *M. rosenbergii* treated with 0.1 and 0.2% anthraquinone extracted from *Rheum officinale* Bail. The results also highlighted that anthraquinone extract promote growth and prevent stress induced by high temperature conditions.

28.5.5 The Role of Medicinal Plants on Pigmentation in M. rosenbergii Aquaculture

The colour of crustaceans is essential in controlling their market price and quality. However, they cannot produce pigments internally and rely on diet for pigmentation, where β-carotene and astaxanthin are main pigments that determine their body colour. Carotenoids from natural or artificial sources are generally utilised as colourants in feeds, however, synthetic carotenoids are expensive and complex (Ruangdej and Laohavisuti 2014; Ashokkumar et al. 2023). Culture of *M. rosenbergii* contributes around 5% of global prawn and shrimp aquaculture (Food and Agriculture Organization (FAO) 2022). Mostly, prawns are sold live, so the colour after cooking is essential (Yang et al. 2012). The colour during cooking prawn could be increased using synthetic astaxanthin (200mgkg^{-1}), however it needs to be imported from different countries (Kumar et al. 2009; Wade et al. 2015). Therefore, natural local products having carotenoids play crucial role in pigmentation of *M. rosenbergii.*

Chilli pepper has distinctive flavours, nutritional value (vitamins A and C) and carotenoids which make them desirable in medical industries (Mohd Hassan et al. 2019; Dave et al. 2020). Aquatic organisms particularly crab and shrimp have the ability of conversion of ß-carotene from consumed carotenoids into astaxanthin (3,3'-dihydroxy-ß, ß-carotene-4,4'-dione) by binding of formed compounds such as carotenoproteins and crustacyanin (Benjakul 2011; Parisenti et al. 2011). Wongphonprateep and Kasirak (2023) studied the effect of crude extract from different concentrations (0, 25 and 50 $mgkg^{-1}$) of chilli peppers viz. *C. frutescens* (Guinea pepper), *C. annuum* (red chilli pepper) and Korean chilli on *M. rosenbergii* for 60 days. The study showed non-significant alterations in growth, feed utilisation and survival among experimental groups feed with 0 and 25 mg/kg of chilli peppers. However, freshwater prawns fed 50 mg/kg of extracts indicated storage of pigmentation in muscles and displayed ability to convert ß-carotene into xanthophyll derivative (470–500 nm) that accumulate in flesh of prawns. Summary of other studies showing notable beneficial roles of medicinal plants in *M. rosenbergii* are depicted in the Table 28.1.

28.6 General Mode of Action of Secondary Metabolites of Medicinal Plants

The antimicrobial, antibacterial and immunostimulatory effects of herbs due to the presence of secondary metabolites including flavonoids, saponins, terpenoids, steroids, polysaccharides, tannins, phenols, essential oils, alkaloids and pigments, and that confer them interesting immunostimulant and antimicrobial activities (Kumar et al. 2012; Tadese et al. 2022b), were extensively studied in shrimp and prawn culture (Chang et al. 2012). Flavonoids, secondary metabolites having polyphenolic structures, act as anti-bacterial agents by destroying their cell membrane, respiratory chain and impeding synthesis of nucleic acids (Yuan et al. 2021). They also act as biofilm inhibiters of foodborne pathogens like *Vibrio parahaemolyticus* (Faleye et al. 2023). They also impair iron acquisition mechanism which is required by *Vibrio* for survival, redox processes, growth and virulence activity (Kustusch et al. 2011). In addition, evidences suggest that tannins form irreversible complexes with proline-rich proteins which further inhibits bacterial cell wall formation. Similarly, the cell wall is devastated upon terpenoids exposure (Kothari et al. 2011). Similarly, saponins display different anti-bacterial actions by binding to their cholesterol membranes leading to lysis of cells (Arabski et al. 2012). Alkaloids also alter the membrane permeability and decline metabolism, nucleic acid synthesis, protein content and wall synthesis of bacteria. Along with this, vitamin C in medicinal plants is responsible for collagen formation, iron absorption, collagen formation, activity of digestive enzymes, antioxidation and metabolism of cholesterol, tyrosine

Table 28.1 A summary of beneficial roles of different medicinal herbs in *M. rosenbergii*

Medicinal plants	Dose	Duration	Medicinal roles	References
Diplazium esculentum (fiddlehead ferns)	0.5, 1, 2, 4 and 10 g/kg methanolic extract	60 days	• Increase in survival, moulting rate and moulting frequency	Hajar-Azira et al. (2023)
A. sativum (garlic)	0, 5, 10 and 20 g/kg (w/w) water extract	60 days	• Significant increase in feed conversion ratio, specific growth rate and weight gain • Activities of GPx, CAT and SOD significantly increased in hepatopancreas • Activities of AST and ALT, and content of proteins and albumin significantly increased in haemolymph	Jahanbakhshi et al. (2022)
Musa sp. (banana)	10 and 20 g/kg blossom powder (BP) and hot-blossom extract (HPE)	56 days	• Significant increase in percentage length gain and weight gain • Both diets displayed higher survival rate, phagocytic activity, clearance efficiency against *L. garvieae*; and significant increase in respiratory bursts and PO activity • Anti-hypothermal (18 °C) stress was observed in prawns fed both diets • At 18 °C, norepinephrine was significantly lesser in haemolymph of HPE fed prawns compared to 28 °C	Mapanao et al. (2022)
Musa sp. (banana)	6 µg/g 50% v/v of methanol (100 °C) peel extract	6 days	• 82% phagocytic activity increased, whereas the cumulative mortality declined to 20% after 6 days	Naksing et al. (2022)
Melaleuca alternifolia (tea tree)	0 and 100 mg/Kg oil	–	• Increase in growth and survival rate and activities of T-AOC, SOD and CAT • Decrease in vacuolation, lipids droplets (hepatopancreas) and hemolymph lipids • Decrease in ALT, AST and MDA	Liu et al. (2022)

(continued)

Table 28.1 (continued)

Medicinal plants	Dose	Duration	Medicinal roles	References
Clostridium butyricum (probiotic butyric acid bacterium)	0, 125, 250, 500 and 1000 mg/kg *C. butyricum* against ammonia stress	8 weeks	• Increase in SOD, MDA and SOA • Increase in the expression of HSPs (HSP60, HSP70 and HSP90)	Sun et al. (2022)
Salvia rosmarinus (rosemary)	0, 2, 5 and 15% extract	90 days	• Decrease in glutathione levels whereas significant increase was observed in LPO, GR and GST	Leite et al. (2021)
Isaria cicadae Miquel (cicada)	0, 100, 200, 400 and 800 mg/Kg	9 weeks	• Significantly increase in specific growth ratio and protein efficiency ratio, whereas feed conversion ratio and feed intake decreased • Significant increase in albumin, urea nitrogen, globulin, ALT and protein • Significant alterations in oxidative enzymes	Yan et al. (2021)
Emodin and *C. butyricum* (probiotic butyric acid bacterium)	0, 25 and 50 mg/kg emodin and 0, 250 and 500 mg/kg *C. butyricum*	8 weeks	• Significant increase in proteins, AST and mRNA gene expression • Contents of Interleukin-1, interferon-γ, Interleukin-6 and tumour necrotic factor-α increased	Tadese et al. (2020)
Lippia alba (bushy matgrass)	0, 1 and 2 ml/kg oil	60 days	• At 1 ml/kg, glutathione-*S*-transferase (GST) and GPx were lesser compared to control • At 2 ml/kg, antioxidant enzymes increased compared to control • Both doses improved growth, survival and decreased LPO	Cagol et al. (2020)
Psidium guajava (guava)	0.5 and 1.0% leaf extract	15 days	• Significantly reduced total bacteria accumulation • Increased storage to 10 days compared to 7 days in untreated group	Karim et al. (2018)

(continued)

Table 28.1 (continued)

Medicinal plants	Dose	Duration	Medicinal roles	References
Scutellaria radix	1, 5 and 10 μg/μl *S. radix* water extract	3, 6, 12and 24 h	• At 5 μg/μl, PO significantly increased • Duration-dependent increase in proPO at all concentrations • Highest survival rate observed at 5 μg/μl against *L. garvieae*	Andriawan et al. (2018)
Morindacitrifolia (noni)	0.6, 4 and 6 g/kg *M. citrifolia* leaves extract	7, 21, 35, 49 and 63 days	• After 63 days, significant increase in the activity of GPx • Upregulation of gene expressions of α2-macroglubulin	Halim and Prajitno (2017)
Ananas comosus (pineapple)	0, 3, 6, 9, 12, 15 and 18 mg/kg dry feed manganese-oxide nanoparticles (Mn-oxide NPs) using *A. comosus* peel extract	90 days	• Significant increase in weight, growth, muscle biochemical compositions, digestive enzymes activity and weight gain • Significant difference in feed conversion ratio • No significant alterations in oxidative enzymes (CAT and SOD) in hepatopancreas and muscle	Asaikkutti et al. (2016b)

and tryptophan which further promotes growth performance and maintains physiological functions (Asaikkutti et al. 2018). Medicinal plant extracts also have an ability to reverse resistance developed against anti-bacterial drugs (Yan et al. 2021). Furthermore, they act as anti-viral agents by stimulating body's innate immune system, inactivating viral particles and limiting their replication (Howlader et al. 2020).

28.7 Limitations in the Application of Medicinal Plants

There are few limitations in using herbs as they are required to be added in right concentrations due to their dose-dependence which significantly activate the immune response (Bulfon et al. 2015). Many studied indicated the incorporation of 7% herbal powder or extracts for *M. rosenbergii* post larvae (Rebecca 2011). Studies reported that utilising different herbal extracts at higher levels might affect animals due to the existence of deterrents including saponins (Kawamura et al. 2019). Additionally, medicinal feed supplements are species-specific, so further research on their dose, extraction methods, exposure duration, combinations and toxicological effects is necessary. The current understanding regarding the mode of action of medicinal plants in prawns is also scarce and requires further study. Further, few medicinal plants are endangered or threatened, so their sustainable utilisation is required to conserve biodiversity (Janke 2018). The overproduction of medicinal plants can also lead to rapid deforestation and destruction of forest species (Bodeker et al. 1997). Therefore, government organisations should prioritise on enticing private companies to cultivate medicinal plants without compromising forest health by offering incentives and constructive support.

28.8 Conclusion

Medicinal plants and their extracts (aqueous or non-aqueous extract) have an ability to improve growth performance, immune parameters, oxidative parameters, appetite and survival of prawns. The literature highlighted different infectious agents (bacteria, virus and fungi) and environmental stressors (water temperature, pH, salinity, ammonia and C/N ratio) impacting survival, growth and heath of prawns. Additionally, potential mode of action and beneficial roles of medicinal plants including growth performance, proper pigmentation, upregulating gene expression, appetite promotion, immunostimulatory effects and feed attraction in prawns especially *M. rosenbergii* were also enlisted.

28.9 Future Perspectives

Future research should focus on the long-term studies to evaluate the safety of medicinal plants. In addition, field trials can also be done to validate the effectiveness of medicinal plants. Furthermore, studies should focus on the mechanisms by which medicinal plants exert immunostimulatory, antimicrobial, anti-stressor or growth-promoting effects in prawns.

References

Adigüzel A, Güllüce M, Şengül M, Öğütcü H, Şahin F, Karaman I (2005) Antimicrobial effects of *Ocimum basilicum* (Labiatae) extract. Turk J Biol 29(3):155–160

Ahmed A, Lodhi S (2024) Effect of heavy metals, cadmium and copper on freshwater prawn, *Macrobrachium lamarreilamarrei* (Crustacea-Decapoda). Int J Entomol Res 9(5):32–36

Amiruddin WM, Sukri SAM, Al-Amsyar SM, Rusli ND, Mat KB, Mohd M, Harun HC (2021) Application of herbal plants in giant freshwater prawn: a review on its opportunities and limitation. In: IOP conference series: earth and environmental science, vol. 756(1). IOP Publishing, p 012022

Andriawan S, Bao HT, Cheng PT, Yuniarti A (2018) Investigation of *Scutellaria radix* on phenoloxidase and prophenoloxidase of giant freshwater prawn, *Macrobrachium rosenbergii*, via reverse-gavage feeding. Res J Life Sci 5(3):195–205

Arabski M, Węgierek-Ciuk A, Czerwonka G, Lankoff A, Kaca W (2012) Effects of saponins against clinical *E. coli* strains and eukaryotic cell line. BioMed Res Int 2012(1):286216

Arockiaraj J, Easwvaran S, Vanaraja P, Singh A, Othman RY, Bhassu S (2012) Molecular cloning, characterization and gene expression of an antioxidant enzyme catalase (MrCat) from *Macrobrachium rosenbergii*. Fish Shellfish Immunol 32(5):670–682

Asaduzzaman M, Rahman MM, Azim ME, Islam MA, Wahab MA, Verdegem MCJ, Verreth JAJ (2010) Effects of C/N ratio and substrate addition on natural food communities in freshwater prawn monoculture ponds. Aquaculture 306(1–4):127–136

Asaikkutti A, Bhavan PS, Vimala K, Karthik M, Cheruparambath P (2016a) Effect of different levels dietary vitamin C on growth performance, muscle composition, antioxidant and enzyme activity of freshwater prawn, *Macrobrachium malcolmsonii*. Aquacult Rep 3:229–236

Asaikkutti A, Bhavan PS, Vimala K, Karthik M, Cheruparambath P (2016b) Dietary supplementation of green synthesized manganese-oxide nanoparticles and its effect on growth performance, muscle composition and digestive enzyme activities of the giant freshwater prawn *Macrobrachium rosenbergii*. J Trace Elem Med Biol 35:7–17

Asaikkutti A, Bhavan PS, Vimala K, Karthik M (2018) Effect of different levels of dietary vitamin C on growth performance, muscle composition, antioxidant and enzyme activity of *Macrobrachium rosenbergii*. Proc Natl Acad Sci India Sect B Biol Sci 88:477–486

Ashokkumar V, Flora G, Sevanan M, Sripriya R, Chen WH, Park JH, Kumar G (2023) Technological advances in the production of carotenoids and their applications–a critical review. Biores Technol 367:128215

Awad E, Awaad A (2017) Role of medicinal plants on growth performance and immune status in fish. Fish Shellfish Immunol 67:40–54

Bachère E (2000) Shrimp immunity and disease control. Aquaculture 191(1–3):3–11

Balasubramanian G, Sarathi M, Venkatesan C, Thomas J, Hameed AS (2008) Oral administration of antiviral plant extract of *Cynodon dactylon* on a large scale production against white spot syndrome virus (WSSV) in Penaeus monodon. Aquaculture 279(1–4):2–5

Banerjee K, Mahakur T, Bindhani PK, Khemundu GR (2024) Assessment of different grass species extracts as source of micronutrients in prawn *Macrobrachium rosenbergii* feeds. Aquaculture and Fisheries 9(6):937–947

Beaulieu M, Costantini D (2014) Biomarkers of oxidative status: missing tools in conservation physiology. Conserv Physiol 2(1):cou014

Beltrán JMG, Espinosa C, Guardiola FA, Esteban MÁ (2018) In vitro effects of *Origanum vulgare* leaf extracts on gilthead seabream (*Sparus aurata* L.) leucocytes, cytotoxic, bactericidal and antioxidant activities. Fish Shellfish Immunol 79:1–10

Benjakul S (2011) Chemical and quality of aquatic animal. Odeon store Part., Ltd. Bangkok, Thailand, p 376

Bermúdez-Almada MC, Espinosa-Plascencia A (2012) The use of antibiotics in shrimp farming. In: Health and environment in aquaculture, pp 199–214

Bhardwaj S, Borana K, Sujad N, Manohar S (2014) Effect of mercuric chloride on protein content of muscles and hepatopancreas of fresh water prawn '*Macrobrachiumlamarreilamarrei*'. Indo Am J Pharmaceut Res 4:2313–2315

Bhavan PS, Devi VG, Shanti R, Radhakrishnan S, Poongodi R (2010) Basic biochemical constituents and profiles of amino acids in the post larvae of *Macrobrachium rosenbergii* fed with Spirulina and yeast enriched Artemia. J Sci Res 2(3):539–549

Bhavan PS, Devi NN, Muralisankar T, Manickam N, Radhakrishnan S, Srinivasan V (2013a) Effects of *Myristica fragrans*, *Glycyrrhiza glabra* and *Quercus infectoria* on growth promotion in the prawn *Macrobrachium rosenbergii*. Int J Life Sci Biotech Pharm Res 2:169–182

Bhavan PS, Saranya C, Manickam N, Muralisankar T, Radhakrishnan S, Srinivasan V (2013b) Effects of *Piper longum*, *Piper nigram* and *Zingiber officinale* on survival, growth, activities of digestive enzymes and contents of total protein, vitamins and minerals in the freshwater prawn *Macrobrachium rosenbergii*. Elixir Bio Tech 58:14824–14828

Bhavan PS, Mohammedsiddiq S, Srinivasan V, Muralisankar T, Manickam N (2014) Effects of seeds of medicinal plants, *Syzygium cumini*, *Phylanthus emblica*, *Azadirachta indica* and *Ricinus communis* on growth promotion in *Macrobrachium malcolmsonii* early juveniles. Int J Res Stud Biosci 2:95–106

Binh VN, Dang N, Anh NTK, Thai PK (2018) Antibiotics in the aquatic environment of Vietnam: sources, concentrations, risk and control strategy. Chemosphere 197:438–450

Bir J, Sarker H, Mita FS, Noor MI, Kumar R, Islam SS, Das M, Huq KA (2024) The impact of salinity and temperature stress on survival, behaviour, immune responseand proximate composition of giant freshwater prawn *Macrobrachium rosenbergii*. Aquacult Int 1–20

Bodeker G, Ghat KKS, Burley J, Vantomme P (eds.) (1997) Medicinal plants for forest conservation and health care. 11:158

Boyd CE, Tucker CS (2012) Pond aquaculture water quality management. Springer Science and Business Media

Bricknell I, Dalmo RA (2005) The use of immunostimulants in fish larval aquaculture. Fish Shellfish Immunol 19(5):457–472

Briggs MR (1991) The performance of juvenile prawns, *Macrobrachium rosenbergii*, fed a range of carbohydrate sources in semi-purified diets. J World Aquacul Soc 22:16A

Bulfon C, Volpatti D, Galeotti M (2015) Current research on the use of plant-derived products in farmed fish. Aquac Res 46(3):513–551

Cagol L, Baldisserotto B, Becker AG, Souza CDF, Heinzmann BM, Caron BO, Leone FA, dos Santos LD, Ballester ELC (2020) Essential oil of *Lippia alba* in the diet of *Macrobrachium rosenbergii*: Effects on antioxidant enzymes and growth parameters. Aquac Res 51(6):2243–2251

Cai J, Chan HL, Yan X, Leung P (2023) A global assessment of species diversification in aquaculture. Aquaculture 576:739837

Chaitanawisuti N, Santhaweesuk W, Wattayakorn G (2013) The combined effects of temperature and salinity on survival of postlarvae tiger prawn *Penaeus monodon* under laboratory conditions. Agric Sci 53–56

Chand BK, Trivedi RK, Dubey SK, Rout SK, Beg MM, Das UK (2015) Effect of salinity on survival and growth of giant freshwater prawn *Macrobrachium rosenbergii* (De Man). Aquacult Rep 2:26–33

Chang YP, Liu CH, Wu CC, Chiang CM, Lian JL, Hsieh SL (2012) Dietary administration of zingerone to enhance growth, non-specific immune response, and resistance to *Vibrio alginolyticus* in Pacific white shrimp (*Litopenaeusvannamei*) juveniles. Fish Shellfish Immunol 32(2):284–290

Chen B, Lin L, Fang L, Yang Y, Chen E, Yuan K, Zou S, Wang X, Luan T (2018) Complex pollution of antibiotic resistance genes due to beta-lactam and aminoglycoside use in aquaculture farming. Water Res 134:200–208

Chen-Fei L, Chou-Min C, Jiun-Yan L (2020) Feasibility of vaccination against *Macrobrachium rosenbergii* nodavirus infection in giant freshwater prawn. Fish Shellfish Immunol 104:431–438

Cheng W, Chen JC (1998) Isolation and characterization of an Enterococcus-like bacterium causing muscle necrosis and mortality in *Macrobrachium rosenbergii* in Taiwan. Dis Aquat Org 34(2):93–101

Cheng W, Chen JC (2000) Effects of pH, temperature and salinity on immune parameters of the freshwater prawn *Macrobrachium rosenbergii*. Fish Shellfish Immunol 10(4):387–391

Choi CY, Shin HS, Choi YJ, Kim NN, Lee J, Kil GS (2012) Effect of LED light spectra on starvation-induced oxidative stress in the cinnamon clownfish *Amphiprionmelanopus*. Comp Biochem Physiol A Mol Integr Physiol 163(3–4):357–363

Citarasu TK, Venkata Malingam MM, Babu R, Sekar RJR, Marian MP (2003) Influence of the antibacterial herbs, *Solanum trilobatum* and *Andrographis paniculata* on the bacterial load of *Penaeus monodon* post larvae. Aquacult Int 11:583–595

Conklin DE (1997) Vitamins: Crustacean nutrition D'Abramo LR, Conklin DE, Akiyama DM (eds). World Aquaculture Society, Baton Rouge, Louisiana, pp 123–149

Coyle SD, Tidwell JH, Yasharian DK, Caporelli A, Skudlarek NA (2005) The effect of biomass density, temperature, and substrate on transport survival of market-size freshwater prawn, *Macrobrachium rosenbergii*. J Appl Aquac 17(4):61–71

Dave D, Liu Y, Pohling J, Trenholm S, Murphy W (2020) Astaxanthin recovery from Atlantic shrimp (*Pandalus borealis*) processing materials. Bioresour Technol Rep 11:100535

Dawood MA, Koshio S, Abdel-Daim MM, Van Doan H (2019) Probiotic application for sustainable aquaculture. Rev Aquac 11(3):907–924

Desai J, Tank SK (2008) Effect of $HgCl_2$ on the protein content of prawn *Macrobrachium rosenbergii* (De Man). J Environ Res Dev 3:62–67

Devi G, Harikrishnan R, Paray BA, Al-Sadoon MK, Hoseinifar SH, Balasundaram C (2019) Effects of aloe-emodin on innate immunity, antioxidant and immune cytokines mechanisms in the head kidney leucocytes of *Labeo rohita* against *Aphanomyces invadans*. Fish Shellfish Immunol 87:669–678

Diab AM, Al-Khefa BT, Khalafallah MM, Salah AS, Farrag FA, Dawood MA (2023) Research article dietary methanolic extract of fenugreek enhanced the growth, haematobiochemical, immune responses, and resistance against *Aeromonas hydrophila* in Nile Tilapia, *Oreochromis niloticus*. Aquac Res 2023:1–13

Dinakaran GK, Soundarapandian P (2009) Biochemical status of edible palaemonid prawn *Macrobrachiumidellaidella* (Hilgendorf, 1898). Adv J Food Sci Technol 1(1):19–26

Direkbusarakom S (2004) Application of medicinal herbs to aquaculture in Asia. Walailak J Sci Technol (WJST) 1(1):7–14

El-Desouky H, El-Asely A, Shaheen AA, Abbass A (2012) Effects of *Zingiber officinalis* and *Cyanodon dactylon* on the growth performance and immune parameters of *Macrobrachium rosenbergii*. World J Fish Mar Sci 4(3):301–307

Elshafie MM, Nawar IA, Algamal MA, Ahmad SM (2012) Evaluation of the biological effects for adding cinnamon volatile oil and TBHQ as antioxidant on rats' lipid profiles. Asian J Plant Sci 11(3):100–108

Faleye OS, Lee JH, Lee J (2023) Selected flavonoids exhibit antibiofilm and antibacterial effects against *Vibrio* by disrupting membrane integrity, virulence and metabolic activities. Biofilm 6:100165

Farook MA, Vimal S, Madan N, Taju G, Abdul Majeed S, Nambi KS, Balasubramanian G, Sahul Hameed AS (2016) Immunomodulatory effect of *Cynodon dactylon* against white tail disease of giant freshwater prawn, *Macrobrachium rosenbergii* (De Man, 1879). Aquac Res 47(11):3421–3431

Farook MA, Mohamed HM, Tariq NPMM, Shariq KM, Ahmed IA (2019) Giant freshwater prawn, *Macrobrachium rosenbergii* (De Man, 1879): a review. Int J Res Analyt Rev 6(1):571–584

Fegan D (2004) Larval shrimp nutrition. In: Global aquaculture advocate. pp 64–66

Figueroa-Lucero G, Hernández-Rubio MC, Gutiérrez-ladrón de guevara MDJ (2012) Acute toxicity of ammonia on *Macrobrachiumtenellum* (Smith) larvae. Revistainternacional de Contaminaciónambiental 28(2):145–150

Food and Agriculture Organization (FAO) (2012) The state of world fisheries and aquaculture: yearbook of fishery statistics. Italy, Rome

Food and Agriculture Organization (FAO) (2018) The state of world fisheries and aquaculture 2018: meeting the sustainable growth. Italy, Rome

Food and Agriculture Organization (FAO) (2020) The state of world fisheries and aquaculture 2020: sustainability in action. Italy, Rome

Food and Agriculture Organization (FAO) (2022) The state of world fisheries and aquaculture 2022: towards blue transformation. Italy, Rome

Food and Agriculture Organization (FAO) (2004) Future prospects for fish and fishery products: mediumterm projections to the years 2010 and 2015. Italy, Rome

Food and Agriculture Organization (FAO) (2009) The state of world fisheries and aquaculture 2008. Italy, Rome, p. 196

Food and Agriculture Organization (FAO) (2010) Technical guidelines for responsible fisheries, vol. 5(4). Rome, Italy, pp 1020–5292

Gimenez AVF, Diaz AC, Velurtas SM, Fenucci JL (2009) Partial substitution of fishmeal by meat and bone meal, soybean meal, and squid concentrate in feeds for the prawn, *Artemesia longinaris*: Effect on digestive proteinases. Israeli J Aquacult-Bamidgeh 61(1):48–56

Glencross B (2006) The nutritional management of barramundi, *Lates calcarifer*–a review. Aquac Nutr 12(4):291–309

Guo JJG, Her BihYueh HB, Chou RueyLing CR, Chen TzyyIng CT (2011) Screening of modern herbal medicines in white shrimp (*Litopenaeusvannamei*) against *Vibrio harveyi* infection. Israeli J Aquacult Bamidgeh 63

Hajar-Azira Z, Aaqillah-Amr MA, Rasdi NW, Ma H, Ikhwanuddin M (2023) Preliminary investigation on the effect of fiddlehead fern, *Diplazium esculentum*, extract to the growth performance of giant freshwater prawn, *Macrobrachium rosenbergii*, postlarvae. Aquacult Int 31(1):81–101

Halim AM, Prajitno A (2017) Aqueous*Morindacitrifolia* leaves extract enhancing glutathione peroxidase activity and α2-macroglobulin gene expression on *Macrobrachium rosenbergii*. Res J Life Sci 4(1):25–33

Harikrishnan R, Jawahar S, Heo MS (2012) Effect of an *Andrographis paniculata* Enriched Diet on Immunomodulation of *Macrobrachiummalcolmsonii* against *Lactococcus garviae*. Israeli J Aquacult-Bamidgeh 64

Hegazi MM, Attia ZI, Ashour OA (2010) Oxidative stress and antioxidant enzymes in liver and white muscle of Nile tilapia juveniles in chronic ammonia exposure. Aquat Toxicol 99(2):118–125

Holthuis LB, Ng PKL (2010) Nomenclature and taxonomy. In: New MB, Valenti WC, Tidwell JH, D'Abramo LR, Kutty (eds) Freshwater Prawns: biology and farming. Wiley-Blackwell, Oxford, pp 12–17

Hossain MY, Ohtomi J, Jaman A, Jasmine S, Vadas RL Jr (2012) Life history traits of the monsoon river prawn *Macrobrachiummalcolmsonii* (Milne-Edwards, 1844) (Palaemonidae) in the Ganges (Padma) river, northwestern Bangladesh. J Freshw Ecol 27(1):131–142

Houssou AM, Lokonon TF, Koussovi G, Adjahouinou DC, Cocan D, Montchowui E (2024) Biological effects of glyphosate and emamectinbenzoate based pesticides on the freshwater prawn *Macrobrachiummacrobrachion* juveniles. Water Air Soil Pollut 235(12):1–14

Howlader P, Ghosh AK, Islam SS, Bir J, Banu GR (2020) Antiviral activity of *Cynodon dactylon* on white spot syndrome virus (WSSV)-infected shrimp: an attempt to mitigate risk in shrimp farming. Aquacult Int 28:1725–1738

Hsu JP, Liu CI (1994) Studies on yeast infection in cultured giant freshwater prawn (*Macrobrachium rosenbergii*). Fish Dis Res 15:55

Huang YH, Zhang M, Li YM, Wu DL, Liu ZQ, Jiang QC, Zhao YL (2019) Effects of salinity acclimation on the growth performance, osmoregulation and energy metabolism of the oriental river prawn, *Macrobrachiumnipponense* (De Haan). Aquac Res 50(2):685–693

Jahanbakhshi A, Pourmozaffar S, Adeshina I, Vayghan AH, Reverter M (2022) Effect of garlic (*Allium sativum*) extract on growth, enzymological and biochemical responses and immune-related gene expressions in giant freshwater prawn (*Macrobrachium rosenbergii*). J Anim Physiol Anim Nutr 106(4):947–956

Janke T (2018) From smokebush to spinifex: Towards recognition of Indigenous knowledge in the commercialisation of plants. Int J Reg Rural Remote Law Policy 1:1–27

Jayamanne SC (1986) A preliminary study of hydrogen sulphide toxicity on juveniles of *Macrobrachium rosenbergii* (De Man). NACA/WP 42:20

Johansson MW, Söderhäll K (1989) A cell adhesion factor from crayfish haemocytes has degranulating activity towards crayfish granular cells. Insect Biochem 19(2):183–190

Kahl R, Kappus H (1993) Toxicology of the synthetic antioxidants BHA and BHT in comparison with the natural antioxidant vitamin E. Zeitschrift Für Lebensmittel-Untersuchung und Forschung 196:329–338

Kanagarasu R, Bhavan PS, Rajkumar G, Nathiya V, Satgurunathan T, Manjula T (2017) Phytochemical characterization of *Alternanthera sessilis* and assessment of its growth promoting potential on the freshwater prawn *Macrobrachium rosenbergii*. Int J Res Stud Zool 3(4):25–38

Kang J, Matsuda O, Imamura N (1995) Effects of hypoxia and hydrogen sulfide on survival of the prawn *Macrobrachiumnipponense* in Lake Kojima, Japan. Nippon SuisanGakkaishi 61(6):821–826

Karim NU, Izzah Rakinah AR, Mohamad Khairi MZ, Nurul Afza Karim NAK, Mhd Ikhwanuddin MI (2018) Effects of guava, *Psidium guajava* leaves extract coating on giant freshwater prawn, *Macrobrachium rosenbergii* during chilled storage. J Sustain Sci Manag 13(1):159–167

Karplus PA (2015) A primer on peroxiredoxin biochemistry. Free Radical Biol Med 80:183–190

Kawamura G, Yong ASK, Au HL, Doison A, Ooi SY, Lim LSLS (2019) Malaysian herbs as feeding attractants and enhancers for the giant freshwater prawn (*Macrobrachium rosenbergii*) and the whiteleg shrimp (*Litopenaeusvannamei*). Borneo J Mar Sci Aquacult (BJoMSA) 3(2):57–67

Kiran, Kumar N, Sindhu S, Gawaria B, Ansari A, Dhariwal A, Sharma P, Kumar R (2025) GWAS approaches in medicinal plants for enhanced secondary metabolite production. In: Chen JT (ed) Biotechnology, multiple omics, and precision breeding in medicinal plants. CRC Press, pp 119–134. https://doi.org/10.1201/9781003475491-10

Kothari V, Naraniwal M, Gupta A (2011) Effect of certain phytochemicals on *Aeromonas hydrophila*. Res Biotechnol 2(4):20–25

Kumar R, Singh D (2022) Insecticide-tolerant bacterial population in *Eisenia fetida*'s Gut and Vermicast exposed to Chlorantraniliprole and Fipronil. Appl Biol Res 24(3):273–279. https://doi.org/10.5958/0974-4517.2022.00033.7

Kumar GS, Nayaka H, Dharmesh SM, Salimath PV (2006) Free and bound phenolic antioxidants in amla (*Emblica officinalis*) and turmeric (*Curcuma longa*). J Food Compos Anal 19(5):446–452

Kumar V, Pillai BR, Sahoo PK, Mohanty J, Mohanty S (2009) Effect of dietary astaxanthin on growth and immune response of Giant freshwater prawn *Macrobrachium rosenbergii* (De Man). Asian Fish Sci 22(1):61–69

Kumar JK, Devi-Prasad AG, Richard SA (2012) In vitro antioxidant activity and preliminary phytochemical analysis of medicinal legumes. J Pharm Res 5:3059–3062

Kumar R, Gupta R, Dahiya T, Yadav R, Saifi R, Yodha K, Kumari B, Singh P, Shefali (2022a) Surveillance of *Heteropneustes fossilis* (Stinging Catfish) diseases and health management practices in Haryana. Appl Ecol Environ Sci 10(4):261–267. https://doi.org/10.12691/aees-10-4-12

Kumar R, Gupta RK, Yadav R, Saifi R, Yodha K, Pooja (2022b) Eisenia fetida as protein source for growth enhancement of Heteropneustes fossilis. Egypt J Aquat Biol Fisher 26(2):577–588. https://doi.org/10.21608/ejabf.2022.234553

Kumar R, Yadav R, Kumar Gupta R, Pooja (2022c) A sustainable way for fish health management by replacement of chemical and drugs by earthworm. In: Environmental degradation in Asia: land degradation, environmental contamination, and human activities. Springer International Publishing, Cham, pp 329–352. https://doi.org/10.1007/978-3-031-12112-8_16

Kumar R, Yadav R, Gupta RK, Pal A, Yodha K, Kumar A (2024) Antioxidant enzyme activities and markers of oxidative stress in the life cycle of different Earthworm species. Indian J Exp Biol 62(07):578–583. https://doi.org/10.56042/ijeb.v62i07.12084

Kumar R, Singh D, Yadav R (2025) Toxic effects of chlorantraniliprole and fipronil on *Eisenia fetida*: from growth and reproductive parameters to avoidance behavior. Environ Eng Manag J 24(7):1499–1508. https://doi.org/10.30638/eemj.2025.117

Kustusch RJ, Kuehl CJ, Crosa JH (2011) Power plays: iron transport and energy transduction in pathogenic vibrios. Biometals 24:559–566

Lalitha KV, Surendran PK (2004) Bacterial microflora associated with farmed freshwater prawn *Macrobrachium rosenbergii* (De Man) and the aquaculture environment. Aquac Res 35(7):629–635

Langer S, Bakhtiyar Y, Lakhnotra R (2011) Replacement of fishmeal with locally available ingredients in diet composition of *Macrobrachiumdayanum*. Afr J Agric Res 6(5):1080–1084

Lardner A (2001) The effects of extracellular pH on immune function. J Leukoc Biol 69(4):522–530

Leite K, Kurosaki JKDAR, Retcheski MC, Tormen L, Romão S, Pinto VZ, Cazarolli LH (2021) Effect of rosemary (*Rosmarinus officinalis*) extract on the antioxidant status and proximate composition of prawn (*Macrobrachium rosenbergii*) meat. J Aquat Food Prod Technol 30(6):683–693

Liew HJ, Rahmah S, Tang PW, Waiho K, Fazhan H, Rasdi NW, Hamin SIA, Mazelan S, Muda S, Lim LS, Chen YM (2022) Low water pH depressed growth and early development of giant freshwater prawn *Macrobrachium rosenbergii* larvae. Heliyon 8(7):e09989

Lim KC, Yusoff FM, Shariff M, Kamarudin MS (2018) Astaxanthin as feed supplement in aquatic animals. Rev Aquac 10(3):738–773

Lio-Pio GD, Inui Y (2010) Health management in aquaculture, 2nd edn. Southeast Asian Fisheries Development Center/Aquaculture Department (SEAFDEC/AQD), Tigbauan, Iloilo

Liu B, Xie J, Ge X, Xu P, Wang A, He Y, Zhou Q, Pan L, Chen R (2010) Effects of anthraquinone extract from *Rheum officinale* Bail on the growth performance and physiological responses of *Macrobrachium rosenbergii* under high temperature stress. Fish Shellfish Immunol 29(1):49–57

Liu J, Gao S, Dong Y, Lu C, Liu Y (2020) Isolation and characterization of bacteriophages against virulent *Aeromonas hydrophila*. BMC Microbiol 20:1–13

Liu M, Sun C, Zheng X, Zhou Q, Liu B, Zhou Y, Xu P, Liu B (2022) Comparative proteomic analysis revealed the mechanism of tea tree oil targeting lipid metabolism and antioxidant system to protect hepatopancreatic health in *Macrobrachium rosenbergii*. Front Immunol 13:906435

Maji SK (2024) Hydrogen sulphide effect in shrimp culture. Chron Aquat Sci 1(10):130–133

Makkar HPS, Francis G, Becker K (2007) Bioactivity of phytochemicals in some lesser-known plants and their effects and potential applications in livestock and aquaculture production systems. Animal 1(9):1371–1391

Malkinson AM (1999) Lung tumorpromation by BHT. University of Colorado, Denver, Crisp Data Base National Institutes of Health

Mallasen M, Valenti WC (2005) Larval development of the giant river prawn *Macrobrachium rosenbergii* at different ammonia concentrations and pH values. J World Aquacult Soc 36(1):32–41

Mapanao R, Rangabpai T, Lee YR, Kuo HW, Cheng W (2022) The effect of banana blossom on growth performance, immune response, disease resistance, and anti-hypothermal stress of *Macrobrachium rosenbergii*. Fish Shellfish Immunol 124:82–91

Mariappan P, Balamurugan P, Balasundaram C (2003) Freshwater prawn, *Macrobrachiumnobilii* a promising candidate for rural nutrition. Curr Sci 8:13–14

Martinez JL (2009) Environmental pollution by antibiotics and by antibiotic resistance determinants. Environ Pollut 157(11):2893–2902

Mather PB, De Bruyn M (2003) Genetic diversity in wild stocks of the giant freshwater prawn (*Macrobrachium rosenbergii*): implications for aquaculture and conservation. World Fish Center Quart 26(4):4–7

Meyer O, Hansen E (1980) Behavioural and developmental effects of butylated hydroxytoluene dosed to rats in utero and in the lactation period. Toxicology 16(3):247–258

Mitra G, Mukhopadhyay PK, Chattopadhyay DN (2005) Nutrition and feeding in freshwater prawn (Macrobrachium rosenbergii) farming. Aqua Feeds Formul Beyond 2(1):17–19

Mohd Hassan N, Yusof NA, Yahaya AF, Mohd Rozali NN, Othman R (2019) Carotenoids of capsicum fruits: pigment profile and health-promoting functional attributes. Antioxidants 8(10):469

Moretti VM, Mentasti T, Bellagamba F, Luzzana U, Caprino F, Turchini GM, Giani I, Valfrè F (2006) Determination of astaxanthin stereoisomers and colour attributes in flesh of rainbow trout (*Oncorhynchus mykiss*) as a tool to distinguish the dietary pigmentation source. Food Addit Contam 23(11):1056–1063

Mugnier C, Zipper E, Goarant C, Lemonnier H (2008) Combined effect of exposure to ammonia and hypoxia on the blue shrimp *Litopenaeusstylirostris* survival and physiological response in relation to molt stage. Aquaculture 274(2–4):398–407

Mukhopadhyay PK, Rangacharyulu PV, Mitra G, Jana BB (2003) Applied nutrition in freshwater prawn, *Macrobrachium rosenbergii*, culture. J Appl Aquac 13(3–4):317–340

Mulyasari, Setijaningsih L, Samsudin R, Taufik I, Kusmini II (2022) The use of local raw materials and fermented feeds for the growth of giant prawns (*Macrobrachium rosenbergii*). IOP Conf Ser Earth Environ Sci 1036:012075

Munoz M, Cedeno R, Rodríguez J, van der Knaap WP, Mialhe E, Bachere E (2000) Measurement of reactive oxygen intermediate production in haemocytes of the penaeid shrimp, *Penaeus vannamei*. Aquaculture 191(1–3):89–107

Muralisankar T, Bhavan PS, Radhakrishnan S, Santhanam P, Jayakumar R (2017) Growth performance, muscle biochemical constituents, amino acid and fatty acid compositions of the giant freshwater prawn, *Macrobrachium rosenbergii*, fed with herb-incorporated diet. Aquac Nutr 23(4):766–776

Muralisankar T, Bhavan PS, Radhakrishnan S, Santhanam P (2018) Dietary supplement of medicinal herbal leaf powder on growth performance, digestive enzymes activities, energy utilization and vitamin levels of the freshwater prawn *Macrobrachium rosenbergii*. In: Proceedings of the zoological society, vol 71. Springer, India, pp 265–271

Mustafa G, Arif R, Atta A, Sharif S, Jamil A (2017) Bioactive compounds from medicinal plants and their importance in drug discovery in Pakistan. Matrix Sci. Pharma 1(1):17–26

Musthafa MS, Ali ARJ, Ali ARH, Mohamed MJ, War M, Naveed MS, Al-Sadoon MK, Paray BA, Rani KU, Arockiaraj J, Balasundaram C (2016) Effect of Shilajit enriched diet on immunity, antioxidants, and disease resistance in *Macrobrachium rosenbergii* (De Man) against *Aeromonas hydrophila*. Fish Shellfish Immunol 57:293–300

Naksing T, Rattanavichai W, Teeka J, Kaewpa D, Borthong J, Areesirisuk A (2022) Biological activities and potential of organic banana (*Musa acuminata*) peel extract in enhancing the immunity of giant freshwater prawn (*Macrobrachium rosenbergii*). Aquac Res 53(7):2645–2656

New MB, Nair CM (2012) Global scale of freshwater prawn farming. Aquac Res 43(7):960–969

New MB (2000) History and global status of freshwater prawn farming. In: Freshwater prawn culture: the farming of Macrobrachium rosenbergii, pp 1–11

Ng WK, Koh CB, Sudesh K, Siti-Zahrah A (2009) Effects of dietary organic acids on growth, nutrient digestibility and gut microflora of red hybrid tilapia, *Oreochromis* sp., and subsequent survival during a challenge test with *Streptococcus agalactiae*. Aquacult Res 40(13):1490–1500

Nikoskelainen S, Bylund G, Lilius EM (2004) Effect of environmental temperature on rainbow trout (*Oncorhynchus mykiss*) innate immunity. Dev Comp Immunol 28(6):581–592

Nisha A, Rani S, Kumar S, Kumar R, Kumar A (2025) Recent technologies and strategies for bioremediation of toxic and hazardous waste. In: Dubey KK, Shah MP, Yadav A (eds) Challenges and sustainable solutions in bioremediation. CRC Press, pp 1–16. https://doi.org/10.1201/978 1003407317-1

Niu J, Chen X, Lu X, Jiang SG, Lin HZ, Liu YJ, Huang Z, Wang J, Wang Y, Tian LX (2015) Effects of different levels of dietary wakame (*Undaria pinnatifida*) on growth, immunity and intestinal structure of juvenile *Penaeus monodon*. Aquaculture 435:78–85

Onunkwor BO, Oguntade R, Babayemi DO, Ademolu KO, Dele-OshinbanJo JS (2022) Toxicology impacts of heavy metal pollution of Ogun and Lagos fishing waters by sewages on the antioxidant status and nutritional value of prawn (*Macrobrachiummacrobrachion*). Niger J Anim Prod 49(1):194–214

Pandey G, Sharma M, Mandloi AK (2012) Immunostimulant effect of medicinal plants on fish. Int Res J Pharm 3(3):112–114

Parameswaran S (1994) A freshwater prawns farming in India. In: Proceedings of the workshop on status of freshwater prawn farming in India CIFE, Mumbai, India, pp 37–41

Parisenti J, Beirão LH, Tramonte VL, Ourique F, da Silveira Brito CC, Moreira CC (2011) Preference ranking of colour in raw and cooked shrimps. Int J Food Sci Technol 46(12):2558–2561

Paul M, Chanda M, Maity J, Gupta S, Patra B, Dash G (2011) The use of antibiotics and disinfectants in the freshwater prawn (*Macrobrachium rosenbergii*) farms of north and south 24 Parganas districts of West Bengal. Chronic Young Sci 2(1):55–55

Pelinson da Fonseca AP, Retcheski MC, Tormen L, Romão S, Cazarolli LH (2025) Metabolic and stress resistance effects in *Macrobrachium rosenbergii* juveniles supplemented with *Curcuma longa* extract. Vet Res Commun 49(1):1–16

Pillai D, Bonami JR (2012) A review on the diseases of freshwater prawns with special focus on white tail disease of *Macrobrachium rosenbergii*. Aquac Res 43(7):1029–1037

Poongodi R, Bhavan PS, Muralisankar T, Radhakrishnan S (2012) Growth promoting potential of garlic, ginger, turmeric and fenugreek on the freshwater prawn *Macrobrachium rosenbergii*. Int J Pharm Bio Sci 3(4):914–926

Pourmozaffar S, Reverter M, Jahromi ST, Harikrishnan R, Pazir MK, Barzkar N, Mozanzadeh MT, Sarvi B, Abolfathi M, Adeshina I, Behzadi S (2024) An overview of the biological functions and mechanisms of action of medicinal plants and seaweeds in the shrimp culture. J Anim Physiol Anim Nutr 1–47

Prakash M, Karmagam N (2013) A study on bacterial flora associated with fresh water prawn, *Macrobrachium rosenbergii*. Int J Curr Res Acad Rev 1(1):1–16

Pu H, Li X, Du Q, Cui H, Xu Y (2017) Research progress in the application of Chinese herbal medicines in aquaculture: a review. Engineering 3(5):731–737

Radhakrishnan EV, Deshmukh VD, Maheswarudu G, Josileen J, Dineshbabu AP, Philipose KK, Sarada PT, Pillai SL, Saleela KN, Chakraborty RD, Dash G (2012) Prawn fauna (Crustacea: Decapoda) of India-an annotated checklist of the Penaeoid, Sergestoid, Stenopodid and Caridean prawns. J Mar Biol Assoc India 54(1):50–72

Radhakrishnan S, Saravana Bhavan P, Seenivasan C, Shanthi R, Poongodi R (2014) Influence of medicinal herbs (*Alterantherasessilis*, *Eclipta alba* and *Cissus quadrangularis*) on growth and biochemical parameters of the freshwater prawn *Macrobrachium rosenbergii*. Aquacult Int 22:551–572

Ragaza JA (2016) Challenges and opportunities in the culture of giant freshwater prawn in the Philippines. World Aquacult 47:59–61

Rajkumar G, Bhavan PS, Srinivasan V, Udayasuriyan R, Karthik M, Satgurunathan T (2017) Partial replacement of fishmeal with marine algae *Turbinaria ornata* and *Gracilariacorticata*for sustainable culture of the freshwater prawn *Macrobrachium rosenbergii*. Int J Res Stud Zool 3:32–44

Rangacharyulu PV (1999) Studies on the nutrition and diet development of the giant freshwater prawn Macrobrachium rosenbergii. Doctoral thesis. West Bengal, India, University of Kalyani

Rani S, Sharma S, Singh D, Sheoran R, Kumar R (2025) Improved production of anti-cancer compounds from medicinal plants through molecular and biotechnological mutagenesis strategies. In: JT Chen (ed.), Biotechnology, multiple omics, and precision breeding in medicinal plants. CRC Press, pp 287–297. https://doi.org/10.1201/9781003475491-21

Rebecca AA (2011) Growth performance of the freshwater prawn *Macrobrachium rosenbergii* post larvae fed with *Ocimum sanctum* (tulsi) and *Withaniasomnifera*(ashwagandha) incorporated feeds. Int J Biol Res Dev 1(1):34–53

Rebecca AA, Bhavan PS (2014) Growth promotion and survival enhancement of the freshwater prawn *Macrobrachium rosenbergii* post larvae fed with *Allium sativum*, *Zingiber officinale* and *Curcuma longa*. Int J Pure Appl Zool 2(2):138–149

Reverter M, Bontemps N, Lecchini D, Banaigs B, Sasal P (2014) Use of plant extracts in fish aquaculture as an alternative to chemotherapy: current status and future perspectives. Aquaculture 433:50–61

Rodrıguez J, Le Moullac G (2000) State of the art of immunological tools and health control of penaeid shrimp. Aquaculture 191(1–3):109–119

Rose KV, Sunil SC, Chatterjee NS, Rangasamy A (2023) Assessment of larval health and metamorphosis of the giant freshwater prawn *Macrobrachium rosenbergii* larvae during acute exposure to Imidacloprid. Int J Bio-Resour Stress Manag 14(6):891–899

Rőszer T (2014) The invertebrate midintestinal gland ("hepatopancreas") is an evolutionary forerunner in the integration of immunity and metabolism. Cell Tissue Res 358:685–695

Ruangdej U, Laohavisuti N (2014) The use of synthetic and natural carotenoid in diet for color enhancement on red cherry shrimp *Neocaridinaheteropoda*. J Fisher Environ 38(1):30–34

Sahoo PK (2008) Diseases in freshwater prawns and their management. National training course manual on seed production and grow out culture of freshwater prawns, CIFA, Kausalyagang, Bhubaneswar, pp.61–67.ts of fresh water prawn farming in India. Fish Farm 26:26–27

Shailender M, Krishna PV, Suresh Babu C, Srikanth B (2012) Impact of diseases on the growth and survival of giant freshwater prawn, *Macrobrachium rosenbergii* (De Man) larvae in the hatchery level. World J Fish Mar Sci 4:620–625

Shanthi R, Bhavan PS, Radhakrishnan S (2012) Influence of medicinal herbs, (*Andrographis paniculata*, *Cissus quadrangularis* and *Eclipta alba*) on growth, digestive enzymes, biochemical constituents and protein profile of the freshwater prawn *Macrobrachium rosenbergii*. Elixir Bio Tech 42:6478–6484

Sharma SR, Jayaprakash S, Philipose KK, Radhakrishnan EV 2(009) Effect of salinity and pH on selected immune functions of the Indian white shrimp, *Fenneropenaeus indicus* (H. Milne Edwards, 1837). Indian J Fisher 56(3):183–187

Sharma MMM, Kapoor D, Rohilla R, Loyal A, Sharma P, Husen A (2024) Role of essential minerals and rhizosphere in plant under changing environmental conditions. In: Husen A (eds) Essential minerals in plant-soil systems. Elsevier, pp. 235–252. https://doi.org/10.1016/B978-0-443-16082-0.00002-3

Sharma P, Kapoor D, Kumar R, Husen A (2025) Rhizosphere and osmolytes coordination in plant systems under adverse situations. In: Husen A (ed) Roles of osmolytes in changing environment. Academic Press, pp 101–117. https://doi.org/10.1016/B978-0-443-26549-5.00020-5

Shefali, Kumar R, Sankhla MS, Kumar R, Sonone SS (2021) Impact of pesticide toxicity in aquatic environment. Biointerface Res Appl Chem 11(3):10131–10140. https://doi.org/10.33263/BRIAC113.1013110140

Silva MCN, Frédou FL, Rosa Filho JS (2007) Estudo do crescimento do camarão*Macrobrachiumamazonicum* (Heller, 1862) da Ilha de Combú, Belém, Estado do Pará. Amazônia, Ciência and Desenvolvimento 2(4):85–104

Simtoe AP, Lugendo BR, Mgaya YD (2024) Efficacy of aromatic plants *Zingiber officinale* and *Syzygium aromaticum* extracts as natural immunostimulants on survival, growth performance, and immune response of farmed prawn *Penaeus monodon* juveniles. Aquacult Int 4499–4513

Smith VJ, Chisholm JR (1992) Non-cellular immunity in crustaceans. Fish Shellfish Immunol 2(1):1–31

Stara A, Machova J, Velisek J (2012) Effect of chronic exposure to simazine on oxidative stress and antioxidant response in common carp (*Cyprinus carpio* L.). Environ Toxicol Pharmacol 33(2):334–343

Stentiford GD, Sritunyalucksana K, Flegel TW, Williams BA, Withyachumnarnkul B, Itsathitphaisarn O, Bass D (2017) New paradigms to help solve the global aquaculture disease crisis. PLoS Pathog 13(2):e1006160

Stratev D, Odeyemi OA (2016) Antimicrobial resistance of *Aeromonas hydrophila* isolated from different food sources: a mini-review. J Infect Public Health 9(5):535–544

Sui J, Luan S, Yang G, Xia Z, Luo K, Tang Q, Lu X, Meng X, Kong J (2019) Genetic parameters and selection response for the harvest body weight of the giant freshwater prawn (*Macrobrachium rosenbergii*) in a breeding program in China. PLoS ONE 14(8):e0218379

Sun C, Tadese DA, Wangari MR, Zhou Q, Zheng X, Liu B, Tamiru M, Dagne A, Janssens GPJ, Zhao Y (2022) Amelioration of ammonia-induced intestinal oxidative stress by dietary *Clostridium butyricum* in giant freshwater prawn (*Macrobrachium rosenbergii*). Fish Shellfish Immunol 131:1173–1181

Tadese DA, Sun C, Liu B, Muritu RW, Kevin NT, Zhou Q, Zhu L, Zhang H, Bo L, Liu M, Xu H (2020) Combined effects of emodin and *Clostridium butyricum* on growth and non-specific immunity of giant freshwater prawns, *Macrobrachium rosenbergii*. Aquaculture 525:735281

Tadese DA, Song C, Sun C, Liu B, Liu B, Zhou Q, Xu P, Ge X, Liu M, Xu X, Tamiru M (2022a) The role of currently used medicinal plants in aquaculture and their action mechanisms: a review. Rev Aquac 14(2):816–847

Tadese DA, Tesfaye G, Tamiru M, Alemu L, Geda F (2022b) Effect of emodin on growth performance, antioxidant status, intestinal morphology and heat shock protein gene expressions of freshwater prawn (*Macrobrachium rosenbergii*). J Fisher 10(3):103205

Tayamen MM (2005) Freshwater prawn program of Bureau of Fisheries and Aquatic Resources (BFAR). Southeast Asian Fisheries Development Center/Aquaculture Department (SEAFDEC/AQD) Institutional Repository (SAIR)

Tibbetts SM, Lall SP, Anderson DM (2000) Dietary protein requirement of juvenile American eel (*Anguilla rostrata*) fed practical diets. Aquaculture 186(1–2):145–155

Tonguthai K (1997) Diseases of the freshwater prawn, *Macrobrachium rosenbergii*. AAHRI Newslett Art 4(2):1–4

Tsai ML, Chen TC, Chou ZY, Wu WY (2025) Rapid determination of the physiological temperature range for the giant freshwater prawn (*Macrobrachium rosenbergii*). Aquacult Rep 40:102628

Uddin SA, Kader MA (2006) The use of antibiotics in shrimp hatcheries in Bangladesh. J Fish Aquat Sci 1(1):64–67

Urmila, Gupta RK, Sharma S, Kumar R (2025) Earthworms: the environmentally preferred bioengineers for managing heavy metals. In: Dey S, Bhattacharya S (eds) Biotechnological interventions in the removal of emerging pollutants. Springer Nature Singapore, Singapore, pp 365–384

Vaid V, Hundal SS (2019) Light microscopic studies to evaluate fish scales as non-invasive indicators of heavy metal–contaminated waters. Environ Monit Assess 191:1–11

Van Hai N (2015) The use of medicinal plants as immunostimulants in aquaculture: a review. Aquaculture 446:88–96

Van Boeckel TP, Pires J, Silvester R, Zhao C, Song J, Criscuolo NG, Gilbert M, Bonhoeffer S, Laxminarayan R (2019) Global trends in antimicrobial resistance in animals in low-and middle-income countries. Science 365:6459

Wade NM, Budd A, Irvin S, Glencross BD (2015) The combined effects of diet, environment and genetics on pigmentation in the giant tiger prawn, *Penaeus monodon*. Aquaculture 449:78–86

Wang M, Wang L, Jia Z, Yi Q, Song L (2018) The various components implied the diversified Toll-like receptor (TLR) signaling pathway in mollusk*Chlamys farreri*. Fish Shellfish Immunol 74:205–212

Wang Y, Wei J, Hong K, Zhou N, Liu X, Hong X, Li W, Zhao J, Chen C, Wu L, Yu L (2022) Transcriptome analysis reveals the molecular response to salinity challenge in larvae of the giant freshwater prawn *Macrobrachium rosenbergii*. Front Physiol 13:885035

Wei J, Tian L, Wang Y, Yu L, Zhu X (2021) Effects of salinity, photoperiod, and light spectrum on larval survival, growth, and related enzyme activities in the giant freshwater prawn Macrobrachium Rosenbergii. Aquaculture 530:735794

White DA, Page GI, Swaile J, Moody AJ, Davies SJ (2002) Effect of esterification on the absorption of astaxanthin in rainbow trout, *Oncorhynchus mykiss* (Walbaum). Aquac Res 33(5):343–350

Wongphonprateep S, Kasirak A (2023) Effect of carotenoid extract from dried chili pepper (*Capsicum annuum*) on growth performance and pigmentation in cooked giant river prawn (*Macrobrachium rosenbergii*). J Fisher Environ 47(2):133–141

Yadav R, Gupta RK, Kumar R, Kaur T, Moun R (2022a) Biochemical response of earthworm, *Eisenia fetida* to heavy metals toxicity. J Appl Nat Sci 14(3):990–998

Yadav R, Gupta RK, Kumar R, Kaur T (2022b) Assessment of toxicity of lead and nickel on the biochemical and immunological parameters of earthworm, *Eudrilus Eugeniae*. Agric Sci Digest 44(5):922–929

Yadav R, Kumar R, Gupta RK, Kaur T, Yodha K, Kour A, Kaur S, Rajput A (2023) Heavy metal toxicity in earthworms and its environmental implications: a review. Environ Adv 100374

Yamuna A, Bhavan PS, Geraldine P (2009) Ultrastructural observations in gills and hepatopancreas of prawn *Macrobrachium malcolmsonii* exposed to mercury. J Environ Biol 30(5):693

Yan WJ, Zhou QL, Liu B, Sun CX, Zhang HM, Song CY (2021) *Isaria cicadae* Miquel improves the growth performance, physiological response and meat quality of giant freshwater prawn, *Macrobrachium rosenbergii*. Pakistan J Zool 1–10

Yang G, Frinsko M, Chen X, Wang J, Hu G, Gao Q (2012) Current status of the giant freshwater prawn (*Macrobrachium rosenbergii*) industry in China, with special reference to live transportation. Aquac Res 43(7):1049–1055

Yin G, Ardó LÁSZLÓ, Thompson KD, Adams A, Jeney Z, Jeney G (2009) Chinese herbs (*Astragalus radix* and *Ganoderma lucidum*) enhance immune response of carp, *Cyprinus carpio*, and protection against *Aeromonas hydrophila*. Fish Shellfish Immunol 26(1):140–145

Yuan G, Guan Y, Yi H, Lai S, Sun Y, Cao S (2021) Antibacterial activity and mechanism of plant flavonoids to gram-positive bacteria predicted from their lipophilicities. Sci Rep 11(1):10471

Zhong Y, Chen ZF, Dai X, Liu SS, Zheng G, Zhu X, Liu S, Yin Y, Liu G, Cai Z (2018) Investigation of the interaction between the fate of antibiotics in aquafarms and their level in the environment. J Environ Manage 207:219–229

Chapter 29
Molluscs as Economic Agents: Biological Mechanisms and Pearl Production

Soumyadyuti Goswami, Jiban Kumar Behera, Bhaskar Behera, and Manojit Bhattacharya

Abstract Pearl cultivation, a subject that is frequently disregarded in agricultural discourse, represents a special fusion of sustainability and innovation in contemporary farming. Rare and valuable jewels, pearls are known for their inherent beauty and the fact that they don't need to be manually carved or processed. Since ancient times, their profound emotional, cultural, and historical importance has been treasured. Pearl-producing molluscs are hence highly valued and economically significant. It is important to remember that not all molluscs that produce pearls are oysters. The evolution and distribution of the several kinds of pearl-producing molluscs in Indian seas are examined in this review. In order to explore the physical and biological mechanisms behind pearl formation and the variables affecting it, we also compile the body of previous research. There is substantial scientific and societal value in researching the development and rearing of molluscs that produce pearls. We can support the preservation and promotion of pearl-producing molluscs by advancing scientific research and putting protective measures in place. The bright future of pearl farming is encouraging as it might raise fishermen's standard of living.

Keywords Pearl · Molluscs · Pearl formation · Fishermen · Economic growth

29.1 Introduction

Soft-bodied creatures, molluscs may live happily in freshwater and saltwater environments. They are known for their capacity to generate pearls and can have either a single (univalve) or two (bivalve) shells (Gökoğlu and Gökoğlu 2021). Invertebrate molluscs do not have backbones or spines. It is estimated that over 100,000 species of molluscs are found in the oceans worldwide (MH 2023). However, only

S. Goswami · J. K. Behera · M. Bhattacharya (✉)
Department of Zoology, Fakir Mohan University, Balasore, Odisha, India
e-mail: mbhattacharya09@gmail.com

B. Behera
Department of Biosciences and Biotechnology, Fakir Mohan University, Balasore, Odisha, India

R. Kumar et al. (eds.), *Sustainability in Aquaculture and Fisheries*,
Springer Water, https://doi.org/10.1007/978-3-032-16605-0_29

about half of these species are utilized in pearl farming, and only a few hundred can produce pearls. Molluscs are divided into two primary categories: freshwater and saltwater. Saltwater molluscs known for pearl production include oysters, while those in freshwater that can produce pearls are commonly called mussels (Lydeard and Cummings 2019). Cultured pearls are created by implanting a small nucleus and a tissue fragment from a donor oyster into a recipient oyster. The mollusc subsequently forms layers around the nucleus, producing a cultured pearl (Gervis and Sims 1992; Taylor and Strack 2008). As the donor's mantle tissue expands, it forms a "pearl sac" around the nucleus (Kishore and Southgate 2015). A cultured pearl forms when nacre gradually accumulates on a nucleus over approximately two years from the secretory cells within the pearl sac. This two-year timeframe is crucial as it allows for the natural and gradual development of the pearl, contributing to its distinct beauty and value (Dix 1972; Taylor and Strack 2008; Cochennec-Laureau et al. 2010).As methods for cultivating *Pinctada fucata* pearl oysters and producing cultured pearls matured in India (Alagarswami 1974), the artificial breeding of this species became necessary due to the considerable variations in their natural populations (Alagarswami et al. 1983). Coastal areas and bays produce only limited amounts of spat, with a low incidence of *P. fucata* (Alagarswami 1977). Despite the established methods for the controlled spawning of pearl oysters, the larval rearing process was limited to achieving only the straight-hinge phase (Alagarswami et al. 1983). Four main species of molluscs are mainly utilised for cultivating cultured pearls, and each species is renowned for yielding exquisite specimens. The molluscs known for producing pearls include Akoya, Tahitian, South Sea, and various freshwater types. Numerous research studies have explored how different culture systems affect pearl oysters' growth and survival rates (Kishore and Southgate 2016), with significant emphasis on the black-lip pearl oyster *P. margaritifera* (Southgate and Beer 2000). The evaluation of pearls is a critical process that significantly influences their market value. Within the pearl industry, two predominant grading systems are employed to assess pearls based on several key criteria, namely lustre, surface quality, shape, colour, and size (Gauthier et al. 2018). These grading systems provide jewellers with detailed and specific guidelines for evaluating pearls, helping ensure transparency and quality within the pearl market.

29.2 Pearl-Producing Molluscs

Freshwater mussels and saltwater oysters produce pearls as a natural defence against irritants, such as parasites that can infiltrate their shells or harm their delicate bodies. The layers of aragonite and conchiolin, which make up the shell of the oyster or mussel, are secreted gradually over time (Farn 2013). The first actual pearl-producing molluscs appeared around 530 million years ago when molluscs evolved shells (Shen 2022). With approximately 128,000 species, molluscs comprise the second-largest phylum in the animal kingdom (Gökoğlu and Gökoğlu 2021). However, pearl-producing species are found in only one of the two subphyla that make up this phylum:

Conchifera. Notable examples of pearl-producing molluscs include the freshwater mussel (Unionidae) and the saltwater oyster (*Pinctada*) (Katupotha 2019; Zhu et al. 2019).

29.2.1 Historical Development of Pearl Formation

The evolutionary tree for the Phylum Mollusca is depicted in the image below (Fig. 29.1). Molluscs that produce pearls are not all members of the same family or group. The evolutionary tree has a variety of pearl producers. Classes of Conchifera that possess the chemical that produces pearls, such as nacre, are indicated by the shaded boxes. The epithelial tissue lining the shell produces nacre, which gives pearls their distinctive sheen.

29.2.1.1 The Extensive Occurrence of Nacre Suggests One of Two Possibilities

Nacre is a fundamental characteristic that has been lost and reacquired throughout the evolutionary history of molluscs (McDougall and Degnan 2018). The composition of nacre is distinctive among specific categories, and certain types foster pearl formation. Most molluscs that generate pearls are bivalves, which are defined by clam-like shells with two halves joined by a hinge (Pabian 1970). A relatively small number of the approximately 20,000 species of bivalve molluscs are used in pearl culture to produce commercial pearls (Southgate 2021). Furthermore, "pearls" found in edible oysters are undesirable for dental health. The Pteriomorpha subclass of bivalves, which includes "scallops," "pen shells," and "hammer shells," is more closely related to pearl oysters than to ordinary oysters.

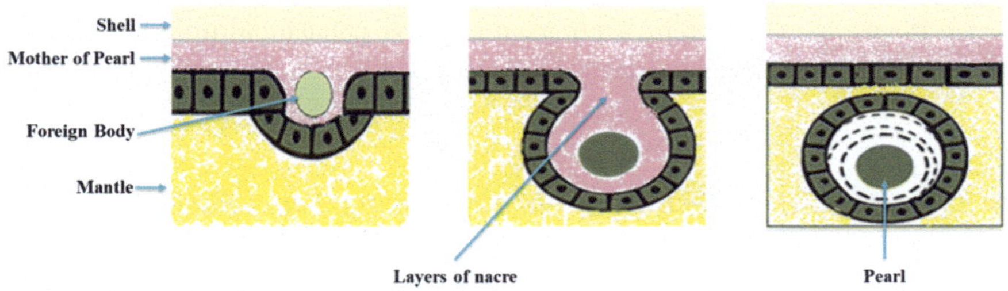

Fig. 29.1 The schematic diagram illustrates the process of Pearl formation within a mollusc

29.2.1.2 History and Exploration of Pinctada

Avicula, meleagrina, and margaritifera are some historical names for Pinctada (Behera 2017). "Margaritifera" means "bearer of pearls" and many oysters that produce pearls are derived from terms related to birds (Streeter 2021). For instance, the Latin word "avis," meaning "bird," is the source of the name Avicula. D'Argenville named the genus after the French word "la pintade," which describes the guinea fowl's grey and white speckled plumage, as he thought the grey color of the pinctada shells resembled that of a guinea fowl. Pearls of a specific species can often be recognized primarily by their shells, but their nacreous lining makes them unique.

29.2.2 Distribution

Pearl oysters are members of the Pteriidae family, specifically the genus *Pinctada* (Roding) (Cunha et al. 2011). Nearly every sea in the tropical and subtropical belt lives on a seafloor that lives from low tide to a depth of 80 m. Despite this, there are only 28 known species of pearl oysters (Southgate and Lucas 2011).Three species generate pearls with commercial worth and gem qualityare *P. margaritifera* (Linnaeus), *P. fucata* (Gould), and *Pinctada maxima* (Jameson) (Table 29.1) (Southgate 2021). Pearl oysters are prevalent in various regions such as Venezuela, the Red Sea, Panama, the Persian Gulf, Mexico, the Philippines, the Gulf of California, Japan, Australis, China, the Cook Islands, South Korea, French Polynesia, Myanmar, Papua New Guinea, India and Indonesia (Zhu et al. 2019).The Gulf of Mannar and the Gulf of Kutch represent the most prolific regions exhibiting the highest concentrations of pearl oysters. The seas surrounding India are home to six distinct varieties of pearl oysters (Venkataraman 2007). The species in concern are *Pinctada fucata* (Gould), *P. margaritifera* (Linnaeus), *P. chemnitzii* (Philippi), *P. sugillata*(Reeve), *P. anomioides* (Reeve), and *P. atropurpurea* (Dunker) (Venkataraman 2007). The black-lip pearl oyster, *Pinctada margaritifera*, is observed in sparse populations in certain regions of the Andaman and Nicobar Islands.

29.2.3 Innovative Techniques in Freshwater Pearl Farming

Recent data from the FAO (2020) shows that freshwater pearl output is still declining due to urbanisation, pollution, climate change, longer cycles of pearl production, habitat deterioration, sedimentation, a lack of competent labour, and pesticide problems (He 2019). For the long-term development of the pearl industry, fisheries departments, policymakers, planners, research organisations, finance agencies, and export agencies must enhance the freshwater production of pearls. The oyster species *Pinctada maxima* and *Pinctada margaritifera* produce most of the pearls used in the

Table 29.1 This table summarizes each species' scientific names, common names, regional distribution, habitat preferences, and IUCN status, along with relevant references

Sl no	Species	Common name	Distribution (Regions)	Habitat	IUCN status	References
1	*Pinctada fucata* (Gould)	Akoya pearl oyster	The Western Pacific Ocean, the Red Sea, the Persian Gulf, the coastal seas of China, Korea, Japan, India, and Venezuela, as well as the Indo-Pacific region	Tropical rocky shores, lagoons and coral reefs	Vulnerable	(Costello, Emblow et al. 2001, Costello, Bouchet et al. 2013)
2	*Pinctada margariti-fera* (Linnaeus)	Black-lip pearl oyster	Persian Gulf, Red Sea, Sudan, Papua New Guinea, Australia, French Polynesia, Cook Islands, Indonesia, Andaman and Nicobar Islands, the Southwestern part of the Indian Ocean, Japan and the Pacific Ocean, and various locations on the coast of India	Oligotrophic coral reef waters with low turbidity	Not evaluated	(Doroudi and Southgate 2003)
3	*Pinctada albina* (Lamarck)	Sharks Bay pearl oyster	Philippines, China, Korea, Vietnam, Indonesia, Micronesia, and northern Australia are all part of the Western Pacific. In Australia, from Perth, Western Australia, to the northern region, and then south to Sydney	Lives in shallow water down to a minimum of 22 m) and is affixed to solid objects on seagrass flats and reefs by a byssus. Not assessed	Not evaluated	(Reeve and Reeve 1857; Beu 1973)
4	*Pinctada maculata* (A. Gould)	Spotted pearl oyster	Indo-West Pacific; In Australia from Dampier Archipelago, North West Australia, around northern Australia, to Clarence River	Lives attached to solid substrates in the shallow subtidal	Not evaluated	(Gould 1850; Reeve and Reeve 1857; Spencer, Marshall et al. 2009)

(continued)

Table 29.1 (continued)

Sl no	Species	Common name	Distribution (Regions)	Habitat	IUCN status	References
5	*Pinctada maxima* (Jameson)	South Sea pearl or Philippine pearlgold-lipped oyster or silver-lipped oyster	Central Indo-West Pacific: Myanmar to Polynesia to the east, and northern Australia to the south. Solitary Island (near Coffs Harbour) in New South Wales, Shark Bay, Western Australia, and occasionally as far south as Sydney	A byssus holds it to a hard substrate while it is young, but as it grows, it separates and can live freely on sand, gravel, or coral debris for up to 80 m, though usually only 30 to 40 m	Not evaluated	(ng Pilipinas 2010, Pangilinan 2014)
6	*Pinctada radiata (Leach)*	Atlantic pearl-oyster or the Gulf pearl oyster	throughout the Indo-Pacific and Mediterranean Sea through the Suez Canal	Attaches itself to hard surfaces like rocks and shipwrecks. It can be found in the littoral, sublittoral, and shelf zones	Not evaluated	(Di Monterosato 1884; Costello, Emblow et al. 2001)
7	*Pinctada chemnitzii* (Philippi)	Cape Pearl Oyster or Tiled Pearl Oyster	Indo-West Pacific: from the Northern Territory to Gove, Queensland, northern New South Wales, and infrequently, Sydney	Lower intertidal and shallow subtidal, attached by byssus to solid substrates	Not evaluated	(Barnard 1927; Macnae 1958; Liu 2008; Tëmkin 2010)

marine industry (Zhu et al. 2019). Innovative techniques in freshwater pearl farming include: advanced grafting methods like using genetically selected donor mussels for better pearl quality, non-nucleated pearl cultivation, utilizing tissue culture to produce large quantities of mantle grafts, employing sophisticated water quality monitoring systems, and integrating technology for precise environmental control and data analysis (Alagarswami 1987).

29.3 Mechanism of Pearl Formation

Pearls come in three different varieties: natural, cultivated, and fake. A natural pearl is produced without the help of humans when an irritant penetrates a specific type of mussel, clam, or oyster. As a protective response to this irritant, the mollusc secretes a fluid that envelops the irritant and then layers and solidifies around it (Taylor and Strack 2008). A shiny pearl forms as a result of the irritant layer being applied consecutively. Together with calcite, aragonite is a polymorph of calcium carbonate ($CaCO_3$) that makes up the majority of these strata. The two minerals are held together by conchiolin, an organic material that has a horn-like shape. Nacre refers to the mixture of conchiolin and argonite that constitutes mother-of-pearl (Zhang et al. 2024) (Fig. 29.1).

29.3.1 Biomineralization

The many processes involved in the "chemistry of pearls" are interesting. Mollusc biomineralization is central. Pearls are produced by this technique, which mixes organic substances such matrix proteins with calcium carbonate (Xu and Xuan 2023). The biomineralization process of molluscs is an interesting combination of organic and inorganic materials. The majority of these shells are composed of calcium carbonate forms called aragonite and calcite. Mollusc shells are distinguished by their robust and long-lasting composition, which is composed of 1–5% organic matrix and around 95% $CaCO_3$ (Azarian and Sutapun 2022).

Mollusca displays a diverse range of biomineral morphologies, including shell epithelial spicules observed in the basal mollusc *Wirenia argentea* (Todt and Wanninger 2010), scales and plates found in bivalves, the operculum of the gastropod *Rapana venosa* (Hashimoto et al. 2012), intracellular detoxifying granules have been identified in the common garden snail *Helix aspers* (Howard et al. 1981) as well as in the egg capsules of the *Patagonian neogastropod* and *Odontocymbiol magellanica* (Bigatti et al. 2010). Furthermore, pearls produced by pearl oysters, statoconia in *Aplysia califonica* (Kondrachuk and Wiederhold 2004) and statoliths in *Nassarius reticulatus* exemplify the remarkable diversity inherent within this group of organisms (Galante-Oliveira et al. 2014). Mollusc shells predominantly consist of

$CaCO_3$ and are characterized by biomineral microstructures that typically incorporate three distinct layers, as evidenced in species such as nautilus, mussels, abalones and oysters. The periostracum, the outermost layer, has a leathery feel and is thin and organic. The two categorized strata are located underneath the periostracum. Long, prismatic-shaped calcitic crystals orientated perpendicular to the periostracum produce the prismatic layer (Marin et al. 2012). Long prismatic-shaped calcitic crystals oriented perpendicular to the periostracum make up the prismatic layer. The inner glossy shell layer, also known as the nacreous layer, is composed of aragonite crystals arranged in a laminar pattern that resembles a brick wall (Kakisawa and Sumitomo 2012). Molluscan shells, which are made of proteins, peptides, lipids, and carbohydrates, are secreted by the mantle epithelium (Addadi et al. 2006; Addadi and Weiner 2014). The organic matrix acts as the structural framework, starts crystal nucleation, and controls the growth of extracellular crystals, according to the matrix-mediated theory, which is generally acknowledged by scientists. The unique crystal morphologies found in the different layers of mollusc shells are the consequence of this process (Addadi et al. 1987; Lowenstam and Weiner 1989; Addadi et al. 2006).

29.3.1.1 Ecological Factors

The biological conditions and water quality of the aquatic ecosystem have a significant impact on pearl formation. Seasonal variations, water temperature, and the types and quantity of planktonic animals in the ocean all have an impact on the physiological metabolism and pearl-making process of molluscs (Collado et al. 2023). One of the main factors that promotes pearl production is consistent water temperatures, which are also present in regions that create pearls, such the South China Sea, due to the presence of planktonic organisms and the clear waters. These environmental elements provide molluscs with a healthy habitat and a plenty of nutrients, which encourages the growth of nacre within their bodies.

29.3.1.2 Environmental Factors

The pH of the aquatic environment, the amount of calcium ions and organic matter in the water, and other variables all have a substantial impact on pearl production (Zhang et al. 2024). Pearl synthesis requires the presence of calcium ions. By collecting calcium ions from the water, molluscs produce nacre. Additionally, the pH of the water affects the solubility and availability of calcium ions, which in turn affects the pace and quality of pearl production (Kumar et al. 2022a, b, c; Siddique, Barman et al. 2022). Environmental instability can cause mollusca to create abnormal or incomplete pearls. Therefore, stable weather conditions are necessary for pearl development.

29.3.1.3 Biological Factors

Pearl creation is directly linked to mollusc species and genetic variations. Various mollusc species have unique traits and rates of pearl production. For instance, during the pearl creation process, Tahitian black pearl oysters and South Sea pearl oysters exhibit distinct biological characteristics, leading to differences in the nacre they produce in terms of both look and quality.

29.4 Growth and Culture of Pearl Producing Molluscs

29.4.1 Spawning and Fertilization

A minor fluctuation in salinity can initiate the spawning process in fully developed gonads of pearl oysters. Once these oysters are reintroduced to seawater, some individuals may reproduce even after undergoing the trauma associated with shell cleaning. Under optimal conditions, pearl oysters harvested from natural beds can reproduce over several months, especially with seawater collected from the surface (Southgate 2021). Under typical circumstances, the spawning process is initiated by males. The sperm in the water stimulates the females, prompting them to respond within thirty minutes (Saucedo and Southgate 2008). The majority of eggs are released during the reproductive process. The initial egg exhibits a pyriform shape, with an average breadth of 45.2 μm and a length of 73.9 μm. It has a prominent and transparent germinal vesicle, measuring 24.7 μm in diameter. The cytoplasm of the yolk displays a somewhat opaque and granular texture. This egg is encapsulated in a vitelline membrane that remains intact until the trochophore developmental stage. Fertilization of the eggs occurs shortly after discharge. The germinal vesicle disintegrates, and they take on a spherical form. The diameter of the fertilized egg is 47.5 μm.

29.4.2 Larval Stages

Blastula: This stage is attained five hours after fertilization. Reoriented cells create blastopores and blastocoels (Løvtrup 1965). ***Gastrulation***: Gastrulation occurs through epiboly. The cells separate into many levels and convolute. The archenteron that has developed at this point communicates with the outside world through the blastopore. Phototropism is present in the embryo. The trip to the stage takes seven hours (Muhr and Ackerman 2023). ***Trochophore larva***: Early trochophore larvae designate the anteroposterior area of the embryo by developing preoral and postoral tufts of cilia. The normal trochophore stage develops a single apical flagellum. The tiny cilia that were seen during the blastula stage vanish. In ten hours, this step is reached (Nielsen 2004). ***Veliger***: The indices characterzing the veliger stage include the formation of a straight hinge line and mantel, the reorganization of preoral cilia into a velum and cessation of the apical flagellum and the preoral and postoral ciliary bands. Along the dorsoventral axis, the average length of the straight hinge larva is 52.5 μm; along the anteroposterior axis, its average length is 67.5 μm. This stage reaches completion approximately 20 h thereafter (Jagadis 2009). ***Umbo stage***: Straight-hinge larvae grow gradually to the umbo stage. The typical clam-shaped umbo stage, which measures 135 × 130 μm, is attained in 10–12 days (PL II B). Mantle folds form and the shell valves are equal (Fig. 29.2) (Alagarswami et al. 1987). ***Eye spot stage***: On the fifteenth day, the eye spot forms when the larva is 190 × 180 μm in size. The foot primordium's base is where the eye spot is located. An eye spot may be seen in a 3.9 mm spit. Ctenidial ridges are developed by the larva (Fig. 29.2) (Victor, Chellam et al. 1995). ***Pediveliger stage***: On the eighteenth day, the foot develops to 200 × 190 μm. Both velum and foot are present throughout the transition from swimming to crawling. Later, as the velum goes away, the foot starts to work. Currently, 2–4 gill filaments are visible (Fig. 29.2) (Vikas and Subramannian 2020). ***Plantigrade***: The stage, 220 by 200 μm, is visible on the twentieth day. Additional gill filaments and labial palps form (Fig. 29.2) (Mathew 2002). ***Spat***: The plantigrade undergoes transformation into a juvenile spat. Distinct morphological features arise from the byssal notch, the anterior and posterior auricles and hinge lines (Fig. 29.2). Notably, the left valve exhibits greater concavity than the right valve. The spat may adhere to the substrate utilizing basal threads. The typical spat is typically observed by the 24th day of development (Carriker 2001).

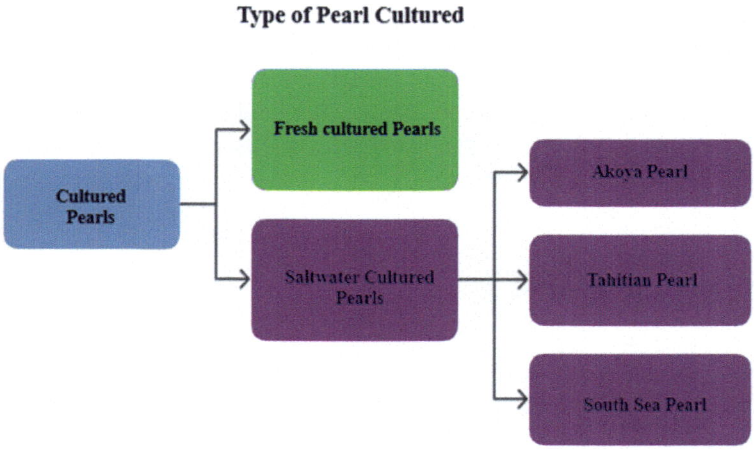

Fig. 29.2 The figure illustrates bivalves' sequential larval development stages, progressing from the umbo stage to the spat stage

29.4.3 *Cultured Pearls*

Freshwater and saltwater cultured pearls are the two types of cultivated pearls. There are three varieties of saltwater farmed pearls (Fig. 29.3).

29.4.3.1 Akoya Pearl

Akoya cultured pearls are widely recognized as the most prominent saltwater grown pearls in global markets. They are cultivated from a specific oyster species renowned for their exceptional quality and lustre (Fig. 29.3). In the early 1900s, Kokichi Mikimoto, a Japanese pioneer in pearl culture, utilised this variety for the first time (Kiefert et al. 2004; Kripa et al. 2007). The filter feeder *Pinctada fucata* is similar to other bivalve molluscs. Water enters the shell through a mantle aperture traverses past the gills, where it filters out food particles and facilitates gas exchange and subsequently exits through a second aperture (Bieler et al. 2014). With a roughly 6 to 9 mm diameter, Akoya cultivated pearls are regarded as some of the smallest ever made. Even if Akoya's origin is after returning to Japan, cultivated pearls are also grown in China, Australia, India, and Mexico (Kiefert et al. 2004; Kripa et al. 2007; Otter et al. 2017).

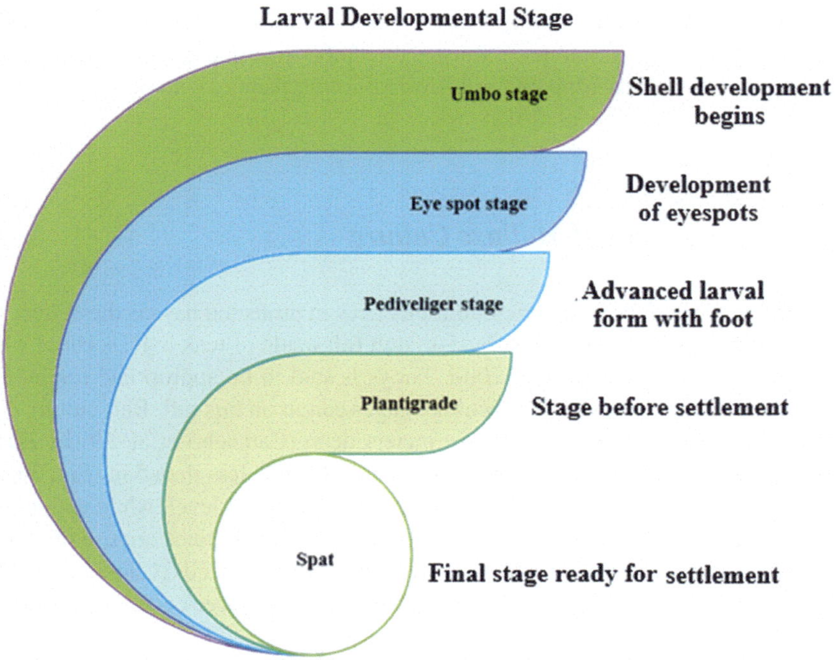

Fig. 29.3 The chat represents the classification of cultured pearls based on their cultivation environment. Cultured pearls are divided into two main categories

The pearls formed are typically white or creamy, although occasionally they require colour treatment.

29.4.3.2 Tahitian Pearl

One type of pearl oyster is *Pinctada margaritifera.* It is a mollusc found in seawater (Fig. 29.3). Tahitian cultured pearls are made from this kind of oyster. It is also referred to as the pearl oyster with a black lip (Kishore and Southgate 2016). The species is found in New Guinea, the Red Sea and the Arabian Gulf among other regions. Tahitian polished pearls have been available on the market since the 1970s. They are offered in various colours exhibiting diverse undertones. Furthermore, these pearls are produced in various dimensions and shapes, including geometric and non-geometric forms. The diameter of these pearls typically ranges from 8 to 14 mm (Hänni 2006).

29.4.3.3 South Sea Pearl

The *Pinctada maxima*, recognized as one of the largest oyster species globally, is responsible for producing a specific type of pearl (Fig. 29.3). This particular oyster yields pearls that measure a minimum of 15 mm in diameter. It is predominantly found in regions such as Indonesia, Burma, the Philippines and Australia. The South Sea variety of this oyster is noted for its diverse colour palette, with most commonly encountered hues (Otter et al. 2017).

29.4.4 Techniques of Mollusc Culture

Raft culture: One of the best agricultural practices in protected bays is the culture of oysters from rafts (Fig. 29.4). A 6-by-5-m unit raft made of teak logs, secured with coir rope, and supported by four 200-L buoys is said to be appropriate for Indian circumstances. Approximately 100 culture cages can fit on this raft. Raft culture can be used when the water is at least five meters deep (Camacho et al. 1991). ***Rack culture***: The rack system can be used if the water depth is less than 5 m (Fig. 29.4). In this, horizontal poles are coir rope-lashed above seawater level, while teak poles are driven into the sea floor at 1-m intervals. The horizontal poles are used to hang culture cages in the water. The rack may be expanded to meet needs (Choi et al. 2017). ***Onshore tank culture***: It is also possible to raise oysters in large concrete tanks built on the beach that can store 75 to 150 tons of saltwater (Fig. 29.4). The water's ideal depth is three meters for oysters to develop normally. Under this approach, oysters and juveniles must be fed phytoplankton cultures. Several systems, such as long-line

Fig. 29.4 The figure represents various methods for cultivating molluscs based on the location and structure used for growth in traditional settings area

and on-bottom cultures, can also be used depending on the sea bottom and other factors (Claus et al. 1983).

29.5 Pearl Grading

The nacreous pearl of pinctada has the following composition: body colour, overtone, and orient are the three primary components determining pearl colour (Table 29.2). The pearl's primary colour is its body colour, which considers saturation, tone, and colour. An obvious transparent hue that seems to cover the body colour is called an overtone. The pearl's orient is a medley of colours that shimmer just under its surface (El-Refi et al. 2022).

Table 29.2 The table presents the relative proportions of water, organic matter, mineral content and unaccounted loss

Sl. no.	Matter	Value (%)
1	Water	3.97
2	Organic matter	3.83
3	Calcite and aragonite	91.53
4	Loss	0.67

Table 29.3 The table outlines the typical size variations observed in Akoya, Tahitian, South Sea and freshwater pearls, indicating both average and commonly occurring diameters in millimetres

S. no	Pearl	Average size (mm)	Common size (mm)
1	Akoya Pearl	1.0–9.5	6.0–6.5
2	Tahitian Pearl	8.0–16.0	12.0–13.0
3	South Sea Pearl	8.0–21.0	16.0–17.0
4	Freshwater Pearl	2.0–12.0	6.0–9.0

29.5.1 Factors Influencing Pearl Quality

29.5.1.1 Size

Pearl measurements are expressed in millimetres, rounded to the nearest 0.5 mm. Measurements are taken perpendicular to the pierced hole to determine the size of spherical pierced pearls. Notably, the 29.65 mm South Sea baroque pearl holds the world's largest farmed pearl. Each type of cultivated pearl possesses distinct dimensions that facilitate effective categorisation and measurement (Table 29.3) (Taylor and Strack 2008).

29.5.1.2 Shape

Pearl shapes are categorized into five primary forms. Round pearls are the rarest shape, as certain forms are inherently complex to produce in nature. Additionally, semi-round or near-round pearls closely resemble round ones, making it challenging for an untrained eye to distinguish between them (Zhong 2023).The striking likeness of drop-shaped pearls to teardrops is the source of their name. If these pearls have a perfect pear form, they can be very precious and symmetrically expanded. The form of baroque pearls is uneven; no two are the same. A subgroup of pearls with baroque shapes is called semi-baroque pearls. They are quite distinctive and easy to identify because of the rings that surround the pearl.

29.5.1.3 Colour

Three primary categories can be utilized to characterize pearl colour: neutral, grey and black. The Gemological Institute of America (GIA) has established a pearl colour grading system. Further, it classifies nineteen pearl colours into overarching groups (Yazawa and Zhou 2018).

29.5.1.4 Lustre

The lustre of a pearl is a crucial factor that influences the quantity and quality of light reflected from its surface and the layers beneath. Several elements contribute to the brightness of a pearl, including the thickness, quality and composition of the nacre (Knoth, Miehe et al. 2005). Although dense nacre doesn't always result in a high-quality pearl, it should have sufficient layers and thickness to interact effectively with light. With all other factors being equal, lustre is greater when nacre is predominantly or entirely aragonite, even if it includes calcite, conchiolin, and aragonite (El-Refi et al. 2022). Conversely, as the amount of calcite increases, brightness decreases. The sheen of pearl is typically associated with the slow deposition of nacre, as opposed to rapid deposition. Pearl producers tend to harvest mature pearls either during cold water periods, as such conditions effectively decelerate the rate of nacre deposition.

29.5.1.5 Surface

The pearl's surface quality can be rated as clean, mildly spotted, moderately spotted, or highly speckled, as its surface is spotless (El-Refi et al. 2022). The moderately speckled surface has many defects, most of which may not be apparent to the unaided eye.

29.5.1.6 Nacre Quality

Pearls are composed of nacre, which significantly influences their brightness and overall quality, thereby affecting their value. Nacre quality can be categorized into three groups. ***Acceptable***: This category denotes that the pearl does not belong to the chalky group. Additionally, the uniform thickness of the nacre renders the nucleus imperceptible. One may "blink" or catch a glimpse of the nucleus (Rinju et al. 2024). ***Nucleus Visible***: When the pearl is turned, it shows intense blinking, with darker and lighter shades flickering. The nucleus, the flash of the dark, can be seen in thin nacre pearls (Antoinette Matlins 2012). ***Chalky Appearance***: This type of pearl exhibits a pale and dull appearance. The minute layers of nacre on these stones do not yield the remarkable lustre typically associated with high-quality pearls (El-Refi et al. 2022).

29.5.2 Improvement of Pearl Quality

Any procedure that improves a pearl's look, other than polishing, is called pearl treatment. This could involve shining, irradiation, and colouring.

29.5.2.1 Dying

Dyeing enhances the depth of the pearl nacre by utilizing silver nitrate. Silver nitrate undergoes a chemical reaction with light and hydrogen sulphide gas upon penetrating the nacre layers, resulting in a relatively dark and black colouration (Ito 2009). Both organic and inorganic materials can create colours other than black. This is a widely used technique for freshwater pearls, as the lower value of these pearls allows manufacturers more flexibility in their processing. Similarly, Akoya pearls are frequently bleached and drilled. In this process, the pearls are immersed in an aniline or a comparable red dye and allowed to soak until the drill hole takes on the colour.

29.5.2.2 Irradiation

Gamma-ray radiation induces darkening of the freshwater nacre layers and the pearl nucleus in Akoya pearls. Conversely, gamma rays result in discolouration of the core in saltwater-grown pearls (ElShazali et al. 2011). Saltwater pearls exposed to radiation will appear grey or blue because of manganese, which is altered by this process. Therefore, only the nucleus undergoes a colour change. However, when exposed to gamma rays, the nacre of freshwater pearls can become quite dark. Moreover, their surface is iridescent and possesses a strong metallic lustre. This occurs because manganese dissolves in freshwater and becomes a typical component of freshwater nacre.

29.5.2.3 Lustre Treatment

Enhancing a pearl's brilliance is the goal of lustre treatment. Maeshoriis the popular technique of thermally treating pearls by heating them first and then cooling them. All akoya, freshwater, and some South Sea pearls are treated with maeshori. The nacre crystals are tightened and moisture is drawn out using maeshori procedures, which differ from factory to factory, but have the drawback of making the pearl's nacre brittle (Ito 2009).

29.6 Socio-Economics of Pearl-Producing Molluscs

Isolation of marine communities, and pearl production may open up additional economic opportunities (Johnston et al. 2019). Pearls benefit these civilizations because they are easily storable and highly valued relative to their weight. These two factors reduce the challenges of transportation. The decline in pearl prices had the most significant impact on medium-sized farms with paid staff; small family farms could continue since they did not have payroll, and large farms managed to survive due to their lower unit costs compared to medium-sized farms. The initial step in determining whether pearl oyster production would be profitable in a developing nation is to prove that it is technically feasible. The mollusc industry confronts various hurdles such as climate change, pollution, habitat destruction, and shortages of skilled workers, infrastructure, and market reach (Table 29.4). Among other considerations, cultural methods must be successfully transmitted to local communities. According to 2004 data from the Centre français du commerce exterior, cited in ISPF (2006), French Polynesia was the world's top exporter with a 27% market share (Tisdell and Poirine 2008). Indonesia with 412 million US dollars in sales of unworked cultivated pearls globally (Shefali et al. 2021; Tisdell and Poirine 2008). The global pearl jewellery market was valued at 11 billion USD in 2023, growing at a CAGR of 12% from 2024 to 2033. Moreover, the market is expected to reach 34.16 billion USD by 2033.

29.7 Conclusion

The pearl market has a bright future due to increasing interest in and respect for pearls. The demand for pearls is expected to grow as Asian economies expand and people's quality of life improves. Additionally, a growing number of individuals are becoming aware of pearls and taking an interest in them due to globalization and the expansion of the tourism sector. This will further drive the growth and development of the pearl oyster market. In the future, new mollusc farming methods may be developed to enhance pearl yield and quality due to technological advancements. Furthermore, pearl design and processing will continue to evolve, creating various pearl products to meet diverse customer demands. The mollusc market will encounter new growth opportunities due to these advancements. Although the pearl farming industry has a lot of promise, sustainability issues exist. Depletion of resources and environmental damage can result from the overexploitation of pearl oyster resources. Therefore, to guarantee their sustainable usage, pearl oyster resources need to be effectively managed and maintained. Furthermore, a number of conventional pearl-growing techniques have the potential to degrade the environment by contaminating water, upsetting the ecology, and producing pollution. In order to reduce environmental effects and attain sustainable growth in the pearl farming industry, green farming methods must be promoted. There are still a lot of unsolved issues with the

Table 29.4 This table outlines key environmental, biological, regulatory and economic issues that affect pearl culture along with their consequences for industry sustainability and growth

Sl no	Challenge	Impact on industry	References
1.	Environmental pollution	The plastic used in pearl farming can generate microplastics and pollute landfills. Regulations on waste disposal from farms can add significant costs to operations, requiring investments in filtration systems and monitoring	(Ram 1997; Saurabh et al. 2014)
2.	Climate change	Climate change can cause harmful red tides and new diseases that can wipe out pearl oysters	(Saurabh et al. 2014)
3.	Overfishing and habitat destruction	Overexploiting lagoons can cause imbalances in the ecosystem	(Latchere et al. 2018)
4.	Diseases and parasites	Organisms weaken oyster shells and increase disease risk Microorganisms compromise the integrity, and predators that consume oysters make them more susceptible	(Rao 1989; Bannister et al. 2019)
5.	Market fluctuations	Pearl growers may see significant income reductions if the market price of pearls declines sharply Shifts in consumer preferences towards alternative pearls can decrease demand for less popular types, resulting in unsold inventory and financial issues Simultaneous increase in production from multiple farms can saturate the market, causing oversupply and price drops Adapting to market changes can make it difficult for growers to invest in new methods, as the risk of price declines may discourage substantial capital commitments Smaller pearl farms, with limited production and product variety, are more vulnerable to market fluctuations	(Saurabh et al. 2014; Mohamed and Sasikumar 2016)

(continued)

Table 29.4 (continued)

Sl no	Challenge	Impact on industry	References
6.	Water temperature and quality	Increased mortality rate due to the rise in temperature Rapid nacre deposition is caused by temperature stress Warm water can increase the infection rate Slow development due to low temperature conditions Low DO and excessive nutrients can cause stress and degrade pearl quality Environmental pollutants can affect pearl growth and quality Fluctuation of pH level can hinder nacre synthesis	(Han et al. 2017)
7.	Regulatory challenges	Strict environmental rules, limitations on harvesting methods and poor waste management may reduce production quality and volume, leading to financial losses and hindering industry growth	(Underwood et al. 2023)

pearl oyster producing process that need more research. The molecular and cellular mechanisms behind the biomineralization and crystal growth processes involved in pearl creation, for example, require more investigation. Additionally, researching the interactions between different biological and environmental elements that result in pearl production might provide light on the intricate regulatory mechanisms at work. Furthermore, we may learn more about the biological characteristics and ecological adaptability of pearl oysters by utilizing contemporary technologies like proteomics and genomics. A more thorough comprehension and useful uses of the pearl manufacturing process will be made possible by these investigations. Molluscs that produce pearls have bright futures overall, but there are certain obstacles. By supporting scientific research, encouraging the use of green farming technologies, and improving resource conservation and management, we can preserve and advance pearl culture, protect pearl oyster resources, and achieve sustainable development in the mollusc farming industry—all of which will contribute more significantly to the advancement of human society. Additionally, research on the pearl-formation process will provide fresh perspectives and chances for advancement in contemporary medicine and technology, expanding its uses and promoting both technical innovation and societal advancement. Research into pearl formation is therefore extremely important from a scientific and sociological standpoint.

29.8 Future Prospects

The economic and ecological significance of molluscs is poised to grow substantially in the coming decades due to the increasing global interest in sustainable aquaculture, biotechnology, and eco-friendly luxury goods such as pearls. The interdisciplinary focus of this book chapter, combining biological mechanisms with economic applications, positions it at the frontier of research and innovation in molluscan studies. The following future directions can be envisioned:

1. **Expansion of Sustainable Pearl Aquaculture**

With the demand for eco-conscious luxury items on the rise, sustainable pearl farming is likely to experience significant growth. Advances in molluscan physiology and breeding technologies may lead to improved pearl yield, quality, and reduced environmental impact.

2. **Biotechnological Innovations**

Understanding the biomineralization processes in molluscs could drive innovations in biomaterials, nanotechnology, and bioengineering, especially in the synthesis of biomimetic materials inspired by nacre (mother-of-pearl).

3. **Molluscs in Blue Economy Models**

As nations increasingly embrace the concept of the Blue Economy, molluscs are expected to play a pivotal role—not only in pearl production but also in biofiltration, coastal ecosystem restoration, and livelihood enhancement in coastal communities.

4. **Genomic and Molecular Research**

Future studies will likely delve deeper into the genomic regulation of nacre formation, growth patterns, and immune responses of molluscs, facilitating selective breeding programs for economically desirable traits.

5. **Climate Change Resilience**

Research on molluscs' adaptability to ocean acidification and warming waters will be crucial in making aquaculture resilient to climate change, ensuring long-term viability and economic returns.

6. **Diversification of Products**

Beyond pearls, molluscs offer potential in cosmetics, nutraceuticals, biomedicine, and environmental biosensors. This diversification enhances their status as economic agents beyond traditional roles.

7. **Policy and Socioeconomic Integration**

Future interdisciplinary collaborations involving policymakers, economists, and marine biologists can facilitate inclusive and sustainable development policies that integrate molluscan aquaculture into rural economies, especially in the Global South.

Acknowledgements Authors are thankful to the Research and Development section of Fakir Mohan University for providing the financial assistance by Seed Money Project Grant.

References

Addadi L, Moradian J, Shay E, Maroudas N, Weiner S (1987) A chemical model for the cooperation of sulfates and carboxylates in calcite crystal nucleation: relevance to biomineralization. Proc Natl Acad Sci 84(9):2732–2736

Addadi L, Joester D, Nudelman F, Weiner S (2006) Mollusk shell formation: a source of new concepts for understanding biomineralization processes. Chem–A Europ J 12(4):980–987

Addadi L, Weiner S (2014) Biomineralization: mineral formation by organisms. Phys Scr 89(9):098003

Alagarswami K, Dharmaraj S, Velayudhan T, Chellam A, Victor A, Gandhi A (1983) Larval rearing and production of spat of pearl oyster Pinctada fucata (Gould). Aquaculture 34(3–4):287–301

Alagarswami K, Dharmaraj S, Velayudhan T, Chellam A (1987) Hatchery technology for pearl oyster production. CMFRI Bull-Pearl Cult 39:62–71

Alagarswami K (1974) Development of cultured pearls in India. Curr Sci 43(7):205–207

Alagarswami K (1977) Larval transport and settlement of pearl oysters (Genus Pinctada) in the Gulf of Mannar. In: Proceedings of the symposium on warm water zooplankton. National Institute of Oceanography, Goa

Alagarswami K (1987) Cultured pearls-production and quality. CMFRI Bull-Pearl Cult 39:107–111

Antoinette Matlins P (2012) The pearl book: the definitive buying guide. Turner Publishing Company

Azarian MH, Sutapun W (2022) Biogenic calcium carbonate derived from waste shells for advanced material applications: a review. Front Mater 9:1024977

Bannister J, Sievers M, Bush F, Bloecher N (2019) Biofouling in marine aquaculture: a review of recent research and developments. Biofouling 35(6):631–648

Barnard K (1927) South African nudibranch Mollusca, with descriptions of new species, and a note on some specimens from Tristan d'Acunha. Ann South Afric Museum 25(1):171–215

Behera PR (2017) Pinctada margaritifera (Linnaeus, 1758)

Beu A (1973) Descriptions of new species and notes on taxonomy of New Zealand Mollusca, No. 2. J R Soc N Z 3(3):307–332

Bieler R, Mikkelsen PM, Collins TM, Glover EA, González VL, Graf DL, Harper EM, Healy J, Kawauchi GY, Sharma PP (2014) Investigating the Bivalve Tree of Life–an exemplar-based approach combining molecular and novel morphological characters. Invertebr Syst 28(1):32–115

Bigatti G, Giraud-Billoud M, Vega IA, Penchaszadeh PE, Castro-Vazquez A (2010) The calcareous egg capsule of the Patagonian neogastropod Odontocymbiola magellanica: morphology, secretion and mineralogy. J Molluscan Stud 76(3):279–288

Camacho AP, Gonzalez R, Fuentes J (1991) Mussel culture in Galicia (NW Spain). Aquaculture 94(2–3):263–278

Carriker MR (2001) Embryogenesis and organogenesis of veligers and early juveniles. Dev Aquacult Fish Sci Elsevier 31:77–115

Choi Y-J, Tri NT, Lee J-M, Kang S-J, Choi B-D (2017) Evaluation of nutrients during rack and bag culture or suspended culture of Pacific oyster Crassostrea gigas. Korean J Fish Aquat Sci 50(3):263–269

Claus C, Maeckelberghe H, de Pauw N (1983) Onshore nursery rearing of bivalve molluscs in Belgium. Aquacult Eng 2(1):13–26

Cochennec-Laureau N, Montagnani C, Saulnier D, Fougerouse A, Levy P, Lo C (2010) A histological examination of grafting success in pearl oyster Pinctada margaritifera in French Polynesia. Aquat Living Resour 23(1):131–140

Collado GA, Valladares MA, Suárez C, Seguel M, Cabello-Guzmán G (2023) Shape, microstructure, and chemical composition of pearls from the freshwater clam Diplodon chilensis native to South America. Animals 13(13):2231

Costello MJ, Emblow C, White RJ (2001) European Register of Marine Species: a check-list of the marine species in Europe and a bibliography of guides to their identification

Costello MJ, Bouchet P, Boxshall G, Fauchald K, Gordon D, Hoeksema BW, Poore GC, van Soest RW, Stöhr S, Walter TC (2013) Global coordination and standardisation in marine biodiversity through the World Register of Marine Species (WoRMS) and related databases. PLoS One 8(1):e51629

Cunha RL, Blanc F, Bonhomme F, Arnaud-Haond S (2011) Evolutionary patterns in pearl oysters of the genus Pinctada (Bivalvia: Pteriidae). Mar Biotechnol 13:181–192

Di Monterosato TA (1884) Nomenclatura generica e specifica di alcune conchiglie mediterranee. (No Title)

Dix TG (1972) Biology of evechinus chloroticus (Echinoidia: Echinometridae) from different localities: 4. Age, growth, and size. New Zealand J Mar Freshwater Res 6(1–2):48–68

Doroudi MS, Southgate PC (2003) Embryonic and larval development of Pinctada margaritifera (Linnaeus, 1758). Molluscan Res 23(2):101–107

El-Refi A, Gharib I, Abdul Fatah N (2022) Pearls: a literature review. Int Des J 12(4):243–251

ElShazali A, Nahid A, Salma H, Elfadil E (2011) Effect of radiation process on antinutrients, protein digestibility and sensory quality of pearl millet flour during processing and storage. Int Food Res J 18(4):1401

Farn AE (2013) Pearls: natural, cultured and imitation. Elsevier

Galante-Oliveira S, Marçal R, Guimarães F, Soares J, Lopes JC, Machado J, Barroso C (2014) Crystallinity and microchemistry of *Nassarius reticulatus* (Caenogastropoda) statoliths: towards their structure stability and homogeneity. J Struct Biol 186(2):292–301

Gauthier J-P, Fereire J, Bui TN (2018) An explanation of a specific type of circling as observed on Ming cultured pearls. In: Bumble bee stone from Indonesia tracking and traceability in the gem industry circling on: 240

Gervis MH, Sims NA (1992) The biology and culture of pearl oysters (Bivalvia pteriidae). WorldFish

Gökoğlu N, Gökoğlu N (2021) Molluscan shellfish. In: Shellfish processing and preservation. pp 129–250

Gould A (1850) Descriptions of shells from the United States exploring expedition. Proc Boston Soc Nat Hist

Han T, Shi R, Qi Z, Huang H, Liang Q, Liu H (2017) Interactive effects of oyster and seaweed on seawater dissolved inorganic carbon systems: implications for integrated multi-trophic aquaculture. Aquaculture Environ Interact 9:469–478

Hänni HA (2006) Keshi Perlen, ein erklärungsbedürftiger Begriff. Verlag nicht ermittelbar

Hashimoto N, Kurita Y, Wada H (2012) Developmental role of dpp in the gastropod shell plate and co-option of the dpp signaling pathway in the evolution of the operculum. Dev Biol 366(2):367–373

He F (2019) Diversity and risk patterns of freshwater megafauna: a global perspective

Howard B, Mitchell PC, Ritchie A, Simkiss K, Taylor M (1981) The composition of intracellular granules from the metal-accumulating cells of the common garden snail (Helix aspersa). Biochem J 194(2):507–511

Ito M (2009) Improving pearl quality by grafting and husbandry methods. Aqua Tips 20(1):1–8

Jagadis I (2009) Seed production and farming of Indian pearl oyster Pinctada fucata-Winter School on recent advances in breeding and larviculture of marine finfish and shellfish

Johnston W, Hine D, Southgate PC (2019) Overview of the development and modern landscape of marine pearl culture in the South Pacific. J Shellfish Res 38(3):499–518

Kakisawa H, Sumitomo T (2012) The toughening mechanism of nacre and structural materials inspired by nacre. Sci Technol Adv Mater 12(6):064710

Katupotha J (2019) Pearl fishery industry

Kiefert L, Moreno D, Arizmendi E, Hanni H, Elen S (2004) Cultured pearls from the Gulf of California, Mexico. Gems Gemol 40(1):26–38

Kishore P, Southgate PC (2015) Does the quality of cultured pearls from the black-lip pearl oyster, Pinctada margaritifera, improve after the second graft? Aquaculture 446:97–102

Kishore P, Southgate PC (2016) The effect of different culture methods on the quality of round pearls produced by the black-lip pearl oyster Pinctada margaritifera (Linnaeus, 1758). Aquaculture 451:65–71

Knoth M, Miehe G, Fuess H (2005) Characterization of pearl luster pigments. Adv Eng Mater 7(10):928–931

Kondrachuk AV, Wiederhold ML (2004) On generation of statoconia in gravireceptors of mollusks. Hear Res 197(1–2):24–34

Kripa V, Mohamed K, Appukuttan K, Velayudhan T (2007) Production of Akoya pearls from the Southwest coast of India. Aquaculture 262(2–4):347–354

Kumar R, Gupta R, Dahiya T, Yadav R, Saifi R, Yodha K, Kumari B, Singh P, Shefali (2022a) Surveillance of *Heteropneustes fossilis* (Stinging Catfish) diseases and health management practices in Haryana. Appl Ecol Environ Sci 10(4):261–267. https://doi.org/10.12691/aees-10-4-12

Kumar R, Gupta RK, Yadav R, Saifi R, Yodha K, Pooja (2022b) *Eisenia fetida* as protein source for growth enhancement of *Heteropneustes fossilis*. Egypt J Aquat Biol Fish 26(2):577–588. https://doi.org/10.21608/ejabf.2022.234553

Kumar R, Yadav R, Kumar Gupta R, Pooja (2022c) A sustainable way for fish health management by replacement of chemical and drugs by earthworm. In: Environmental degradation in Asia: land degradation, environmental contamination, and human activities. Springer International Publishing, Cham, pp 329–352. https://doi.org/10.1007/978-3-031-12112-8_16

Latchere O, Mehn V, Gaertner-Mazouni N, Le Moullac G, Fievet J, Belliard C, Cabral P, Saulnier D (2018) Influence of water temperature and food on the last stages of cultured pearl mineralization from the black-lip pearl oyster Pinctada margaritifera. PLoS One 13(3):e0193863

Liu L (2008) Checklist of marine biota of China seas. Institute of Oceanography, Chinese Academy of Science, China

Løvtrup S (1965) Morphogenesis in the amphibian embryo fertilization and blastula formation. Wilhelm Roux'archiv Für Entwicklungsmechanik der Organismen 156:204–248

Lowenstam HA, Weiner S (1989) On biomineralization. Oxford University Press

Lydeard C, Cummings KS (2019) Freshwater mollusks of the world: a distribution atlas. JHU Press

Macnae W (1958) The geographic and oceanographic setting of Inhaca. In: Macnae W, Kalk M (eds.) A natural history of Inhaca Island, Mecambique. Johannesburg, pp 1–4

Marin F, Le Roy N, Marie B (2012) The formation and mineralization of mollusk shell. Front Biosci 4(3):1099–1125

Mathew A (2002) Nursery rearing of pearl oyster Pinctada fucata (Gould, 1850) under onshore conditions. Central Marine Fisheries Research Institute, Cochin

McDougall C, Degnan BM (2018) The evolution of mollusc shells. Wiley Interdiscipl Rev Dev Biol 7(3):e313

MH V (2023) Taxonomic identification of molluscan species. St Teresa's College (Autonomous), Ernakulam

Mohamed KS, Sasikumar G (2016) Overview of bivalve fisheries of India

Muhr J, Ackerman KM (2023) Embryology, gastrulation. StatPearls [Internet], StatPearls Publishing

ng Pilipinas R (2010) Prehistory (pre–900). Ethnic Groups 5:33–37

Nielsen C (2004) Trochophora larvae: CEll-lineages, ciliary bands, and body regions. 1. Annelida and Mollusca. J Exp Zool B Mol Dev Evol 302(1):35–68

Otter LM, Agbaje OB, Huong LT-T, Häger T, Jacob DE (2017) Akoya cultured pearl farming in Eastern Australia. Gems Gemol 53:423–437

Pabian RK (1970) Record in rock: a handbook of the invertebrate fossils of Nebraska. Conserv Surv Div 1

Pangilinan JL (2014) In focus: 9 facts you may not know about philippine national symbols. National Commission for Culture and the Arts

Ram KJ (1997) Freshwater pearl culture in India
Rao NS (1989) Handbook, freshwater molluscs of India, Zoological Survey of India
Reeve L, Reeve L (1857) Monograph of the genus Avicula. Conchol Icon 10:1–160
Rinju RM, Anil M, Williams ES (2024). Studies on the Biometry of hatchery - produced and farmed Asian pearl oyster Pinctada imbricata fucata (Gould 1850) along the southwest coast of India.. 12. 123-130.
Saucedo PE, Southgate PC (2008) Reproduction, development and growth. Elsevier Amsterdam, the Netherlands
Saurabh S, Mohanty U, Mohanty J, Jayasankar P (2014) Pearl culture technology in freshwater environment. Aquaculture and fisheries environment. Discovery Publishing House Pvt. Ltd., New Delhi, India, pp 51–78
Shefali, Kumar R, Sankhla MS, Kumar R, Sonone SS (2021) Impact of pesticide toxicity in aquatic environment. Biointerface Res Appl Chem 11(3):10131–10140. https://doi.org/10.33263/BRIAC113.1013110140
Shen FL (2022) Pearl: nature's perfect gem. Reaktion Books.
Siddique MF, Barman AC, Haque MA, Tanu M, Shahjahan M, Uddin MJ (2022) Effects of pH on image pearl production in freshwater mussels (Lamellidens marginalis) under controlled temperature. Int J Nat Soc Sci 9(3):74–83
Southgate PC, Beer AC (2000) Growth of blacklip pearl oyster (Pinctada margaritifera) juveniles using different nursery culture techniques. Aquaculture 187(1–2):97–104
Southgate P, Lucas J (2011) The pearl oyster. Elsevier
Southgate PC (2021) Pearl oyster culture and pearl production. Molluscan shellfish aquaculture: a practical guide. CABI GB, pp 205–231
Spencer H, Marshall B, Willan R (2009) Checklist of New Zealand living Mollusca. Canterbury University Press Christchurch, pp 196–219
Streeter EW (2021) Pearls and pearling life. Read Books Ltd.
Taylor J, Strack E (2008) Pearl production. Pearl Oyster 273–302
Tëmkin I (2010) Molecular phylogeny of pearl oysters and their relatives (Mollusca, Bivalvia, Pterioidea). BMC Evol Biol 10:1–28
Tisdell C, Poirine B (2008) Economics of pearl farming. Pearl Oyster 473:495
Todt C, Wanninger A (2010) Of tests, trochs, shells, and spicules: development of the basal mollusk Wirenia argentea (Solenogastres) and its bearing on the evolution of trochozoan larval key features. Front Zool 7:1–17
Underwood LH, van der Reis A, Jeffs AG (2023) Diet of snapper (*Chrysophrys auratus*) in green-lipped mussel farms and adjacent soft-sediment habitats. Aquacult Fish Fisheries 3(3):268–286
Venkataraman K (2007) Marine ecosystems of India. Indian J Environ Educ 7:7–26
Victor A, Chellam A, Dharmaraj S, Velayudhan T (1995) Manual on pearl oyster seed production, farming and pearl culture. CMFRI Specl Publ 63:1–53
Vikas P, Subramannian S (2020) Accelerated seed production of pearl spot using modified hatchery method. Train Manual-Aquacult Work 57–67
Xu X, Xuan J (2023) Exploration of the mechanism and process of pearl formation. Int J Aquacult 13(11):1
Yazawa E, Zhou C (2018) Colored stones and pearls. Gems Gemol 54(3)
Zhang Y, Geng S, Yu G, Hong Y, Hu B (2024) Research progress on formation mechanism of pearl. Heliyon
Zhong Y (2023) Analysis of the morphology, optical properties and internal structure of pearls. Int J Aquacult 13(9)
Zhu C, Southgate PC, Li T (2019) Production of pearls. In: Goods and services of marine bivalves, pp 73–93

Chapter 30
Carangids in Indian Waters: Importance, Knowledge Gaps, and Pathways to Sustainable Fisheries

Sonal Rajendra Kalbande, Kusum Komal Karati, and Ajit Antony

Abstract Carangids, also known as pompanos, scads, trevallies, runners, amberjacks, queenfish, and jacks, are a significant group of Indian marine pelagic fisheries. They perform a vital role in both the fisheries and marine ecosystems. In Indian waters approximately 60 species of carangids have been reported, which exhibit wide-ranging adaptations from coastal to oceanic pelagic habitats, and play critical roles both as predators and prey in marine food webs. This study at first provides a comprehensive overview of Indian carangids, emphasizing their distribution, species diversity, spawning and recruitment success, their ecological significance and addresses the gap area which requires to achieve sustainable carangid fisheries. Despite their ecological and economic importance, carangid population face threats from overexploitation, habitat degradation, pollution, and climatic changes, necessitating urgent sustainable management measures. By integrating ecological research with modern fisheries practices, the study underscores the potential for sustainable fisheries and resource management strategies that ensure the long-term viability of carangid stocks. This approach supports coastal livelihoods and promotes conservation of the carangid diversity, offering a pathway for the responsible development and utilization of India's carangid resources.

Keywords Carangids · Distribution and habitat · Spawning · Sustainable fishery

30.1 Introduction

India has diverse fishery resources, from hill streams to deep-sea areas, and is home to four major biodiversity hotspots. Globally, a total of about 36,272 fish species are recorded (Fricke et al. 2022), and among them around 3496 of these species are

S. R. Kalbande · K. K. Karati (✉) · A. Antony
Centre for Marine Living Resources and Ecology, Kochi, Kerala, India
e-mail: karatikusum@cmlre.gov.in

S. R. Kalbande · A. Antony
Kerala University of Fisheries and Ocean Studies, Kochi, Kerala, India

R. Kumar et al. (eds.), *Sustainability in Aquaculture and Fisheries*,
Springer Water, https://doi.org/10.1007/978-3-032-16605-0_30

found in India, contributing 9.6% of the world's total fish diversity (Banerjee et al. 2022). Considering the total fish diversity in India, marine fish represent about 70% of the total, consisting of 2443 species across 230 families and 40 orders (Jayakumar et al. 2022). Notably, the order Perciformes is the most diverse, and includes 1367 species, accounting for 56% of the identified marine fish species in this region.

Among these, the family Carangidae is an important group within the large order Perciformes, which includes spiny-rayed fishes (Acanthopterygii), making it one of the most prominent families of bony fishes (Joshi et al. 2011). Globally, these fishes are found across tropical and subtropical marine environments, whereas some species are extended upto temperate waters. Carangids are important as they are commercially valuable, widely distributed, and exhibit diverse ecological roles, leading to their recognition as vital components of marine ecosystems (Rakhunde et al. 2023).

Carangidae were considered to have approximately 140 species under 32 genera worldwide (Abdussamad et al. 2013). The community composition in this group has been revised further and Nelson (2016) reported 147 species across 30 genera. Recently, Froese and Pauly (2023) has identified 148 species within 30 genera, making Carangidae the largest family in the order Carangiformes. Considering the surrounding seas of Indian subcontinent, Carangidae is represented by 60 species, and thus they form a highly diverse group in this water (Abdussamad et al. 2013). The complex morphological and meristic traits of fishes in this group make species identification quite challenging.

A variety of commercially important species supports India's carangid fishery, such as Indian threadfish (*Alectis indicus*), Threadfinned trevally (*A.ciliaris*), shrimp scad (*Alepes djedaba*), Razorbelly scad (*A. para*), Indian scad *(Decapterus russellii),* Redtail scad *(D. kurroides),* Torpedo scad *(Megalaspis cordyla),* Double spotted queenfish *(Scomberoides lysan),* Talang queenfish (*S. commersonianus*), Barred queenfish (*S. tala*), Leather jacket (*S. tol*), Snubnose pompano *(Trachinotus blochii),* Small spotted Pompano *(T. baillonii),* Largespotted dart *(T. botla),* Oxeye scad *(Selar boops),* Bigeye scad *(S. crumenophthalmus),* Kuweh trevally *(Atropus atropos),* Bigeye trevally *(Caranx sexfasciatus),* Giant kingfish *(C. ignobilis),* Longfin trevally *(Carangoides armatus),* Malabar trevally *(C. malabaricus),* Coachwhip trevally *(C. oblongus),* Yellowtail scad *(Atule mate),* Black-banded trevally *(Seriolina nigrofasciata),* Black pomfret *(Parastromateus niger),* and Yellow-stripe scad *(Selaroides leptolepis)* (Nair and Radhakrishnan 2000; Kasim 2003) (Fig. 30.1). All these species of fishes in the Carangidae family contribute to its diversity and economical importance. Especially, queenfishes belonging to the Scomberoid genera and jacks under both *Naucrates* and *Uraspis* genera are valuable because of their large size, and thus having high economic importance (Kasim 2003; Kumar et al. 2022a, b). Considering the overall carangid community, they are of great commercial importance due to their size, abundance, and high economic value. As a result, these fishes are widely harvested and both artisanal and industrial fisheries play an important role in their catch particularly in tropical and subtropical regions.

In Indian waters, carangids have a crucial part of total marine fish landings, with 14 commercially significant species identified along the coastline (Abdussamad et al. 2013). The carangid community in Indian waters include large and medium-sized

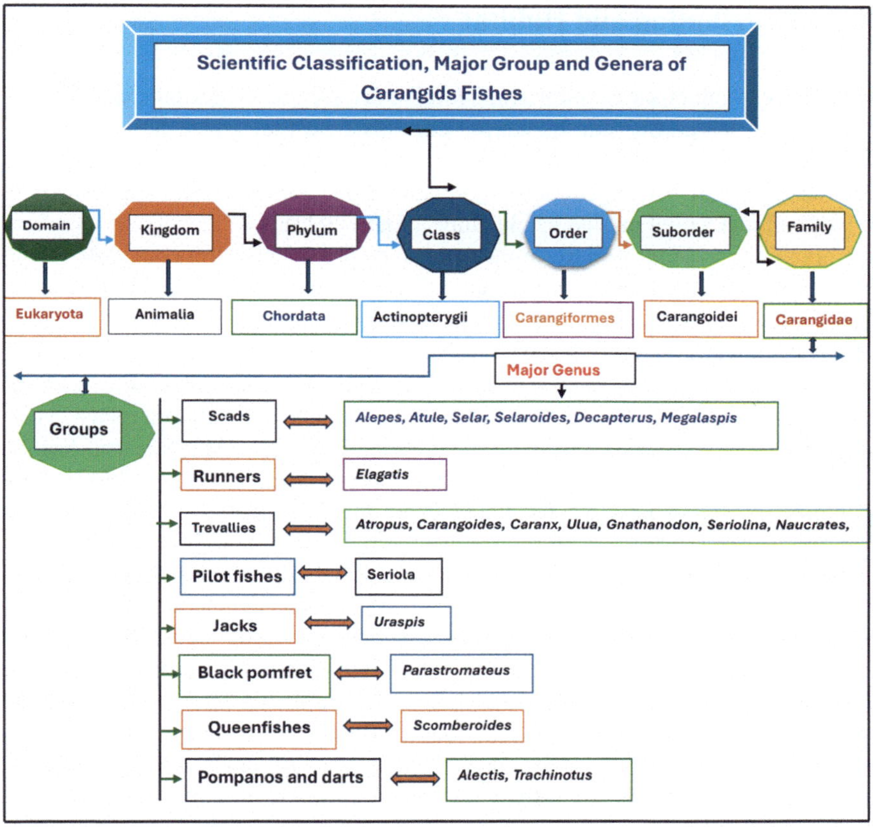

Fig. 30.1 Scientific classification, major groups and genera of Carangids *Source* Author

fishes, including *Caranx* spp., *Alectis* spp., *Trachinotus* spp., and *Selar* spp. All these species are highly valued in the market and thus have economic importance. Carangid catches are most prevalent during the monsoon season, especially along the northeastern and southwestern coasts of India (Rohit et al. 2024). Fishermen use different methods to catch these fishes such as boat seines, purse seines, trawls, hooks and lines, drift gillnets (Kumar et al. 2022c; Bandkar et al. 2022). Understanding the fishery, biology, and life cycle of carangid species are essential due to their ecological significance and economical importance. This research not only offers insights into the sustainable management and conservation of these resources but also emphasizes their potential for further development in sustainable fishing practices.

30.2 Distribution and Habitat

Although, carangid fishes exhibit a high level of diversity and contribute significantly to global marine production, they represent only 0.56% of global finfish aquaculture production (FAO 2022). Considering capture fishery, in Indian waters, carangid fisheries have recently experienced significant growth, contributing approximately 6–10% to India's total marine fish landings (Anon 2022).

Carangids can adapt well to various aquatic environments across the globe, especially in the Pacific, Indian, and Atlantic Oceans. (Poojary et al. 2015; Rakhunde et al. 2023). These fishes primarily inhabit tropical and subtropical marine environments; however, certain species are also found in temperate regions (Joshi et al. 2011). Carangid species are found across the Indian Ocean, inhabiting both the eastern and western coasts of the Indian subcontinent and support robust, year-round fisheries, particularly in the coastal states of Gujarat, Maharashtra, Kerala, Tamil Nadu, Karnataka, and Andhra Pradesh (Nair and Radhakrishnan 2000). They primarily inhabit shallow waters, typically up to 60 m depth; although they can also be located at depths of up to 100 meters in certain areas (Nair and Radhakrishnan 2000). While most carangid species are widely distributed globally, certain species, such as *Alectis* spp., are generally solitary and tend to inhabit specific coastal habitats, including coral reefs, rocky outcrops, and mangrove areas. Their solitary behavior may result in a more localized distribution within these broader geographical zones. Some species, such as the Indian Threadfish (*Alectis indicus*) and the Indian Scad (*Decapterus russellii*), predominantly inhabit continental shelf regions during their juvenile stages, where they thrive in brackish environments (Sreepada 2005). In contrast, species like Bigeye trevally (*Caranx sexfasciatus*) (Shefali et al. 2021; Froese and Pauly 2024) are oceanic pelagics, commonly found in open waters.

30.3 Carangid Landing Trends in India

According to Najmudeen (2024), from 2000 to 2023, India's average annual landings of carangids reached 213,131 tonnes, constituting 8.73% of the nation's total finfish landings. Among the diverse species of carangids, scads, particularly *Decapterus russelli*, emerged as the most prevalent, accounting for 39% of the overall carangid landings. The landing trends for scads indicate a consistent increase over the years. Notably, the coastal zone of southwest India is responsible for the majority (>75%) of scad landings, with Kerala and Karnataka being the primary contributors. Kerala accounts for 36.5% of the total scad landings, while Karnataka contributes 29.6%.

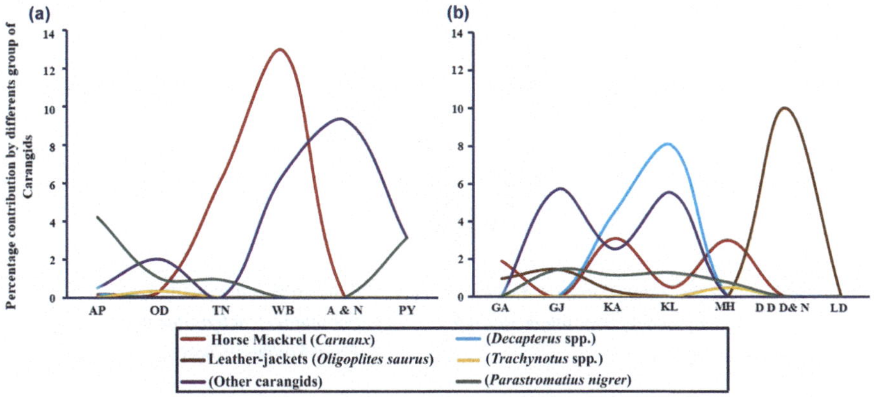

Fig. 30.2 Contribution of different group of Carangids to the total marine fish landing in the coastal states of India along **a** western Bay of Bengal **b** eastern Arabian Sea during 2020–2021 prepared based on data of Handbook of Fisheries Statistics (2022). (AP-Andhra Pradesh; OD-Odisha; TN-Tamil Nadu; WB-West Bengal; A &N-Andaman & Nicobar Islands; PY- Puducherry; GA-Goa; GJ-Gujrat; KA-Karnataka, KL-Kerala; MH-Maharashtra; D D D & N-Daman, Diu, Dadra & Nagar Haveli; LD-Lakshadweep)

30.4 Species Contribution of Carangids

In Indian waters, Pelagic finfish accounted for 57% of the total landing of marine fishes in 2022 (Najmudeen 2023). Based on the Handbook of Fisheries Statistics (Anon 2022), carangids in India are classified into six groups: horse mackerel (*Caranx* and *Trachurus* spp.), scads (*Decapterus* spp.), leatherjackets (*Oligoplites saurus*), pompano (*Trachinotus* spp.), other carangids, and black pomfret (*Parastromateus niger*). Figure 30.2 shows the percentage composition of these major groups across various states. During 2020–2021, the national carangid catch was contributed by: horse mackerel 2.3%, scads 1.3%, leatherjackets 0.4%, pompano 0.1%, other carangids 2.6%, and black pomfret 1.5% (Fig. 30.3). Along the west coast, the species composition closely followed the all-India trend, whereas notable variations were observed along the east coast.

30.5 Feeding Habit, Growth, and Life History

Research on the feeding habit of carangid species reveals that they are predominantly carnivores, preying on a diverse range of smaller crustaceans and fishes (Shetkar and Nirmale 2023). These fishes exhibit fast growth rates and relatively short lifespans, typically ranging from 2 to 6.6 years. Smaller species, such as *Selaroides leptolepis*, reach a maximum size of 185 mm, while larger species like queenfish and jacks can grow to lengths of 1–2 m. The smaller species frequently occur in huge groups in the

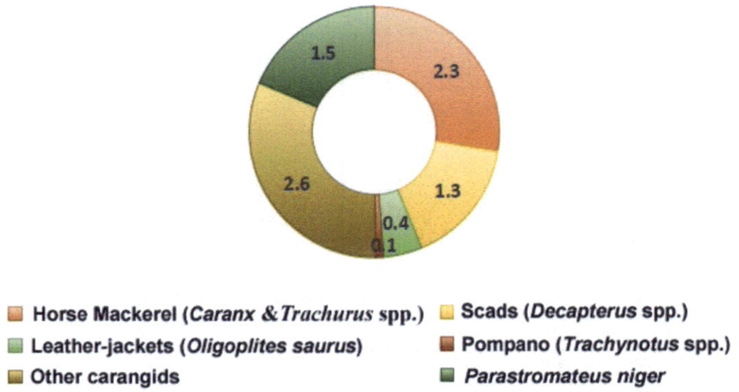

Fig. 30.3 Percentage contribution of different groups of carangids to the total marine fish landing along the total Indian coast

shallower coastal waters. Due to their rapid growth, prolonged spawning seasons, and continuous recruitment, carangid species are well-adapted to withstand significant fishing pressure, maintaining their populations despite current levels of exploitation (CMFRI Report 2022).

30.6 Spawning Details in Carangids

There are limited studies on various carangid species spawning behavior along the Indian coast (Bandkar et al. 2022). These investigations are mainly concentrated on factors such as gonadal development, ova diameter, distribution, and the size and timing of recruitment. (Laroche et al. 2006). These helped in the preliminary understanding of the spawning behavior of this important group. The details of the case study information available in Indian waters for the period of 1995–2023 are tabulated in (Table 30.1).

30.7 Knowledge Gap in Carangid Research in Indian Waters

Although carangids are economically significant, research on the developmental biology of this family is relatively limited, particularly regarding initial development of life stages (Rathnasuriya et al. 2021). Most of these studies have focused on adult specimens (Costa et al. 2005). Research on carangid species in Indian waters also has advanced, yet several knowledge gaps persist:

Table 30.1 The spawning case study information of some Indian carangid fish species

Carangid species	Regions of report	Estimated Season of Spawning	Estimated season of peak spawning	Size and age at first maturity	Fecundity	Sex Ratio (Male: Female)	Ova diameter	References
Megalaspis cordyla	Cochin	April–February	May–august	251–260 mm	91,854 to 324,292	–	–	Sivakami (1995)
Decapterus russelli	Vizhinjam,	March–may and November–December	June–July and February	150 mm (male) and 140 mm (female); 0.91 years	–	–	–	Balasubramanian and Natarajan (2000)
Decapterus macrosoma	Vizhinjam,	February, June, October and December	May and august	(male 158 mm and females 157 mm); 1.05 years	–	–	–	Balasubramanian and Natarajan (2000)
Decapterus russelli	Malabar	–	April—May; august–October	–	–	–	–	Manojkumar (2007)
Megalaspis cordyla	Ratnagiri	May–January	May–October	210 mm	92,268 -, 5, 49,900 average 2,47,671	1:1.3	–	Jadhav and Mohite (2014)
Decapterus russelli	Maharashtra waters, northwest coast of India	December–march and June–October	January and September	153 mm	29,986 - 1,52,123	1:0.43 (140–149 mm) to 1:1 (110–119) mm	–	Poojary et al. (2015)
Atropus atropos	Mangalore coast, Karnataka	September–December and march–may	–	160–180 mm (male) and 180–200 mm (female)	33,298–1, 88,675 average 94,083	1:1.3	0.113 mm to 0.787 mm	Rajesh et al. (2018)

(continued)

Table 30.1 (continued)

Carangid species	Regions of report	Estimated Season of Spawning	Estimated season of peak spawning	Size and age at first maturity	Fecundity	Sex Ratio (Male: Female)	Ova diameter	References
Seriolina nigrofasciata	South-west coast	–	March–may; October–December	25 cm (male) and 27 cm (female)	–	1:1.15	–	Rajesh et al. (2019)
Alepes djedaba	Cochin coast, Kerala	June to September	July.	193 mm (male) and 174 mm (female)	59,744 to 3,30,190 average 1,94,965.	1:1		Sajana (2019)
Alepes djedaba	Ratnagiri	–	–	149 mm	15,004–640,800 average 309,916	1: 0.85	–	Bandakar et al. (2022)
Atropus atropos	Ratnagiri	May–December	–	171 mm	16,458–138,688 average 70,097	–	–	Rakhunde et al. (2023)
Alepes kleinii	Northeastern Arabian Sea		January–march	12.25 cm(female) 12.57 cm (male)	30,653 to 70,275	1:1.45	230–720 μm	Ghosh (2023)

30.7.1 Life History and Ecological Role

Understanding about the life cycles, breeding patterns, and ecological roles of various carangid species is still limited. This is crucial for developing effective management and conservation strategies. While around 60 species are found in Indian waters, comprehensive biological data is available only for a few commercially important species (Joshi et al. 2011).

30.7.2 Abundance and Distribution

The overall distribution and abundance of carangids within India's exclusive economic zone (EEZ) is mostly known and few areas have been identified with higher densities. However, there is lack of comprehensive data across various seasons and regions which poses challenges for effective management and conservation of fisheries planning (Kasim 2003; Joshi et al. 2011; Vijith et al. 2023).

30.7.3 Taxonomy and Species Identification

The carangid family comprises approximately 60 species in Indian waters, many exhibiting morphological variations that complicate accurate identification. A comprehensive taxonomic review and the development of reliable field identification keys are essential to address these challenges. (Abdussamad et al. 2013; Ray et al. 2021).

30.8 Spawning and Hatching Development

Carangid eggs are pelagic, spherical, and vary in size between 0.7 mm and 1.5 mm (Ikeda and Mito 1988) while Honebrink and Randy (2000) recorded a slightly narrower range of 0.7 and 1.3 mm. These eggs possess narrow perivitelline space, generally transparent, with presence of a segmented yolk, one or more oil globules, and commonly accompanied by melanophores (Fig. 30.4a). Their appearances make them challenging to differentiate from the eggs of several other marine fish families. Hatching typically happens within 24–48 h after spawning, depending on size of eggs and temperature of surrounding water (Honebrink and Randy 2000).

Carangid larvae generally measure between 2 and 4.3 mm in length at the time of hatching (Leis and Trnski 2000). They possess a yolk sac which is generally large in size, unpigmented eyes, and an undeveloped mouth. The pigmentation of the larvae changes as the yolk is absorbed, as reported by Silva et al. (2024) (Fig. 30.4b). Leis

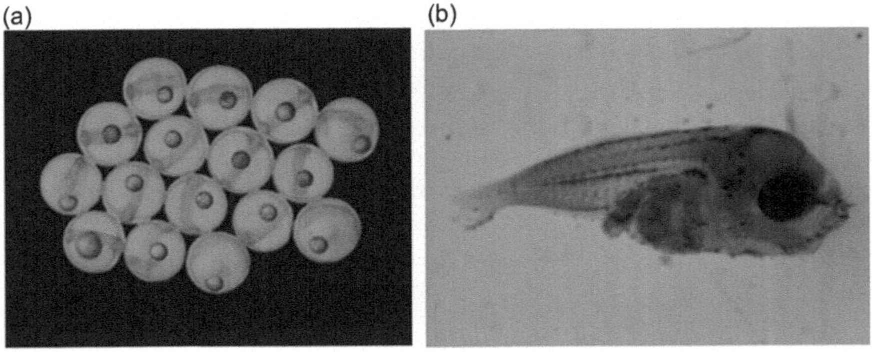

Fig. 30.4 Photos of **a**: Carangid fish eggs, **b**: Carangid fish larva (*Source* Author)

and Trnski (2000) reported that carangid larvae typically measure notochord length (NL) between 1-and 2-mm. Notable characteristics of larvae in the yolk sac stage include an oil globule at the frontal region of the sac, positioned at the ventral region of the head, and possessing 24–27 myomeres. However, the identification of newly hatched larvae can be difficult as differentiation of them from other marine species of fishes is quite challenging.

30.9 Conservation and Management for Sustainable Fisheries

The adoption of maximum sustainable yield (MSY) strategies for overfished stocks could increase yields by 10.6 megatons, representing 12% of total catches and 6% of aquatic animal production for the year 2022 (Elleby et al. 2025). A considerable number of fish populations are currently suffering from overexploitation, primarily due to the absence of sustainable management practices. Some are overexploited, while others are underexploited. In India, stock assessments indicate that many fish stocks are being exploited at or near MSY levels, with some stocks already overexploited (Pillai 2006). To develop effective conservation strategies for pelagic fisheries, it is essential to consider biological and socio-economic factors contributing to their vulnerability. The coastal fisheries sector faces challenges such as overfishing, habitat degradation, and pollution, threatening resource sustainability and impacting local communities (James 2010).

To address this issue, it is essential to implement comprehensive management and conservation strategies. Considering the long-term sustainability of carangid fishery, a multifaceted approach is required in any marine ecosystem (Joshi et al. 2011). This approach needs to consider factors like consistent monitoring, regulation of fish stocks, protection of habitats and spawning grounds. The enforcement of catch limits, reduction of bycatch, and the improvement of ecosystem health through habitat restoration efforts will further enhance ecosystem health through habitat restoration efforts (FAO 2022).

30.10 Sustainable Conservation Strategies

30.10.1 Marine Protected Areas (MPAs)

MPAs play an important role for the protection of the habitat of adult fishes as well as to safeguard their spawning and rearing grounds (FAO 2019). Considering the importance of carangid population, suitable areas need to designate as MPA where carangids are known to have prime spawning habitats so that in case of the overexploitation of any species also rebuilding of fish stock can be possible and thus the diversity as well as their ecological importance is maintained.

30.10.2 Regulations on Catch Limits and Quota Systems

The overfishing is a vital threat to the fishery having both ecological and economic impact. Hence, to maintain the sustainable yield, a few simple regulations are needed to consider. The regulation on the total allowable catch and quota system are the common control which needs to be considered for the maximum utilization of the potential yield of carangids in a sustainable way. The direct regulation on fishing effort can be done by controlling number of licenses, and closing areas during particular period of time. The instruction on minimum size of catch for different species of fishes and restrictions on types of fishing gear can aid further in the maintenance of the catch limit.

30.10.3 Seasonal Closures

Policy makers consider seasonal closures during breeding seasons can help to reduce fishing pressure and hence it can be a potential measure in the maintenance of carangid population. The seasonal closure allows the adult populations to reproduce and replenish without the chance of getting overexploited. The Implementation of seasonal closure during breeding period is an effective strategy to protect the early stages of the carangids and thus provides a perfect mechanism to maintain their populations (George and Shanmugam 2005). In Indian waters, important carangid species such as Bluefin Trevally (*Caranx melampygus*) and Giant Trevally (*Caranx ignobilis*), are known to have particular spawning seasons, with peak period of May to July (Pardee et al. 2023). In general, it has been recorded that carangid species are abundant along the coasts of Kerala, Tamil Nadu, and Andhra Pradesh of India, with peak seasons from January to June and further in September to November and southwest monsoon has been observed to be their main spawning period (Premlatha 1988). The implementation of seasonal closures during the peak spawning period can significantly contribute to the sustainable yield and management of carangid populations in Indian waters.

30.10.4 Bycatch Reduction

The development of fishing gear that helps to reduce bycatch of non-target species is crucial for the conservation of the species diversity of fishes. Hence, improvements of technologies to minimize bycatch are important for protecting both carangid stocks and overall marine biodiversity (Kasim and Hamsa 2003). The soft bycatch reduction devices for trawl fisheries have improved significantly in India in recent years, and thus playing vital role to decrease the capture of juveniles and unintended species. Continuous monitoring and collaboration with the fishing industry and fishermen can significantly help to reduce the negative impacts of bycatch on the marine ecosystem. Implementing suitable bycatch reduction devices along with proper monitoring can be vital for ensuring sustainable management of carangid resources and conserving ecological diversity.

30.10.5 Ecosystem-Based Fisheries Management (EBM)

The EBM deserves importance due to its consideration of the entire ecosystem, including species interactions and habitat conditions and thus promotes sustainable fisheries management. This approach thus not only helps to prevent overfishing; it ensures a healthier marine ecosystem. Hence, this approach can be employed for the management of the carangid fishery and this in turn will support a sustainable fishery with overall ecological balance (Rani et al. 2025; Nisha et al. 2025; Kiran et al. 2025; Muthiah and Shanmugam 2014).

30.10.6 Public Awareness and Community Involvement

For the effective implementation of sustainable fisheries management, public awareness and active involvement of local communities are crucial. Although, involving coastal communities in managing the marine resources can be difficult, educational and outreach programs play an instrumental role in encouraging fishermen to adopt sustainable practices. This will help in fostering an appreciation for the importance of conserving the fishes. Considering the importance of carangids, these initiatives will aim to bridge the gap between conservation objectives and community participation, and thus will ensure long-term success (Suresh and Sivakumar 2012).

30.10.7 Monitoring Targeted Fish Species

Monitoring targeted fish species is crucial because they are widely available, easy to catch, and have high consumer demand. Both small-scale (artisanal) and large-scale (industrial) fishing can negatively impact these fish populations. One effective way to protect these species is through continuously monitoring and regulating their individual species stocks (Dehadrai and Yadava 2004).

30.10.8 Enhancing Fisherwomen's Livelihoods

To empower fisherwomen, it is essential to provide financial support, equitable policies, and opportunities for skill development. To enhance their income, both small and large-scale financing and training in sustainable processing and market development are crucial. Access to direct marketing can further accelerate their earnings growth. Inclusive policies that promote gender equality, livelihood security, and opportunities for visionary leadership play a significant role in advancing sustainability in fisheries. Supporting fisherwomen not only uplifts communities but also fosters sustainable fisheries practices (Das 2023).

30.11 Conclusion

Carangids play an important role in global marine fisheries both economically and ecologically, and in Indian waters also, they have crucial significance. However, habitat degradation, intense fishing pressures, and changing climatic conditions necessitate urgent action to conserve and sustainably manage carangid stocks. This action should focus on improving stock assessments, protecting vital habitats, implementing responsible fishing practices, and adopting ecosystem-based management strategies. For effective conservation, it is crucial to engage the public, raise awareness, strengthen regulatory frameworks, and promote cooperation. These practical steps are essential for ensuring the long-term viability of carangid fisheries while supporting livelihoods and protecting carangid fisheries and marine biodiversity.

30.12 Future Prospects

The carangid fishery in Indian waters exhibits an important future prospect to due to its high market demand and export potential. This will be a vital step to achieve Sustainable Development Goal 14, Life Below Water. The required long-term monitoring of target species, technological advancement in gear, employment of suitable improved management policies, and public awareness can help us to achieve this goal.

Acknowledgements The authors thank the Secretary, Ministry of Earth Sciences (MoES), Government of India for the support of this study. We are grateful to the Director, Centre for Marine Living Resources & Ecology (CMLRE), Kochi for allowing us to complete the work. The first author conveys gratitude to Desk IITM, Pune for the support. This is a CMLRE Contribution No 218.

References

Abdussamad EM, Rohit P, Koya KP, Mohamed OMMJ, Jayabalan K (2013) Carangids (family: Carangidae) in the seas around Indian subcontinent with description of macro-taxonomic characters for the field identification of genera and species. Indian J Fish 60(2):21–36

Anon (2022) Handbook on fisheries statistics. Ministry of Agriculture (Department of Agriculture and co-operation, fisheries division). Government of India, New Delhi, p 153

Balasubrahmanian NK, Natarajan P (2000) Studies on the biology of the scads, *Decapterus russelli* and *Decapterus macrosoma* at Vizhinjam, southwest coast of India. Indian J Fish 47(4):291–300

Bandkar D, Nirmale VH, Gurjar UR, Metar SY, Pawar RA, Chogale ND (2022) Reproductive biology and feeding dynamics of the shrimp scad, *Alepes djedaba* (Forsskal, 1775) from Ratnagiri, central west coast of India. Indian J Fish 69(1):37–44

Banerjee D, Raghunathan C, Rizvi AN, Das D (2022) Animal discoveries 2021: New Species and New Records. Director, Zool. Surv, India, Kolkata, p 232

CMFRI (2022) Annual Report 2021. Central Marine Fisheries Research Institute, Kochi, p 300

Costa MR, Albieri RJ, Araujo FG (2005) Size distribution of the jack *Chloroscombrus chrysurus* (Linnaeus) (Actinopterygii, Carangidae) in a tropical bay at southeastern Brazil. Rev bras Zool 22:580–586

Das J (2023) Blue economy, blue growth, social equity and small-scale fisheries: A global and national level review. Stud Soc Sci Res 4(1):38–82

Dehadarai PV, Yadava YS (2004) Fisheries development. In: State of the Indian farmer–A millennium study, vol 13. Publ. By Dept. of agriculture and co-operation, Ministry of Agriculture, Govt. of India, New Delhi, p 173

Elleby C, Domínguez IP, Nielsen R, Nielsen M, Hoff A (2025) Introducing maximum sustainable yield targets in fisheries could enhance global food security. Commun Earth Environ 6(1):33

FAO (2019) Marine protected areas and fisheries. *Food and Agriculture Organization of the United Nations*. www.fao.org

Food and Agricultural Organization (2022) Fishery and aquaculture statistics. In: Global aquaculture production 1950–2020 (fish stat J). FAO Fisheries and Aquaculture Division, Rome, Italy. www.fao.org/fishery/statistics/software/fishstatj/en

Fricke R, Eschmeyer WN, VanderLaan R (2022) Eschmeyer's catalog of fishes: generaspecies. http://researcharchive.calacademy.org/research/ichthyology/catalogue/fishcatmain.asp

Froese R, Pauly D (2024) *Caranx sexfasciatus* (bigeye trevally). FishBase. www.fishbase.org

Froese R, Pauly D (eds) (2023). FishBase. World Wide Web electronic publication. https://www.fishbase.org

George MD, Shanmugam P (2005) Status of marine fishery resources and sustainable management practices in Indian waters. Indian J Mar Sci 34(4):319–328

Ghosh S (2023) Reproductive biology and feeding habits of Razorbelly Scad, Alepes kleinii (Bloch, 1793), in the northeastern Arabian Sea. J Coast Res 39(6):1094–1104

Honebrink RR (2000) A review of the biology of the family Carangidae, with emphasis on species found in Hawaiian waters. Resources, Hawaii Department of Land and Natural

Ikeda T and Mito S 1988 Pelagic fish eggs. In: Okiyama M (ed) An atlas of the early stage fishes in Japan. pp 999–1083

Jadhav TD, Mohite SA (2014) Reproductive biology of horse mackerel *Megalaspis cordyla* (Linnaeus, 1758) along Ratnagiri coast of Maharashtra. India J Mar Biol Ass India 55(2):35–40

James PSBR (2010) Taxonomic status of marine pelagic fishes of India, research priorities and conservation strategies for the sustainability of their fisheries. Indian J Anim Sci 80(4(Supp):39–45

Jayakumar TT, Bineesh KK, Kumar TA, Lal KK (2022) The New Finfi sh Genetic Resource Discoveries from Indian Waters–Need for Enhancing Taxonomy Capacity. Indian J Plant Genet Resour 35(03):293–297

Joshi KK, Nair RJ, Samad EMA, Thomas S, Kakati VS, Jasmine S, Varghese M, Miriam Paul S, Sukumaran S, George RM, Manisseri MK (2011) The carangids of India: A monograph. Central Marine Fisheries Research Institute, Indian Council of Agricultural Research

Kasim HM, Hamsa KMSA (2003) Conservation and management strategies for pelagic fisheries in India. J Mar Biol Assoc India 45(1–2):29–35

Kiran, Kumar N, Sindhu S, Gawaria B, Ansari A, Dhariwal A, Sharma P, Kumar R (2025) GWAS approaches in medicinal plants for enhanced secondary metabolite production. In: Chen JT (ed) Biotechnology, multiple omics, and precision breeding in medicinal plants. CRC Press, pp 119–134. https://doi.org/10.1201/9781003475491-10

Kumar R, Gupta R, Dahiya T, Yadav R, Saifi R, Yodha K, Kumari B, Singh P, Shefali (2022a) Surveillance of *Heteropneustes fossilis* (stinging catfish) diseases and health management practices in Haryana. Appl Ecol Environ Sci 10(4):261–267. https://doi.org/10.12691/aees-10-4-12

Kumar R, Gupta RK, Yadav R, Saifi R, Yodha K, Pooja (2022b) *Eisenia fetida* as protein source for growth enhancement of *Heteropneustes fossilis*. Egypt J Aquat Biol Fish 26(2):577–588. https://doi.org/10.21608/ejabf.2022.234553

Kumar R, Yadav R, Kumar Gupta R, Pooja (2022c) A sustainable way for fish health management by replacement of chemical and drugs by earthworm. In: Environmental degradation in Asia: land degradation, environmental contamination, and human activities. Springer International Publishing, Cham, pp 329–352. https://doi.org/10.1007/978-3-031-12112-8_16

Laroche JG, Ditty JT, Lamkin SR (2006) Carangidae: jacks. In: Richard WJ (ed) Early stages of Atlantic fishes: an identification guide for the Western Central North Atlantic, vol II. CRC press, Boca Raton, Florida, pp 1439–1510

Leis JM, Trnski T (2000) Carangidae. In: Leis JM, Carson-Ewart BM (eds) The larvae of indo-Pacific coastal fishes: an identification guide to marine fish larvae. Brill, Leiden, pp 499–517

Manojkumar PP (2007) Stock assessment of Indian scad, *Decapterus russelli* (Ruppell, 1830) off Malabar. J Mar Biol Assoc India 49(1):79–80

Mohamad Kasim H (2003) Carangids. Status of Exploited Marine Fishery Resources of India. CMFRI, Cochin, In, pp 66–75. isbn:81-901219-3-6

Muthiah S, Shanmugam P (2014) Ecosystem-based management for carangid fisheries: challenges and strategies in Indian waters. Indian J Mar Sci 43(4):455–463

Nair PN, Radhakrishnan (2000) Carangid resources of India. In: Marine fisheries research and management. CMFRI Kochi, Kochi, pp 317–348

Najmudeen TM 2023. Overview of marine fisheries of India. In: International workshop cum training on fisheries and aquaculture: African-Asian rural development organization (AARDO). CMFRI training manual series no. 35/2023 (35). ICAR-Central Marine Fisheries Research Institute, Kochi, pp 12–22. ISBN 978-93-82263-62-3

Najmudeen TM (2024) Marine finfish fisheries of India. In: Training manual on fisheries management practices and techniques. CMFRI training manual series no. 43/2024. ICAR-Central Marine Fisheries Research Institute, Kochi, pp 10–29

Nelson JS, Grande TC, Wilson MV (2016) Fishes of the world. John Wiley & Sons

Nisha A, Rani S, Kumar S, Kumar R, Kumar A (2025) Recent technologies and strategies for bioremediation of toxic and hazardous waste. In: Dubey KK, Shah MP, Yadav A (eds) Challenges and sustainable solutions in bioremediation. CRC Press, pp 1–16. https://doi.org/10.1201/978 1003407317-1

Pardee C, Wiley J, Fendrick T (2023) Working with community fishers to determine the spawning seasonality of two commonly targeted jack species. Mar Coast Fish 15(5):e10264

Pillai NGK (2006) Pelagic fisheries of India. In: Handbook of fisheries and aquaculture, published by Directorate of Information and Publications of agriculture. ICAR, New Delhi, pp 56–77

Poojary N, Tiwari LR, Sundaram S (2015) Reproductive biology of the Indian scad, *Decapterus russelli* (Ruppell, 1830) from Maharashtra waters, northwest coast of India. J Mar Biol Assoc India 57(1):71–77. ISSN 2321-7898

Premlatha P (1988) Studies on the distribution and abundance of carangid fish larvae along the southwest coast of India. J Mar Biol Assoc India 30(1&2):75–82

Rajesh DP, Anjanayappa HN, Benakappa S (2018) Maturation and spawning frequencies of *Atropus atropos* from Mangalore coast, Karnataka. India J Fish Life Sci 3:46–53

Rajesh KM, Rohit P, Mini KG, Sathyavathi T, Abdul Hakeem MM (2019) *Seriolina nigrofasciata* (Ruppel, 1829), its fishery and biological aspects off south-west coast of India. Indian J Fish 66(1):9–16

Rakhunde AV, Nirmale VH, Pawar RA, Metar SY, Bhosale BP (2023) Biology of the cleftbelly trevally *Atropus atropos* (Bloch and Schneider, 1801) from Ratnagiri coast of India. Indian J Fish 70(2)

Rani S, Sharma S, Singh D, Sheoran R, Kumar R (2025) Improved production of anti-cancer compounds from medicinal plants through molecular and biotechnological mutagenesis strategies. In: Chen JT (ed) Biotechnology, multiple omics, and precision breeding in medicinal plants. CRC Press, pp 287–297. https://doi.org/10.1201/9781003475491-21

Rathnasuriya MIG, Mateos-Rivera A, Skern-Mauritzen R, Wimalasiri HBU, Jayasinghe RPPK, Krakstad JO, Dalpadado P (2021) Composition and diversity of larval fish in the Indian Ocean using morphological and molecular methods. Mar Biodivers 51:1–15

Ray D, Mishra SS, Mohapatra A, Ghorai N (2021) Notes on carangids (Carangiformes: Carangidae) from West Bengal coast with new records. Rec Zool Surv India:511–520

Rohit P, Abdussamad EM, Rathinam MM, A., Ganga, U., Ghosh, S., Rajesh, K.M., Koya, K.P., Koya, M., Chellappan, A., Nakhawa, A.D. and Surya, S. (2024) Pelagic fisheries of India–an overview. Indian J Fish 71(1):12–28

Sajana N, Nandan SB, Radhakrishnan CK (2019) Feeding behaviour and reproductive biology of the shrimp scad *Alepes djedaba* (Forsskal, 1775) off Cochin coast, Kerala, South India. Indian J Fish 66(3):32–40

Shefali, Kumar, R., Sankhla, M.S., Kumar, R., and Swaroop S. Sonone (2021). Impact of pesticide toxicity in aquatic environment. Biointerface Res Appl Chem,11(3):10131–10140. https://doi.org/10.33263/BRIAC113.1013110140

Shetkar M, Nirmale V (2023) A review of food and feeding habits of carangids occurring along the Indian coast. J Indian Fish Assoc 50(3)

Silva-Segundo CA, Funes-Rodríguez R, Anaya-Godínez E, Gómez-Gutiérrez J (2024) Molecular and morphological identification of larvae of Carangidae (Teleostei, Carangiformes) species from southern Gulf of California. ZooKeys 1212:195

Sivakami S (1995) Fishery and biology of the carangid fish *Megalaspis cordyla* (Linnaeus) off Cochin. J Mar Biol Assoc India 37(1–2):237–248

Sreepada RA (2005) Marine fishes of the Western coast of India. Daya Publishing House, New Delhi, India

Suresh KP, Sivakumar K (2012) Overfishing and its consequences on fishery resources of the Indian Ocean region. Mar Pollut Bull 64(10):2130–2135

Vijith V, Anil AC, Shankar D, Antony SP (2023) Assessing the geographical distribution of carnivorous and planktivorous fishes in the western bay of Bengal. Eco Inform 74:101961

Chapter 31
White Button Mushroom: A Promising and Sustainable Alternative as a Feed Ingredient for Fish

Priya Singh, Rajender Kumar Gupta, Amrendra Kumar, Smruthi Hareendran, Shubham Janbandhu, and Nehal A. Tandel

Abstract Growing demand for sustainable aquaculture practices has spurred interest in alternative feed sources for finfish and shellfish. Mushrooms, particularly *Agaricus bisporus* from the Basidiomycetes family, are recognized as a good source of vital nutrients and physiologically active components having potential therapeutic properties. As the most widely cultivated mushroom globally, *A. bisporus* offers a nutritional profile that includes carbohydrates, proteins, lipids, fiber, minerals, and vitamins, making it a valuable dietary option. Its inclusion in fish diets can partially replace traditional fishmeal, resulting in various nutritional and physiological benefits across multiple species. These advantages of feeding fish diets supplemented with mushrooms include improved growth performance, improved immunological function and higher antioxidant activity. Additionally, the bioactive compounds found in white button mushrooms may improve disease resistance, promoting more resilient and sustainable aquaculture systems. This chapter investigates the possibilities of white button mushroom uses and its bioactive byproducts as effective feed ingredients for fish farming.

P. Singh (✉)
Department of Aquaculture, College of Fisheries Science, Chaudhary Charan Singh Haryana Agricultural University, Hisar, Haryana, India
e-mail: priya222jnv@gmail.com

R. K. Gupta
Department of Zoology, College of Basic Sciences & Humanities, Chaudhary Charan Singh Haryana Agricultural University, Hisar, Haryana, India

A. Kumar
The International Graduate Course of Veterinary Science and Technology (VST), Faculty of Veterinary Science, Chulalongkorn University, Bangkok, Thailand

S. Hareendran
Kerala University of Fisheries and Ocean Studies, Kochi, Kerala, India

S. Janbandhu · N. A. Tandel
Department of Fisheries Resource and Management, College of Fisheries, Mangalore, Karnataka Veterinary Animal and Fisheries Sciences University, Bidar, India

R. Kumar et al. (eds.), *Sustainability in Aquaculture and Fisheries*,
Springer Water, https://doi.org/10.1007/978-3-032-16605-0_31

Keywords Aquaculture · *Agaricus bisporus* · Growth promoters · Immunostimulants

31.1 Introduction

Production from aquaculture and fisheries worldwide reached a historic peak, yielding 223.2 million tons (MT), valued at USD 472 billion, and providing about 20.7 kg of aquatic food per person. Aquaculture is the most economical source of animal protein and the fastest-growing food production industry in the world. In 2024, 130.9 MT were produced from this sector worldwide, with a value of USD 312.8 billion (FAO 2024). It has been established in several areas across developing countries, including those in Asia and Africa, that offer rural people's opportunities to enhance their standard of living and escape poverty through additional family income (Moisa et al. 2022; Tran et al. 2023). Aquaculture has more than doubled its percentage of intake of fish oil and fish meal over the past ten years due to its rapid expansion. As a result, these resources have become increasingly limited and expensive. Therefore, a major current concern in aquaculture is to decrease its impact on environment through providing farm-raised fish with a well-balanced nutritional diet, cost-effective, and eco-friendly feed alternatives. The development of artificial feeds, made with diverse ingredients, was a significant advancement for the global aquaculture industry in the last century. However, challenges remain, including the accessibility and price of non-conventional feed sources, intersectoral rivalry, maintaining demand and supply balance while safeguarding quality of environment, as well as ensuring social acceptance and economic sustainability (Mitra 2021).

The continuous growth in the global population is placing increasing strain on existing agricultural systems. Consequently, numerous nations are focusing on boosting food production to satisfy the demands of a population that is predicted to grow to 10 billion people by 2050 (Shefali et al. 2021). In view of rising demands, aquaculture is becoming a sustainable potential source of animal-based protein for the global population (Dawood and Koshio 2019; FAOSTAT 2018). However, this growth has brought significant challenges to the aquaculture sector in terms of expansion, sustainability and maintaining environmental integrity (Mitra 2021). In order to meet these growth expectations, the aquaculture industry must address not only present issues, but also prepare for future demands that are likely to arise by global challenges, like climate change and growing threat of superbugs, as well as by the intensification of aquaculture systems (Burgos-Aceves et al. 2018). Farmers struggle to increase fish density while lowering disease risks in order to meet the rising demand for fish (Dawood and Koshio 2016). More intense and super-intensive culture techniques have replaced large, low-productivity systems in aquaculture practices worldwide in recent years. However, fish are becoming more susceptible to stressful situations as a result of this intensification, which raises the possibility of disease outbreaks (Chen et al. 2014). The use of antibiotics and chemotherapeutic medicines has been a common response to disease outbreaks (Kennedy et al. 2016).

However, the widespread use of these medications has led to complications such as immune suppression, nature-related problems and microorganisms resistant to antibiotics (Allameh et al. 2016; Dawood and Koshio 2016).

Alternative methods are being investigated to enhance fish immunity, disease resistance and general health. Prebiotics, probiotics and other immune-modulating feed additives have shown especially encouraging results (Hoseinifar et al. 2018; Iswarya et al. 2018; Van Doan et al. 2018; Wu et al. 2015). Natural herbal items have been successfully used in aquaculture to improve blood parameters or strengthen immune function in a number of studies. These comprise essential oils from *Thymus vulgaris* (Zargar et al. 2019), *Cupressus macrocarpa* leaves (Kesbic et al. 2020) and *Punica granatum* seed oil (Acar et al. 2018). Other plants that are commonly utilized in aquaculture include *Zingiber officinale*, *Cynodon dactylon*, *Allium sativum*, *Punica granatum* and *Withania somnifera* (Reverter et al. 2017).

Since the quality of the raw materials used to formulate feed determines its effectiveness, feed is essential to the aquaculture value chain. It affects the rates of growth of different fish species raised in particular environmental conditions. In order to sustain successful aquaculture farm operations, feed components must be both affordable and suitably nutrient-dense (Chaitanawisuti et al. 2011). Research on the immunomodulatory effects of herbal extracts in aquatic animal cultures suggests that they could be useful substitutes for vaccinations, antibiotics or chemotherapeutic medications, providing a cost-effective and efficient therapeutic option without the potential for toxicity (FAO 2016). Developing feed that uses locally available, low-cost by-products is essential since feed prices have a significant impact on overall operating expense (Langer et al. 2011). Fishmeal, a highly prized ingredient and the primary source of protein in artificial feeds, is gradually declining while its price keeps rising. This situation emphasizes the vital need to identify other protein sources. Fishmeal can be effectively substituted with alternatives like Soy-based protein and poultry waste products (Hernández et al. 2007; Kumar et al. 2022a, b). Also, mushroom by-products appear to be a promising substitute for fishmeal protein in fish diets. Notably, mushroom stalks are rich in protein, polysaccharides, and antioxidants. Feeding Nile tilapia fingerlings (*Oreochromis niloticus*) with the mushroom stalks of *Pleurotus sajor-caju* as a substitute for fishmeal protein has been shown to promote significant growth (Muin et al. 2015). Dietary fibers and high-fiber foods are gaining attention for their health benefits, especially in the nutraceutical industry. Edible mushrooms, valued as the "elixir of life" since ancient times, are recognized as a potential reservoir of fiber. They provide essential amino acids and polysaccharides, offering advantages over animal and plant protein sources. Mushrooms' ability to grow efficiently in controlled settings further enhances their value as a food source. For very long time they have been ingested by humans and used in homeopathic medicine, mushrooms' global consumption has risen. Recently, mushroom-based products have attracted interest in aquaculture (Kumar et al. 2022a, b, c, d; Dias et al. 2012; Sartori et al. 2015).

Over the past few years, several researches have proven that edible mushrooms are abundant in bioactive substances, such as fibers, terpenes, peptides, glycoproteins, minerals, unsaturated fatty acids, antioxidants (such as phenolic compounds), vitamins E and C and functional polysaccharides (Aida et al. 2009; Rathore et al. 2017; Wasser 2002). They provide several health advantages such as improved antioxidant defenses, illness prevention, immunological support and antiviral and anticancer properties (Chen et al. 2006; Harikrishnan et al. 2012a, b; Jeong et al. 2012). Although mushrooms have been used by humans for many generations, their variety and quantity have recently risen. Edible mushrooms have many nutritional values and health advantages, such as their high protein, vitamins, minerals, low calorie count, minimal fat and essential fatty acids. The common edible mushrooms include *Agaricus bisporus* (Button Mushroom), *Calocybe indica* (Milky Mushroom), *Pleurotus ostreatus* (Oyster Mushroom), and *Ganoderma lucidum* (Reishi Mushroom). Because of its robust market demand and extensive cultivation, *A. bisporus* has the highest economic value of all the others (Liu et al. 2013). According to Tian et al. (2012), it is also recognized as a medicinal herb that is abundant in bioactive compounds such as vitamins, minerals, ergothioneine, polyphenols, and polysaccharides. This mushroom, which is well-known for both its flavor and health advantages, is highly valued and widely available worldwide. Along with fat, it has high levels of protein, moisture, fiber, carbs, and ash. It contributes to its overall nutritional value by providing vital vitamins and minerals. Additionally, it contains fat that is high in linoleic acid, glycogen (4.5–5.0%), and essential amino acids like lysine and tryptophan (Harikrishnan et al. 2018). In addition, the white button mushroom also displays pharmaceutical properties such as antibacterial, antifungal, immunomodulating, glucose and cholesterol-lowering qualities. Research on button mushrooms is still limited, despite its potential (Rajni et al. 2006). The first button mushrooms to be used in aquaculture was Harikrishnan et al. (2018), who found that extracts at 5 and 10% might improve immunological and hematological parameters in *Clarias gariepinus* against *Flavobacterium columnare*.

The edible basidiomycete fungus *A. bisporus*, commonly referred to as the white button mushroom (WBM) (Have et al. 2003), is native to grasslands in North America, Europe and India (Hood et al. 2004). About 90% of all mushrooms consumed in the United States come from this one, which is one of the most widely grown mushrooms in the world (Grove 1981). Originally having a brownish cap and dark brown gills, the most popular variety now has brown gills and a white cap, stalk and flesh (Parslew et al. 1999). Numerous studies have examined the health advantages of *A. bisporus* mushrooms in both humans and animals, including how they can strengthen the immune system. However, the use of WBM in aquaculture has not received much attention. This book chapter's goal is to analyze the current state of conventional aquafeeds while discussing the growing significance of alternative feeds and how they address crucial sustainability issues for the creation of future policies.

31.2 The Antimicrobial Agents and Its Impacts on Aquaculture Practices

Antibiotics, which come from either synthetic or natural origins, are described as agents able to cause destruction or suppress the development of microbes (Romero et al. 2012). Host safety is a crucial consideration while selecting an antibiotic. A "safe" antibiotic is one that successfully cures bacterial infections while causing minimal adverse effects. Besides therapeutic use, antibiotics are also applied in prophylactic and metaphylactic purposes, which are considered as preventive and control methods (Cabello 2006). In aquaculture, antibiotics require approval from regulatory bodies, which establish rules for administration techniques, doses and withdrawal periods (Dawood et al. 2018). Feed is the most often utilized administration route in aquaculture; however, infected animals frequently lose their appetite, resulting in lower feed consumption and reduced metabolism. This may cause antibiotics to be excreted into the environment. Burridge et al. (2010) stated that over 75% of antibiotics given to fish are released into surrounding waters. Defoirdt et al. (2011) and Hoseinifar et al. (2018) have expressed worries over the overuse of antibiotics in aquaculture, particularly in intensive systems, which can lead to more frequent disease outbreaks. Estimating global antibiotic use in aquaculture is challenging due to differences in national registration systems (Dawood et al. 2018); nonetheless, differences in antibiotic application worldwide are considered substantial (Burridge et al. 2010). For instance, shrimp farms in Thailand used antibiotics at a rate of 500–600 g per metric ton, Vietnam reported 700 g per metric ton, while Norway only used 1 g per MT (Defoirdt et al. 2011).

Beyond disease control, prophylactic antibiotic use has been shown to enhance host growth but also increases the likelihood of antibiotic resistance (Hoseinifar et al. 2017; Kumar and Singh 2022). Resistance genes have been detected in aquatic bacteria, terrestrial animals, and humans (Cabello 2006). Consequently, the use of antibiotics in aquaculture, both as treatment and prevention, is tightly regulated (Hoseinifar et al. 2017). Additionally, antibiotics used in human medicine, including non-biodegradable compounds, enter aquatic ecosystems and exert selective pressure over time (Iwashita et al. 2015). Overuse in aquaculture further leads to environmental accumulation of unmetabolized antibiotic residues. To cater to expanding market for aquaculture products free of antibiotics, alternatives like probiotics, prebiotics, with medicinal plants are being explored (Doan et al. 2019).

31.3 Performance-Improving Dietary Supplements in Aquaculture

During diet formulation, feed supplements are included to improve fish health outcomes, to increase feed quality and optimize feeding efficiency. Most of the supplements that used are non-nutritive which include substances such as antioxidants, immune boosters, probiotics and antibiotics, all are meant to promote growth and improve water quality in aquaculture systems. But these supplements also increase aquaculture production costs. To cope with this challenge, feed manufacturers have switched to functional feed ingredients (Bharathi et al. 2019). These functional additives not only promote growth and immunity but also support physiological processes and general fish health, offers benefits beyond conventional additives. These are thought to be effective substitutes to antibiotics and chemical therapies (Yousefi et al. 2018). Furthermore, as a potent alternative for vaccines functional additives can enhance disease resistance by activating the innate immune system. Some examples of these additives are plant-based substances, organic acids, mycotoxin binders, immune boosters, yeast derivatives, probiotics, prebiotics, enzymes, fungi, seaweeds and microalgae (Alemayehu et al. 2018; Bharathi et al. 2019). Fungi and their extracts are the most promising and potent functional feed supplements among them (Van Doan et al. 2019). Table 31.1 gives an overview of recent studies about the use of fungus and their by-products as immune-stimulating agents and growth enhancers in aquaculture.

31.4 Replacement of Fishmeal with Other Conventional Protein Source

Whole fish or fish byproducts are dried and ground to create fishmeal, a high-protein flour. For farmed fish, it is generally considered to be one of the most nutrient-dense and easily digestible foods. Additionally, it serves as a key source of omega-3 long-chain polyunsaturated fatty acids—specifically eicosapentaenoic acid (EPA) and docosahexaenoic acid (DHA)—in animal nutrition (FAO 2024). Fishmeal accounts for 50–70% of the ingredients in fish feed (Jannathulla et al. 2019). In 2022, fish by-products accounted and 53% and 34%of the overall production of fish oil and fish meal, respectively (FAO 2024). Feed costs constitute more than half of the total operating expenses for an aquaculture farm. However, decreasing fish stocks (Macusi et al. 2022) and increasing fishmeal demand cast a challenging outlook for the aquaculture sector-unless a shift occurs toward incorporating non-fish ingredients in the production of feed. Over the next two decades, aquaculture will be anticipated to grow, driving greater fishmeal consumption. The rising demand for fish oil and fishmeal may put additional pressure on the dwindling marine resource stocks (Aladetohun and Sogbesan 2013; Oliva-Teles et al. 2022). Recently, fishmeal prices have more than doubled globally (Lin et al. 2022; Bansemer et al. 2023). Nowadays,

Table 31.1 Studies on the effect of mushrooms and their by-products on immune systems, physiological reactions and growth performance of aquatic species

Finfish/ shellfish species	Mushrooms	Compound used	Doses and duration	Effect	References
Rainbow trout (*Oncorhynchus mykiss*) 20 g	*Pleurotus ostreatus*	Extract	0, 10, 20 g/kg 6 weeks	PA (↑) LA (↑) Gross Ig level (↑) Resistance to *L. gravieae* (↑)	Uluköy et al. (2016)
Labeo rohita 60–80 g	*Pleurotus florida*	Glucan	0.5, 1.0 g/ kg 2 weeks	SOA (↑) PA (↑) BA (↑) ACH50 (↑) LA (↑)	Kamilya et al. (2008)
Kuruma shrimp (*Marsupenaeus japonicus*) 10–20 g	*Schizophyllum commune*	Polysaccharide	50 and 100 mg/ kg 7 days	PA (↑) Resistance against virus (↑)	Itoh (1997)
Litopeneaus vannamei 13 g	*Ganoderma lucidum*	Beta-glucan	0.5, 1 g/ kg 4 weeks	THC (↑) SGC (↑) SOA (↑)	Yang ChihChiu et al. (2014)
Orange-spotted grouper (*Epinephelus coioides*) 6.4 ± 0.65 g	*Coriolus versicolor*	Beta-glucan	1, 2 g/ kg 2 weeks	LA (↑) PA (↑) RBA (↑) Resistance to *V. alginolyticus* (↑)	Chang et al. (2013)
Rainbow trout (*O. mykiss*) 20 g	*Lentinula edodes*	Extract	10 and 20 g/ kg 42 days	THC (↑) HCL (↑) PA (↑) LA (↑) SMA (↑) IgM (↑) Survivability against *L. garvieae* (↑)	Baba et al. (2015)

PA: Phagocytic activity; LA: Lysozyme activity; BA: Bactericidal activity; SOA: Superoxide anion; ACH50: serum alternative complement activity; THC: Total hemocyte count; SGC: Semi granular cell; HCL: Hematocrit level; RBA: Respiratory burst activity; SMA: Serum myeloperoxidase activity TLC: Total leucocytes count

reducing fish oil and fishmeal in feeds is crucial for easing global marine resource pressures. Substituting fishmeal from feed can also reduce production expenses (De Francesco et al. 2007). Additionally, combining complementary amino acids with other protein sources is strongly advised, including lysine, methionine, threonine and tryptophan (Ogunji and Wirth 2001). Fishmeal is inadequate to meet the increasing demand in aquaculture practices as its consumption rises since alternative protein sources usually do not provide same essential amino acid profile as fishmeal in feed (Wang et al. 2023). Therefore, it is vital to find substitute protein sources to fishmeal (Wang et al. 2023; Galkanda-Arachchige 2020). Due to its scarcity, fishmeal raises concerns regarding sustainability and the environment. Therefore, it is crucial to look into alternative protein sources for aquafeed.

Alternatives to fish meal that have been investigated for use in aquafeed include insects, animal waste, microbial biomass, plant-based proteins such as micro and macroalgae and food waste. These decisions have been carefully considered. They are valued for their affordability and environmental friendliness and include plant-based proteins such as soybean, pea, canola, groundnut oilcake, sunflower, cottonseed and leaf meals. However, by employing specific processing methods, these plant proteins drawbacks can be minimized. Particularly soybean meals excellent nutritional value and simplicity of digesting have attracted a lot of interest. It is thus the most often used plant protein in aquafeed to replace and cut fish meal. Though contradicting results, many studies agree that soybean meal can effectively replace fish meal in large quantities and offers both financial and nutritional benefits. Still, the rising cost of soybean meal has driven demand for less costly substitutes. Studies indicate that substituting soybean meal for fish meal yields the best growth, feed efficiency, body composition and cost results. Moreover, mixed with soybean meal, mushroom stems show promise as a partial replacement for fish meal. These components taken together offer a workable approach to create more reasonably priced fish feed (Muin et al. 2015).

Fish meals including mushrooms are more fit and preferred than fishmeal, claims Deborah Paripuranam et al. (2011), which promotes the development of *Labeo rohita* and *Hemigrammus caudovittatus* fingerlings. When Zhang et al. (2017) replaced mushroom fermented bran hydrolysate for up to 80% of the fishmeal in the diet, significantly enhanced *Carassius Carassius* weight gain, protein utilization, antioxidant levels and digestive enzyme performance. A study by Eswari et al. (2019) showed that substituting 75% of fishmeal led to marked improvements in survival and growth, muscle protein content, amino acid profiles, carbohydrates, lipids, ash content, and enzyme activities (protease, amylase and lipase) in contrast to the diet under control.

31.5 Beneficial Attributes of Mushroom in Boosting Fish Immune Response

The fish immune system exhibits many similarities when compared to vertebrates, though several key differences exist. Unlike vertebrates, fish are self-sufficient organisms from early developmental stages, relying exclusively on survival through innate immunity (Rombout et al. 2005). In dealing with immunological threats, non-specific factors play an important part of disease resistance. While fish's adaptive immune response tends to be slower, it is important for sustained immunity and significantly contributes to effective vaccination. Furthermore, adaptive immunity is crucial for controlling immunological responses and using receptor proteins for preserving homeostasis. The immune response that is not specific is influenced by various internal and external influences, such as temperature, stress and population density, which may impair innate immunity. However, certain dietary supplements and immune boosters can improve immune response efficiency (Magnadottir 2010). In fish, the innate immune system plays an especially important role in fighting off infections, because of their adaptive immune defenses, weaker antibody response and slow lymphocyte proliferation and maturation (Whyte 2007). Overall, the innate immune system in fish is composed of three primary parts: a mucosal barrier that includes the skin, gills and digestive system; cellular components such as non-specific cytotoxic cells (NCCs), phagocytic cells (granulocytes, monocytes/macrophages), and NK cells (Neumann et al. 2001); as well as humoral components such complement peptides, lysozyme, antimicrobial peptides and α-2 macroglobulin (Nakao et al. 2011).

Fungi of higher order are important providers of a broad range of essential natural substances. Bioactive polysaccharides are found in a variety of Basidiomycete mushrooms in their culture broth, mycelium and fruiting bodies, such as *Saccharomyces cerevisiae*, *Lentinula edodes*, *Schizophyllum commune*, *Coriolus versicolor*, *Pleurotus ostreatus* and *Ganoderma lucidum* (Wasser 2002). Mushrooms are abundant in chitin, making them a significant dietary fiber source. The proteins and bioactive substances present in these fungi are associated with numerous medicinal benefits, such as anticancer, anti-inflammatory, antidiabetic, antihyperlipidemic, antioxidant, antiviral, and antimicrobial effects (Muszyńska et al. 2017). Among the most studied proteins in WBM are often called polyphenol oxidase, PPO (tyrosinase) and lectin (Weijn et al. 2013; Yu et al. 1999). Lectins, which are sugar-binding proteins, are known for their ability to agglutinate foreign particles and function as humoral defense factors in crustaceans, exhibiting antitumor, antiproliferative, and immunomodulatory activities (Santos et al. 2014). PPO catalyzes the transformation of L-tyrosine into L-DOPA, which is further converted into dopaquinone, a precursor in melanin biosynthesis (Weijn et al. 2013). Melanins inhibit microbial growth by suppressing extracellular proteinase and chitinase activities. Additionally, PPO enhances phagocytosis, nodule formation, encapsulation, and promotes hemolymph protein production. Mushrooms possess tiny bioactive substances like statins, indoles, phenolic compounds and additional secondary metabolites. Antioxidant properties are exhibited by ergothioneine that enhances monocyte function, indoles exhibit anticancer and anti-aging effects, while ergosterol and ergocalciferol aid in preventing vitamin D deficiency (Muszyńska et al. 2017; Kiran et al. 2025). Fish diet containing mushrooms enhance growth, strong the immune system and make disease resistance (Fig. 31.1) because of its nutritional benefits and presence of bioactive substances like β-glucans.

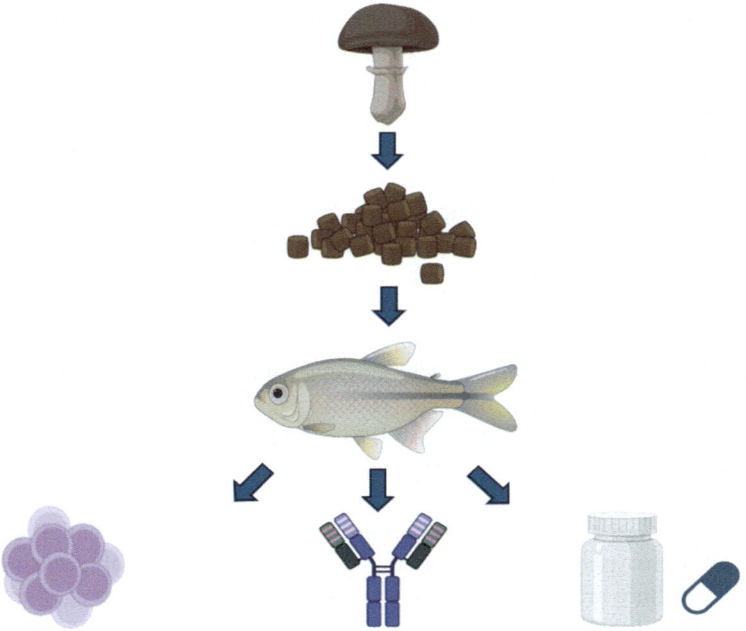

Fig. 31.1 This figure illustrates the effect of mushroom incorporated diet on fish

31.5.1 Potential Prebiotics

Polysaccharides found in mushrooms include α- and β-glucans, hemicellulose, chitin, mannans, galactans as well as xylans. These substances act as potential prebiotics, supporting the growth and function of beneficial gut microbiota while inhibiting harmful pathogens, thereby boosting immunity (Cheung 1996; Kalač 2009). Chitin, an indigestible polysaccharide, functions as dietary fiber (Cheung 1996; Kalač 2009). On the other hand, water-soluble polysaccharides generally demonstrate greater biological activity compared to their insoluble counterparts (Tao et al. 2006). In mushrooms, the majority of the polysaccharides are found in both straight and branching glucan forms, with glycosidic bonds like (1–3)-glucans, (1–3)-α-glucans and (1–6)-β-glucans. Additionally, certain polysaccharides, such as arabinose, mannose, fructose, galactose, xylose, glucose and glucuronic acid, are considered true heteroglycans (Wasser 2002). While they differ chemically, the majority of mushroom polysaccharides are categorized as β-glucans (Wasser 2002). Three principal polysaccharides, including α- and β-glucan and galactomannan, are mostly found in the cell walls of white button mushrooms (Smiderle et al. 2011), with β-glucan identified as a prebiotic. The β-glucosidic bonds in β-glucans are resistant to breakdown by digestive enzymes in vertebrates' pancreas or brush border, allowing these polysaccharides to survive stomach acid and remain undigested. This resistance to digestion supports their role as prebiotics. In addition to serving as prebiotics, polysaccharides found in mushrooms have a number of biological characteristics, including immune-modulating, antiviral, antioxidant, anticancer and antibacterial activities (Lindequist et al. 2005; Thatoi and Singdevsachan 2014). Mushrooms have recently gained popularity as natural prebiotics in aquaculture (Lindequist et al. 2005; Aida et al. 2009). Growing amounts of research have been conducted on the advantages of mushrooms in aquaculture and their use as natural prebiotics (Zou et al. 2016; Van Doan et al. 2016; Harikrishnan et al. 2012a, b).

31.6 Using White Button Mushrooms Instead of Traditional Feed Sources

As macrofungi, mushrooms are among the most sustainable crops. They consume less water, energy, space and other resources than many vegetables (Rosenberg 2020). *A. bisporus*, the white button mushroom, belongs to the class Basidiomycetes and the family Agaricaceae. WBM is the most widely grown and eaten mushroom worldwide, it comes from grasslands of North America and Europe. It is also renowned for its numerous health benefits. China is the world's largest producer of mushrooms, followed by the United States and the Netherlands. In 2018, global production reached approximately 8.99 million tons (FAOSTAT 2020). WBM is the most widely used type, commanding high prices in India (Mehta et al. 2011). Initially cultivated only during winter, advancements in farming techniques have made year-round production possible. This is supported by favorable climatic conditions, plentiful agricultural waste, affordable labor, and diverse fungal biodiversity in Indian region. India's total mushroom production is around 0.13 MT., with major contributors being Himachal Pradesh, Uttar Pradesh, Punjab, and Haryana. The mushroom industry in India expanded at an average annual rate of 4.3% from 2010 to 2017. WBM constitutes 73% of the entire production, with oyster mushrooms 16%, paddy straw mushrooms 7% and milky mushrooms 3%. However, in India consumption per capita remains low at under 100 g per year. In 2016–2017, India exported 1,054 quintals of canned and frozen WBM, earning Rs. 7,282.26 lakh. The country's spawn demand is estimated to be 8,000–10,000 tons annually. Most of this demand is met by private producers, while public sector contributions remain limited to 10% (Sharma et al. 2017).

There is a growing focus on balanced diets for daily health and managing chronic diseases. Mushrooms are widely acknowledged as nutrient-dense and restorative foods. Their therapeutic properties are being researched in clinical studies by organizations like the World Health Organization (WHO). As a food and medication, mushrooms are categorized by the FDA as "healthy foods." Histidine, arginine methionine and lysine, are among the nine essential amino acids found in mushrooms. They are abundant in high-quality protein, minimal in fat, and cholesterol-free, making them valuable health supplements. Mushrooms also have potential to address food security challenges in developing nations (Dutta 2007). Despite these benefits, research on the nutritional properties of various mushroom species remains limited (Crisan and Sands 1978; Didukh et al. 2004). In aquaculture, feed expenses account for 50–80% of total production costs. Ensuring proper nutrition is critical for achieving cost-effective results. The cost of feed depends on the quality and price of its components, which include a mix of organic and inorganic materials. In shrimp farming, feed alone represents 50–60% of total expenses (Samocha et al. 2004). Protein, the most expensive nutrient, is often sourced from fishmeal (FM).

Fishmeal is preferred because it provides essential amino acids, fatty acids, minerals, and other nutrients (Oujifard et al. 2012). Despite its benefits, fishmeal is costly due to limited availability and high demand. To reduce costs, the aquaculture industry is replacing fishmeal with affordable, high-quality plant-based protein sources. This shift not only reduces expenses but also increases profitability in shrimp farming (Naylor et al. 2009; Amaya et al. 2007). Currently, replacing fishmeal with more affordable animal- or plant-based ingredients is a key research priority in global aquaculture due to the increasing scarcity of quality fishmeal and rising costs (Fontainhas-Fernandes et al. 1999). Various protein sources are available that have the potential to replace fishmeal in aquaculture feeds without compromising fish growth performance (Tacon 2004). WBM holds potential as a dietary component, either as an additive or alternative, owing to its rich nutrients and immune-enhancing benefits. It may act as a valuable, underutilized resource for finfish and shellfish, contributing to improved health and addressing malnutrition challenges. The nutritional value of *A. bisporus* is shown in Table 31.2, which highlights its high moisture content, high-quality protein, dietary fiber, essential vitamins and minerals, all of which contribute to its potential as a functional food.

31.6.1 Basic Nutrient Profile

Mushrooms are characterized by high moisture content, which varies based on factors like harvest timing, growth conditions, culinary use and their storage methods (Guillamón et al. 2010). *A. bisporus* samples contain moisture levels ranging from 91 to 92 g per 100 g, with ash content at 0.9 to 1 g per 100 g and energy values between 29 and 31 kcal per 100 g (Reis et al. 2012). The proximate composition of *A. bisporus* fruiting bodies, reveals that when compared to other mushrooms, they have higher protein level (29.14%), fat (1.56%) and carbohydrates (51.05%) (Ahlawat et al. 2016). In dried button mushroom fruit bodies, carbohydrates range from 48.9 to 38.3%, fiber from 23.3 to 17.7%, ash from 11.00 to 7.77% and fat from 3.92 to 2.53% (Tsai et al. 2007).

Table 31.2 Nutritional value of *Agaricus bisporus*

Nutritional analysis of Agaricus bisporus	
Crude protein	34.30%
Crude fiber	9.81%
Carbohydrates	59.4%
Fat	3.1%
Moisture	92.5%
Ash	12.31%
Essential minerals and vitamins (in mg)	
Potassium	2850
Sodium	106
Iron	8.8
Phosphorus	912
Calcium	71
Thiamine (B1)	8.9
Riboflavin (B2)	3.7
Niacin (B3)	42.5
Ascorbic acid (C)	26.5

31.6.2 Amino Acid and Protein Profile

Mushrooms are recognized as a superior source of protein. According to Corrêa et al. (2016), while mushrooms contain less protein than animal meats, they surpass most other foods, including milk, in crude protein content. A number of variables may affect the chemical makeup and nutritional content of mushrooms that are cultivated, including growth substrates (Gothwal et al. 2012), developmental stages, as well as post-harvest circumstances (Guillamón et al. 2010). As a result, studies report varying protein content for mushrooms, even for the alike species. For instance, the amount of protein present in *A. bisporus* has been reported at 11.01% (Saiqa et al. 2008), 25% (Muszyńska et al. 2011), 17.7–24.7% (Mohiuddin et al. 2015) and 29.14% by Ahlawat et al. (2016), depending on the growing substrate used. Mushroom proteins have an amino acid profile similar to that of animal proteins, which can help in balancing high intake of animal protein, particularly in developed nations (Guillamón et al. 2010). According to Kakon et al. (2012), mushrooms are a viable meat substitute since their proteins provide all nine of the essential amino acids needed by humans. Alanine, arginine, glutamic acid, serine, aspartic acid, lysine, leucine, phenylalanine, proline, threonine and tyrosine are the amino acids that are most prevalent in *A. bisporus* (Muszyńska et al. 2013). Additionally, *A. bisporus* contains vital amino acids that are beneficial for public health, such as methionine, cystine, phenylalanine, valine, leucine, isoleucine, lysine, tyrosine and threonine (Muslat et al. 2014).

31.6.3 Dietary Fiber and Carbohydrates

Humans do not primarily obtain their energy from the carbs in mushrooms. The amount of digestible carbohydrates, such as glucose and mannitol, are found in minimal quantities (dry weight below 1%), while glycogen constitutes 5–10% of dry weight. Mushroom carbohydrates are mostly composed of non-digestible carbohydrates, including oligosaccharides like trehalose and non-starch polysaccharides (NSPs) like β-glucans, chitin and mannans, which make up the majority of the carbs in mushrooms (Kumar et al. 2022c; Cheung 2010). Mannitol and trehalose are typical sugars that are present in cultivated edible mushrooms, with mannitol being especially prominent in both white as well as brown button mushrooms (Reis et al. 2012). The health advantages of mushrooms are attributed to dietary fiber, which includes hemicelluloses, mannans, β-glucans and chitin (Maftoun et al. 2015), which make up the cell walls of fungi (Cheung 2009). According to Nitschke et al. (2011), *A. bisporus* contains 9.60 g of chitin for every 100 g of dry matter. Compared to oyster mushroom, WBM contains twice as much chitin (Cherno et al. 2016). In the same way, *A. bisporus* has more chitin than *L. edodes* and *P. ostreatus* (Vetter 2007).

31.6.4 Mineral Make-Up

The ability of mushrooms to absorb and retain minerals from their surroundings is well recognized. According to Owaid (2015), *A. bisporus* is a significant source of K, Fe, Zn, Cu, Na, Se, Co and Mn. According to Mohiuddin et al. (2015), the mineral concentrations of *A. bisporus* fruiting bodies from various regions of Bangladesh such as K, Na, Fe, Cu, Zn and Mn contents ranged from 0.54 to 1.58%, 37.2 to 61.9 μg/g, 143.6 to 396 μg/g, 54.6 to 163.4 μg/g, and 36.6–58.0 μg/g, respectively. Zinc ranged from 8.1 to 7.0 mg/kg, iron from 7.4 to 7.9 mg/kg, phosphorus from 7.4 to 7.9 mg/kg, magnesium from 88.0 to 76.3 mg/kg, sodium from 2652 to 2500 mg/kg, potassium from 213.3 to 238.8 mg/kg and calcium 534.2 to 554.8 mg/kg (Caglarirmak 2009). Additionally, potassium around 4.21%, sodium about 500.8 mg/kg and selenium 1.34 mg/kg were found in WBM fruiting bodies (Ahlawat et al. 2016).

31.6.5 Nutrient Cofactors

Mushrooms are an excellent reservoir of vitamins. In *Agaricus* species, niacin is reported to be the most abundant vitamin, succeeded by the vitamin riboflavin, along with more vitamins such as thiamine, vitamin C and E (Bernaś and Jaworska 2016). According to Çağlaırmak (2009) brown *A. bisporus* (portobello mushroom) provides higher amounts of vitamin B9 (0.08–0.09 mg/kg), B2 (0.27–0.29 mg/kg), B3 (2.9–3.6 mg/kg), and B1 (0.085–0.09 mg/kg), but not a lot of vitamin C. Mushrooms are also recognized as a natural reservoir of vitamin D. *A. bisporus* has a vitamin D concentration of 984 IU/g (Ahlawat et al. 2016), while wild mushrooms generally having higher levels than cultivated ones (Simon et al. 2011). Growing in low light levels may be the cause of the cultivated white button mushrooms reduced vitamin D content (Reis et al. 2012). Fungal cell membranes contain ergosterol, which is a pioneer to vitamin D_2, which is crucial for bone health.

31.6.6 Essential Fatty Components

Although *A. bisporus* has little fat, it still contains linoleic acid and other essential fatty acids. According to Barros et al. (2008), wild *Agaricus* species exhibited greater polyunsaturated fatty acid content and decreased levels of monounsaturated fatty acids. This was mostly because wild species had a higher proportion of linoleic acid than farmed types. The total fatty acid contents of *A. bisporus* strains varied between 180 and 5,818 mg/kg of dry weight, with linoleic acid comprising nearly 90% of the total fatty acids on average (Baars et al. 2016). Saiqa et al. (2008) state that the primary fatty acids found in *A. bisporus* are linoleic acid, which makes up 44.19% of the total, along with palmitic, linoleic, stearic, caprylic, oleic, erucic and eicosanoic acids. Öztürk et al. (2011) discovered that the most prevalent fatty acid types in *A. bisporus* oil were palmitic (12.67–14.71%) and linoleic (61.82–67.29%). Similarly, linoleic, stearic and palmitic acids were identified as the primary fatty acids by Shao et al. (2010). Saiqa et al. (2008) found that human health depends on linoleic acid and has several advantages, including reducing atherosclerosis by interacting with HDL in the blood profile. According to Hossain et al. (2007), *A. bisporus* had a linoleic acid concentration that was five times greater than that of Pleurotus ostreatus and twenty times greater than that of *Ganoderma lucidum*. Mushrooms are valued as gourmet foods all over the world because of their unique, delicate flavor and healthy nutritional value. They are necessary orthomolecules with high K and P levels, low Na content that helps those with hypertension, and a high mannitol content and low glycemic index that helps diabetics. Furthermore, phenolics, lipids, hydrolytic and oxidative enzymes, glycoproteins and other biologically active substances found in mushrooms enhance disease resistance in aquaculture species, increase non-specific immunity, and alter the immune system (Ullah et al. 2018; Zhang et al. 2007; Raa 1996; Dalmo and Seljelid 1995).

31.6.7 Medicinal Value

The extraction of bioactive compounds from mushrooms, especially *A. bisporus*, has drawn increasing attention to create functional foods. WBM has long been used in traditional medicine, and because of its anti-inflammatory, anti-cancer and antioxidant properties, its extracts and bioactive chemicals are being used more and more all over the world. These substances are used to treat a variety of human illnesses, including cancer, fungal, bacterial, and diabetic infections, coronary heart disease and immune-related problems (Dhamodharan and Mirunalini 2010). Although direct human intervention trials are currently scarce, involving the consumption of mushrooms, completed research shows that mushrooms and its extracts have few negative effects and are generally well tolerated (Volman et al. 2010). Because *A. bisporus* extracts can reduce the activity of acetylcholinesterase and butyrylcholinesterase, they may also be utilized to treat Alzheimer's disease (Öztürk et al. 2011). Mohamed (2012) employed GC/MS to discover 174 important metabolites in ethanolic extracts of *A. bisporus*, with concentrations ranging from less than 1 to 83% (w/w), and categorized them into twelve groups. Numerous therapeutic effects of these metabolites promote immunological function and improve overall health. Additionally, mushrooms contain many therapeutic compounds, including glycoproteins, natural antibiotics, triterpenoids and enzymes (Beelman et al. 2003). Zinc, is an essential component of numerous enzyme systems, which is crucial for regulating various metabolic processes. Due to their biological capacity to accumulate nutrients, mushrooms are said to be a rich zinc source (Bano 1981).

31.6.8 Pathogen-Inhibiting Properties

Previous studies have shown that methyl alcohol extracts of *A. bisporus* have antimicrobial activity against specific bacterial and fungal species (Akyuz et al. 2010). The stipes of *A. bisporus* may act as spontaneous antimicrobials, as suggested by Ndungutse et al. (2015), who also reported microbial suppression (Sharma et al. 2024, 2025) by *A. bisporus* extracts. Strong antimicrobial activity is demonstrated by the aqueous total protein extracts of *A. bisporus*, especially against Methicillin-Resistant *Staphylococcus aureus* (MRSA) and *S. aureus*. According to Owaid et al. (2017), *A. bisporus* is essential to the biosynthesis of silver nanoparticles (AgNPs), which are well known for their strong antibacterial and antifungal properties and it trailed second to the oyster mushroom by about 11% in terms of nanoparticle synthesis (Owaid and Ibraheem 2017). AgNPs were successfully prepared from *A. bisporus* extract by Sudhakar et al. (2014), highlighting its advantage over conventional antibiotics, as no microbial resistance has been observed. Using UV–visible spectroscopy, TEM and FT-IR Manzoor-ul-Haq et al. (2015) described these biosynthesized AgNPs and found that AgNPs generated from *A. bisporus* exhibited a larger inhibitory zone against MRSA than AgNPs derived from *Fomes fomentarius*, *Helvella lacunosa*, *Pleurotus florida* and *Ganoderma appalanatum*. Additionally, they examined how *A. bisporus*-AgNPs worked in concert with the antibiotics Ceftriaxone and Gentamicin, finding that the inhibition zones were 25 mm and 26 mm, respectively. The antibacterial activity of produced *A. bisporus*-AgNPs was further examined by Mirunalini et al. (2012) against gram-positive bacteria such as *S. aureus*, and tested against *S. typhi*, *Proteus* species, *Klebsiella* species and *Enterobacter* species (Dhanasekaran et al. 2013).

31.6.9 Anti-tumor Effects

A popular edible fungus in Western nations, *A. bisporus*, has a lectin that, when applied to human epithelial cancer cells, has potent antiproliferative actions without appearing to be cytotoxic. This lectin, known as ABL (*A. bisporus* lectin), demonstrates antitumoral activity across various cell types. It inhibits the proliferation of cells and lattice contraction in a dose-dependent way without generating significant toxicity. When subtle healing regulation is required, like in glaucoma surgery of eye, ABL may be very helpful (Yu et al. 1999; Irazoqui et al. 1997). A vital trace element like selenium has a great role in cancer prevention in humans as well as animals. More selenium is found in mushrooms than in the majority of fruit and vegetable meals, making them a significant selenium source, especially for vegetarians (Clark et al. 1996). Selenium may help prevent cancer through its antioxidant properties and enhance immune function. Human studies indicate that higher selenium intake may reduce cancer incidence, and intervention trials suggest that selenium can decrease cancer risk, especially in the lung, colon, liver and prostate, with the most pronounced benefits in individuals with low selenium levels (Spolar et al. 1999). White button mushrooms have potential as chemopreventive agents against breast cancer due to their ability to suppress aromatase, an enzyme involved in estrogen synthesis. Aromatase converts androgens to estrogen, and elevated aromatase levels are regarded as a risk factor for breast cancer in breast tissue. Mushroom extract was the most successful in reducing the activity of this enzyme out of the seven vegetable extracts that were examined (Sun et al. 1997).

31.6.10 The Effect of Genoprotective

The most commonly grown and consumed mushroom is button mushroom. Fruit bodies of *A. bisporus* have demonstrated protective properties preventing cellular DNA oxidative damage caused by H_2O_2. This species contains a heat-sensitive protein referred to as FIIb-1, which is identified as tyrosinase and is present in the fruit body. The genoprotective impacts of cold-water extracts from WBM correlated with the tyrosinase activity in its fruit bodies. The protective effect of WBM tyrosinase relies on the enzymatic conversion of tyrosine to L-DOPA, followed by the conversion of this metabolite into dopaquinone (Soler-Rivas et al. 1999).

31.6.11 Effect of WBM on Genes Linked to Immunity

When fish are exposed to environmental stress, feed additives can effectively reduce immunosuppression, oxidative damage, development retardation and inflammation (Dawood et al. 2018, 2020a). Due to its numerous physiological advantages for animals, WBM stands out among these feed additives (Aida et al. 2009; Hoseinifar et al. 2019; Liu et al. 2013). Dietary WBM has been shown to improve intestinal barrier integrity (Zou et al. 2016), boost antioxidant capacity (Hoseinifar et al. 2019), modulate immune response (Safari et al. 2018, 2019) and aid fish species in growing (Amiri et al. 2018) when used in place of antibiotics. Many nutritional techniques, such as adding functional feed additives, have been employed to lessen the detrimental effects of heat stress on fish (Abass et al. 2018; El Basuini et al. 2020).

31.6.12 Immunostimulant

Intensive aquaculture operations often result in an increasing number of pathogenic organisms. To reduce the incidence of these microbial diseases, aquaculture companies have primarily relied on vaccines and antibiotics. However, these methods have the potential to produce a large no. of bacteria that are resistant to antibiotics, highlighting the need for innovative approaches to pathogen control. As an alternative, immunostimulants are increasingly recommended for inclusion in feeds to control diseases and enhance immune responses in aquatic farming (Anderson 1992). In the aquaculture practices, various immunostimulants kinds have been employed. Table 31.3 presents findings from past studies examining the influence of white button mushrooms and its by-products on the growth, physiological conditions and immunological variables of different fish species.

31.6.13 Cultivation of Mushrooms

Mushrooms, as a group, are found in diverse regions across the globe, from arctic zones to tropical areas. Some species are restricted to specific regions, while others are distributed across distant geographical locations. Most species, however, tend to favor particular habitats. For instance, some like *Pleurotus tuber-regium* are commonly found in upland forests, others in swampy areas, and still others in open spaces like gardens, lawns, or pastures. Numerous varieties of mushroom, especially mycorrhizal forms, are linked with specific kind of vegetation. Mushrooms may also show a predilection for particular substrates within a particular ecosystem. Some basidiocarps are known as terrestrial forms because they usually develop on soil. Other mushrooms, like *Cortinarius melliolens* and *Tricholoma lobayensis*, can grow on dead leaves (follicolous) or litter; others, like *Lentinus edodes*, can grow on wood (lignicolous); and still more, like *Coprinus lagopus*, can grow on dung (coprophilous). Some are categorized as fungicolous and they can even grow on the fruiting bodies of other mushrooms (Abulude and Muhammed 2013). Mushrooms commonly grow on waste materials including sawdust, compost and garbage (Gbolagade 2006). Stamets (2001) observed that by using fermented horse manure

Table 31.3 Investigations on the influence of white button mushrooms and its by-products on the growth, physiological conditions and immunological variables of different fish species

Finfish/shellfish species	Compound used	Doses and duration	Effect	References
Oreochromis niloticus	Powder	0.5, 1, 2 and 4% 60 days	Mitigate the effects of heat stress, Growth (↑), Antioxidative defense (↑), Immunological response (↑)	Dawood et al. (2020b)
Persian sturgeon (*Acipenser persicus*)	Powder	0.5 and 1% Eight weeks	Detrimental impacts on certain stress indicators and blood parameters	Monsef Shokri et al. (2023)
Cyprinus carpio	Powder	0.5, 1 and 2% Eight weeks	Antioxidative defense gene expression (↑), immunological gene expression (↑), growth parameters (↑)	Hoseinifar et al. (2019)
Oreochromis niloticus	Powder	2, 4 and 6% Six weeks	Lysozyme activity (↑), ACH50 levels (↑), WBC (↑), RBC (↑), Hb (↑), Hct (↑), differential leukocytes (↑)	Habib et al. (2021)
Rainbow Trout (*Oncorhynchus mykiss*)	Powder	0.5, 1 and 2% 8 weeks	Expression of interleukin (IL)-1β (↑), Intestinal IL-8 expression (↑)	Amiri et al. (2018)
Litopenaeus vannamei	Powder	1, 1.5, 2, 2.5 and 3% 8 weeks	Growth (↑), SGR (↑), FCR (↑), PER (↑), Improved resistance against *Vibrio parahaemolyticus*	Singh et al. (2022, 2023)
Zebra fish (*Danio rerio*)	Powder	5, 10, and 20 g/kg 11 days	Immunological functions (↑)	Ezatrahimi et al. (2024)
Ctenopharyngodon Idella, Hypophthalmichthys molitrix and Hypophthalmichthys nobilis	Button mushroom compost and chemical fertilizer	NA 2 months	Significant difference in hemoglobin levels among both fishes, Interstitial connective tissue (↑)	Asad Seftjani et al. (2021)

(continued)

Table 31.3 (continued)

Finfish/shellfish species	Compound used	Doses and duration	Effect	References
Goldfish (*Carassius auratus*)	*A. bisporus* + *Lactobacillus casei*	Diet containing *L. casei* (Pro), 1% mushroom powder (MP) and *L. casei* + 1% mushroom powder (MP + Pro) 8 weeks	In MP + Pro Immune-related gene expression (↑), Ig level (↑), Lysozyme activity (↑), ACH50 levels (↑), Protease (↑)	Safari et al. (2019)
Zebrafish (*Danio rerio*)	WBM + *Lactobacillus casei*	*L. casei* (Pro), 1% mushroom powder (MP) and *L. casei* + 1% mushroom powder (MP + Pro) 60 days	In MP + Pro Growth related gene expression (↑), Mucosal immunological gene expression (↑), Antioxidant gene expression (↑), Skin mucosal immune related parameters (↑)	Safari et al. (2018)
Zebrafish (*Danio rerio*)	Synbiotics containing button mushroom extract + *Lactobacillus acidophilus* and *L. bulgaricus*	*L. acidophilus* in combination with 1% WBM extract as synbiotic1, *L. bulgaricus* with 1% WBM extract as synbiotic2 and *L. acidophilus* and *L. bulgaricus* with 1% WBM extract as synbiotic3 60 days	Growth performance (↑), Digestive enzymes (↑)	Zakariaei et al. (2022)
Clarias gariepinus	Powder	1, 5 and 10% 4 weeks	Promote better growth, non-specific immune system performance enhanced, Resistance against *Flavobacterium columnare* (↑)	Harikrishnan et al. (2018)

(continued)

Table 31.3 (continued)

Finfish/shellfish species	Compound used	Doses and duration	Effect	References
Cyprinus carpio	Powder	0, 0.5, 1 and 2% 8 weeks	TNF-alpha (↑), IL1b gene expression (↑)	Zou et al. (2016)
Macrobrachium rosenbergii post-larvae	Powder	25, 50, 75 and 100% 90 days	Up to 75% of fishmeal can be substituted with *A. bisporus* for the sustainable rearing of *M. rosenbergii* seeds in the nursery	Eswari et al. (2019)

and damp litter, garden mushrooms can be cultivated year-round. Miles and Chang (2004) based on early writings, summarized that mushroom farming began in France nearby 1630. After that, numerous investigations have monitored various growth parameters in both natural and controlled environments. The composting process was assessed by observing changes in temperature, pH, and the compost's ability to support mushroom mycelium growth. Mushroom growth was found to occur across a wide pH range, from pH 3.4 to pH 9.0. Studies indicated that lower pH levels promoted *Coprinus cinereus* growth, whereas pH 7 (neutral) favored *Volvariella volvacea.* Several factors, such as pH, temperature, light, substrate properties and the ratio of nutritional components, affects both mycelial growth and fruiting body production.

In the course of mycelia extension on substrates, mushrooms generate various enzymes that break down complex organic substances like lignin and cellulose into soluble products, that are taken up by the hyphae. Several mushroom species have been evaluated qualitatively due to their capacity to hydrolyze starch. *Lentinus edodes* mycelia can develop on a variety of liquid cultures and agar media (Kuforiji and Fasidi 2008). Composts enhanced with organic materials that include B-complex vitamins may also increase the yield of mushrooms, according to certain theories. The overall pattern of carbon source use in white button mushroom is consistent throughout studies, despite significant differences in strains, timing and research locations. Typically, xylose, fructose and glucose are identified as effective carbon suppliers. Like every fungus, *A. bisporus* also requires a nitrogen source. Kuforiji and Fasidi (2007) conducted investigations into the effects of various atmospheric parameters on the growth of mushroom, individually as well as under diverse situations. They discovered that factors like temperature, pH, physical properties of the substrate, and proportions of nutrients influence mycelial growth. Mycelia, which result from spore germination, are essential for mushroom development. However, the fruiting body of the mushroom is what attracts mushroom hunters because of its fleshy texture and conspicuous appearance.

31.6.14 Mushroom Storage

After being harvested, mushrooms need to be preserved in fresh conditions to preserve their flavor and quality. At 0 °C, mushrooms can be kept for up to five weeks, by storing them in dense polyethylene bags (0.02–0.03 mm) in nitrogen. At 5 °C, they can be kept for up to four weeks and up to two weeks at 15 °C under similar conditions. For up to 10 days, mushrooms can be preserved by gamma radiation or can be stored in a controlled environment with 9% oxygen and 25% carbon dioxide. It's also possible to freeze-dry mushrooms. It works well to dehydrate, grind, and store in an airtight container; later the powder can be used to prepare mushroom soup (Abulude and Muhammed 2013).

31.6.15 Other Uses

Due to their nutritional and therapeutic advantages, these have been prized as necessary foods since ancient times. They were utilized as food and medicine in ancient China, where they were thought to improve human health and longevity (Safwat and Al Kholi 2006). While the Egyptians viewed it as a gift from their god Osiris, the Greeks believed it would give warriors strength in combat (Maihara et al. 2012). As "Food of the Gods," edible mushrooms were revered by the Romans and saved for joyous occasions (Rahi and Malik 2016). In Mayan culture, psychoactive mushrooms were mostly used in religious events, a tradition still practiced in some areas (Matsushima et al. 2009).

31.7 Conclusion

A. bisporus may significantly help in eliminating malnutrition, especially in developing and underdeveloped countries, due to its high nutritional value. Beyond nutritional advantages, *A. bisporus* also offers therapeutic properties, mainly in cancer prevention, cardiovascular health, diabetes management and as an antioxidant and antimicrobial agent. In recent decades, edible mushrooms have increasingly been utilized as treatments or health food supplements. Research suggests that nutraceutical therapy has potential as a novel approach in curing a number of serious diseases. Further clinical study is vital to validate the medical effects of WBMs bioactive chemicals, even though they potentially represent a significant advancement in medication development. More widespread usage of mushrooms as a feed additive or substitute in fish feed could be a useful strategy to harness their potential on a global scale.

31.8 Future Prospects

The use of WBM in aquafeeds offers promising future directions in sustainable aquaculture. As the demand for fish protein rises worldwide, there is an urgent need for alternative feed ingredients that are not only nutritionally but also ecologically and economically feasible. Rich in protein, fiber and bioactive substances like β-glucans, it provides not only nutritional advantages but also immunomodulating properties that can enhance fish health and reduce disease risk. As the industry seeks alternatives to fishmeal and soy, white button mushroom presents a cost-effective and eco-friendly option, especially when utilizing mushroom byproducts. WBM has a promising role in aquaculture for enhancing growth rate and preventing disease incidences in fish.

Acknowledgements The authors would like to express their profound gratitude to the Vice Chancellor of Chaudhary Charan Singh Haryana Agricultural University, Hisar, Haryana and the Dean of the College of Fisheries Science, Hisar for providing the necessary facilities for the study.

References

Abass DA, Obirikorang KA, Campion BB, Edziyie RE, Skov PV (2018) Dietary supplementation of yeast (*Saccharomyces cerevisiae*) improves growth, stress tolerance, and disease resistance in juvenile Nile tilapia (*Oreochromis niloticus*). Aquac Int 26:843–855

Abulude FO, Muhammed NM (2013) Fungi: a review on mushrooms. Knowl Glob Dev 1:18–31

Acar Ü, Parrino V, Kesbiç OS, Lo Paro G, Saoca C, Abbate F, Fazio F (2018) Effects of different levels of pomegranate seed oil on some blood parameters and disease resistance against *Yersinia ruckeri* in rainbow trout. Front Physiol 9:596

Ahlawat OP, Manikandan K, Singh M (2016) Proximate composition of different mushroom varieties and effect of UV light exposure on vitamin D content in *Agaricus bisporus* and *Volvariella volvacea.* Mushroom Res 25(1)

Aida FMNA, Shuhaimi M, Yazid M, Maaruf AG (2009) Mushroom as a potential source of prebiotics: a review. Trends Food Sci Technol 20(11–12):567–575

Akyuz M, Onganer A, Erecevıt P, Kırbag S (2010) Antimicrobial activity of some edible mushrooms in the eastern and southeast Anatolia region of Turkey. Gazi Univ J Sci 23(2):125–130

Aladetohun NF, Sogbesan OA (2013) Utilization of blood meal as a protein ingredient from animal waste product in the diet of *Oreochromis niloticus.* Int J Fisheries Aquac 5(9):234–237

Alemayehu TA, Geremew A, Getahun A (2018) The role of functional feed additives in tilapia nutrition. Fish Aquac J 9(2):1g–1g

Allameh SK, Yusoff FM, Ringø E, Daud HM, Saad CR, Ideris A (2016) Effects of dietary mono-and multiprobiotic strains on growth performance, gut bacteria and body composition of Javanese carp (*Puntius gonionotus*, B leeker 1850). Aquac Nutr 22(2):367–373

Amaya EA, Davis DA, Rouse DB (2007) Replacement of fish meal in practical diets for the Pacific white shrimp (*Litopenaeus vannamei*) reared under pond conditions. Aquaculture 262(2–4):393–401

Amiri O, Miandare HK, Hoseinifar SH, Shabni A, Safari R (2018) Skin mucus protein profile, immune parameters, immune-related gene expression, and growth performance of rainbow trout (*Oncorhynchus mykiss*) fed white button mushroom (*Agaricus bisporus*) powder. Int J Med Mushrooms 20(4)

Anderson DP (1992) Immunostimulants, adjuvants, and vaccine carriers in fish: applications to aquaculture. Annu Rev Fish Dis 2:281–307

Asad Seftjani A, Abdi R, Salari-Aliabadi MA, Basir Z (2021) Comparative study of impact of button mushroom compost and chemical fertilizer on head kidney and some blood biochemical parameters in warm water aquaculture. Iranian Vet J 17(2):5–14

Baars JJP, Sonnenberg ASM, Mumm R, Stijger I, Wehrens, H. R. M. J. (2016). *Metabolites contributing to taste in Agaricus bisporus* (No. 2016–1). Plant Research International.

Baba E, Uluköy G, Öntaş C (2015) Effects of feed supplemented with Lentinula edodes mushroom extract on the immune response of rainbow trout, *Oncorhynchus mykiss*, and disease resistance against *Lactococcus garvieae.* Aquaculture 448:476–482

Bano Z (1981) Mineral and heavy metal contents in the sporophores of *Pleurotus* species. Mushroom Newsl. Trop. 2:3–6

Bansemer MS, Salini MJ, Nankervis L, Stone DA (2023) Reducing dietary wild derived fishmeal inclusion levels in production diets for large yellowtail kingfish (*Seriola lalandi*). Aquaculture 572:739487

Barros L, Cruz T, Baptista P, Estevinho LM, Ferreira IC (2008) Wild and commercial mushrooms as source of nutrients and nutraceuticals. Food Chem Toxicol 46(8):2742–2747

Beelman RB, Royse DJ, Chikthimmah N (2003) Bioactive components in button mushroom *Agaricus bisporus* (J. Lge) Imbach (Agaricomycetideae) of nutritional, medicinal, and biological importance. Int J Med Mushrooms 5(4)

Bernaś E, Jaworska G (2016) Vitamins profile as an indicator of the quality of frozen *Agaricus bisporus* mushrooms. J Food Compos Anal 49:1–8

Bharathi S, Antony C, Cbt R, Arumugam U, Ahilan B, Aanand S (2019) Functional feed additives used in fish feeds. Int J Fish Aquat Stud 7(3):44–52

Burgos-Aceves MA, Cohen A, Smith Y, Faggio C (2018) MicroRNAs and their role on fish oxidative stress during xenobiotic environmental exposures. Ecotoxicol Environ Saf 148:995–1000

Burridge L, Weis JS, Cabello F, Pizarro J, Bostick K (2010) Chemical use in salmon aquaculture: a review of current practices and possible environmental effects. Aquaculture 306(1–4):7–23

Cabello FC (2006) Heavy use of prophylactic antibiotics in aquaculture: a growing problem for human and animal health and for the environment. Environ Microbiol 8(7):1137–1144

Çağlarırmak N (2009) Determination of nutrients and volatile constituents of *Agaricus bisporus* (brown) at different stages. J Sci Food Agric 89(4):634–638

Chaitanawisuti N, Choeychom C, Piyatiratitivorakul S (2011) Effect of dietary supplementation of brewers yeast and nucleotide singularly on growth, survival and vibriosis resistance on juveniles of the gastropod spotted babylon (*Babylonia areolata*). Aquac Int 19:489–496

Chang CS, Huang SL, Chen S, Chen SN (2013) Innate immune responses and efficacy of using mushroom beta-glucan mixture (MBG) on orange-spotted grouper, *Epinephelus coioides*, aquaculture. Fish Shellfish Immunol 35(1):115–125

Chen S, Oh SR, Phung S, Hur G, Ye JJ, Kwok SL, Williams D (2006) Anti-aromatase activity of phytochemicals in white button mushrooms (*Agaricus bisporus*). Can Res 66(24):12026–12034

Chen Y, Fei S, Zhang K (2014) Stabilisation for switched linear systems with time-varying delay and input saturation. Int J Syst Sci 45(3):532–546

Cherno N, Osolina S, Nikitina A (2016) Chemical composition of *Agaricus bisporus* and *Pleurotus ostreatus* fruiting bodies and their morphological parts. Food Environ Saf J 12(4)

Cheung PCK (1996) Dietary fiber content and composition of some cultivated edible mushroom fruiting bodies and mycelia. J Agric Food Chem 44(2):468–471

Cheung PCK (2009) Mushrooms as functional foods (ed P. C. K. Cheung), John Wiley & Sons, Inc., Hoboken, NJ, USA

Cheung PC (2010) The nutritional and health benefits of mushrooms. Nutr Bull 35(4):292–299

Clark LC, Combs GF, Turnbull BW, Slate EH, Chalker DK, Chow J, Taylor JR (1996) Effects of selenium supplementation for cancer prevention in patients with carcinoma of the skin: a randomized controlled trial. JAMA 276(24):1957–1963

Corrêa RCG, Brugnari T, Bracht A, Peralta RM, Ferreira IC (2016) Biotechnological, nutritional and therapeutic uses of *Pleurotus* spp. (Oyster mushroom) related with its chemical composition: a review on the past decade findings. Trends Food Sci Technol 50:103–117

Dalmo RA, Seljelid R (1995) The immunomodulatory effect of LPS, laminaran and sulphated laminaran [β (l,3)-D-glucan] on Atlantic salmon, *Salmo salar* L., macrophages in vitro. J Fish Dis 18(2):175–185

Dawood MA, Koshio S (2016) Recent advances in the role of probiotics and prebiotics in carp aquaculture: a review. Aquaculture 454:243–251

Dawood MAO, Koshio S (2019) Application of fermentation strategy in aquafeed for sustainable aquaculture. Rev Aquac 12(2): 987–1002

Dawood MA, Koshio S, Esteban MÁ (2018) Beneficial roles of feed additives as immunostimulants in aquaculture: a review. Rev Aquac 10(4):950–974

Dawood MA, Abo-Al-Ela HG, Hasan MT (2020a) Modulation of transcriptomic profile in aquatic animals: probiotics, prebiotics and synbiotics scenarios. Fish Shellfish Immunol 97:268–282

Dawood MA, Eweedah NM, El-Sharawy ME, Awad SS, Van Doan H, Paray BA (2020b) Dietary white button mushroom improved the growth, immunity, antioxidative status and resistance against heat stress in Nile tilapia (*Oreochromis niloticus*). Aquaculture 523:735229

De Francesco M, Parisi G, Pérez-Sanchez J, Gomez-Réqueni P, Médale F, Kaushik SJ, Poli BM (2007) Effect of high-level fish meal replacement by plant proteins in gilthead sea bream (*Sparus aurata*) on growth and body/fillet quality traits. Aquac Nutr 13(5):361–372

Defoirdt T, Sorgeloos P, Bossier P (2011) Alternatives to antibiotics for the control of bacterial disease in aquaculture. Curr Opin Microbiol 14(3): 251–258

Deborah Paripuranam T, Divya VV, Ulaganathan P, Balamurugan V, Umamaheswari S (2011) Replacing fish meal with earthworm and mushroom meals in practical diets of *Labeo rohita* and *Hemigrammus caudovittatus* fingerlings. Indian J Anim Res 45(2):115–119

Dhamodharan G, Mirunalini S (2010) A novel medicinal characterization of *Agaricus bisporus* (white button mushroom). Pharmacol Online 2:456–463

Dhanasekaran D, Latha S, Saha S, Thajuddin N, Panneerselvam A (2013) Extracellular biosynthesis, characterisation and in-vitro antibacterial potential of silver nanoparticles using *Agaricus bisporus*. J Exp Nanosci 8(4):579–588

Dias DA, Urban S, Roessner U (2012) A historical overview of natural products in drug discovery. Metabolites 2(2):303–336

Didukh M, Nevo E, Volz PA, Volz PA (2004) Impact of the family 'Agaricaceae' (Fr.) cohn on nutrition and medicine. ARG Gantner

Dutta R (2007) Advances in mushroom science, vol 14. Satish Serial Publishing House, Delhi, p 240

El Basuini MF, Teiba II, Zaki MA, Alabssawy AN, El-Hais AM, Gabr AA, Dossou S (2020) Assessing the effectiveness of CoQ10 dietary supplementation on growth performance, digestive enzymes, blood health, immune response, and oxidative-related genes expression of Nile tilapia (*Oreochromis niloticus*). Fish Shellfish Immunol 98:420–428

Eswari S, Bhavan PS, Kalpana R, Dharani C, Manjula T, Sarumathi K, Rajkumar G (2019) Phytochemical Characterization of the mushroom *Agaricus bisporus* and assessment of its nutritional ability in the place of fishmeal for survival and growth of the freshwater prawn *Macrobrachium rosenbergii* post-larvae. Integrative food. Nutr Metab 6:1–11

Ezatrahimi N, Soltanian S, Hoseinifar SH (2024) Skin mucosal immune parameters and expression of the immune-relevant genes in *Danio rerio* treated by white button mushroom (*Agaricus bisporus*). Fish Physiol Biochem 1–14

FAO (2016) Agriculture Organization, 2014. Livestock Primary. Food and Agriculture Organization of the United Nations

FAO (2024) The State of World Fisheries and Aquaculture 2024. Blue Transformation in Action. Food and Agriculture Organization of the United Nations, Rome, Italy, pp 7

FAOSTAT (2018) Food and agriculture organization of the United Nations. Rome, p 403. http://faostat.fao.org

FAOSTAT (2020) Food and agriculture data. http://www.fao.org/faostat/en/#home.

Fontainhas-Fernandes A, Gomes E, Reis-Henriques MA, Coimbra J (1999) Replacement of fish meal by plant proteins in the diet of Nile tilapia: digestibility and growth performance. Aquac Int 7:57–67

Galkanda-Arachchige HS, Wilson AE, Davis DA (2020) Success of fishmeal replacement through poultry by-product meal in aquaculture feed formulations: a meta-analysis. Rev Aquac 12(3):1624–1636

Gbolagade JS (2006) Bacteria associated with compost used for cultivation of Nigerian edible mushrooms *Pleurotus tuber-regium (Fr.) Singer*, and *Lentinus squarrosulus* (Berk.). Afr J Biotechnol 5(4):338–342

Gothwal R, Gupta A, Kumar A, Sharma S, Alappat BJ (2012) Feasibility of dairy waste water (DWW) and distillery spent wash (DSW) effluents in increasing the yield potential of *Pleurotus flabellatus* (PF 1832) and *Pleurotus sajor-caju* (PS 1610) on bagasse. 3 Biotech 2:249–257

Grove JF (1981) Volatile compounds from the mycelium of the mushroom *Agaricus bisporus*. Phytochemistry 20(8):2021–2022

Guillamón E, García-Lafuente A, Lozano M, Rostagno MA, Villares A, Martínez JA (2010) Edible mushrooms: role in the prevention of cardiovascular diseases. Fitoterapia 81(7):715–723

Habib SS, Naz S, Khalid S, Kanwal R, Ameer I, Khan SNA, Nazir N (2021) Effect of white button mushroom (*Agaricus bisporus*) on immunity and haematological parameters of Oreochromis niloticus. Pak J Zool 7:1–4

Harikrishnan R, Balasundaram C, Heo MS (2012a) Effect of *Inonotus obliquus* enriched diet on hematology, immune response, and disease protection in kelp grouper, *Epinephelus bruneus* against *Vibrio harveyi*. Aquaculture 344:48–53

Harikrishnan R, Kim MC, Kim JS, Balasundaram C, Heo MS (2012b) Effect of Coriolus versicolor supplemented diet on innate immune response and disease resistance in kelp grouper *Epinephelus bruneus* against *Listonella anguillarum*. Fish Shellfish Immunol 32(2):339–344

Harikrishnan R, Naafar A, Musthafa MS, Ahamed A, Arif IA, Balasundaram C (2018) Effect of *Agaricus bisporus* enriched diet on growth, hematology, and immune protection in *Clarias gariepinus* against *Flavobacterium columnare*. Fish Shellfish Immunol 73:245–251

Hernández MD, Martínez FJ, Jover M, García BG (2007) Effects of partial replacement of fish meal by soybean meal in sharpsnout seabream (*Diplodus puntazzo*) diet. Aquaculture 263(1–4):159–167

Hood IA, Beets PN, Kimberley MO, Gardner JF, Oliver GR, Pearce S (2004) Colonisation of podocarp coarse woody debris by decomposer basidiomycete fungi in an indigenous forest in the central North Island of New Zealand. For Ecol Manag 196(2–3):311–325

Hoseinifar SH, Dadar M, Ringø E (2017) Modulation of nutrient digestibility and digestive enzyme activities in aquatic animals: the functional feed additives scenario. Aquac Res 48(8):3987–4000

Hoseinifar SH, Yousefi S, Capillo G, Paknejad H, Khalili M, Tabarraei A, Faggio C (2018) Mucosal immune parameters, immune and antioxidant defence related genes expression and growth performance of zebrafish (*Danio rerio*) fed on *Gracilaria gracilis* powder. Fish Shellfish Immunol 83:232–237

Hoseinifar SH, Zou HK, Paknejad H, Hajimoradloo A, Van Doan H (2019) Effects of dietary white-button mushroom powder on mucosal immunity, antioxidant defence, and growth of common carp (*Cyprinus carpio*). Aquaculture 501:448–454

Hossain MS, Alam N, Amin SR, Basunia MA, Rahman A (2007) Essential fatty acid contents of *Pleurotus ostreatus*, *Ganoderma lucidum* and *Agaricus bisporus*. Bangladesh J Mushroom 1(1):1–7

Irazoqui FJ, Zalazar FE, Nores GA, Vides MA (1997) *Agaricus bisporus* lectin binds mainly O-glycans but also N-glycans of human IgA subclasses. Glycoconj J 14:313–319

Iswarya A, Vaseeharan B, Anjugam M, Gobi N, Divya M, Faggio C (2018) β-1, 3 glucan binding protein-based selenium nanowire enhances the immune status of *Cyprinus carpio* and protection against *Aeromonas hydrophila* infection. Fish Shellfish Immunol 83:61–75

Itoh W (1997) Augmentation of protective immune responses against viral infection by oral administration of schizophyllan. Mediators Inflamm 6(4): 267–269

Iwashita MKP, Nakandakare IB, Terhune JS, Wood T, Ranzani-Paiva MJT (2015) Dietary supplementation with *Bacillus subtilis, Saccharomyces cerevisiae and Aspergillus oryzae* enhance immunity and disease resistance against *Aeromonas hydrophila* and *Streptococcus iniae* infection in juvenile tilapia *Oreochromis niloticus*. Fish Shellfish Immunol 43(1):60–66

Jannathulla R, Rajaram V, Kalanjiam R, Ambasankar K, Muralidhar M, Dayal JS (2019) Fishmeal availability in the scenarios of climate change: Inevitability of fishmeal replacement in aquafeeds and approaches for the utilization of plant protein sources. Aquac Res 50(12):3493–3506

Jeong SC, Koyyalamudi SR, Jeong YT, Song CH, Pang G (2012) Macrophage immunomodulating and antitumor activities of polysaccharides isolated from *Agaricus bisporus* white button mushrooms. J Med Food 15(1):58–65

Kakon AJ, Choudhury MBK, Saha S (2012) Mushroom is an ideal food supplement. J Dhaka Natl Med College Hosp 18(1):58–62

Kalač P (2009) Chemical composition and nutritional value of European species of wild growing mushrooms: a review. Food Chem 113(1):9–16

Kamilya D, Joardar SN, Mal C, Maiti K (2008) Effects of a glucan from the edible mushroom (*Pleurotus florida*) as an immunostimulant in farmed indian major carp (*Catla catla*). Israeli J Aquac 60(1–4):9

Kennedy DA, Kurath G, Brito IL, Purcell MK, Read AF, Winton JR, Wargo AR (2016) Potential drivers of virulence evolution in aquaculture. Evol Appl 9(2):344–354

Kiran, Kumar N, Sindhu S, Gawaria B, Ansari A, Dhariwal A, Sharma P, Kumar R (2025) GWAS approaches in medicinal plants for enhanced secondary metabolite production. In: Chen JT (ed) Biotechnology, multiple omics, and precision breeding in medicinal plants. CRC Press, pp 119–134

Kuforiji O, Fasidi I (2008) Compositional studies on Pleurotus tuber-regium Sclerotia. Adv Food Sci 30(1):2–5

Kuforiji OO, Fasidi IO (2007) Factors affecting the yields of fruitbody and sclerotia in *Pleurotus tuber-regium*. Adv Food Sci 29(4):211–215

Kumar R, Singh D (2022) Insecticide-tolerant bacterial population in *Eisenia fetida*'s Gut and Vermicast exposed to Chlorantraniliprole and Fipronil. Appl Biol Res 24(3):273–279

Kumar R, Gupta R, Dahiya T, Yadav R, Saifi R, Yodha K, Kumari B, Singh P, Shefali (2022a) Surveillance of *Heteropneustes fossilis* (Stinging Catfish) diseases and health management practices in Haryana. Appl Ecol Environ Sci 10(4):261–267. https://doi.org/10.12691/aees-10-4-12

Kumar R, Gupta RK, Yadav R, Saifi R, Yodha K, Pooja (2022b) Eisenia fetida as protein source for growth enhancement of Heteropneustes fossilis. Egypt J Aquat Biol Fisheries 26(2):577–588. https://doi.org/10.21608/ejabf.2022.234553

Kumar R, Yadav R, Kumar Gupta R, Pooja (2022c) A sustainable way for fish health management by replacement of chemical and drugs by earthworm. In: Environmental degradation in Asia: land degradation, environmental contamination, and human activities. Springer International Publishing, Cham, pp 329–352. https://doi.org/10.1007/978-3-031-12112-8_16

Kumar H, Bhardwaj K, Kuča K, Sharifi-Rad J, Verma R, Machado M, Cruz-Martins N (2022d) Edible mushrooms' enrichment in food and feed: a mini review. Int J Food Sci Technol 57(3):1386–1398

Langer S, Bakhtiyar Y, Lakhnotra R (2011) Replacement of fishmeal with locally available ingredients in diet composition of *Macrobrachium dayanum*. Afr J Agric Res 6(5):1080–1084

Lin Z, Yoshikawa S, Hamasaki M, Koyama T, Kikuchi K, Hosoya S (2022) Effects of low fishmeal diets on growth performance, blood chemical composition, parasite resistance, and gene expression in the tiger pufferfish, *Takifugu Rubripes*. Aquaculture 560:738484

Lindequist U, Niedermeyer TH, Jülich WD (2005) The pharmacological potential of mushrooms. Evid Based Complement Alternat Med 2(3): 285-299

Liu J, Wu YC, Kan J, Wang Y, Jin CH (2013) Changes in reactive oxygen species production and antioxidant enzyme activity of *Agaricus bisporus* harvested at different stages of maturity. J Sci Food Agric 93(9):2201–2206

Macusi ED, Liguez CGO, Macusi ES, Liguez AKO, Digal LN (2022) Factors that influence small-scale Fishers' readiness to exit a declining fishery in Davao Gulf, Philippines. Ocean Coastal Manag 230:106378

Magnadottir B (2010) Immunological control of fish diseases. Mar Biotechnol 12:361–379

Maihara VA, Moura PLDC, Catharino MGM, Moreira EG, Castro LP, Figueira RCL (2012) Cadmium determination in *Lentinus edodes* mushroom species. Food Sci Technol 32:553–557

Manzoor-ul-Haq VR, Singh D, Singh AK, Ninganagouda S, Hiremath J (2015) Dried mushroom *Agaricus bisporus* mediated synthesis of silver nanoparticles from Bandipora District (Jammu and Kashmir) and their efficacy against methicillin resistant Staphylococcus aureus (MRSA) strains. Nanosci Nanotechnol Int J 5:1–8

Matsushima Y, Eguchi F, Kikukawa T, Matsuda T (2009) Historical overview of psychoactive mushrooms. Inflamm Regen 29(1):47–58

Mehta BK, Jain SK, Sharma GP, Doshi A, Jain HK (2011) Cultivation of button mushroom and its processing: an techno-economic feasibility. Int J Adv Biotechnol Res 2(1):201–207

Miles PG, Chang ST (2004) Mushrooms: cultivation, nutritional value, medicinal effect, and environmental impact. CRC Press

Mirunalini S, Arulmozhi V, Deepalakshmi K, Krishnaveni M (2012) Intracellular biosynthesis and antibacterial activity of silver nanoparticles using edible mushrooms. Notulae Scientia Biologicae 4(4):55–61

Mitra A (2021) Thought of alternate aquafeed: conundrum in aquaculture sustainability?. Proc Zool Soc 74(1):1–18 (Springer India, New Delhi)

Mohamed EM (2012) Chemical profile, agaritine and selenium content of *Agaricus bisporus*. Braz Arch Biol Technol 55:911–920

Mohiuddin KM, Alam MM, Arefin MT, Ahmed I (2015) Assessment of nutritional composition and heavy metal content in some edible mushroom varieties collected from different areas of Bangladesh. Asian J Med Biol Res 1(3):495–501

Moisa MB, Tufa CA, Gabissa BT, Gurmessa MM, Wedajo YN, Feyissa ME, Gemeda DO (2022) Integration of geospatial technologies with multi-criteria decision analysis for aquaculture land suitability evaluation: the case of Fincha'a River Sub-basin, Western Ethiopia. J Agric Food Res 10:100448

Monsef Shokri M, Yousefi S, Yousefi Jourdehi A, Nazemroaya S, Hossenifar SH, Sayed Hassani M (2023) Growth performance, hematological and serum biochemical parameters of juvenile Persian sturgeon (*Acipenser persicus*) fed with white button mushroom (*Agaricus bisporus*) powder. J Fisheries 76(4):565–577

Muin H, Taufek NM, Abiodun RA, Yusof HM, Razak SA (2015) Effect of partial and complete replacement of fishmeal with mushroom stalk meal and soy bean meal on growth performance of Nile tilapia, Oreochromis Niloticus Fingerlings. Sains Malaysiana 44(4):511–516

Muslat MM, Al-Assaffii IAA, Owaid MN (2014) *Agaricus bisporus* product development by using local substrate with bio-amendment. Int J Environ Global Clim 2(4):176–188

Muszyńska B, Sułkowska-Ziaja K, Ekiert H (2011) Indole compounds in fruiting bodies of some edible Basidiomycota species. Food Chem 125(4):1306–1308

Muszyńska B, Sułkowska-Ziaja K, Wojcik A (2013) Levels of physiologically active indole derivatives in the fruiting bodies of some edible mushrooms (Basidiomycota) before and after thermal processing. Mycoscience 54(5):321–326

Muszyńska B, Kała K, Rojowski J, Grzywacz-Kisielewska A, Opoka W (2017) Composition and biological properties of *Agaricus bisporus* fruiting bodies: a review

Nakao M, Tsujikura M, Ichiki S, Vo TK, Somamoto T (2011) The complement system in teleost fish: progress of post-homolog-hunting researches. Dev Comp Immunol 35(12):1296–1308

Naylor RL, Hardy RW, Bureau DP, Chiu A, Elliott M, Farrell AP, Nichols PD (2009) Feeding aquaculture in an era of finite resources. Proc Natl Acad Sci 106(36):15103–15110

Ndungutse V, Mereddy R, Sultanbawa Y (2015) Bioactive properties of mushroom (*A Garicus bisporus*) stipe extracts. J Food Process Preserv 39(6):2225–2233

Neumann NF, Stafford JL, Barreda D, Ainsworth AJ, Belosevic M (2001) Antimicrobial mechanisms of fish phagocytes and their role in host defense. Dev Comp Immunol 25(8–9):807–825

Nitschke J, Altenbach HJ, Malolepszy T, Mölleken H (2011) A new method for the quantification of chitin and chitosan in edible mushrooms. Carbohyd Res 346(11):1307–1310

Ogunji JO, Wirth M (2001) Alternative protein sources as substitutes for fishmeal in the diet of young tilapia *Oreochromis niloticus* (Linn.)

Oliva-Teles A, Enes P, Couto A, Peres H (2022) Replacing fish meal and fish oil in industrial fish feeds. In: Feed and feeding practices in aquaculture, pp 231–268

Oujifard A, Seyfabadi J, Kenari AA, Rezaei M (2012) Fish meal replacement with rice protein concentrate in a practical diet for the Pacific white shrimp, *Litopenaeus vannamei* Boone, 1931. Aquac Int 20:117–129

Owaid MN (2015) Mineral elements content in two sources of *Agaricus bisporus* in Iraqi market. J Adv Appl Sci 3(2):46–50

Owaid MN, Ibraheem IJ (2017) Mycosynthesis of nanoparticles using edible and medicinal mushrooms. Eur J Nanomed 9(1):5–23

Owaid MN, Barish A, Ali Shariati M (2017) Cultivation of *Agaricus bisporus* (button mushroom) and its usages in the biosynthesis of nanoparticles. Open Agric 2(1):537–543

Öztürk M, Duru ME, Kivrak Ş, Mercan-Doğan N, Türkoglu A, Özler MA (2011) In vitro antioxidant, anticholinesterase and antimicrobial activity studies on three *Agaricus* species with fatty acid compositions and iron contents: a comparative study on the three most edible mushrooms. Food Chem Toxicol 49(6):1353–1360

Parisa Maftoun PM, Helmi Johari HJ, Mohammad Soltani MS, Roslinda Malik RM, Nor Zalina NZ (2015) The edible mushroom *Pleurotus* spp.: I. biodiversity and nutritional values

Parslew R, Jones KT, Rhodes JM, Sharpe GR (1999) The antiproliferative effect of lectin from the edible mushroom (*Agaricus bisporus*) on human keratinocytes: preliminary studies on its use in psoriasis. Br J Dermatol 140(1):56–60

Raa J (1996) The use of immune stimulatory substances of fish and shellfish farming. Rev Fish Sci 4: 229-288

Rahi DK, Malik D (2016) Diversity of mushrooms and their metabolites of nutraceutical and therapeutic significance. J Mycol 2016(1):7654123

Rajni G, Grewal RB, Goya RK (2006) Nutritional attributes of Agaricus bisporus and Pleurotus sajor caju mushrooms. Nutr Hlth 18: 179–184

Rathore H, Prasad S, Sharma S (2017) Mushroom nutraceuticals for improved nutrition and better human health: a review. PharmaNutrition 5(2):35–46

Reis FS, Martins A, Barros L, Ferreira IC (2012) Antioxidant properties and phenolic profile of the most widely appreciated cultivated mushrooms: a comparative study between in vivo and in vitro samples. Food Chem Toxicol 50(5):1201–1207

Reverter M, Tapissier-Bontemps N, Sasal P, Saulnier D (2017) Use of medicinal plants in aquaculture. In:Diagnosis and control of diseases of fish and shellfish, pp 223–261

Rombout JHWM, Huttenhuis HBT, Picchietti S, Scapigliati G (2005) Phylogeny and ontogeny of fish leucocytes. Fish Shellfish Immunol 19(5):441–455

Romero J, Feijoó CG, Navarrete P (2012) Antibiotics in aquaculture—Use, abuse and alternatives. Health Environ Aquac 159(1):159–198

Rosenberg L (2020) Mushroom are considered to be the most sustainable vegetable-here's why? Green matters. https://www.greenmatters.com/p/why-are-mushrooms-sustainable

Safari R, Hoseinifar SH, Dadar M, Khalili M (2018) Powder of the white bottom mushroom, *Agaricus bisporus* (Agaricomycetes), improved immunomodulatory and health-promoting effects of *Lactobacillus casei* in zebrafish (*Danio rerio*). Int J Med Mushrooms 20(7):695–704

Safari R, Hoseinifar SH, Nejadmoghaddam S, Dadar M (2019) Combined administration of the white button mushroom *Agaricus bisporus* (agaricomycetes) and *lactobacillus casei* modulated immune-related gene expression and mucosal and serum immune parameters in a goldfish (*Carassius auratus*) model. Int J Med Mushrooms 21(5):503–513

Safwat MSIA, Al Kholi MAJ (2006) Recent trends, reality and future in the production, manufacture and marketing of medicinal and aromatic plants. In: Egyptian association of producers, manufacturers and exporters of medical and aromatic plants esmab, Giza, Egypt, p 76

Saiqa S, Haq NB, Muhammad AH, Muhammad AA, ATA UR (2008) Studies on chemical composition and nutritive evaluation of wild edible mushrooms

Samocha TM, Davis DA, Saoud IP, DeBault K (2004) Substitution of fish meal by co-extruded soybean poultry by-product meal in practical diets for the Pacific white shrimp, Litopenaeus Vannamei. Aquaculture 231(1–4):197–203

Santos AF, Da Silva MDC, Napoleão TH, Paiva PMG, Correia MDS, Coelho LCBB (2014) Lectins: function, structure, biological properties and potential applications

Sartori SB, Ferreira LFR, Messias TG, Souza GD, Pompeu GB, Monteiro RTR (2015) *Pleurotus* biomass production on vinasse and its potential use for aquaculture feed. Mycology 6(1):28–34

Shao S, Hernandez M, Kramer JK, Rinker DL, Tsao R (2010) Ergosterol profiles, fatty acid composition, and antioxidant activities of button mushrooms as affected by tissue part and developmental stage. J Agric Food Chem 58(22):11616–11625

Sharma VP, Annepu SK, Gautam Y, Singh M, Kamal S (2017) Status of mushroom production in India. Mushroom Res 26(2):111–120

Sharma MMM, Kapoor D, Loyal A, Kumar R, Sharma P, Husen A (2024) Environmental toxicity of engineered carbon nanoparticles. In: Bachheti AJ, Bachheti RK, Husen A (eds) Carbon-based nanomaterials: synthesis, agricultural, biomedical, and environmental interventions. Springer, pp 337–353. https://doi.org/10.1007/978-981-97-0240-4_16

Sharma MMM, Kapoor D, Loyal A, Kumar R, Sharma P, Husen A (2025) Positive and negative impact of carbon-based nanomaterials on the plant growth performance. In: Husen A (ed) Emerging carbon nanomaterials for sustainable agricultural practices: synthesis, plant growth, performance, production and protection. Springer Nature Singapore, pp 63–83

Shefali, Kumar R, Sankhla MS, Kumar R, Sonone SS (2021) Impact of pesticide toxicity in aquatic environment. Biointerface Res Appl Chem 11(3):10131–10140. https://doi.org/10.33263/BRIAC113.1013110140

Simon RR, Phillips KM, Horst RL, Munro IC (2011) Vitamin D mushrooms: comparison of the composition of button mushrooms (*Agaricus bisporus*) treated postharvest with UVB light or sunlight. J Agric Food Chem 59(16):8724–8732

Singh P, Tank PR, Janbandhu S, Motivarash Y (2022) Effect of supplementation of white button mushroom, *Agaricus bisporus* (Imbach, 1946) on growth performance and survival in white leg shrimp, *Litopenaeus vannamei* (Boone, 1931). J Exp Zool 25(2):2521–2528

Singh P, Tank PR, Janbandhu S, Jatav SK, Qureshi QA, Patel ND (2023) Effect of supplementation of white button mushroom, *Agaricus bisporus* (Imbach, 1946) on disease resistance in white leg shrimp, Litopenaeus vannamei (Boone 1931). Pharma Innov 12(10):1229–1233

Smiderle FR, Ruthes AC, van Arkel J, Chanput W, Iacomini M, Wichers HJ, Van Griensven LJ (2011) Polysaccharides from *Agaricus bisporus* and *Agaricus brasiliensis* show similarities in their structures and their immunomodulatory effects on human monocytic THP-1 cells. BMC Complement Altern Med 11:1–11

Soler-Rivas C, Jolivet S, Arpin N, Olivier JM, Wichers HJ (1999) Biochemical and physiological aspects of brown blotch disease of *Agaricus bisporus*. FEMS Microbiol Rev 23(5):591–614

Spolar MR, Schaffer EM, Beelman RB, Milner JA (1999) Selenium-enriched *Agaricus bisporus* mushrooms suppress 7, 12-dimethlybenz [a] anthracene bioactivation in mammary tissue. Cancer Lett 138(1–2):145–150

Stamets P (2001) A novel approach to farm waste management. Mushroom J Winter 30:28–32

Sudhakar T, Nanda A, Babu SG, Janani S, Evans MD Markose TK (2014). Synthesis of silver nanoparticles from edible mushroom and its antimicrobial activity against human pathogens. Int J Pharmtech Res 6(5): 1718-1723

Sun XZ, Zhou D, Chen S (1997) Autocrine and paracrine actions of breast tumor aromatase. A three-dimensional cell culture study involving aromatase transfected MCF-7 and T-47D cells. J Steroid Biochem Mol Biol 63(1–3):29–36

Tacon AG (2004) Use of fish meal and fish oil in aquaculture: a global perspective

Tao Y, Zhang L, Cheung PC (2006) Physicochemical properties and antitumor activities of water-soluble native and sulfated hyperbranched mushroom polysaccharides. Carbohyd Res 341(13):2261–2269

ten Have R, Wijngaard H, Ariës-Kronenburg NA, Straatsma G, Schaap PJ (2003) Lignin degradation by *Agaricus bisporus* accounts for a 30 increase in bioavailable holocellulose during cultivation on compost. J Agric Food Chem 51(8):2242–2245

Thatoi H, Singdevsachan SK (2014) Diversity, nutritional composition and medicinal potential of Indian mushrooms: a review. Afr J Biotechnol 13(4)

Tian Y, Zeng H, Xu Z, Zheng B, Lin Y, Gan C, Lo YM (2012) Ultrasonic-assisted extraction and antioxidant activity of polysaccharides recovered from white button mushroom (*Agaricus bisporus*). Carbohyd Polym 88(2):522–529

Tran TQ, Van Vu H, Nguyen TV (2023) Aquaculture, household income and inequality in Vietnam's coastal region. Marine Policy 153:105634

Tsai SY, Wu TP, Huang SJ, Mau JL (2007) Nonvolatile taste components of *Agaricus bisporus* harvested at different stages of maturity. Food Chem 103(4):1457–1464

Ullah MI, Akhtar M, Awais MM, Anwar MI, Khaliq K (2018) Evaluation of immunostimulatory and immunotherapeutic effects of tropical mushroom (*Lentinus edodes*) against eimeriasis in chicken. Trop Anim Health Prod 50:97–104

Uluköy G, Baba E, Öntaş C (2016) Effect of oyster mushroom, Pleurotus ostreatus, extract on hemato-immunological parameters of Rainbow trout, Oncorhynchus mykiss. J World Aquac Soc 47(5):676–684

Van Doan H, Doolgindachbaporn S, Suksri A (2016) Effects of Eryngii mushroom (*Pleurotus eryngii*) and *Lactobacillus plantarum* on growth performance, immunity and disease resistance of Pangasius catfish (*Pangasius bocourti*, Sauvage 1880). Fish Physiol Biochem 42:1427–1440

Van Doan H, Hoseinifar SH, Faggio C, Chitmanat C, Mai NT, Jaturasitha S, Ringø E (2018) Effects of corncob derived xylooligosaccharide on innate immune response, disease resistance, and growth performance in Nile tilapia (*Oreochromis niloticus*) fingerlings. Aquaculture 495:786–793

Van Doan H, Hoseinifar SH, Esteban MÁ, Dadar M, Thu TTN (2019) Mushrooms, seaweed, and their derivatives as functional feed additives for aquaculture: an updated view. Stud Nat Prod Chem 62:41–90

Vetter J (2007) Chitin content of cultivated mushrooms *Agaricus bisporus*, *Pleurotus ostreatus* and *Lentinula edodes*. Food Chem 102(1):6–9

Volman JJ, Mensink RP, Van Griensven LJLD, Plat J (2010) Effects of α-glucans from *Agaricus bisporus* on ex vivo cytokine production by LPS and PHA-stimulated PBMCs; a placebo-controlled study in slightly hypercholesterolemic subjects. Eur J Clin Nutr 64(7):720–726

Wang X, Luo H, Zheng Y, Wang D, Wang Y, Zhang W, Shao J (2023) Effects of poultry by-product meal replacing fish meal on growth performance, feed utilization, intestinal morphology and microbiota communities in juvenile large yellow croaker (*Larimichthys crocea*). Aquac Rep 30:101547

Wasser SJAMB (2002) Medicinal mushrooms as a source of antitumor and immunomodulating polysaccharides. Appl Microbiol Biotechnol 60:258–274

Weijn A, Bastiaan-Net S, Wichers HJ, Mes JJ (2013) Melanin biosynthesis pathway in *Agaricus bisporus* mushrooms. Fungal Genet Biol 55:42–53

Whyte SK (2007) The innate immune response of finfish-a review of current knowledge. Fish Shellfish Immunol 23:1127–1151

Wu ZQ, Jiang C, Ling F, Wang GX (2015) Effects of dietary supplementation of intestinal autochthonous bacteria on the innate immunity and disease resistance of grass carp (*Ctenopharyngodon idellus*). Aquaculture 438:105–114

Yang ChihChiu YC, Chen ShiuNan CS, Lu ChungLun LC, Chen S, Lai KamChiu LK, Liao WenLiang LW (2014) Effect of mushroom beta glucan (MBG) on immune and haemocyte response in Pacific white shrimp (*Litopenaeus vannamei*)

Yousefi S, Hoseinifar SH, Paknejad H, Hajimoradloo A (2018) The effects of dietary supplement of galactooligosaccharide on innate immunity, immune related genes expression and growth performance in zebrafish (*Danio rerio*). Fish Shellfish Immunol 73:192–196

Yu LG, Fernig DG, White MR, Spiller DG, Appleton P, Evans RC, Grierson I, Smith JA, Davies H, Gerasimenko OV, Peterson OH, Milton JD, Rhodes JM (1999) Edible mushroom, *Agaricus bisporus* lectin, which reversibly inhibits epithelial cell proliferation, blocks nuclear localization sequence-dependent nuclear protein import. J Biol Chem 274(8):4890–4899

Zakariaei H, Sudagar M, Hoseini SS, Paknejad H, Baruah K (2022) The effect of using synbiotics containing button mushroom (*Agaricus bisporus*) extract in combination with two species of lactic acid bacteria on digestive enzyme activities, body composition, growth and intestinal microbial flora in zebrafish (*Danio rerio*). J Mar Sci Technol 21(2):63–75

Zargar Z, Rahimi-Afzal Z, Soltani E, Mirghaed TA, Ebrahimzadeh-Mousavi AH, Soltani M, Yuosefi P (2019) Growth performance, immune response and disease resistance of rainbow trout (*Oncorhynchus mykiss*) fed *Thymus vulgaris* essential oils. Aquac Res 50:3097–3106

Zhang D, Zhang Y, Liu B, Jiang Y, Zhou Q, Wang J, Kuang Q (2017) Effect of replacing fish meal with fermented mushroom bran hydrolysate on the growth, digestive enzyme activity, and antioxidant capacity of allogynogenetic crucian carp (*Carassius auratus gibelio*). Turk J Fish Aquat Sci 17(5):1039–1048

Zhang M, Cui SW, Cheung PCK, Wang Q (2007) Antitumor polysaccharides from mushrooms: a review on their isolation process, structural characteristics and antitumor activity. Trends Food Sci Technol 18(1):4–19

Zou HK, Hoseinifar SH, Miandare HK, Hajimoradloo A (2016) *Agaricus bisporus* powder improved cutaneous mucosal and serum immune parameters and up-regulated intestinal cytokines gene expression in common carp (*Cyprinus carpio*) fingerlings. Fish Shellfish Immunol 58:380–386

Chapter 32
Tinospora Cordifolia: A Natural Remedy for Sustainable Aquaculture

Chinmoyee Bhattacharyya, Pavitra Chutia, and Susmita Dey

Abstract *Tinospora cordifolia* (TC), a prominent medicinal plant often used in traditional medicinal practices has recently emerged as a good potential for enhancing productivity and health in aquaculture. This chapter reviews recent research developments explaining TC's profound benefits in fish and shrimp aquaculture. Studies revealed potent antioxidant properties of TC extracts which effectively mitigate oxidative stress induced by anthropogenic factors in fish species. TC supplements effectively enhanced antioxidant enzyme activities and reduced pro-inflammatory cytokine levels, thereby maintaining cellular activities and boosting overall health. Various extracts of TC also showed noticeable immunomodulatory activities, improving both specific and non-specific immune responses. Studies also revealed that TC extracts fed oraly enhanced antibody titers, neutrophil activity, and phagocytic efficiency, which resulted in improved disease resistance against some common pathogens of aquaculture. Furthermore, TC-enriched diets promote growth performance in aquatic organisms, as evidenced by significant improvements in overall weight, Specific Growth Rate (SGR), and Feed Conversion Efficiency (FCR). Enhanced hematological and serum biochemical parameters reflect improved physiological functions and overall health status in TC-supplemented fish. Moreover, in shrimp aquaculture, TC supplementation augments non-specific immune responses, leading to increased survival rates and disease resistance against prevalent pathogens. *Tinospora cordifolia* emerges as a natural and effective adjunct in aquaculture practices, offering sustainable solutions for enhancing aquatic organism health and productivity. Future research should explore optimal dosages and delivery

---

C. Bhattacharyya (✉)
Department of Life Sciences, Dibrugarh University, Dibrugarh, Assam, India
e-mail: rs_chinmoyeebhattacharyya@dibru.ac.in

P. Chutia
Barpathar College, Golaghat, Assam, India
e-mail: pavitra.chutia@gmail.com

S. Dey
Department of Biomedical Engineering and Interventional Radiology, Lerner Research Institute, Cleveland, USA
e-mail: susmitadedas@gmail.com

R. Kumar et al. (eds.), *Sustainability in Aquaculture and Fisheries*,
Springer Water, https://doi.org/10.1007/978-3-032-16605-0_32

mechanisms to maximize TC's beneficial effects while ensuring environmental sustainability and economic viability in aquaculture operations.

Keywords Aquafeed · Shrimp culture · Mechanism · *Guduchi* · Pisciculture

32.1 Introduction

The current global population stands at eight billion and is anticipated to surge to ten billion by 2050, prompting a heightened demand for cost-effective, readily available, and nutrient-rich food sources. Aquaculture and fisheries are pivotal in meeting this demand, as their products offer essential macro and micronutrients for human consumption. Between 1990 and 2016, global per capita fish consumption rose significantly from 14.4 to 20.3 kg, with fish contributing to nearly 6.5% of total protein and 17% of animal protein intake in 2009 (Mishra 2020; Kumar et al. 2022a, b). In 2021, aquaculture products contributed nearly 15% of animal protein intake and 6% of total protein intake globally with at least 20% of per capita protein supply from animal sources. As per the report of FAO global annual per capita apparent consumption of aquatic foods reached 20.7 kg with a projected value of 21.3 kg in 2032 (FAO 2024). Consequently, there has been a gradual escalation in global fish production over the past five decades (Reverter et al. 2014). Aquaculture, in particular, has emerged as one of the rapidly growing animal production industries, boasting growth rate of 5.3% per annum from 2001 to 2018. In 2018, global fish production soared to 179 million tonnes (Mt), a significant increase from 62 Mt. in 2011, with 96.4 Mt. sourced from fisheries and 82.1 Mt. from aquaculture (FAO 2020). In 2022, global fisheries and aquaculture production had surged to 223.2 Mt. of which 130.9 Mt. was from aquaculture sector, setting a new high for the first time in history with an increase of 4.4% from 2020. This sector was valued at USD 312.8 billion in 2022, which employed approximately 61.8 million people (FAO 2020).

Shrimp and shellfish culture stand out as significant sectors within the aquaculture industry, with India playing a major role. Asian countries collectively dominate approximately 85% of the global shellfish production. In 2018, global shrimp production experienced a 3–5% rise compared to 2017, reaching nearly 4 Mt. (FAO 2021). India currently ranks as the third-largest shrimp producer globally, with export values witnessing a remarkable 31.71% increase in rupees in 2021–22. During the same financial year, India exported 7,28,123 MT of shrimp valued at USD 5828.59 Million, with the United States serving as the largest importer. Notably, frozen shrimps constituted the majority of the exported seafood, comprising approximately 75% (MPEDA 2022). However, as of January–September 2024, the global shrimp import amounted to 2.6 Mt., indicating a decline by 3.4% than that of 2023. Ecuador ranked top in export of shrimps at 918.13 Mt. showing 0.90% increase than previous year, followed by India who faced a remarkable decline by 6.69% compared to 2023 (FAO 2025).

However, the burgeoning demand has led to the intensification of finfish and shellfish culture, resulting in mass mortalities attributed to various environmental stresses (Bondad-Reantaso et al. 2005; Shefali et al. 2021). Overcrowding, malnutrition, poor water quality, and unhygienic conditions compromise the immune systems of fish, facilitating the proliferation of opportunistic pathogens and subsequent mass mortalities. Similarly, the shrimp culture industry encounters comparable challenges stemming from increased production and disease outbreaks. Over the past 15 years, disease outbreaks have inflicted losses totalling nearly 15 billion US dollars on the industry, with Asia bearing the brunt, accounting for 80% of these losses (Kumar and Singh 2022; Ghosh et al. 2021).

To mitigate economic losses due to several common diseases, aquaculture relies heavily on synthetic veterinary drugs for growth promotion and medication (Rico et al. 2013). However, their widespread use poses significant health risks to aquatic organisms and humans which contributes to contamination of the aquatic environment and antibiotic resistance. Moreover, the bioaccumulation of drugs in different tissues and water poses hazards to the environment as well as human health (Cabello 2006; Cherukumudi et al. 2024; Romero et al. 2012). Though vaccines available in the market are often successful in disease control, their affordability and specificity limits its widespread use (Reverter et al. 2014). As a result, there is a growing interest in natural alternatives like plant secondary metabolites to promote sustainable aquaculture practices.

Recently, medicinal plants which are commonly used for human and other veterinary practices, are gaining attention in aquaculture particularly in Asia which highlights their potential in aquatic health management (Kumar et al. 2022c; Bulfon et al. 2015). Traditional medicinal plants like ginger, garlic, cinnamon, amla, and guava are being effectively integrated into feeds for their diverse health benefits (Van Hai 2015). The phytochemicals present in these plants offer cost-effective, readily available substitutes with varied range of efficiency against different pathogens. These plants offer enhanced growth, improve immunity, and mitigate stress in fish due to the presence of bioactive compounds such as alkaloids, flavonoids, and phenols (Citarasu 2010; Mbokane and Moyo 2024). Among these medicinal plants, *T. cordifolia* is well known for its medicinal properties in Ayurveda and also shows promising potential in mitigating several problems of aquaculture by enhancing growth, boosting immunity, and increasing disease resistance in freshwater fishes (Bulfon et al. 2015). Thus, this book chapter aims to explore the reported benefits of *T. cordifolia* in promoting growth, health, and enhancing disease resilience in both freshwater fish and shellfish.

32.2 *Tinospora Cordifolia*: A Miraculous Plant

32.2.1 Morphology and Habitat

Tinospora cordifolia (TC) belongs to the Menispermaceae family and is native to the Indian subcontinent. It is also found Bangladesh, Sri Lanka, Pakistan, Malaysia, and Indonesia (Saha and Ghosh 2012). It is known by various vernacular names such as amrita, giloy, guduchi, and madhuparni (Modi et al. 2021). TC is characterized by large deciduous, woody, climbing shrubs. It has heart shaped alternate leaves, unisexual flowers, grey-colored fibrous stems, and fleshy single-seeded fruits (Joshi and Kaur 2016; Tripathi et al. 2015). TC can grow in a variety of soil types, ranging from acidic to alkaline, with very little soil moisture (Spandana et al. 2013).

32.2.2 Chemical Composition

Tinospora cordifolia contains high protein, low fat and several essential micronutrients which are considered beneficial for human health. Several medicinally important phyto-chemicals have been isolated from various parts of *TC* which belongs to several classes such as steroids, phenols, diterpenoid lactones, glycosides, alkaloids, terpenoids, polysaccharides, lignans, sesquiterpenoids and aliphatic compounds(Choudhary et al. 2014; Modi et al. 2021).

32.2.3 Nutrient Contents

The shoots of TC boast a nutritional value of 292.54 calories per 100 g, with high protein, low fat, and abundant micronutrients like nitrogen, phosphorus, potassium, and calcium(Nile and Khobragade 2009). Ripened TC fruit exhibits a slightly lower nutritional value at 132.8 calories per 100 g, yet with higher macronutrient content compared to shoots, along with tocopherol, niacin, and ascorbic acid. The study also reported the presence of tocopherol, niacin, and ascorbic acid and minerals such as iron, Potassium, Copper, Zinc, Magnesium, and Sodium in the fruit (Imtiyaj Khan et al. 2011).

32.2.4 Phytochemicals

Alkaloids mostly isolated from the stem and root of TC are aporphine, berberine, choline, palmetin, jatrorrhizine, isocolumbin, tembetarine, tinosporin, magnoflorine, tetrahydropalmatine (Singh and Devi 2017). Prominent steroids isolated from the stem are β-sitosterol, δ-sitosterol, Ecdysterone, Giloinsterol, 20 β-hydroxyecdysone, and Makisterone A (Mckeown et al. 2012). Other TC stem phytoconstituents are sesquiterpene compound tinocordifolin, and Glycosides such as tinocordiside, cordioside, palmatosides, tinocordifolioside (Arunachalam et al. 2022; Kapil and Sharma 1997; Maurya and Handa 1998). The whole plant extract reported the presence of diterpenoid lactones such as tinosporin, Furanolactone, tinosporides, clerodane derivatives, jateorine, columbin, and aliphatic compounds such as heptacosanol Nonacosan-15-one dichloromethane, octacosanol (Dhanasekaran et al. 2009;

Sriramaneni et al. 2010; Thippeswamy et al. 2008). Tinosporofuranol, tinospora-clerodanol, tinosporafurandiol, tinosporaclerodanoid, β-sitosterol are some steroids reported from the bark of TC (Spandana et al. 2013). Isolation of phytochemicals from roots of TC reported the presence of compounds like Tinosporic acid, Giloin, and N-trans-feruloyl tyramine, which are known to have various medicinal properties (Patel and Mishra 2011).

32.2.5 Biological Activity

TC exhibits a diverse array of biological activities, including anticancer, antidiabetic, neuro-regenerative, antibacterial, and immunomodulatory properties(Joshi and Kaur 2016; Modi et al. 2021; Sharma et al. 2019). Its traditional use in Ayurveda encompasses treatments for jaundice, diabetes, rheumatism, and various other ailments, highlighting its significance in traditional medicine(Goel et al. 2004; Sinha et al. 2004). TC is revered for its role as an appetite stimulant, energy booster, and immunomodulator, reflecting its multifaceted medicinal value (Salkar et al. 2017).

32.3 Potential Use of *Tinospora Cordifolia* in Freshwater Fish Culture

Tinospora cordifolia, a long-valued medicinal herb is often used in fish culture by different communities as a traditional medicine for various ailments. Extracts of different parts of TC possesses strong free radical scavenging, antioxidant, growth promoting and immunomodulatory properties which make it more relevant in current scenarios (Fig. 32.1). Several scientific research highlights that the bioactive compounds such as polyphenols, alkaloids, and flavonoids contribute in enhancing cellular functions by mitigating oxidative stress (Table 32.1). Also, it influences overall metabolism that ultimately results in providing resistance to different diseases and promoting good health, resilience and sustainability in aquaculture

32.3.1 Antioxidant Potential

Antioxidant enzymes are important biomarkers of stress in aquatic environments, which shows shift from its normal values under various anthropogenic stress (Kumar et al. 2024; Liu et al. 2021; Livingstone 2003). Incorporation of bioactive compounds extracted from the medicinal plants offers a promising strategy for managing stress induced alterations by enhancing antioxidant defence, scavenging free radicals and minimizing ROS generation. Phytochemicals are known to bind to the metal ions that

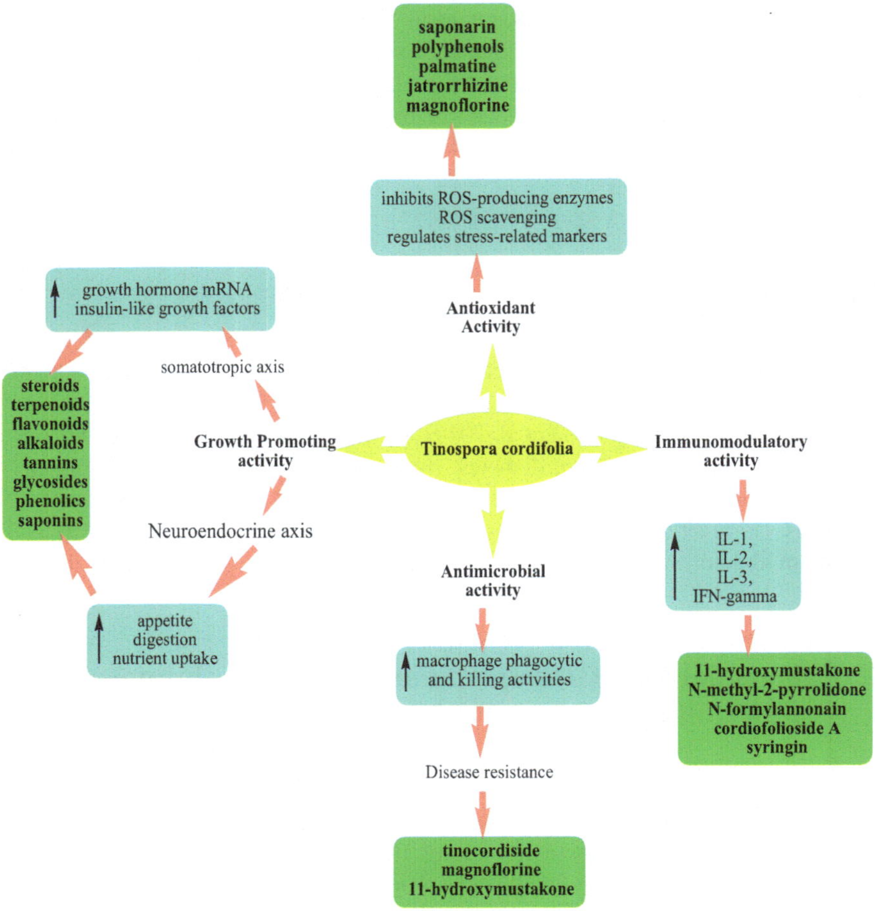

Fig. 32.1 Effect of phytochemicals isolated from Tinospora cordifolia

are involved in oxidative reaction. Secondary metabolites like flavonoids not only inhibit ROS-generating enzymes but also acts as scavengers of reactive molecules, protecting the cells from lipid peroxidation (Faheem et al. 2022).

In a study by El-Basuini et al.(2022), dietary supplementation of TC extracts against hypoxia stress in Nile tilapia in varying dose showed significant improvement in antioxidant enzyme activities, particularly in liver and intestinal tissues. Fishes fed with supplemented feed exhibited reduced ROS production and helped in mitigating the damage to immune cells and DNA, resulting in more robust defense system. Reduced circulating cortisol and glucose, which are identified as stress related markers, in treated fishes further indicated lowered physiological stress. In another study by Vineetha et al. (2021) reported the antioxidative effects of an ethanolic extract of *T. cordifolia* (TCE) in Nile tilapia exposed to titanium dioxide nanoparticle (TNP) toxicity. Exposure to TNP resulted in suppressed activity of

Table 32.1 *Tinospora cordifolia* in freshwater aquaculture

Extract	Species	Route	Doses	Duration	Against	Results	Reference
Petroleum ether Ethanol and extracts of the leaves	*Oreochromis mossambicus*	Intraperitoneal (IP) injection	0, 0.8, 8 and 80 mg/kg	–	*Aeromonas hydrophilla*	Ethanolic extract improved primary and secondary immune response, enhanced neutrophil activity, improved disease resistance Petroleum ether extract significantly improved secondary response	(Sudhakaran et al. 2006)
ImmuPlus (Mixture of *T. cordifolia*, *Ocimum sanctum*, *Withania somnifera*, and *Emblica officinalis*)	*Labeo rohita* larvae	Oral	0, 0.25, 0.5 and 0.75g/kg	10 days	*Aeromonas hydrophila*	Improved growth, survival and disease resistance	(Kumari et al. 2007)
	Advanced larvae		0, 0.5, 1, and 2g/kg	45 days		Increased body weight, survival	
	Fingerlings		0 and 1g/kg	30 days		Increased growth, reduced mortality	
	Juveniles		0 and 1g/kg	15 and 30 days		Increased serum total protein, albumin and globulin. Enhanced non-specific and specific immunity	
Water soluble fraction of leaves	*Oreochromis mossambicus*	IP injection	0, 6, 60 and 600 mg/kg	–	*Aeromonas hydrophila*	Enhanced Non-specific humoral and cellular immunity and disease resistance	(Alexander et al. 2010)
Stem Powder	*Labeo rohita*	Oral	0%, 0.25%, 0.5% and 0.75%	90 days	–	Improved hemato-biochemical parameters such as TLC, TEC, Hemoglobin, total protein, albumin and globulin.	(Anita et al. 2015)

(continued)

Table 32.1 (continued)

Extract	Species	Route	Doses	Duration	Against	Results	Reference
Leaf powder	*Cyprinus carpio haematopterus*	Oral	0%, 0.25%, 0.5% and 0.75%	90 days	–	Improved growth, hematological and serum biochemical parameters like total protein, albumin and globulin	(Anita et al. 2016)
Methanolic stem extract	*Labeo rohita*	Oral	0, 100, 200, 300 mg/kg	60 days	*Aeromonas hydrophila*	Improved disease resistance and non-specific immune response	(A. Sharma et al. 2017)
Leaf powder	*Catla catla* (post larvae)	Oral	0, 0.25, 0.5, 0.75, 1%	21 days	–	Improved survival and growth of the post larvae	(Upreti and Chauhan 2018)
Methanolic extract of leaves	*Pangasius sutchi*	Oral	0, 100, 200, 400, 800 mg/kg	45 days	–	Improvement in growth parameters, hematological parameters and serum biochemical parameters	(Latha et al. 2020)

(continued)

Table 32.1 (continued)

Extract	Species	Route	Doses	Duration	Against	Results	Reference
Ethanolic extract	*Oreochromis niloticus*	Oral	0, 5, 10, 15 g/kg	14 days	Titanium dioxide nanoparticle (TNP)	Improved growth performance, antioxidant enzyme activity, histopathological damages in liver, gill and kidney Reduced liver marker enzymes, pro inflammatory cytokines and excessive ROS production Regulated expression of HSPs, antioxidant, apoptotic and antiapoptotic proteins	(Vineetha et al. 2021)
Water extract	*Oreochromis niloticus*	Oral	0, 2, 4, 6, and 8 g/kg of diet	56 days	Hypoxia stress for 72 hours	Improvement in growth performance and increase in digestive enzymes, serum and mucosal immunological parameters, antioxidant enzyme activity in liver and intestine with increasing level of doses. Lowered the level of stress markers such as glucose and cortisol	(El Basuini et al. 2022)

antioxidant enzymes which were effectively elevated by TCE supplementation by enhancing enzyme levels in the liver, kidney, and gills of Nile tilapia (El Basuini et al. 2022). The subsequent reduction in ROS production mitigated cellular damage from macromolecular oxidation. Notably, TCE exhibited dose-dependent regulation of elevated protein carbonyl, malondialdehyde (MDA), and ROS levels induced by TNP toxicity. Moreover, TCE supplementation dose-dependently attenuated elevated levels of AST, ALT, ALP in serum preserving liver integrity. The most pronounced effects were observed with the highest TCE supplementation dose.

Tinospora cordifolia is recognized for its potent antioxidant properties attributed to its abundant polyphenolic content, extensively studied across various organisms, including humans (Reddy and Reddy 2015). The substantial ROS scavenging property of *T. cordifolia* may be attributed to specific compounds such as saponarin, known for its α-glucosidase inhibitory activity.The high polyphenolic content in *T. cordifolia* extracts likely contributes to the regulation of stress-related markers by modulating glycogenesis.

32.3.2 Immunostimulant and Antimicrobial Properties

Plants represent a rich source of diverse immunomodulatory compounds within their secondary metabolites. These compounds not only bolster the plants' own defense mechanisms but also augment primary and secondary immune responses in other organisms (Sharma et al. 2024; Sharma et al. 2025).

Innate immune system of fishes comprises a complex interplay of macrophages, phagocytic cells, lysozymes, granulocytes, and monocytes, collaborating to eliminate various pathogens, including bacteria, viruses, and parasites, through mechanisms such as bacterial cell wall rupture, reactive oxygen species (ROS) production, and phagocytosis (Harikrishnan et al. 2011; Van Hai 2015). Plant-derived secondary metabolites have shown promise in enhancing nonspecific immune responses in fish, likely by modulating MAPKs and NF-κB pathways at the Toll-like Receptor (TLR) level, thereby upregulating genes involved in immune response pathways (Faheem et al. 2022).

Jinish (2002) demonstrated strong antimicrobial activity of TC and several other medicinal plants against Vibrio species isolated from the gut of *Epinephelus tauvina*. Another study on *Oreochromis mossambicus* revealed that administration of ethanolic extracts of TC leaves improved antibody titers and neutrophil activity against *Aeromonas hydrophila*, leading to enhanced disease resistance (Sudhakaran et al. 2006). Furthermore, ImmuPlus, a polyherbal mixture containing TC, has shown remarkable improvement in both specific and nonspecific immunity in different growth stages of *Labeo rohita* against *A. hydrophila*. The fishes fed with it showed improved total protein, albumin and globulin, natural hemolysin titre, neutrophil and lysozume activity. An optimal dose of 1 g/kg was found to be effective for the fingerlings (Kumari et al. 2007). Again, *Oreochromis mossambicus, challenged with A. hydrophila when fed with aqueous extracts of the leaves showed enhanced nonspecific* *immune response (*Alexander et al. 2010). Another study by A. Sharma et al. (2017) revealed that methanolic extracts of TC can significantly improve immune parameters in *Labeo rohita.* El-Basuini et al.(2022) suggested an optimal dose of 5.49 g/kg for enhancing the immune system of *Oreochromis niloticus* against hypoxia stress.

Different TC extracts improved serum and mucus bactericidal and lysozyme activity, nitroblue tetrazolium (NBT) activity. In TNP-intoxicated fish, regulation of pro-inflammatory cytokine secretion supports the immunomodulatory activity of the extracts (Vineetha et al. 2021). Presence of polysaccharides and polyphenols may contribute to the immunomodulatory activities of the extracts (Saeed et al. 2020). Secondary metabolites such as N-methyl-2-pyrrolidone, cordifolioside A, tinocordiside,, N-formylannonain, syringin,, magnoflorine, and 11-hydroxymustakone present in TC may help in enhancing phagocytic and killing activities of macrophage (More and Pai 2012; Sharp and Secombes 1993).

32.3.3 Growth Promotion

One of the prominent problems in aquaculture industry is consistent growth of the fishes for good yield. Researches have suggested integration of plant-based products into aquafeeds as a sustainable approach to promote growth rates (Rani et al. 2025; Nisha et al. 2025; Kiran et al. 2025). Several investigations have demonstrated the growth-promoting effects of *Tinospora cordifolia* (TC) dietary supplementation in freshwater fish species. Kumari et al. (2007), tested ImmuPlus supplementation on *Labeo rohita.* The study found that it exhibits positive effects on total body weight and growth at different developmental stages of *Labeo rohita.* Results suggested that incorporating ImmuPlus at 0.5 g/kg and at 1 g/kg in advanced larval and fingerling stage diets can improve body weight.

Anita et al. (2016) observed growth enhancement in *Cyprinus carpio haematopterus* fingerlings that were fed with an experimental diet (0.75% TC leaf powder) for 90 days. There were significant improvements in Specific Growth Rate (SGR), Feed Conversion Efficiency (FCR), and Gross Conversion Efficiency (GCE) compared to the control group. Upreti and Chauhan (2018) reported that when *Catla catla* were fed with 1% TC leaf supplemented diet for 21 days, it showed enhanced SGR, FCR and survival. These experiments indicated the potential of TC to promote healthy growth and increase survival rates.

Methanolic extracts of TC leaves can enhance growth in *Pangasius sutchi* fingerlings Latha et al. (2020). Fingerlings fed with a diet supplemented with 400 mg/kg TC extract showed the highest increase in SGR, Protein Efficiency Ratio, survival rates, and decreased FCR. Vineetha et al. (2021) demonstrated that dietary supplementation of ethanolic extract of TC can improve growth performance in *Oreochromis niloticus* exposed to titanium dioxide nanoparticle toxicity.

While the precise molecular mechanisms underlying these effects remain elusive, secondary metabolites from plants probably can influence the somatotropic axis of fish, that ultimately leads to elevation of growth hormone mRNA and various insulin-like growth factors (Sruthi et al. 2018). Also, modulation of the neuroendocrine axis by plant secondary metabolites may stimulate fish growth by enhancing appetite, improving digestion, and promoting nutrient uptake from the intestine (Faheem et al. 2022). The growth promoting activity of TC may be attributed to its ability to stimulate appetite, enhance digestibility, and improve nutrient utilization, due to its rich array of biologically active compounds, including steroids, terpenoids, flavonoids, alkaloids, tannins, glycosides, phenolics, and saponins (Awad and Awaad 2017; Saeed et al. 2020).

32.3.4 *Enhancement of Hemato-Biochemical Parameters*

Serum hematological and biochemical parameters are important markers for assessing fish health, immune response, and stress levels. Regular monitoring of this is vital for understanding the impacts of environmental, dietary changes and disease conditions. Medicinal plants have been proved to be beneficial for supporting better survival in fishes.

ImmuPlus, a plant-based supplementation, showed significant improvements in serum biochemical parameters such as total protein, albumin, and globulin in *Labeo rohita* Kumari et al. (2007). Anita et al. (2015, 2016) found that incorporation of TC leaf powder in the diet of *Cyprinus carpio haematopterus* can improve RBC, WBC count, hemoglobin percentage, serum albumin, globulin, and total protein after 90 days of feeding trials. The highest alteration observed in the group supplemented with 0.75% TC leaf powder.

Another study by Latha et al. (2020), tested methanolic extract of TC leaves on *Pangasius sutchi* fingerlings with a dose of 400 mg/kg and it marked significant improvements in all hematological and serum biochemical parameters compared to the control group, reinforcing the idea that such plant based diet can remarkably improve fish health.

32.4 *Tinospora Cordifolia* in Shrimp Aquaculture

In crustaceans, the non-specific immune response constitutes the primary defense mechanism against invading pathogens, wherein lysozyme and haemocytes play pivotal roles (Sarathi et al. 2007). Haemocytes and plasma proteins initiate the immune cascade by recognizing invading pathogens, while lysozyme aids in effectively eliminating both gram-positive and gram-negative bacteria(Saurabh and Sahoo 2008; Vargas-Albores and Yepiz-Plascencia 2000). Activation of putative phenol oxidase system components is integral to cellular defense mechanisms, further stimulating nodule formation, phagocytosis, and haemocyte locomotion (Söderhäll et al. 1986). The drawbacks associated with antibiotics and synthetic immunostimulants have sparked interest in plant-based immunostimulants in recent years. Phytochemicals have been found to enhance immunomodulatory properties, activating non-specific immune responses such as phagocytic, respiratory burst, and lysozyme activities in shrimp and other crustaceans (Table 32.2) (Kim et al. 1999).

A polyherbal mixture containing *T. cordifolia*, along with other medicinal herbs, has demonstrated beneficial effects on various shrimp species. Dietary supplementation of TC, along with *Cyanodon dactylon, Picrorhiza kurooa, Aegle marmelos*and *Eclipta alba*, exhibited protective action against white spot syndrome virus (WSSV) in *Penaeus monodon* (black tiger shrimp). Supplementation at doses of 100, 200, 400, and 800 mg/kg significantly improved shrimp survival following WSSV challenge, with the highest survival (74%) observed at the 800 mg/kg dose. PCR diagnosis indicated reduced WSSV infection with increasing herbal mix concentration. Significant changes in haematological parameters, such as reduced haemolymph coagulation time, increased total haemocyte count, and oxyhemocyanin levels, further supports the protective effect of the supplementation. The diet also altered muscle protein, carbohydrate, and glucose levels significantly, as well as increased prophenol oxidase activity and intracellular superoxide anion production (Citarasu et al. 2006).

The results of another study revealed that a blend of *Acalypha indica, Hygrophila spinosa, Zingiber officinale, Picrorhiza kurooa*, and *T. cordifolia* can provide protection against infection of *Vibrio harveyi* in *Fenneropenaeus indicus* (Indian white shrimp). Methanolic extracts of these five medicinal herbs significantly improved growth, survival, haematological parameters, and disease resistance against *V. harveyi*. Increased weight, specific growth rate, reduced haemolymph coagulation time, enhanced total haemocyte count, phagocytic activity, haemagglutinin activity, phenol oxidase activity, and bacterial clearance were observed with increasing concentrations of formulated diet. In vitro screening of different extracts against *V. harveyi* revealed that the methanolic extract can be beneficial in inhibiting bacterial growth (Rajeswari et al. 2012).

Commercially formulated Herbal Food Additive (HFA), a mixture of *T. cordifolia, Terminalia chebula*, and *Emblica officinalis*, had been reported to improve immune responses in various animals. Incorporation of HFA into the diet of *Penaeus semisulcatus* significantly altered protein and moisture levels, with increased phenol oxidase activity in the 1% diet group. Phagocytic activity was highest in the 2% HFA diet group; however, overall growth was not markedly influenced by HFA supplementation (Al-Musalam et al. 2014).

Incorporation of AquaImmu, a polyherbal mixture containing *Ocimum sanctum, T. cordifolia, Withania somnifera*, and *Emblica officinalis*, at a concentration of 300 mg/100 g, improved growth and immune modulation in *Penaeus monodon* in a dose-dependent manner. Haematological and immunological parameters were significantly enhanced, with the 300 mg/100 g supplementation group exhibiting the highest values, indicating the immunomodulatory potential of the polyherbal mix (Chandran et al. 2016).

Table 32.2 Tinospora cordifolia in shrimp culture

Plant extract	Species	Route	Doses	Duration	Against	Result	References
Mixture of *C. dactylon, A. marmelos, TC, P. kurooa and E. alba*	*Penaeus monodon*	Oral	0, 100, 200, 400, 800 mg/kg	25 days	White spot syndrome virus	Increased survival, decreased viral infection Reduced haemolymph coagulation time, improved oxyhaemocyanin level, haemocyte count Increased protein, carbohydrate, glucose in muscle Increased prophenol oxidase activity, intracellular superoxide anion production	(Citarasu et al. 2006)
Methanolic extract of *A. indica, H. spinosa, P. kurooa, T. cordifolia andZ. officinale*	*Fenneropenaeus indicus*	Oral	0, 500, 1,000, 1,500, 2,000 mg/kg	60 days	*Vibrio harveyi*	Suppressed bacterial growth in vitro. Decreased mortality rate. Enhanced growth parameters. Decreased haemolymph coagulation, increased total haemocyte count. Increased phenol oxidase activity, phagocytosis	(Rajeswari et al. 2012)

(continued)

Table 32.2 (continued)

Plant extract	Species	Route	Doses	Duration	Against	Result	References
HFA (mixture of *T. cardiofolia, T. chebula* and *E. officianalis*	*Penaeus semisulcatus*	Oral	0%, 1%, 2%	190 days	–	Higher phenol oxidase activity, phagocytic activity. Lower crude protein and higher moisture content	(Al-Musalam et al. 2014)
AquaImmu (Mixture of *O. sanctum, W.somnifera, T. cordifolia*and *E. officinalis*)	*Penaeus monodon*	Oral	100, 200, 300, 400 mg/ 100 g	90 days	–	Improved survival and growth performance. Increased total haemocyte count Increased phenol oxidase, respiratory burst, lysozyme activities, bacterial clearance efficiency	(Chandran et al. 2016)

32.5 Conclusion

In the realm of aquaculture, where aquatic organisms face myriad environmental and anthropogenic stressors, finding sustainable solutions to mitigate these challenges remains elusive. While aquaculture practitioners employ various strategies to combat adverse conditions using antibiotics and other synthetic chemicals, enhancing the immune system through natural supplements has emerged as a crucial approach. In rural settings, where homemade remedies are prevalent due to their minimal adverse effects on organisms and ecosystems, *T. cordifolia* has long served as a beacon of hope. Known for its bioactive compounds it has proved to improve overall health of the aquatic species, minimizing the need for reliance on conventional drugs contributing more towards eco-friendly and sustainable aquaculture practices.

32.6 Future Prospects

Scientific validation has underscored its utility in improving aquaculture practices, with a significant number of rural communities integrating this phytomedicine into their small-scale aquatic endeavours. However, for widespread commercial utilization, rigorous scientific investigations for understanding the precise molecular mechanism of the secondary metabolites, are imperative to optimize dosages and formulate effective preparations. Looking ahead, a better understanding of the mechanisms of action could lead to development of cost-effective and sustainable alternatives to conventional drugs which not only enhance productivity and resilience but will also support the achievement of United nation's Sustainable Development Goals (SDGs) ensuring that aquaculture can thrive in harmony with sustainability while meeting the food security needs worldwide

Acknowledgement The authors are highly thankful to the parents' university for the valuable guidance.

References

Alexander CP, Kirubakaran CJW, Michael RD (2010) Water soluble fraction of Tinospora cordifolia leaves enhanced the non-specific immune mechanisms and disease resistance in Oreochromis mossambicus. Fish Shellfish Immunol 29(5):765–772

Al-Musalam LS, Al-Ameeri AA, Saheb AS, Al-Yaqout A (2014) Effect of herbal feed additive on the growth, survival and immune response of green tiger prawn (penaeus semisulcatus). Pak J Nutr 13(7):366–371. https://doi.org/10.3923/pjn.2014.366.371

Anita, Dubey MK, Khati A, Chauhan RS (2016) Immunostimulatory and growth promoting potential of Tinospora cordifolia (Thunb.) miers on fingerlings of Amur carp. Indian J Exp Biol 54(10):659–663

Anita, Dubey MK, Khati A, Kanya P (2015) Study of Haematological profile of Labeo Rohita (Rohu) fed with Tinospora Cordifolia (Giloy) as Neutraceutical. J Environ Bio Sci 29(2):1–4
Arunachalam K, Yang X, San TT (2022) Tinospora cordifolia (Willd.) Miers: protection mechanisms and strategies against oxidative stress-related diseases. J Ethnopharmacol 283:114540
Awad E, Awaad A (2017) Role of medicinal plants on growth performance and immune status in fish. Fish Shellfish Immunol 67:40–54
Bondad-Reantaso MG, Subasinghe RP, Arthur JR, Ogawa K, Chinabut S, Adlard R, Tan Z, Shariff M (2005) Disease and health management in Asian aquaculture. Vet Parasitol 132(3–4 SPEC. ISS):249–272. https://doi.org/10.1016/j.vetpar.2005.07.005
Bulfon C, Volpatti D, Galeotti M (2015) Current research on the use of plant-derived products in farmed fish. Aquac Res 46(3):513–551. https://doi.org/10.1111/are.12238
Cabello, FC (2006) Heavy use of prophylactic antibiotics in aquaculture: a growing problem for human and animal health and for the environment. In: Environmental microbiology Vol 8, Issue 7, pp 1137–1144 https://doi.org/10.1111/j.1462-2920.2006.01054.x
Chandran MN, Moovendhan S, Suganya AM, Tamilselvi A, Bebin I, G., & Palavesam, A. (2016) Influence of polyherbal formulation (AquaImmu) as a potential growth promotor and immunomodulator in shrimp Penaeus monodon. Aquac Rep 4:143–149. https://doi.org/10.1016/j.aqrep.2016.10.002
Cherukumudi K, Augustian HT, Kausar N, Biju J, Rao KG, Gopalraaj J, Velayudhannair K, Radhakrishnan DK, Tomson M (2024) Exploring botanical innovations in fish aquaculture: A review of biological impacts and industry prospects: exploring botanical innovations in fish aquaculture. Fish Technol 61(4)
Choudhary N, Siddiqui MB, Khatoon S (2014) Pharmacognostic evaluation of Tinospora cordifolia (Willd.) Miers and identification of biomarkers. Indian J Tradit Knowl 13(3):543–550
Citarasu, T. (2010). Herbal biomedicines: a new opportunity for aquaculture industry. In Aquaculture international (Vol. 18 Issue 3, pp. 403–414) https://doi.org/10.1007/s10499-009-9253-7
Citarasu T, Sivaram V, Immanuel G, Rout N, Murugan V (2006) Influence of selected Indian immunostimulant herbs against white spot syndrome virus (WSSV) infection in black tiger shrimp, Penaeus monodon with reference to haematological, biochemical and immunological changes. Fish Shellfish Immunol 21(4):372–384
Dhanasekaran M, Baskar A-A, Ignacimuthu S, Agastian P, Duraipandiyan V (2009) Chemopreventive potential of epoxy clerodane diterpene from Tinospora cordifolia against diethylnitrosamine-induced hepatocellular carcinoma. Investig New Drugs 27(4):347–355
El Basuini MF, Teiba II, Shahin SA, Mourad MM, Zaki MAA, Labib EMH, Azra MN, Sewilam H, El-Dakroury MF, Dawood MAO (2022) Dietary Guduchi (Tinospora cordifolia) enhanced the growth performance, antioxidative capacity, immune response and ameliorated stress-related markers induced by hypoxia stress in Nile tilapia (Oreochromis niloticus). Fish Shellfish Immunol 120:337–344
Faheem M, Abbas RZ, Liaqat I, Hoseinifar SH, Maneepitaksanti W, Van Doan H (2022) Bioactive components in medicinal plants: a mechanistic review of their effects on fish growth and physiological parameters. Annals of Animal Science
FAO (2020) The state of food and agriculture 2020. Sustainability in action
FAO (2021) GLOBEFISH Highlights—A quarterly update on world seafood markets. In: GLOBEFISH Highlights—A quarterly update on world seafood markets (Issue April 2019 ISSUE, with Jan.—Dec. 2018 Statistics). https://doi.org/10.4060/cb6414en
FAO (2024) In brief to the state of world fisheries and aquaculture 2024. In: Blue transformation in action. https://doi.org/10.4060/cd0690en
FAO (2025) GLOBEFISH | quarterly shrimp analysis—February 2025
Ghosh AK, Panda SK, Luyten W (2021) Anti-vibrio and immune-enhancing activity of medicinal plants in shrimp: A comprehensive review. Fish Shellfish Immunol 117(April):192–210. https://doi.org/10.1016/j.fsi.2021.08.006
Goel HC, Prasad J, Singh S, Sagar RK, Agrawala PK, Bala M, Sinha AK, Dogra R (2004) Radioprotective potential of an herbal extract of Tinospora cordifolia. J Radiat Res 45(1):61–68

Harikrishnan R, Heo MS, Balasundaram C (2011) Impact of plant products on innate and adaptive immune system of cultured finfish and shellfish. Aquaculture 317(1–4):1–15. https://doi.org/10.1016/j.aquaculture.2011.03.039

Imtiyaj Khan M, Sri Harsha PSC, Giridhar P, Ravishankar GA (2011) Pigment identification, antioxidant activity, and nutrient composition of Tinospora cordifolia (willd.) Miers ex hook. F & Thoms fruit. Int J Food Sci Nutr 62(3):239–249. https://doi.org/10.3109/09637486.2010.529069

Jinish, L. J. G. (2002). Ayurvedic phytotherapeutics against vibrio infections in grouper, Epinephelus tauvina. M. Sc. Dissertation, MS University, Tirunelveli, Tamil Nadu, India

Joshi G, Kaur R (2016) Tinospora cordifolia: a phytopharmacological review. Int J Pharm Sci Res 7(3):890. https://doi.org/10.13040/IJPSR.0975-8232.7(3).890-97

Kapil A, Sharma S (1997) Immunopotentiating compounds from Tinospora cordifolia. In. J Ethnopharmacol 58

Kim KH, Hwang YJ, Bai SC (1999) Resistance to vibrio alginolyticus in juvenile rockfish (Sebastes schlegeli) fed diets containing different doses of aloe. Aquaculture 180(1–2):13–21

Kiran, Kumar N, Sindhu S, Gawaria B, Ansari A, Dhariwal A, Sharma P, Kumar R (2025) GWAS approaches in medicinal plants for enhanced secondary metabolite production. In: Chen JT (ed) Biotechnology, multiple Omics, and precision breeding in medicinal plants. CRC Press, pp 119–134. https://doi.org/10.1201/9781003475491-10

Kumar R, Gupta R, Dahiya T, Yadav R, Saifi R, Yodha K, Kumari B, Singh P, Shefali. (2022a) Surveillance of *Heteropneustes fossilis* (stinging catfish) diseases and health management practices in Haryana. Appl Ecol Environ Sci 10(4):261–267. https://doi.org/10.12691/aees-10-4-12

Kumar R, Gupta RK, Yadav R, Saifi R, Yodha K, Pooja (2022b) *Eisenia fetida* as protein source for growth enhancement of *Heteropneustes fossilis*. Egypt J Aquat Biol Fish 26(2):577–588. https://doi.org/10.21608/ejabf.2022.234553

Kumar R, Singh D (2022) Insecticide-tolerant bacterial population in *Eisenia fetida*'s gut and Vermicast exposed to Chlorantraniliprole and Fipronil. Appl Biol Res 24(3):273–279. https://doi.org/10.5958/0974-4517.2022.00033.7

Kumar R, Yadav R, Gupta RK, Pal A, Yodha K, Kumar A (2024) Antioxidant enzyme activities and markers of oxidative stress in the life cycle of different earthworm species. Indian J Exp Biol 62(07):578–583. https://doi.org/10.56042/ijeb.v62i07.12084

Kumar R, Yadav R, Kumar Gupta R, Pooja (2022c) A sustainable way for fish health management by replacement of chemical and drugs by earthworm. In: Environmental degradation in Asia: land degradation, environmental contamination, and human activities. Springer International Publishing, Cham, pp 329–352. https://doi.org/10.1007/978-3-031-12112-8_16

Kumari J, Sahoo PK, Giri SS (2007) Effects of polyherbal formulation "ImmuPlus" on immunity and disease resistance of indian major carp, labeo rohita at different stages of growth. Indian J Exp Biol 45(3):291–298

Latha C, Krishnakumar V, Munuswamy N (2020) Growth, survival and Haemato-biochemical profiles of the freshwater catfish, Pangasius sutchi (fowler, 1937) fingerlings fed on Tinospora cordifolia leaf extract supplemented diet. Egypt J Aquat Biol Fish 24(1):249–262

Liu, B, Li, Y, Mehmood, K, Nabi, F, Ahmed, S, Rehman, T, Faheem, M, AshraF, M, Tang, Z, & Zhang, H (2021) Role of oxidative stress and antioxidants in thiram-induced tibial dyschondroplasia

Livingstone DR (2003) Oxidative stress in aquatic organisms in relation to pollution and aquaculture. Rev Médecine Vétérinaire (Fr)

Maurya R, Handa SS (1998) Tinocordifolin, a sesquiterpene from Tinospora cordifolia. Phytochemistry 49(5):1343–1345. https://doi.org/10.1016/S0031-9422(98)00093-4

Mbokane EM, Moyo NAG (2024) A systematic review and meta-analysis of the potential effect of medicinal plants on innate immunity of selected freshwater fish species: its implications for fish farming in southern Africa. Aquac Int 32(1):315–335

Mckeown E, Bykerk VP, De Leon F, Bonner A, Thorne C, Hitchon CA, Boire G, Haraoui B, Ferland DS, Keystone EC, Pope JE (2012) Quality assurance study of the use of preventative therapies in glucocorticoid-induced osteoporosis in early inflammatory arthritis: results from the CATCH cohort. Rheumatology (United Kingdom) 51(9):1662–1669. https://doi.org/10.1093/rheumatology/kes079

Mishra SP (2020) Significance of fish nutrients for human health. Int J Fish Aquat Res 5(3):47–49

Modi B, Shah KK, Shrestha J, Shrestha P, Basnet A, Tiwari I, Aryal SP (2021) Morphology, biological activity, chemical composition, and medicinal value of Tinospora cordifolia (willd.) Miers. Adv J Chem Sect B 3(1):36–53. https://doi.org/10.22034/ajcb.2020.243751.1058

More P, Pai K (2012) In vitro NADH-oxidase, NADPH-oxidase and myeloperoxidase activity of macrophages after Tinospora cordifolia (guduchi) treatment. Immunopharmacol Immunotoxicol 34(3):368–372

MPEDA (2022) EXPORT PERFORMANCE 2021–22

Nile SH, Khobragade CNN (2009) Determination of nutritive value and mineral elements of some important medicinal plants from western part of India. J Med Plants 8(SUPPL. 5):79–88

Nisha A, Rani S, Kumar S, Kumar R, Kumar A (2025) Recent technologies and strategies for bioremediation of toxic and hazardous waste. In: Dubey KK, Shah MP, Yadav A (eds) Challenges and sustainable solutions in bioremediation. CRC Press, pp 1–16. https://doi.org/10.1201/9781003407317-1

Patel MB, Mishra S (2011) Hypoglycemic activity of alkaloidal fraction of Tinospora cordifolia. Phytomedicine 18(12):1045–1052

Raja, Rajeswari P, Velmurugan S, Michael Babu M, Albin Dhas S, Kesavan K, Citarasu T (2012) A study on the influence of selected Indian herbal active principles on enhancing the immune system in Fenneropenaeus indicus against Vibrio harveyi infection. Aquac Int 20(5):1009–1020. https://doi.org/10.1007/s10499-012-9525-5

Rani S, Sharma S, Singh D, Sheoran R, Kumar R (2025) Improved production of anti-cancer compounds from medicinal plants through molecular and biotechnological mutagenesis strategies. In: Chen JT (ed) Biotechnology, multiple Omics, and precision breeding in medicinal plants. CRC Press, pp 287–297. https://doi.org/10.1201/9781003475491-21

Reddy NM, Reddy RN (2015) Tinospora cordifolia chemical constituents and medicinal properties: a review. Sch Acad J Pharm 4(8):364–369

Reverter M, Bontemps N, Lecchini D, Banaigs B, Sasal P (2014) Use of plant extracts in fish aquaculture as an alternative to chemotherapy: current status and future perspectives. In: Aquaculture, vol 433. Elsevier, pp 50–61. https://doi.org/10.1016/j.aquaculture.2014.05.048

Rico A, Phu TM, Satapornvanit K, Min J, Shahabuddin AM, Henriksson PJG, Murray FJ, Little DC, Dalsgaard A, Van den Brink PJ (2013) Use of veterinary medicines, feed additives and probiotics in four major internationally traded aquaculture species farmed in Asia. Aquaculture 412–413:231–243. https://doi.org/10.1016/j.aquaculture.2013.07.028

Romero J, Feijoó CG, Navarrete P (2012) Antibiotics in aquaculture–use, abuse and alternatives. Health Environ Aquac 159:159–198

Saeed M, Naveed M, Leskovec J, Kakar I, Ullah K, Ahmad F, Sharif M, Javaid A, Rauf M, Abd El-Hack ME (2020) Using Guduchi (Tinospora cordifolia) as an eco-friendly feed supplement in human and poultry nutrition. Poult Sci 99(2):801–811

Saha S, Ghosh S (2012) Tinospora cordifolia: one plant, many roles. Anc Sci Life 31(4):151–159. https://doi.org/10.4103/0257-7941.107344

Salkar K, Chotalia C, Salvi R (2017) Tinospora cordifolia: an antimicrobial and immunity enhancer plant. Int J Sci Res 6:1603–1607

Sarathi M, Ahmed VPI, Venkatesan C, Balasubramanian G, Prabavathy J, Hameed ASS (2007) Comparative study on immune response of Fenneropenaeus indicus to vibrio alginolyticus and white spot syndrome virus. Aquaculture 271(1–4):8–20

Saurabh S, Sahoo PK (2008) Lysozyme: an important defence molecule of fish innate immune system. Aquac Res 39(3):223–239

Sharma A, Chadha NK, Das SK, Sen A, Roy SD, Chanu TI, Sawant PB, Prakash C (2017) Tinospora cordifolia extract induced effects on cellular immune reactions of labeo rohita (Hamilton) challenged against aeromonas hydrophila. Int J Pure Appl Biosci 5:765–775

Sharma MMM, Kapoor D, Rohilla R, Loyal A, Sharma P, Husen A (2024) Role of essential minerals and rhizosphere in plant under changing environmental conditions. In: Husen A (ed) Essential minerals in plant-soil systems. Elsevier, pp 235–252. https://doi.org/10.1016/B978-0-443-16082-0.00002-3

Sharma P, Dwivedee BP, Bisht D, Dash AK, Kumar D (2019) The chemical constituents and diverse pharmacological importance of Tinospora cordifolia. Heliyon 5(9):e02437

Sharma P, Kapoor D, Kumar R, Husen A (2025) Rhizosphere and osmolytes coordination in plant systems under adverse situations. In: Husen A (ed) Roles of Osmolytes in changing environment. Academic, pp 101–117. https://doi.org/10.1016/B978-0-443-26549-5.00020-5

Sharp GJE, Secombes CJ (1993) The role of reactive oxygen species in the killing of the bacterial fish pathogen Aeromonas salmonicida by rainbow trout macrophages. Fish Shellfish Immunol 3(2):119–129

Shefali, Kumar R, Sankhla MS, Kumar R, Sonone SS (2021) Impact of pesticide toxicity in aquatic environment. Biointerface Res Appl Chem 11(3):10131–10140. https://doi.org/10.33263/BRIAC113.1013110140

Singh S, Devi P (2017) Pharmacological potential of Tinospora cordifolia (Willd.) Miers ex hook. & Thoms.(Giloy): a review. J Pharmacogn Phytochem 6(6):1644–1647

Sinha, K, Mishra, NP, Singh, J, & Khanuja, SPS (2004) Tinospora cordifolia (Guduchi), a reservoir plant for therapeutic applications: a review

Söderhäll K, Smith VJ, Johansson MW (1986) Exocytosis and uptake of bacteria by isolated haemocyte populations of two crustaceans: evidence for cellular co-operation in the defence reactions of arthropods. Cell Tissue Res 245(1):43–49. https://doi.org/10.1007/BF00218085

Spandana, U., Ali, S. L., Nirmala, T., Santhi, M., & Sipai Babu, S. D. (2013). A review on Tinospora cordifolia. In International journal of current pharmaceutical review and research (Vol. 4, Issue (2), pp. 61–68). www.ijcpr.com

Sriramaneni RN, Omar AZ, Ibrahim SM, Amirin S, Zaini AM (2010) Vasorelaxant effect of diterpenoid lactones from Andrographis paniculata chloroform extract on rat aortic rings. Pharm Res 2(4):242

Sruthi MV, Nair AB, Arun D, Thushara VV, Sheeja CC, Vijayasree AS, Oommen OV, Divya L (2018) Dietary curcumin influences leptin, growth hormone and hepatic growth factors in tilapia (Oreochromis mossambicus). Aquaculture 496:105–111

Sudhakaran, DS, Srirekha, P, Devasree, L. D, Premsingh, S, & Michael, RD (2006) Immunostimulatory effect of Tinospora cordifolia Miers leaf extract in Oreochromis mossambicus

Thippeswamy G, Sheela ML, Salimath BP (2008) Octacosanol isolated from Tinospora cordifolia downregulates VEGF gene expression by inhibiting nuclear translocation of NF-< kappa> B and its DNA binding activity. Eur J Pharmacol 588(2–3):141–150

Tripathi BM, Singh DC, Chaubey S, Kour G, Arya R (2015) A critical review on guduchi (Tinospora cordifolia (willd.) miers) and its medicinal properties. Int J Ayur Pharma Research 3(5):6–12

Upreti P, Chauhan RS (2018) Effect of leaf powder of giloy (Tinospora cordifolia) in fish feed on survival and growth of post larvae of Catla catla. J Appl Nat Sci 10(1):144–148

Van Hai N (2015) The use of medicinal plants as immunostimulants in aquaculture: a review. In: Aquaculture, vol 446. Elsevier, pp 88–96. https://doi.org/10.1016/j.aquaculture.2015.03.014

Vargas-Albores F, Yepiz-Plascencia G (2000) Beta glucan binding protein and its role in shrimp immune response. Aquaculture 191(1–3):13–21

Vineetha VP, Devika P, Prasitha K, Anilkumar TV (2021) Tinospora cordifolia ameliorated titanium dioxide nanoparticle-induced toxicity via regulating oxidative stress-activated MAPK and NRF2/Keap1 signaling pathways in Nile tilapia (Oreochromis niloticus). Comp Biochem Physiol Part C Toxicol Pharmacol 240:108908

Chapter 33
Unrecognized Role of Fisherwomen: A Barrier to Achieving Sustainable Fisheries Development

Khushbu Sharma, Rachna Gulati, Amit Kour, Nikki, Parveen Kumar Nimbrayan, Nisha Kumari, and Mayur Mukut Murlidhar Sharma

Abstract Fish is the most abundant food source in the world. For many fish families in the developing countries, fish provide a substantial supply of protein as well as their main source of income. In order to support their function as home managers in the majority of households, women play a critical role in the development of the fishing industry. In addition to handling, processing, and preservation, the fisherwomen serve as a conduit between producers and customers. The active involvement of fisherwomen can play a pivotal role in advancing sustainable development within the fisheries sector. Their work is more tough and hard due to less resources, challenging tasks, and more competition. Women are required to labor in the markets' dirtiest spaces due to gender discrimination, which is a widespread problem in the fisheries. They lack basic amenities like toilets, a place to change clothes, clean drinking water, etc. The exploiters view the fisherwomen's passive suffering as a weakness, and this is becoming unbearable. Fisherwomen need to be given more social and economic influence if India's fisheries industry is to be preserved. Only education in nutrition,

K. Sharma (✉) · A. Kour · Nikki
Department of Zoology, Chaudhary Charan Singh Haryana Agricultural University, Hisar, Haryana, India
e-mail: khushbu181997@gmail.com

R. Gulati
Department of Aquatic Animal Health Management, Chaudhary Charan Singh Haryana Agricultural University, Hisar, Haryana, India
e-mail: jdccshau@gmail.com

P. K. Nimbrayan
Department of Agricultural Economics, Chaudhary Charan Singh Haryana Agricultural University, College of Agriculture, Bawal, Haryana, India

N. Kumari
Department of Commerce, Gurugram University, Gurugram, India

M. M. M. Sharma
Department of Agriculture and Life Industry, Kangwon National University, Chuncheon, South Korea

R. Kumar et al. (eds.), *Sustainability in Aquaculture and Fisheries*, Springer Water, https://doi.org/10.1007/978-3-032-16605-0_33

hygiene, health, and child care, as well as instruction in modern technologies and best practices, can achieve this.

Keywords Discrimination · Education · Economic · Fisherwomen

33.1 Introduction

Women constitute 50% of the world's population, 30% of the labor force, 60% of working hours, 10% of global revenue, yet hold less than 10% of global property (Dutta and Naidu 2025; Singh 2022a, b). Despite women's participation in the economy through the production of goods and services, which is accounted for in national income figures, their contribution is not recognized as economically productive. It is estimated to be 26% for women and 52% for men (Ceballos-Concha et al. 2025; Singh and Sukhwinder, 2022a, 2022b). Market-oriented conception dismisses women's contributions as non-economic, non-market, and domestic labor. The rural population in India accounts for 68.8 percent of the total (83.3 crores). Agriculture and related sectors such as horticulture, animal husbandry, and fishing support a substantial share of the rural population. According to the National Commission for Enterprises in the Unorganized Sector (NCEUS) Report 2009, the unorganized sector employed 93 percent of India's total employment in 2017 (Maurya and Mohanty 2019). The informal sector would account for 86–87 percent of all workers, with the remaining 6 percent working in formal sectors. These employees lack job security, income security, and social security, rendering them highly vulnerable (Agarwal 2010). Despite the absence of incentives in the unorganized sector, women seek employment in it due to their lack of awareness, illiteracy, ease of entry and exit, lack of skills, shorter waiting period, and irregular nature of employment (Bhatt 2006). Furthermore, poverty and economic necessity have compelled women to work in this industry. These women are predominantly from marginalized groups such as scheduled castes, scheduled tribes, the most backward classes, and economically disadvantaged segments of society. Women from low-income households accept any type of employment, regardless of its nature. The rising cost of living has driven impoverished women to take up such occupations. They are burdened by family expenses and seek to fulfill the family's responsibilities (Choudhary 2001). India's population has been increasing at an alarming rate, resulting in a scarcity of land. Due to population growth, agricultural land is now being utilized for domestic purposes. Consequently, fisheries, poultry, and related industries are becoming increasingly significant (Dhawan 2017). The value of fisheries is well recognized as a source of income for millions of fishermen, a primary driver of the coastal economy, and a substantial source of foreign exchange. The development of technology, along with extension operations, has increased Indian fish production by a factor of a thousand, propelling India to become the second largest inland aquaculture fish producer and resulting in a thousand-fold increase in the value of marine product exports (Campling et al. 2012; Kumar et al. 2022a). According to the UNDP's most recent Human Development Report, women constitute 70% of the poor. This is attributed to the fact that women face numerous complex cultural, social, traditional, economic, and legal constraints that impoverished men do not (Malik 2013). They have significantly fewer opportunities than men to obtain education and training, and consequently possess very few skills required in the contemporary environment. The feminization of poverty appears to be increasing. In impoverished and oppressed communities, women often bear a disproportionate burden. Fisherwomen's husbands may be unable to work due to age or alcoholism (Rimstead 2016). Fisherwomen purchase fish wholesale or retail for various reasons, including reduced catch sizes. Fisherwomen typically work more than 8 to 11 h daily. They must wait for the boats to return and subsequently participate in competitive auctions. When the catch is low, demand increases (Sarkar and Bhuimali 2005). The highest bidder purchases the fish and later sells it for profit. Financially stronger fisherwomen are the primary beneficiaries. Larger catches allow for participation of the poorest segments of the population in the market. A social hierarchy exists among fisherwomen whose male relatives or husbands have brought in the catch (D'SOUZA 2020). Women in the middle of the hierarchy purchase fish from the first group of women and sell it at the market. Women engage in retail sales ranging from Rs 500 to Rs 2000. This indicates that the majority of fisherwomen operate as small-scale fish merchants. The fisherwomen must then transport the catch and sell it at the market. They utilize ice to preserve the seafood (Mathew 1991; Kusakabe and Thongprasert 2022). The majority of fisherwomen work 8–12 h daily acquiring and selling fish. Fisherwomen play a crucial role in the overall life of fishing families. They not only manage day-to-day family and domestic tasks but also make significant contributions to the majority of fisheries and aquaculture activities. Thus, the contribution of women to societal evolution is substantial. However, women are frequently excluded from lucrative sectors of fisheries and marine-related activities. Unequal power relations and access to resources have perpetuated poverty in coastal communities (Kathirvel et al. 2007). Furthermore, the Liberalization, Privatization, and Globalization (LPG) economic reforms did not include any specific agricultural package. With a 5.15 percent contribution to India's agricultural GDP alone, the potential for individuals is considerable. Additionally, the industry employs approximately 14 million people. While women's roles in this sector were traditionally limited to post-harvest operations such as processing and marketing and were deemed negligible, their position as active participants is increasingly being recognized (Immanuel et al. 2016). Transportation of fish, sales of fish and dry fish, cleaning and drying of fish, development of fish feed, ornamental fish culture, fin-fish culture, crab fattening in cages and pens, shrimp processing, shrimp, crab and fish farming and hatchery management, development of value-added fish products, aquaculture by-products business, sea weed culture and drying etc. (Shah and Bhukhari 2019; Kumar et al. 2022b). The revenue generated thus serves as a complement to their family's income. However, there are numerous impediments to ensuring the economic development of women working in this sector, including illiteracy, a lack of awareness of government programs, social taboos and conservative norms, income disparities, and other inequalities resulting

Table 33.1 Fish production (MMT) and fisheries export per year

S.no	Year	Fish production (MMT)	Value of fisheries export (Crores)
1	2019–2020	14.16	46,662.85
2	2018–2019	13.57	46,589.37
3	2017–2018	12.7	45,106.89
4	2016–2017	11.43	37,870.90
5	2015–2016	10.76	30,420.83

Source Department of Fisheries, States Government / UTs Administration

from unaddressed gender disparities, which should become the focus of addressing the development of these communities (Gopal et al. 2020).The contributions of aquaculture, which has acquired the status of "second biggest producer" internationally, account for the majority of this steady performance, with an overall expected growth rate of 6% according to the Twelfth Five Year Plan as summarized in Table 33.1. With around 2200 species of fish – warm fresh water (24.7%), cold water (3.3%), estuaries (6.5%), and sea (65.5%) – the huge and rich diversity accessible for sustainable exploitation is great. This sector's contribution to GDP is expected to be at 1.07 percent, with yearly export revenues alone amounting to approximately 33,441.61 crores (Salih 2019).

Despite efforts in aquaculture to meet growing demand by incorporating culture-based capture fisheries, marine fishery production has been steadily declining. This decline is primarily due to overexploitation, reaching maximum sustainable yields, and the absence of advanced technological innovations and effective management frameworks aimed at promoting sustainable livelihoods.

33.2 Need of Study

Women play a vital role within the family structure, contributing to its overall development through their participation in employment, adoption of family planning methods, and financial support, especially in households of fisherwomen.

33.3 Study Plan

This entire study is based on research studies in relevant fields and on secondary data available on various governmental and private institutes site. All the information was summarized for reviewing the role of fisherwomen in fishery sector.

33.4 Women in Fisheries in India

India has an 8,118-km coastline with 3,638 fishing communities (Chandravanshi et al. 2020). India is the world's seventh largest producer of fish. Exports of fish and fish products have risen, and the monetary worth of exports has more than fivefold. As a result, India's fishing industry has expanded significantly during the last few decades (Emam et al. 2021).

Women have an important part in the fishing industry. They provide assistance. Mechanization in fisheries has enhanced fishing product exports (Matovu et al. 2024). Five lakh women work in the post-harvest sector of marine fisheries. During 2002–2003, it contributed to earnings of almost 7000 crores (Rajan 2024). There is a scarcity of quantitative data on women's employment, production, and wages in many areas. Tamil Nadu leads the four southern states with 13 coastal districts, followed by Andhra Pradesh and Kerala, each with nine, and Karnataka with three.

These states have a total of 2864 km of coastline, with 1465 fish landing stations and an equal number of settlements (Fig. 33.1). The distinct fisheries communities found in each state, some of them are shown in Table 33.2 (Hapke 2001). Tamil Nadu is the second-largest maritime state with 422 inland fishing hubs, after Andhra Pradesh (508). There is a rare chance for aquaculture expansion since rivers encompass 31,026 km^2, reservoirs cover 5.63 lakh ha, ponds cover 16.52 lakh ha, and brackish water covers 3.71 lakh ha. There are 27.60% of fishermen living in Tamil Nadu overall. Fish eaters make up as little as 27% of the population in Tamil Nadu, while the vast bulk of potential fish is exported to neighboring states (Vijayan et al. 2025). State-wise fishermen population in India (2019–2020) are shown in Table 33.3.

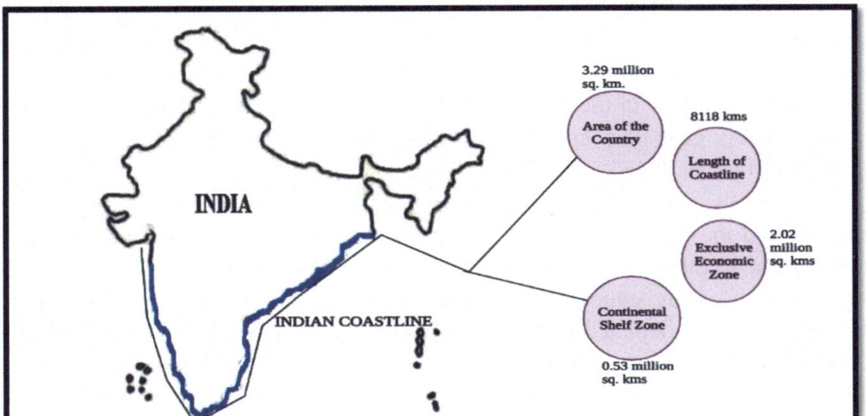

Fig. 33.1 Show Indian coastline *Source* Author

Table 33.2 Fishing communities of India State fishing communities

S. No	State	Fish communities
1	Odisha	Kaibartas, Jalaris, Rajbhansis Vadabalijas, and Khandayats
2	West Bengal	Kaibartas
3	Andhra Pradesh	Palles, Neyyala, Balijas, Vada Jalaris, Pattapu
4	Tamil Nadu	Mukkuvars, Pattinavars, and Paravas

33.5 Women's Role in the Aquaculture Sector

33.5.1 Pre-harvest and Harvest Activities

According to studies conducted on the east and west coastlines of India, fisherwomen are important to the pre-harvest sector. Before men travel to sea, they are involved in tasks like mending and repairing nets, getting ready hooks, gathering bait, and buoys, among other things. More males than in any other state on India's east coast are involved in net production and maintenance, according to a study done in West Bengal (Biswas and Rao 2014). Working offshore, fisherwomen engage in tasks including collecting mussels and oysters, harvesting seaweed, and more (Fig. 33.2). As a result, the proverb "Shells are for women, fish are for men" was coined (Klieber et al. 2015). This remark has some reality in the context of India, as gleaning is undertaken by fisherwomen closer to home. It also features cheap investment, which increases women's engagement (Klieber et al. 2015).

A middle-aged married woman in the Rameswaram the village is shown gathering seaweed in a My Nation film while wearing very little protective gear. After thereafter, the catch is sold for INR 100. These women gather crabs and a few huge fish near the ocean to augment their income, which they then sell at the neighborhood market (MyNation 2019). In a similar vein, fishermen, the most of whom are married, are seen hand-picking seaweed in the Gulf of Munnar hamlet of Ramanathapuram without any basic safety gear. This sport has been enjoyed for generations despite similar challenges (Tanuku 2013).

Consequently, similar actions by fisherwomen have been documented by governments around the west coast of India. For instance, a group of middle-aged tribal women from Kerala's Vedar tribe were shown in a Desert Story video fishing in a river while wearing sarees (Desert Story 2017). In contrast, tribal women from the Dangs tribe are depicted in the Magik India documentary (2018) catching fish in contaminated river water using a cloth. In both cases, the fish harvested is used for domestic purposes. Women, for instance, have traditionally been associated with unluck for a number of reasons. According to Philomena, an activist and fisherwoman, "whenever someone met with danger at sea, it was thought to be because the ladies at home had been unfaithful' (Program for Social Action 2017) or "If we went and stood on the beach with our hair undone, it was supposed to bring ill luck to those who ventured out to sea.

Table 33.3 State wise fishermen population in India (2019–2020)

State/UT	Inland Fishermen Population		Marine Fishermen Population	
	Female	Male	Female	Male
Andhra Pradesh	288,293	333,558	413,125	461,712
Arunachal Pradesh	8,868	15,147	**NA**	**NA**
Assam	900,047	1,624,059	**NA**	**NA**
Bihar	2,899,642	3,127,733	**NA**	**NA**
Chhattisgarh	73,384	146,971	**NA**	**NA**
Goa	**NA**	**NA**	5,272	5,273
Gujarat	110,085	120,900	159,848	167,858
Haryana	521	117,934	**NA**	**NA**
Himachal Pradesh	1,067	10,739	**NA**	**NA**
Jharkhand	46,400	94,497	**NA**	**NA**
Karnataka	180,924	465,352	131,031	196,970
Kerala	117,819	122,377	372,718	431,447
Madhya Pradesh	999,865	1,232,957	**NA**	**NA**
Maharashtra	529,433	602,536	188,499	197,760
Manipur	44,433	3,278	**NA**	**NA**
Meghalaya	4,822	11,745	**NA**	**NA**
Mizoram	961	5,328	**NA**	**NA**
Nagaland	2,062	5,896	**NA**	**NA**
Orissa	371,613	549,806	282,086	314,069
Punjab	573	7,018	**NA**	**NA**
Rajasthan	25,380	31,880	**NA**	**NA**
Sikkim	**93**	488	**NA**	**NA**
Tamil Nadu	110,220	125,694	512,500	535,337
Telangana	379,127	483,094	**NA**	**NA**
Tripura	**43**	7,718	**NA**	**NA**
Uttarakhand	1,092	7,260	**NA**	**NA**
Uttar Pradesh	1,798,622	2,101,383	**NA**	**NA**
West Bengal	1,204,093	1,629,481	156,462	246,225
Andaman Nicobar island	946	2,807	10,164	12,024
Chandigarh	**NA**	524	**NA**	**NA**
Dadra & Nagar Haveli, Daman & Diu	**NA**	**NA**	19,487	20,529
Delhi	30	3,316	**NA**	**NA**
Jammu and Kashmir	**NA**	17,396	**NA**	**NA**
Ladakh	**NA**	**22**	**NA**	**NA**
Lakshadweep	**NA**	**NA**	1,380	5,138

(continued)

Table 33.3 (continued)

State/UT	Inland Fishermen Population		Marine Fishermen Population	
	Female	Male	Female	Male
Puducherry	**3384**	**5084**	41,494	57,310
All India	**10,103,842**	**13,013,978**	**22,94,066**	**26,51,652**

Source Department of Fisheries, States Government / UTs Administration

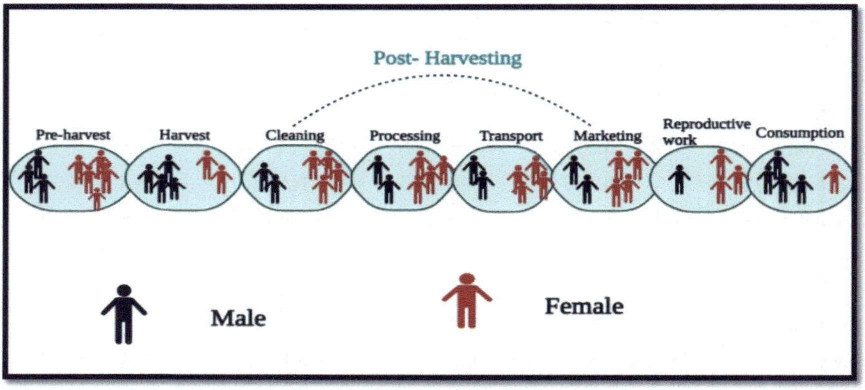

Fig. 33.2 Show various activities performed by male and female in fishery sector *Source* Author

33.5.2 Post-harvest

After the fish is harvested, post-harvest operations start and continue until the fish is delivered to the clients. As soon as the boats arrive in the harbors and on the beach, women are involved in every activity until the fish reaches its destination. For instance, fisherwomen in Mumbai, Maharashtra, are not involved in fish harvesting; instead, they may be seen in wholesale markets, fish landing facilities, and locations where fish is sorted, cleaned, packaged, and sold (Peke 2013). Fish processing needs to be done with the utmost care because it is one of the most edible commodities. A large quantity of waste happens in small-scale fishing, which is often conducted by artisanal fisherwomen, due to a lack of basic facilities like sanitary landing centers, storage facilities, transportation, and market facilities. Women's post-harvest activities become even more crucial in this circumstance (Bhatt and Rao, 2003; Lekshmi et al. 2022).

33.5.3 *Fish Cleaning*

The engagement of fisherwomen as fish dealers and prawn cleaning in Kerala was captured in Klein and Davis' edited book "To Work and to Weep: Women in Fishing Economies" (Shakir 2022). These married women share the difficulty of obtaining fish at fish auctions where they have to compete with males. These auctions were known to be attended by 10% of women (Gulati 1984). Philomena, the campaigner, stated that women had several health difficulties while working in prawn cleaning or peeling because they had to kneel for 2–3 h on ice that may results in dermatological, musculoskeletal, and gynecological disorders (Programme for Social Action 2017).

33.5.4 *Processing*

Generations of Indian fisherwomen were involved in the traditional process of fish processing and preserving (Fig. 33.3). The preserved/processed fish might have been caught by their family or acquired from other fish dealers. Fish (fresh or processed) is an essential element of the cuisine in the north-eastern state of Manipur and is usually in great demand, particularly in markets due to unique or traditional taste and flavors. As a result, women preserve the fish by drying, smoking, fermenting, and other methods (Meetei et al. 2016). Women in the majority perform vendering when compared to men. The fisherwomen collected the fish from big contractors as early as 3 am in the morning. This routine hard work in the early morning disturbs their sleep cycle may cause serious health issues (Shah 2010).

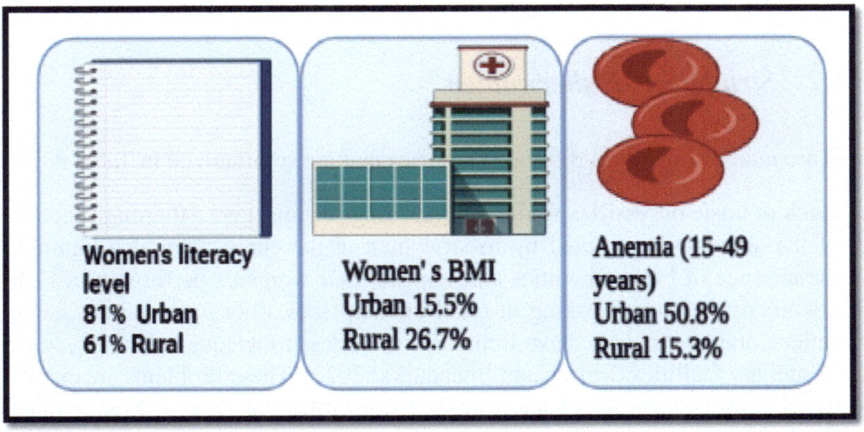

Fig. 33.3 Show health status and literacy ratio fisherwomen

33.5.5 Women as Vendors

Door-to-door fisherwomen buy their catch in groups or individually at fish auctions. Those that purchase from auction houses divide the cost of the fish as well as the cost of transportation. These ladies have pre-determined selling zones and regular clients to whom they sell their fish to avoid confrontations. These ladies, however, are experiencing fresh competition from 'bhayias' who do not obey the 'unspoken rule,' which harms their company. Furthermore, due to a lack of identity documents, women have additional problems (i) restricted access to buildings (ii) bribing officials. Furthermore, women suffer from health issues as a result of their physically demanding jobs (carrying fish baskets on their heads and travelling long distances) (Sindhu 2022).

33.5.6 The Plight of Women

Women's roles are confined to second grade workers or assisting with processing, maintaining equipment, and marketing fish. However, their importance is frequently overlooked, both socially and economically. They are typically considered as low-wage employees. Their troubles have been exacerbated by a lack of work stability and a contract system; they were paid a poor pay. They were compelled to undertake their occupations because they lack the necessary training to pursue alternative opportunities (Chattrjee et al. 2022). They were frequently susceptible to intermediaries who charge exorbitant interest rates, making it nearly hard for them to break free from the cycle of poverty

33.5.7 Struggle of Fisherwomen

There are many struggles in different countries that are summarized in Table 33.4.

A. **Lack of basic necessities** While the preceding sections give a thorough account of the activities performed by fisherwomen in the chain, they also highlight the absence of basic amenities that impede their women's performance. Fisherwomen who were working in retail marketplaces, door-to-door sales, street sellers, and auctioneers, have limited or no access to adequate drinking water or sanitary facilities (Rajeev and Bhandarkar 2022). These problems are exacerbated, especially for menstrual women. These violate Sustainable Development Goal (SDG) 6, which mandates that everyone should have access to safe drinking water and sanitary services.

Furthermore, amenities such as toilet access (for breakfast), lighting facilities, particularly for shops and street sellers are not available (Peke 2013). There have

Table 33.4 Show challenges faced by fisher-women at different level

S.no	Challenges	Pre-harvest	Harvest	Post harvest
1	Lack of basic infrastructure	✔	✔	✔
2	Lack of safe drinking water	✔	✔	✔
3	Health issues	✔	✔	✔
4	Use of poor gears	✔	✔	✔
5	Low wages	✔	✔	✔
6	Less opportunity	✔	✔	✔
7	Abuse and harassments	✔	✔	✔
8	Labor Intensive work	✔	✔	✔
9	Lack of license	✔	✔	✔

been reports of fisherwomen in Mumbai failing to get clean water, which has reduced the number of women engaged in the business over time (Peke 2013).

B. **Lack of Transportation services**: While some fisherwomen sell fish near harbours, others must haul the seafood to marketplaces. In many cases, fisherwomen on India's west and east coasts face various hurdles in this phase of the chain. Because fisherwomen are sometimes denied access to public transportation (buses and auto-rickshaws), they must rely on alternative private modes of transportation, making them exposed to intimidation and harassments (ICSF 2010).
C. **Laborious work**: Fisherwomen in India's local value chain are seen conducting strenuous labor tasks whether it is any stage of harvesting that has a direct impact on their health. For example, oysters, picking shells, seaweed in adverse weather conditions, physically wading through water with their hands to catch shrimps, fishing with plastic hooks, ropes, and so on. Women were observed working from dawn to night in the post-harvest, washing and sorting fish without any safety equipment (Sharma et al. 2022).
D. **Low wages and less opportunity**: In underprivileged areas, women work two jobs, taking care of their families at home and taking care of their careers (Agarwal 2010). Over 45% of women globally between the ages of 15 and 64 make major economic contributions, according to the International Labor Organization (ILO). Two categories of discrimination target women. Pre-market discrimination refers to the denial of access to resources that aid in the growth of capital, such as training, education, experience, and so on. Market discrimination, on the other hand, refers to the differential payment of equivalent work. Despite receiving lower earnings and less opportunities than males, women make greater contributions to the fishing industry.

Table 33.5 Summarized rights of fisherwomen

S.no	Rights of fisherwomen
1	• Carrying facility has to be given to all the fish selling women in all villages
2	• Provide special bogey in all trains for fish vending women for the particular route
3	• Monetary root has to provide to fish vending women if any accidents happen during their work
4	• Unlawful collection of money should be stopped in local markets
5	• Fund should be given for the education of girls studying in private colleges from pre degree onwards
6	• Fish loads by themselves should be excused from market tax
7	• Local government should supervise the operation of markets. Market tax should be directly collected from fish vendors and it should be limited to Rs. 1/-
8	• Set up savings cum relief scheme for women too
9	• Set up residential education system for the girls from fish community
10	• Representation of fisherwomen should be ensured in head load worker's welfare boards

33.6 Rights of Fisherwomen

The profession of fish vending is ignored due to the development process. The money earned by the fisherwomen has been the key source of livelihood in the families of the fishing communities. There has been a rapid boost in practices in this field, neglecting the privileges of fisherwomen. The exploiters think the silent pain of the fisherwomen as weak point and this has now become intolerable (Kuruvilla 2008). Their rights as working fisherwomen are summarized in Table 33.5.

33.7 Conclusion

Women have higher participation rates in unorganized sectors compared to men, yet they continue to face unequal pay. Government policies and initiatives can play a key role in empowering women by ensuring employment opportunities, increasing income levels, and reducing the influence of intermediaries who exploit the illiteracy of rural women. To guarantee that rural women benefit from these efforts, the administrative system must enhance its vigilance, particularly concerning the safety of employed women in rural areas. The first and most essential step toward empowering women is enabling them to recognize their fundamental right to education. Sustainable economic growth can only be achieved if gender equality is realized through the elimination of discrimination, especially for marginalized, unprotected, and informal female workers. True empowerment will be achieved when women, regardless of their social status, possess full decision-making power and economic independence.

33.8 Future Perspectives

Empowering fisherwomen requires gender-sensitive policies, access to education, skill development, and modern technologies. Ensuring proper sanitation, healthcare, safe transportation, and market infrastructure is equally important. Organizing women into cooperatives and ensuring their representation in policy-making can enhance their bargaining power and livelihood security. A focus on climate-resilient practices like seaweed farming and value addition can further strengthen their role in sustainable fisheries development.

References

Agarwal B (2010) Gender and green governance: the political economy of women's presence within and beyond community forestry. OUP Oxford

Bhatt ER (2006) We are poor but so many: The story of self-employed women in India. Oxford University Press on Demand.

Bhatta R, Rao KA (2003) Women's livelihood in fisheries in coastal Karnataka. India. Indian J Gender Stu 10(2):261–278

Biswas MP, Rao MRM (2014) Fisherwomen of the east coastal India: a study. Internat J Gender and Women's Stud 2(2):297–308

Campling L, Havice E, McCall Howard P (2012) The political economy and ecology of capture fisheries: market dynamics, resource access and relations of exploitation and resistance. J Agrar Chang 12(2–3):177–203

Ceballos-Concha A, Asche F, Cárdenas-Retamal R (2025) Salmon aquaculture in Chile: production growth and socioeconomic impacts. Rev Aquac 17(1):e12993

Chandravanshi R, Neelmani VM, Kumar M, Devi BN (2020) A review on coastal tourism in India. Indian J Pure App Biosci 8(4):138–171

Chatterjee R, Roy D, Kumari G, Mondal P (2022) Gender mainstreaming and women empowerment in small holdings. Innov Small-Farm Agric: Improv Livelihoods Sustain

Choudhary AR (2001) Common property resource management: gender, equity and participation. A case study of the fish workers of Kerala. Indian Social Institute

D'souza S (2020) Unmasking the silent role fisherwomen play in the small-scale fisheries local value chain in India

Department of Fisheries (2020) National Fisheries Policy, 2020. New Delhi, India

Desert Story (Director) (2017) ആദിവാസികള് മീന് പിടിക്കുന്ന് ടിക്കുകള്Kerala Tribal Fishing Tricks and traps. Fishing Live [Motion Picture]. Accessed 14 Aug 2020. https://www.youtube.com/watch?v=9sXiSRTZlk

Dhawan V (2017) Water and agriculture in India. In: Background paper for the south Asia expert panel during the global forum for food and agriculture, vol 28

Dutta M, Naidu SC (2025) Women workers. In: The Elgar companion to decent work and the sustainable development goals. Edward Elgar Publishing, pp 188–199

Emam MA, Leibrecht M, Chen T (2021) Fish exports and the growth of the agricultural sector: the case of south and southeast Asian countries. Sustainability 13(20):11177

Gopal N, Hapke HM, Kusakabe K, Rajaratnam S, Williams MJ (2020) Expanding the horizons for women in fisheries and aquaculture. Gend Technol Dev 24(1):1–9

Gulati L (1984) Technological change and women's work participation and demographic behaviour: a case study of three fishing villages. Econ Polit Wkly 2089–2094

Hapke HM (2001) Gender, work, and household survival in South Indian fishing communities: a preliminary analysis. Prof Geogr 53(3):313–331

ICSF (2010) Women Fish Vendors in India: An Information Booklet. (Kumar, K ed) Chennai, Tamil Nadu, India: Chandrika Sharma

Immanuel S, Pillai VN, Vivekanandan E, Kurup KN, Srinath M (2003) A preliminary assessment of the coastal fishery resources in India-socioeconomic and bioeconomic perspective. In: Assessment, management and future directions for coastal fisheries in Asian countries WorldFish Cent Conf Proc 67:439–478

Kathirvel M, Gokul A, Thirumilu P (2007) Biodiversity and utilization of brachyuran crabs of the exclusive economic zone of India. Conserv Valuat Mar Biodivers 145:145

Khushbu Gulati R, Sushma A, Sharma, P (2022) Ecological impact of heavy metals on aquatic environment with reference to fish and human health 14(4):1471–1484

Kleiber D, Harris LM, Vincent AC (2015) Gender and small-scale fisheries: a case for counting women and beyond. Fish Fish 16(4):547–562

Kumar R, Gupta R, Dahiya T, Yadav R, Saifi R, Yodha K, Kumari B, Singh P, Shefali (2022a) Surveillance of Heteropneustes fossilis (Stinging Catfish) diseases and health management practices in Haryana. Appl Ecol Environ Sci 10(4):261–267

Kumar R, Gupta RK, Yadav R, Saifi R, Yodha K, Pooja (2022b) Eisenia fetida as protein source for growth enhancement of Heteropneustes fossilis. Egypt J Aquat Biol Fish 26(2):577–588

Kuruvilla C (2008) Fish vending women–a human rights perspective. Doctoral dissertation, Loyola College of Social Sciences

Lekshmi PS, Radhakrishnan K, Narayanakumar R, Vipinkumar VP, Parappurathu S, Salim SS, Johnson B, Pattnaik P (2022) Gender and small-scale fisheries: contribution to livelihood and local economies. Mar Policy 136:104913

Magik India (Director) (2018) Dang tribal women fishing near Ahwa village–Gujarat [Motion Picture]. Accessed 15 Aug 2020. https://www.youtube.com/watch?v=LTiBgmRW5MU&feature=emb_logo

Malik K (2013) Human development report 2013. The rise of the South: human progress in a diverse world. The Rise of the South: human progress in a diverse world (March 15, 2013). UNDP-HDRO Human Development Reports

Matovu B, Bleischwitz R, Alkoyak-Yildiz M, Arlikatti S (2024) Invigorating women's empowerment in marine fishing to promote transformative cultures and narratives for sustainability in the blue economy: a scoping literature review from the Global South. Mitig Adapt Strat Glob Change 29(8):83

Maurya P, Mohanty PC (2019) What restricts credit to women enterprises? Evidence from India's informal sector. Int J Soc Econ

Meetei WT, Saha B, Pal P (2016) Participation of women in fisheries: a study on gender issues in Manipur, India. Int J Bio-Resour Stress Manag 7(4):906–914

MyNation (Director) (2010) Tamil Nadu fisherwomen risk their lives; use traditional method to catch shrimps [Motion Picture]. India. Accessed 13 Aug 2020. https://www.youtube.com/watch?v=QQq1KlqS2Ew

Peke S (2013) Women fish vendors in Mumbai: a study report

Programme for Social Action (2017) Occupation of the coast–blue economy in India. The research collective, New Delhi. Accessed 20 Oct 2020. https://updatecollective.files.wordpress.com/2018/01/occupationofthecoast_trc2017.pdf

Rajan SI (2024) Migration and development: new evidence from the Kerala Migration Survey 2023. Migr Dev 13(2):139–175

Rajeev M, Bhandarkar S (2022) Unravelling supply chain networks of fisheries in India: The Transformation of Retail. Springer Nature

Rimstead R (2016) Remnants of Nation. University of Toronto Press, In Remnants of Nation

Sarkar S, Bhuimali A (2005) Disadvantaged women in informal sector: a study ofDomestic Help and Bidi Workers. Abhijeet Publications

Shah RH, Bukhari R (2019) Strengthening the role of women in fisheries and aquaculture. J Entomol Zool Stud 7(3):648–654

Shah D (2010) Women in fisheries case studies: India

Shakir SA (2022) Theeramythri Project and Fisherwomen Empowerment: a Case of Malappuram District, Kerala

Sindhu L (2022) Women Empowerment in 21st Century in India: Some Issues and Challenges. IJNRD-Int J Nov Res Dev (IJNRD) 7(3):104–114

Singh S (2022) Women workers in unorganized sector: a study of Patiala City. J Women Empower Stud (JWES) 2(01)14–20. ISSN: 2799-1253

Singh S (2022) Women workers in unorganized sector: a study of Patiala City. J Women Empower Stud (JWES) 2(01):14–20. ISSN: 2799-1253

Tanuku U (Producer), Khosla M (Writer), Banerji R (Director) (2013) Shifting undercurrents–women seaweed collectors of Gulf of Mannar [Motion Picture]. Accessed 15 Oct 2020. https://www.youtube.com/watch?v=mqIWFUfLn4Y&feature=emb_title

Vijayan V, Suresh A, Sabu SS, Padi HP (2025) Fish for food and nutrition security in India: a comprehensive framework analysis. Food Security 1–17

Index

R. Kumar et al. (eds.), *Sustainability in Aquaculture and Fisheries*, Springer Water, https://doi.org/10.1007/978-3-032-16605-0

N

O

P

R

The manufacturer's authorised representative in the EU is Springer Nature Customer Service Centre GmbH, Europaplatz 3, 69115 Heidelberg, Germany. If you have any concerns regarding our products, please contact ProductSafety@springernature.com

Printed and bound by CPI Group (UK) Ltd, Croydon, CR0 4YY
07/10/2026
02229357-0020